Catalysis and Zeolites

Springer

Berlin
Heidelberg
New York
Barcelona
Hong Kong
London
Milan
Paris
Singapur
Tokyo

J. Weitkamp · L. Puppe (Eds.)

Catalysis and Zeolites

Fundamentals and Applications

With 211 Figures and 48 Tables

Prof. Dr.-Ing. Jens Weitkamp
Universität Stuttgart
Institut für Technische Chemie I
D-70550 Stuttgart
Germany

Dr. Lothar Puppe
Bayer AG
Geschäftsbereich Chemikalien
CH-F-FC, Gebäude G 8
D-51368 Leverkusen
Germany

ISBN 3-540-63650-1 Springer-Verlag Berlin Heidelberg New York

Library of Congress Cataloging-in-Publication Data

Catalysis and zeolites : fundamentals and applications / J. Weitkamp, L. Puppe (eds.).
includes bibliographical references and index.
ISBN 3540636501 (hardcover : acid-free paper)
1. Zeolites. 2. Catalysis. I. Weitkamp, J. (Jens) II. Puppe, L. (Lothar)
TP245.S5 C39 199 660'.2995-dc21 98-48578

This work is subject to copyright. All rights are reserved, whether the whole or part of the material is concerned, specifically the rights of translation, reprinting, reuse of illustrations, recitation, broadcasting, reproduction on microfilm or in other ways, and storage in data banks. Duplication of this publication or parts there of is permitted only under the provisions of the German Copyright Law of September 9, 1965, in its current version, and permission for use must always be obtained from Springer-Verlag. Violations are liable for prosecution act under German Copyright Law.

© Springer-Verlag Berlin Heidelberg 1999
Printed in Germany

The use of general descriptive names, registered names, trademarks, etc. in this publication does not imply, even in the absence of a specific statement, that such names are exempt from the relevant protective laws and regulations and therefore free for general use.

Typesetting: Fotosatz-Service Köhler GmbH, Würzburg
Cover layout: design & production, Heidelberg

SPIN: 10076794 02/3020 - 5 4 3 2 1 - Printed on acid-free paper

Preface

Zeolites occur in nature and have been known for almost 250 years as aluminosilicate minerals. Examples are clinoptilolite, mordenite, offretite, ferrierite, erionite and chabazite. Today, most of these and many other zeolites are of great interest in heterogeneous catalysis, yet their naturally occurring forms are of limited value as catalysts because nature has not optimized their properties for catalytic applications and the naturally occurring zeolites almost always contain undesired impurity phases.

It was only with the advent of synthetic zeolites in the period from about 1948 to 1959 (thanks to the pioneering work of R.M. Barrer and R.M. Milton) that this class of porous materials began to play a role in catalysis. A landmark event was the introduction of synthetic faujasites (zeolite X at first, zeolite Y slightly later) as catalysts in fluid catalytic cracking (FCC) of heavy petroleum distillates in 1962, one of the most important chemical processes with a worldwide capacity of the order of 500 million t/a. Compared to the previously used amorphous silica-alumina catalysts, the zeolites were not only orders of magnitude more active, which enabled drastic process engineering improvements to be made, but they also brought about a significant increase in the yield of the target product, viz. motor gasoline. With the huge FCC capacity worldwide, the added value of this yield enhancement is of the order of 10 billion US $ per year. No wonder therefore, that the introduction of zeolitic FCC catalysts is often referred to as a true revolution in petroleum refining.

Given this enormous success, it is not surprising that, in the period after 1962, zeolite catalysts rapidly conquered additional processes, at first in the field of petroleum refining, somewhat later in the field of basic petrochemistry. Examples of industrial processes (besides FCC) where zeolite catalysts nowadays have a firm place are hydrocracking of heavy petroleum distillates, octane number enhancement of light gasoline by isomerization, dewaxing of heavy petroleum distillates and lubricating oils, the synthesis of ethylbenzene (the precursor chemical of styrene and polystyrene) from benzene and ethene, the isomerization of xylenes (to produce *para*-xylene, the precursor chemical for terephthalic acid and polyesters derived from it) and the disproportionation of toluene into benzene and xylenes. Today, catalysis is the single most important application of zeolites in terms of financial market size (not in terms of tons produced per year) with an estimated turnover in the order of 1 billion US $ per year worldwide.

In these more traditional fields, zeolite catalysis has without doubt reached a certain degree of maturity. However, new applications for zeolites as catalysts are emerging, especially in the broad area of manufacturing organic intermediates and fine chemicals. In addition, the potential of zeolite catalysts for the purification of exhaust gases, e.g., from diesel engines, is under intense scrutiny all over the world. Here, the environmental aspect of zeolite catalysis becomes obvious, yet the search for zeolitic catalysts in organic syntheses is often driven by environmental aspects as well, especially if less benign catalyst systems like hydrofluoric acid, aluminum trichloride etc. are to be replaced by zeolites.

Altogether, zeolite catalysis has become a most important sub-field of heterogeneous catalysis. Nobody who is active in heterogeneous catalysis, either in industry or at a research institution, can afford to ignore zeolites and the remarkable progress that is still being made in their synthesis, post-synthesis modification, pysico-chemical characterization and testing.

This book therefore starts, in Chapter 1 written by *J.-L. Guth* and *H. Kessler*, with a state-of-the-art review of the synthesis methods for aluminosilicate zeolites and related silica-based materials. The term „zeolite" is nowadays used in a much broader sense than the one traditionally used by mineralogists, and it encompasses numerous zeolite-like, cristalline microporous solids with elements other than silicon and aluminum on tetrahedral framework positions. Many of these materials do have a potential as catalysts, and the most important family of such materials, the aluminophosphates and their derivatives, are dealt with in Chapter 2 by *J.A. Martens* and *P.A. Jacobs*. An overview of the available post-synthesis techniques for the transformation of a zeolite with the desired framework into a catalyst with optimal acidic, shape-selective etc. properties is given in Chapter 3 written by *G. Kühl*. An in-depth treatment of the most relevant techniques for zeolite characterization, viz. X-ray powder diffraction, infrared spectroscopy and nuclear magnetic resonance spectroscopy, has been contributed by *H.G. Karge, M. Hunger,* and *H.K. Beyer* in the central Chapter 4. Chapter 5, written by *J. Weitkamp, S. Ernst,* and *L. Puppe*, deals with the principles and fundamentals of shape-selective catalysis which is unique for catalysts with pore widths in the order of molecular dimensions. A critical review of the recent literature on the use of zeolite catalysts in organic chemistry is presented in Chapter 6 by *P. Espeel, R. Parton, H. Toufar, J.A. Martens, W. Hölderich,* and *P.A. Jacobs*. The final Chapter 7 by *P.M.M. Blauwhoff, J.W. Gosselink, E.P. Kieffer, S.T. Sie,* and *W.H.J. Stork* discusses the industrial processes which are currently in operation using zeolitic catalysts and the specific effects of the zeolites in those processes. The editors sincerely thank the authors for having contributed to this book and for having devoted a large amount of their precious time to the respective chapters.

Catalysis and Zeolites is meant to provide comprehensive and valuable information to all those who already are working in the field or who plan to use zeolites as catalysts in their future research. It is our hope that the book will fully meet the readers' expectations.

Jens Weitkamp
Lothar Puppe

April 1999

Contents

Authors

Dr. Hermann K. Beyer
X-Ray Diffraction Department
Central Research Institute for Chemistry
Hungarian Academy of Sciences
P. O. Box 17
H-1525 Budapest, Hungary
E-mail: H6040BEY@ELLA.HU

Dr. P. M. M. Blauwhoff
Shell-Raffinaderiet Fredericia
P.O. Box 106
DK-Fredericia, Denmark

Prof. Stefan Ernst
Department of Chemistry
Chemical Technology
University of Kaiserslautern
P.O. Box 3049
D-67653 Kaiserslautern, Germany

Dr. Patrick Espeel
Exxon Chemical International, Inc.
Hermeslaan 2
B-1830 Machelen, Belgium

Dr. J. W. Gosselink
Shell International Oil Products
Shell Research and Technology Centre,
Amsterdam
Badhuisweg 3
NL-1031 CM Amsterdam, The Netherlands

Prof. Jean-Louis Guth
Laboratoire de Matériaux Minéraux
URA-CNRS 428
Ecole Nationale Supérieure
de Chimie de Mulhouse
3, rue Alfed Werner
F-68093 Mulhouse Cedex, France

Prof. Wolfgang Hölderich
Institute for Chemical Technology
and Heterogeneous Catalysis
RWTH Aachen
Worringerweg 1
D-52074 Aachen, Germany

Priv.-Doz. Dr. Michael Hunger
Institute of Chemical Technology I
University of Stuttgart
D-70550 Stuttgart, Germany
E-mail: michael.hunger@po.uni-stuttgart.de

Professor Pierre A. Jacobs
Centrum voor Oppervlaktechemie
en Katalyse
Katholieke Universiteit Leuven
Kardinaal Mercierlaan 92
B-3001 Heverlee, Belgium
E-mail: Pierre.Jacobs@agr.kuleuven.ac.be

Dr. Hellmut G. Karge
Faradayweg 8
D-14195 Berlin, Germany

Dr. Henri Kessler
Laboratoire de Matériaux Minéraux
Ecole Nationale Supérieure
de Chimie de Mulhouse
3, rue Alfred Werner
F-68093 Mulhouse Cedex, France
E-mail: H.Kessler@univ-mulhouse.fr

Dr. E.P. Kieffer
Shell Research and Technology
Centre, Amsterdam
Shell International Chemicals B.V.
Badhuisweg 3
NL-1031 CM Amsterdam, The Netherlands

Prof. Günter H. Kühl
Department of Chemical Engineering
311A Towne Building
University of Pennsylvania
220 South 33rd Street
Philadelphia, PA 19104-6393, USA
E-mail: kuehl@seas.upenn.edu

Prof. Johan A. Martens
Centrum voor Oppervlaktechemie
en Katalyse
Katholieke Universiteit Leuven
Kardinaal Mercierlaan 92
B-3001 Heverlee, Belgium
E-mail: Martens@agr.kuleuven.ac.be

Dr. Rudy Parton
DSM Research
New Polymers
EP/NP
P. O. Box 18
NL-6160 MD Geleen, The Netherlands
E-mail: R.F.M.J.Parton@research.dsm.nl

Dr. Lothar Puppe
Bayer AG
GB Chemikalien, CH-F-FC
Gebäude G 8
Bayerwerk
D-51368 Leverkusen, Germany
E-mail: lothar.puppe.lp@bayer-ag.de

Prof. S. T. Sie
Faculty of Chemical Technology &
Materials Science
Delft University of Technology
Julianalaan 136
NL-2628 BL Delft, The Netherlands
E-mail: S.T. Sie@stm.tudelft.nl

Dr. Wim H. J. Stork
Shell Research and Technology Centre,
Amsterdam
Shell International Chemicals B.V.
Badhuisweg 3
NL-1031 CM Amsterdam, The Netherlands

Dr. Helge Toufar
Fachbereich Chemie
Institut für Technische Chemie
Martin-Luther-Universität Halle-Wittenberg
Schloßberg 2
D-06108 Halle, Germany

Prof. Jens Weitkamp
Institute of Chemical Technology I
University of Stuttgart
D-70550 Stuttgart, Germany
E-mail: jens.weitkamp@po.uni-stuttgart.de

CHAPTER 1

Synthesis of Aluminosilicate Zeolites and Related Silica-Based Materials

Jean-Louis Guth and Henri Kessler

1.1 Scope

Five important monographs are devoted to the synthesis of aluminosilicate zeolites. The first one, by Breck [1], is the most general in its content, since as well as syntheses which are reviewed until 1973, modifications and properties (ion exchange, adsorption) are also described. Barrer's book [2] is fully devoted to the synthesis and transformation, with a complete survey up to 1981, of the results related to zeolites with low or medium Si/Al ratios. The third one, by Jacobs and Martens [3], presents exhaustive information about the preparation and identification of silica-rich zeolites. The two latest ones, by Szostak [4, 4a], contain in addition an important part on non-aluminosilicate molecular sieves.

This chapter completes these monographs in three domains, viz.

- formation mechanisms and crystallogenesis,
- updated reviews of synthetic results not only from the (Si, Al) system but also from binary (or ternary) systems in which the main element Si is associated with divalent (Be, Cu...), trivalent (B, Ga, Fe...), tetravalent (Ge, Sn, Ti...) and pentavalent (V...) elements and
- recent progress concerning zeolites which have reached industrial development or which have been most widely studied.

1.2 Introduction

1.2.1 Structure, Composition, Nomenclature

Aluminosilicate zeolites comply with the following definition: "Crystallized solids characterized by a structure which comprises

- a three-dimensional and regular framework formed by linked TO_4 tetrahedra ($T = Si, Al...$), each oxygen being shared between two T elements,
- channels and cavities with molecular sizes which can host the charge-compensating cations, water or other molecules and salts.

Fig. 1.1. Hexagonal crystals of a zeolite with the novel EMT-type structure (Si/Al = 4, hexagonal polytype of the cubic faujasite). The overall breadth of the photograph corresponds to ca. 8.5 μm

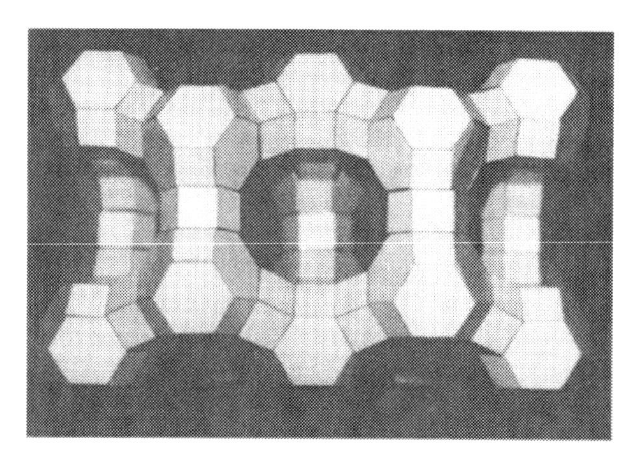

Fig. 1.2. Schematic representation of the framework of the EMT structure type ([001] direction: ↑)

The microporosity must be "open", and the framework must have enough stability to allow the transfer of matter between the interior of the crystals and the exterior.

This definition corresponds to an ideal structure. According to the structure type and the composition, and also to the synthesis method or post-synthesis treatments the framework may present defects (non-bridging oxygen, vacant sites, mesopores), and the coordination of the T elements may be modified by species present in the micropores.

The size of the synthetic crystals is generally between a fraction of a micrometer and several micrometers, but may reach several hundred micrometers (Fig. 1.1). The diameter of the channels and cavities varies according to the structure from 0.3 to 1.3 nm (Fig. 1.2), the highest values of the internal surface area and pore volume being, respectively, 800 $m^2 g^{-1}$ and 0.35 $cm^3 g^{-1}$.

The topology of the framework defines a structure type symbolized by a group of three letters, e.g., MFI for the structure type of zeolite ZSM-5. A list of the 98 currently accepted structure types, with their type materials, can be found in the Atlas of Zeolite Structure Types [5] which contains an additional list of zeolite material designations related to known structure types (e.g., Omega, ZSM-4 and LZ-202 are materials with the structure type MAZ derived from the material type Mazzite, which is a natural zeolite). However, there are numerous other zeolites with silicon as a main T element of which the structure is not yet known or is only hypothetical, or without a structure code (e.g., SSZ-27, SSZ-35, ZSM-47, CJS-1 etc.). Updated data can be found on the world wide web in Europe (http://www.iza-sc.ethz.ch/IZA-SC/) or in North America (http://www.iza-sc.csb.yale.edu/IZA-SC/).

The composition is best described by the general formula:

$$x_1M_1^{n_1+}; x_2M_2^{n_2+}; [(y_1T_1; y_2T_2\ldots)O_{2(y_1+y_2+\ldots)}]^{x-} z_1A_1; z_2A_2\ldots$$

in which the expression in square brackets represents the framework composition, the other terms corresponding to species in the micropores.

- M_1, M_2, cations with charge n_1, n_2 which compensate the negative charge of the framework ($x_1n_1 + x_2n_2 + \ldots = x$)
- $T_1, T_2\ldots$, elements (Si, Al ...) in the tetrahedra
- $A_1, A_2\ldots$, water, molecules, ion pairs.

Silicon may be considered as the principal or key element of the framework. Elements which substitute for silicon must accept a tetrahedral coordination with oxygen. The first Pauling rule ($0.225 < T_T/R_O < 0.414$) is useful, but it must not be applied in too strict a manner, above all when the element enters a framework with elements which accept the tetrahedral coordination easily. Besides the coordination state, chemical factors such as condensation possibilities or the oxidation state in relation with the overall framework charge must also be taken into account when selecting possible T elements (see Sect. 1.3.2.2).

Aluminum is the element which replaces silicon most easily. The limits of the Si/Al ratio are 0.5 (e.g., bicchulite) and infinity (e.g., silicalite-1). However, up to now, no structure is known which can be synthesized in this whole composition range. Zeolites are usually classified into three classes, namely

- zeolites with low Si/Al ratios (< 5),
- zeolites with medium Si/Al ratios (5 to 10) and
- zeolites with high Si/Al ratios (> 10).

The formation of the framework charges is due to the replacement of the tetravalent silicon by trivalent or divalent elements. These charges contribute to the stabilization of the structure, through their interaction with the compensating cations which makes the synthesis easier. After synthesis, the cations present are normally alkali, alkaline-earth or R_4N^+ (R = H, alkyl, aryl) cations. The latter transform into protons during calcination, and the former can be exchanged.

The cations are generally hydrated, especially when they are strongly polarizing. According to the composition of the reaction mixture, and to the structure type, ionic pairs (e.g., NaCl, $BaCl_2$, Pr_4NF etc.) or molecules (e.g., amines, alcohols etc.) may also be present. Removal of these different species creates the microporous volume ready for use.

1.2.2
History of Zeolite Synthesis [6]

The aim of the first experiments on zeolite synthesis, which go back to the middle of the 19th century, was to reproduce the natural zeolites. However, the absence of unquestionable identification methods, such as XRD, did not allow a conclusion of any certitude. According to the synthetic conditions used, quite dense phases (e.g., analcite) were usually obtained. Meanwhile, the remarkable molecular sieve properties of certain zeolites were recognized, but industrial application was impossible because of the scarceness of these zeolites.

The merit is due to Milton (Union Carbide) who, in the early 1950s, found an easy method which allowed the preparation of zeolites with large apertures and high adsorption capacity. This method was based on the use of reactive alkaline aluminosilicate gels and of low crystallization temperatures (between 80 and 150 °C). Higher Si/Al ratios were obtained by adding finely divided amorphous silica to the gel.

At the same time Barrer and co-workers realized systematic syntheses at temperatures between 60 and 450 °C from reaction mixtures with the composition

$$aM_{2/n}O; Al_2O_3; nSiO_2; bH_2O$$

by varying the nature of the base (hydroxide of Li, Na, K, Rb, Cs, Ca, Sr, Ba, NH_4, etc.) and the parameters a, n and b of the mixture.

Thus, over about twenty years, the majority of the aluminosilicate zeolites with Si/Al ≤ 5 now known were found. Among these materials, the best known are: zeolite A with no natural counterpart, zeolites X and Y which have the same framework topology as natural faujasite, zeolites L, erionite, offretite and mordenite [1, 2].

In 1961 Barrer and Denny carried out pioneering work in the field. They were the first to replace inorganic bases with organic ones, and they observed that this resulted in an increase in the Si/Al ratio in the zeolites. Thus, with the association of Na^+ and $(CH_3)_4N^+$ they obtained zeolite A with Si/Al ratios up to 3, whereas with Na^+ alone the ratio was 1. The limited uptake possibility of bulky cations resulted in a limited substitution of Si by Al. Continuing in this way, researchers at Mobil Oil Corp. worked with quaternary ammonium ions with longer alkyl chains, such as tetraethyl- or tetrapropylammonium, and thus obtained the first zeolites with Si/Al ratios largely over 5. These were zeolites Beta and ZSM-5 which also had new structures. This race for high Si/Al ratios ended in 1978 with the synthesis of silicalite-1 which is the pure-silica end-member of ZSM-5 [3]. The absence of framework charges and, as a consequence, the absence of compensating cations conferred new properties to this molecular sieve, the most remarkable being the hydrophobic and organophilic character of the internal surface.

The first attempt to substitute other T-elements in the (Si, Al) pair dates from 1952 with a partial replacement of Al by Ga in a thomsonite-type zeolite with the formula $Ca_4Ga_3Al_5Si_8O_{32}$, xH_2O [2]. In 1959, Barrer and co-workers studied the hydrothermal crystallization of alkaline gels based on the (Ge, Al), (Si, Ga) and (Ge, Ga) pairs. They obtained several zeolites, among them zeolite A and faujasite [2]. The introduction of Be in place of Al has been the object of more recent work, the first zeolitic beryllosilicate (analcite-type) being reported in 1971. Since then, other zeolites containing Be, such as beryllo-faujasite, have been synthesized [2].

Incorporation of boron beside silicon presents more difficulties, and to date Si/B ratios below 10 have not been attained. Other elements such as Ti, Sn, Zn, Fe could be substituted for Si but always with a low degree of substitution. Some elements like Cr or V were claimed to be in the framework but without complete characterization.

All previously mentioned zeolites contain a tetravalent element (e.g., Si, Ge), generally in association with a trivalent element (e.g., Al, Ga, B etc.) and more seldom with a divalent element (e.g., Be). A new class of materials based on the structural model of zeolites was reported in 1982. By associating a trivalent element (Al) with a pentavalent element (P) in a framework of TO_4, Union Carbide researchers crystallized several aluminophosphates with molecular sieve properties and with an overall zero framework charge. Negative charges arise if P or Al are partly replaced by tetravalent or divalent elements [4].

1.3 Theoretical Part

1.3.1 Crystallogenesis [7, 8]

As shown below, zeolites crystallize in a solution from soluble species. The different principles and laws of crystallogenesis which apply will shortly be described and discussed.

1.3.1.1 Nucleation

Separate nucleation and growth mechanisms occur each time the appearance of a new phase involves the formation of an interface separating this from the mother phase.

A new solid phase nucleates in a solution provided that the free enthalpy diminution due to the creation of the bulk (volume) surpasses the free enthalpy augmentation due to the formation of the interface. The variation of the free enthalpy G is given by the function

$$G = V\Delta G_V + S\Delta G_S \qquad (1.1)$$

where $\Delta G_V = -\frac{kT}{v} \ln \alpha$ and $\Delta G_S = \sigma$ represent the free enthalpy variation per unit volume and unit surface formation, respectively, k is the Boltzmann constant, v the volume of the "molecular" unit, α the ratio between the concentration Css of the solutes in the supersaturated solutions (out of equilibrium) and the concentration Cs in the saturated solution (in equilibrium with the solid having a surface negligible with regard to the bulk) and σ is the free enthalpy per surface unit.

If the nucleus has a spherical shape with a radius r, Eq. 1.1 becomes:

$$G_r = -\frac{4\pi^3 kT \ln \alpha}{3v} + 4\pi r^2 \sigma \tag{1.2}$$

Figure 1.3 represents the variation of G_r as a function of r and shows the possible evolutions of a nucleus: if $r < r^*$, the nucleus is only an "embryo" which may dissolve, and if $r > r^*$, the nucleus may grow. The values of r^* (critical nucleus radius) and of ΔG_r^* are calculated from Eq. 1.2 for $\partial G_r/\partial r = 0$:

$$r^* = \frac{2\sigma v}{kT \ln \alpha} \tag{1.3}$$

$$\Delta G_r^* = \frac{16\pi\sigma^3 v^2}{3(kT \ln \alpha)^2} \tag{1.4}$$

These values change with α (Fig. 1.4): the size of the critical nucleus decreases with the increase of α (supersaturation degree). This makes its formation easier through a local fluctuation mechanism.

ΔG_r^* can be considered as an activation energy, and the nucleation rate J (number N of nuclei formed per unit volume and unit time) may be calculated from Eq. 1.5:

$$J = \frac{dN}{dt} = J_O e^{-\Delta G^*/RT} \quad \text{with} \quad J_O = J' S \nu \tag{1.5}$$

S is the number of moles of species involved in the crystallization per unit volume and ν the frequency at which the critical nuclei transform into crystals. The term J′ is specific to the chemical reaction and will be discussed below.

An inspection of Eqs. 1.4 and 1.5 giving ΔG^* and J clearly shows that the nucleation rate depends on at least three parameters: α, σ and S.

The variation in J as a function of α (Fig. 1.5a) shows that for a low supersaturation (A to B) the nucleation rate remains close to zero and then increases strongly to a practical limit because the spontaneous nucleation rate becomes so large that the supersaturation drops immediately. A maximum in the nucleation rate may also be observed, however, especially in complex systems (Fig. 1.5b). An increase in the viscosity with α, which results in a variable diffusion activation energy added up to ΔG^* in Eq. 1.5, may sometimes be the explanation. But in the case of complex systems such as those involved in zeolite synthesis,

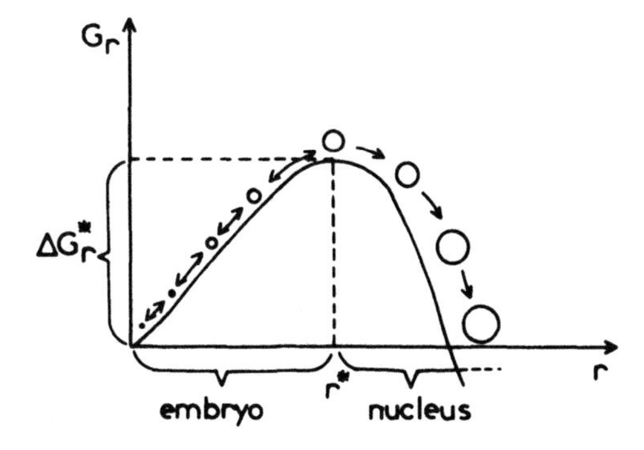

Fig. 1.3. Evolution of the free enthalpy G_r versus the radius r during the formation of a nucleus

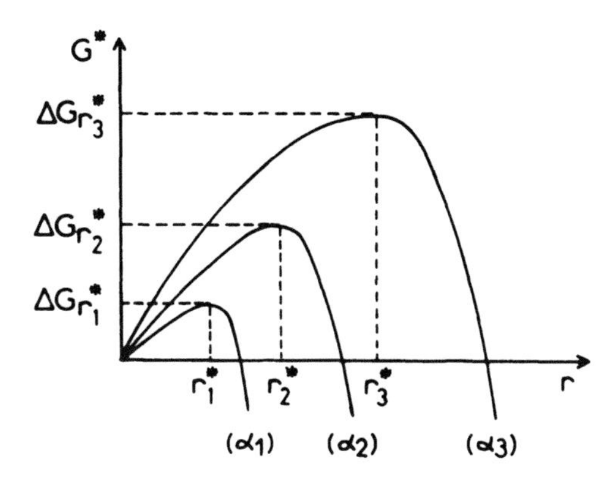

Fig. 1.4. Decrease of ΔG_r^* and r* with the increase of the supersaturation α ($\alpha_1 > \alpha_2 > \alpha_3$)

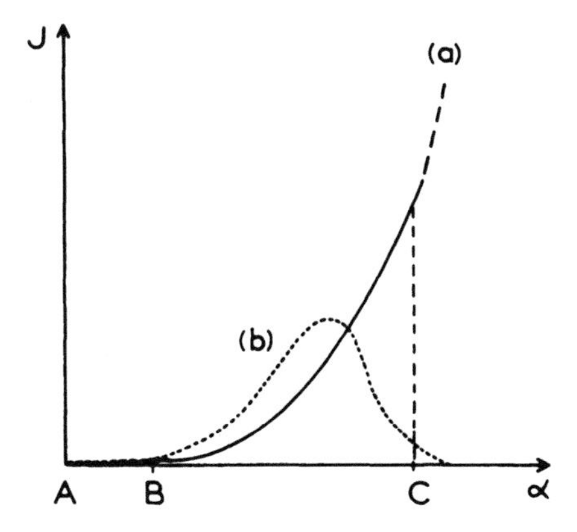

Fig. 1.5. (**a**) Variation of the nucleation rate J as a function of the supersaturation α. AB is the metastable domain of the system (J ~ o) (**b**) The nucleation rate passes through a maximum in systems with increasing viscosity or with high r_{min}^*

another explanation can be proposed [9]. The decrease in the critical nucleus size, related to the increase in the supersaturation, must have a limit corresponding to a minimal size. Below this size, the nucleus does not contain enough atoms allowing, through their structural information, the growth of a crystal with a given structure. A phase with more "simplexity" [10], characterized by a critical nucleus with a smaller minimal size, may then appear (see Sect. 1.3.1.4).

All things being equal, nucleation is faster in solutions where the solubility is high. This means that if in two different solvents the same supersaturation can be achieved, the nucleation rate will be highest in the solvent where the concentration of the crystallizing species is highest. This is easy to understand since there are more species encountered when they are more numerous and closer to each other [8a].

Another important parameter is the interfacial free energy σ which represents the work necessary to generate a solid/solution interface when the crystal is cut parallely to a given plane. It corresponds to the energy of the broken bonds minus the energy which is recovered when the atoms of the fresh surface interact with the solution. Therefore, σ will be more important as the "cohesion" of the crystal along this plane is strong and as the affinity of the crystal towards the solvent is low.

Thus when the solubility increases, by changing the solvent or the solution composition, the interfacial free energy decreases [8b]. In other words, the greater the affinity of the solvent for the crystal, the larger the solubility S, the smaller s and the higher the nucleation rate.

Conversely, when the material is sparingly soluble, nucleation is more difficult and occurs only when a high supersaturation is reached. This occurs, for instance, when a $AgNO_3$(1 M) solution is mixed with a NaCl (1 M) solution. The supersaturation vs. AgCl is $\sim 10^{10}$ and the phase precipitates instantaneously in the form of crystalline AgCl. Four factors favor the direct precipitation of a crystalline phase from very highly supersaturated solution: high "simplexity" of the structure, presence of only one kind of species which can form the crystal, ionic or metallic bonds formation and bond strength. Like AgCl, zeolites are sparingly soluble solids. Unfortunately, these favorable factors are not met during their crystallization. Thus if a strong supersaturation is created, for instance by mixing concentrated silicate and aluminate solutions, there results precipation of an amorphous gel with a drastic drop of the supersaturation, which is now determined by the solubility of the amorphous gel.

It can be noticed that this problem does not arise in the preparation of the precursor of organized mesoporous solids. Indeed, in that case the precursor can be directly precipitated from a highly supersaturated solution [8c]. These solids are closer to gels than to zeolites. There is no organized arrangement and periodicity in the framework. Thus the nucleation needs a cooperative organization of surfactant molecules into micelles (supramolecular organization involving van der Waals interactions) and a polycondensation of monomers into a disorganized, and generally interrupted, framework forming walls around the micelles. It can be assumed that the arrangement of the micelles into a more or less perfect hexagonal array is related to the degree of supersaturation.

Before discussing the means which will allow the crystallization of the precursors of microporous solids like zeolites it is also important to bear in mind that the nucleation of such precursors is not only a side by side juxtaposition of units from the solution. It implies chemical condensation reactions between some of the units to form a completely connected framework. Such reactions [8d] have their own activation energy ΔG_C which is included in the pre-exponential term J′of the nucleation rate equation with:

$$J' = J'_O e \frac{-\Delta G_C}{RT}$$

Besides the above-mentioned homogeneous nucleation, heterogeneous nucleation occurs on solid surfaces (reactor walls, solid particles in suspension or crystals): the surface creation for a nucleus is smaller, which makes its formation easier.

The rate of the primary nucleation (homogeneous and heterogeneous) may be modified by

- adsorption of foreign species present in the reaction mixture which changes σ and leads generally to a decrease of the rate owing to the activation energy needed for the desorption
- secondary nucleation by introduction of seeds and breeding which corresponds to the formation of small seeds (splinters) when crystals collide.

If the reaction mixture is not supplied with new species the overall nucleation rate no longer remains constant and decreases rapidly with time due to the fall in the degree of supersaturation. But if, as in the case of zeolite synthesis, the dissolution of a solid source of reactants (e.g., reactive gel) maintains the supersaturation almost constant, the nucleation rate may remain constant and even increase due to a development of heterogeneous and secondary nucleation (autocatalytic process).

1.3.1.2
Crystal Growth

The second step of crystal formation is the growth of the nucleus. A nucleus is growing until it is thermodynamically in equilibrium with the solution. The Gibbs–Thomson relationship gives the solubility of a spherical particle of radius r:

$$\frac{kT \ln (Cr/Cs)}{2v} = \frac{\sigma}{r} \tag{1.6}$$

Cr and Cs are, respectively, the solubility of particles with radius r and ∞. From Fig. 1.6, which shows the variation of Cr/Cs as a function of r, one can infer that if the supersaturation $\alpha = Css/Cs$ of the solution is higher than the ratio $(Cr/Cs)_i$ corresponding to a particle of radius r_i, this particle will grow until α reaches the new value of the $(Cr/Cs)_f$ ratio. The final size r_f of the particle then depends on the slope of the decrease in α with increasing r.

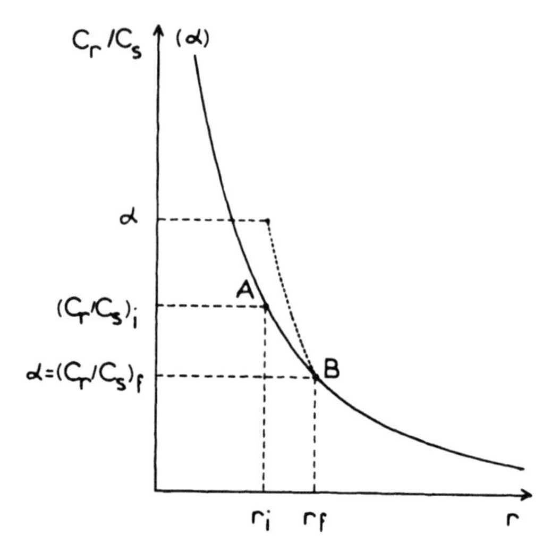

Fig. 1.6. Cr/Cs ratio as a function of the particle radius r. (The *dashed curve* shows the growth of a particle in a solution with $\alpha > (Cr/Cs)_i$, from the radius r_i to r_f for which $\alpha = (Cr/Cs)_f$)

Thus, the different factors determining the crystal sizes and their distribution are:

- the initial supersaturation,
- the volume of the solution,
- the number of nuclei initially formed or present,
- the overall nucleation rate during crystal growth and
- Ostwald ripening.

Ostwald ripening, which corresponds to the dissolution of the smallest crystals in favor of the largest, can be explained by considering the relationship between r and Cr/Cs (Fig. 1.6). If a particle is in a solution characterized by an α value below the value of Cr/Cs in equilibrium with the radius r, then the particle dissolves. Thus, in a solution containing crystals with different sizes the large crystals are responsible for the drop in the concentration to a value consistent with their size. But such a concentration is below the value needed for smaller crystals which then disappear by dissolution.

The morphology of a crystal at equilibrium can theoretically be calculated using Wulff's relationship:

$$\frac{\sigma_i}{h_i} = \frac{kT \ln \alpha}{2v} \tag{1.7}$$

where h_i its the distance of a face i from the center of the crystal and σ_i is its free enthalpy per unit surface. This relationship is an extension of Eq. 1.6 to solids for which σ is not constant on all the surface and results from the fact that at equilibrium the free enthalpy of the surface ($G_s = \sum_i \sigma_i s_i$) must be minimal.

However, such an equilibrium form is generally not reached because of the influence of the kinetic factors which govern the growth rate of the different

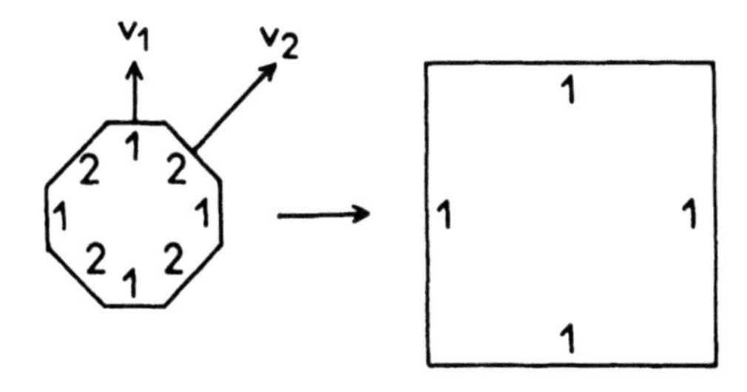

Fig. 1.7. Schematic representation of the disappearance of the faces with the highest growth rate

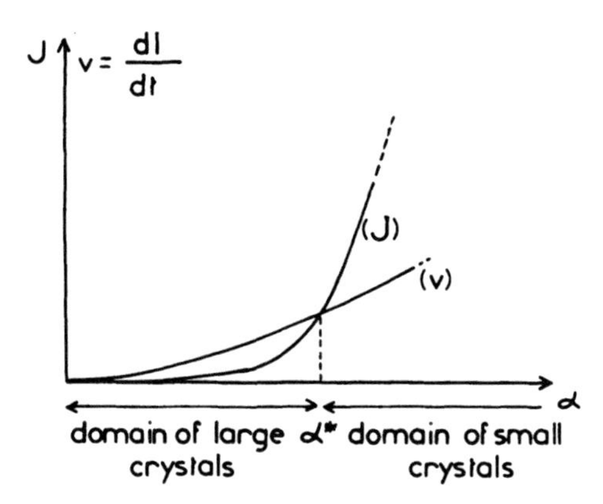

Fig. 1.8. Comparison of the nucleation rate with the growth rate as a function of α. Domains of large (v > J) and small (v < J) crystals formation

faces. Before describing these factors one may recall that during crystal growth the face with the highest growing rate will disappear in favor of the others (Fig. 1.7).

When the nucleation rate J and growth rate V as a function of α are compared, it can be seen that nucleation needs a higher degree of supersaturation than growth does (growth may continue even at low α values). Through the control of α, large ($\alpha < \alpha^*$) or small ($\alpha > \alpha^*$) crystals can be produced (Fig. 1.8).

For high values of α, the growth rate is generally under diffusional control (see Fig. 1.9, Css >> Cc, Cc being the concentration on the crystal surface):

$$V \sim D\,(Css - Cc) \tag{1.8}$$

Factors which then affect the growth rate are, besides those related to diffusion:

- convection (solution flow due to thermal heterogeneities) and
- agitation (mechanically produced solution flow).

For lower α values (Css ~ Cc), the growth rate is controlled by the building units fixing mechanism:

$$V \sim k\,(Cc - Cs) \tag{1.9}$$

This process depends on the structure and bonds on the crystal faces. According to the P.C.B. (Periodic Chain Bond) theory, the different faces of a crystal belong

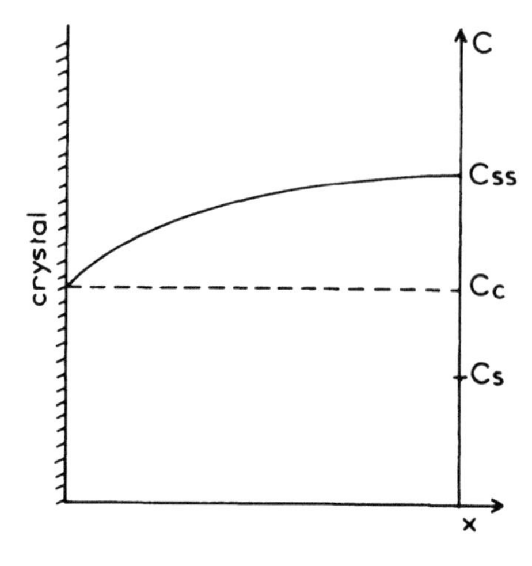

Fig. 1.9. Variation of Css in the solution as a function of the distance x from the crystal (Cc is the concentration on the crystal face and Cs is the concentration at equilibrium for a crystal with $r = \infty$)

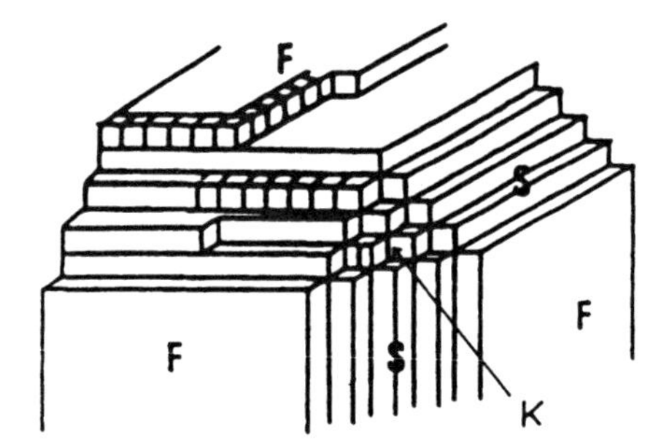

Fig. 1.10. Schematical representation of the K-, S- and F-type faces of a crystal

to three types: F (flat faces), S (stepped faces) and K (kinked faces) (Fig. 1.10). The bonding of the building units on the faces K and S, which are faces of high surface energy, is promoted by the presence of kinks and steps, and may be without any relationship with the neighboring units (formation of curved faces):

$$V_K \sim V_S \sim k \ln \alpha \tag{1.10}$$

For F faces three different mechanisms are possible (Fig. 1.11). If α is very low, growth occurs with a low rate due to the presence of screw dislocation steps, which lead to spiral growth (B.C.D. theory of Burton, Cabrera and Frank). For mean values of α, a bidimensional nucleation mechanism on the surface of the face is responsible for the increase in the growth rate (K.S. theory of Koessel and Stransky). For still higher α values, the growth proceeds through a mechanism comparable to that to the K and S faces.

On a more microscopic scale the growth mechanism also depends on the nature of the building units and their bonding mechanism. Although few relevant data are generally available, the different parameters involved are: possible

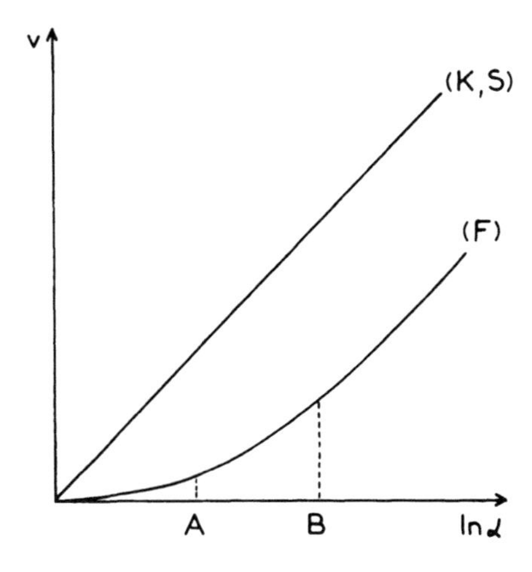

Fig. 1.11. The three different growth regimes for the F-type faces compared to the unique growth regime for the K- and S-type faces ($\ln \alpha < A$: spiral growth; $A < \ln \alpha < B$: growth by bidimensional nucleation; $\ln \alpha > B$: "continuous" growth as for K- and S-type faces)

building units in relation to the structure and composition, orientation and diffusion on the surface, desolvation, bonding reactions.

Finally, the growth rate of certain faces may be strongly modified through the specific adsorption of foreign species which change the morphology of the crystal.

The accumulation of defects on the faces during the growth can lead to a stop in growth: the generated disorder increases the free enthalpy of the crystal; this results in a supersaturation α of the solution relative to the faulty crystal which decreases to the saturation value ($\alpha = 1$). The crystallization can then proceed with the formation of new nuclei, principally on the faces of the crystal, leading to polycrystalline oriented aggregates. This is often observed in highly supersaturated media with high growth rates.

1.3.1.3
Advancement of the Crystallization with Time

The S-shaped curve (Fig. 1.12) which represents the advancement of the crystallization with time is described by the Kholmogorov equation:

$$\frac{M_t}{M_f} = X_t = 1 - e^{-Bt^n}$$

where M_t and M_f are, respectively, the mass of crystals in the product at time t and in the final product. n is a constant depending on the nucleation and growth conditions. B is only a constant if homogeneous nucleation occurs. However, such a curve does not contain any information on the basic kinetic parameters of crystallization: nucleation rate and crystal growth rate. More accurate information on the crystallization kinetics can be provided when, based on crystal size and size distribution, the linear crystal growth rate and the rate of nucleation as a function of time can be determined [11].

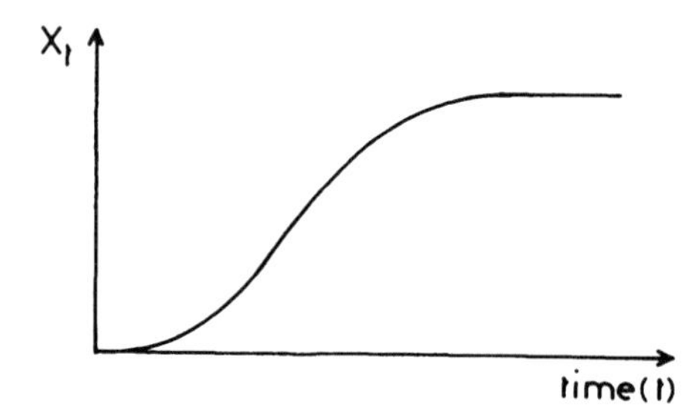

Fig. 1.12. Advancement of the crystallization X_t as a function of time t

1.3.1.4
Ostwald's Rule

"If a system is far from equilibrium, intermediate metastable phases crystallize generally before the thermodynamically stable phase".

In the case of zeolite synthesis, Ostwald's rule is illustrated by the following experiment. Starting from clear aluminosilicate solutions (NaOH 3 M, Si/Al = 1, 60 °C) zeolites, A, X, losod and sodalite crystallize in that order when the concentrations of SiO_2 and AlO_2^- decrease from 90 mmol l^{-1} to 40 mmol l^{-1}. Above 90 mmol l^{-1} a gel forms by mixing the silicate and aluminate solutions. This observation can be related to the solubility of the zeolite, measured under the same conditions, and whose values decrease in the same order: 25.5, 22.8, 21 and 18.8 mmol l^{-1} SiO_2 or AlO_2^- [9]. These results are illustrated in Fig. 1.13 which shows schematically the different nucleation rates as a function of the solution concentration and the different solubility levels of two metastable phases (a gel and a zeolite A) and one stable phase (a zeolite B). The drop in the concentration in the highly supersaturated solution, due to the metastable phase formation, allows the nucleation and growth of a less metastable phase and, finally, of the stable phase.

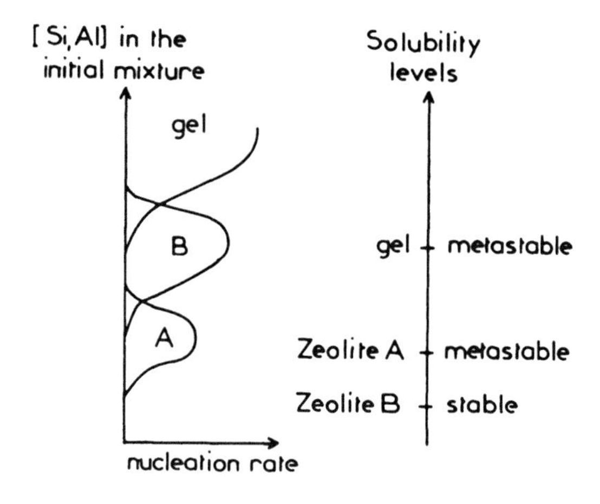

Fig. 1.13. Nucleation domains and rates as a function of the [Si, Al] species concentration for the gel and for two zeolites B and A. Comparison with the solubility levels

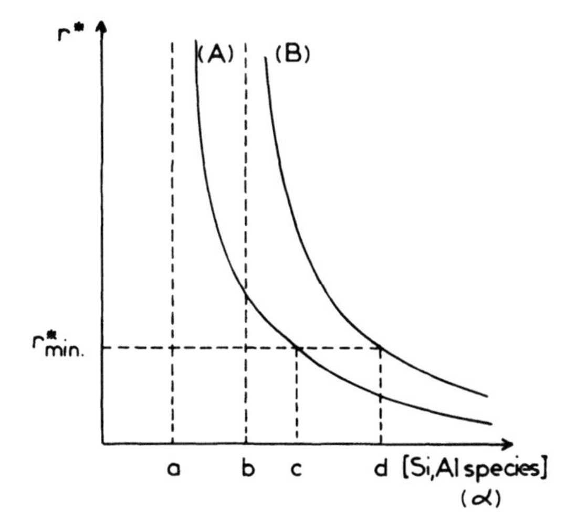

Fig. 1.14. Variation of the critical nucleation radius r* as a function of [Si] of two zeolites A and B with solubilities a and b, but with the same minimal critical nucleation radius r^*_{min} ([Si] > d: no crystallization; b < [Si] < d: nucleation and growth of B; a < [Si] < c: nucleation and growth of A)

Two remarks can be made about Ostwald's rule:

- Ostwald's rule applies in most cases to systems leading to rather complex structures from rather complex media. Thus, no metastable phases appear during the crystallization of NaCl.
- The metastable phases exhibit generally higher "simplexity" than the stable ones [109].

A hypothesis based on a minimal critical radius (r^*_{min}) of the nucleus (see Sect. 1.3.1.1) can be proposed as an explanation of Ostwald's rule [9]. From the curves (Fig. 1.14) representing the variation of the critical radius r* allowing nucleation as a function of [Si, Al species] for two zeolites A and B with different solubilities but with the same r^*_{min}, it can be shown that if [Si, Al species] > d, neither A nor B crystallize: the r* value of the nuclei which are formed is below r^*_{min} needed for the two zeolites. In this case, a phase with a lower r^*_{min} (a more metastable phase or with higher "simplexity", i.e., a gel) will appear. Between c and d, only zeolite B can crystallize and below c, A is probably the only phase which will form, the supersaturation for B nucleation being too low.

Other situations are possible. Thus, two solids having the same solubility (same stability) but with different r^*_{min} for instance due to a difference in the complexity of the structure may crystallize separately (or together), according to the supersaturation domain of the solution.

1.3.2
Zeolite Synthesis Mechanism and Chemistry

1.3.2.1
Presentation of the Synthesis System

1.3.2.1.1
Solubility and Crystallization from the Solution

The framework of each zeolite crystal is an iono-covalent macromolecule which is sparingly soluble in aqueous solutions like those used for its synthesis. For instance, solubility values [9, 12] obtained at 60 °C in 0.5 mol l^{-1} NaOH solution are for:

Zeolite A	$Na_{12}[AlSiO_4]_{12} \cdot 27\,H_2O$	0.43 g l^{-1} SiO_2 and 0.37 g l^{-1} Al_2O_3
Analcite	$Na_{16}[AlSi_2O_4]_{16} \cdot 16\,H_2O$	0.14 g l^{-1} SiO_2 and 0.06 g l^{-1} Al_2O_3

It is possible to crystallize directly zeolites from such solutions if their supersaturation is appropriately adjusted [9, 13, 14]. However, only small amounts are collected because of the dilution of the solution, the narrow allowed concentration range and a pH increase due to hydrolysis of the soluble species during their condensation onto the crystal.

1.3.2.1.2
The Role of the Gel and Synthesis Mechanism

When the different constituents of the framework are mixed in their soluble form (e.g., concentrated silicate and aluminate solutions) no crystalline material generally precipitates. Too high supersaturation does not allow the formation of a complex organized crystal. The rapid polycondensation is too anarchic, and the radius of the nucleus is too small for the development of a crystal: an amorphous solid forming a gel with the aqueous phase appears. The solubility of this gel is, however, high enough to give a solution with a supersaturation adapted to the crystallization of a zeolite if the other required species such as the templates are available in the solution. According to the supersaturation level, metastable and/or stable zeolites may crystallize (Fig. 1.13). As the framework-forming species in the solution are bound to the crystals, they are replaced by new ones resulting from the dissolution of the gel and this until its disappearance. This mechanism which is presently accepted [15–18], is illustrated in Fig. 1.15. Numerous experimental facts which are in agreement with such a mechanism are listed in [6].

However, besides its role as a reservoir for the reactants, the gel is also a regularizer of the concentration levels in the solution. If this function is no longer ensured (e.g., modification of the solubility of the gel, dissolution rate lower than the crystallization rate) other zeolites, which are less soluble or which possess other compositions, may appear. Finally the first formed zeolite can also play the role of a reactants reservoir and dissolve in favor of a more stable one.

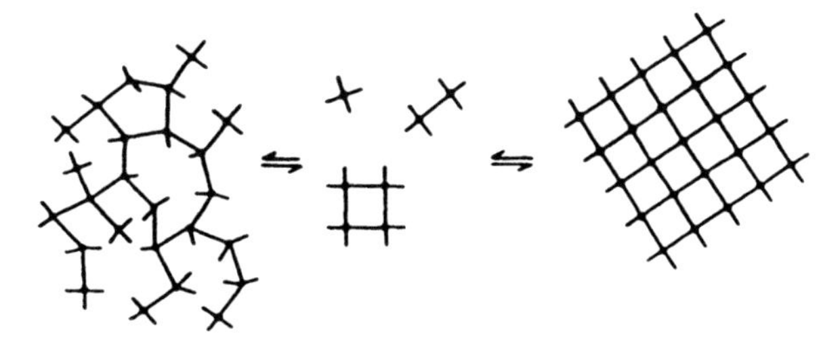

Fig. 1.15. Schematic representation of zeolite formation through the dissolution of the gel and solution-mediated crystallization

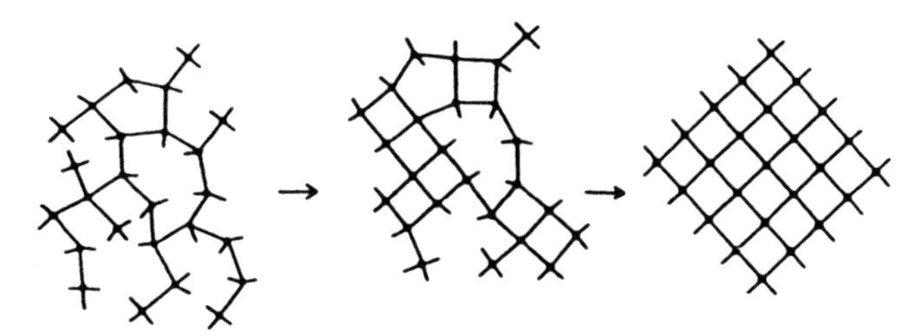

Fig. 1.16. Schematic representation of zeolite formation by "*in situ*" rearrangement of the gel

It has been suggested by some authors [19, 20] that the gel could transform "*in situ*" into crystalline zeolites by solid–solid reactions (Fig. 1.16). But Barrer [2] has noted that even if the crystal nucleated in the solid phase of the gel, its lower density relative to the crystalline phase would give rise to gaps between the growing crystals and the gel, as the latter is incorporated into the crystals. The remaining gel would need to dissolve to reach the crystal surface.

Nuclei of zeolite crystals probably appear inside the smallest pores of the gel in which supersaturation will be highest (Fig. 1.17) but a decondensation of the constituents of the gel with a passage through the solution is necessary. It can be noticed that the presence of a very small amount of liquid phase (e.g., an adsorbed layer) may be sufficient [21]. Crystalline solids are also used as a source of framework-forming elements, and their transformation shows clearly a passage through the solution.

1.3.2.1.3
The Synthesis System

The synthesis system involves three main constituents:

- the source of the framework T elements (generally in the solid state),
- the template(s) (mostly in the liquid phase) and
- the mobilizer or mineralizer (mostly in the liquid phase),

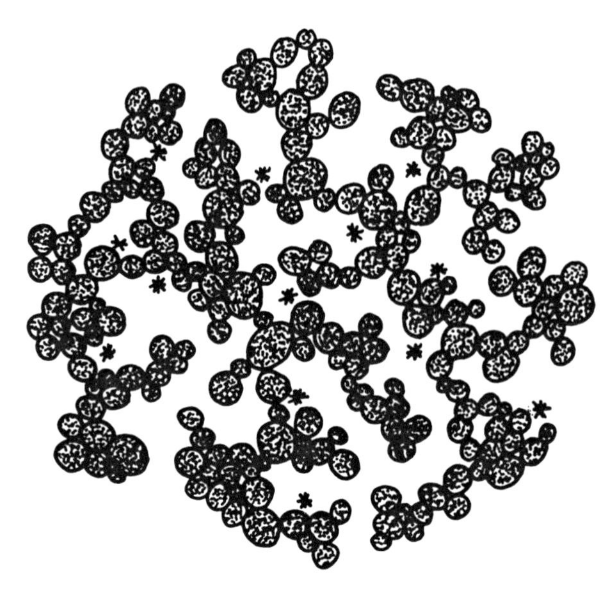

Fig. 1.17. Schematic representation of a gel particle formed by the concatenation of micelles with different sizes. The zeolite nuclei (*) appear in the pores in which the supersaturation of the solution is higher than in the outer solution

in a liquid phase which is also the solvent. In practice, water is the most common solvent but mixtures with other solvents or non-aqueous solvents have also been proposed.

1.3.2.2
Framework T Elements

1.3.2.2.1
Selection Rules

The selection of the T elements with their oxidation state, which are compatible with a framework of linked TO_4 tetrahedra, can be based on several criteria, some of which are empiric. Pauling's first rule (tetrahedral coordination of T^{n+} if $0.225 < R_{T^{n+}}/R_{O^{2-}} < 0.414$), which is often used, applies normally to ionic bonds in coordination polyhedra and structures with maximum compactness. This is not quite the case for zeolites and related oxide materials, for which the T–O bond has a mixed iono-covalent character. Indeed, the compromise between a directed and rigid covalent bond and a non-directed flexible ionic bond makes the building of complex frameworks from tetrahedra with common oxygen atoms easier. Furthermore, elements with high or low oxidation states are not easily incorporated into such frameworks, one of the reasons probably being the high formal charge (n) which would result on the "TO_2" units and which would be difficult to compensate. But other reasons, related to the composition (e.g., absence of polycondensable OH groups) and to the stability (e.g., no need for better "screening" of the T element [22] in the form of monomeric species) may be invoked to explain their non-incorporation. Thus, ClO_4^- or ClO_3OH are more stable as monomeric species in the solution than in a framework in the form of "ClO_2^{3+}" units. At the opposite, Li^+, which can accept tetrahedral coordination with oxygen, is present in an aqueous solution in the form of a hydrated cation, the low need for screening being ensured by the water molecules. Generally, an element will be incorporated into the frameworks in the form of "TO_2" units if this ensures a better "screening" of the elements than in the species present in the solution. As an illustration, solutions containing $Si(OH)_4$ species transform more easily into silica than those containing highly ionized silicate species such as $SiO_3(OH)^{3-}$. These considerations are useful in order to define the synthesis medium characteristics (concentration of the species, nature of the ligand, pH etc.) for a given element.

Another empiric rule, also related to the nature of the elements in the framework, concerns its overall chemical composition which must be such that the mean charge reduced to one TO_2 unit lies between -1 and 0. Positively or too highly negatively charged frameworks are not known.

There are some examples of zeolite frameworks which accept elements with a coordination state higher or lower than 4. But in these cases the additional ligands are not used to establish supplementary linkages to the framework. This occurs mainly with elements which can exhibit another coordination state besides the tetahedral coordination (e.g., B^{3+}, Al^{3+}, Ga^{3+}) or elements which are not normally in a fourfold coordination (e.g., Ti^{4+} for a limited substitution degree).

Materials with elements which are not tetrahedrally coordinated to the framework oxygens form a new class of crystalline microporous solids (see Sect. 1.5.1).

In summary, incorporation of T elements is favored if the following conditions are fulfilled:

- $R_{T^{n+}}/R_{O^{2-}}$ is between 0.225 and 0.414 or near these limits,
- electronegativity allows a balanced iono-covalent bonding with oxygen,
- the oxidation state is between +2 and +5,
- an improvement of the "screening" by polycondensation and "TO_2" units formation and
- a resulting framework with a mean charge/tetrahedron value between -1 and 0.

1.3.2.2.2
Elements Associated with Si in the Framework

Elements which are found in zeolite materials along with their oxidation states are listed in Table 1.1. Elements for which the localization in the framework is not completely proven or which exhibit non-tetrahedral coordination or which substitute for silicon only to a small extent are given in parentheses.

Table 1.1. Elements, with Their Oxidation state, Which Can Substitute for Silicon in the Tetrahedral Framework (Elements for which localization, coordination or oxidation states are uncertain are in parentheses)

Oxidation state	II	III	IV	V
Elements	Be	Al	Ge	P
	(Co)	B	Sn	(V)
	Cu	(Cr)	Ti	
	(Fe)	Fe	(V)	
	Zn	Ga	(Zr)	

1.3.2.3
Mineralizer and T Element Species in the Solution

1.3.2.3.1
Definition of a Mineralizer

A mineralizer is a chemical species which makes possible the formation of a more stable solid phase from a less stable solid phase, through a dissolution-precipitation or crystallization process. Thus, it allows the formation of a supersaturated solution, generally from a gel. In the case of zeolite synthesis mixtures, the solution contains, besides the polycondensable (useful) silicate and aluminosilicate species, other species which are in equilibrium with the former. One of the roles of the mineralizer is to increase the overall solubility, i.e., the concentration of all the species. It may be assumed that the mineralizer not only increases the dissolution (decondensation) and condensation rates but also the conversion rates between the different species in the solution. It thus maintains the availability of the useful species at the level needed for nucleation and crystal growth. Like a catalyst, the mineralizer is normally not consumed during the transformation: the mineralizer used during the dissolution to form soluble species is recovered during the crystallization. A complete discussion of the role of the mineralizer can be found in [22a].

The mineralizer also has an effect on the nature of the different species which are in the solution and, for instance, an increase of its concentration can cause as a consequence a change in the resulting zeolite composition and a decrease of the concentration of the polycondensable species which may lead to the formation of a more stable phase (e.g., direct formation of hydroxysodalite in place of zeolite A when the OH^- concentration is too high).

1.3.2.3.2
Solution Chemistry with OH^- as Mineralizer

In the presence of OH^- oxygenated species of the T elements are formed in aqueous solution. But the OH^- concentration needed has to be higher if the oxides of the T elements are less acidic. The main T elements in zeolites, i.e., Si and Al, are present as silicate and aluminosilicate anions in alkaline solutions [23, 24].

OH^- increases the solubility of silica by ionizing the silanol groups and breaking siloxane bonds

$$\equiv SiOH + OH^- \rightarrow \equiv SiO^- + H_2O$$
$$\equiv Si{-}O{-}Si\equiv + OH^- \rightarrow \equiv SiO^- + HO{-}Si\equiv$$

Thus, the monomeric silicic acid $Si(OH)_4$, which is sparingly soluble at pH 7 ($\cong 2 \cdot 10^{-3}$ mol l^{-1}) forms highly soluble silicate anions at high pH, but with fewer silanol groups

$$Si(OH)_4 + nOH^- \rightarrow Si(OH)_{4-n}\,O_n^{n-} + nH_2O \; (n = 1-4)$$

These anions are in equilibrium with oligomeric anions which have chain, cyclic or polyhedric structures. High concentrations and not too akaline pH values favor these oligomeric species. The presence of organic cations such as Me_4N^+ or Pr_4N^+ modifies the distribution of these species and may lead to the formation of specific species (e.g., double four- or five-membered rings).

OH^- solubilizes alumina as $Al(OH)_4^-$ anions which do not undergo further ionization or oligomerization.

The condensation of the silicate and aluminate anions to form aluminosilicate anions and then the zeolite framework needs the presence of hydroxide groups. When the concentration of OH^- increases, the silicate species will be less and less able to condense (the $\equiv SiO^-/\equiv SiOH$ ratio increases), the ability of aluminate anions to condense remaining constant. This fact has several consequences:

- alumina-rich zeolites crystallize preferentially at a higher pH, a lower pH favoring silica-rich zeolites;
- during crystallization from a gel, the pH increases generally (OH^- is liberated by the transformation of non-bridged $\equiv SiO^-$ groups in the gel into bridged $\equiv Si{-}O{-}Si\equiv$ groups in the zeolite), thus Si/Al in the crystals may decrease from the core to the outer shell;
- at high pH, the low concentration of polycondensable silicate species (lower supersaturation) may favor the formation of a stable and dense phase;
- a high $\equiv SiO^-/\equiv SiOH$ ratio on the silicate and aluminosilicate anions prevents complete $\equiv Si{-}O{-}Si\equiv$ bridging in the framework ($\equiv SiO^-M^+$ defects are formed).

1.3.2.3.2
Solution Chemistry with F^- as Mineralizer

F^- becomes "active" generally at pH lower than 11 – 12. It increases the solubility of silica and alumina by substituting for OH^-. This provokes the transition from tetrahedral to octahedral coordination, i.e.:

$$Si(OH)_4 + 6\,F^- \rightarrow SiF_6^{2-} + 4\,OH^-$$

When the pH becomes acidic, H_2O can also act as a ligand to Si and Al.

As for OH^-, there is an optimum for the F^- concentration, depending on the stability of the fluoro complexes. This can be used to control the incorporation of the substituting T element [25]. The stability of the fluoro complexes TF_6^{n-} of some selected T elements decreases in the following order:

$$Al^{III} > Fe^{III} > Ga^{III} \geq Si^{IV}$$

Thus the Si/Al ratio in a MFI-type zeolite increases with the F/(Si+Al) ratio in the reaction mixture. For a small substitution degree of a T^{III} element, such as Fe^{III}, the concentration gradient of T^{III} in the crystal can be strongly reduced with high F^- concentration or if Si^{IV} and T^{III} are combined in the same reactive solid source. The contrary can also be obtained (T^{III}-rich core and pure Si^{IV} outer shell in the crystals) with low F^- concentrations or with separated Si^{IV} and T^{III} sources (e.g., silica gel and ferric chloride solution). When several trivalent elements are competing during the crystallization of the same zeolite, the elements whose fluoro complexes are the least stable are incorporated first (e.g., Ga^{III} in the core and Al^{III} in the outer shell of the crystals) [25, 26].

Other differences may be noticed when F^- is used in place of OH^-:

- according to the stability of the different fluoro complexes it may be concluded that F^- is a mineralizer suitable for the synthesis of silica-rich zeolites;
- F^- allows the direct incorporation of cations which are sparingly soluble (e.g., Co^{2+}) or unstable (e.g., NH_4^+) at high pH;
- by working at a pH below 10–11, non-bridging $\equiv SiO^-$ defects are avoided;
- organic templates such as quaternary ammonium ions are generally more stable in a neutral medium than in an alkaline medium;
- the supersaturation obtained for the framework-forming species is generally lower. This allows a better control of the crystallization but needs longer crystallization times and optimized templates;
- on account of the neutral or slightly acid pH, framework elements such as Al^{III} may also be incorporated as extra-framework species (e.g., compensating cations) during synthesis.

1.3.2.4
Templates

The incorporation of a template during the synthesis of the solid:

- contributes, by the bonds which are set up, to the stability of the solid and allows its formation and
- controls, by a templating effect, the formation of a potentially microporous framework structure.

The removal of the template after synthesis provides the microporous voids (channels and cavities).

According to the nature of the T elements the charge density of the framework and the shape of the microporous spaces, different stabilizing interactions can be produced by template–template bonds and/or template–framework bonds (cation-anion, dipole-ion, hydrogen and/or van der Waals bonds).

Thus the type of the inorganic or organic template species, viz.

- framework-charge compensating cations, e.g., Na^+, Ca^{2+}, alkylammonium,
- molecules, e.g., H_2O, amines, alcohols or
- ion pairs, e.g., NaCl, Pr_4NF

will also depend on the same parameters, and the selection of a template has to take into account these parameters in order to provide the best stabilizing interactions.

For silica-based zeolites one may distinguish two cases according to the degree of substitution of the T^{III} elements (T = Al, Ga etc.).

- $Si/T^{III} < 5-10$: the high framework negative charge density is best compensated by small cations (alkali, alkaline-earth, $(CH_3)_4N$ etc.) which are more or less hydrated according to the available space and the charges of the cations;
- $Si/T^{III} > 5-10$: the low framework negative charge density is best neutralized by large organic cations constituted by many atoms providing van der Waals interactions which compensate the deficiency of stabilization due to the small number of Coulombic interactions. Even in the absence of trivalent T elements, negative charges can be present on non-bridging groups, such as $\equiv SiO^-$, if the material has been prepared at high pH. Templates are then present in the form of ion pairs or molecules only when the framework bears no excess negative charge (zero-charged or autocompensated frameworks) or as void fillers between the charge-compensating cations.

The example of faujasite synthesis, which is made possible by the stabilizing interactions between the hydrated sodium cation and the negative framework charges, may be discussed to illustrate these considerations. The synthesis of zeolite Y is more difficult than that of zeolite X: the decrease in the number of Al in the framework needs more reactive silica and alumina sources in order to compensate for the loss of stability due to the smaller number of sodium cations. A higher Si/Al ratio than that of zeolite Y could be achieved by using, as templates, sodium cations complexed by large crown ether molecules allowing van der Waals interactions in addition to the smaller number of Coulombic interactions. Moreover, the use of a crown ether (18-crown-6) adapted to a hexagonal symmetry led to the formation of the hexagonal polymorph of cubic faujasite [26].

1.4 Experimental Part

1.4.1 Experimental Factors

Despite its apparent simplicity, zeolite synthesis is a difficult art, chiefly if one is looking for a material with a specific structure and composition and not contaminated by amorphous or crystalline impurities. All the experimental factors must be taken into account [27], through a mode of procedure whose description must be as complete, detailed and precise as possible, following the recommendations of the Synthesis Commission of the International Zeolite Association (IZA). Some synthesis examples are commented on in Sect. 1.4.2 and 1.4.3.

Nevertheless, owing to the numerous factors, which can be varied, it is possible to obtain a very large number of zeolites differing in structure and composition but also in crystal size and morphology.

1.4.1.1
Nature of the Reactants

1.4.1.1.1
The Solvent

Until now water remains the most widely used solvent in the reaction mixture. Its properties are well suited for the dissolution, with the help of the mineralizer, of all species needed for the crystallization and the thermal conditions of the transformation. Moreover, in numerous applications water also plays the role of a template, often in association with other templating species (e.g., charge-compensating cations). Non-toxicity, low cost, good thermal stability and conductivity are other qualities of water. There are only a few cases where the use of another solvent has led to original results justifying the replacement of water (e.g., use of ethylene glycol for the synthesis of silicasodalite).

1.4.1.1.2
Sources of the Framework T Elements

The T elements are generally present in the reaction mixture as amorphous hydroxides, hydrous oxides or related solids such as alkaline aluminosilicate. These solids may be introduced with different textures (e.g., precipitated gels, ground glasses, volcanic ashes, colloidal suspensions or fumed silicas), and primary reactants may be used to prepare such solids (e.g., alkaline silicate solution, halides, fluoro complexes or alkoxides for Si; aluminate solution or various salt solutions for Al). Different T elements may occur in the same source (e.g., calcined clays for Al and Si). In some cases crystalline solids such as natural zeolites can be recrystallized into other zeolites depending on the templates present in the mixture. The choice of the reactants of the T elements is critical, since reactive sources are necessary to obtain the most metastable zeolites.

1.4.1.1.3
Mineralizer

OH^-, the most common mineralizer for silica- and aluminosilica-based zeolites is often present in the source of the T elements (e.g., hydrolized silicate solution). When this is not the case, or when its concentration is not high enough, adjustment is made with the inorganic or organic base which is also used as template.

According to the pH required for the reaction mixture, fluoride salts or acids are added as sources of the F^- mineralizer. It may also be combined in the T element reactants and liberated through hydrolysis [e.g., $(NH_4)_2SiF_6$; $AlF_3 \cdot H_2O$].

1.4.1.1.4
Templates

Besides water, the inorganic templates are cations and/or anions of bases (e.g., LiOH, NaOH), salts (e.g., NH_4F, NaCl) or acids (e.g., HF) which are added or completed as such when they are not already present in the sources of the T element (e.g., Na^+ in silicate or aluminate solutions used to prepare the gel).

Organic templates, chosen from among the large variety of alkyl- or arylammonium hydroxides or salts, amines, alcohols etc. are directly introduced into the reaction mixture.

1.4.1.2
Composition of the Reaction Mixture

This is, together with the nature of the reactants, the second important factor which will determine the nature of the zeolite formed. But generally, a given zeolite crystallizes not only from one single composition: there exists a composition domain inside which only more or less important variations are observed for the composition of the zeolite, its structure remaining the same (Fig. 1.18).

The overall reaction mixture composition is often calculated in the form of molar ratios of the constituents expressed as oxides [e.g., SiO_2/Al_2O_3; Na_2O/Al_2O_3; $(Me_4N)_2O/Al_2O_3$; H_2O/Al_2O_3 etc.]. However it is sometimes more convenient to formulate the composition by quoting the exact nature of the reactants (e.g., a SiO_2; b $Al(OH)_3$; c NaOH; d Me_4NCl; e H_2O) and indicating the starting molar fraction with the pH of the mixture.

1.4.1.3
Preparation Procedure of the Reaction Mixture

The preparation conditions will strongly influence the dissolution rates of the gels, the nature and concentration of the species in the liquid phase and hence

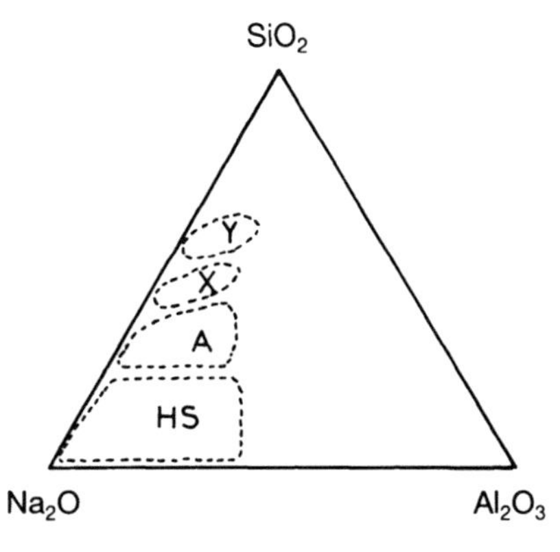

Fig. 1.18. Formation domains of zeolites Y, X, A and HS in the ternary Na_2O–SiO_2–Al_2O_3 diagram

the result of the synthesis. These conditions include the physical state of the reactants, the way and the order of introducing them, the homogenization of the mixture, etc. It is the lack of control of this type of factors which is often responsible for the non-reproducibility or for the difficulties which arise in the scale-up step. Indeed, these factors have an effect on the texture and composition homogeneity through the entire volume of the mixture and are thus responsible for the constancy of the solution characteristics during the crystallization.

1.4.1.4
Aging

It is sometimes necessary to "age" the reaction mixture during a given time at a temperature below the crystallization temperature and generally near room temperature. During this period a chemical and structural reorganization occurs which affects both the solid and the liquid phases. For instance, with the same reaction mixture, depending on whether an aging period is applied or not, zeolite Y or zeolite P will be produced. It can be assumed that during the aging, the supersaturation in the solution rises enough to allow the crystallization of the more metastable zeolite Y when the mixture is heated.

1.4.1.5
Seeding

Addition of seeds of the desired zeolite (crushed crystals or "nucleation solutions") allows, through a reduction of the crystallization time, the formation of a pure metastable phase competing with more stable phases (kinetic aspect). Sometimes, however, it is the only way to obtain a metastable zeolite in a reaction mixture with a supersaturation degree high enough for crystal growth but not for nucleation (thermodynamic aspect). Moreover, the crystal size may be adjusted by varying the amount of added seeds.

1.4.1.6
Nature of the Reactor

The nucleation rate can be modified by the walls of the reactor, in particular if it has been previously used for synthesis (increasing heterogeneous nucleation due to the roughness, seeding due to the presence of small crystal particles). A non-controlled pollution can also arise from a corrosion or a dissolution of the walls. Thus, synthesis performed in glass containers will be all the more polluted when the temperature is high and the pH alkaline. PTFE walls are at present the best solution to such problems when the synthesis temperature is below 250 °C, but their low thermal conductivity may be a disadvantage when high heating rates are needed. The shape and volume of the reactor have to be taken into consideration for a good temperature homogeneity of the reaction mass, especially in the absence of agitation.

For vapor pressures below atmospheric pressure, open reactors are normally employed, but care should then be taken to (i) avoid the reaction of the alkaline solution with CO_2 from the air resulting in the formation of CO_3^{2-} or HCO_3^- and the concomitant lowering of the pH and (ii) a loss of solvent. When the pressure is higher than 1 bar, the crystallization is carried out in a closed reactor (autoclave) but the filling must be below a given limit which depends on the nature of the solvent and the temperature. A pressure increase due to the decomposition of the organic template has to be controlled.

1.4.1.7
Crystallization Temperature

The crystallization temperature domain of zeolites ranges between room temperature and about 300 °C. But according to the zeolite, the temperature domain will be more or less broad and located at a more or less high temperature. The most "open" zeolites are preferably obtained at lower temperature, especially if water is a template.

The rate of the temperature increase is sometimes critical. For example, in the synthesis of silicalite-1 at 170°C and 20 h heating, the latter may be obtained when 1 °C min^{-1} heating rate is used, whereas magadiite is formed when the heating rate is 10°C min^{-1}.

1.4.1.8
Pressure

Crystallization is generally carried out under autogenous pressure. The influence of the pressure has been little studied except in the case of natural zeolite formation where it has been shown that a strong pressure increase favors the densest structures.

1.4.1.9
Agitation

In order to get and maintain a good temperature homogeneity, a synthesis under agitation is preferred. However, technical problems are not always easy to resolve (working in a closed reactor) and in some cases agitation is not recommended. Indeed, there are zeolites which form only under static conditions, other products crystallizing under stirring. This fact can probably be related to a decrease of the high supersaturation inside the small pores of the gel particles due to the movement of the liquid phase which prevents the nucleation of the metastable phase in these small pores.

1.4.1.10
Heating Time

According to the desired zeolite and the chosen operating conditions (e.g., temperature), the crystallization lasts from a few minutes to several months. When the synthesis medium is allowed to evolve, other phases will frequently succeed from the less stable to the most stable, according to Ostwald's rule (Fig. 1.19).

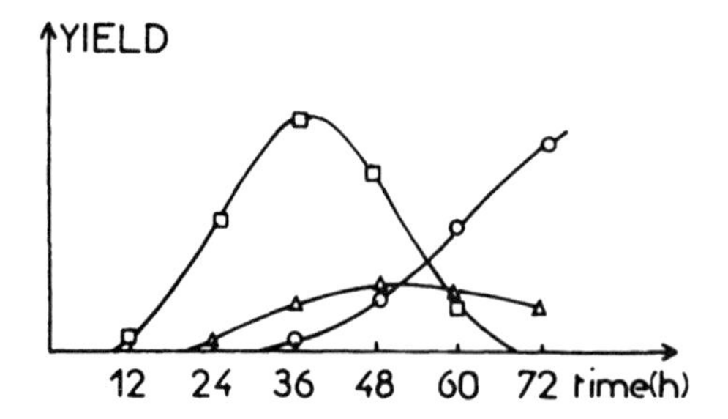

Fig. 1.19. Evolution of the yield of zeolites Omega (□), sodalite (Δ) and analcite + cristobalite (○) in dependence of time in the mixture $5\,Na_2O$; $0.11\,(TMA)_2O$; $1\,Al_2O_3$; $13.7\,SiO_2$; $220\,H_2O$ at 150 °C

This is a further difficulty in obtaining pure phases, since a second or a third phase may form before the first has finished crystallizing.

1.4.2
Review of the Zeolites Obtained from Various Reaction Systems

1.4.2.1
All-Silica Molecular Sieves (T = Si)

Several all-silica molecular sieves, which are generally hydrophobic in their calcined form, have already been reported; they were all prepared in the presence of an organic template (Table 1.2). All, except silicasodalite and silica-AFI, which is the silica form of $AlPO_4$-5, correspond to high-silica zeolite structures, i.e., zeolites for which Si/Al is larger than about 5. In Table 1.2, an example of an organic template is given for each structure type. Some of the materials, such as silica-ferrierite, silicalites-1 and -2, silica-ZSM-22 and -48 can be prepared from several organics; the others have been obtained so far from one single specific template. The fluoride route has also been successful for the synthesis of the siliceous end-members of ZSM-5 [33, 34], ZSM-11 [36] and ZSM-48 [36].

It can be expected that more silica forms of other porous tectosilicate structure types are likely to be obtained, perhaps by finding new templates with efficient host–guest interactions.

1.4.2.2
(Si, Al) Systems with Inorganic Cations

An extensive list of about 30 synthetic zeolites, some of which are counterparts of natural zeolites, and the preferred inorganic cations (alkali or alkaline-earth) for their synthesis, have been given by Barrer [2, 40].

Table 1.2. All-Silica Molecular Sieves Obtained To Date by Direct Synthesis

Material	Example of organic template	Ref.
silica-AFI (SSZ-24)	*N,N,N*-trimethyl-1-adamantammonium	[28]
silica-Beta	dibenzyldimethylammonium	[29a]
silica-ferrierite	n-propylamine + pyridine + HF	[30]
silica-sodalite	ethylene glycol or trioxane	[31a,b]
silica – UTD – 1	bis(pentamenthylcyclopentadienyl)cobalt(III) hydroxide	[31c]
silica-ZSM-5 (silicalite-1)	tetrapropylammonium	[32–34]
silica-ZSM-11 (silicalite-2)	tetrabutylammonium	[35,36]
silica-ZSM-12	4,4′-trimethylenedipiperidine	[37]
silica-ZSM-22	diethylamine	[38]
silica-ZSM-23	di-*n*-propylamine	[39]
silica-ZSM-48	1,3-diaminopropane	[29b]

1.4.2.3 (Si, Al) Systems with Inorganic and Organic Templates

A list of zeolite syntheses in the presence of inorganic cations and organic species, essentially quaternary ammonium bases, is given by Barrer [2]. An updated list has been published by Lok et al. [41]. It contains numerous other zeolite structures obtained with tetraalkylammonium bases, tri-, di- and mono-alkylamines, cyclic amines, alkyldiamines, cyclic diamines, polyamines and alcohols. A few novel materials since reported and the corresponding structure-directing species are given in Table 1.3.

1.4.2.4 (Si, Al) Systems with Organic Templates

Only a few (Si, Al)-based zeolites have been prepared in the presence of organic templates alone. Indeed, the presence of an inorganic base in the starting gel is usually required. The advantage of the absence of alkali cations in the as-synthesized material is that a simple calcination yields the H-form of the zeolite directly. No ion exchange or acid treatment is necessary in order to remove the alkali cations. The Al present in the framework will be charge-compensated by the organic cations, therefore the Al content will be dependent on the number of cations that can be accommodated in the voids and hence on their size and their charge. Examples of syntheses with organic templates alone are given in Table 1.4.

Table 1.3. Examples of Recently Reported (Si, Al) Zeolites with the Inorganic and Organic Species Used in Their Synthesis

Zeolite	Inorganic + organic species	Ref.
High-silica faujasite	Na^+ + 15-crown-5	[42]
	$Na^+ + (HOCH_2CH_2)_2Me_2N^+$	[43]
	$Na^+ + Pr_4N^+$ and/or Bu_4N^+	[43]
Hexagonal analogues of faujasite	Na^+ + 18-crown-6	[42]
	$Na^+ + MeEt_3N^+$	[43]
MCM-22	Na^+ + hexamethyleneimine	[44a]
Nu-86	$Na^+ + Me_3N^+(CH_2)_nN^+Me_3$ (n = 8, 9)	[44b]
Nu-87	$Na^+ + Me_3N^+(CH_2)_nN^+Me_3$ (n = 8–12)	[44c]
ZSM-18	Na^+ + tris-quaternary ammonium	[45]
ZSM-57	$Na^+ + Et_3N^+(CH_2)_5N^+Et_3$	[46]
ZSM-58	Na^+ + methyltropinium	[47]

Table 1.4. Examples of (Si, Al) Systems with Organic Templates Alone

Zeolite	Template	Ref.
Beta	DABCO (+ HF) (+ $MeNH_2$)	[48]
ferrierite	$H_2N(CH_2)_nNH_2$, n = 3, 4	[49]
gismondine	Me_4N^+	[50]
sodalite	Me_4N^+	[51]
ZSM-5	Pr_4N^+ (+ F^-)	[25]
	Pr_3NH^+ (+ F^-)	[25]
	Pr_2NH_2+ (+ F^-)	[25]
	$H_2N(CH_2)_nNH_2$, n = 5, 6	[49]
ZSM-11	$H_2N(CH_2)_nNH_2$, n = 7, 10	[49]
ZSM-12	DABCO (+ HF) (+ $MeNH_2$)	[48]
ZSM-22	$CH_3(CH_2)_4NH_2$ (+ HF)	[25]
ZSM-23	d-*n*-propylamine (+ HF)	[25]

1.4.2.5
(Si, T^{II}) Systems, T^{II} = Be, Co, Cu, Zn

Only a few microporous materials with Be, Co, Cu and Zn as framework elements beside Si have been synthesized.

A synthetic analogue [52] of the beryllosilicate mineral lovdarite $K_4Na_{12}Be_8Si_{28}O_{72} \cdot 18\,H_2O$ [53] has been described recently. Beryllium [54] and cobalt [55, 56] have been introduced in very limited amounts into the MFI structure. By chemical analysis, the Si/Be ratio was found to be equal to 30–60, and a tetrahedral coordination for Be was derived from ^{9}Be NMR spectroscopy [54]. The amount of the Co content in Co-ZSM-5 was even smaller with Si/Co ≈ 200 [55]. On the other hand (Cu, Si)-analcime $Cs_2CuSi_5O_{12}$ has been made by high temperature synthesis, and its crystal structure has been refined by the Rietveld technique [57]. Several new zincosilicates have been reported [58a]. For example, a microporous sodium zincosilicate named VPI-7 with a novel structure [58b] was synthesized in the presence of tetraethylammonium as a crystallization support species. The peculiarity of the structure of VPI-7 is that it contains three-membered rings beside larger rings as does lovdarite [53].

1.4.2.6
(Si, T^{III}) Systems, T^{III} = B, Fe, Ga

In Table 1.5 the T^{III}-containing materials (T^{III} = B, Fe, Ga) obtained so far by direct synthesis are given.

The synthesis of the boron, iron and gallium analogues of ZSM-5 has been the most investigated. They were generally obtained by a procedure derived from the one patented by Argauer and Landolt [111] for the synthesis of Al-ZSM-5. The fluoride route was also successfully employed for B-ZSM-5 [112], Fe-ZSM-5 [113] and Ga-ZSM-5 [114, 115]. The Si/T^{III} values are very close to those observed for Si/Al in the MFI-type aluminosilicate.

The evidence for B incorporation into the tetrahedral framework was mostly based on X-ray diffraction data (unit cell variation with B-content), IR and ^{11}B NMR spectroscopy [90, 91]. For iron(III), techniques such as ESR, EXAFS, UV-visible diffuse reflectance and Mössbauer spectroscopy were used to demon-

Table 1.5. Structure Types or Materials Obtained from (Si, T^{III}) Systems. (The numbers in parentheses refer to the literature)

Structure type or material	B	Fe	Ga
ABW			[59-61]
AFI	[62, 63]		
analcime	[64a]		[64b]
Beta	[63, 65]	[66]	[67]
erionite			[68a]
EU-1			[68b]
faujasite			[69-74]
ferrierite	[75]		[76]
L			[77-79]
NU-1	[65, 80]	[80]	[80]
mordenite		[81]	[82]
natrolite			[64]
offretite			[83, 84]
omega			[64]
sodalite	[85]	[86]	[87, 88]
thomsonite			[89]
ZSM-5	[65, 90-93]	[94-99]	[100-103]
ZSM-11	[65, 92]	[104]	
ZSM-12	[65, 105, 106]	[106]	
ZSM-22		[107]	[108]
ZSM-23		[109]	
ZSM-48	[110]		

strate the tetrahedral coordination of the element; for Ga, EXAFS, infrared and Ga NMR spectroscopy were the most frequently employed.

For other structure types, the synthesis route was generally that used for the (Si, Al) analogue by replacing the aluminum source by a boron, iron or gallium source. In the case of iron in particular, care must be taken to avoid the formation of insoluble brown-colored ferric hydroxides on mixing the reactants [104]. The techniques used to study the location of the T^{III} element were essentially the same as for the MFI-type materials.

1.4.2.7
(Si, T^{IV}) Systems, T^{IV} = Ge, Ti, Zr

The substitution of Si by Ge when a trivalent element is present has been reviewed by Barrer [2, 116]. (Ge, Al)-faujasite and (Ge, Al)-phillipsite corresponding to a total replacement of Si by Ge were obtained. The partial replacement of Si in the MFI-type strcture, i.e., in the absence of a trivalent element, was reported by Gabelica and Guth [117]. A substitution degree Ge/(Ge + Si) as high as 0.33 could be reached.

Two types of microporous titanosilicates where the transition element is, respectively, in TiO_6 and TiO_4 groups are known. The first type comprising a mixed tetrahedral-octahedral framework is discussed in Sect. 1.5. Concerning the tetrahedral frameworks, Ti incorporation into the structure-types *BEA [118] and MFI [119–123] has been reported. Ti substitution in titanosilicalite-1 (TS-1) has been shown by techniques such as X-ray diffraction and infrared spectroscopy [120, 123] and more recently by EXAFS [124]. The substitution degree of Ti turns out to be small, i.e., $Ti/(Si + Ti) \cong 2.5$ mol%.

The synthesis of Zr-silicalite-1 has been reported [125, 126]. The evidence for partial isomorphous substitution of Zr for Si was based on powder X-ray diffraction data (unit cell volume increase).

1.4.2.8
Other Si-Based Systems – V, Cr and Mo in Tetrahedral Frameworks

The direct synthesis of microporous metallosilicates based on vanadium, chromium and molybdenum and corresponding to high-silica structures has been reported. The amount of element present in the materials is generally very small, therefore, it is difficult to get sound evidence for its substitution into the zeolite framework.

A vanadium silicate TRS-48 has been claimed by Taramasso et al. [127] in a patent. More recently, Kornatowski et al. [128] have described the synthesis of the MFI-type vanadium silicate KVS-5. The bulk analysis showed that approximately 1 wt% vanadium was present in the solid. The ESR results were consistent with V^{4+} in the as-synthesized form and V^{5+} in the calcined form. The synthesis of V-ZSM-48 was also reported [129]. Here too, it was concluded that vanadium was in the +4 state in the as-synthesized material.

A microporous chromosilicate AMS-1 Cr has been claimed by Klotz [130], and Rao et al. [131] published the synthesis of Cr(III)-silicalite-1. Cr-ZSM-48 has been made by Pang et al. [129] in the presence of diaminohexane as a template. Chromium was found to be in the + 3 state in the framework.

As to molybdenum substitution, Pang et al. [132] have recently published a study on the factors affecting the synthesis of Mo-ZSM-5. Molybdenum was found to be present in the framework.

1.4.3 Synthesis of Some Selected Important Zeolites

1.4.3.1 *Zeolites with the LTA-Type Structure*

The synthesis of zeolite A is very well documented. The crystallization which is rapid can be performed in the temperature range from 20° to 175 °C [1], 100 °C being particularly advantageous [133]. Zeolite A is usually prepared in the Na form (Na-A), Na^+ being the preferred cation in the synthesis. In a typical preparation, a reaction mixture with the composition

$$5.03\,Na_2O:Al_2O_3:1.35\,SiO_2:229.6\,H_2O$$

was heated at 86 °C. Pure zeolite A was formed after 2 h [134]. Zeolite A has the overall composition $Na_2O, Al_2O_3, 2\,SiO_2 \cdot 4.5\,H_2O$ [1], the corresponding unit cell formula (per pseudo cell) being $Na_{12}Al_{12}Si_{12}O_{48} \cdot 27\,H_2O$ [5].

On prolonged heating, it may convert to the more stable sodalite, all the more if the alkalinity of the mixture is high. If zeolite A is left in the mother liquor it may transform slowly into zeolite P [1].

The direct synthesis of the Na, K [134–136] and Na, Ca [137] forms of zeolite A, which is of interest for special purposes, is also possible. A subsequent ion exchange of Na^+ by K^+ and Ca^{2+} may thus be spared.

Table 1.6. FAU-Type and Related Materials: Structure-Directing Species and Si/Al Ranges in the Solids

Material	Structure-directing species	Si/Al	Ref.
CSZ-1	$Na^+ + Cs^+$	1.5–3.5	[152]
CSZ-3	$Na^+ + Cs^+$	2.5–3.5	[147]
ECR-4	$Na^+ + (HOEt)_2Me_2N^+$	2.5–4	[144]
ECR-30	$Na^+ + Et_3MeN^+$	3–5.14	[151]
ECR-32	$Na^+ + Pr_4N^+$	4.3–5.6	[145]
EMC-1	Na^+ + 15-crown-5	3–5	[146]
EMC-2	Na^+ + 18-crown-6	3–5	[146]
VPI-6	Na^+	1–1.5	[149]
X	Na^+	1–1.5	[142]
Y	Na^+	1.5–3	[143]
ZSM-3	$Na^+ + Li^+$	1.4–2.25	[148]
ZSM-20	$Na^+ + Et_4N^+$	3.5–5	[150]

The Si/Al ratio of zeolite A can be increased to above 1 by using gels of high silica and sodium hydroxide content [138] or by adding Me_4N^+ cations to the starting mixture in place of part of the Na^+. A Si/Al ratio equal to 1.24 was reached by the first procedure [138]. ZK-4 with Si/Al up to about 3 was obtained with $Me_4N^+ + Na^+$ [139, 140]. In such a silica-rich material, essentially all sodalite cages contain a Me_4N^+ ion [141]. ZK-4 is isotypic with the LTA-type materials N-A and α [5].

1.4.3.2
Zeolites with FAU-Type Structures and Polytypes

Several materials comprising a cubic and hexagonal packing of faujasitic sheets are known. Pure cubic (FAU), quasi-pure hexagonal (EMT) and intergrowth phases have been reported. The FAU-type phases are zeolites X [142], Y [143], ECR-4 [144], ECR-32 [145], EMC-1 [146] and CSZ-3 [147]. The cubic-hexagonal intergrowth materials are ZSM-3 [148], VPI-6 [149], ZSM-20 [150] and ECR-30 [151]. EMC-2 [146] is the quasi-pure hexagonal analogue of faujasite. For the additional material CSZ-1, several descriptions of the crystal structure have been given. Treacy et al. [152] proposed a rhombohedrally distorted faujasite structure, whereas Martens et al. [153] had evidence from the n-decane cracking test that CSZ-1 is an intergrowth of the FAU and EMT structures.

The structure-directing species used in the synthesis of these materials are reported in Table 1.6 together with the Si/Al ranges in the solids.

1.4.3.2.1
Synthesis of Zeolites X and Y

Zeolites X and Y can be obtained from starting mixtures differing essentially in the Si/Al ratio (larger for Y than for X) in the temperature ranges 20–120 °C and 20–175 °C [1]. A crystallization temperature around 100 °C is preferred.

For example [142b], zeolite X was prepared by mixing sodium aluminate, sodium silicate, sodium hydroxide and water at room temperature. The mixture with a composition

$$4.2\,Na_2O : Al_2O_3 : 3\,SiO_2 : 180.6\,H_2O$$

was aged at room temperature for 16 h and then heated at 100 °C for 3 to 8 h, both without stirring.

Zeolite P (GIS type) is formed as an impurity, if the mixture is not aged at room temperature or if it is stirred during the heating at the crystallization temperature.

Zeolite Y was prepared [154] from a mixture with the composition

$$3.4\,Na_2O : Al_2O_3 : 9.5\,SiO_2 : 136\,H_2O.$$

The alumina and silica sources were, respectively, sodium aluminate and colloidal silica. The mixture was aged at 20 °C for 24 h and heated without stirring at 95 °C for 24 h. Pure zeolite Y with Si/Al = 2.1 was obtained.

As for zeolite X, zeolite P (GIS type) is formed as a second phase when the reaction mixture is not aged at room temperature or when it is stirred during the crystallization.

1.4.3.2.2
Synthesis of Silica-Rich Cubic FAU-Type Materials

From Wholly Inorganic Media. A possible way to increase the Si/Al ratio of zeolite Y above 3 from wholly inorganic mixtures (Na^+ cations) by decreasing the pH to about 11 and improved seeding with a nucleation liquor was discussed by Robson [155].

In the Presence of Na^+ and a Crown Ether. Delprato et al. [146] reported the synthesis of silica-rich faujasite using the crown ether 15-crown-5 as a structure-directing agent. A weakly basic aluminosilicate hydrogel with the composition

$$2.2\,Na_2O:Al_2O_3:10\,SiO_2:1\,(15\text{-crown-}5):140\,H_2O$$

was aged for 24 h at 20°C, and heated for 8 days at 110°C. Pure faujasite with Si/Al = 4.5, higher than in zeolite Y was obtained. It appears that the presence of the template allowed such a low alkalinity and led to an increase of Si/Al.

1.4.3.2.3
Synthesis of the Silica-Rich Hexagonal EMT-Type Material

Delprato et al. [146] also reported the synthesis of a silica-rich hexagonal EMT-type material by using the crown ether 18-crown-6 as a template. A mixture with the composition

$$1.8\,Na_2O:Al_2O_3:10\,SiO_2:1\,(18\text{-crown-}6):140\,H_2O$$

was aged for 24 h at 20°C, and heated at 110°C during 40 days. A well-crystallized pure EMT-type phase with Si/Al = 4.9 was obtained.

1.4.3.2.4
Synthesis of the FAU-EMT Intergrowth Phase ZSM-20

Ernst et al. [156] published an optimized synthetic procedure for ZSM-20. They obtained a pure product with a crystal size of ca. 0.5 μm from a gel composition

$$1.25\,Na_2O:Al_2O_3:22.2\,SiO_2:22.6\,Et_4OH:258\,H_2O$$

after 14 days at 100°C.

Sodium aluminate and a tetraalkylorthosilicate were used as the alumina and silica source, respectively. The Et_4OH solution must be as free of K^+ cations as possible, otherwise zeolite Beta and other phases are obtained. Zeolite Beta also forms, even in the absence of K^+, when the synthesis time is too long or when the crystallization temperature is increased to 120°C for example.

The authors found that the crystallization time can be considerably shortened, either by aging the gel at room temperature for 1 d or seeding with 10 wt% ZSM-20 crystallites, or by replacing tetraethylorthosilicate by tetramethylorthosilicate.

Arhancet and Davis [156a] reported the synthesis of other FAU-EMT intergrowths by using a mixture of 15-crown-5 and 18-crown-6 templates. However, it should be mentioned that with the same mixture of crown ethers, Dougnier et al. [156b] observed an overgrowth of both cubic and hexagonal structures.

1.4.3.3
Synthesis of Zeolite Beta

1.4.3.3.1
Synthesis in Conventional Media

Zeolite Beta is a large pore (12-membered ring apertures), high-silica zeolite (Si/Al from ca. 10 to 100) which was first synthesized in 1967 [157] from alkaline aluminosilicate gels in the presence of sodium and tetraethylammonium cations. A natural counterpart of zeolite Beta, tschernichite, was discovered recently [5]. Many patents have been published since the first synthesis, but in the literature there are only a few studies on the influence of the synthetic conditions on the crystallization of zeolite Beta. Whereas for zeolite ZSM-20, which is also synthesized with Na^+ and Et_4N^+ as structure-directing cations, tetramethyl- or tetraethylorthosilicate is the preferred silica source, various sources were used for the synthesis of zeolite Beta, i.e., tetraethylorthosilicate, colloidal or amorphous silica [158]. However, when the reactivity of the silica source decreased, zeolites ZSM-12 or ZSM-5 formed as well as zeolite Beta [159].

Camblor et al. [160] observed that the addition of K^+ to the synthesis mixture increased the average crystal size of the final product. A typical starting composition they used was

$$0.53\,Na_2O:0.47\,K_2O:Al_2O_3:50\,SiO_2:25\,Et_4NOH:750\,H_2O.$$

Amorphous silica (Aerosil) was the SiO_2 source. The mixture was heated at 135 °C in the static mode. After 24 h the amount of zeolite formed was maximum. Pure zeolite Beta with Si/Al in the zeolite equal to 13.3 and a mean crystal size of ca. 0.2 µm was obtained. The authors observed that agitation of the gel during the heating decreased the induction period and the crystallization time. On the other hand, Di Renzo et al. [161] found that in moderately silicic systems (starting Si/Al = 25–50) zeolite Beta formed before ZSM-12, which appeared when the mixture has been partly depleted in aluminum. Indeed, the Si/Al ratio in zeolite Beta is smaller than in the gel, e.g., 13.3 vs. 50 [160].

1.4.3.3.2
Synthesis in Fluoride Medium

Caullet et al. [162] succeeded in preparing zeolite Beta from near neutral aqueous aluminosilicate gels in the presence of fluoride as the mineralizing agent and 1,4-diazabicyclo[2,2,2]octane (DABCO) and methylamine as the organic species. With this route, the zeolite obtained is less siliceous (Si/Al = 9 – 22) than when prepared in alkaline medium with Na^+ and Et_4N^+ cations. Pure zeolite Beta (2 – 3 μm crystals) was typically obtained when a gel with the composition,

$$Al_2O_3:(5-7)\,SiO_2:DABCO:CH_3NH_2:2\,HF:10\,H_2O$$

and containing crushed zeolite Beta crystals as seeds was heated at 170 °C for 15 d. Indeed the full transformation of the gel into the zeolite requires the presence of Beta seeds. DABCO was occluded in the zeolite as a polymeric compound identified as polyethylenepiperazine whereas methylamine was not detected in the solid. However, its presence together with DABCO in the starting mixture was found to be necessary for the complete crystallization of the gel to be achieved.

Zeolite Beta can be obtained in unstirred or stirred systems (decrease of the crystallization time). For stirred systems, zeolite Beta crystallized at 170 °C when Si/Al in the gel was smaller than ca. 12, above this ZSM-12 was formed. At 200 °C, ZSM-12 was obtained solely for all Si/Al values.

1.4.3.4
Synthesis of Zeolite LTL

Zeolite Linde type L, whose natural counterpart is perlialite [163], is a large-pore zeolite with a one-dimensional 12-membered ring channel system. It was first synthesized by Breck [164]. K^+ and Ba^{2+} are the preferred cations, but they can be partly replaced by others such as Na^+ [2] and Cs^+ [165]. The formula of the K/Na form is $K_6Na_3Al_9Si_{27}O_{72} \cdot 21\,H_2O$ [5]. The synthesis of zeolite L with inorganic and organic cations such as $Ba^{2+} + Me_4N^+$ or $K^+ + Na^+ +$ tetraalkylammonium cations was reported by Barrer et al. [166] and Vaughan [167], respectively.

More recently, Joshi et al. [168] systematically investigated the synthesis over a range of $K_2O/(K_2O + Na_2O)$ ratios in the temperature range 120 – 180 °C using the typical Breck formulation

$$8\,[x\,K_2O + (1-x)\,Na_2O]:Al_2O_3:20\,SiO_2:200\,H_2O.$$

The silica source was fused silica. At 150 °C, for $x = 1$, well-crystallized zeolite L in the K-form was obtained after 12 h. On prolonged heating a transformation to zeolite W of the MER-type structure was observed. For $0.33 \le x < 1$, zeolite L was again obtained but it converted to zeolite T on further heating. For $x < 0.33$ only zeolite Y was obtained.

1.4.3.5
Synthesis of Zeolites with the MAZ-Type Structure

Zeolite Linde Omega and ZSM-4 are the synthetic counterparts of natural mazzite whose formula is $(Na_2, K_2, Ca, Mg)_5\,Al_{10}Si_{26}O_{72} \cdot 28\,H_2O$ [5]. They were first synthesized by Flanigen [169] and Ciric [170], respectively, from aluminosilicate gels containing Na^+ and Me_4N^+ cations. The crystal structure shows a one-dimensional 12-membered ring channel system with 8-membered ring openings. The synthesis temperature is usually in the range 80–150 °C for a heating time of 2–8 d. The most suitable silica source is fumed silica or a colloidal silica sol. The Si/Al ratio in the zeolite is in the range 2.5–3.6 [1]. The use of other organics such as choline or pyrrolidine [171] and DABCO [172] was also reported.

Recently, Fajula et al. [173] examined the influence of synthesis parameters such as temperature (100 and 135 °C) and reactivity of the alumina and silica sources on the crystallization of zeolite Omega. It was obtained in a pure form by heating a gel with the composition

$$5.03\,Na_2O:1.1\,(Me_4N)_2O:Al_2O_3:13.8\,SiO_2:210\,H_2O.$$

for ca. one day at 135 °C. The silica source used was a low-surface-area silica gel SiO_2 175 (175 $m^2 g^{-1}$), and sodium aluminate was the aluminum source. After more than one day of heating, sodalite appeared, and later both zeolites, Omega and sodalite, dissolved, and cristobalite formed.

The same authors [174] performed an almost organic-free synthesis by using a structure-directing mixture which was prepared by heating a mixture whose composition was

$$3.2\,Na_2O:0.8\,(Me_4N)_2O:Al_2O_3:8.2\,SiO_2:160\,H_2O$$

at 50 °C for 20 d. About 25–30 vol% of this seeding mixture was then added at room temperature to the Me_4N-free crystallization mixture

$$3.2\,Na_2O:Al_2O_3:8\,SiO_2:160\,H_2O.$$

The autoclave was then heated at the desired temperature with stirring. The observed conversion as a function of time at various temperatures is given in Fig. 1.20. It can be seen that the induction period can thus be drastically reduced.

Several advantages were found for such a procedure:

- The induction period decreased with respect to that of the traditional synthesis procedure.
- The consumption of Me_4N template was reduced.
- The structure-directing mixture allowed a separation of the nucleation and the growth steps; therefore, a very homogeneous distribution of the crystal size was achived and well-shaped zeolite crystals were obtained.

1.4.3.6
Synthesis of Zeolites with the MFI-Type Structure

ZSM-5 is a synthetic high-silica zeolite (Si/Al larger than ca. 7) first reported in 1973 by Argauer and Landolt [111]. The all-silica end member silicalite-1 was synthesized by Flanigen et al. in 1977 [32]. The MFI-type topology shows a three-

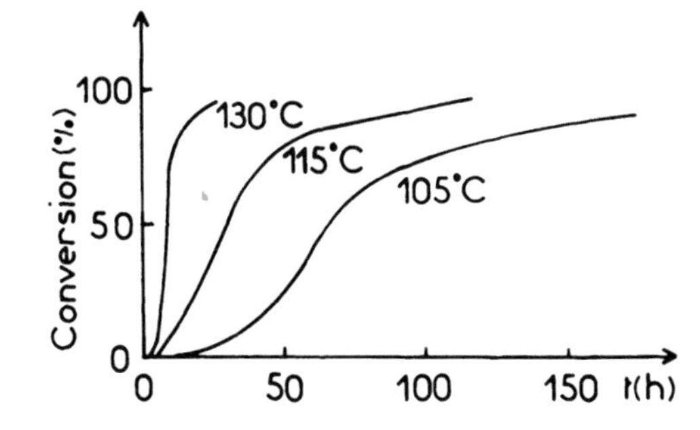

Fig. 1.20. Synthesis of zeolite Omega with a structure-directing mixture: crystallization as a function of time at various temperatures [174]

dimensional 10-membered channel system. Due to its sorption and catalytic properties, ZSM-5 aroused great interest. A large number of patents and publications were devoted to its synthesis, in particular to the factors governing the crystallization [3]. The first syntheses were performed in the presence of Na^+ and organic additives. About 25 organic species were claimed to be useful [41], among which the tetrapropylammonium cations are the most structure-directing. Indeed, ZSM-5 is easily synthesized with this template in a very broad range of experimental conditions. The organic-free synthesis in the presence of Na^+ cations only has also been reported [175–177].

1.4.3.6.1
Synthesis in the Presence of an Organic Additive

The synthesis of ZSM-5 in alkaline medium in the presence of tetrapropylammonium cations or other organic species has been thorougly discussed by Jacobs and Martens [3]. They report three optimized recipes for the synthesis of ZSM-5 with tetrapropylammonium cations. In the three procedures, the silica source was pyrogenic silica (Aerosil) or sodium silicate (water glass), and the alumina source was sodium aluminate or aluminum nitrate. Tetrapropylammonium bromide was used as the tetrapropylammonium source; it is much less expensive than the hydroxide used by Argauer and Landolt [111]. As an example, in their recipe (a), a gel with the starting molar composition

$$90\,(Pr_4N)_2O : 14.4\,Na_2O : Al_2O_3 : 72\,SiO_2 : 10500\,H_2O : 1430\ \text{glycerol}$$

with Aerosil and aluminum nitrate, respectively, as the silica and alumina sources, was heated at 150 °C for 3 d. High dilution together with the presence of glycerol led to large elongated crystals (ca. 8×6 µm), the Si/Al ratio in the zeolite was 35.

Erdem and Sand [178] studied the influence of $r = Na/(Na + Pr_4N)$ on the phases obtained at 175 °C from a gel with the composition

$$10\,(Na, Pr_4N)_2O : Al_2O_3 : 28\,SiO_2 : 750\,H_2O.$$

When r was in the range 0.7–1.0, only mordenite was formed. For $0.4 < r < 0.7$ mordenite formed first and dissolved to recrystallize into analcime and

ZSM-5, analcime transformed finally into ZSM-5. For a low Na^+ content ($r<0.3$), pure ZSM-5 was formed directly. This shows that for intermediate r values, the crystallization time has to be large enough to allow a complete transformation into ZSM-5. On the other hand when the heating time of a ZSM-5 synthesis is too long, quartz impurities may form [179]. Likewise, quartz crystallizes easily when the temperature is too high, e.g., ca. 200 °C [179].

1.4.3.6.2
Organic-Free Synthesis of ZSM-5

Synthesis in the Absence of Seeds. Several organic-free syntheses were reported in the patent literature [175–177] some years after Argauer and Landolt's discovery [111]. Recently, there have also been reports in the open literature [180–182]. Such a synthesis is indeed very attractive because it is less expensive, and the calcination steps to turn the zeolite into a sorbent or a potential catalyst by removal of the template can be avoided. There is agreement [180–182] that pure ZSM-5 can be obtained at 190 °C, 24–40 h, in the absence of seeds, from organic-free gels having the composition

$$(5-14)\,Na_2O:Al_2O_3:(50-70)\,SiO_2:(1000-3200)\,H_2O.$$

Thus, Dai et al. [181] carried out their syntheses in PTFE-lined autoclaves which were carefully cleaned with HF prior to each run, in order to eliminate any seeds left from a preceding experiment. By heating a gel with the composition

$$11.2\,Na_2O:Al_2O_3:70\,SiO_2:3213\,H_2O$$

at 190 °C for 24 h without stirring, they obtained pure ZSM-5, the Si/Al ratio in the zeolite being equal to 22. On increasing the crystallization time, ZSM-5 transformed into mordenite which converted to quartz. For $SiO_2/Al_2O_3 = 20-25$ or $Na_2O/Al_2O_3 > 14$, mordenite is the major phase, and for $SiO_2/Al_2O_3 \sim 200$ essentially quartz is obtained [181–182].

Thus, it appears that only ZSM-5 containing sufficient framework-Al (Si/Al = 20–25) can be prepared from organic-free gels [180–182]. This presumably corresponds to the composition for which enough interactions are still present between the framework charges and the Na^+ cations to stabilize the structure.

Synthesis in the Presence of Seeds. Numerous recent patents [182a] and publications in the open literature [183–184] report on the synthesis of ZSM-5 in the presence of seeds of the same zeolite.

The seeds are usually finely crushed ZSM-5 crystals, either prepared in the presence of an organic or recycled from an organic-free synthesis. In these wholly inorganic syntheses, silica sol appears to be the best silica source. In the presence of seeds the crystallization time is reduced with respect to that in the absence of seeds.

1.4.3.7
Synthesis of Zeolites with the MOR-Type Structure

1.4.3.7.1
Small- and Large-Port Mordenite

Mordenite is a natural high-silica zeolite whose typical unit cell formula is $Na_8Al_8Si_{40}O_{26} \cdot 24H_2O$. The crystal structure [5] shows a 12-membered ring channel system parallel to $\vec{c}$ (opening 6.5 Å × 7.0 Å). These channels are interconnected through small side channels parallel to $\vec{b}$ and circumscribed by 8-membered rings (opening 2.6 × 5.7 Å). Thus, mordenite can be considered as having an essentially one-dimensional channel system.

The first extensive investigation of the synthesis of mordenite was performed by Barrer [185] by using the starting composition

$$Na_2O : Al_2O_3 : (8.2-12.3)\,SiO_2 : mH_2O$$

with sodium aluminate and silica gel as the reactants. The best yields of mordenite were obtained at rather high temperatures, viz. in the range 265–295 °C. In another study, Barrer and White [186] found that the starting SiO_2/Al_2O_3 ratio, n, was a key factor for obtaining mordenite. Analcime was produced for n = 4–5, mordenite for n ~ 10 and quartz was formed for n larger than 12. Domine and Quobex [186a] investigated the influence of various factors such as the nature of the silica and alumina sources, pH, temperature and heating time. The best reactants source was found to be an amorphous sodium silicoaluminate (SiO_2/Al_2O_3 = 1–12), and the best pH values were between 12 and 13 (adjusted by NaOH addition). Pure mordenite crystallized at 300 °C after 7–10 h heating. Various groups also have synthesized other cationic forms (Ca, Sr) and mixed forms (Na–Ca). Most of the above materials had openings of about 4 Å in diameter, which corresponds approximately to the openings of most natural mordenites.

Such mordenites with reduced adsorption diameter are called small-port type mordenites. Large-port-type mordenite, which adsorbs molecules of about 7 Å in diameter, was obtained over a large range of temperatures (100–260 °C) for 12–168 h with various starting compositions. For example, large-port-type mordenite crystallized from the molar starting composition

$$3.5\,Na_2O : Al_2O_3 : 10\,SiO_2 : 219\,H_2O$$

by heating at 135–165 °C for 12–24 h [187]. The reactants were sodium silicate, silica gel, aluminum and sodium hydroxide.

Three suggestions have been made to explain the differences of sorption properties between small-port and large-pot mordenites (see, for example, [188]), particular localization of the cations in the channels, stacking faults on planes perpendicular to the main channels or the presence of amorphous material in the pores.

A recent study by Raatz et al. [189] accredited the hypothesis of planar structural defects resulting from the synthetic conditions. Sanders [190] came to the conclusion that natural mordenite contains linear faults in the direction of the main set of mordenite chains, whereas synthetic mordenite contains a much lower concentration of faults.

1.4.3.7.2
Synthesis of Silica-Rich Mordenite

Acid-leaching of the structural alumina or steam treatment results in an increased Si/Al ratio. The direct synthesis of silica-rich mordenite with increased thermal stability is also possible in the absence or in the presence of an organic additive. For example, a recent publication on the synthesis of silica-rich mordenite in wholly inorganic media is that of Kim and Ahn [191]. They synthesized a range of mordenite samples with Si/Al from 6.3 to 15.75 at 170 °C by increasing the starting SiO_2/Al_2O_3 and keeping the Na_2O/SiO_2 and Na_2O/H_2O ratios constant. In particular, by heating a gel with the molar composition

$$6\,Na_2O : Al_2O_3 : 30\,SiO_2 : 780\,H_2O$$

at 170 °C for 20 h, pure mordenite with Si/Al = 8.6 was obtained.

Several organic additives, among them the tetraethyl- and benzyltrimethylammonium cations, ethylpyridinium and neopentylamine, have been used for the synthesis of silica-rich mordenite. Thus, with tetraethylammonium cations, Chumbhale et al. [192] obtained at 170 °C large-port silica mordenite samples with Si/Al ranging from 7.40 to 16.18 for SiO_2/Al_2O_3 varying between 19.6 and 37.1 in the starting mixture.

1.4.3.8
Synthesis of Zeolites with the OFF- and/or the ERI-Type Structure

1.4.3.8.1
Composition and Structure of Offretite and Erionite

Offretite and erionite are two natural zeolites with the unit cell formulae $(Ca, Mg, K_2)_{2.5}Al_5Si_{13}O_{36} \cdot 15\,H_2O$ and $(Ca, Mg, Na_2, K_2)_{4.5}Al_9Si_{27}O_{72} \cdot 27H_2O$, respectively [2]. The Al content of the synthetic materials is generally lower (Si/Al higher) than in natural samples because bivalent cations compensate part of the framework charges in the latter.

Their crystal structures are closely related [5]. Offretite shows a 12-membered ring channel system parallel to $\vec{c}$ (6.7 Å in diameter), linked by a set of sinusoidal 8-membered ring channels perpendicular to $\vec{c}$ (3.6 × 4.9 Å openings). In erionite, the 12-membered ring channels are blocked by 6-membered rings at 15.2 Å intervals, erionite-type cages are thus generated. They are interconnected three-dimensionally by 8-membered ring apertures (3.6 × 5.1 Å).

1.4.3.8.2
Synthesis of Offretite

Offretite was first synthesized by Aiello and Barrer [193] by using a mixture of NaOH, KOH and tetramethylammonium hydroxide (Me_4NOH). It turned out that the Me_4N^+ cation played a key role for obtaining pure offretite.

Moudafi et al. [194] studied the influence of the concentration of Me_4NOH in the case where Me_4NOH and KOH alone are used as bases. They found that pure offretite was obtained at 110 °C, 72 h with the starting composition

$$6.2\,(Me_4N,K)_2O:Al_2O_3:15.3\,SiO_2:280\,H_2O$$

when $0.04 \le Me_4N/(Me_4N + K) \le 0.5$.

They also investigated the role of sodium by varying separately the two parameters x and b/(a + b) in the cation distribution $xMe_4N + aK + bNa$ with $x + a + b = 1$. Offretite was only isolated in a pure form when x was lower than 0.2. As an example, for $x = 0.04$, up to 75% of the K^+ cations could be replaced by Na^+ cations in the starting mixture to produce pure offretite. When Na/(Na + K) was equal to 0.9, a mixture of offretite and zeolite Omega crystallized. For larger values of x, a mixture of offretite and sodalite ($x = 0.3-0.5$) or pure sodalite ($x = 0.8$) was obtained.

The same group used the nucleation method to reduce the Me_4N^+ consumption in the synthesis of pure offretite [195]. A nucleation gel was prepared by aging a mixture with the composition

$$6\,K_2O:2.5\,(Me_4N)_2O:Al_2O_3:15.3\,SiO_2:280\,H_2O$$

over 3 d at room temperature. 2–15% of this nucleation gel was added to the crystallization gel of the same composition but without Me_4N^+. Pure offretite was obtained by heating the resulting mixtures at 150 °C for 15 h. The crystals exhibited high Si/Al ratios ($\cong 4.3$), and their Me_4N^+ content per unit cell was reduced to 0.17 compared to 1.20–1.96 for the offretite obtained by the direct route [193].

1.4.3.8.3
Synthesis of Erionite

The synthesis of pure erionite has been much less investigated than that of pure offretite. Aiello and Barrer [193] reported its preparation from gels containing the mixed bases Me_4NOH and NaOH [$Me_4N/(Me_4N + Na) = 0.8$]. In an investigation of the crystallization of erionite, offretite and erionite–offretite intergrowths in the base system NaOH–KOH–Me_4NOH, Ueda et al. [196] obtained pure erionite with $Si/Al = 4$ from a gel with the composition

$$5\,K_2O:15\,Na_2O:0.1\,(Me_4N)_2O:Al_2O_3:60\,SiO_2:350\,H_2O$$

which was heated at 100 °C for 6 d.

1.4.3.8.4
Synthesis of Offretite-Erionite Intergrowths

Owing to the close relationship between their structures, intergrowth of erionite in offretite and vice versa is not rare [197]. Zeolite T [198] and zeolite ZSM-34 [199] are two such intergrowth phases. Zeolite T was prepared typically at 100 °C in the absence of any organic additive from a mixture containing only Na^+ and K^+ cations ($Na^+/K^+ \cong 3$; $SiO_2/Al_2O_3 \cong 25$). A typical unit cell formula of the intergrowth was $Na_{1.2}K_{2.8}Al_4Si_{14}O_{36} \cdot 14.4\,H_2O$ [1]. In the synthesis of ZSM-34, choline chloride $(CH_3)_3NCH_2CH_2OH \cdot HCl$ was used as the organic additive together with Na^+ and K^+ cations [199].

The differentiation between pure erionite and erionite–offretite intergrowths by powder X-ray diffraction is not clear cut. Therefore, the cyclohexane adsorption capacity of the materials is generally used to estimate the extent of erionite intergrowth in offretite [195].

The synthesis of offretite–erionite type materials from mixtures containing neither Me_4NOH nor choline chloride as organic species was investigated by Occelli et al. [200]. Benzyltrimethylammonium, benzyltriethylammonium and mono- or dibasic substituted 1,4-diazabicyclo[2,2,2]octane cations were added to a hydrogel with the molar composition

$$2.2\,Na_2O:0.64\,K_2O:Al_2O_3:12\,SiO_2:200\,H_2O.$$

Heating at 150 °C for 5 d without stirring resulted in pure offretite when benzyltrimethylammonium cations were used, whereas with the three others a ZSM-34-type intergrowth material was produced.

1.5
Synthesis of Other Selected Materials

1.5.1
Titanosilicates with Mixed Octahedral-Tetrahedral Frameworks

A novel microporous titanosilicate named GTS-1 was prepared by Chapman and Roe [201] by heating strongly alkaline titanium silicate gels. It is a structural analogue of the mineral pharmacosiderite, and its ideal formula would be $M_4Ti_4Si_3O_{16}$ (M = K, Cs). The framework consists of TiO_6 and SiO_4 units whose arrangement generates a 3-dimensional channel system circumscribed by 8-membered rings of alternating octahedra and tetrahedra.

Other types of microporous titanosilicates, ETS-4 and ETS-10 were reported by Kuznicki et al. [202]. They were obtained by autoclaving aqueous mixtures of sodium silicate, $TiCl_3$ or $TiCl_4$, NaOH and KF (pH ~ 10). Thus, ETS-10 crystallized from a gel with the molar composition

$$3.72\,Na_2O:1.46\,KF:1\,TiCl_4:4.85\,SiO_2:182\,H_2O \text{ (with seeds)}$$

after heating at 200°C for 18 h. Valtchev [203] reported a synthetic route using organic amines or quaternary ammonium cation additives. Tetramethylammonium was found to be most efficient. According to adsorption measurements [202b] the pore sizes are ca. 4 and 8 Å for ETS-4 and ETS-10, respectively. Composition analysis showed that titanium is octahedrally coordinated in the framework requiring two charge-compensating cations per titanium which are readily exchanged. This results in a change in the effective pore size. The crystal structure of ETS-10 has been reported [204] and the presence of SiO_4 tetrahedra and TiO_6 octahedra confirmed. The composition of the framework is $Si_5TiO_{13}^{2-}$. The pore system contains 12-membered rings and presents a considerable degree of disorder.

Recently, more novel microporous titanosilicates have been reported, among which are monoclinic $Na_{2.7}K_{5.3}Ti_4Si_{12}O_{36} \cdot 4\,H_2O$ [205] and orthorhombic $K_2TiSi_3O_9 \cdot H_2O$ [206]. Both show 8-membered ring channels with exchangeable cations and water molecules.

1.5.2
Synthesis of Mesoporous Aluminosilicates

In 1991, researchers at Mobil reported the synthesis of a new family of mesostructured aluminosilicate molecular sieves called M41S [207]. This family includes (a) the type MCM-41 displaying a one-dimensional hexagonal arrangement of uniformly open channels (15 to 100 Å), (b) the type MCM-48 with a three-dimensional pore system with ~ 30 Å openings, and (c) other types among which are lamellar phases. These solids can be synthesized from surfactants such as quaternery ammonium cations like cetyltrimethylammonium cations ($CTMA^+$).

As an example, a material with a specific surface area as high as 1000 $m^2\,g^{-1}$ (when calcined) was obtained at 150°C for 48 h from a gel with the molar composition

$$6.1\,(CTMA)_2O:5.2\,(Me_4N)_2O:0.23\,Na_2O:Al_2O_3:33.2\,SiO_2:789\,H_2O$$

Such materials may be prepared from mixtures with Si/Al in the range 15–∞ and over a wide temperature range, even at room temperature.

A liquid-crystal templating mechanism was proposed by the Mobil Group for the M41S-type materials [208]. The mechanism suggested was further investigated by Monnier et al. [209] and Chen et al. [210]. It was concluded that it is a silicated surfactant assembly that is responsible for the formation of the M41S structures.

Another type of material named FSM-16 with uniform mesopores was prepared by Inagaki and co-workers [211] via hydrothermal treatment of the layered sodium silicate kanemite $NaHSi_2O_5 \cdot 3\,H_2O$ in the presence of alkyltrimethylammonium surfactants.

Chen et al. [212] found that both MCM-41 and FSM-16 show narrow pore-size distributions. However, the latter showed higher thermal and hydrothermal stability owing to a higher degree of condensation in the silicate walls.

Numerous other mesophase materials prepared via other synthetic routes have since been reported. A review of recent advances in the field has been published recently by Beck and Vartuli [213].

1.6 Activation of Zeolites

As-synthesized zeolites which contain inorganic cations only have to be dehydrated before being used in adsorption or catalytic processes. This has to be done carefully in order to avoid a steaming resulting in various effects such as siloxane bond hydrolysis and/or dealumination of the framework.

Zeolites which are prepared in the presence of organic additives generally trap these organic species in the cages and/or channels of their pore system. In order to turn the solids into open materials, the organic species have to be removed. This is mostly performed by calcination of the as-synthesized zeolite. Care has to be taken to avoid undesired effects such as local overheating which may result in a partial destruction of the crystal structure, or dealumination of the framework by the water vapor already present in the solid or generated during the combustion of the organics. The steaming effect will be all the stronger as a deep-bed calcination is performed. Therefore, the calcination should be carried out under controlled conditions depending on the type of zeolite and the organic species. A procedure using a low heating rate (~ 1 °C min^{-1}) with intermediate steps up to a maximum temperature in the range 400 - 600 °C under a dry oxygen-poor gas flow followed by a burning of the residual organics under pure oxygen at 400 - 600 °C is generally appropriate.

References

1. Breck DW (1974) Zeolite molecular sieves. Structure chemistry and use. John Wiley, New York
2. Barrer RM (1982) Hydrothermal chemistry of zeolites. Academic Press, London
3. Jacobs PA, Martens JA (1987) Synthesis of high-silica aluminosilicate zeolites. Elsevier, Amsterdam (Studies in Surface Science and Catalysis, vol 33)
4. Szostak R (1989) Molecular sieves, principles of synthesis and identification. Van Nostrand Reinhold, New York

4a. Szostak R (1992) Handbook of molecular sieves. Van Nostrand Reinhold, New York

5. Meier WM, Olson DH, Baerlocher Ch (1996) Atlas of zeolites structure types. (Fourth Revised Edition) Elsevier, London
6. Guth JL, Caullet P (1986) Journal de Chimie Physique 83(3):155
7. Strickland-Constable RF (1968) Kinetics and mechanism of crystallization. Academic Press, London
8. Tiller WA (1991) The science of crystallization: Macroscopic phenomena and defect generation. Academic Press, London

8a. Boistelle R, Astier JP (1988) J Crystal Growth 90:14

8b. Bennema P, Söhnel O (1990) J Crystal Growth 102:547

8c. Huo Q, Margolese Di, Ciesla U, Demuth DG, Feng P, Gier TE, Sieger P, Firouzi A, Chmelka BF, Schüth F, Stucky GD (1994) Chem Mater 6:1176

8d. Jolivet JP (1994) De la Solution à l'Oxyde. Interéditions/CNRS Éditions, Paris

9. Caullet P, Guth JL, Hurtrez G, Wey R (1981) Bull Soc Ch France 7-8:I 253
10. Goldsmith JR (1953) J Geol 61:349
11. Zhdanov SP, Feoktistova NN, Vtjurina LM (1991) In: Öhlmann G, Pfeifer H, Fricke R (eds) Catalysis and adsorption by zeolites. Elsevier, Amsterdam, pp 287–296
12. Caullet P, Guth JL, Wey R (1980) C R Acad Sci. Paris 291 D:117
13. Guth JL, Caullet P, Wey R (1975) Bull Soc Chim. France 11-12:2375

14. Pang W, Ueda S, Koizumi M (1986) In: Murakami Y, Iijima A, Ward JW (eds) New developments in zeolite science and technology. Kodansha - Elsevier, Tokyo, pp 177-184
15. Barrer RM, Baynham JW, Bultitude FW, Meyer WM (1959) J Chem Soc 195
16. Kerr GT (1966) J Phys Chem 70:1047
17. Kühl GH (1971) Second international conference on Molecular Sieve Zeolites. American Chemical Society, Washington DC, pp 63-74 (Advanc Chem Ser 101)
18. Zhdanov SP, Samulevitch NN (1980) In: Rees LV (ed) Proceedings of the fifth international conference on zeolites. Heyden, London, pp 75-84
19. Breck DW, Flanigen EM (1968) Molecular sieves. Society of Chemical Industry, London, pp 47-61
20. McNicol BD, Pott GT, Loos KR (1972) J Phys Chem 76:338
21. Xu W, Dong J, Li J (1990) J Chem Soc, Chem Commun 131
22. Weyl WA, Marboe EC (1962) The constitution of glasses. A dynamic interpretation Vol I Fundamental of the structure of inorganic liquids and solids. John Wiley and Sons, New York
22a. Guth JL (1995) In: Aiello R (ed) Proceedings III Convegno Nazionale Scienza e Tecnologia delle Zeoliti. Associazione Italiana Zeoliti, Universita degli Studi della Calabria
23. Mc Cormick AV, Bell AT (1989) Catal Rev - Sci Eng, 31 (1 & 2):97
24. Caullet P, Guth JL (1989) In: Occelli ML, Robson HE (eds) Zeolite synthesis. ACS, Washington DC, pp 83-97 (ACS Symposium Series 398)
25. Guth JL, Kessler H, Higel JM, Lamblin JM, Patarin J, Seive A, Chézeau JM, Wey R (1989) In: Occelli ML, Robson HE (eds) Zeolite synthesis. American Chemical Society, Washington DC, pp 176-195 (ACS Symposium Series 398)
26. Guth JL, Caullet P, Seive A, Patarin J, Delprato F (1990) In: Barthomeuf D, Derouane EG, Hölderich W (eds) Guidelines for mastering the properties of molecular sieves. Plenum Press, New York, pp 69-86 (NATO ASI Series B: Physics Vol 221)
27. Jansen JC (1991) In: Van Bekkum H, Flanigen EM, Jansen JC (eds) Introduction to zeolite science and practice. Elsevier, Amterdam, pp 77-136
28. (a) Zones SI (1987) US Pat 4665110; (b) Bialek R, Meier WM, Davis ME, Annen MJ (1991) Zeolites 11:438
29. (a) Van der Waal JC, Rigutto MS, Van Bekkum H (1994) J Chem Soc Chem Comm 1241; (b) Gies H (1991) In: Atwood JL, Davies JED, MacNicol DD (eds) Inclusion compounds. Oxford University Press, pp 1-36 (Inorganic and physical aspects of inclusion, vol 5)
30. Kupermann, A, Nadimi S, Oliver S, Ozin GA, Garcès JM, Olken MM (1993) Nature (London) 365:239
31. (a) Bibby DM, Dale MP (1985) Nature 317:157; (b) Keijsper J, den Ouden CJJ, Post MFM (1989) In: Jacobs PA, van Santen RA (eds) Zeolites: facts, figures, future. Elsevier, Amsterdam, pp 237-247 (Studies in surface science and catalysis, vol 49); (c) Balkus KJ Jr, Gabrielov AG, Zones SI, Chan IY (1997) In: Occelli ML, Kessler H (eds) Synthesis of porous materials, Zeolites, clays, and nanostructures. Marcel Decker, New York, pp 77-92
32. Flanigen EM, Bennett JM, Grose RW, Cohen JP, Patton RL, Kirchner RM, Smith JV (1978) Nature 271:512
33. Price GD, Pluth JJ, Smith JV, Bennett JM, Patton RL (1982) J Am Chem Soc 104:5971
34. Patarin J, Soulard M, Kessler H, Guth JL, Baron J (1989) Zeolites 9:397
35. Bibby DM, Milestone NB, Aldridge LP (1979) Nature 280:664
36. Mostowicz R, Nastro A, Crea F, Nagy JB (1991) Zeolites 11:732
37. Gies H, Marler B, Fyfe CA, Kokotailo GT, Cox DE (1990) J Phys Chem 94:3718
38. Marler B (1987) Zeolites 7:393
39. Franklin KR, Lowe BM (1989) In: Jacobs PA, van Santen RA (eds) Zeolites: facts, figures, future. Elsevier, Amsterdam, pp 179-188 (Studies in surface science and catalysis, vol 49)
40. Barrer RM (1981) Zeolites 1:130
41. Lok BM, Cannan TR, Messina CA (1983) Zeolites 3:282

42. Delprato F, Delmotte L, Guth JL, Huve L (1990) Zeolites 10:546
43. Vaughan DEW (1991) In: Öhlmann G, Pfeifer H, Fricke R (eds) Catalysis and adsorption by zeolites. Elsevier, Amsterdam, pp 275-286 (Studies in surface science and catalysis, vol 65)
44. (a) Casci JL, Stewart A (1990) Eur Pat Appl 337291; (b) Casci JL, Shannon MD, Cox PA, Andrews SJ (1992) In: Occelli ML, Robson HE (eds) Molecular sieves. Van Nostrand Reinhold, New York, pp 359-372 (Synthesis of microporous solids, vol 1)
45. Lawton SL, Rohrbaugh WJ (1990) Science 247:1319
46. (a) Valyocsik EW, Page NM (1986) Eur Pat Appl 174121; (b) Ernst S, Weitkamp J (1991) In: Öhlmann G, Pfeiffer H, Fricke R (eds) Catalysis and adsorption by zeolites. Elsevier, Amsterdam, pp 645-652 (Studies in surface science and catalysis, vol 65)
47. (a) Rodewald G, Valyocsik EW (1987) US Pat 4665264; (b) Ernst S, Weitkamp J (1991) Chem Ing Tech 63:748
48. Caullet P, Hazm J, Guth JL, Joly JF, Lynch J, Raatz F (1992) Zeolites 12:240
49. Valyocsik EW, Rollmann LD (1985) Zeolites 5:123
50. Baerlocher C, Meier WM (1970) Helv Chim Acta 53:1285
51. Baerlocher C, Meier WM (1969) Helv Chim Acta 52:1853
52. Ueda S, Koizumi M, Baerlocher C, McCusker LB, Meier WM (1986) In: Preprints of poster papers of the 7th International Zeolite Conference, Japan association of zeolite. Tokyo, poster paper 3C-3
53. (a) Merlino S (1984) In: Olson DH, Bisio A (eds) Proceedings of the 6th International Zeolite Conference. Butterworths, Guildford, pp 747-759; (b) (1990) Eur J Mineral 2:809
54. Romannikov VN, Chumachenko LS; Mastikhin VM, Ione KG (1985) J Catalysis 94:508
55. Rossin JA, Saldarriaga C, Davis ME (1987) Zeolites 7:295
56. Mostowicz R, Dabrowski AJ, Jablonski JM (1989) In: Jacobs PA, van Santen RA (eds) Zeolites: facts, figures, future. Elsevier, Amsterdam, pp 249-259 (Studies in surface science and catalysis, vol 49)
57. Heinrich AR, Baerlocher C (1991) Acta Cryst C47:237
58. (a) Camblor MA, Yoshikawa M, Zones SI, Davis ME (1997) In: Occelli ML, Kessler H (eds) Synthesis of porous materials, Zeolites, clays, and nanostructures, Marcel Decker, New York, pp 243-261; (b) Annen MJ, Davis ME, Higgins JB, Schlenker JL (1991) In: Bedard RL, Bein T, Davis ME, Garcés J, Maroni VA, Stucky GD (eds) Synthesis/characterization and novel applications of molecular sieves materials. Materials Research Society, Pittsburgh, pp 245-254
59. Newsam JM (1986) J Chem Soc Chem Comm 1295
60. Newsam JM (1988) J Phys Chem 92:445
61. Yang J, Xie D, Yellon WB, Newsam JM (1988) J Phys Chem 92:3586
62. Zones SI, Holtermann DL, Innes RA, Santilli DS, Yuen LT, Ziemer JN (1991) PCT Int Appl WO 91-844
63. Van Nostrand RA, Santilli DS, Zones SI (1992) In: Occelli ML, Robson HE (eds) Molecular sieves. Van Nostrand Reinhold, New York, pp 359-372 (Synthesis of microporous solids, vol 1)
64. (a) Millini R, Montanari L, Bellussi G (1993) Microporous Mater. 1:9; (b) Yelon WB, Xie D, Newsam JM, Dunn J (1990) Zeolites 10:553
65. Kuehl GH (1987) US Pat 4661467
66. Kumar R, Thangaraj A, Bhat RN, Ratnasamy P (1990) Zeolites 10:85
67. Camblor MA, Perez-Pariente J, Fornés V (1992) Zeolites 12:280
68. (a) Occelli ML (1990) US Pat 4931266; (b) Rao GN, Shiralkar VP, Kotasthane AN, Ratnasamy P (1992) In: Occelli ML, Robson HE (eds) Molecular sieves. Van Nostrand Reinhold, New York, pp 153-166 (Synthesis of microporous solids, vol 1)
69. Selbinaud J, Mason RB (1961) J Inorg Nucl Chem 20:222
70. Kühl GH (1971) J Inorg Nucl Chem 33:3261
71. Suzuki K, Kiyozumi Y, Shin S, Ueda S (1985) Zeolites 5:11
72. Newsam JM, Jacobson AJ, Vaughan DEW (1986) J Phys Chem 90:6858
73. Occelli ML (1989) US Pat 4803060

74. Delprato F, Guth JL, Zivkov C (1991) Fr Demande FR 2650582
75. Shawki MS, Schimandle JJ (1987) Eur Pat Appl 234766
76. Sulikowski B, Klinowski J (1989) J Chem Soc Chem Comm 1289
77. Xu R, Liu X (1984) Chimica Sinica 42:227
78. Newsam JM (1986) J Chem Soc Chem Comm 1295
79. Occelli ML (1990) US Pat 4919907
80. Bellussi G, Millimi R, Carati A, Madinelli G, Gervasini A (1990) Zeolites 10:642
81. Chandwadkar AJ, Bhat RN, Ratnasamy P (1991) Zeolites 11:42
82. Barri SAI, Young D (1985) Eur Pat Appl 130013
83. Liu X (1986) Thesis, University of Cambridge
84. Union Oil Co of California (1988) Jpn Kokai Tokkyo Koho 88129014
85. Hoelderich W, Lermer H (1987) DE 3537998
86. Szostak R, Thomas TL (1986) J Chem Soc Chem Comm 113
87. McCusker LB, Meier WM, Suzuki K, Shin S (1986) Zeolites 6:388
88. Newsam JM, Jorgensen JD (1987) Zeolites 7:569
89. Goldsmith JR (1952) Miner Mag 29:952
90. Coudurier G, Vedrine JC (1986) In: Murakami Y, Iijima A, Ward JW (eds) New development in zeolite science and technology. Kodansha-Elsevier, Tokyo, pp 643-652 and references cited therein
91. Kutz NA (1988) In: Flank WH, Whyte TE Jr (eds) Perspectives in molecular sieve science. ACS, Washington DC, pp 532-541 and references cited therein (ACS symposium series, vol 368)
92. Perego G, Bellussi G, Carati A, Millini R, Fattore V (1989) In: Occelli ML, Robson HE (eds) Zeolite synthesis. ACS Washington, DC, pp 360-373 (ACS symposium series, vol 398)
93. Cichocki A, Parasiewicz-Kaczmarska J, Michalik M, Bus M (1990) Zeolites 10:577
94. Kotasthane AN, Shiralkar VP, Hegde SG, Kulkarni SB (1986) Zeolites 6:253
95. Calis G, Frenken P, De Boer E, Swolfs A, Hefni MA (1987) Zeolites 7:319
96. Borade RB (1987) Zeolites 7:398
97. Meagher A, Nair V, Szostak R (1988) Zeolites 8:3
98. Meagher A (1989) Zeolites 9:87
99. Ulan JG, Gronski R, Szostak R (1991) Zeolites 11:466, 472
100. Ione KG, Vostrikova LA, Petrova AV, Mastikhin VM (1984) In: Jacobs PA, Jaeger NI, Jirû P, Kazansky VP, Schulz-Ekloff G (eds) Structure and reactivity of modified zeolites. Elsevier, Amsterdam, pp 151-156
101. Weitkamp J, Beyer HK, Borbély G, Cortés-Corbéran V, Ernst S (1986) Chem Ing Techn 58:969
102. Kentgens APM, Bayense CR, Van Hoof JHC, De Haan JW, Van de Ven LJM (1991) Chem Phys Lett 176:399
103. Challoner R, Harris RK, Barri SAI, Taylor MJ (1991) Zeolites 11:827
104. Reddy JS, Reddy KR, Kumar R, Ratnasamy P (1991) Zeolites 11:553
105. Yanan Z, Shouhe X, Hexuan L (1990) Cuihua Xuebao 11:323
106. Yanan Z, Hexuan L (1990) Shiyon Xuebao, Shiyon Jiagong 6:33
107. Borade RB, Adnot A, Kaliagine S (1991) Zeolites 11:710
108. Barri SAI, Young D (1984) Eur Pat Appl 106478
109. Kumar R, Ratnasamy P (1990) J Catal 121:89
110. Quinzhu J, Pang W (1990) Huaxue Xuebao 48:761
111. Argauer RJ, Landolt GR (1972) US Pat 3702886
112. Kessler H, Chézeau JM, Guth JL, Strub H, Coudurier G (1987) Zeolites 7:360
113. Patarin J, Kessler H, Guth JL (1990) Zeolites 10:674
114. Seive A, Guth JL, Raatz F, Petit L (1989) Eur Pat Appl 342075
115. Axon SA, Huddersman K, Klinowski J (1990) J Chem Phys Letters 172:398
116. Barrer RM (1984) In: Olson D, Bisio A (eds) Proceedings of the 6th International Zeolite Conference. Butterworth, Guildford, pp 870-886
117. Gabelica Z, Guth JL (1989) In: Jacobs PA, van Santen RA (eds) Zeolites: facts, figures, future. Elsevier, Amsterdam, pp 421-430 (Studies in surface science and catalysis, vol 49)

118. Chen NY, McCullen SM (1989) Eur Pat Appl 325053
119. Taramasso M, Perego G, Notari B (1983) US Pat 4410501
120. Perego G, Bellussi G, Corno C, Taramasso M, Buonomo F, Esposito A (1986) In: Murakami Y, Iijima A, Ward JW (eds) New developments in zeolite science and technology. Kodansha-Elsevier, Tokyo, pp 129-136
121. Popa JM, Guth JL, Kessler H (1988) Eur Pat Appl 2923673
122. Kraushaar B, Van Hooff JHC (1988) Catal Lett 1:81
123. Qiu S, Pang W, Yao S (1989) In: Jacobs PA, van Santen RA (eds) Zeolites: facts, figures, future. Elsevier, Amsterdam, pp 133-142 (Studies in surface science and catalysis, vol 49)
124. Lopez A, Tuilier MH, Kessler H, Guth JL, Popa JM (1991) In: Hasnain SS (ed) X-ray absorption fine structure. Ellis Harwood, New York, p 549
125. Pang W, Yu L, Wu Y (1989) Geodeng Xuexiao Huaxue Xuebao 19:951
126. Dongare MK, Singh P, Moghe PP, Ratnasamy P (1991) Zeolites 11:690
127. Taramasso M, Manara G, Fattore V, Notari B (1980) UK Pat Appl 2024790
128. (a) Kornatowski J, Sychev M, Goncharnk V, Baur WH (1991) In: Öhlmann G, Pfeiffer H, Fricke R (eds) Catalysis and adsorption by zeolites. Elsevier, Amsterdam, pp 581-590 (Studies in surface science and catalysis, vol 65); (b) Kornatowski J, Wichterlova B, Jirkovsky J, Löffler E, Pilz W (1996) J Chem Soc, Faraday Trans 92:1067
129. Jiao Q, Pang W (1990) Gaodeng Xuexiao, Huaxue Xuebao 11:797
130. Klotz MR (1983) US Pat 4405502
131. Prasado Rao TSR, Borade RB, Halgeri AB (1986) Chem Express 1:709
132. Yu L, Pang W (1989) Jilin Daxue Ziran Kexue Xuebao (1977)
133. Milton RM (1959) US Pat 2882243
134. Warzywoda J, Thompson RW (1991) Zeolites 11:577
135. Vaughan DEW (1985) US Pat 4534947
136. Puppe L, Reiss G (1987) Ger Offen 3617840
137. Puppe L, Mengel L (1989) Eur Pat Appl 310916
138. Guth JL, Collin P, Wey R (1970) Bull Soc Fr Minéral Cristallogr 93:59
139. Kerr GT (1966) Inorg Chem 5:1537
140. Jarman RH, Melchior MT; Vaughan DEW (1983) In: Stucky GD, Dwyer FG (eds) Intrazeolite chemistry. Am Chem Soc, Washington DC, pp 267-281 (ACS symposium series, vol 218)
141. Jarman RH, Melchior MT (1984) J Chem Soc Chem Comm 414
142. (a) Milton RM (1959) US Pat 2882244; (b) Union Carbide (1962) French Patent 1286136
143. Breck DW (1964) US Pat 3130007
144. Vaughan DEW (1988) US Pat 4714601
145. Vaughan DEW, Strohmaier KG (1990) US Pat 4931267
146. Delprato F, Delmotte L, Guth JL, Huve L (1990) Zeolites 10:546
147. Vaughan DEW, Barret MG (1982) US Pat 4333859
148. Kokotailo GT, Ciric J (1971) Adv Chem Ser 101:109
149. Davis ME (1992) In: Occelli ML, Robson HE (eds) Molecular sieves. Van Nostrand Reinhold, New York, pp 60-69 (Synthesis of microporous materials, vol 1)
150. Ciric J (1976) US Pat 3972983
151. Vaughan DEW (1989) Eur Pat Appl 315461
152. Treacy MMJ, Newsam JM, Beyerlein RA, Leonowicz ME, Vaughan DEW (1986) J Chem Soc Chem Comm 1211
153. Martens JA, Jacobs PA, Cartlidge S (1989) Zeolites 9:423
154. Delprato F (1989) Nouvelles zéolithes de la famille de la faujasite synthétisées en présence d'éther-couronnes. Thesis, Université de Haute Alsace, Mulhouse, France
155. Robson HE (1989) in: Occelli ML, Robson HE (eds) Zeolite synthesis. ACS, Washington, DC, pp 436-447 (ACS symposium series, vol 398)
156. Ernst S, Kokotailo GT, Weitkamp J (1987) In: Grobet PJ, Mortier WJ, Vansant EF, Schulz-Ekloff G (eds) Innovation in zeolite materials science. Elsevier, Amsterdam, pp 29-36 (Studies in surface science and catalysis, vol 37)
156a. Arhancet JP, Davis ME (1991) Chem Materials 3:567

156b. Dougnier F, Patarin J, Guth JL, Anglerot D (1992) Zeolites 12:160
157. Wadlinger RL, Kerr GT, Rosinski EJ (1967) US Pat 3308069
158. Gabelica Z, Dewaele N, Maistriau L, B Nagy J, Derouane E (1989) In: Occelli ML, Robson HE (eds) Zeolite synthesis. ACS, Washington, DC, pp 518-543 (ACS symposium series, vol 398)
159. Bhat RM, Kumar R (1990) J Chem Techn Biotechnol 48:453
160. Camblor MA, Mifsud A, Pérez-Pariente J (1991) Zeolites 11:792, 202
161. Di Renzo F, Albizane A, Nicolle MA, Fajula F (1991) In: Öhlmann G, Pfeiffer H, Fricke R (eds) Catalysis and adsorption by zeolites. Elsevier, Amsterdam, pp 603-612 (Studies in surface science and catalysis, vol 65)
162. Caullet P, Hazm J, Guth JL, Joly JF, Lynch J, Raatz F (1992) Zeolites 12:240
163. Artioli G, Kvick A (1990) Eur J Mineralogy 2:749
164. Breck DEW (1965) US Pat 3216789
165. Verduijn JP, Gellings EP (1988) Eur Pat Appl 280513
166. Barrer RM, Mainwaring DE (1972) J Chem Soc Dalton 1259
167. Vaughan DW (1985) Eur Pat Appl 142348
168. Joshi PN, Kotasthane AN, Shiralkar VP (1990) Zeolites 10:598
169. Flanigen EM (1968) Dutch Pat 6710729
170. Ciric J (1968) Br Pat 1117568
171. Rubin MK, Plank CJ, Rozinski EJ (1977) US Pat 4021447
172. Rubin MK, Rozinski EJ (1982) US Pat 4331643
173. Fajula F, Vera-Pacheco M, Figueras F (1987) Zeolites 7:203
174. Fajula F, Nicolas S, Di Renzo F, Gueguen C, Figueras F (1989) In: Occelli ML, Robson HE (eds) Zeolite synthesis. ACS, Washington, DC, pp 493-505 (ACS symposium series, vol 398)
175. Plank CJ, Rosinski EJ, Rubin MK (1979) US Pat 4175114
176. Grose RW, Flanigen EM (1981) US Pat 4257885
177. Bale WJ, Stewart DG (1981) Eur Pat Appl 30811
178. Erdem A, Sand LB (1979) J Catalysis 60:241
179. Delmotte L (1985) Etude des conditions de formation des zéolithes très riches en silice. Thesis, Université de Haute Alsace, Mulhouse, France
180. Van Santen RA, Keijsper J, Oms G, Kortbeek AGTG (1986) In: Murakami Y, Iijima A, Ward JW (eds) New developments in zeolite science and technology. Kodansha-Elsevier, Tokyo, pp 169-175
181. Dai FY, Suzuki M, Takeahashi H, Saito Y (a) (1986) In: Murakami Y, Iijima A, Ward J (eds) New developments in zeolite science and technology. Kodansha-Elsevier, Tokyo, pp 223-230; (b) (1989) In: Ocelli ML, Robson HE (eds) Zeolite synthesis. ACS Washington, DC, pp 244-256 (ACS symposium series, vol 398)
182. Shiralkar VP, Clearfield A (1989) Zeolites 9:363
182a. See for example: Joly JF, Caullet P, Guth JL, Faust AC, Brunard N, Kolenda F (1990) Fr Appl 90/16529
183. Mravec D, Riecanova, Ilavsky J, Majling J (1991) Chem Papers 45:27
184. Tissler A, Polanek P, Girrbach U, Müller U, Unger KK (1989) In: Karge HG, Weitkamp J (eds) Zeolites as catalysts, sorbents and detergent builders. Elsevier, Amsterdam, pp 399-408 (Studies in surface science and catalysis, vol 46)
185. Barrer RM (1948) J Chem Soc 2158
186. Barrer RM, White EAD (1952) J Chem Soc 1561
186a. Domine D, Quobex J (1968) In: Barrer RM (ed) Molecular sieves. Society of Chemical Industry, London, pp 78-84
187. (a) Bajpai PK, Rao MS, Gokhale KVGK (1978) Ind Eng Chem Prod Res Develop 17: 223; (b) Bajpai PK (1986) Zeolites 6:2
188. Sand LB (1968) In: Barrer RM (ed) Molecular Sieves. Society of Chemical Industry, London, pp 71-77
189. Raatz F, Marcilly C, Freund E (1985) Zeolites 5:329
190. Sanders JV (1985) Zeolites 5:81

191. Kim GJ, Ahn WS (1991) Zeolites 11:745 and references cited therein
192. Chumbhale VR, Chandwadkar AJ, Rao BS (1992) Zeolites 12:63
193. Aiello R, Barrer RM (1970) J Chem Soc A 1470
194. Moudafi L, Massiani P, Fajula F, Figueras F (1987) Zeolites 7:63
195. Moudafi L, Dutartre R, Fajula F, Figueras F (1986) Applied Catalysis 20:189
196. Ueda S, Nishimura M, Koizumi M (1985) In: Drzaj S, Hocevar S, Pejovnik S (eds) Zeolites: synthesis, structure, technology and application, Elsevier, Amsterdam, pp 105–110 (Studies in surface science and catalysis, vol 24)
197. Millward GR, Thomas JM (1984) J Chem Soc Chem Comm 77
198. Breck DW, Acara NA (1960) US Pat 2950952
199. Rubin MK, Rosinski EJ, Plank CJ (1978) US Pat 4086186
200. Occelli ML, Innes RA, Pollak SS, Sanders JV (1987) Zeolites 7:265
201. Chapman DM, Roe AL (1990) Zeolites 10:730
202. (a) Kuznicki SM (1989) US Pat 4853202; (b) Kuznicki SM, Trush KA, Allen FM, Levine SM, Hamil MM, Hayhurst DT, Manson M (1992) In: Occelli ML, Robson HE (eds) Synthesis of microporous materials. Van Nostrand Reinhold, New York, pp 427–453 (Molelcular sieves, vol 1)
203. Valtchev VP (1994) J Chem Soc Chem Comm 261
204. Anderson MW, Terasaki O, Oshuna T, Philippou A, MacKay SP, Ferreira A, Rocha J, Lidin S (1994) Nature (London) 367:347
205. Liu X, Shang M, Thomas JK (1997) Microporous Mater 10:273
206. Dadachov MS, Le Bail A (1997) Eur J Solid State Inorg Chem 34:381
207. Beck JS, Chu CTW, Johnson ID, Kresge CT, Leonowicz ME, Roth WJ, Vartuli JC (1991) PCT Int Appl WO 91-11390
208. Beck JS, Vartuli JC, Roth WJ, Leonowicz ME, Kresge CT, Schmitt KD, Chu CTW, Olson DH, Sheppard EW, McCullen SB, Higgins JB, Schlenker JL (1992) J Am Chem Soc 114:10834
209. Monnier A, Schüth F, Huo Q, Kumar D, Margolese D, Maxwell RS, Stucky GD, Kirshnamurty M, Petroff P, Firouzi A, Janicke M, Chmelka BF (1993) Science 261:1299
210. Chen CY, Burkett SL, Li HX, Davis ME (1993) Microporous Mater 2:27
211. (a) Inagaki S, Fukushima Y, Okada A, Kuranchi T, Kuroda K, Kato C (1993) In: von Ballmoos R, Higgins JB, Treacy MMJ (eds) Proceedings of the 9th International Zeolite Conference, Montreal 1992. Butterworth-Heinemann, London, pp 305–311; (b) Inagaki S, FukushimaY, Kuroda K (1994) In: Weitkamp J, Karge HG, Pfeifer H, Hölderlich W (eds) Proceedings of the 10th International Zeolite Conference, Garmisch-Partenkirchen, 1994. Elevier, Amsterdam, pp 125–132 (Studies in surface science and catalysis, vol 49B)
212. Chen CY, Xiao SQ, Davis ME (1995) Microporous Mater 4:1
213. Beck JS, Vartuli JC (1996) Current Opinion in Solid State and Materials Science 1:76

CHAPTER 2

Phosphate-Based Zeolites and Molecular Sieves

Johan A. Martens and Pierre A. Jacobs

2.1 Introduction

Material scientists tend to subdivide the family of crystalline, microporous three-dimensional oxide structures into silicate-based and phosphate-based specimens. The distinction has a historical origin, poro-tecto-phosphates having been discovered after the poro-tecto-silicates. This subdivision is sometimes arbitrary. Some materials contain both silicon and phosphorus. The first publication on aluminophosphate molecular sieves was published in 1982 [1]. Since that time, there has been substantial research activity in the area of phosphate-based zeolites and molecular sieves. Presently, the number of phosphate-based structures and compositions exceeds even that of silicate-based materials.

A large number of poro-tecto-aluminophosphate structures synthesized by Wilson et al. [1] are denoted with the acronym '$AlPO_4$-n' referring to three-dimensional oxide frameworks with P/Al ratios equal to 1:1. The number 'n' refers to the specific crystallographic structure. The numbering is arbitrary. Analogous materials containing other elements besides Al and P are conveniently thought of as being derived from an imaginary, isostructural $AlPO_4$, in which Al, or P, or both, have undergone partial or even complete isomorphic substitution. The nature of the isomorphic substitution is currently indicated in the acronyms [2, 3], e.g., SAPO-5 is a silicoaluminophosphate with structure type No. 5, CoAPO-11 is a cobaltoaluminophosphate with structure type No. 11, etc. For the naming of materials, several inventors have used codes derived from the name of their institute or company. For example, the code name 'TAMU' stands for 'Texas A & M University', 'VPI' for 'Virginia Polytechnic Institute' and 'JDF' for 'Jilin David Faraday'.

The purpose of this chapter is to provide the reader with a feeling for the versatility of physicochemical properties that can be generated in phosphate-based zeolites and molecular sieves. Structural and physicochemical concepts are explained that allow the behavior of individual, as well as families of species, to be rationalized and predicted.

2.2
Structural, Synthetic and Physicochemical Concepts Relevant to Poro-tecto-phosphates

Based on their pioneering work in the field of poro-tecto-phosphates, Bennett et al. and Flanigen et al. discovered a number of structural, synthetic and physicochemical rules in these materials [3–5]. Based on the wealth of structural and physicochemical data that are actually available, these rules are now firmly established and additional concepts could be developed [6]. In the first part of this chapter, the main concepts in poro-tecto-phosphates are developed and illustrated. In the second part, ways in which these rules can be handled to rationalize and predict material properties such as for instance the relative siting in the oxide frameworks of the individual elements, the generation of framework charges, susceptibility to isomorphic substitutions, sorption behavior, pore size and thermal stability are explained and illustrated.

2.2.1
$AlPO_4$ ($GaPO_4$) Topological Concept

In all presently known phosphate-based crystalline microporous three-dimensional oxides, phosphorus is four-coordinated and has a valency of five. Aluminum is trivalent, and its coordination number can be IV, V or VI. Usually, four ligands of aluminum are oxo bridges with framework phosphorus atoms, the eventual fifth and sixth ligand positions being occupied with non-framework guest molecules such as hydroxo, fluoro, phosphate, aquo or other species. The '$AlPO_4$' topological concept arises from the observation that, when neglecting these secondary coordinations of aluminum, $AlPO_4$ structures can be represented by an idealized framework composed of corner sharing, strictly alternating AlO_4 and PO_4 tetrahedra [3, 4]. For substituted specimens, this basic $AlPO_4$ structure with Al, P alternation is obtained after appropriate replacement of the third elements with Al and/or P [3, 5].

Because each oxygen of the framework is two-connected, a three-dimensional network of linked AlO_4 and PO_4 tetrahedra can conveniently be represented by a line drawing, representing only the branches between adjacent Al and P atoms. In the line drawing, the centers of the oxygen tetrahedra lie at the nodes, each node representing an Al or P atom. Topological line drawings of most of the phosphate-based structures can, e.g., be found in the Atlas of Zeolite Structure Types, issued by the International Zeolite Association [7].

The strict alternation of Al and P in the tetrahedral nodes of the idealized framework precludes the occurrence of odd-membered rings of corner sharing tetrahedra in poro-tecto-phosphates. In poro-tecto-silicates, five rings are, e.g., very common odd-membered rings.

The crystal chemistry of gallium in microporous crystalline phosphates is analogous to that of aluminum, and a '$GaPO_4$' topological concept is convenient when treating gallium containing materials.

Zeolites and molecular sieves are often classified according to their pore size [7]. The pore size is conveniently measured as the number of nodes in the smallest ring of the topological framework model giving access to microvoid volumes. Small, medium and large pore materials contain 8-ring, 10-ring and 12-ring apertures, respectively, while openings larger than 12-rings are indicated as 'extra-large' [8]. In all presently known structures with 10 and 12 rings, the '$AlPO_4$' concept is obeyed (Table 2.1). Exceptions to this simple topological concept are found in some structures with extra-large pores and in small pore

Table 2.1. Poro-tecto-phosphates and the '$AlPO_4$' ('$GaPO_4$') Concept

Window size[a]	Framework structure	'$AlPO_4$' Concept	Type species	Ref.
20	-CLO	framework interrupted at Ga and P	cloverite	[9]
	'JDF-20'[b]	framework interrupted at P		[10]
18	VFI	obeyed	VPI-5	[11]
			$AlPO_4$-54	[12]
16	'ULM-5'[b]	framework interrupted at Ga and P		[13]
14	AET	obeyed	$AlPO_4$-8	[14, 15]
			MCM-37	[16]
12	AFI	obeyed	$AlPO_4$-5	[17]
	AFR	obeyed	SAPO-40	[18, 19]
	AFS	obeyed	MAPSO-46	[20]
	AFY	obeyed	CoAPO-50	[20]
	ATO	obeyed	$AlPO_4$-31	[21]
	ATS	obeyed	MAPO-36	[22]
	BPH	obeyed	beryllophosphate-H	[23]
	'DAF-1'[b]	obeyed		[24]
	FAU	obeyed	SAPO-37	[2]
10	AEL	obeyed	$AlPO_4$-11	[25, 26]
	AFO	obeyed	$AlPO_4$-41	[27]
	'$Ga_9P_9O_{36}OH \cdot HNEt_3$'[b]	obeyed		[28]
	'ULM-3'[b]	obeyed		[13]
	'ULM-4'[b]	obeyed		[13]
8	ABW	obeyed	ZnPO-ABW	[29]
	AEI	obeyed	$AlPO_4$-18	[30]
	AFT	obeyed	$AlPO_4$-52	[31]
	ANA	obeyed	$AlPO_4$-24	[1]
	APC	obeyed	$AlPO_4$-C	[32]
			$AlPO_4$-H3	[33, 34]
			MCM-1	[35, 36]
	APD	obeyed	$AlPO_4$-D	[32]
	ATN	obeyed	MAPO-39	[37]
	ATT	obeyed	$AlPO_4$-33	[38]
			$AlPO_4$-12-TAMU	[39]
	ATV	obeyed	$AlPO_4$-25	[40]

Table 2.1 (continued)

Window size[a]	Framework structure	'$AlPO_4$' Concept	Type species	Ref.
8	AWW	obeyed	$AlPO_4$-22	[41]
	CHA	obeyed	SAPO-34	[2]
			CoAPO-44,	[20]
			CoAPO-47	[20]
			ZYT-6	[42]
	ERI	obeyed	$AlPO_4$-17	[43]
	GIS	obeyed	MAPSO-43	[3]
	LEV	obeyed	SAPO-35	[2]
	LTA	obeyed	SAPO-42	[44]
	RHO	obeyed	BeAsPO-RHO	[29]
	'$AlPO_4$-12'[b]	framework interrupted at P contains Al-O-Al bonds		[45]
	'$AlPO_4$-14'[b]	obeyed		[46,47]
	'$AlPO_4$-14A'[b]	contains Al-O-Al bonds		[48]
	'$AlPO_4$-15'[b]	obeyed		[49,50]
	'$AlPO_4$-21'[b]	obeyed		[51,52]
	'$AlPO_4$-H2'[b]	obeyed		[31,53]
	'$AlPO_4$-EN3'[b]	obeyed		[54]
	'$GaPO_4$-C_7'[b]	obeyed		[55]
	'$AlPO_4$-JDF'[b]	obeyed		[56]
	'$AlPO_4$-CJ2'[b]	obeyed		[57]
	'ZnPO-dab'[b]	framework interrupted at Zn and P		[29]
6	AST	obeyed	$AlPO_4$-16	[58]
	LOS	obeyed	BePO-LOS	[59]
	SOD	obeyed	$AlPO_4$-20	[1]

[a] Number of T-atoms in smallest ring circumscribing channels.
[b] IZA structure code not assigned.

structures (Table 2.1). In these instances, framework elements at specific crystallographic positions have only three instead of four oxo bridges to neighboring framework elements, giving rise to a systematic interruption of the three-dimensional network (Table 2.1). At the framework interruptions at P, Al and Ga elements, the coordination is saturated with a hydroxo ligand; the coordination of lower valent elements such as Zn is saturated typically with an aquo ligand.

2.2.2 Al(Ga) Coordination Concept

Aluminum exhibits a rich coordination chemistry in $AlPO_4$s. Neutral molecules (e.g., water) and anionic species (e.g., fluoride, hydroxide or phosphate anions) adsorbed in the micropores can interact with aluminum atoms thus generating

Table 2.2. Distribution (%) of Al-centered Oxygen, Fluorine Polyhedra in As-synthesized $AlPO_4$ Materials

Coordination:	Al^{IV}	Al^{V}	Al^{VI}			Ref.
Ligands:	O_4	$O_4(OH)$	$O_4(OH)_2$	$O_4(OH_2)(OH)$	$O_4(OH_2)_2$	
Four-coordinated $AlPO_4$s						
$AlPO_4$-5	100	-	-	-	-	[60, 61]
$AlPO_4$-11	100	-	-	-	-	[60, 62]
$AlPO_4$-12 TAMU	100	-	-	-	-	[39]
JDF-20	100	-	-	-	-	[10]
$AlPO_4$-hydrates						
$AlPO_4$-H_3	50	-	-	-	50	[33, 63, 64]
$AlPO_4$-8	66.7	-	-	-	33.3	[65]
$AlPO_4$-H2	66.7	-	-	-	33.3	[53]
VPI-5	66.7	-	-	-	33.3	[66, 67]
$AlPO_4$-hydroxides						
$AlPO_4$-12	33.3	66.7	-	-	-	[45]
$AlPO_4$-21	33.3	66.7	-	-	-	[51, 52, 68, 69]
$AlPO_4$-EN3	33.3	66.7	-	-	-	[54]
$AlPO_4$-14	50	25	25	-	-	[46]
$AlPO_4$-17	55.56	44.44	-	-	-	[9, 60, 70]
$AlPO_4$-18	66.7	33.3	-	-	-	[30]
$AlPO_4$-14A	71.43[a]	-	-	28.57[b]	-	[48, 71]
$AlPO_4$-31	x[c]	x[c]	-	-	-	[60]
$AlPO_4$-20	x[c]	x[c]	-	-	-	[72]
$AlPO_4$-15	-	-	50	50	-	[50]
$Al_2(PO_4)_2$-$(OH)(NH_4)(H_2O)_2$	-	-	-	100	-	[49]
$AlPO_4$-phosphates						
$AlPO_4$-22	66.7	33.3	-	-	-	[41]
$AlPO_4$-fluorides						
$AlPO_4$-CJ2	-	x[c,d]	x[c,d]	-	-	[57, 73]
$AlPO_4$-CHA	x[c]	-	x[c,e]	-	-	[74]

[a] $O_2(OH)_2$ and O_4.
[b] $O_5(OH)$.
[c] Percentages not available.
[d] OH partially substituted with F.
[e] O_4F_2.

Al^{V} and Al^{VI} coordinations. In most instances, only a fraction of the Al atoms located at specific crystallographic positions exhibit this interaction with guest molecules. It will be shown later in this chapter that the physicochemical properties of an $AlPO_4$ material are determined primarily by the coordination chemistry of aluminum and, therefore, it is useful to classify the materials according to the coordinations adopted by the framework aluminum atoms.

A convenient reference state to determine the Al coordinations in an $AlPO_4$ phase is the as-synthesized form. After their crystallization in hydrothermal conditions, the microporous crystals are filled with organic structurizing agents (templates), water and other cationic or anionic molecular species. Negative electronic charges of the framework-bound anions are balanced by the positive charges of organic and/or inorganic cations that are encapsulated in the micropores. For the majority of $AlPO_4$ phases, the Al coordination has been determined either by structure refinement based on diffraction data or using high-resolution ^{27}Al solid state NMR. The as-synthesized $AlPO_4$ materials can be subdivided according to the Al coordination into (i) *four-coordinated* $AlPO_4s$, (ii) $AlPO_4$*-hydrates*, (iii) $AlPO_4$*-hydroxides*, (iv) $AlPO_4$*-fluorides* and (v) $AlPO_4$*-phosphates*. Literature data on the distribution of Al coordinations in as-synthesized $AlPO_4$ materials are collected in Table 2.2. It has to be emphasized that the Al coordination does not depend on the nature of the organic structurizing agent that is encapsulated during the crystallization, but is an intrinsic property, as demonstrated, e.g., with $AlPO_4$-5 [17, 60, 61] and $AlPO_4$-17 [9, 60, 70, 77]. The different Al coordinations are explained in Fig. 2.1. In *four coordinated* $AlPO_4s$, all framework Al atoms are four-coordinated (Al^{IV}). In $AlPO_4$ *hydrates*, Al atoms

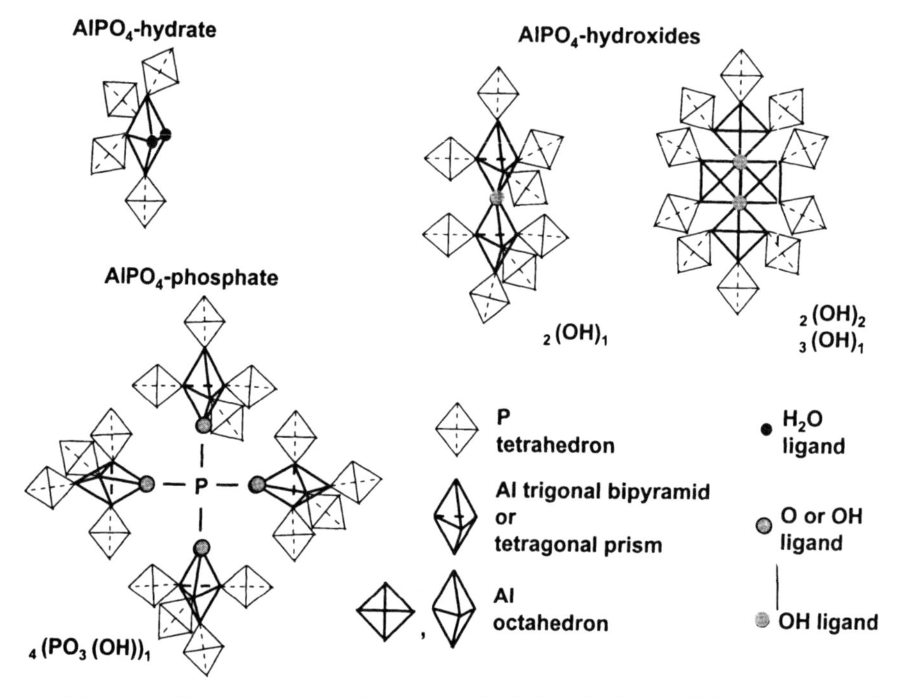

Fig. 2.1. Al coordinations occurring in as-synthesized $AlPO_4$-hydrates, $AlPO_4$-hydroxides and $AlPO_4$-phosphates. The terminology $_x(X)_y$ defines the number, x, of framework elements bridged by a single X species and the number, y, of bridges of type X between these framework elements

at specific crystallographic framework positions are six-coordinated (Al^{VI}) and have two aquo ligands in addition to the four oxo bridges to framework P atoms. This type of aluminum is denoted as 'aluminum dihydrate' [67]. In *$AlPO_4$-hydroxides*, a number of specific Al atoms are five- or six-coordinated depending on whether these atoms are linked either to one hydroxo ligand (Al^V), one hydroxo ligand and one aquo ligand or to two hydroxo ligands (Al^{VI}). Fluorine exhibits a similar chemistry in the coordination sphere of Al atoms of *$AlPO_4$-fluorides*. In *$AlPO_4$-phosphates* a phosphate anion not belonging to the framework is trapped in a systematic way in specific cages. The oxygens of these phosphate anions occupy a fifth coordination site of framework Al^V atoms. Whereas aquo ligands bind to one Al atom, hydroxo, fluoro and phosphate ligands are always bound to two Al atoms at least (Fig. 2.1).

A classification of $GaPO_4$ materials according to Ga coordination chemistry in their as-synthesized form is provided in Table 2.3. The as-synthesized $GaPO_4$s are either *$GaPO_4$-hydroxides* or *$GaPO_4$-fluorides*, with Ga coordinations of IV, V and VI. $GaPO_4$-hydrates and four coordinated $GaPO_4$s have not been reported.

Table 2.3. Distribution (%) of Ga-centered Oxygen, Fluorine Polyhedra in As-synthesized $GaPO_4$ Materials

Coordination:	Ga^{IV}	Ga^V	Ga^{VI}		Ref.
Ligands:	O_4	$O_4(OH)$	$O_4(OH)_2$	$O_4(OH_2)(OH)$	
$GaPO_4$-hydroxides					
$Ga_9P_9O_{36}OH \cdot HNEt_3$	66.7	33.3	–	–	[28]
$GaPO_4$-14	50	25	25	–	[46]
$GaPO_4$-21	33.3	66.7	–	–	[46]
$GaPO_4$-12	–	100	–	–	[46]
$GaPO_4$-C_7	–	–	50	50	[55]
$GaPO_4$-fluorides					
ULM-5	x[a,b]	x[a,b]	x[a,b]	–	[13]
cloverite	–	100[c]	–	–	[9,75]
$GAPO_4$-LTA	–	100[c]	–	–	[76]
ULM-3	–	66.7[b]	33.3[b]	–	[13]
ULM-4	–	66.7[b]	33.3[b]	–	[13]
$GaPO_4$-CJ2	–	x[a,b]	x[a,b]	–	[57]

[a] Percentages not available.
[b] F instead of OH.
[c] 75% O_4F and 25% $O_3F(OH)$.

2.2.3
Template: Framework Stoichiometry Concept

The organic amines and quaternary ammonium compounds that are encapsulated in the intracrystalline cavities during crystallization can exhibit different types of interactions with the host framework [78]. At one extreme, the encapsulated molecule acts as a space-filler. The channels and cages in the structure are approximately filled by uncharged encapsulated organic molecules positioned at van der Waals bonding distances to framework atoms. The space filling organic is often an undissociated ion pair such as a tetraalkylammonium hydroxide molecule or an unprotonated amine. This type of host-guest interaction occurs in four-coordinated $AlPO_4$s and $AlPO_4$-hydrates. At the other extreme, the encapsulated molecule carries a positive charge (organic ammonium cation) and is electrostatically coupled with a negative charge on the $AlPO_4$ framework, generated by addition of a hydroxo, fluoro or phosphate ligand to a framework aluminum (gallium) atom. For this type of interaction, there generally exists a simple stoichiometric relationship between template molecules and structural elements, such as cavities and repeat distances in channel directions (Table 2.4). $AlPO_4$-17 synthesized with piperidine is an illustrative example of the arrangement of charged template molecules in cages [43, 78] (Fig. 2.2). Two piperidinium ions are present in each ERI cavity. Each positive charge introduced by the template in the ERI cavity is balanced by an OH group, crosslinking a specific pair of Al^V atoms. In $AlPO_4$-18, a tetraethylammonium cation is located in the widest part of the AEI cage, while the hydroxyl ion is located in the narrow part (Fig. 2.2). For the first category of uncharged templates in van der Waals interaction with $AlPO_4$ frameworks, there is no such strict stoichiometric relationship. The number of template molecules per structural unit depends on their nature, as illustrated for the AFI-type of material in Table 2.4.

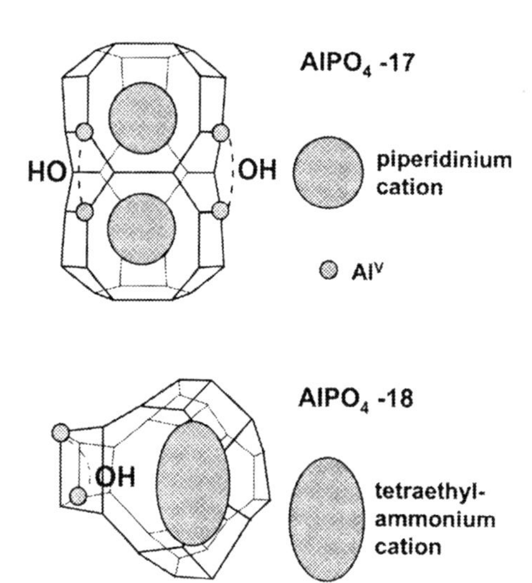

Fig. 2.2. Examples of the siting of organic cations and pairs of Al^V linked through a common hydroxo ligand in as-synthesized $AlPO_4$-17 [43] and $AlPO_4$-18 [30] materials. In $AlPO_4$-17, pairs of Al^V occupy only two-thirds of symmetrically related sites

Table 2.4. Examples for Stoichiometries of Template Incorporation[a] in $AlPO_4$-based Microporous Materials [5, 74, 78–80]

Framework	Template (R)	Stoichiometry[b]
AEI	tetraethylammonium	1/*aei* cage
AEL	dipropylamine	1/c-repeat in channel
	dipropylammonium	1/c-repeat in channel
AFI	tetrapropylammonium hydroxide	1/c-repeat in channel
	tetrapropylammonium	1/c-repeat in channel
	tripropylammine	1/c-repeat in channel
	tetraethylammonium	1/c-repeat in channel
	N,N′-dimethylpiperazine	1.3/c-repeat in channel
	1,4-diazabicyclo[2,2,2]octane	1.8/c-repeat in channel
	N,N-diethylethanolamine	1.3/c-repeat in channel
	triethylamine	1.2/c-repeat in channel
	2-methylpyridine	1.3/c-repeat in channel
	N-methylpyridine	1.3/c-repeat in channel
	tropine	1.4/c-repeat in channel
AFR	tetrapropylammonium	1/c-repeat in channel
AFS	dipropylamine	4/*afs* cage
AST	quinuclidine	1/*trd* cage
CHA	triethylamine	2/*cha* cage
	tetraethylammonium	1/*cha* cage
	morpholinium	2/*cha* cage
CHA	cyclohexylamine	2/*cha* cage
	N,N-diethylethanolamine	1.5/*cha* cage
ERI	piperidinium	2/*eri* cage
	1-azabicyclo[2,2,2]octane	2/*eri* cage
	2,2-dimethylpropylammine	2/*eri* cage
	cyclohexylamine	2/*eri* cage
	quinuclidine	2/*eri* cage
FAU	tetrapropylammonium	2/*fau* cage
	tetramethylammonium	1/*toc* cage
LEV	quinuclidine	1/*lev* cage
LTA	ethanolamine	3/*grc* cage
	tetramethylammonium	1/*toc* cage
	dipropylammonium	3/*grc* cage
SOD	tetramethylammonium	1/*toc* cage

[a] Templates in cationic form are charge-balanced with anions coordinated to framework Al atoms. For the amines, it is not established whether they are protonated.

[b] Topologic and geometric descriptions of the channels and cages can be found in [81].

2.3 Rationalization of Properties of Poro-tecto-phosphates with Structural, Synthetic and Physicochemical Concepts

The three concepts in poro-tecto-phosphates developed in the previous section readily explain a number of material properties such as pore size, thermal stability, adsorption properties and isomorphic substitution.

2.3.1 Pore Size

For fully four-coordinated frameworks, the upper limit of the number of tetrahedrally coordinated T atoms in a ring is 12, limiting the maximum free diameter to ca. 0.8 nm. Increasing the window size necessitates an unfavorable enlargement of the O–T–O angles in the ring far beyond the ideal tetrahedral angle of 109°. One way to facilitate the formation of O–T–O angles approaching 180° is by introducing framework elements with a coordination number of VI, such as dihydrate aluminum. The two water molecules occupy *cis*-positions in the coordination sphere of aluminum and protrude into the channel. Examples of this class of phosphates with extra-large pores are the materials with VFI topology, possessing 18-ring pore openings, and the molecular sieve $AlPO_4$-8, having a pore system with 14-ring windows (Fig. 2.3). This combination of tetrahedral and octahedral coordinations is an example of the recent tendency in zeolite material science to develop octahedrally coordinated frameworks (Fig. 2.4).

The alternative approach to avoid strain in the O–T–O angles is to interrupt the framework at positions in the wall of the extra-large pore. This situation is encountered in the JDF-20 structure. ULM-5 and cloverite are extra-large pore materials in which higher coordination numbers and systematic framework interruption are combined. In the interrupted framework structures, hydroxyl groups protrude into the channels, giving rise to pore appertures with special shapes (like the 4-parted clover leaf in cloverite, Fig. 2.3).

In the early literature on $AlPO_4$s [1, 4, 26], it was thought that aluminum with a coordination number of four was essential in order to obtain molecular sieving properties. Meanwhile, it has become clear that there is no such correlation of Al coordination and pore size.

2.3.2 Thermal Stability

For the purpose of sorptive and catalytic applications, the pores of the as-synthesized molecular sieves need to be evacuated. This is currently done by calcination. The high temperature stability of $AlPO_4$s ($GaPO_4$s) is dependent on the Al (Ga) coordination. When the interactions of Al (Ga) atoms with their fifth and sixth ligand are removed, displacive transformations (involving changes of relative atom positions), topotactic and disruptive transformations (involving bond breaking and re-linking) or structural collapse may be ob-

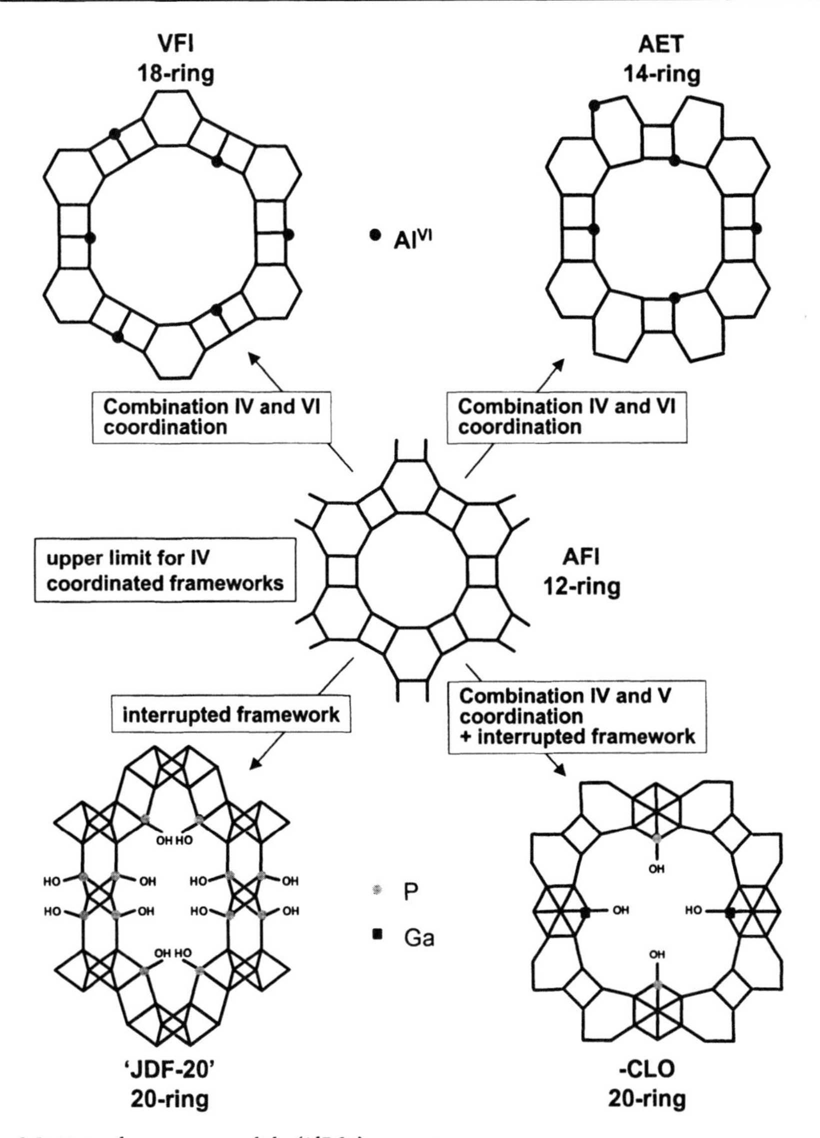

Fig. 2.3. Extra-large pores and the '$AlPO_4$' concept

served. An overview of the behavior of the different $AlPO_4$-type materials is given in Table 2.5.

The *four-coordinated* $AlPO_4$*s* are thermally and hydrothermally very stable structures. These structures withstand heating at temperatures exceeding 1300 K. Only minor structural changes (typically alterations of the crystallographic space group) are observed upon calcination and rehydration [26, 60, 62, 84, 85].

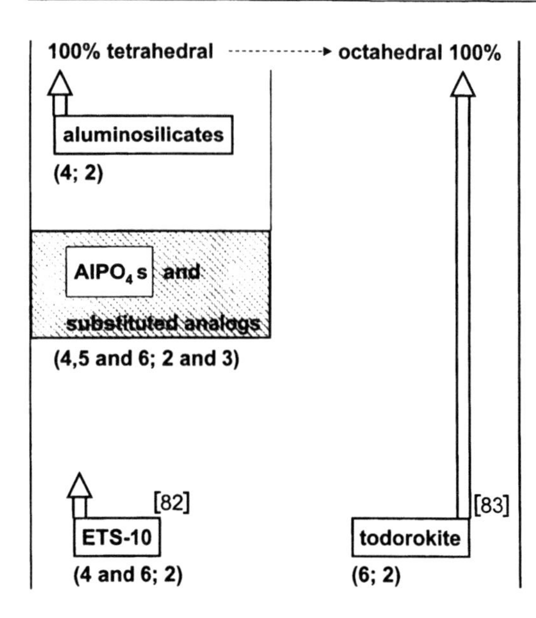

Fig. 2.4. Evolution from tetrahedral to octahedral coordination in molecular sieve frameworks. In the topological description of frameworks, the (*x*; *y*) designation specifies that the framework elements occupy *x*-connected vertices of the three-dimensional network; oxygen atoms occupy *y*-connected positions between the *x*-connected vertices [84]

In the *$AlPO_4$-hydrates*, water molecules play an important structural role. After the crystallization of $AlPO_4$-H3 [33] and VPI-5 [66, 86] only water molecules are contained in their micropores. Upon dehydration, the coordination number of Al is decreased gradually from VI over V to IV by removal of the aquo ligands, resulting in dramatic changes of the bond angles in the framework. $AlPO_4$-H3 exhibits complex topotactic and displacive phase transitions upon dehydration-rehydration treatments, related to the removal and reinsertion of the two aquo ligands of aluminum [32].

A VPI-5 structure with aluminum monohydrate sites can be obtained by careful dehydration [87]. The phase transitions of VPI-5-dihydrate, VPI-5-monohydrate, and anhydrous VPI-5 upon dehydration and rehydration are of the displacive type and are reversible [87]. Depending on the temperature and

Table 2.5. Thermal Stability of $AlPO_4$s

$AlPO_4$ type	Stability
Four coordinated $AlPO_4$s	very stable; minor structural changes
$AlPO_4$-hydrates	very unstable; disruptive (sometimes topotactic) phase transitions
$AlPO_4$-hydroxides	stability depending on Al^{IV} content high Al^{IV} content: displacive phase transitions low Al^{IV} content: disruptive (sometimes topotactic) phase transitions or collapse

atmosphere, a disruptive transition occurs, yielding another $AlPO_4$ hydrate structure with AET topology [88–92]. The VFI into AET transition reduces the pore openings from 18 to 14 rings and is sometimes reversible [88]. In both structures (VPI-5 and $AlPO_4$-8), one-third of the framework aluminum is aluminum dihydrate (Table 2.2).

The thermal and hydrothermal stability of *$AlPO_4$-hydroxides* ($GaPO_4$-hydroxides) depends on the structure type, and especially on the share of Al (Ga) atoms adopting a coordination number higher than IV. The hydroxo bridging between two or three Al^{V} or Al^{VI} atoms distorts the framework. This distortion vanishes upon removal of the hydroxo ligands on heating. For example, in $AlPO_4$-21, one third of the framework Al is four-coordinated; two thirds are five-coordinated due to the presence of a hydroxo ligand (Table 2.2). Calcination of $AlPO_4$-21 leads to a topotactic transformation and yields the $AlPO_4$-25 material, which is a four-coordinated $AlPO_4$ [40]. In structures in which the majority of Al atoms are four-coordinated, only displacive transformations are observed, for example, in $AlPO_4$-17 (with ERI topology) [77].

2.3.3
Adsorption Properties

Adsorption of hydrocarbons on $AlPO_4$s follows a micropore volume-filling mechanism, resulting in type I isotherms [90–92], as illustrated in Fig. 2.5 for benzene and hexane adsorption on $AlPO_4$-5. The polarity of the $AlPO_4$s is situated between that of aluminum-rich aluminosilicate zeolites (such as NaX) and silica molecular sieves (such as SiO_2-FAU) (Fig. 2.6).

The adsorption isotherm of polar molecules in $AlPO_4$ materials is of type V or VI, depending on the $AlPO_4$-type. When water is adsorbed on an evacuated $AlPO_4$, at low relative pressures, only small amounts are adsorbed (Figs. 2.5 and 2.6). At a specific higher pressure, there is a sudden rise of the adsorbed amount. This first adsorption step may be followed by an additional uptake step before saturation of the adsorbent, depending on the structure type. In $AlPO_4$-5 there is only one step, whereas in VPI-5, there are two (Fig. 2.6). With water and other polar adsorbates, $AlPO_4$s exhibit hystereses extending to very low pressures, as illustrated with $AlPO_4$-5 in Fig. 2.5 [88, 92–95]. This sorption behavior of polar molecules in $AlPO_4$s is explained by coordinative adsorption of the polar molecule on specific framework aluminum sites. The polar molecule enters the coordination shell of Al only after a sufficient amount is adsorbed. The reason is that the Al^{IV} to Al^{V} and Al^{VI} transitions require a bulk crystallographic phase transition. The hysteresis observed in the isotherms of polar molecules is not due to capillary condensation, but to the persistance of the structure with Al^{V} and/or Al^{VI} coordinations up to very low desorption pressures.

The sorption behavior of an $AlPO_4$ with respect to polar molecules is predictable based on its aluminum coordination chemistry. In $AlPO_4$s that occur as *hydrates*, the formation of Al^{V} and Al^{VI} upon coordinative adsorption of polar molecules occurs typically in two consecutive steps, as illustrated in Fig. 2.6 for the adsorption of water on VPI-5. The Al atoms involved in the coordinative

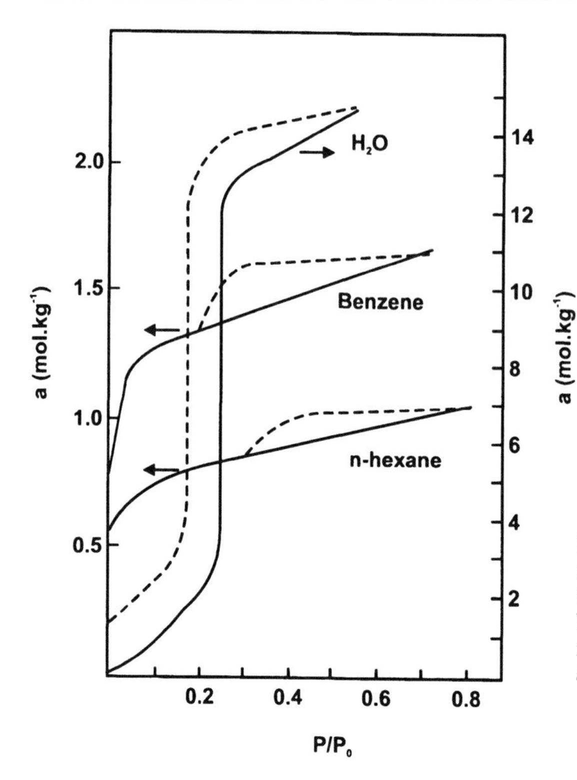

Fig. 2.5. Adsorption (*full lines*) and desorption (*dotted lines*) isotherms at 301 K for hexane, benzene and water on $AlPO_4$-5 molecular sieve (data from [88 and 89])

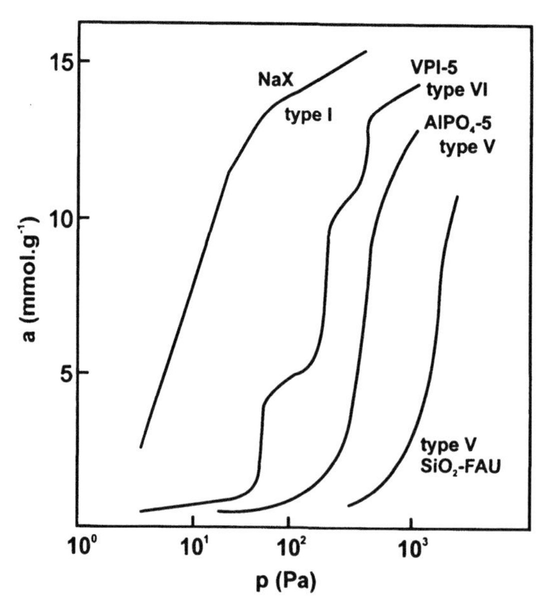

Fig. 2.6. Adsorption isotherms of water on NaX, VPI-5, $AlPO_4$-5 and silica-FAU (adapted from [93])

Table 2.6. Al Coordinations in Calcined, Four-Coordinated $AlPO_4$s upon Chemisorption of Polar Molecules [96–98]

Adsorbate	Dipole moment (Debye)	Molecular sieve	
		$AlPO_4$-5	$AlPO_4$-11
water	1.8	40% (Al^V + Al^{VI})	20% Al^{VI}
ammonia	1.3	50% (Al^V + Al^{VI})	50% (Al^V + Al^{VI})
methanol	1.7	no Al^V; no Al^{VI}	no Al^V; no Al^{VI}
ethanol	1.7	no Al^V; no Al^{VI}	no Al^V; no Al^{VI}
acetonitrile	3.9	no Al^V; no Al^{VI}	no Al^V; no Al^{VI}

adsorption processes and in aluminum dihydrate formation during the crystallization are the same [93].

In $AlPO_4$s that are *four-coordinated*, there is only one uptake step in the adsorption isotherm of polar molecules. Small molecules with high dipole moment (water and ammonia) can penetrate into the coordination shell of part of the Al atoms, generating Al^V and Al^{VI} coordinations, detectable with high-resolution ^{27}Al solid-state NMR (Table 2.6), whereas bulkier polar molecules do not adsorb coordinatively. The coordinative adsorptions of water and ammonia are accompanied by minor structural changes only.

When polar compounds are adsorbed in calcined *$AlPO_4$-hydroxides*, they can occupy the coordination sites of the hydroxo ligands. Either one or two polar adsorbate molecules are coordinatively adsorbed per Al site, depending on their nature and the $AlPO_4$ framework type (Table 2.7). For example, methanol creates Al^V environments in $AlPO_4$-18 (AEI topology), whereas water and ammonia create Al^{VI} coordinations (Table 2.7). Methanol starts generating Al^V coordinations only after ca. one molecule of methanol is sorbed per cavity (Fig. 2.7). Jänchen et al. suggested that the sorbate structure effectively interacting with the framework aluminum atoms and causing the phase transition is a methanol dimer [95]. Based on ^{27}Al DOR NMR, it could be shown that the Al atoms that adopt the Al^V coordination in the as-synthesized material and after

Table 2.7. Al Coordinations in $AlPO_4$-hydrates and -hydroxides upon Adsorption of Polar Molecules [93, 95]

Adsorbate	Dipole moment (Debye)	Molecular sieve	
		VPI-5	$AlPO_4$-18
water	1.8	33% Al^{VI}	20% Al^{VI}
ammonia	1.3	n.d.[a]	Al^{VI} [b]
methanol	1.7	33% Al^{VI}	33% Al^V; no Al^{VI}
ethanol	1.7	33% Al^{VI}	n.d.[a]

[a] Not determined.
[b] Not quantified.

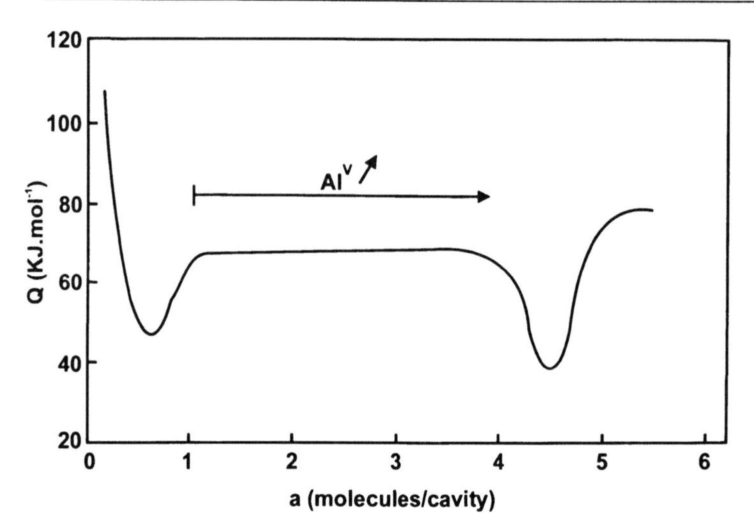

Fig. 2.7. Differential molar heat of adsorption of methanol on $AlPO_4$-18 against the amount adsorbed (adapted from [95])

methanol adsorption are the same atoms [95]. Their position in the narrow part of the AEI cage is shown in Fig. 2.2.

The incorporation of other elements into the $AlPO_4$ framework alters the sorption characteristics. The introduction of framework charges in substituted specimens generates strong adsorption centers. The substituted analogs encompass a range of moderately to highly hydrophilic surface properties, comparable to the variability found in silicate based materials [2, 3].

2.3.4
Isomorphic Substitution

Per definition, isomorphic substitution corresponds to a replacement of an element in a crystalline lattice by another element with similar cation radius and coordination requirements. Isomorphic substitution in $AlPO_4$s is currently achieved during crystallization by adding the element to be incorporated into the synthesis mixture. In poro-tecto-phosphates, many types of isomorphic substitutions are possible. The complexity arises from the presence of two types of atoms, Al and P, which are both susceptible to substitution, and from the diversity of coordinations of the framework Al atoms (IV, V or VI).

The first $AlPO_4$-based materials reported in 1982 contained Al and P only [1, 99]. The addition of Si to the Al and P framework elements resulted in 1984 in the synthesis of silicoaluminophosphate (SAPO) molecular sieves [2, 100]. Later, in 1985 and 1986, microporous metal aluminophosphates with frameworks containing Mg, Mn, Fe, Co or Zn were successfully prepared [3, 101], and the incorporation of other elements including Li, B, Be, Ti, Mn, Ga, Ge and As was achieved [3, 102]. The list of possible framework elements in $AlPO_4$-based materials is still growing to include, e.g., Cr, Ni, V and Sn. Furthermore, there seems to be no upper limit to the number of elements that can be incorporated simultaneously. The synthesis of multi-element phases containing up to six different framework elements has been claimed [103].

For phosphates obeying the $AlPO_4$ concept or having interrupted frameworks with strict Al, P alternation, three mechanisms of isomorphic substitution have been observed: substitutions of Al atoms (SM I), substitutions of P atoms (SM II), and substitutions of pairs of adjacent Al and P atoms (SM III) (Fig. 2.8).

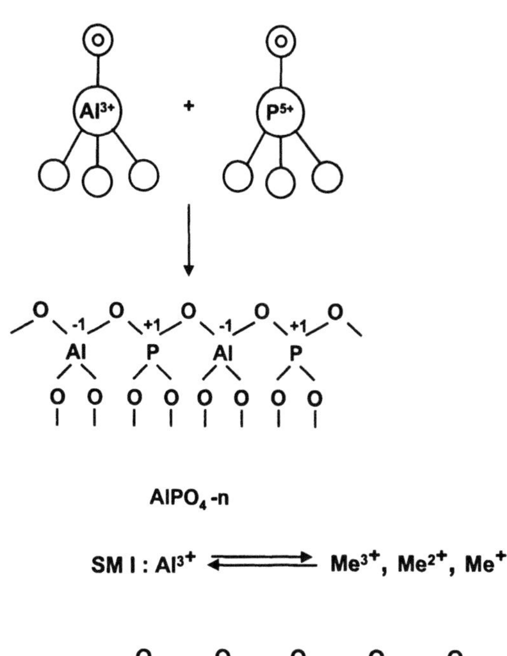

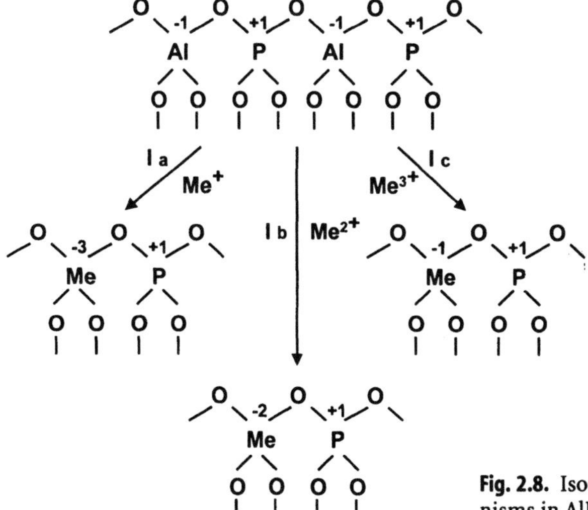

Fig. 2.8. Isomorphic substitution mechanisms in $AlPO_4$ molecular sieves

SM II : $P^{5+} \rightleftharpoons Me^{5+}, Me^{4+}$

SM III : $Al^{3+} \rightleftharpoons 2Me^{4+}$

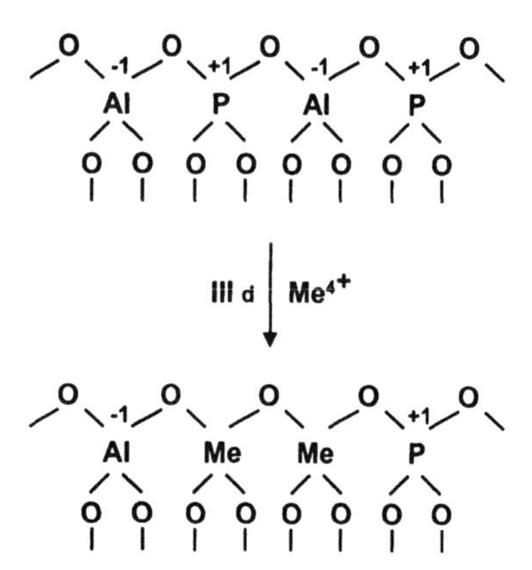

Fig. 2.8 (continued)

SM III has been observed with silicon only. The SM I and SM II mechanisms can be subdivided according to the formal charge of the element introduced (Fig. 2.8). Other types of substitutions not mentioned in Fig. 2.8 are unlikely [5]. They would lead either to positively charged frameworks or to a too high negative framework charge density.

Pauling's minimum radius ratio, R, concept [104], useful for predicting the potential incorporation of an element into silicate-based lattices [105] is of limited value for predicting the substitution in $AlPO_4$s (Fig. 2.9). Elements not normally predisposed to tetrahedral coordination with oxygen and situated largely outside the preferred R range for tetrahedral coordination ($0.225 < R < 0.414$) can be incorporated. Their successful incorporation is ascribed to the flexibility of phosphate frameworks [3] and the unique possibility offered in $AlPO_4$s to extend the coordination sphere of the element with additional ligands

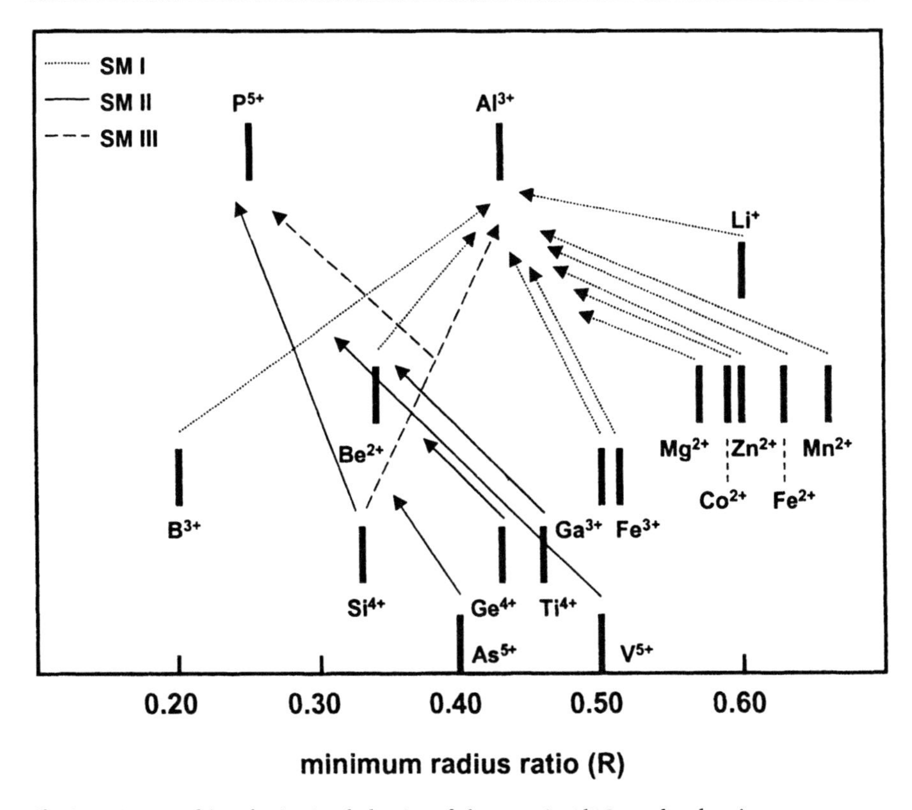

Fig. 2.9. Isomorphic substitution behavior of elements in $AlPO_4$ molecular sieves

besides the oxo bridges of the framework. In $AlPO_4$s, the formal charge rather than the ionic radius determines the substitution behavior (Fig. 2.9). With respect to isomorphic substitution, $GaPO_4$s have received much less attention compared to $AlPO_4$s.

2.3.5 Isomorphic Substitutions SM Ia, Ib and IIa Generating Framework Charges

Isomorphic substitutions of the types SM Ia, Ib and IIa generate negative framework charges and have received the most attention in the literature. Specimens with one of these types of substitution have the potential of Brønsted acidity and cation exchange provided that the organic templates can be evacuated with preservation of the crystallinity and provided the substituted elements are stable in the framework.

The behavior of the different classes of $AlPO_4$s with respect to isomorphic substitutions generating framework charges (SM Ia, Ib and IIa) is summarized in Table 2.8.

Table 2.8. Isomorphic Substitutions Generating Framework Charges in $AlPO_4$-based Materials [6]

Mechanism:	SM Ia	SM Ib	SM IIa
Elements:	Li(I)	Co(II), Fe(II) Mg(II), Mn(II) Zn(II)	Si(IV)
Four-coordinated $AlPO_4$s			
Framework topologies:	probably all	all	all
Substitution degree:[a]	n.d.[b]	0…S…ES	0…S
$AlPO_4$-hydrates			
Framework topologies:	none	reported for VFI	reported for VFI
Substitution degree:	–	traces	traces
$AlPO_4$-hydroxides, -fluorides and -phosphates			
Framework topologies:	specific	specific	specific
Substitution degree:[a]	n.d.[b]	0…S	0…<S or 0…S
No $AlPO_4$ analog			
Framework topolgies:	none	specific	specific
Substitution degree:[a]	–	S	S

[a] S: stoichiometric substitution level; ES: substitution level exceeding stoichiometric substitution.
[b] Not determined.

The *four-coordinated $AlPO_4$s* are susceptible to the three types of substitutions which introduce framework charges [3, 5]. In these $AlPO_4$s, the templates are either unprotonated amines, protonated amines or tetraalkylammonium cations, encapsulated as ion pairs, e.g., as organic ammonium hydroxide and tetraalkylammonium hydroxide molecules. The negative framework charge generated upon SM Ia, SM Ib or SM IIa substitution can effectively be balanced by the organic template, which becomes positively charged by protonation (in the case of amines), or omission of the anion (in the case of ion pairs). The template/framework stoichiometry (Table 2.4) dictates the maximum degree of substitution. Specimens in which the framework charge corresponds to the maximum charge tolerated by the template-framework stoichiometry are called 'stoichiometrically' substituted specimens [6].

The template/framework stoichiometry is less stringent with respect to the incorporation of bivalent elements according to SM Ib in four-coordinated $AlPO_4$s (Table 2.8). When the stoichiometric degree of substitution is exceeded, extra-framework Me^{2+} cations function as additional charge compensating species [106]. With Co(II), Fe(II) and Mn(II), part of the metal occurs frequently in extra-framework positions readily from low levels of substitution on [107–113].

The incorporation of Li(I) according to the SM Ia mechanism creates two formal negative framework charges at each Li site. The framework charge is balanced partly by the monovalent organic cations, partly by extra-framework Li^+ cations [3, 4].

In their as-synthesized form, *$AlPO_4$-hydroxides, -fluorides and -phosphates* have negatively charged frameworks due to the binding of anionic ligands with framework Al atoms. Isomorphic substitutions of the type SM Ia, Ib and IIa disturb the electrostatic charge coupling of the negative framework charges of the anionic framework ligands and encapsulated organic cations. With the increasing degree of incorporation of the foreign element, the anionic framework ligands are progressively removed such that the framework-template charge balance remains satisfied. The success of the substitution depends on the stability of the framework in the absence of interactions with the hydroxo, fluoro and phosphate species (Table 2.8).

$AlPO_4$-hydrates are rather unreactive towards isomorphic substitutions generating framework charges (Table 2.8). Several explanations can be advanced to explain this behavior [6]. (i) For some materials like VPI-5 and $AlPO_4$-H3, there is almost no incorporation of organic molecules in the micropores during crystallization. Isomorphic substitutions generating framework charges are unlikely since there is no species in the micropores capable of balancing negative framework charges. (ii) The introduction of charged species in the micropores is expected to disturb the water structures. (iii) The aluminum dihydrate sites are essential for the structural integrity. A replacement of framework aluminum atoms coordinated with two aquo ligands with elements not capable of adopting such coordination is unlikely. (iv) With respect to Si incorporation according to SM IIa, the generation of Si(4Al) environments in which some of the Al atoms are hexa-coordinated is probably unfavored given the analogy with aluminosilicate zeolites, in which Al^{VI} coordinations do not occur.

Several multi-element compounds do not have an $AlPO_4$ analog [5, 20] (Table 2.8). Most of the time these compounds have stoichiometric compositions and the framework Al atoms are four-coordinated. Lowering of the degree of substitution is difficult to achieve, as it would disturb the charge balance or the template/framework stoichiometry.

Microporous crystalline beryllophosphates (BePOs) and zincophosphates (ZnPOs) are 1:1 combinations of Be(II) or Zn(II) with P(V) in tetrahedral oxide frameworks [29, 59, 106, 107]. The large negative framework charge is balanced with Li^+, Na^+ and tetramethylammonium cations. Stability problems arise when these compounds are heated at moderate temperatures.

2.3.6
Isomorphic Substitutions Ic and IIb Generating Electroneutral Frameworks

Literature on SM Ic type isomorphic substitution according to which Al is replaced by other trivalent elements is rather scarce [5]. However, this type of chemistry may lead to interesting catalytic properties. Cr(III) has been incorporated into the framework of $AlPO_4$-14, an *$AlPO_4$-hydroxide* [108]. The siting of chromium in the framework could be determined. Cr adopts a sixfold coordination and replaces selectively the aluminum at $AlO_4(OH)_2$ sites in the framework.

Due to its small cation radius, B(III) cannot mimick Al^{V} and Al^{VI} coordinations and substitutes Al^{IV} selectively. Boron incorporation is typically found in *four-coordinated $AlPO_4$s* [5].

Partial replacement of P with As(V) has been reported to be possible for $AlPO_4$s with AEL and AEI topologies [5]. The incorporation of traces of V(V) in $AlPO_4$-5 seems possible. Its siting in the framework is complex and is probably not in agreement with the SM IIb concept [109, 110].

2.3.7
Si Incorporation According to SM III Generating Si and AlP Domains

Silicon is the only element exhibiting SM III. By replacing systematically all Al and P atoms with Si in a certain section of the crystal starting from the external surface, the formation of framework charges can be avoided. A transition from an aluminophosphate into a silicate compositional domain in an AFI type of framework is illustrated in Fig. 2.10. An electroneutral framework is generated, comprising an $AlPO_4$ core (AlP domain) and topotactic SiO_2 overlayer(s) (Si domains). At the boundary of the two crystal domains, Si (3Si, 1Al) environments are present (Fig. 2.10). It has to be stressed that formal framework charges must not necessarily be associated with these Si environments.

SM III occurs in four-coordinated $AlPO_4$s. In AEL frameworks, extensive and virtually pure SM III has been realized in a large part of the crystals with levels of Si of up to 46% of the T atoms [111–113]. In many SAPO materials, Si is incorporated according to a combination of SM IIa and SM III mechanisms.

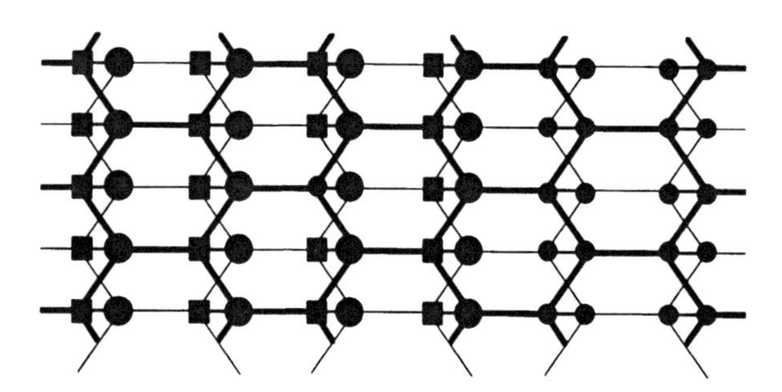

Fig. 2.10. Example of heterogeneous framework composition in a silicoaluminophosphate with AFI topology, generated through SM III substitutions

2.3.8
Si Incorporation According to Combinations of SM IIa and SM III Generating SiAl and SiAlP Domains

The majority of reported SAPO materials can be rationalized based on a combination of SM IIa and SM III types of substitutions [6]. The combination of two types of substitutions gives rise to a whole domain of possible silicoaluminophosphate compositions in the Si, Al, P ternary framework composition diagram (Fig. 2.11). The chemical composition in individual SAPO crystals is heterogeneous. One part of a SAPO crystal corresponds chemically to a silicate with SiO_2 composition (Si domain). Situations may be encountered where some Si atoms are replaced by Al atoms, thus generating negative framework charges in an aluminosilicate (SiAl domain). The remaining part combines the Al, P and Si elements, is generated by a SM IIa mechanism, and contains isolated Si atoms having four Al neighbors in the framework (SiAlP domain).

The *four-coordinated $AlPO_4$s* are susceptible to Si incorporation via a combination of SM IIa and SM III mechanisms [111, 114]. SAPO-5 crystals containing large layers of SiO_2 composition, representing, e.g., up to 25% of the oxide lattice have been obtained [115]. The silicon is always concentrated at the surface of the crystals [116, 117]. SAPO-11 samples with SM IIa and SM III combinations have been prepared [118].

Silicon incorporation according to SM IIa and SM III combinations is also possible in specific *$AlPO_4$-hydroxides* and *-fluorides*, SAPO-31 [119] and SAPO-34 [120], but the amount of silicon incorporated is lower. Individual crystals

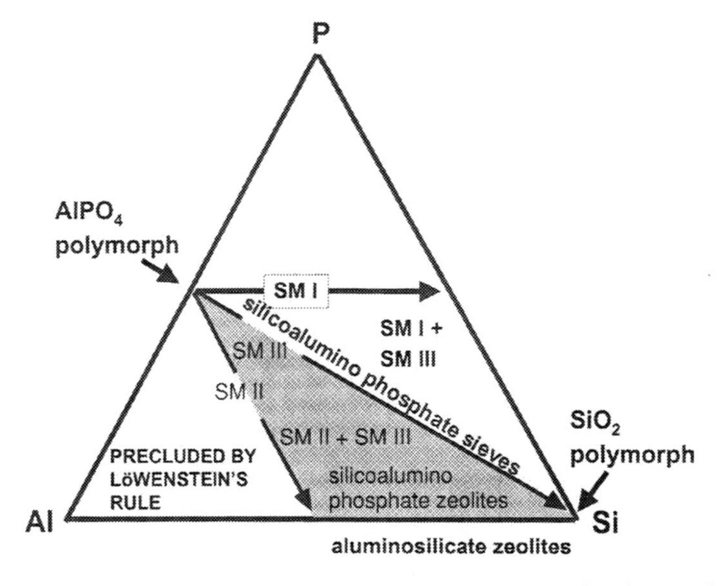

Fig. 2.11. Si, Al, P compositional diagram for silicoaluminophosphate zeolites and molecular sieves. Actually observed compositions are situated in the shaded area

of SAPO-42 and SAPO-20 (derived from $AlPO_4$ phosphates) contain a small amount of SiAlP intergrown with SiAl moieties [72, 121]. The sodalite cages of these SAPO-20 and SAPO-42 materials contain occluded phosphate anions.

The stoichiometric SAPO-37 species according to SM IIa with silicon has a framework T-atom composition of 12.5% Si, 50.0% Al and 37.5% P [122]. Each type of framework atom has only one type of environment. Each Si tetrahedron is linked to 4Al tetrahedra [Si(4Al) environment], each P tetrahedron to 4Al tetrahedra [P(4Al) environment]) and each Al tetrahedron to 3P and 1Si tetrahedra [Al(3P, 1Si) environment]. The negative framework charge is balanced with two tetrapropylammonium cations in the supercages, and one tetramethylammonium cation in each truncated octahedral cage [122]. In specimens with higher levels of Si, the framework can be subdivided into SiAlP regions with the Si, Al, P composition of stoichiometric SAPO-37, and SiAl regions [115, 123].

From the discussion on the mechanisms of Si incorporation in SAPOs, it follows that attempts to rationalize the catalytic activity of SAPOs by concepts such as overall Si content or lattice charge are meaningless. In SAPO crystals containing Si or SiAl and SiAlP domains, Brønsted acid sites are located in the SiAlP domains, in the SiAl domains and at the domain interfaces. The contribution to the catalytic turnovers of the different types of acid sites is strongly dependent on the structure type [115].

In SAPO-5, the SiAl domains have essentially a SiO_2 composition and do not contain catalytically active Brønsted acid sites [115]. The catalytic activity of SAPO-5 is situated in the SiAlP domains, where Si atoms give rise to the Brønsted acidity. The Brønsted acid strength in the SiAlP domains is very homogeneous, and the catalytic activity per acid site does not depend on the presence of SiO_2 domains [124]. Similar acid strengths were found for samples with pure SM IIa and samples with SM III and SM IIa combinations [125, 126].

In SAPO-11 with pure SM III, the catalytic activity is due to acid sites located at SiAlP-Si interfaces [115]. In SAPO-37, Brønsted acid sites are present in the SiAl and SiAlP domains as well as on the phase boundaries, where the sites exhibit the highest catalytic turnover numbers [115].

Acknowledgments. This work has been sponsored by the Belgian Federal Ministry of Science Policy in the frame of an IUAP-PAI programme and by the Flemish FWO.

References

1. Wilson ST, Lok BM, Messina CA, Cannan TR, Flanigen EM (1982) J Am Chem Soc 104:1146
2. Lok BM, Messina CA, Patton RL, Gajek RT, Cannan TR, Flanigen EM (1984) J Am Chem Soc 106:6092
3. Flanigen EM, Lok BM, Patton RL, Wilson ST (1986) In: Murakami Y, Iijima A, Ward JW (eds) New Developments in Zeolite Science and Technology, Proceed. 7th Int. Zeolite Conf, Kodansha-Elsevier, Tokyo 103

4. Bennett JM, Dytrych WJ, Pluth JJ, Richardson JW Jr, Smith JV (1986) Zeolites 6:349
5. Flanigen EM, Patton RL, Wilson ST (1988) In: Grobet PJ, Mortier WJ, Vansant EF, Schulz-Ekloff G (eds) Innovation in Zeolite Materials Science. Stud Surf Sci Catal 37:13
6. Martens JA, Jacobs PA (1994) In: Jansen JC, Stöcker M, Karge HG, Weitkamp J (eds) Advanced Zeolite Science and Applications. Stud Surf Sci Catal 85:653
7. Meier WM, Olson DH (1992) Zeolites 12:449
8. Li H-X, Davis ME (1994) Catal. Today 19:61
9. Estermann M, McCusker LB, Baerlocher C, Merrouche A, Kessler H (1991) Nature 352:320
10. Huo Q, Xu R, Li S, Xu Y, Ma Z, Thomas JM, Chippendale AM (1992) J Chem Soc Commun 875
11. Davis ME, Saldarriaga C, Montes C, Garces J, Crowder C (1988) Nature 331:698
12. Richardson JW Jr, Smith JV, Pluth JJ (1989) J Phys Chem 93:8212
13. Férey G, Loiseau T, Riou D (1994) Materials Science Forum 125:152-153
14. Dessau RM, Schlenker JL, Higgins JB (1990) Zeolites 10:522
15. Vogt ETC, Richardson JW (1990) J Solid State Chem 87:469
16. Chu CTW, Schlenker JL, Lutner JD, Chang CD (1992) US Patent 5,091,073
17. Bennett JM, Cohen JP, Flanigen EM, Pluth JJ, Smith JV (1983) In: Intrazeolite Chemistry. Am Chem Soc Symp Ser 218:109
18. Meier WM, Olson DH (1992) Zeolites 12:30
19. Dumont N, Gabelica Z, Derouane EG, McCusker LB (1993) Microporous Materials 1:149
20. Bennett JM, Marcus BK (1988) In: Grobet PJ, Mortier WJ, Vansant EF, Schulz-Ekloff G (eds) Innovation in Zeolite Materials Science. Stud Surf Sci Catal 37:269
21. Bennett JM, Kirchner RM (1992) Zeolites 12:338
22. Smith JV, Pluth JJ, Andries KJ (1993) Zeolites 13:166
23. Harvey G (1988) Z Kristallogr 182:123
24. Wright PA, Jones RH, Natarajan S, Bell RG, Chen J, Hursthouse MB, Thomas JM (1993) J Chem Soc Chem Commun 633
25. Bennett JM, Smith JV (1985) Z Kristallogr 171:65
26. Bennett JM, Richardson JW Jr, Pluth JJ, Smith JV (1987) Zeolites 7:160
27. Meier WM, Olson DH (1992) Zeolites 12:477
28. Yang G, Feng S, Xu R (1987) J Chem Soc Chem Commun 1254
29. Gier TE, Stucky GD (1991) Nature 349:508
30. Simmen A, McCusker LB, Baerlocher Ch, Meier WM (1991) Zeolites 11:654
31. Bennett JM, Kirchner RM, Wilson ST (1989) In: Jacobs PA, van Santen RA (eds) Zeolites: Facts, Figures, Future. Stud Surf Sci Catal 49:731
32. Keller EB, Meier WM, Kirchner RM (1990) Solid State Ionics 43:93
33. Pluth JJ, Smith JV (1985) Nature 318:165
34. d'Yvoir F (1961) Bull Soc Chim 372:1762
35. Derouane EG, Valyocsik EW, von Ballmoos R (1984) Eur Pat Appl 146:384
36. Martens JA, Verlinden B, Mertens M, Grobet PJ, Jacobs PA (1989) In: Occelli ML, Robson HE (eds) Zeolite Synthesis. ACS Symp Ser No 398, Washington 305
37. McCusker LB, Brunner GO, Oju AF (1990) Acta Crystallogr A46:C55
38. Patton RL, Gajek RT (1984) US Patent 4,473,663
39. Rudolf PR, Saldarriaga-Molina C, Clearfield A (1986) J Phys Chem 90:6122
40. Richardson JW Jr, Smith JV, Pluth JJ (1990) J Phys Chem 94:3365
41. Richardson JW Jr, Pluth J, Smith JV (1989) Naturwissenschaften 76:467
42. Ito M, Shimoyama Y, Saito Y, Tsurita Y, Otake M (1985) Acta Crystallogr C41:1698
43. Pluth JJ, Smith JV, Bennett JM (1986) Acta Crystallogr C42:283
44. Kessler H, Patarin J, Schott-Darie C (1994) In: Jansen JC, Stöcker M, Karge HG, Weitkamp J (eds) Advanced Zeolite Science and Applications. Stud Surf Sci Catal 85:75
45. Parise JB (1984) J Chem Soc Chem Commun 1449
46. Parise JB (1985) J Chem Soc Chem Commun 606

47. Zibrowius B, Lohse U, Szulzewsky K, Fichtner-Schmittler H, Pritzkow W, Richter-Mendau J (1991) In: Öhlmann G et al. (eds) Catalysis and Adsorption by Zeolites. Stud Surf Sci Catal 65:549
48. Pluth JJ, Smith JV (1987) Acta Crystallogr C43:866
49. Parise JB (1984) Acta Crystallogr C40:1641
50. Pluth JJ, Smith JV, Bennet JM, Cohen JP (1984) Acta Crystallogr C40:2008
51. Bennett JM, Cohen JP, Artioli G, Pluth JJ, Smith JV (1985) Inorg Chem 24:188
52. Parise JB, Day CS (1985) Acta Crystallogr C41:515
53. Li H-X, Davis ME (1993) J Chem Soc Faraday Trans 89:952
54. Parise JB (1985) In: Drzaj B et al. (eds) Zeolites: Synthesis, Structure, Technology and Applications. Stud Surf Sci Catal 24:271
55. Wang T, Yang G, Feng S, Shang C, Xu R (1989) J Chem Soc Chem Commun 948
56. Xu R, Huo Q, Pang W (1993) In: von Ballmoos R, Higgins JB, Treacy MMJ (eds) Proceed. 9th Int Zeolite Conf. Vol I, Butterworth-Heinemann, Stoneham 271
57. Ferey G, Loiseau T, Lacorre P, Taulelle F (1993) J Solid State Chem 105(1):179
58. Bennett JM, Kirchner RM (1991) Zeolites 11:502
59. Harrison WTA, Gier TE, Stucky GD (1993) Zeolites 13:242
60. Blackwell CS, Patton L (1984) J Phys Chem 88:6135
61. Müller D, Jahn E, Fahlke B, Ladwig G, Haubenreisser U (1985) Zeolites 5:53
62. Goepper M, Guth F, Delmotte L, Guth JL, Kessler H (1989) In: Jacobs PA, van Santen RA (eds) Zeolites: Facts, Figures, Future. Stud Surf Sci Catal 49B:857
63. Pluth JJ, Smith JV (1986) Acta Crystallogr C42:1118
64. Grobet PJ, Geerts H, Martens JA, Jacobs PA (1989) In: Klinowski J, Barrie PJ (eds) Recent Advances in Zeolite Science. Stud Surf Sci Catal 52:193
65. Martens JA, Geerts H, Grobet PJ, Jacobs PA (1992) In: Derouane EG, Lemos F, Naccache C, Ribeiro FR (eds) Zeolite Microporous Solids: Synthesis, Structure and Reactivity. NATO ASI Ser C, Vol 352, Kluwer Academic Publishers, Dordrecht 477
66. McCusker LB, Baerlocher Ch, Jahn E, Bülow M (1991) Zeolites 11:308
67. Grobet PJ, Martens JA, Balakrishnan I, Mertens M, Jacobs PA (1989) Appl Catal 56:L21
68. Jelinek R, Chmelka BF, Wu Y, Grandinetti PJ, Pines A, Barrie PJ, Klinowski J (1991) J Am Chem Soc 113:4097
69. Alemany LB, Timken HKC, Johnson ID (1988) J Magn Res 80:427
70. Zibrowius B, Lohse U (1992) Solid State Nuclear Magnetic Resonance 1:137
71. Goepper M, Guth JL (1991) Zeolites 11:477
72. Hasha D, de Saldarriaga LS, Saldarriaga C, Hathaway PE, Cox DF, Davis ME (1988) J Am Chem Soc 110:2127
73. Taulelle F, Loiseau T, Maquet J, Livage J, Ferey G (1993) J Solid State Chem 105(1):191
74. Klock E, Delmotte L, Soulard M, Guth JL (1993) In: von Ballmoos R, Higgins JB, Treacy MMJ (eds) Proceed. 9th Int Zeolite Conf. Vol I, Butterworth-Heinemann, Stoneham 633
75. Merrouche A, Patarin J, Kessler H, Soulard M, Delmotte L, Guth JL, Joly JF (1992) Zeolites 12:226
76. Simmen A, Patarin J, Bearlocher Ch (1992) In: von Ballmoos R, Higgins JB, Treacy MMJ (eds) Proceed. 9th Int Zeolite Conf. Vol I, Butterworth-Heinemann 433
77. Lohse U, Löffler E, Kosche K, Jänchen J, Parlitz B (1993) Zeolites 13:549
78. Pluth JJ, Smith JV (1989) In: Jacobs PA, van Santen R (eds) Zeolites: Facts, Figures, Future. Stud Surf Sci Catal 49B:835
79. Wilson ST, Lok BM, Messina CA, Flanigen EM (1984) In: Olson D, Bisio A (eds) Proceed. 6th International Zeolite Conf. Butterworths, Guildford 97
80. Wilson ST, Flanigen EM (1989) In: Occelli ML, Robson HE (eds) Zeolite Synthesis. ACS Symp Ser No 398, Washington 329
81. Smith JV (1989) In: Jacobs PA, van Santen RA (eds) Zeolites: Facts, Figures, Future. Stud Surf Sci Catal 49:29
82. Anderson MW, Terasaki O, Oshuna T, Philippou A, Mackay SP, Ferreira A, Rocha J, Lidin S (1994) Nature 367:347

83. Shen Y-F, Zeger RP, DeGuzman RN, Suib SL, McCurdy L, Potter DI, O'Young C-L (1993) Science 260:511
84. Khouzami R, Coudurier G, Lefebvre F, Vedrine JC, Mentzen BF (1990) Zeolites 10:183
85. Meinhold RH, Tapp NJ (1990) J Chem Soc Chem Commun 219
86. Rudolf PR, Crowder CE (1990) Zeolites 10:163
87. Martens JA, Feijen E, Lievens JL, Grobet PJ, Jacobs PA (1991) J Phys Chem 95:10025
88. Malla PB, Komarneni S (1991) Mat Res Soc Symp Proceed. Vol 233, Material Research Society 237
89. Thamm H, Stach H, Jahn E, Fahlke B (1986) Adsorption Sci Techn 3:217
90. Wilson ST, Lok BM, Messina CA, Cannan TR, Flanigen EM (1983) In: Stucky GD, Dwyer FG (eds) Intrazeolite Chemistry. Am Chem Soc Symp Ser No 218:79
91. Jänchen J, Stach H, Grobet PJ, Martens JA, Jacobs PA (1992) Zeolites 12:9
92. Stach H, Thamm H, Fiedler K, Grauert B, Wieker W, Jahn E, Öhlmann G (1986) In: Murakami Y, Lijima A, Ward JW (eds) New Developments in Zeolite Science and Technology, Proceed. 7th Int Zeolite Conf, Kodansha-Elsevier, Tokyo 539
93. Jänchen J, Stach H, Grobet PJ, Martens JA, Jacobs PA (1993) Proceed. 9th Int Zeolite Conf. Vol II, von Ballmoos R, Higgens JB, Treacy MMJ (eds) Butterworth-Heinemann, Stoneham 21
94. Kenny MB, Sing KSW, Theocharis CR (1991) J Chem Soc Chem Commun 974
95. Jänchen J, Peeters MPJ, de Haan JW, van de Ven LJM, van Hooff JHC, Girnus I, Lohse U (1993) J Phys Chem 97(46):12042
96. Kustanovich I, Goldfarb D (1991) J Phys Chem 95:8818
97. Peeters MPJ, de Haan JW, van de Ven LJM, van Hooff JHC (1993) J Phys Chem 97:5363
98. Peeters MPJ, de Haan JW, van de Ven LJM, van Hooff JHC (1993) J Phys Chem 97:8254
99. Wilson ST, Lok BM, Flanigen EM (1982) US Patent 4,310,440
100. Lok BM, Messina CA, Patton RL, Gajek RT, Cannan TR, Flanigen EM (1984) US Patent 4,440,871
101. Messina CA, Lok BM, Flanigen EM (1985) US Patent 4,544,143
102. Wilson ST, Flanigen EM (1985/86) Eur Patent Appl 132,708; US Patent 4,567,029
103. Lok MB, Marcus BK, Flanigen EM (1985) Eur Patent Appl 158,350
104. Pauling L (1960) The Nature of the Chemical Bond. 3rd ed, Cornell University Press 545
105. Tielen M, Geelen M, Jacobs PA (1985) In: Fejes P, Kallo D (eds) Proceedings International Symposium on Zeolite Catalysis. Acta Physica et Chemica Szegedensis 1
106. Harvey G, Meier WM (1989) In: Jacobs PA, van Santen RA (eds) Zeolites: Figures, Facts, Future. Stud Surf Sci Catal 49A:179
107. Nenoff TM, Harrison WTA, Gier TE, Nicol JM, Stucky GD (1992) Zeolites 12:770
108. Helliwell M, Kaucic V, Cheetham GMT, Harding MM, Kariuki BM, Rizkallah PJ (1992) 9th International Zeolite Conference. Montreal, Recent Research Report No 203
109. Montes C, Davis ME, Murray B, Narayana M (1990) J Phys Chem 94:6431
110. Rigutto MS, van Bekkum H (1993) J Mol Catal 81:77
111. Mertens M, Martens JA, Grobet PJ, Jacobs PA (1990) In: Barthomeuf D, Derouane EG, Höldrich W (eds) Guidelines for Mastering the Properties of Molecular Sieves - Relationship between the Physicochemical Properties of Zeolitic Systems and Their Low Dimensionality. NATO ASI, Ser B, Vol 221. Plenum Press, New York, London 1
112. Jahn E, Müller D, Becker K (1990) Zeolites 10:151
113. Martens JA, Grobet PJ, Jacobs PA (1990) J Catalysis 126:299
114. Martens JA, Mertens M, Grobet PJ, Jacobs PA (1988) In: Grobet PJ, Mortier WJ, Vansant EF, Schulz-Ekloff G (eds) Innovation in zeolite materials science. Stud Surf Sci Catal 37:97
115. Martens JA, Grobet PJ, Jacobs PA (1990) J Catalysis 126:299
116. Rajic N, Stojakovic D, Hocevar S, Kaucic V (1992) Zeolites 13:384
117. Young D, Davis ME (1991) Zeolites 11:277
118. Yang L, Aizhen Y, Qinhua X (1991) Appl Catal 67:169
119. Zubowa H-L, Alsdorf E, Fricke R, Neissendorfer F, Richter-Mendau J, Schreier E, Zeigan D, Zibrowius B (1990) J Chem Soc Faraday Trans 86(12):2307

120. Xu Y, Maddox P, Couves JW (1990) J Chem Soc Faraday Trans 86(2):425
121. Kühl GH, Schmitt KD (1990) Zeolites 10:2
122. Sierra de Saldarriaga L, Saldarriaga C, Davis M (1987) J Am Chem Soc 109:2686
123. Martens JA, Janssens C, Grobet PJ, Beyer HK, Jacobs PA (1989) In: Jacobs PA, van Santen RA (eds) Zeolites, Facts, Figures and Future. Stud Surf Sci Catal 46:215
124. Halik C, Chaudhuri SN, Lercher JA (1989) J Chem Soc Faraday Trans I 85(11):3879
125. Halik C, Lercher JA (1988) J Chem Soc Faraday Trans I 84:4457
126. Meusinger J, Vinek H, Dworeckow G, Goepper M, Lercher JA (1991) In: Jacobs PA, Jaeger NI, Kubelkova L, Wichterlova B (eds) Zeolite Chemistry and Catalysis. Stud Surf Sci Catal 69:373

CHAPTER 3

Modification of Zeolites

Günter H. Kühl

3.1
Ion Exchange of Zeolites

3.1.1
Introduction and Theory

Zeolites, in a narrow definition, are porous crystalline aluminosilicates having a uniform pore structure and exhibiting ion-exchange behavior. The framework of zeolites consists of SiO_4 and AlO_4 tetrahedra sharing oxygen ions located at their apices, thus resulting in the general framework formula $(AlO_2)_x(SiO_2)_{(n-x)}$, where n is the number of tetrahedra per unit cell, and $x \leq n/2$. Since aluminum is trivalent, every AlO_2 unit carries a negative charge, which is compensated by a positive charge associated with a cation. Therefore, the ion-exchange capacity of a zeolite depends on the chemical composition, i.e., a higher ion-exchange capacity is observed in zeolites of low SiO_2/Al_2O_3 ratio [1a]. The specific ion-exchange capacity varies with the structure of the zeolite and the exchange cation.

Ion exchange is generally carried out in aqueous systems. A limited amount of work has been published where a non-aqueous co-solvent was used along with water. Solid-state ion exchange using fused salts has been the subject of some recent work.

The study of ion exchange in zeolites has been less extensive than in ion-exchange resins, largely because of the few applications that zeolites have found in this field. Major applications are in the use as water softeners in laundry detergents (low-cost and non-polluting substitute for phosphate), and in the preparation of various types of catalysts. One advantage of zeolites for ion exchange is the availability of a great variety of zeolites with different, but uniform, pore sizes, so that "ion sieving" becomes possible. A disadvantage is the frequently lower capacity and the lower stability at extreme pH values.

3.1.2
Aqueous Ion Exchange

In the as-crystallized form of the zeolites, the cations are usually alkali and/or quaternary ammonium ions. Whereas the quaternary ions are hydrophobic and not hydrated, the alkali ions are surrounded by a hydration shell. The size of this hydration shell depends on the size and the charge of the cation as well as on the temperature. Hydrated ions are mobile within the zeolite channels and do not necessarily occupy specific positions. However, if the charge density of the zeolite framework is high, i.e., if the SiO_2/Al_2O_3 ratio is low, ions may have to shed part of their hydration shell so that the zeolite can accommodate a sufficient number of cations within the pore system. In such a case, cations may complete their coordination sphere with framework-O atoms and occupy specific sites even in the hydrated form of the zeolite. Usually, three water molecules are replaced by three framework oxygens, and water molecules occupy the remaining coordination sites. Since the framework oxygens are relatively fixed in their positions, the proper cation-oxygen distances are established by the cations assuming positions on an axis normal to the plane defined by the three oxygen atoms. The distance from the three-oxygen plane depends on the size of the bare ion. Thus, Li^+ is located close to the plane, Na^+, K^+ etc. are displaced along the axis into one of the cavities.

When the zeolite crystals are immersed in an aqueous electrolyte, the zeolitic ions communicate with the zeolite-external solution, resulting in an exchange of ions between the solid phase and the solution. Since the anionic charges of the framework are fixed, the number of cationic charges within the zeolite pores is constant. Depending on the particular zeolite, the pores may consist of one type of channel having essentially the same width throughout the unit cell; there may be two types of channels with different diameters, and the channels may or may not intersect, or the intracrystalline void may consist of one or more types of cages connected by smaller openings.

Anhydrous cations, e.g., quaternary ammonium ions, have a definite size. If they are larger than the pore openings of the zeolite structure, they are unable to enter or leave the pore system, and ion exchange cannot take place. Large hydrated cations, on the other hand, can vary their size by temporarily losing some water molecules, and the smaller, less hydrated ions may be able to penetrate the pore apertures. A fully hydrated cation is in equilibrium with partial hydration states of the ion and the bare ion. The equilibrium depends mainly on the temperature, but also on the concentration, and the influence of a co-solvent, if present. The size of the hydration shell at a particular temperature depends on the size and charge of the anhydrous ion. Generally, the higher the charge and the smaller the ion, the larger is the hydration shell. As an example, a hydrated potassium ion may be smaller than a hydrated sodium ion at the same temperature.

The rate of ion exchange depends on the concentration of ions of a size capable of penetrating the pores of the zeolite. At ambient temperature, a solution of a large hydrated ion may contain very few partially hydrated ions of a size smaller than the pore opening of the zeolite, so that the exchange is very slow. The rate of exchange increases with rising temperature, as water is stripped from the ions and the hydration equilibrium shifts toward less hydrated ions.

The anions associated with the cations in solution are usually excluded from the zeolite channels because of the repulsion exerted by the negative charges of the pore apertures. Salt imbibition becomes significant only when the ion exchange is carried out at concentrations of 0.5 molar or higher [2].

The theory of zeolitic ion exchange has recently been updated [3], and the reader is referred to this very thorough paper. For the present purpose, the established theory, although deficient in parts, will serve the reader better to follow 35 years of literature in this field.

Usually, ion exchange is performed by simply contacting the zeolite with a salt solution of a different cation at ambient temperature, or at elevated temperature if an accelerated exchange rate is desired. The exchange reaction, in which one type of cation is replaced with another, assumes an equlibrium state that is unique for the particular zeolite and the particular cations. Exchange between ion A^{a+}, initially in solution, and ion B^{b+}, initially in the zeolite, may be expressed as

$$bA_s^{a+} + aB_z^{b+} \rightleftarrows bA_z^{a+} + aB_s^{b+} \tag{3.1}$$

where a and b are the valencies of the exchanging cations A^{a+} and B^{b+}, and s and z designate the solution and zeolite phases. The definition of the thermodynamic equilibrium constant follows from Eq. 3.1:

$$K = (f_A^b Z_A^b \gamma_B^a m_B^a)/(f_B^a Z_B^a \gamma_A^b m_A^b) \tag{3.2}$$

where f_A and f_B are the rational single-ion activity coefficients of ions A and B in the zeolite phase, γ_A and γ_B the molal single-ion activity coefficients in the solution phase, Z_A and Z_B the equivalent fractions of ions A and B in the zeolite phase, and m_A and m_B the molarities of ions A and B in the solution phase. If

$$^{N}K_B^A = (m_B^a Z_A^b)/(m_A^b Z_B^a)$$

is the rational selectivity coefficient or concentration quotient, Eq. 3.2 changes to

$$K = {}^{N}K_B^A (f_A^b \gamma_B^a)/(f_B^a \gamma_A^b) \tag{3.3}$$

The corrected rational selectivity coefficient, K_C, is defined as

$$K_{cB}^A = {}^{N}K_B^A (\gamma_B^a/\gamma_A^b) = K(f_B^a/f_A^b) \tag{3.4}$$

For a binary reaction, the standard free energy per equivalent of exchange, defined as

$$\Delta G^0 = -(RT/ab) \ln K, \tag{3.5}$$

is the free energy change per equivalent associated with the reaction of Eq. 3.1, when the reactants proceed completely to products, all of the reactants and products being in their standard states. The standard states are defined as the homoionic forms of the exchanger immersed in an infinitely dilute solution of the corresponding ions [4].

3.1.2.1
Ion-Exchange Isotherms

The exchange equilibrium for ions A^{a+} and B^{b+} (Eq. 3.1) can be characterized conveniently by the ion-exchange isotherm, which is obtained by plotting the equivalent fraction of the entering ion in the zeolite over the equivalent fraction of the same ion in solution (the order is usually reversed in the British literature [5, 6]) at a specific temperature and concentration of the salt in the solution phase.

The equivalent fractions of A^{a+} in solution and in the zeolite, respectively, are given by

$$S_A = am_A/(am_A + bm_B) \tag{3.6}$$

$$Z_A = aM_A/(aM_A + bM_B) \tag{3.7}$$

where m_A, m_B are the molalities of the ions in solution and M_A, M_B are the concentrations (mol/kg) of the respective ions in the hydrated zeolite. It is sometimes more convenient to express the concentrations in the zeolite phase in terms of the ion-exchange capacity of the exchanger,

$$Z_A = bM_A/Q \tag{3.8}$$

where Q is the ion-exchange capacity, expressed as the number of charges per 100 g of zeolite after equilibration over saturated NaCl solution.

Since $S_A + S_B = 1.0$ and $Z_A + Z_B = 1.0$, the ion-exchange isotherms can be plotted from the equilibrium values S_A and Z_A. In a perfect zeolite structure, every framework-Al carries a negative charge. Therefore, Z_A and Z_B may be expressed as the ratios of eq. A^{a+}/Al and eq. B^{b+}/Al, respectively, provided that hydrolytic reactions of the zeolite or the solute do not lead to incidental hydronium ion exchange.

The preference of the zeolite phase for one of the two ions is expressed as the selectivity coefficient (or separation factor), α^A_B, where

$$\alpha^A_B = Z_A S_B / S_A Z_B \tag{3.9}$$

If A^+ is preferred by the zeolite over B^+, then α^A_B is greater than 1.0.

Five types of isotherms have been observed for binary ion exchange (Fig. 3.1) [1b].

When the logarithm of the corrected rational selectivity coefficient, $K_{cB}^{\ A}$, is plotted as a function of the equivalent fraction of the cation in the zeolite, Z_A, three types of curves may be obtained (Fig. 3.2) [1c]. For univalent exchange, $a^+ = b^+ = 1$, such a plot is linear and

$$\log K_{cB}^{\ A} = 2\,CZ_A + \log K^A_B \tag{3.10}$$

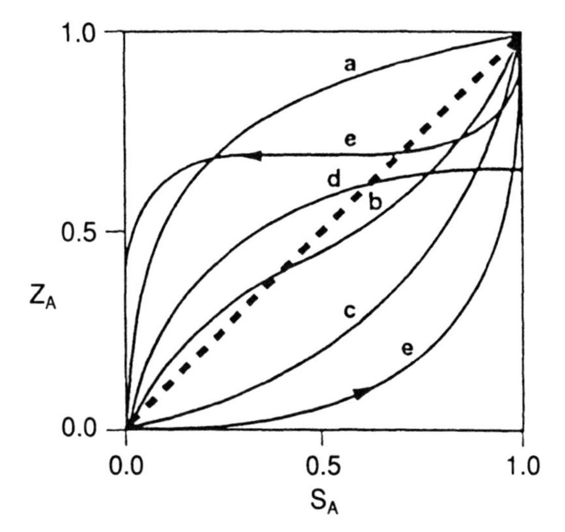

Fig. 3.1. Types of isotherms for binary ion exchange [1b], **a**) Selectivity for the entering cation over the entire range of zeolite compositions, **b**) Selectivity reversal with increasing equivalent fraction in the zeolite, **c**) Selectivity for the leaving cation over the entire range of zeolite compositions, **d**) Ion exchange does not go to completion although the entering cation is initially preferred, **e**) Hysteresis effects may result from formation of two zeolite phases

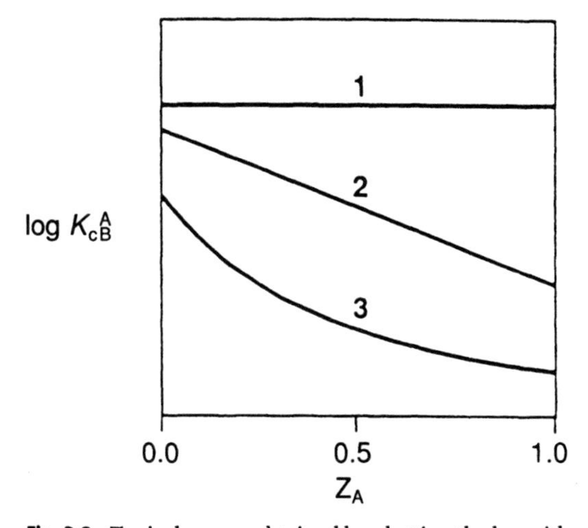

Fig. 3.2. Typical curves obtained by plotting the logarithm of the corrected rational selectivity coefficient, K_{cB}^{A}, as a function of the equivalent fraction of the entering cation in the zeolite, Z_A [1c]

The slope, 2C, is named the Kielland coefficient. For curve 1, $d \log K_{cB}^{A}/dZ_A$ is o, and $K = K_{cB}^{A}$ for $\gamma_A/\gamma_B = 1$. If the relationship is linear as shown by curve 2, $d \log K_{cB}^{A}/dZ_A$ is a constant, and K is given by

$$\log K = \log K_{cB}^{A} + C(1 - 2Z_A) \tag{3.11}$$

In this case,

$$\log \gamma_A = C(Z_B)^2$$
$$\log \gamma_B = C(Z_A)^2$$

and $K = K_{cB}^{A}$ at $Z_A = 0.5$. When $d \log K_{cB}^{A}/dA_z$ is not constant, then the curve may have many forms, e.g., curve 3. These curves may have maxima, minima or inflection points [7, 8]. Kielland plots of this type have been related to the presence of exchange sites of different types [9], but false maxima may be caused by non-equilibrium conditions [7, 8, 10, 11].

The standard free energies, standard enthalpies, and standard entropies of exchange can be calculated from the corrected selectivity data by the method of Gaines and Thomas [4]. For simplification, a modified form of their equation that neglects imbibement of electrolyte by the crystals and assumes the water activity to be 1.0 may be used. The equation is then

$$\ln K = (b - a) + \int_0^1 \ln K_c dZ_A \int_0^1 \ln K_c dZ_A \tag{3.12}$$

The first term on the right-hand side of the equation is the difference between the charges of the two ions, the second term is graphically evaluated by calculating the area under the curves of the Kielland plots. The standard free energies, enthalpies, and entropies of reaction per equivalent of exchange are then computed using the relationships

$$\Delta G_T^0 = -RT \ln K \tag{3.13}$$

$$d \ln K/dT = \Delta H^0/RT^2 \tag{3.14}$$

$$\Delta S^0 = (\Delta H^0 - \Delta G_T^0)/T \tag{3.15}$$

Instead of using the graphical method, Barri and Rees [12] calculated the corrected selectivity coefficient for the calcium exchange of NaA from the equation

$$K_c = (Z_{Ca}/Z_{Na}^2) \cdot (S_{Na}^2/S_{Ca}) \cdot 2N(\gamma_{\pm NaCl}^4)/(\gamma_{\pm CaCl_2}^3) \tag{3.16}$$

The activity coefficient ratios, Γ, where $\Gamma = \gamma_{\pm NaCl}^4/\gamma_{\pm CaCl_2}^3$, were obtained from the experimental data of Moore and Ross [13].

3.1.2.2
Experimental

In order to establish the ion-exchange isotherm for a zeolite, it is necessary to determine the equilibrium distribution of the cations between zeolite and solution phases. This is done by contacting equal amounts of homoionic zeolite, equilibrated over saturated NaCl solution for at least 1 week to ensure a constant water content (which must be accurately determined), with metal chloride or nitrate solutions of known, and varying, proportions of the competing ions at the same total normality. All equilibrations should be carried out in shaken polyethylene containers immersed in a water bath maintained at constant temperature, usually 298 ± 0.1 K. The sample should be agitated until equilibrium has been reached. Usually, a week is adequate for uni-univalent exchange in zeolites, but, e.g., multivalent transition metal ions may hydrolyze if the solutions are in contact with the zeolite for extended periods. Even if no precipitation occurs, hydronium ions formed are selectively exchanged into the zeolite. For this reason, it should not be assumed that $(A^{a+} + B^{b+})/Al = 1$, and both ions should be determined. If this is done, a deficiency in the total ion content of the zeolite, due to incidental hydronium ion exchange, is frequently observed. After equilibration, the phases are separated by filtration and washing with water, making sure that the wash water is collected along with the filtrate in order to determine the total ion content in solution. Alternatively, since the volume of the aqueous phase is known, the ion concentration can be determined in the filtrate prior to washing of the solid. The exchanging metal ions are determined in both solid and aqueous phases. The displacement of a metal ion, e.g., Na^+, labeled with a radioisotope may be studied radiochemically. This method provides an easy analysis of the distribution of metal ions between two phases [14].

In order to assure that the results obtained are equilibrium data, it is necessary to determine points on the isotherm by back-exchange. For this purpose, a completely exchanged material is prepared by multiple-step exchange with the ion to be introduced, e.g., Ca^{2+}. Prior to determination of SiO_2, Al_2O_3, Na_2O, metal oxide, e.g., CaO, and water, the ion-exchanged zeolite should be dried at ambient temperature only (drying at elevated temperature can cause a redistribution of ions within the cavities and may lead to irreversible isotherms [15, 16]) and equilibrated, as above, over saturated NaCl solution. The product should ideally have an atomic ratio of eq. $Ca^{2+}/Al = 1.00$. The same procedure as described above is used for the back-exchange with Na^+. If the points obtained lie on the curve established by forward-exchange, it is certain that equilibrium has been attained.

It is good practice to verify, by X-ray diffraction and sorption measurements, that the structure is intact after ion exchange. This is particularly advisable for low Si/Al zeolites, hydrolyzable cations, and elevated temperatures. Even if the structure does not collapse, incidental hydronium ion exchange (see 3.1.3.1) affects the ion exchange and may cause framework-Al to be hydrolyzed (see 3.3.2). It is, therefore, absolutely necessary to ascertain that the exchange isotherm is reversible. If a hysteresis is observed, the reason can be found in one or more of these causes or in the kinetics of a particularly slow reverse reaction, as observed, e.g., in the Ca^{2+} exchange of low-silica zeolite X [17], probably accompanied by a phase change, as found in the Na-Sr-X system [18].

Further information on the plotting and interpretation of ion-exchange isotherms is available [6], and for the design of ion-exchange processes see [19].

The kinetics of ion exchange can be monitored by means of ion-selective electrodes [20]. An earlier method rapidly separates, at the reaction temperature, solid and solution of small slurry samples by means of a fritted disk funnel attached to the reaction vessel [21]. The degree of exchange is followed by determining the metal ion content in the solution using conventional procedures. Still earlier, Stamires [22] calculated self-diffusion coefficients from electrical conductivity data. Alternatively, the ion content can be measured radiochemically using, e.g., ^{22}Na as a radiotracer in the zeolite phase [14, 20, 23–25]. This method is particularly useful for the study of cation self-diffusion and yields diffusion and exchange parameters unimpeded by the different rate of exchange of a competing cation [26–29]. Energy barriers against self-diffusion can be assessed from the Arrhenius equation.

3.1.2.3
Thermochemistry of Ion Exchange

The heat of exchange can be determined by contacting a hydrated zeolite in a calorimeter with the exchange solution and measuring the heat evolved [8]. Although the zeolite is hydrated by sorbing water vapor at 25 °C prior to the reaction, the heat of wetting and the heat of equilibration are to be considered in the interpretation of the data.

The standard heat of partial exchange, ΔH_x^0 (kJ/g), varies with the degree of exchange, x, and can be plotted as a function of x. The degree of exchange can be measured advantageously by employing ^{22}Na as a tracer either in the zeolite, when the pure sodium form is used, or in the mixed Na/Me exchange solution. The differential heat of partial exchange, $\partial \Delta H_x^0 / \partial x$, can be obtained from these curves and is plotted as a function of x.

The standard heat of exchange, ΔH^0, cannot be measured directly because complete exchange cannot be obtained in the calorimeter. However, the heat of exchange can be measured for the exchange of both pure ionic forms of the zeolite to the same intermediate composition, if complete exchange can be obtained. ΔH^0 can then be calculated from these data. The result can be checked by extrapolating the curve of ΔH_x^0 over x to $x = 1$. The accuracy of the latter method depends, of course, on the extent of extrapolation required, but if the zeolite cannot be completely exchanged with a particular ion, such extrapolation yields the only approximation attainable.

Since the standard free energy, ΔG^0, can be derived from the isotherm (Eqs. 3.6 and 3.14), knowledge of ΔH^0 now permits the standard entropy, ΔS^0, to be calculated (Eq. 3.15). For further details, see [8].

After this general treatment, two zeolite structures that are most widely used as catalysts – faujasite (zeolites X and Y), and ZSM-5 – will be discussed in detail as examples for ion exchange. Publications on ion exchange of other zeolites, e.g., zeolite A, mordenite, chabazite, clinoptilolite, erionite, etc., can be found in the literature. Ternary ion exchange exceeds the scope of this book. It is referred to the work by Townsend and co-workers [3, 30–37].

3.1.3
Ion Exchange of Zeolites X and Y

These zeolites have the structure of the mineral faujasite and differ from each other in composition. Zeolite X is arbitrarily defined as having a SiO_2/Al_2O_3 molar ratio in the range of 2.0 to 3.0, while synthetic faujasite with SiO_2/Al_2O_3

of 3.0 to 6.0 is designated as zeolite Y. Because of the ordering of the tetrahedral silicon and aluminum in the framework, zeolite X should be more properly limited to the range of 2.0–2.8 and zeolite Y to 4.0–6.0, while a transition form with mixed ordering exists in the range of 2.8–4.0 [38–40].

The framework structure of faujasite consists of truncated octahedra (β-cages or sodalite cages) connected through six-membered rings (6R) to form double-six rings (D6R, also called hexagonal prisms) in a tetrahedral arrangement. The α-cages or supercages formed in this manner are accessible through four nearly planar 12-membered rings (12R) with a crystallographic aperture of 7.4 Å and an effective diameter of about 9 Å. The aperture of the 6R of the sodalite cages is 2.2 Å, and the effective diameter is 2.5–2.6 Å.

The 12R are shared by adjacent supercages, and the channels thus formed intersect in every supercage. The cubic unit cell has a lattice parameter, a_o, of 24.61–25.13 Å, depending on the SiO_2/Al_2O_3 ratio. For zeolite Y, the range is 24.85 ($SiO_2/Al_2O_3 = 3$) to 24.61 Å ($SiO_2/Al_2O_3 = 6$), unit cell parameters above 24.85 Å define zeolite X. The upper limit of 25.02 Å [1d] may have been estimated, and $a_o = 25.13$ Å was later measured for zeolite X of $SiO_2/Al_2O_3 \sim 2.0$ [39]. The unit cell contains 8 sodalite cages, 8 supercages, and 16 hexagonal prisms. Since adjacent rings are always shared, the total number of tetrahedra in the framework is 192/unit cell. The maximum number of Al is 96/u.c. because adjacent Al tetrahedra are not allowed (Loewenstein's rule) [41]. Therefore, the maximum number of monovalent cations is also 96, so that the unit cell formula in the anhydrous state is $Na_{96}[AlO_2)_{96} \cdot (SiO_2)_{96}]$. Commercial zeolite X usually has a SiO_2/Al_2O_3 molar ratio in the range of 2.45 ± 0.2. Some differences found in structural work, especially in the cation position occupancies, may be attributable to the different SiO_2/Al_2O_3 of the zeolite examined.

Zeolites X and Y possess a variety of cation positions. The more common ones are site SI in the hexagonal prism (octahedral coordination by 6 framework-O), site SI′ in the center of the 6R just outside the D6R in the sodalite cage, site SII in the center of the 6R just outside the sodalite cage in the supercage, site SII′ similarly in the center of the 6R, but displaced just inside the sodalite cage, and site SIII in the supercage on a 4R. If SI is occupied, then SI′ contains no cations, and if site SII is occupied, there are no cations in SII′. A hydrated zeolite NaX with a SiO_2/Al_2O_3 molar ratio of 2.8 (80 Na = 80 Al/u.c.) contains 16 Na^+ in SI and 32 Na^+ in SII, while the remaining 32 ions have not been located and are believed to be hydrated and mobile [42]. The 48 located cations may be required for the faujasite structure to crystallize, thus imposing an upper limit of $SiO_2/Al_2O_3 = 6$ for an anhydrous unit cell formula of $Na_{48}[AlO_2)_{48}(SiO_2)_{144}]$. In order to accommodate more Na^+ ions, a higher sodium content ($SiO_2/Al_2O_3 = 2.36$; 88 Na = 88 Al per unit cell) may cause Na^+ to be placed in the more plentiful SI′ sites (15 Na^+ in SI, 2 in SI′) [43], although it is stated in this paper that the number of 17 cations in the small cages is within experimental error of 16. No crystallographic study of NaY has been reported, but Baur's work on hydrated mineral faujasite ($SiO_2/Al_2O_3 = 4.62$) indicates that the number of cations located in the small-cage network does not decrease with increasing SiO_2/Al_2O_3 ratio [44].

In order to replace the ions in the small-cage network by ion exchange, the entering ions must pass through the 6-membered rings. The largest univalent ions that can completely replace all Na^+ in zeolite X are expected to be K^+ (r_c = 1.33 Å) and Ag^+ (r_c = 1.26 Å), since the 6R appears to have an effective aperture of about 2.6 Å [45]. However, Tl^+ (r_c = 1.47 Å) has also been found to penetrate the 6R of zeolite X (*vide infra*).

Since zeolites X and Y have the same framework topology, their ion-exchange isotherms will be discussed simultaneously.

3.1.3.1
Univalent Ion Exchange

Isotherms for the exchange of NaX with alkali-metal ions at 25 °C were determined contemporaneously by Sherry [46] and by Barrer et al. [47]. The former also includes isotherms for silver and thallium. The zeolites used by these two groups had similar SiO_2/Al_2O_3 (2.46 ± 0.02 and 2.43) and Na/Al (1.00 ± 0.02 and 1.00) ratios and were equilibrated at a water vapor partial pressure provided by saturated NH_4Cl solution. The exchange was carried out at 0.1 total normality, and the results were in essential agreement. The corresponding isotherms, including Ag^+ and Tl^+, for NaY were determined by Barrer et al. [45] and by Sherry [46].

The isotherms for Li^+ and K^+ exchange are similar to those obtained for these cations with zeolite A. Again, the selectivity of NaX for lithium is low and, for potassium, it is high at low coverage and low at high coverage (Fig. 3.3). Zeolite X is more selective for Li^+ than is zeolite Y whereas it is less selective for K^+, Rb^+, and Cs^+ than zeolite Y (Figs. 3.3 and 3.4). The low selectivity of both zeolites for Li^+ is attributed to the high energy required to remove water of hydration from the ion. Therefore, Li^+ remains in the zeolites preferentially in the hydrated form, whereas the other alkali ions tend to be coordinated by framework oxygens. A hydrated ion can be more readily accommodated by the high charge density of the zeolite X framework than by the less hydrophilic zeolite Y. The potassium ion migrates easily through the 6R, and even a slightly smaller 6R, as in zeolite Y, is no obstacle (Fig. 3.4). The explanation may be found in the relatively high selectivity of the supercage sites for potassium and the resulting high concentration of K^+ in the α-cage prior to significant exchange in the sodalite cage. Upon exchange of the β-cage sites, the Na^+ ions need to migrate through the high concentration of K^+ in the supercages, both in blocking positions in site SII and as fully hydrated ions. Like potassium, Rb^+ and Cs^+ show selectivity reversals. Barrer et al. [45] report 65% exchange with these ions, whereas Sherry [46] reasons that a maximum exchange of 82% should be achievable, if only the 16 cations in the small cages are unexchangeable. However, all attempts to attain 82% exchange with Rb^+ and Cs^+ have failed. Perhaps the high concentration of the large ions in the supercages forces the smaller ions (Na^+), capable of penetrating the 6R, into the sodalite cages where they cannot be exchanged with the large alkali ions. Therefore, it appears that up to 32 ions, now probably located in SI′ sites, remain unexchangeable, so that the total exchange may be as low as 62%. This explanation is supported by the finding that all but 16 Na^+ in zeolite Y (50 Na^+/u.c.) can be exchanged with Rb^+ or Cs^+; the lower population density of the large ions in the supercage does not force additional sodium ions into

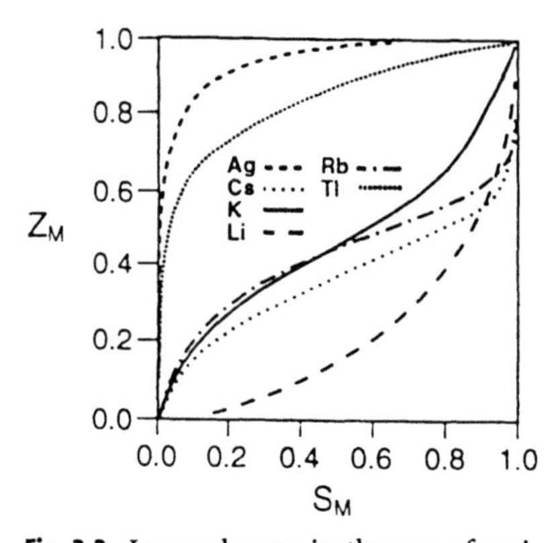

Fig. 3.3. Ion-exchange isotherms of univalent ions with NaX (SiO_2/Al_2O_3=2.46) at 0.1 N, 25°C [46]

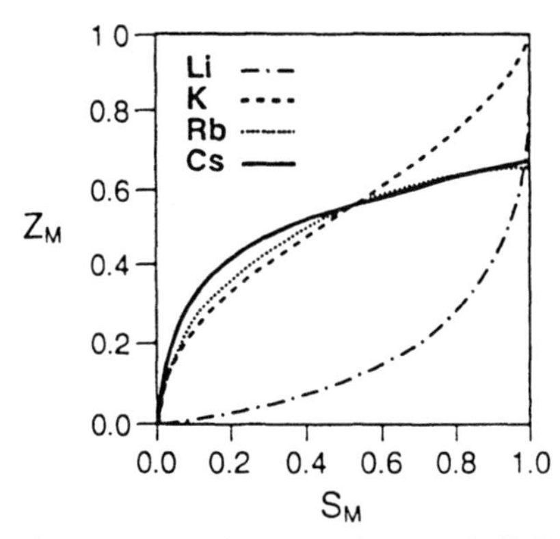

Fig. 3.4. Ion-exchange isotherms of alkali ions with NaY (SiO_2/Al_2O_3 = 5.33) at 0.1 N, 25°C [46]

the sodalite cage. A slightly different explanation is given by Barrer et al. [48]. Zeolite Y of higher Al content, 68 Al/u.c. (Si/Al = 1.82), gave 63% Rb^+ and Cs^+ exchange, leaving 25 Na^+ in the small cages unexchanged [49].

Zeolite X is very selective for Ag^+ and somewhat less so for Tl^+ (Fig. 3.3). Silver easily exchanges with ions in all sites, while Tl^+ does not enter the D6R. The Tl^+ ion does penetrate the 6R into the sodalite cages [46,47] and is found in SI′ positions as well as in SII sites [50]. Zeolite Y does not admit Tl^+ into the sodalite cage at 25°C [45,46] (Fig. 3.5), while complete exchange can be achieved at 100°C [51], indicating that the incomplete exchange at 25°C is attributable to the slow rate of diffusion through the 6R. The Tl^+ ion, although as large as Rb^+, is very polarizable and appears to be sufficiently deformable to enable it to pass through the 6R [48]. Although zeolite Y does admit Ag^+ into the sodalite cage, this zeolite is much less selective for Ag^+ than is zeolite X suggesting that there is much more ion binding in zeolite X than in zeolite Y (Fig. 3.5). This conclusion is supported by the higher selectivity for thallium ions at low coverage of zeolite X compared with zeolite Y.

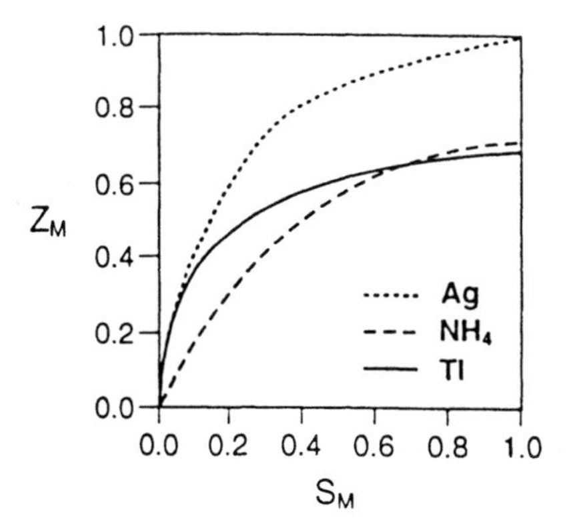

Fig. 3.5. Ion-exchange isotherms of NH_4^+, Ag^+, Tl^+ with NaY (SiO_2/Al_2O_3 = 5.33) at 0.1 N and 25°C [46]

The results discussed above suggest the selectivity series [46], at low coverage,

Ag ≫ Tl > Cs ≥ Rb > K > Na > Li

and, at 50% replacement of Na^+,

Ag ≫ Tl > Na > K > Rb ≥ Cs > Li

for zeolite X, and

Ag ≫ Tl > Cs > Rb > K > Na > Li

for zeolite Y.

Incidental hydronium ion exchange occurs upon exchanging a zeolite, e.g., of faujasite topology, with a less basic cation, e.g., NH_4^+, and decreases with increasing Si/Al ratio, i.e., decreasing charge density of the zeolite framework [52–54]. The degree of hydrolysis of the sodium form of zeolites can be estimated from the pH obtained when the zeolite is slurried in a 0.1 M sodium nitrate solution [52, 53]. In this experiment, the high-silica NaY gave a pH of about 6.5, whereas the pH for NaX was about 9.6, with NaY with lower SiO_2/Al_2O_3 being intermediate. High-silica forms of zeolite Y, Si/Al of about 2.5, gave no evidence of the presence of H_3O^+ upon exchange with NH_4Cl, while almost 20% of the cations in NH_4-exchanged NaX were H_3O^+.

The exchange isotherm of NaY ($SiO_2/Al_2/O_3$ = 5.33, ~52 Al/u.c.) for NH_4^+ at 0.1 total normality and 25 °C was found to terminate at Z_M of about 0.7 (Fig. 3.5) [46], leaving the 16 Na^+ in the sodalite cage unexchanged. Further exchange can be obtained at elevated temperature [55], and complete exchange was achieved at 160 °C [56]. The ternary ion-exchange equilibria involving H_3O^+, NH_4^+, and Na^+ ions in zeolites X and Y of various Si/Al ratios have been studied in greater detail by Franklin et al. [54]. Their work established that NH_4^+ exchange into the sodalite cage increases up to SiO_2/Al_2O_3 ~ 4.0 and then decreases again (Table 3.1).

Binary exchange isotherms of the Na-NH_4-X system found in the early literature neglected this incidental hydronium ion exchange, and only the Na^+ concentration was monitored, as the removal of sodium was of primary importance for the preparation of catalysts. Incidental H_3O^+ exchange occurs with many multivalent ions, e.g., transition metal ions. The degree of hydronium exchange with ammonium salts, the hydronium ion exchange can be reduced considerably by adding ammonium hydroxide in order to increase the pH [57].

Table 3.1. Ion Distribution in Na-NH_4-H_3O-X and -Y System [54]

SiO_2/Al_2O_3	Na^+/u.c.	H_3O^+/u.c.	NH_4^+/u.c.	Al/u.c.
2.52	0.09	13.86	71.05	85.0
3.62	0.50	4.68	63.08	68.3
4.18	1.76	2.22	58.12	62.1
4.88	7.37	0	48.43	55.8

The pores of faujasite-type zeolite are sufficiently large to allow a variety of organic-substituted ammonium ions to exchange with sodium ions in the supercage. Ionic sizes up to tetramethylammonium [58] and triethylammonium [59] have been exchanged, although the latter was found to replace only 2% Na^+ in NaX, but 26% in NaY. Exchange isotherms of alkylammonium ions show higher selectivities and higher capacities for zeolite Y than for zeolite X, probably because of the lower ionic field strength (lower hydrophilicity and higher organophilicity) of the more siliceous zeolite Y [59]. Similarly, selectivity and capacity decrease with increasing size and increasing number of organic groups on the nitrogen. The selectivity sequences were

$$NH_4^+ > CH_3NH_3^+ > Na^+ > C_2H_5NH_3^+ > C_3H_7NH_3^+$$
$$> C_4H_9NH_3^+ > (CH_3)_2NH_2^+ > (C_2H_5)_2NH_2^+$$

for zeolite X, and

$$NH_4^+ > CH_3NH_3^+ > C_2H_5NH_3^+ > C_3H_7NH_3^+ > C_4H_9NH_3^+$$
$$> Na^+ > (CH_3)_2NH_2^+ > (C_2H_5)_2NH_2^+$$

for zeolite Y. Thermodynamic data for the exchange of NaY with $C_3H_7NH_3^+$ have been determined [60].

3.1.3.2
Divalent Ion Exchange

The isotherms of divalent ions for zeolites X and Y are sigmoidal when the cations can exchange with the Na^+ located in the small cages. When the ions are too large to penetrate the 6R, the increase in loading at high S_M does not occur, as the 16 Na^+ in the small cages remain unexchanged.

Alkaline Earth. The magnesium isotherm of zeolite X (85.1 Al/unit cell, $SiO_2/Al_2O_3 = 2.52$) indicates that the zeolite is initially selective for Mg^{2+} and becomes unselective at 40% coverage [61]. The isotherm was extrapolated to about 70% exchange; however, no data higher than about 59% were measured. The endpoint for the Mg^{2+} isotherm of low-silica X (96 Al/u.c., Si/Al = 1.0) was determined to be 67% [62] (Fig. 3.6), otherwise the isotherm is very similar to that above. A 67% exchange corresponds to 32 remaining Na^+ ions for $SiO_2/Al_2O_3 = 2.0$. Apparently the situation is very similar to that observed for the Rb^+ and Cs^+ exchange of zeolite X (*vide supra*). The magnesium ion is highly hydrated and does not shed its water of hydration easily at ambient temperature; therefore, it does not enter the sodalite cage. Instead, the high concentration of ions in the supercage forces the only ions small enough to penetrate the 6R, the Na^+ ions, into the β-cage, so that up to 32 SI′ sites may be occupied by Na^+. The 65 SII sites in the supercage are available for the 32 Mg^{2+} ions present at 67% exchange, but the isotherm indicates that only about 20 Mg^{2+} form ion pairs with framework anions, the rest remain fully hydrated. Wolf et al. [61] determined the Mg^{2+} iso-

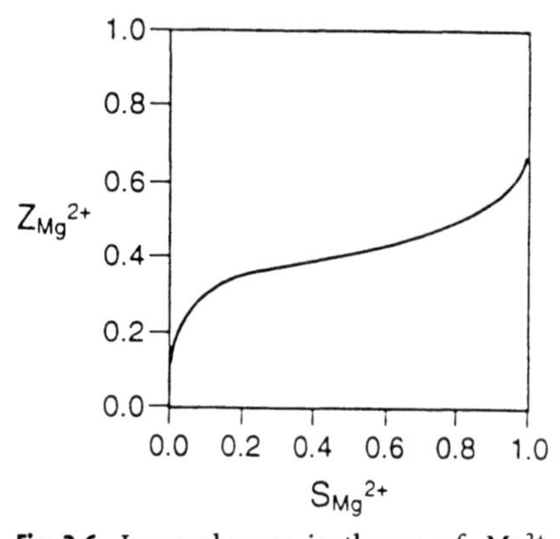

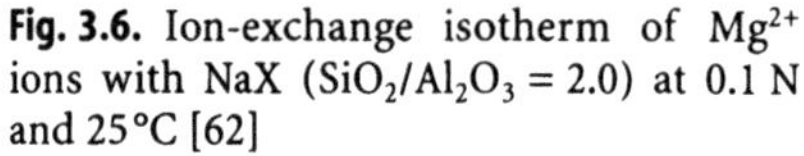
Fig. 3.6. Ion-exchange isotherm of Mg^{2+} ions with NaX (SiO_2/Al_2O_3 = 2.0) at 0.1 N and 25 °C [62]

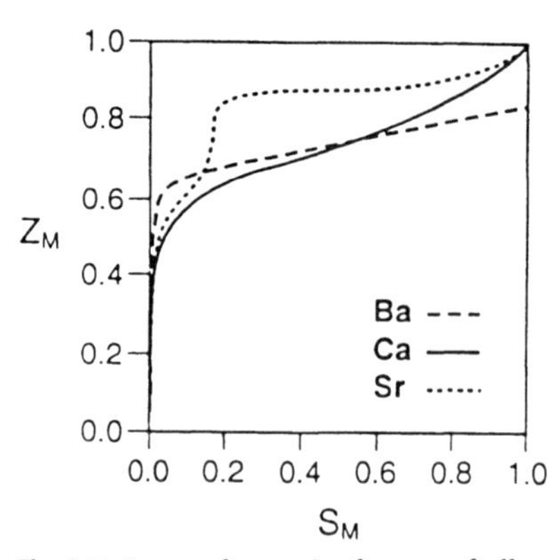

Fig. 3.7. Ion-exchange isotherms of alkaline earth ions with NaX (SiO_2/Al_2O_3 = 2.52) at 0.1 N and 25 °C [64]

therms at 10° and 30 °C and found a slight advantage at the higher temperature. However, the temperature is not sufficiently high to enable the Mg^{2+} to penetrate the 6R. At some higher temperature, Mg^{2+} should be able to exchange with β-cage Na^+. More recently [63] the Mg^{2+} isotherms of NaX (SiO_2/Al_2O_3 = 2.54) and two samples of NaY (SiO_2/Al_2O_3 = 3.62 and 4.88) have been reported, but the incidental H_3O^+ exchange observed was ignored in the calculation of the isotherms.

The sodium ions of NaX can be completely exchanged with Ca^{2+} (Fig. 3.7) [47, 61, 64]. After 24 h equilibration at 25 °C, the maximum exchange was 82%, corresponding to complete exchange of the supercage Na^+, but the 16 cations in the hexagonal prisms were not exchanged [64]. The Na^+ was completely exchanged only after an extended equilibration period. In the Ca/Na low-silica X system (LSX), a hysteresis was found in the back-exchange of CaLSX with Na^+ after one week equilibration [62]. If the isotherm for the forward-exchange, which is similar to that found by others for NaX of SiO_2/Al_2O_3 = 2.39–2.52 [61, 64], is accepted to be correct, then the Ca^{2+} ions migrate through the 6R in the back-exchange even more slowly than they did in the initial exchange. This behavior may be caused by a similar phase change as with SrX [64]. The electroselective effect [65] causes the Ca^{2+} ion specificity to increase with decreasing total normality [66].

The 25 °C isotherm for the Ca-Na-Y system (SiO_2/Al_2O_3 = 5.53, 51 Al/u.c.) terminates at about 70% exchange [45, 64], indicating that again the 16 Na^+ in the β-cages are not exchanged. It is expected that complete exchange can be achieved at some higher temperature, but has not been observed at 50 °C [64]. The slightly larger free diameter of the 6R in zeolite X is seen to contribute little to the different behavior of Ca^{2+} in this zeolite. The preferred interpretation of the results is the crowding of the supercage with hydrated Ca^{2+} ions in zeolite X, which induces partial dehydration and coordination of cations to framework-O in order to accommodate all Ca^{2+}.

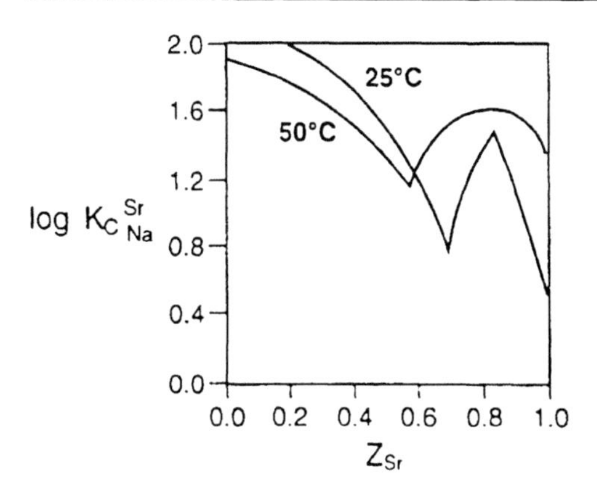

Fig. 3.8. Corrected selectivity coefficients for the Sr-Na-X system at 0.1 N as a function of zeolite composition, 25 and 50°C [64]

Complete exchange of NaX with strontium has been attained, but the strontium isotherm shows a peculiar irregularity in the region of 65–80% Sr^{2+} loading, particularly at 25°C, but also, to a lesser degree, at 50°C [64] (Fig. 3.8). Data reported by Ames [66] show a similar irregularity, although at higher S_{Sr}. This unusual behavior is caused by a region of limited solubility of the end members NaX and SrX. Examination of the region between 70 and 85% exchange by X-ray diffraction reveals one faujasite phase changing into another with an expanded lattice [67]. Simultaneously, the selectivity coefficient, which falls with increasing Sr uptake, begins to increase at about 71% Sr loading and reaches a maximum at about 84% Sr loading when the miscibility gap has been traversed, and then resumes its decrease with increasing Sr loading (Fig. 3.8). The inversion point of the selectivity curve drops to $S_{Sr} < 0.6$ at 50°C [64]. The phase change is caused by the removal of Na^+ from the hexagonal prisms (SI site) and positioning of Sr^{2+} in SI′ sites, which is possible only after dehydration of the Sr^{2+} ions has progressed sufficiently to allow these ions to penetrate the 6R. The results indicate that the hydration energy of a Sr^{2+} ion is smaller than that of Ca^{2+}, a fact expected because of the larger ionic radius of the bare Sr^{2+} ion having the same charge. The isotherms for the Sr-Na-Y system at 25°C and 50°C terminate at about 67% exchange. Not being crowded in the supercage, Sr^{2+}, like Ca^{2+}, does not dehydrate sufficiently to enable it to penetrate the 6R.

The barium isotherm for zeolite X of Si/Al = 1.26 (85 Al/u.c.) terminates at 0.82 eq. Ba^{2+}/Al at 25°C [64, 68]. Clearly, the energy of hydration is lower for Ba^{2+} than for Sr^{2+}, but, despite the crowding in the supercage, Ba^{2+} ions do not penetrate the 6R at ambient temperature. The bare ion, which has a crystal radius of 1.35 Å, is just small enough to migrate through the 6R, and complete removal of the water shell is required. Indeed, a complete exchange is achieved at 50°C [64]. The Ba^{2+} isotherm for the Ba-Na-Y (51 Al/u.c.) system at 25°C terminates at about 67% barium loading [64]. A higher Al content (68 Al/u.c.) gives a maximum of 76% exchange with Ba^{2+} at 25°C [49]. All three results obtained at 25°C leave ~16 Na^+ in the small cages unexchanged.

Transition Metal Ions. Transition metal ion solutions are generally acidic because of hydrolysis. Therefore, at low transition metal ion loading, a considerable degree of hydronium ion exchange is observed; at high loading with transition metal ions, this hydronium ion exchange is partially reversible. Ion exchange with such ions should, therefore, be treated as a ternary exchange. However, the experimental conditions are further complicated by the hydrolysis of NaX and NaY [69] and the interaction of the acidic and basic solutions resulting in further hydrolysis of the zeolite and transition metal ions [70, 71], even to the point where transition metal hydroxide precipitates [71, 72]. The latter is probably the reason for the irreversibility of some exchanges. Hydrolysis of the transition metal salt solution can be reduced or eliminated by pre-exchanging the zeolite with an ammonium salt solution before contacting it with the transition metal salt solution [55]. Exchange with a cationic, e.g., ammine complex also eliminates the problem of hydrolysis (*vide infra*).

It can be hypothesized that the hydrolysis of both zeolite and transition metal ion is aggravated by the inability of the hydrated transition metal ions to migrate through the 6R, thereby generating an ion sieving effect that greatly prefers hydronium ions to exchange into the small cages. This sieving effect generates initially a high concentration of hydronium ions in the small cages which is expected to promote hydrolytic attack on the zeolite framework. Simultaneously, the selective removal of hydronium ions from the solution within the zeolite enhances the tendency of the transition metal ion to hydrolyze. Successful exchange at 180 °C of Ni^{2+} into the sodalite cages of zeolite Y has been reported [73], but the subject of hydrolysis is not discussed in this paper.

Copper exchange of NaX was first reported in 1973 [61]. The isotherm determined at 10 °C terminates at 100% exchange. In contrast, at 30 °C, the 100% exchange point was reached at S_{Cu} of less than 0.7, a result that may have been caused by hydrolysis of the cation, precipitation of, e.g., copper hydroxide, and exchange of Na^+ with H_3O^+. A copper isotherm similar to that determined at 10 °C was obtained at 25 °C with 0.01 total normality [74], and the isotherm was found to be reversible. However, an exhaustive exchange with 0.01 N $Cu(NO_3)_2$ yielded a product with only 0.83 eq. Cu^{2+}/Al and 0.01 Na^+/Al, indicating exchange of 0.16 H_3O^+/Al. It therefore appears that very little, if any, Cu^{2+} migrated into the sodalite cages.

The ion-exchange isotherms of NaX with Mn^{2+}, Co^{2+}, Ni^{2+} at 25 °C and 0.1 total normality are very similar [75, 76], the selectivity for the transition metal ion is somewhat higher at 0.01 total normality due to the electroselective effect [65]. Zeolite X is slightly more selective for Zn^{2+} [74, 76], and complete exchange was found in the Cd-Na-X system [76]. Exchange of NaY with Mn^{2+}, Co^{2+}, Ni^{2+}, Cu^{2+}, and Zn^{2+} again yields very similar isotherms [74, 75], when allowance is made for the different normality used in the two publications. The maximum exchange levels were found to increase for all of these ions with temperature.

All transition metal ions discussed above can penetrate the 6R into the sodalite cage in the anhydrous state. The small crystal radii of these divalent ions ranging from 0.69 to 0.80 Å are responsible for the tightly bound hydration shell. Even more resistance to dehydration was encountered for Mg^{2+} (*vide supra*), $r_c = 0.66$, whereas Cd^{2+} ($r_c = 0.97$) is more readily dehydrated thus enabling it to enter the sodalite cage with ease, even at low temperature, and exchanging all sodium ions. Removal of the hydration shell by calcination causes the ions to migrate into the sodalite cage and/or hexagonal prisms [77], from where they cannot be removed by ion exchange at room temperature, i.e., they are „locked in“ [49]. This behavior can be used to place these ions exclusively in certain positions [78].

Complex Cations. If the ion to be exchanged into the zeolite forms an ammine complex, the presence of ammonia increases the pH of the solution, and hydronium ion exchange can be minimized. Since these complexes are more stable than the aquo complexes, ammonia is not stripped from the solvation shell at normal exchange conditions, and the ion is too large to penetrate the 6R. Tetramminecopper(II) ion, for example, replaces all but the 16 ions in the small cages of NH_4NaY, but 32 monovalent ions are found to be retained upon exchange of NH_4NaX [71]. The explanation appears to be again (*vide supra*) that, in the crowded environment of zeolite X, small ions are displaced into the small cages, and the large $[Cu(NH_3)_4]^{2+}$ ions, unable to penetrate the 6R, assume the positions in the supercages, about 24/supercage. The isotherms obtained were found to be reversible in contrast to those of $Cu(H_2O)_4^{2+}$. Other ammine complexes used in ion exchange reactions were those of Zn^{2+} and Co^{3+} [79].

Exchange of NaY with $[Pt(NH_3)_4]^{2+}$ ions gave an isotherm with ~68% exchange of Na^+, leaving ~20 Na^+ unexchanged [80] (Fig. 3.9). The isotherm of NaX with this cation leveled off initially at ~60% exchange of Na^+, retaining ~34 Na^+; further exchange up to 71% was observed above S_{Pt} of 0.85, leaving ~24 Na^+ unexchanged. The ion exchange with $[Pd(NH_3)_4]^{2+}$ ions shows ~62% replacement of Na in NaY and 68% in NaX, leaving ~24 and ~27 Na^+, respectively, unexchanged [80] (Fig. 3.10). The numbers for NaY being higher than the expected 16 Na^+ may reflect the low SiO_2/Al_2O_3 of the NaY used, 4.24, whereas the data for NaX indicate that other factors in addition to the size of the ions influence the degree of ion exchange [45, 81].

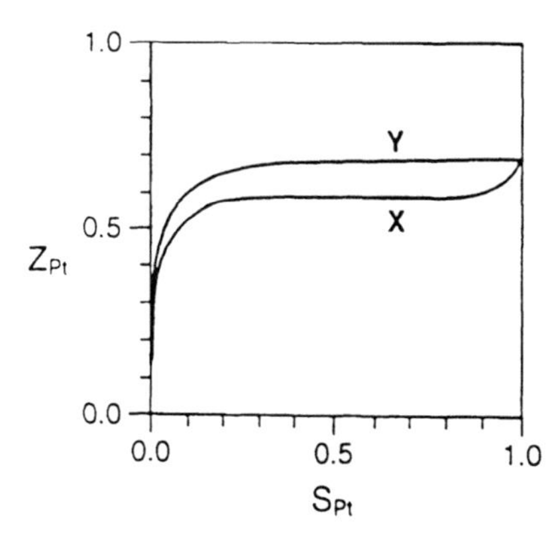

Fig. 3.9. Ion-exchange isotherms of NaX and NaY with $[Pt(NH_3)_4]^{2+}$ at 0.1 N and 25°C [80]

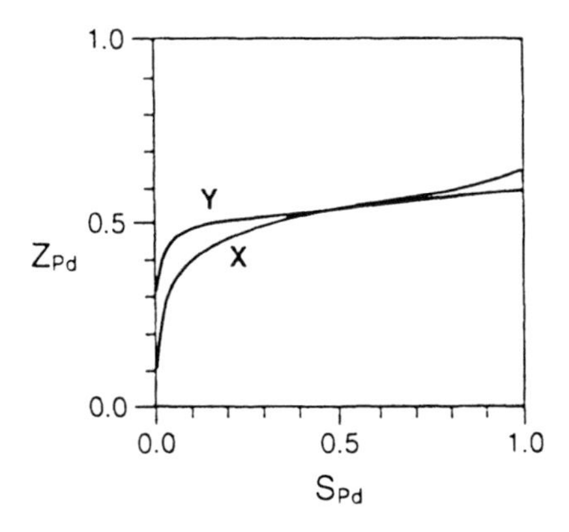

Fig. 3.10. Ion-exchange isotherms of NaX and NaY with $[Pd(NH_3)_4]^{2+}$ at 0.1 N and 25°C [80]

A further decrease in the exchange capacity is expected with monovalent ammine complexes, e.g., diamminesilver(I) ions, because of the space requirement per charge of these ions. Indeed, while aquosilver ions replace Na in NaX and NaY completely, the $(NH_3)_2Ag$-Na-X isotherm terminates at 57% (48 Ag/u.c.), and the $(NH_3)_2Ag$-Na-Y isotherm at 55% (34 Ag/u.c.) loading [82, 83]. The 27 Na^+ left in $(NH_3)_2Ag$-Na-Y may be located in the small cavities, whereas the number of Na^+ in $(NH_3)_2Ag$-Na-X, 36 Na^+, appears to be too high to fit into the small cages, and some of these ions may be located in the supercage but are not replaced by $[(NH_3)_2(Ag]^+$ because of space limitations.

3.1.3.3
Trivalent Ion Exchange

Many trivalent ions undergo hydrolysis reactions in aqueous solution resulting in the formation of hydroxylated and polynuclear species as well as hydronium ions. Results of ion exchange with ions such as Cr^{3+} and Fe^{3+} have to be interpreted with caution, and reversibility needs to be established [84, 85]. The same is true for Al^{3+} exchange of faujasite, where a combined exchange with aluminum and hydronium ions has been reported [86].

At 25 °C and 0.3 total normality, sodium faujasites are initially extremely selective for La^{3+}, up to about 50–60% loading [87]. The isotherms terminate at Z_{La} equal to 0.85 for NaX (SiO_2/Al_2O_3 = 2.52) and 0.69 for NaY (SiO_2/Al_2O_3 = 5.52), leaving 13 and 16 Na^+, respectively, unexchanged (Fig. 3.11). The somewhat low number of 13 probably indicates incidental hydronium ion exchange with NaX. The zeolite becomes less selective for La^{3+} with increasing total normality – the predicted electroselective effect [65] – but the isotherms still terminate at the same point (Fig. 3.12). The isotherms thus generated are completely reversible. The end points indicate that the hydrated La^{3+} ion is too large to migrate through the 6R into the sodalite cages. At an elevated temperature,

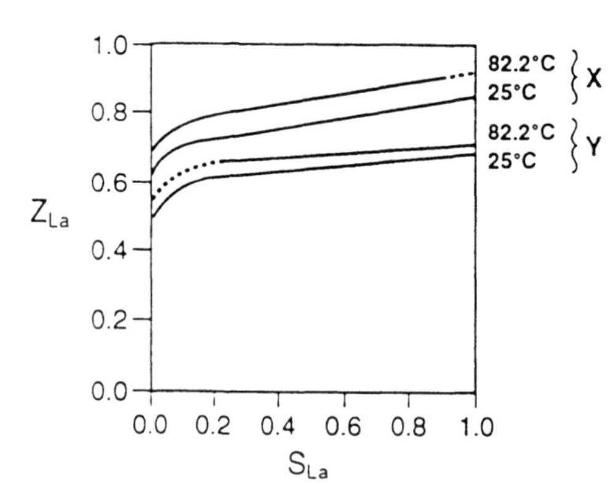

Fig. 3.11. Ion-exchange isotherms for the La-Na-X and La-Na-Y systems at 0.3 N, 25 and 82.2 °C [87]

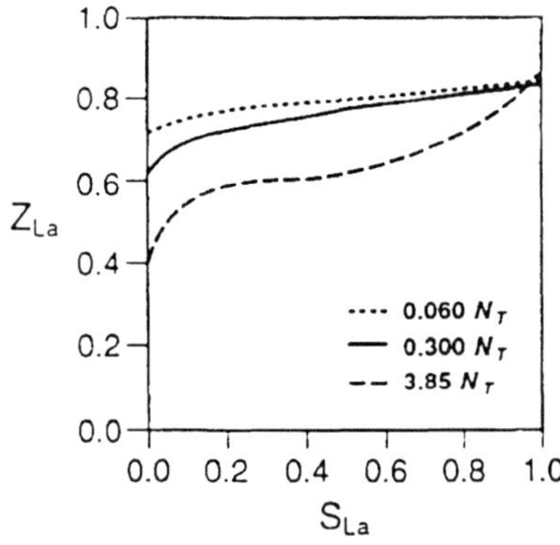

Fig. 3.12. Ion-exchange isotherms for the La-Na-X system at varying total normality, 25 °C [87]

stripping of water of hydration should enable the ion to penetrate the 6R. Indeed, a somewhat higher degree of exchange was achieved at 82 °C, but these isotherms were irreversible [87]. Complete exchange of zeolite NaY with La^{3+} was obtained at 180 °C [88].

The need to dehydrate cations such as La^{3+} to enable them to penetrate the 6R apertures points to calcination as an alternative and frequently more convenient procedure because it does not require the use of pressurized equipment. Since La^{3+} ions, exchanged into the sodalite cages at elevated temperature, become locked in, i.e., the exchange is irreversible at atmospheric conditions, the distribution of La^{3+} between α- and β-cages after calcination can be determined by back-exchange at ambient temperature [89]. The result is a curve that steeply rises as the calcination temperature is increased, and reaches a level of 100% locked-in La^{3+} ions at 220 °C [49]. Any sodium ions replaced by La^{3+} in the sodalite cages migrate into the supercages where they can be exchanged in a second exchange step. If the ion to be introduced is again La^{3+}, a subsequent calcination will again move these ions into the sodalite cage. Repeated exchange and calcination does not only remove all Na^{+}, but causes also the valency of the La^{3+} ions in the β-cage to be reduced to an average of 2.5. This lowering of the ionic charge is most likely caused by the formation of hydroxylated lanthanum ions, or O- or OH-bridged binuclear ions [90]. Further information on lanthanum-exchanged zeolite Y can be gleaned [91–94]. The presence of such hydroxylated species is held responsible for the development of strong acidity [95–97].

The cerium-exchange isotherm of bound NaX extrudate at 25 °C and total normality of 0.5 [66] is very similar to that obtained by Sherry [87] for lanthanum at 0.3 total normality, if the apparently unsupported extrapolation to 100% exchange is disregarded. However, the 70 °C isotherm includes data points at Z_{Ce} equal to 0.90 and 0.99. When Ce^{3+}-exchanged Y is calcined in air, at least part of the Ce^{3+} is oxidized to Ce^{4+}; this reaction can be avoided by calcining in an inert atmosphere, e.g., nitrogen [98].

The behavior of faujasite-type zeolites toward ion exchange with rare earth ions (RE^{3+}) does not change monotonously over the entire range of SiO_2/Al_2O_3 ratios [40]. Constant values of eq. RE^{3+}/Al and Na^{+}/Al were obtained above $SiO_2/Al_2O_3 = 4.0$, and breaks in the uptake of RE^{3+} (removal of Na^{+}) were found at SiO_2/Al_2O_3 molar ratios of slightly below 4.0 (64 Al/u.c.) and at 2.8 (80 Al/u.c.). These discontinuities coincide with those observed in the correlation of the lattice parameter of the sodium form of faujasite-type zeolites plotted as a function of Al/u.c. [39].

Zeolite NaY is somewhat less selective for yttrium, but the isotherm again terminates at 85% loading [87]. With the exception of the terminating point, the Y^{3+} isotherm constructed by Ames [66] is very similar to that of Sherry. The isotherm generated from the exchange of Y^{3+} into strontium zeolite Y at 25 °C and a total normality of 0.5 is also reported [66].

As the Si/Al ratio increases further by dealumination, the zeolite becomes more selective for Cs^{+}, NH_4^{+}, Ca^{2+}, and La^{3+}. Additional enhancement of the selectivity is observed at elevated temperature [99]. Any cationic aluminum species compete with incoming ions [100].

3.1.4
Ion Exchange of ZSM-5

The pore system of ZSM-5 consists of channels formed by puckered 10-membered rings (10R) [101]. Straight channels parallel to the **b** axis, with elliptical openings of 5.1 × 5.5 Å, intersect sinusoidal channels along the **a** direction, with nearly circular 5.4 × 5.6 Å apertures. There is no channel in the **c** direction. Specific cation positions have been identified recently for a Ni-exchanged ZSM-5 [102]. SI is a location in a side pocket near the center of a 6R facing the cavity at the intersection of two channels. It is accessible through an elongated 6R with an aperture of 1.8 × 3.1 Å. SII is located inside the sinusoidal channel about 1.6 Å from the center of the 10R. From these locations it would be expected that site II is accessible to all cations, while site I is inaccessible to cations with radii of >0.9 Å. However, even the large caesium ion is able to exchange with all sodium ions, thus indicating that this site may be displaced into the channel for ions that do not require a strong coordination to framework oxygens (*vide infra*).

Ion-exchange data for ZSM-5 are available at 0.1 total normality for SiO_2/Al_2O_3 molar ratios in the range of 40–200 [103] and at 0.05 total normality for $SiO_2/Al_2O_3 = 78$ [104]. The zeolite samples were crystallized in the presence of tetrapropylammonium (TPA) ions. These ions are located in the channel intersections with one propyl group extending into each channel leading from the intersection. When the as-crystallized zeolite is ammonium ion exchanged, only two-thirds of the original sodium ions can be removed [103]. Therefore, in order to study the ion-exchange behavior of ZSM-5, it is necessary to remove the bulky TPA ions which impede the ionic movement during the ion-exchange process and cannot be exchanged out of the zeolite, e.g., by calcination in an ammonia atmosphere, followed by repeated sodium or ammonium ion exchange to prepare a pure cationic form [103]. The selectivities at 0.1 total normality were found to be independent of the $SiO_2/Al_2/O_3$ ratio and the crystallite size [103].

3.1.4.1
Univalent Ion Exchange

The selectivities for all univalent ions studied except lithium were found to decrease with rising temperature [104]. ZSM-5 has a lower selectivity for lithium than for sodium; the selectivity for the larger alkali ions increases with the ionic radius (Fig. 3.13 and Table 3.2). The Cs^+ ions were found to be located, in their hydrated form, in the channels parallel to the **b** axis, and attach themselves to framework-O upon dehydration [105]. Thermodynamic parameters for the Na/K isotherm were determined for three Si/Al ratios [106]. Ammonium and hydronium ion exchanges yield isotherms similar to that of Cs^+ [103, 104]. Partial isotherms for the exchange of NH_4-ZSM-5 with Li^+, Na^+, and K^+ are shown by Chu and Dwyer [103]. In sharp contrast to the low Si/Al zeolites, Ag^+ is not greatly preferred, and the separation factor at $S_{Ag} = 0.5$ lies between Na^+ and K^+, in agreement with the ionic radius. Thus, the selectivity sequence is

$$Cs^+ > H_3O^+ > NH_4^+ > K^+ > Ag^+ > Na^+ > Li^+$$

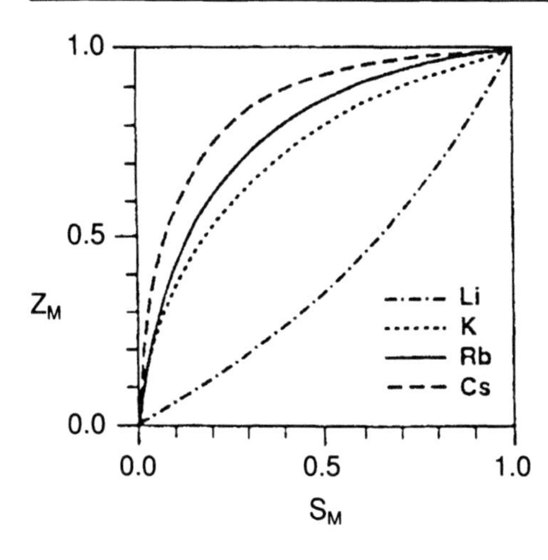

Fig. 3.13. Ion-exchange isotherms for alkali ions in NaZSM-5 (SiO_2/Al_2O_3 = 78) at 25 °C and 0.05 total normality [104]

Table 3.2. Separation Factor (α^X_{Na}) of ZSM-5 at 25 °C [103]

Ions (X)	α^X_{Na} $S_X = S_{Na} = 0.5$	Ionic radius (Å)	Hydrated radius (Å)
Univalent			
Cs	20	1.69	3.29
H_3O	18	1.5	(2.82) [a]
NH_4	13	1.48	3.31
K	5.2	1.33	3.31
Ag	3.3	1.26	3.41
Na	1.0	0.95	3.58
Li	0.6	0.60	3.82
Divalent			
Ba	0.24	1.35	4.1
Ca	0.07	0.99	4.2
Mg	0.11	0.65	4.4
Cu	2.33	0.72	4.2
Zn	1.63	0.74	4.3
Ni	0.59	0.70	4.04
Trivalent			
Al	low	0.50	4.75
La	0.22	1.15	4.52

[a] From molecular diameter of water.

Kinetic studies of ammonium exchange of HZSM-5 have been published by Ikeda et al. [107].

In contrast to the ion-exchange behavior of faujasite-type zeolites [59], for which the maximum extent of exchange with alkylammonium ions decreases with increasing molecular weight and polarizability of the cations, the sodium ions of Na-ZSM-5 can be exchanged completely with most alkylammonium ions studied [108]. The selectivity of the zeolite for mono- and dialkylammonium ions increases with the length of the alkyl group:

$$Na \ll CH_3NH_3^+ < C_2H_5NH_3^+ < C_3H_7NH_3^+ < C_4H_9NH_3^+$$

$$Na \ll (CH_3)_2NH_2^+ < (C_2H_5)_2NH_2^+ < (C_3H_7)_2NH_2^+ < (C_4H_9)_2NH_2^+$$

The same selectivity sequence was found for dealuminized zeolite Y, so that the reversal from low- to high-SiO_2/Al_2O_3 zeolites can be attributed to the ionic field strength [109] and the space available to larger ions when the number of anionic sites decreases [110].

ZSM-5 is particularly selective for tetramethylammonium ions, which easily replace all Na^+ in Na-ZSM-5; the selectivity of methylammonium ions increases in the order

$$CH_3NH_3^+ < (CH_3)_2NH_2^+ < (CH_3)_4N^+$$

Ions larger than TMA^+ (6.4 Å diameter), such as tetraethyl- and tetrapropylammonium ions with ionic sizes of 7.9 and 9 Å, respectively, are excluded from the zeolite structure by ion exchange, but small amounts may exchange with surface sites. Benzyltrimethylammonium ions (critical dimensions of 6.4 × 6.9 Å) can replace only about 65% of the Na^+ in ZSM-5. This behavior is attributed to the presence of two types of channels and the inability of $BTMA^+$ to move through the sinusoidal channels. Support for this explanation is found in the complete exchange of ZSM-11 (which has a similar structure, but only straight channels with near-circular diameter) with $BTMA^+$ ions [108].

3.1.4.2 Divalent Ion Exchange

The isotherms for the exchange of Na-ZSM-5 with alkaline-earth ions are sinusoidal and terminate below 100% loading. The highest degree of exchange attainable increases with the size of the bare ion [104]. These results, along with the selectivities observed for the univalent ions, indicate that large, weakly hydrated ions are preferred by ZSM-5, in agreement with Sherry's prediction for high-silica zeolites [109]. Therefore, when water is stripped off on heating, higher loadings are attained, in contrast to the results obtained with univalent ions. A higher charge density of the zeolite framework, i.e., a lower Si/Al ratio, was found to permit higher loading with Ba^{2+} [106].

Copper, zinc and nickel ions replace all sodium in ZSM-5 (Fig. 3.14 and Table 3.2) [103], and the selectivity does not change when the SiO_2/Al_2O_3 increases from 40 to 206. According to Kucherov et al. [111], Cu^{2+} ions exist in hydrated ZSM-5 ($SiO_2/Al_2O_3 = 69$) with only 10–15% copper loading in an octahedral environment. Upon dehydration in vacuo at 100 °C, the coordination changes to square pyramidal and square planar with all ligands being framework-oxygens. However, it is expected that, similar to mordenite, high copper loading, particularly at low SiO_2/Al_2O_3 ratio, leads to interactions between copper ions and possibly cluster formation.

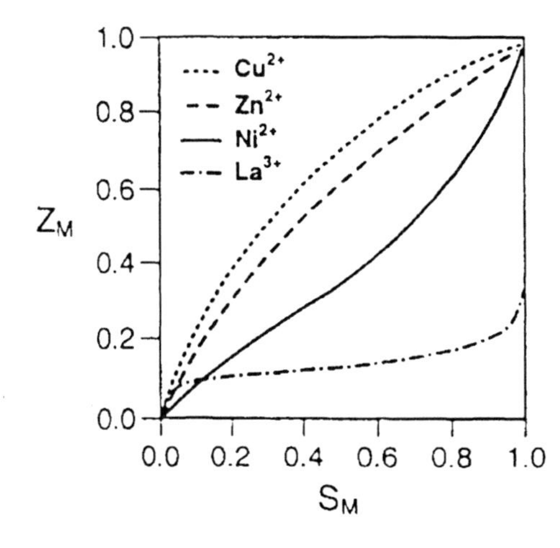

Fig. 3.14. Ion-exchange isotherms of NaZSM-5 at 25 °C with Cu^{2+}, Zn^{2+}, Ni^{2+} (SiO_2/Al_2O_3 = 40–206, 0.1 N) [103] and La^{3+} (SiO_2/Al_2O_3 = 78, 0.05 N) [104]

3.1.4.3
Trivalent Ion Exchange

The large La^{3+} ion is less hydrated than the small Al^{3+} ion, in agreement with the lower charge density. While no significant exchange of ZSM-5 with Al^{3+} has been observed, the La^{3+} isotherm of NaZSM-5 (SiO_2/Al_2O_3 = 78) at 0.05 normality is similar to that of Ca^{2+} (Fig. 3.14) [104]. The degree of exchange is greater at 65 °C, and it seems that the large hydration energy of polyvalent ions becomes the major factor in the ion exchange of high-silica zeolites [104]. The higher degree of exchange reported to be observed at 300 °C can most likely be attributed to hydrolysis of the La^{3+} ion, exchange of lower-valency hydroxylated ions and/or precipitation of $La(OH)_3$ [112].

3.1.4.4
Aluminum-Independent Ion Exchange

Fegan and Lowe [113] suggested and Chester et al. [114] demonstrated that high-silica forms of ZSM-5 and ZSM-11 show ion-exchange capacity in excess over the amount of tetrahedral aluminum in the framework. High-silica ZSM-5, in the as-crystallized form, contains cations (sodium and tetrapropylammonium) in excess over the number required by tetrahedral framework-Al, and the total cation content appears to be relatively independent of the SiO_2/Al_2O_3 ratio. The Na/Al atomic ratio of the higher SiO_2/Al_2O_3 materials was greater than 1.0 and, after thermal decomposition of the tetrapropylammonium ions, the aqueous slurry had a pH of about 10. Exchange with $Pt(NH_3)_4^{2+}$ ions, which are stable in basic solution, yielded products with Pt contents well in excess of that required as counterions for the framework-Al. The ratio of eq. $[Pt(NH_3)_4]^{2+}/Al$ increased with increasing SiO_2/Al_2O_3 ratio. When the calcined zeolites were first exchanged with ammonium ions, all products had NH_4/Al ratios of 1. Subsequent exchange with $[Pt(NH_3)_4]^{2+}$ at pH 7 yielded products with eq.

$[Pt(NH_3)_4]^{2+}/Al$ ratios below 1, indicating that excess exchange capacity exists only at basic pH. The findings are in agreement with the explanation that zeolite-internal framework $\equiv Si-O^-$ groups are formed during synthesis as counterions for the quaternary cations, that these anions are retained in the sodium form on calcination, and that the sodium is exchangeable at high pH. At lower pH, the cation is replaced by H^+ to form a silanol group that does not participate in ion exchange unless the pH is again raised to restore the siloxy cation site.

3.2 Metals Supported on Zeolites

A metal function, e.g., for reduction, hydrogenation, or oxidation, can be combined with the acid function of a zeolite. If the metal is located inside the zeolite cavities, only those molecules capable of entering the zeolite pores can be, e.g., hydrogenated.

There are basically two procedures by which precursors of metals can be introduced into a zeolite, by ion exchange [115, 116] (see Sect. 3.1) if the metal forms cations, and by sorbing neutral metal compounds, e.g., carbonyl complexes [117]. A third method frequently applied to bound catalysts is impregnation, which introduces anions as well as cations. Although impregnation of zeolites, particularly those of high SiO_2/Al_2O_3 ratio, is possible, impregnation of a bound catalyst frequently involves only the binder, from which the cations then migrate into the zeolite by ion exchange. In order to impregnate a zeolite of low Si/Al ratio, a high concentration of the solution is required to overcome the anion-repelling effect of the negative charges around the pore mouth. A convenient method involves evaporation of the impregnating solution in contact with the zeolite. Impregnation of a zeolite is employed when the quantity of metal ion to be introduced is greater than would be obtainable by ion exchange of, e.g., high-silica zeolites, but the distribution is more uniform in the ion-exchanged sample [118].

Finally, large zeolite cages offer themselves as locations for synthesis of metal complexes with dimensions exceeding those of the pore apertures, so that they are trapped like a ship in a bottle. The ship-in-a-bottle method [119] is used to immobilize such complexes within the zeolite cavities [120] providing an opportunity to heterogenize a homogeneous catalyst system. One particular method consists of three steps:

1. Dealumination of the zeolite framework in order to generate OH-groups;
2. fixation of 2-pyridylethyl groups using a silane coupling reagent, e.g., 2-[2-(trichlorosilyl)ethyl]pyridine, in tetrahydrofuran; and
3. formation of the copper complex by reaction with $CuCl_2$ in ethanol.

The thus immobilized Cu^{2+} ions coordinated with 2-pyridylethyl groups are easily reduced to Cu^+ by heating for 1 h at 200 °C in vacuum.

Most hydrated metal cations can easily be stripped temporarily of at least part of their hydration shell, so that they can be ion-exchanged even into small-pore zeolites. On the other hand, elements that do not readily form hydrated cations, e.g., Pt, are usually ion exchanged in the form of their ammine complexes, e.g., $[(NH_3)_4Pt]^{2+}$.

The ion-exchanged zeolite is usually calcined prior to reduction of the metal ion. The reduction of zeolitic metal cations causes zeolitic acid sites to be generated, so that a high loading with such cations may destabilize the zeolite upon reduction, particularly if the SiO_2/Al_2O_3 is low. The reduction of zeolitic cations at low loading or in a high-silica zeolite,

$$M^+Z^- + 1/2\,H_2 \rightleftarrows H^+Z^- + M \tag{3.17}$$

is frequently reversible. Depending on the conditions of calcination and reduction, individual metal atoms or metal clusters can be formed. Reduction to catalytically active lower valence states are included in the discussion.

In spite of some early papers, this field of research is still in the process of maturing and has been rather active for several years. Further work is required to combine the many pieces of information to form a clear picture.

This chapter will concentrate again on zeolites of faujasite and ZSM-5 topologies. A considerable amount of work on metals supported on other zeolites, i.e., types A [121–135], L [136–139], and mordenite [140–147] has been published. Elements of groups VIA [117, 145, 148–166] and VIIA [117, 167] are excluded from the discussion.

3.2.1
Reduction of Metal Ions in Zeolites

In early work, Yates [168] reduced metal ions in zeolite X with hydrogen and observed that Ni^{2+} and Ag^+ formed metal crystallites of 240 and 170 Å average size, respectively, as determined by X-ray diffraction. It is noted that nickel and silver entities must migrate in order to aggregate and that the metal particles, because of their size, must be generated on the external surface of the zeolite crystals, unless the zeolite framework is, at least locally, destroyed. Other reducing agents applicable for some cations include CO and hydrocarbons [169].

3.2.1.1
Group IB

All Ag^+ ions in completely exchanged AgX and AgY are located in positions SI, SI′, SII, SII′ and, for high Al contents, on 4R of the supercage. The number of cations in the sodalite cage does not exceed four, but silver ions seem to be the only ions that can occupy sites SI and SI′ simultaneously. Linear Ag^+ clusters at D6R cannot interact with one another in faujasite, while in zeolite A a red color is observed with two Ag_3 clusters per sodalite cage. Paramagnetic Ag species are produced in faujasite by γ-irradiation at 77 K, and the highest nuclearity of observed clusters is three [170–172]. Isolated Ag^0 atoms can be stabilized in hydrated zeolites or in oxidized zeolites Y with less than 1 Ag/u.c. at 77 K, and hydrated $Ag^0(H_2O)_4$ was found to be located in the center of the sodalite cage.

An excellent summary of prior work on Ag and Cu in A, X, and Y is given by Schoonheydt [172].

The distribution of Cu^{2+} ions in zeolites X and Y between various sites is still a matter of contention. The siting depends on the number of Cu^{2+} ions and of the other ions present. It appears that the preferred site for Cu^{2+} in dehydrated zeolites X and Y is SI′ [173]. The Cu^{2+} ion is coordinated to 6R oxygens, and the spectral properties of Cu^{2+} in SI′ and SII are independent of the structure type and SiO_2/Al_2O_3 ratio.

Autoreduction of Ag^+ in AgY or AgNaY to Ag^0 has been observed at temperatures as low as 120°–150 °C [174]. At higher temperatures, e.g., 420 °C [175], Ag atoms migrate to the surface of the zeolite crystals, where they agglomerate to form silver metal [174, 175]. As the temperature is raised, the intensity of the XPS Ag $3d_{5/2}$ line increases and reaches a limit peculiar for the temperature applied. A similar migration to the surface is observed with copper and other metals. This autoreduction has been explained with formation of dehydroxylated Y and elimination of O_2 [172].

The absorption spectra of fully hydrated AgNaY (8 Ag^+/u.c.), after several stages of dehydration, measured in the form of the Schuster–Kubelka–Munk remission function, $F(R_\infty)$, contains a broad, modulated absorption at 1–3.5 eV, which increases with temperature up to 500 °C and is assigned to Ag_n clusters ($1 < n < 7$) in the zeolite pores. It can be removed by treatment with oxygen at 400 °C. A broad, featureless peak at 3.32 eV is assigned to intrazeolite silver particles and a sharp peak at 4.05 eV to atomic silver located in the center of the hexagonal prism (site SI). This assignment is at variance with the conclusion of Beyer et al. [176] that the Ag^+ ion is not reduced in this location. Finally, a broad multiplet in the range of 4.5–5.5 eV is assigned to Ag^+ ions in one or more of the regular cation sites in zeolite Y [177].

Treatment of fresh AgY with O_2 at 400° or 600 °C generates charged clusters with Ag located in sites I and I′. These clusters are destroyed upon reduction at 75 °C, while charged aggregates stable to 450 °C are formed within the β-cage with Ag in sites I′ and II′. A prolonged reoxidation restores the original site I – site I′ clusters consisting of Ag^0 and Ag^+, which are responsible for the yellow color [178].

The Cu^{2+} ion in ZSM-5 resides essentially in one type of site [179]. It is easily reduced to Cu^+ under the influence of water at elevated temperature [180] and even at ambient temperature by irradiation with X-rays [181]. The reaction probably involves dissociation of water caused by the polarization of Cu^{2+} [182–187], dehydration of the hydroxylated ions [187, 188], and autoreduction to Cu^+:

$$2\,Cu^{2+} + 2\,H_2O \rightleftarrows 2\,Cu(OH)^+ + 2\,H^+ \tag{3.18}$$

$$2\,Cu(OH)^+ \rightleftarrows [Cu\text{-}O\text{-}Cu]^{2+} + H_2O \tag{3.19}$$

$$[Cu\text{-}O\text{-}Cu]^{2+} \rightleftarrows 2\,Cu^+ + 0.5\,O_2 \tag{3.20}$$

Excessively exchanged CuZSM-5 contains CuO after calcination, in addition to Cu^{2+} and $[Cu\text{-}O\text{-}Cu]^{2+}$. CuZSM-5 catalyzes the decomposition of NO to the elements [189–194]. Of particular interest recently has been a super-exchanged CuZSM-5 as catalyst for the removal of NO from exhaust gases [195, 199]. The super-exchange was achieved by first ion exchanging with 0.012 M $Cu(NO_3)_2$ (1 l/15 g) for 24 h, then increasing the pH to 7.5 with 3 M NH_4OH or to 5.8 with saturated $Mg(OH)_2$ solution. Excess loading of ZSM-5 can also be achieved by repeated ion exchange with copper acetate [197]. Polynuclear hydroxylated cations may be formed within the zeolite cavities, $Cu(OH)^+$ cations may have exchanged into the zeolite at the elevated pH [184] or were formed by hydrolysis within the zeolite, along with H^+ [182].

Alkaline-earth co-cations in CuZSM-5 were found to promote the direct decomposition of NO at temperatures above 450 °C. The effect depends on the ion-exchange mode and is more pronounced when Cu^{2+} is exchanged first [200]. Particularly active was a AgCuZSM-5, which was found to be stable for 50 h at 500 °C. Cu^+ ions appear to be more stable in ZSM-5 than in zeolite Y where they are more easily oxidized to Cu^{2+} [201].

Dehydrating 68% exchanged Cu^{2+}NaY above 350 °C causes autoreduction of Cu^{2+} to Cu^+, accompanied by formation of Lewis acid sites [202]. Cu^+ ions migrate gradually from the small cavities to the supercages. The location of the ion can be determined by FTIR after chemisorption of CO. Different sample pretreatment conditions give different cation distributions and different Cu^{2+}/Cu^+ ratios [203].

The Brønsted acid sites formed upon reduction of Cu^{2+}Y with H_2 yield the same bridging OH-groups as in HY, as evidenced by the IR bands at 3550 and 3640 cm^{-1} [204]. Reduction of CuNaY with D_2 generates the same Brønsted acid sites with IR bands at 2690 and 2630^{-1} [205]. When a 32% exchanged AgNaY is calcined and reduced with hydrogen at 450 °C, a high activity for the conversion of cumene is obtained due to the replacement of Ag^+ with H^+. The silver crystallite size has no influence on the catalytic activity in the acid-catalyzed reaction [175].

The consumption of hydrogen in the reduction of AgY has been followed kinetically using IR spectroscopy and X-ray diffraction [176]. Different mechanisms were found for low- and high-temperature reduction (Figs. 3.15 and 3.16). Silver ions are reduced quantitatively at temperatures above 160 °C only, and the number of Ag^+/u.c. not reducible at lower temperatures agrees with the number of Ag^+ located in site SI. Ag^+ ions must move out of this site in order to be reduced by hydrogen, and the migration of Ag^+ from SI sites becomes rate controlling. The high-temperature reduction is independent of the hydrogen pressure and the degree of Ag^+ exchange. The kinetics fit the first-order expression [174].

$$dc/dt = k_2(C_0 - C) \quad (3.21)$$

in which C_0 is the initial concentration of Ag^+, and C the concentration of Ag^+ after time t and also equal to twice the number of moles of H_2 consumed. The activation energy of the high-temperature reduction is 97.5 kJ/mol. The reaction rate calculated from the above becomes negligible at ≤160 °C indicating that a different mechanism applies at lower temperature, where the rate of reaction

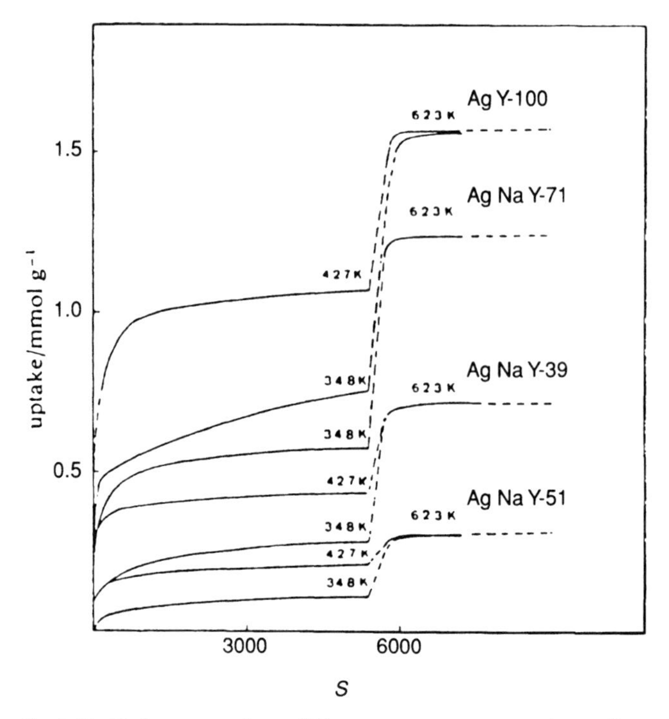

Fig. 3.15. Hydrogen uptake at different temperatures over Ag zeolites with different Ag content

is proportional to the hydrogen pressure and to the concentration of Ag^+, and inversely proportional to the concentration of the reaction product:

$$dc/dt = k'P(C_0' - C)/C \tag{3.22}$$

where C_0' is the initial concentration of Ag^+ ions available for this reaction, C is the concentration of Ag^+ ions reduced after time t, and P is the hydrogen pressure. The linearized version of this expression is:

$$dc/Pdt = K'C_0'/C - k' \tag{3.23}$$

which is represented by Fig. 3.17. The apparent activation energy for the low-temperature reduction is 40 kJ/mol.

Highly dispersed Ag_3^+ clusters are generated when the reduction is performed at low temperature. Any silver particles formed under these conditions must have a diameter of < 3.5 nm, because they are not detectable by XRD [206]. The reaction

$$ZO^- + Ag^+ + 0.5\,H_2 \rightleftarrows Ag^0 + ZOH \tag{3.24}$$

is reversible by thermal degassing at > 300 °C [207–209]. Reversible interconversion is supported by the IR spectrum of CO chemisorbed on Ag^+ in the super-

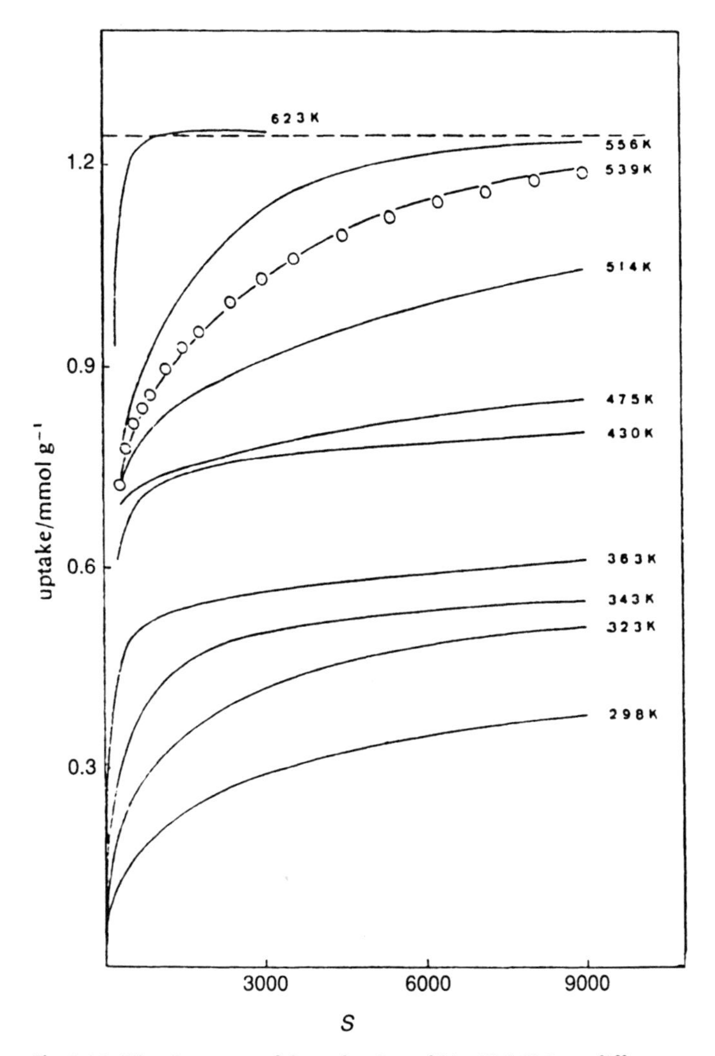

Fig. 3.16. Kinetic curves of the reduction of (Ag, Na)-Y-71 at different temperatures [176]

cage (2174 cm^{-1}). The extent of reversibility decreases with increased degree of Ag^+ reduction.

The reduction of Ag^+ in zeolite AgNaY with H_2 at 350°C is completely reversible by treatment with oxygen [210]:

$$2\,Ag^0 + 2\,H^+ + 0.5\,O_2 \rightarrow 2\,Ag^+ + H_2O \qquad (3.25)$$

Reoxidation is very slow at low temperature, 90–210°C, and only highly dispersed silver in the zeolite cages is reoxidized in this temperature range [211]. There is an induction period at low temperature, which can be eliminated by

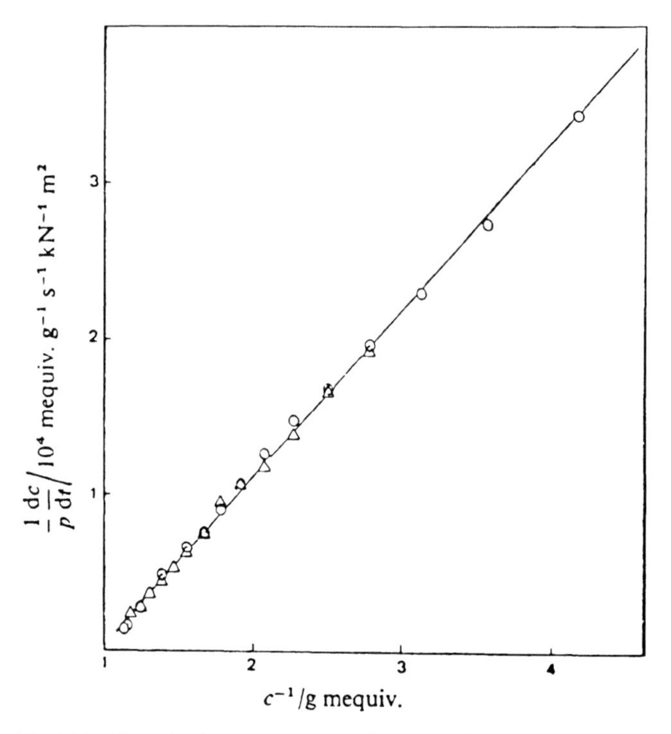

Fig. 3.17. Linearized representation of Eq. 3.23 for the reduction at 343 K of (Ag, Na)-Y-71 at different hydrogen pressures [176]. O: ~20 kN m^{-2}; Δ: ~ 55 kN m^{-2}

introducing water vapor. Reoxidation is rapid and proceeds to completion at temperatures ≥230 °C; it involves oxidation of external silver clusters causing Ag^+ ions to migrate back into the cavities [176]. The reaction rate in this temperature regime depends on the crystallite size as well as on the temperature.

The reduction of Cu^{2+} in H_2 correlates with reduction temperature and time, heating rate and ammonia treatment, which facilitates migration of Cu^{2+} from the sodalite cage to the supercage. The nature of the alkali co-cation strongly influences the degree of Cu^{2+} reduction [212]. Reduction of Cu^{2+} in Cu^{2+} Y proceeds via a two-step mechanism, the first step to Cu^+ is complete at 200 °C within 1 h and is completely reversible [213], as it is for silver. The degree of reduction depends on the hydrogen pressure [210].

$$2\,Cu^{2+}Y + H_2\,(133\text{ Pa}, 360\,°C) \rightarrow 2\,Cu^{+}HY \qquad (3.26)$$

The rate of reduction decreases abruptly at 75% reduction of Cu^{2+}, and this reaction is not reversible at 10^{-3} Pa, 360 °C. Reoxidation with O_2,

$$2\,Cu^{+}HY + 0.5\,O_2\,(1.03\text{ kPa}, 360\,°C) \rightarrow 2\,Cu^{2+} + H_2O$$

remains incomplete, and only 76% of the Cu^+ is reoxidized.

Some of the copper may be present as $Cu(OH)^+$, or its dehydrated form, $[Cu-O-Cu]^{2+}$ [187, 214, 215]. Two types of such exchange-coupled copper ions have been identified by their EPR spectra, a linear or identical pair, observed in CuMgY, CuCaY, CuLaY, but not in CuNaY or CuBaY, and a non-linear pair in all highly Cu^{2+}-exchanged Y [214].

The second step occurs at 400 °C, when Cu^+ is reduced to copper crystallites, which can grow, in the course of 20 h, to a size of 300 Å. These larger crystallites on the external surface of the zeolite crystals are reoxidized slowly to CuO, while small metal particles within the pores of the zeolite are reoxidized readily to Cu^{2+} ions located in ion-exchange sites [213]. When Cu^{2+} Y is reduced at 500 °C, the color changes to grey-blue and finally to red-pink attributable to Cu metal [216], while Cu^+ Y samples containing no other Cu species are white [217].

Temperature-programmed reduction/oxidation can be used to determine the transition metal inside and outside the zeolite [211]. External CuO is reduced at a lower temperature than Cu^{2+} in the zeolite channels [218]. The formation of Cu^+ from Cu^{2+} is determined by the mobility of the Cu^{2+} species. The rate of Cu^0 formation is determined by the diffusion of the reactant through the pores, and Cu^{2+} in the SI site is more difficult to reduce [218]. Super-exchanged or quickly degassed Cu^{2+} Y does not form Cu^+ upon reduction, and Cu^0 accumulates on the surface of the zeolite particles. Cu^+ and small clusters of Cu^0 are reoxidized reversibly, and the mobility of Cu^+ determines the rate of the reaction [218]. Reoxidation of Cu^0-containing zeolite Y gives two maxima in the temperature-programmed oxidation profile; the one at lower temperature corresponds to reversible oxidation of Cu^+ and Cu^0 inside the zeolite cavities [213] to yield Cu^{2+} in zeolitic ion-exchange positions, while the high-temperature maximum corresponds to the oxidation of Cu^0 crystals external to the zeolite to CuO as a second phase [211, 219]. A similar observation is made for zeolite Y containing small silver particles. No dealumination was observed after reduction of CuY with H_2 or CO and reoxidation [187].

In an effort to prepare a catalyst for low-temperature CO oxidation, Matsumoto et al. [220] generated an active Cu species by a sequence of preliminary evacuation, reduction with H_2, and reoxidation with O_2. The catalytic activity increased proportionally with the amount of this species estimated from TPR, while the activation energy remained constant. CO reduced the active species directly and reversibly to Cu^0 clusters at low temperature, while original Cu^{2+} species were reduced to Cu^+ at high temperature. EXAFS suggests that small CuO clusters may function as catalytic centers.

Zeolite Y materials containing Cu^+ and samples containing Cu^0 in the zeolite cages exhibit the same kinetic behavior, whereas the kinetic parameters for catalysts containing large Cu^0 crystals were different. A minimum of 5 Cu^+/u.c. in Cu^{2+}/Cu^+Y or Cu^+Y gave maximum activity, while the activity of Cu^0Y increased with the amount of metal in the cages [217].

Very high dispersion of Cu^0 was obtained by the following procedure [219, 221]: Prior to the reduction process, the Cu^{2+}-exchanged zeolite Y was washed three times with ethanol and anhydrous diethyl ether and then dried at 110 °C. The copper was reduced by slowly heating the dried material at 1 °C/min in H_2 to 250 °C and holding at this temperature for 3 h. The presence of small amounts of Ce^{3+} in CuNaY increased the degree of Cu^{2+} reduction to very small Cu^0 clusters. Addition of Pt^{2+} also increased the degree of copper reduction, but the Cu^0 formed large, 20–30 nm, crystallites [221]. The Cu TPR peaks merge with the Pt TPR peaks, even if all Pt is located initially in the supercages and the Cu in hexagonal prisms. The surface of the bimetallic particles obtained in the reduction is enriched in Cu [222]. Similarly, reduction of Cu in NaY is enhanced by the presence of Pd. Calcination at 500 °C causes the bare ions to migrate into the sodalite cages, and the reduction enhancement is seen only for those Cu^{2+} ions sharing a cage with Pd^{2+} [223].

In contrast to reducing an ion, e.g., Cu^{2+}, with hydrogen, where a Brønsted acid site is generated, reduction with CO gives rise to a Lewis acid site [216]. Reduction of Cu^{2+} Y by treatment with CO at 500 °C for 48 h generates Cu^+ ions, and the color changes from blue-grey to white. Formation of positively charged radicals by reaction of Cu^+ with anthracene is inhibited by rehydration of Cu^+ Y, attributed to poisoning of Lewis acid sites with water. Cu^{2+} ions located in the sodalite cages of dehydrated Cu^{2+} Y [214, 224] are not reduced by CO. Adsorption of NH_3 causes these ions to migrate into the supercages where they form copper(II)-ammine complexes [225].

Therefore, Cu^{2+}Y is quantitatively reduced to Cu^+Y in 1 h at 400 °C in the presence of 1.3 kPa NH_3 via $[Cu(NH_3)_2]^+$ [226]. For the same reason, Cu^+ ions in anhydrous zeolite Y do not react with oxygen at ambient temperature, but are readily oxidized when ammonia is preadsorbed [226]. Cu^{2+} complexed with pyridine in zeolite Y is partially reduced to Cu^+ upon outgassing at temperatures up to 200 °C, and the reduced form is reoxidized completely by heating in 0.27 bar O_2 at 100 °C for 3 h [227]. Others report that Cu^+–CO complexes have been identified by FTIR in SI or SI′, SII′ and SII sites [203].

Reduction with CO produces CO_2, apparently removing oxygen from the framework and generating AlO^+ cations. When a CO-reduced sample is treated with H_2, no Brønsted acid sites are formed. After reduction with CO, the Cu^+ ion in zeolite Y reacts with NO to form the complex Cu^+NO [228]. A CuY reduced with CO is stable at 500 °C through many oxidation (with O_2 or NO)/reduction (with CO) cycles, indicating that CuY is an excellent catalyst for CO oxidation to CO_2 [220] or NO reduction to N_2. The reduced form is always close to Cu^+, i.e., a change of 1 e^-/Cu. Reduction with CO is also reversible at higher temperature, but now the reaction proceeds to Cu^0. The zeolite structure ($SiO_2/Al_2O_3 = 4.5$) is perfectly stable in these reactions, even after treatment at 750 °C. At this temperature, the copper is always present as Cu^0 or CuO after reduction or oxidation cycles, respectively [229].

While Cu^{2+}, as well as other divalent cations, interact with CO by van der Waals forces, Ag^+ and Cu^+ located in SII positions of zeolite Y form rather stable ethylene complexes with $d\pi$ bonding [230], as evidenced by strong IR bands near 1430 cm^{-1}, lowering of the C=C stretching frequency compared with that of the gas phase, and a combination band at 1930 cm^{-1}. The enhanced adsorption of ethylene corresponds to $C_2H_4/Cu = 1$ at 25 °C and >0.1 bar, as Cu^{2+} was reduced to Cu^+.

3.2.1.2
Group VIIIA, Fourth Period

Iron. Ferrous ions exchanged into zeolite Y are reported to be located preferentially in site I. Dehydration leads to fourfold coordination near the hexagonal windows [231]. Zeolite Y of the composition $Na_{19}Fe^{II}_{13}(AlO_2)_{58}(SiO_2)_{134}$ could be reversibly oxidized and reduced between Fe^{2+} and Fe^{3+}. The catalyst gave almost complete conversion of CO and/or NO at temperatures < 500 °C. FeY was examined for suitability as a catalyst for the oxidation of CO with NO, O_2, and N_2O and operated in nearly completely oxidized state when O_2 was present in the feed, but close to the reduced state when NO was the only oxidizing agent [232, 233]. In contrast to conventional FeY, in which most Fe is in site I, Fe^{2+} ion exchange of zeolite Y dealuminated with $(NH_4)_2SiF_6$ places a larger fraction of Fe^{2+} in sites I′, II′, and/or II, where they serve as adsorption centers for water and, in site II, as adsorption center for CO as well.

The Mössbauer spectrum of Fe^{2+}, after reduction with sodium metal vapor, indicated that iron clusters with a diameter of < 13 Å and a narrow particle size distribution were preferentially formed. These clusters were superparamagnetic [234].

Nickel. Similar to copper, nickel ions located in SI of dehydrated zeolite Y are extracted from this site and placed in the supercage by adsorption of a strong complexing agent, especially NH_3; the change in coordination is signified as a color change from pink to green. Slow migration is also observed on adsorption of NO or pyridine, but not with hydrocarbons or CO [235].

Bivalent cations coordinated to three framework-O in the supercages react with CO to form a complex with IR frequencies specific for the cation. The adsorption follows Langmuir isotherms suggesting that a single CO molecule is attached to every cation. Co^{2+} and Ni^{2+} ions adsorb CO much more tenaciously than alkaline-earth ions, signifying the higher ability to form coordination bonds [236]. Ni^{+} ions are generated when Ni^{2+} Y is heated for 1 h at 575 °C with alkali-metal vapor (ESR signal at $g = 2.065$), although most nickel ions are reduced to Ni^{0}, the final stage attained completely after 16 h. Reduction with hydrogen, on the other hand, was found to produce only Ni^{0}.

In contrast, formation of Ni^{+} ions was observed with diffuse reflectance spectroscopy (DRS) when NiCaNaY was reduced with H_2 at 200 °C. The appearance of strong bands at 13500 and 29500 cm^{-1}, a weak one at 4700 and a shoulder at 25000 cm^{-1} is accompanied by a color change to malachite green. These bands decrease in intensity upon adsorption of O_2 at room temperature and disappear completely when the sample is heated in O_2 at 100–200 °C. After reduction, the remaining Ni^{2+} ions are inaccessible to CO. An IR band at 2140 cm^{-1} is assigned to CO chemisorbed on Ni^{+} and a band at 2095 cm^{-1} to CO chemisorption on Ni^{0} in linear configuration [237]. The degree of nickel reduction increases with decreasing Brønsted acidity, i.e., increasing SiO_2/Al_2O_3 ratio, decreasing degree of exchange, and increasing basicity of the cocations [238] (see Eq. 3.17). Addition of acidity by exchanging with strongly polarizing Ce^{3+} as cocation suppresses the reduction process as well as crystal growth during thermal activation, and reduction of Ni^{2+} in NH_4Y is negligible [239, 240]. The reducibility of Ni^{2+} varies with the reducing agent, e.g., at 400 °C on zeolite X, the reducibility decreases in the order H_2 > 1-butene > propene > ethene and, on zeolite Y (SiO_2/Al_2O_3 = 5.2), in the order 1-butene = propene > H_2 = ethene [241].

Reduction with hydrogen at high temperature causes the nickel atoms to migrate towards the external surface of the zeolite crystals, where they aggregate to form nickel crystallites. The activity of these catalysts for the hydrogenolysis of ethane increases with rising reduction temperature, while the poison resistance decreases. Low-temperature reduction yields a catalyst with very little surface nickel and greatly improved nickel dispersion throughout the pores resulting in high poison resistance, but lower activity because of the decreased degree of reduction. Addition of Cr_2O_3 can inhibit migration of Ni^0 and reduce the build-up of surface nickel during the reduction process [169, 242].

Temperature-programmed reduction and oxidation are useful techniques for determining the distribution of transition metal between internal and external locations. TPO is not applicable to the nickel/zeolite system. Instead, Ni^0 is completely reoxidized prior to determining the bidisperse metal distribution by TPR. Like CuO, NiO located on the external surface of the zeolite crystals is more easily reduced than Ni^{2+} in internal cation positions [211]. The reduction characteristics of Ni^{2+} at three different exchange sites, SII (and/or SII'), SI', and SI, can be distinguished by three TPR peaks at 520, 640, and 820 °C, respectively [243]. The reducibility of Ni^{2+} is appreciably enhanced by replacing Na^+ with Ca^{2+} or Sr^{2+}; this observation is explained by the crystallographic positions of Ni^{2+}. The zeolite structure remains intact up to 780 °C when Ni^{2+} corresponding to the first two peaks is reduced, but collapses during the third peak. The latter TPR peak does not appear when the SI sites are occupied by Ca^{2+} or Sr^{2+} and is, therefore, assigned to Ni^{2+} in the SI position. The competition for this site causes some Ni^{2+} to become accessible to CO in the α-cage [237, 244]. The location of cations in the supercage and in the small-cage network can be monitored by Cs^+ back-exchange and conformed by the IR spectra of the CO adducts. Similarly, iodometric titration and Na^+ back-exchange can be utilized to determine the degree of Ni^{2+} reduction. As the Ni^{2+} ion is dehydrated with rising temperature, it tends to migrate from the large into the small cages and enters the hexagonal prism (SI) [245]. Therefore, the degree of reduction is greater at lower temperature, as the Ni^{2+} ions located in the supercages are reduced preferentially [246].

Sintering of elemental nickel appears to occur via a crystallite migration mechanism [247]. The basicity of the alkali cocations strongly influences the dimensions of the supported metal. Thus, at low exchange levels (< 10 Ni^{2+}/u.c.), Ni reducibility is inversely related to the coordination strength of Ni^{2+} to the zeolite framework and increases in the sequence

NiCsNaY < NiRbNaY < NiKY < NiNaY < NiLiY

At higher exchange levels (> 10 Ni^{2+}/u.c.), the degree of Ni^{2+} reduction decreases with increasing Brønsted acidity in the order

NiCsNaY > NiRbNaY > NiKY > NiNaY > NiLiY

If the zeolite contains two reducible cations, e.g., Cu^{2+} and Ni^{2+}, reduction may lead to the formation of alloy crystallites. Alloy formation has a profound influence on the type of reaction, e.g., of n-hexane. Nickel supported on zeolite causes hydrogenolysis with formation of CH_4 and straight chains of shorter length, whereas Ni/Cu zeolite catalysts, while less active, give considerable isomerization to singly branched isoparaffins. Small quantities of 2,2-dimethylbutane were also found. The pattern of reaction thus becomes more similar to that of Pt/zeolite catalysts [248]. The reduction rate of zeolitic Ni^{2+} is greatly increased by the presence of metallic Pd^0 [249] or Pt^0 [250] in the vicinity of these ions.

Nickel can also be introduced by sorbing $Ni(CO)_4$ on the zeolite [251]. The IR spectra indicate that the sorbate is undisturbed in dealuminated zeolite Y, but bridges of the type Na^+–OC-Ni are formed in cationic forms of zeolite Y. Although the nickel carbonyl is more strongly bound in the latter case, the $Ni(CO)_4$ in either configuration can be extracted with tetrahydrofuran. Thermal treatment results in the loss of CO and the formation of a bimodal Ni^0 phase, which consists of small clusters of 5–15 Å size within the cavities and 50–300 Å crystallites on the external zeolite surface. Oxidation of $Ni(CO)_4$/NaY to NiO/NaY occurs at room temperature, and the small NiO particles are stable to sintering up to 400 °C. Reaction of adsorbed $Ni(CO)_4$ with other ligands replaces one CO and, if appropriately sized ligands are employed, a "ship-in-a-bottle" complex can be formed. Even nickel phthalocyanine can be synthesized in the zeolite pores by reacting the carbonyl with phthalonitrile [252].

Cobalt. The behavior of $[Co(NH_3)_6]^{3+}$ at calcination conditions differs markedly from that of $[Co(H_2O)_6]^{2+}$. While the latter loses all ligands rapidly and migrates swiftly into sodalite cages and hexagonal prisms, the $[Co(NH_3)_6]^{3+}$ ion loses five of its six ammine ligands easily while being reduced, even in flowing O_2, to $[CoNH_3]^{2+}$. A stable tetrahedral configuration is attained with three framework-oxygens and the strongly bonded ammine ligand, detectable by strong absorption bands in the UV-VIS DRS spectrum at 16600, 17300, and 18500 cm^{-1}. Probing with ethylenediamine shows that this complex is still located in the supercages. The last ammine ligand is oxidized at 500 °C. The oxidation is promoted in the bimetallic PdCo/NaY system by the simultaneous oxidation of the ammine ligands of $[Pd(NH_3)_4]^{2+}$ at 300 °C. While Pd ions migrate into sodalite cages at calcination temperatures $\geq$300 °C, the monoammine-cobalt(II) ions remain in supercages. Cobalt ions do migrate into the sodalite cages at higher temperature.

The reducibility of Co^{2+} is markedly enhanced by the presence of Pd [134, 253] and Pt [254]. The temperature required to reduce Co^{2+} in NaY was significantly lower when Co^{2+} was located in the vicinity of Pd or Pt. The resulting bimetallic particles in the zeolite supercage impart high activity and selectivity for CO hydrogenation to CH_3OH. The enhancement depends on the calcination temperature and the metal content, which determine the locations of Pt(Pd) and Co in the zeolite. Most of the Co^{2+} is co-reduced with the Pt^{2+} at temperatures $<$450 °C. Formation of $PtCo_xO_4$, which prevents migration of the Co^{2+} into the small cages, seems to be an intermediate for Pt-catalyzed Co^{2+} reduction. If the calcined PdCo/NaY zeolite is exposed to H_2O prior to reduction, large alloy particles detectable by XRD are formed. Simultaneously, a shift in CO hydrogenation toward hydrocarbons is observed [255].

The ferromagnetic resonance band of CoNaY reduced with H_2 for 36 h at 500 °C with intermittent oxygen treatments was twice as intense as that of CoNaY reduced with H_2 only. The result suggests that reduction of CoNaY with H_2–O_2 treatment favors production of zerovalent clusters due to increase in mobility of Co^{2+}. The size of the clusters is so small that H_2 chemisorption, X-ray diffraction, and transmission electron microscopy cannot detect the quantity of zerovalent Co clusters present. Beside FMR spectroscopy, measurement of H_2 consumption during the reduction process, and the activity for CO hydrogenation, proved useful for the characterization of the clusters [256].

Highly dispersed superparamagnetic Co clusters have been obtained in NaX by decomposing adsorbed $Co_2(CO)_8$ with microwave plasma [257]. The clusters have a diameter of 7 Å, as FMR data indicate, and are not observable by transmission electron microscopy [258]. Information on the size and location of Co clusters in NaY can be obtained by measurement of the ^{59}Co spin-echo NMR spectrum. The critical temperature for the superparamagnetic/ferromagnetic transition can be used to estimate the average Co cluster size, and selective treatment with triphenylphosphine, which cannot enter the cavities, can determine the location of the clusters. The methods showed that decomposition of adsorbed $Co_2(CO)_8$, followed by annealing at 200 °C, resulted in the formation of Co clusters with an average diameter of 6–10 Å inside the NaY cavities. In contrast, decomposition and annealing at 500 °C caused clusters > 10 Å to be formed outside the NaY cages [259], while others report that larger particles are formed at temperatures as low as 160 °C [258].

3.2.1.3
Group VIIIA, Fifth Period

Ruthenium. Zeolites X and Y can be ion-exchanged at room temperature with 0.05M $RuCl_3$. In this exchange, the ruthenium displays a charge of 2.1, indicating that the exchanging cation is mainly $RuOH^{2+}$. The products are thermally stable to outgassing in vacuo at 350 °C and to reduction with hydrogen at 450 °C [260].

When RuNaY prepared by ion exchange with $[Ru(NH_3)_6]^{3+}$ is thermally decomposed in vacuo at temperatures increasing to 350 °C, the initially white sample turns successively deep purple, white, brownish and grey. The appearance of N_2 in the decomposition products suggests a reduction of the Ru^{3+} ion [261]. When a fresh sample is contacted with H_2 before the $Ru(NH_3)_6^{3+}$ complex is decomposed and the reduction is continued at temperatures rising to 560 °C, the subsequent TPO shows a minor part of the Ru metal to be oxidized at room temperature and is therefore located in the zeolite cavities. All metal is reoxidized below 500 °C indicating that no external metal is present. The maximum rate at 270 °C suggests that this oxidation is due to particles of intermediate size, which may be accommodated in mesopores of the zeolite generated by dehydroxylation. The various procedures employed demonstrate that only degassing

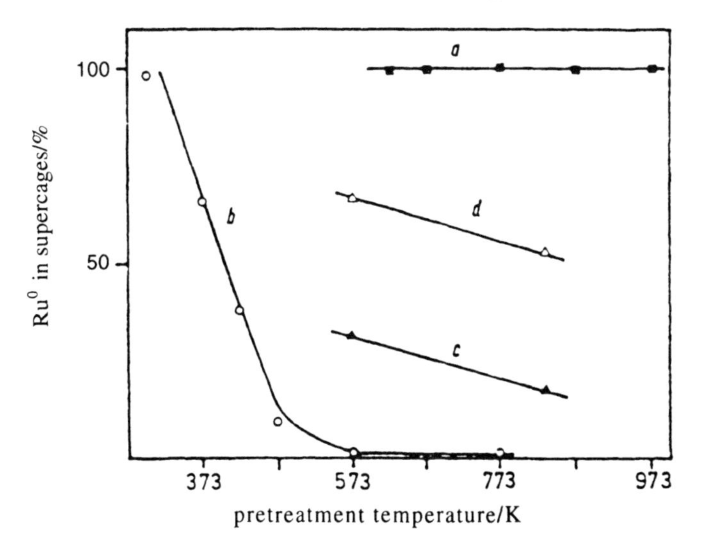

Fig. 3.18. Ruthenium metal inside zeolite cages of RuNaY-40 after different pretreatments [261], **a**) degassed at 350 °C and reduced at increasing temperatures, **b**) outgassed and reduced at 350 °C, oxidized at increasing temperatures, followed by a second reduction at the same temperature, **c**) degassed at room temperature and reduced at different temperatures during TPR experiment, **d**) same as **c**), but degassed at 100 °C prior to TPR run

at 350 °C and reducing at increasing temperatures yields a product with all the metal located in the zeolite cavities (Fig. 3.18). The presence of water or ammonia during the reduction gives a bimodal distribution of Ru^0 particles [261].

The ruthenium-red complex cation can be ion-exchanged into NaY at 75 °C. As shown by ^{129}Xe NMR, very small Ru clusters consisting of ~20 Ru atoms are formed in the supercage by autoreduction of the supported ruthenium species during evacuation under heating to 400 °C. The average number of Ru atoms/cluster gradually increases to 50 on further heating to 550 °C, either in H_2 or in vacuo. Exposure of the 20-atom cluster to O_2 caused a greater increase in the cluster size than exposure to H_2 or vacuum at the same temperature, leading eventually to large Ru agglomerates greater than 1 μm on the external surface of the zeolite crystals [262].

Chemisorption of CO on Ru^{3+} Y at ambient temperature causes formation of monocarbonyl or polycarbonyl complexes of Ru^{3+}, depending on the partial pressure of CO. The monocarbonyl complex adds O_2 at room temperature to form a superoxide ion with CO and O_2 being coordinated simultaneously [263]. Interaction of CO with both oxidized and reduced ruthenium supported on highly dealuminized Y results in formation of well-defined surface carbonyl with IR stretching bands at 2152, 2091, and 2086 cm^{-1}. The clear splitting of the CO stretching band usually found unresolved at ~2080 cm^{-1} with supported Ru allows assignment to a tricarbonyl with slightly distorted C_{3n} structure. Surface tricarbonyl transforms reversibly into a well-defined surface dicarbonyl during evacuation at higher temperature [264]. Dispersion of ruthenium on dealuminated zeolite Y was found to increase with the degree of dealumination. The maximum turnover frequency for CO hydrogenation occurred at a SiO_2/Al_2O_3 molar ratio of about 10 [264a].

Zeolite Y adsorbs $Ru_3(CO)_{12}$ without loss of CO. The complex is located near the supercage aperture. Heating at 300 °C causes all CO to be lost, and the Ru atoms are randomly distributed [117].

Faujasite-type zeolites containing Ru^0 crystallites within the pores are active catalysts for ammonia synthesis from N_2 and H_2. The activity is altered by varying the cocations and the SiO_2/Al_2O_3 ratio, with RuKX being the most active and RuHY the least active catalyst [265]. Stable ruthenium metal clusters were obtained by ion exchange with $[Ru(NH_3)_6]Cl_3$ at 333 K involving controlled formation of a ruthenium complex oligomer upon drying, and activation in synthesis gas by heating at 10 K/min to 844 K [265a]. While Pt/HZSM-5 deactivates during ethane aromatization at temperatures >530 °C, Ru/HZSM-5 is more stable below 700 °C, but is less active due to the lower dehydrogenation activity of Ru [266]. The hydrogenolysis activity of ZSM-5-supported Ru is reduced by the addition of Cu to the catalyst. The TPR peak occurs at 180 °C instead of 240 °C for the monometallic Cu-sample, close to the monometallic Ru/ZSM-5, suggesting alloy formation. This effect is more pronounced with $RuCl_3$ than with $Ru(NO)(NO_3)_3$ [267].

Rhodium. Rhodium can be introduced into zeolite Y by exchanging with $[Rh(NH_3)_5Cl]Cl_2$ at 80 °C, $RhCl_3$ at 90 °C, or $Rh(ClO_4)_3$, after adjusting to pH 3.7, at 70 °C [268]. The complex ion remains stable only at high concentration, whereas hydrolysis at 80 °C causes a limited ligand exchange and formation of the ion $[Rh(NH_3)_5OH]^{2+}$ at the low concentration usually applied for this ion exchange. The chloro complex, which has also been exchanged into NaX, is retained if the exchange is carried out at room temperature [269, 270]. In contrast to exchange with $Rh(ClO_4)_3$, ion exchange of NaY with aqueous $RhCl_3$ yields a product containing ~0.1 Cl/Rh.

Thermal activation of NaY exchanged with $[Rh^{III}(NH_3)_5Cl]^{2+}$ yields $[Rh^{II}(NH_3)_5]^{2+}$ at 200 °C, and complete decomposition of the complex occurs at temperatures >500 °C. The Rh^{2+} ions generated migrate to sites SI (10%), SI′, SII, SII′ [269]. The Rh^{2+} is partially oxidized with O_2 at 500 °C and may form a binuclear $[(O)_3\text{-}Rh\text{-}Rh\text{-}(O)_3]^{4+}$ complex [271]. Slow heating to 350° or 500 °C in flowing O_2, followed by outgassing at the same temperature, retains trivalent Rh [272].

Four different paramagnetic Rh^{II} species have been detected by ESR after activation with O_2 at 200°–475 °C and evacuation at room temperature, and a fifth after evacuation at high temperature [273]. The latter is interpreted as being due to isolated Rh^0 atoms stabilized in the small cages of zeolite Y. This reduction is believed to involve O from the zeolite framework. The species is also formed upon exposure of RhNaY to reducing chemicals, e.g., H_2 and CO, at ambient temperature [274]. Almost no paramagnetic species were detected after activation in O_2 at >500 °C indicating oxidation of all Rh^{II}. All Rh^{II} species are located in the supercages. Similar observations were made with RhNaX [270].

Shape-selective hydrogenation of cyclopentene in preference to 4-methylcyclohexene has been achieved with Rh/ZSM-11 [275].

Rhodium hydride can be deposited on zeolite NaHX or NaHY by first reacting the acidic OH-groups with triallylrhodium ($-C_3H_6$), followed by reduction in H_2 ($-C_3H_8$), reaction with HCl ($-C_3H_6$), and final reduction with H_2 (-HCl), to yield $-O_z$-RhH_2 [276].

Rhodium carbonyl clusters can be entrapped by first ion-exchanging with $[Rh(NH_3)_6]Cl_3$. The dried zeolite, when treated with 80 bar of CO + H_2 (1:1) at 130 °C, gave IR bands at 2095, 2080, and 2060 cm^{-1} (all coordinated terminal carbonyls), as well as a band at 1765 cm^{-1} (bridging carbonyl), assigned to rhodium carbonyl clusters. The product had high activity and selectivity for the hydroformylation of hexene-1 in liquid phase to aldehydes [277].

After carbonylation of Rh^{3+} at atmospheric pressure, two species, Rh(III)(CO) and $Rh(I)(CO)_2$, were identified in the sample initially calcined at 500 °C, while the 350 °C calcined sample contained only monovalent Rh. The trivalent species also disappeared upon rehydration of the sample initially dehydrated at high temperature. Thus, the overall reaction mechanism for the reduction of rhodium in a zeolitic environment appears to be identical to that observed in solution, suggesting the following reactions [272]:

$$Rh(III) + CO \rightarrow Rh(III)\text{-}CO \tag{3.28}$$

$$Rh(III)\text{-}CO + 2CO + H_2O \rightarrow Rh(I)(CO)_2 + CO_2 + 2H^+ \tag{3.29}$$

A catalyst prepared in a similar manner from zeolite X gave superior performance in the carbonylation of methanol to methyl acetate [278], compared with $RhCl_3$/NaX [279]. Rh(III)NaY, prepared by treatment of NaY with $RhCl_3$, was applied as catalyst for hydroformylation of olefins [280]. The complex $[Rh(CO)_2]^+$ was observed to exist in RhNaX after hydroformylation, and the cluster $Rh_6(CO)_{16}$ was found in RhNaY [281]. Further investigation showed that reduction of the Rh to the metal cluster occurs in RhNaY only in the presence of water [282]. Precise conditions for the preparation of RhNaY hydroformylation catalysts have been established in order to assure reproducible activity [283].

The complex $[Rh(CO)_2(acac)]$ can be deposited from the vapor phase in basic NaY, prepared by treating NaY with NaN_3 in methanol and thermal decomposition in N_2 flow at specified conditions. The product catalyzes CO hydrogenation to low-molecular weight olefins and methanol. The complex is stabilized by high partial pressures of CO or CO + H_2. It is believed that zeolite-entrapped Rh carbonyl clusters, e.g., $Rh_6(CO)_{16}$, may be the active species, with added selectivity provided by the zeolite pore structure [284].

Palladium. The $[Pd(NH_3)_4]^{2+}$ ion penetrates into the channels of ZSM-5. Vacuum treatment at 180–500 °C results in destruction of the complex, formation of Pd^{2+}, and reduction to Pd^0. Larger clusters of Pd^0 formed upon reduction in H_2 can be redispersed in O_2 with formation of Pd^{2+} in the zeolite [285]. Pd/HZSM-5 is highly active for the reduction of NO_x with methane in the presence of O_2 in the range 350–550 °C [286]. Agglomeration of Pd particles is the dominant cause of catalyst deactivation. Triatomic palladium particles in a freshly reduced

0.88 wt% Pd/HZSM-5 are stabilized by protons bridging between Pd_3 and the cage wall. During the conversion of methylcyclopentane at 250 °C, the Pd_3 clusters coalesce to generate particles of about 20 Å, thus lowering the conversion from 11.5 to 1.5% in 3 h [286a]. The importance of proper calcination conditions is illustrated by the calcination of $PdNaNH_4Y$ prepared by ion exchange of NaY with $[Pd(NH_3)_4]Cl_2$ [287]. The position of Pd^{2+} in the products obtained varies depending on the temperature applied during the activation with O_2, and the final state of metal dispersion is determined by the location of the ions. The $[Pd(NH_3)_4]^{2+}$ ions are located in the supercages at temperatures up to 250 °C (characteristic absorbance maximum at 33500 cm^{-1} in the UV-VIS DRS [288]) and reduction with H_2 at 20–130 °C (120–220 °C [287]) leads to formation of Pd particles fitting in the supercages [287]. Upon treatment with O_2 at 200–500 °C, these particles are reoxidized to Pd^{2+} cations located in SI′ sites in the sodalite cages [287]. Oxidation of the ammine ligands in $[Pd(NH_3)_4]^{2+}$ is a stepwise process and produces $[Pd(NH_3)_2]^{2+}$ in the supercages while $[Pd(NH_3)_4]^{2+}$ and Pd^{2+} are located in the sodalite cages. The relative abundance and location of these ions is controlled by the calcination program [287, 289, 290]. Calcination at <250 °C leaves palladium ions in the supercages where the coordination to ammine ligands decreases with increasing calcination temperature. At 300 °C, the Pd ions lose the third ammine ligand, the monoammine Pd ions can then migrate into the sodalite cages, where the remaining ligand is destroyed at 400 °C [289], and calcination at ≥ 500 °C places Pd^{2+} quantitatively in the sodalite cages [290], while EXAFS and DRS data indicate that Pd ions coexist in sodalite cages and hexagonal prisms [287]. Reduction in H_2 between 20° and 230 °C of such calcined product forms atomically dispersed Pd^0 particles in the sodalite cages [287]. Their subsequent release into the supercage network is an activated process. The ratio of adsorbed H_2/reduced Pd initially increases with temperature and then passes through a maximum, indicating that isolated Pd atoms are incapable of dissociatively chemisorbing H_2 [290]. Calcination at 500 °C, followed by reduction at 350 °C, causes small Pd clusters to be formed in the supercages with a compressed Pd–Pd distance of 2.68 Å and an average coordination number of 4 corresponding to a nuclearity of 6 Pd atoms [291]. At higher temperatures, the Pd^0 agglomerates to form 20–35 Å particles, too large to fit in the supercages. The atomically dispersed Pd^0 is reoxidized in O_2 to Pd^{2+} cations at 180–480 °C, but reoxidation of the 20–35 Å particles at 230–500 °C leads to formation of PdO_x [287].

Autoreduction during the calcination of the tetramminepalladium ion in inert atmosphere is negligible in zeolite Y, although at less rigorously controlled conditions it may become significant [292]. Rehydration after calcination causes Pd^{2+} ions in SI′ sites to migrate towards the supercages and may become square-planar coordinated with H_2O. When $[Pd(H_2O)_4]^{2+}$ ions are recalcined, large PdO particles are formed in the supercages as a result of hydrolysis. Subsequent reduction always leads to large Pd particles. In situ calcination of $[Pd(NH_3)_4]Y$ at 0.5 °C/min was found to achieve the highest Pd dispersion [293].

Pretreatment conditions have a significant effect on the performance of a metal/zeolite catalyst. A Pd/stabilized HY catalyst reduced at 300 °C showed stable activity and high selectivity for isomerization in the hydroconversion of n-nonane at 200 °C and 0.1 MPa, while that reduced at ≥500 °C initially favored cracking at the expense of isomerization. The reaction mechanism can be correlated with the change in Pd dispersion [294].

The described reactions suggest a procedure for redispersing large palladium aggregates. The large Pd particles are first oxidized in O_2 at a temperature where PdO particles react with zeolite protons to Pd^{2+} and H_2O [295]. This reaction indicates that redispersion is much more effective in a zeolite containing a high concentration of protons. The Pd^{2+} ions, dispersed in the sodalite cages, can be distinguished from PdO by TPR, where their reduction gives rise to a peak in the range of 160–200 °C, while PdO is reduced at ~35 °C. Subsequent gentle reduction of the Pd^{2+} ions in the sodalite cages results in formation of small Pd particles in the supercages. However, the proton concentration in the vicinity of very large Pd particles is insufficient for complete redispersion, even in strongly acidic HY. In such cases, exposure to gaseous ammonia causes $[Pd(NH_3)_4]^{2+}$ and NH_4^+ ions to be formed in the supercages. These ions are highly mobile so that concentration gradients are lowered and, after calcination, protons are again present in the vicinity of PdO particles [296]. Another method for redispersing large Pd particles on Pd/NaY or Pd/HY involves exposure of these zeolites to Cl_2 flow at 350 °C leading to formation of mobile $PdCl_2$, which, upon reduction at 350 °C, is converted to HCl and highly dispersed Pd^0 in the supercages. Thermal activation of chlorinated Pd/NaY and Pd/HY leads to ion exchange of $PdCl_2$ with zeolitic H^+ forming HCl and Pd^{2+} in the sodalite cages. Chlorination and reduction of physical mixtures of Pd with NaY or HY also produce highly dispersed Pd particles. It is proposed, that transport of Pd from the external surface to the interior of zeolite Y occurs through vapor phase transport of volatile $PdCl_2$ [297].

Probing the palladium by chemisorption of CO and characterizing the product with FTIR confirmed that most Pd is inaccessible to CO after calcination at 500 °C, and those Pd that are located in the supercage have various oxidation numbers [298]. After reduction at 350 °C and CO chemisorption, sharp bands appear in the FTIR spectrum suggesting that most Pd atoms are present in similar clusters. Reduction at 450 °C shows little structure in the FTIR spectrum after adsorption of CO, indicating that the reduction temperature determines the microgeometry of the sites adsorbing CO on Pd particles in the supercages.

Admission of CO to reduced Pd/NaY at room temperature induces substantial migration and coalescence of primary Pd clusters. After calcination at 500 °C and reduction at 200 °C, isolated Pd^0 atoms are located in the sodalite cages, where they are inaccessible to CO at temperatures below 120 K [135]. Pd clusters in supercages with Pd coordination numbers of 2.4 or 4 (6 Pd atoms) are generated by calcination at 250 °C/reduction at 200 °C or calcination at 500 °C/reduction at 350 °C, respectively. Adsorption of CO on the latter results in increased cluster size and a relaxed Pd–Pd distance of 2.7 Å. The primary mobile Pd-CO clusters coalesce in the supercages until further transport is sterically hindered, leading to $Pd_{13}(CO)_x$ clusters (CN = 6) with double-bridged CO [299] at ambient temperature as the 8.2 Å Pd_{13} core is unable to traverse the 7.5 Å apertures [135, 291]. The enhanced growth of these clusters indicates that CO adsorption weakens the Pd–Pd bonds in the clusters as well as the interaction of the clusters with the zeolite framework, although no cluster disintegration was observed.

At elevated temperature, e.g., 200 °C, further growth is possible as the supercage windows expand, and $Pd_{40}(CO)_x$ clusters are formed which completely fill the α-cage [300].

The geometry of the faujasite supercage appears to favor formation of Pd carbonyl clusters. As the clusters lose some of their CO ligands by purging with an inert gas, the intensities of IR bands decrease, while new bands with sharp isosbestic points emerge. Readsorption of CO reverses this process. The low activation energy of CO release and concomitant changes in the IR bands characteristic of Brønsted acid OH-groups indicate a chemical interaction between zeolite protons and Pd carbonyl clusters [301]. The $Pd_{13}(CO)_x$ clusters contain singly and multiply bonded CO, possibly including a butterfly configuration. The Pd ions initially located in the α- or the β-cage yield very similar spectra. The clusters formed may consist of cubooctahedral anions of the type $[Pd_{13}(CO)_{24}H_{5-n}]^{n-}$ and are probably located in the center of the cage. The partially decarbonylated cluster interacts with cage walls resulting in an increased stretching frequency of CO ligands in threefold coordination sites [302].

FTIR spectra of adsorbed CO on Pd particles in the supercages of zeolite Y show that positively charged Pd_n^+ clusters are formed when the proton concentration in the supercages is high. It is hypothesized that $[Pd_nH]^+$ adducts are the electron-deficient Pd clusters responsible for catalytic superactivity. Formation of such adducts is favored by high concentrations of divalent cations. CO displaces H^+ from $[Pd_nH]^+$ adducts, causing an increased intensity of the IR band due to supercage O_z-H groups. TPR suggests that divalent ions preferentially populate sodalite cages and hexagonal prisms, thus preventing Pd^{2+} from migrating into these positions. During reduction, both Pd_n clusters and H^+ are formed in supercages, setting the stage for $[Pd_nH]^+$ adducts [303].

The interaction between acid sites and Pd is demonstrated by the conversion of methylcyclopentane over Pd/MY, where M = Li, Na, K. The activity for ring opening increases from Li to K, while that for ring enlargement decreases and the benzene/cyclohexane ratio in the ring enlargement product increases. The results are explained with a decrease in the acid strength from Li through K, while the electron transfer from donor sites to Pd clusters increases in the same order [304]. Isomerization is catalyzed by Pd_n clusters, while ring enlargement to cyclo-C_6 and C_6H_6 requires both metal sites and protons. It is proposed that $[Pd_nH]^+$ adducts are the active sites where MCP is converted to benzene during one residence of the molecule. This novel reaction path compares favorably with the classical "bifunctional" route, which requires shuttling of reaction intermediates between metal and acid sites [305].

Reduction of $Pd^{2+}Cu^{2+}$/NaY with hydrogen yields a PdCu alloy. The protons generated by the reduction selectively reoxidize the Cu^0 at 280 °C in inert atmosphere to Cu^{2+}; complete oxidation takes place at 500 °C. The Cu^{2+} ions generated migrate into the small cages, and the monometallic Pd particles obtained form hydrides more readily than ordinary Pd particles [223]. A PdCo/NaY catalyst (9 Co + 9 Pd/u.c.), calcined and reduced without delay, yields methanol and dimethyl ether as predominant products of CO hydrogenation. If, however, prior to reduction, the calcined precursor is exposed to moist atmosphere for several days, methane and higher hydrocarbons prevail [255]. The explanation for this behavior is that the Co^{2+} ion, upon rehydration, returns to the supercage where it is reduced together with the Pd^{2+} to yield an alloy that catalyzes the formation of hydrocarbons. In contrast, reduction of freshly calcined PdCo/NaY leaves the Co^{2+} ions in the sodalite cages; the Pd^{2+} in the supercages, after reduction, catalyzes the formation of oxygenates. The chemistry of bringing metal ions back to the supercages by offering attractive ligands has potential for other applications, e.g., rejuvenation of aged metal/zeolite catalysts with NH_3 [296]. PdCo alloy particles prevent the formation of palladium carbide. A high Co content of the alloy is obtained by reduction at high temperatures, and the catalysts thus prepared show similarities to Fischer-Tropsch catalysts [306]. Similarly, Pd and Ni in the same zeolite enhance each other's reducibility and form bimetallic particles. Pretreatment conditions have a pronounced effect on the TPR profiles of monometallic and bimetallic Pd/Ni catalysts. At high pH, Ni^{2+} is hydrolyzed in the supercages; at low pH, Ni^{2+} ions migrate to small cages upon calcination. In the first case, NiO is formed in the supercages while Pd^{2+} ions migrate into the sodalite cages upon calcination. Upon reduction, Pd atoms leave the small cages, and some become attached to NiO clusters and enhance their reducibility. Calcination of a sample exchanged at low pH places both ions in the small cages. Reduction at >200°C yields PdNi dimers, which subsequently migrate into supercages, but the activation energy for this migration is higher than for Pd atoms. If the reduction is carried out at a temperature >400°C, unreduced Ni^{2+} ions migrate to Pd or $PdNi_x$ particles in the supercages and are then reduced to form $PdNi_x$. In a reversion of this process in inert atmosphere at elevated temperature, protons generated during reduction selectively reoxidize Ni atoms of $PdNi_x$ particles [307, 308].

3.2.1.4
Group VIIIA, Sixth Period

Osmium. Zeolite-entrapped osmium-carbonyl clusters can be prepared by reacting adsorbed $H_2Os(CO)_4$ in the pores of a basic form of zeolite Y, prepared by treatment with NaN_3, with CO + H_2 at 300°C and 1 bar. The yellow product has a different IR spectrum compared with that prior to CO/H_2 treatment. The new CO bands at 2092 w, 2057 m, 1988 vs, and 1966 s cm^{-1} are suggestive of Os carbonyl clusters. The catalyst had low activity for CO hydrogenation, but was quite stable and gave non-Schulz-Flory distribution of C_1-C_5 hydrocarbons with high olefin/paraffin ratios. After 20 d of operation, the still yellow catalyst gave a new IR spectrum with CO bands at 2097 w, 1987 sh, 1960 vs, and 1870 sh cm^{-1}, suggesting a change in the structure of the carbonyl cluster. In contrast, when HY was used for this preparation, the resulting catalyst was much less stable during CO hydrogenation and gave common Schulz-Flory product distribution. Osmium migrated to the surface, where it formed metal crystallites [309, 310].

Iridium. Iridium can be deposited on zeolites NaX [311] and NaY [312-315] by adsorbing the complex [Ir$(CO)_2$(acac)] from a solution in hexane. By treatment with CO at 1 bar and 70°C, this complex is converted to $[HIr_4(CO)_{11}]^-$/NaX and, with CO/H_2 at 225°C and 20 bar, to $[Ir_6(CO)_{15}]^{2-}$/NaX. In NaY, reductive carbonylation of the adsorbed precursor at 40 °C leads to formation of $[Ir_4(CO)_{12}]$. These complexes are converted in CO at 125 °C to the isomer of $[Ir_6(CO)_{16}]$ with edge-bridging ligands and in CO + H_2 at 250 °C and 20 bar to the isomer of $[Ir_6(CO)_{16}]$ with face-bridging ligands. The zeolites containing the latter catalyze CO hydrogenation at 250 °C and 20 bar with high selectivity for propane. Either isomer of $[Ir_6(CO)_{16}]$ in the zeolite an be decarbonylated in H_2 at 300 °C and 1 bar and recarbonylated to give either isomer. EXAFS shows that the decarbonylated cluster is predominantly Ir_6 with an octahedral structure and an Ir–Ir bond distance of 2.71 Å.

Platinum. An early isomerization catalyst was prepared by loading CaY with 0.5 wt% Pt using ion exchange with $[Pt(NH_3)_4]^{2+}$. The product was pretreated at 350 °C and then reduced in H_2 at 300 °C. The dispersion of platinum can be advantageously measured by the ability to chemisorb hydrogen, expressed as H/Pt. Hydrogen chemisorbs rapidly at 100 °C and 0.93 bar to reach a ratio of H/Pt = 2. Further temperature increase to 200 °C at the same H_2 pressure causes initially rapid desorption of H_2, followed by slow re-adsorption of H_2 to the same ratio of H/Pt. The result was interpreted as weak chemisorption of molecular hydrogen at 100 °C, dissociation and strong bonding in hydride form at 200 °C. It is believed that the Pt remains in atomic dispersion throughout this procedure [316]. EXAFS examination of a commercial catalyst, Linde SK-200, 0.5% Pt/CaY, similar to the above, revealed that all Pt was reduced to Pt^0 and that this Pt is smaller in size than in conventional reforming catalysts. About 60% of the Pt is sufficiently small to be dissolved in acid, the remaining 40% is of ~60 Å size. The acid-soluble particles are estimated to be ~10 Å in diameter, small enough to fit into the supercages. Exposure to H_2 leaves Pt with a negative charge, and the ratio of H/Pt is 0.5 [317]. In contrast, Mashchenko et al. obtained a positively charged surface, $(Pt\text{-}H)^+$, upon thermal decomposition at 350 °C of $[Pt(NH_3)_4]^{2+}$ in zeolite Y, followed by reduction in H_2 at temperatures >300 °C [318].

Platinum is most readily introduced into zeolite Y by ion exchange with $[Pt(NH_3)_4]^{2+}$. Direct reduction of this complex leads to formation of large Pt agglomerates, probably via a neutral mobile Pt hydride, formulated as $Pt(NH_3)_2H_2$ [319]. Therefore, it is necessary to decompose the tetrammine complex prior to reduction with H_2. The temperatures applied in the latter two steps determine the position and the dispersion of the Pt^0. Empirical results show the optimal dispersion after drying in vacuo, followed by calcination in air at 360 °C and reduction in H_2 at 400 °C [320]. Others found that thermal decomposition of $[Pt(NH_3)_4]Y$ in air causes complete reduction of the platinum [292]. In addition, TPD, FTIR and XPS have been used to elucidate the decomposition of the $[Pt(NH_3)_4]^{2+}$ complex in KX and KY [320a]. The degree of agglomeration of Pt from autoreduction of decomposing $Pt(NH_3)_4^{2+}$ in zeolite X depends on a great number of factors, such as the degree of Pt exchange, the degree of hydration, the medium (inert gas, O_2, or vacuum) used in the calcination, the heating rate, and on the constraints imposed by the zeolite lattice [321].

Since simple thermal decomposition of the platinum complex causes some reduction and sintering of the reduced Pt [322], it is important to perform the initial calcination in a rapid flow of O_2 at a heating rate of 1 °C/min to 220 °C and of 0.25 °C/min to 300 °C [323]. It is essential that volatile decomposition products, i.e., H_2O and NH_3, are removed rapidly in order to prevent migration of Pt. This treatment leaves most Pt^{2+} in the supercage. Upon reduction at 300 °C, 6–13 Å agglomerates fitting into the supercages and stable to 800 °C are formed. Chemisorption of H_2 gives H/Pt = 1 corresponding to very small particles [324]. The zeolite structure breaks down at 900 °C, and the agglomerates are transformed into crystallites of 25–30 Å diameter. Initial calcination at 600 °C causes most Pt^{2+} ions to migrate into the sodalite cages. On reducing at 300 °C, the Pt^0 is atomically dispersed in these cages and does not chemisorb H_2. When this material is evacuated at higher temperature, Pt atoms gradually agglomerate and form crystallites, and H_2 chemisorption increases up to 800 °C to give H/Pt = 0.65 [325].

The presence of small quantities of cerium in zeolite Y favors the formation of smaller and more homogeneous Pt particles [326]. After activation in flowing O_2 (0.5 K/min to 630 K) and reduction at 580 K under 0.53 bar H_2 pressure, a zeolite of the composition $Pt_{11}Ce_1Na_{19}H_{12}[Al_{56}Si_{136}O_{384}]$ encaged Pt particles in the range of 7–11 Å. These Pt particles, when free of an adsorbate, were found to have a distorted structure, and the interatomic distances were contracted compared with the normal fcc structure of bulk platinum. Adsorption of H_2 caused complete relaxation, and the aggregates had the normal fcc structure and interatomic distances [327]. More recently, Rh, Ir, and Pt were found to form fcc crystallites in the cavities of zeolite X with their crystallographic axes parallel to the corresponding axes of the zeolite lattice. While Rh and Ir are bounded almost exclusively by {111} surfaces, Pt particles have less well-defined morphology, but the {111} surfaces seem to predominate in these crystallites as well [328].

Platinum and palladium are advantageously introduced into ZSM-5 by ion exchanging the ammonium form at 50 °C with aqueous $[Pt(NH_3)_4]Cl_2$ or the corresponding palladium complex, respectively. The product is then reduced in a stream of hydrogen and olefin:

$$[Pt(NH_3)_4]^{2+} Z^{2-} + H_2 \rightarrow Pt + 4NH_3 + (H^+)_2Z^{2-} \tag{3.30}$$

Alternatively, a shape-selective catalyst can be obtained by calcination in flowing oxygen to 350 °C at 0.5 °C/min, followed by reduction with hydrogen at 500 °C for 0.5 to 2 h [329]. Reduction at 300 °C for a period of 1 h produces an active, but non-shape-selective hydrogenation catalyst as a result of migration and agglomeration of Pt on the surface of the zeolite crystals. This migration during reduction with hydrogen has been attributed to the formation of mobile platinum hydride, $Pt(NH_3)_2H_2$ [319]. A highly selective catalyst containing well-dispersed Pt clusters of < 10 Å was obtained by reduction in a hydrogen/olefin mixture [329]. The olefin may act as a hydride trap decomposing the platinum hydride before it can migrate. Small quantities of external Pt can be poisoned by addition of a phosphine that is larger than the pore opening, e.g., tri-p-tolylphosphine, thus causing further improvement in shape-selectivity. If it is desired to minimize acid-catalyzed reactions, the acidity generated during the reduction of the complex ions to the metals (Eq. 3.30) can be neutralized by injecting ammonia into the reactor at a temperature below 300 °C, or the reduced catalyst may be treated with dilute base.

A bifunctional Pt/HZSM-5, prepared by a method similar to the one described above, proved to be a stable catalyst for the conversion of propane to aromatics [330]. The high activity and particularly the high stability are ascribed to the proximity of the acid function of the zeolite and the dehydrogenation function of the small platinum crystallites highly dispersed in the interior of the zeolite pore network. No migration of Pt to the surface was observed. Pyridine adsorbed on acid sites of ZSM-5 is hydrogenated to piperidine over Pt/HZSM-5 in the presence of gaseous hydrogen, as well as over a physical mixture of Pt/SiO_2 and HZSM-5, whereas this reaction does not occur over Pt/SiO_2 or HZSM-5. The reversible reaction is interpreted with hydrogen spillover from Pt sites to acid sites of the zeolite [330 a].

NH_4ZSM-5 exchanged with $[Pt(NH_3)_4]Cl_2$ and subsequently impregnated with the same salt for a Pt content of 0.5 wt% was calcined in dry air at 1.3 °C/min to 350 °C, held at this temperature for 30 min, the temperature was then raised to 450 °C and held for 3 h. After reduction in hydrogen at 350 or 520 °C, small particles within the zeolite cavities with an average diameter of about 8 Å were deposited. Calcination of a similar catalyst at 520° instead of 450 °C produced a catalyst containing Pt aggregates of 12–15 Å average diameter. EXAFS indicated that the Pt particles were bonded to framework oxygens with Pt–O distances in the range of 1.92–2.03 Å, while the Pt–Pt distances are 2.63–2.68 Å [331, 332]. A significant increase in the constraint index [333] was observed upon loading ZSM-5 with Pt and attributed to the influence of olefins, formed by alkane dehydrogenation, on the cracking mechanism [334]. Pt/ZSM-5 gave higher activity for NO reduction with C_2H_4 than CuZSM-5. Water vapor does not affect the performance of Pt/ZSM-5, but it reduces the activity of CuZSM-5 to almost zero [335].

The position of the stretching vibration of CO linearly chemisorbed on Pt in faujasite shifts to lower wavenumbers as the dispersion of Pt increases [336]. Similarly, a particle size dependent effect is observed in the desorption of O_2 from platinum dispersed in faujasite, as the activation energy increases by ~60 kJ/mol when the particle size decreases from 5 to 1 nm [337]. In contrast to CO adsorption on Pt^0, CO chemisorption on NaY or CsNaY containing $[Pt(NH_3)_4]^{2+}$ causes a bright red-purple color to develop. The characterization by UV-Vis and IR spectroscopy suggests that $[Pt_3(CO)_3(\mu_2CO)_3]_n^{2-}$ complexes of the Chini type may be formed and stabilized in the supercages of basic faujasite [338]. IR spectroscopy of adsorbed CO was also used to characterize the platinum particles generated in 0.6% Pt/MBaL (M = Li, Na, K, Rb, Cs). Low-frequency IR bands which appeared on heating these catalysts containing chemisorbed CO suggest that the CO-covered Pt clusters change their structure during heating [338 a].

A new vapor phase impregnation method with $Pt(acac)_2$ has recently been developed and applied to a variety of zeolites. Specifically, the dehydrated zeolite and $Pt(acac)_2$ are physically mixed in a tube, which is subsequently evacuated to 10^{-2} Pa and sealed. The sample in the sealed tube is slowly heated to 145 °C and held at this temperature for 16–24 h. Since ion-exchange sites are not required for this impregnation, Pt can be loaded into the pore systems of both charged and neutral frameworks. Pt introduced by this method does not migrate to the external surface of the zeolite crystals at reaction conditions of 460–520 °C and atmospheric pressure [339].

3.3
Dealumination of Zeolites

3.3.1
Thermal Treatment

In the as-crystallized form, zeolites are hydrated, i.e., the cavities are filled with water. This water may be present in a physisorbed state and/or in the form of the hydration shell of cations. Because of the equilibrium between the water of hydration of the cations and the physisorbed water, these two types of water are lost over a relatively wide temperature range in one continuous step, as seen by thermal analysis. This dehydration is endothermic and reversible, and is utilized in the application of zeolites as adsorbents. Zeolitic adsorbents are customarily activated by calcination at 350 °C. Several zeolites, e.g., harmotome [340–342] and phillipsite [341, 343], undergo a structural change upon dehydration. Such a reaction is usually associated with a sharp endotherm in the DTA profile and a distinct step in the TGA curve.

For use as acid catalysts, zeolites are frequently applied in the acid form, which can most conveniently be prepared by calcination of the ammonium form. The acid forms of low-silica zeolites are inherently unstable. Some thermal stability can be attained by partial exchange with ammonium ions, leaving most of the cation sites occupied by stable cations; but even mild hydrothermal treatment causes loss of crystal structure [344–346]. However, zeolites X and Y can be stabilized by introducing rare earth ions into the β-cages while hydrogen ions can be present in the α-cages (see 3.1.3.3). Zeolites of higher SiO_2/Al_2O_3 ratio, e.g., mordenite, can be exchanged with dilute acid to yield the hydronium form; but thermal dehydration of the hydronium form does not yield the pure hydrogen form, as hydrolytic reactions interfere [347].

Most zeolites with SiO_2/Al_2O_3 molar ratios above about 20 are crystallized in the presence of an organic cation, usually a quaternary ammonium ion. Frequently these ions are larger than the pore openings, but even if they are not, ion exchange is impractical because of the high selectivity of the siliceous zeolites for such ions. Instead, the organic ions can be removed by thermal and/or oxidative decomposition.

Several examples given below will illustrate the general chemistry involved in the conversion of zeolites to the hydrogen form. The methods applicable for zeolites of a particular topology, SiO_2/Al_2O_3 ratio, and cationic form may be selected or designed from these examples based on the chemistry described.

3.3.1.1 Hydrogen Zeolites

Early attempts to prepare hydrogen zeolites of low Si/Al ratio by ion exchange with an acid caused the structure to collapse [348–350]. The more siliceous zeolites, which may have remained crystalline, yielded at best hydronium zeolites, but most likely aluminum-deficient materials.

The preparation of crystalline H-mordenite and H-chabazite by calcination of the ammonium forms in air was first reported by Barrer [351].

$$4\,NH_4Z + 3\,O_2 \rightarrow 4\,HZ + 6\,H_2O + 2\,N_2 \tag{3.31}$$

Uytterhoeven et al. [352] prepared hydrogen zeolite Y by careful calcination of NH_4Y and found the highest hydroxyl content after calcination at 440 °C, as determined by deuterium exchange. As ammonia was removed by thermal treatment,

$$NH_4Y \rightleftarrows HY + NH_3 \tag{3.32}$$

a new OH band grew at a frequency of 3660 cm^{-1} in the IR spectrum. This band, distinctly different from the 3740 cm^{-1} band of the terminal ≡SiOH groups, was assigned to a ≡SiOH adjacent to a "trigonal Al" and explained by the reaction of the proton formed by deammination with the zeolite framework:

$$\overset{H^+}{(\equiv SiO)_3Al^- - OSi\equiv} \leftrightarrow (\equiv SiO)_3Al^{\delta-} \cdots \overset{H}{O^{\delta+}} - Si\equiv \leftrightarrow (\equiv SiO)_3Al\ HO - Si\equiv \tag{3.33}$$

The conclusion by Uytterhoeven et al. [352] that the Brønsted acid shown on the left-hand side of Eq. 3.33 is not present in substantial quantity in the hydrogen form of zeolite Y was challenged by Benesi [353], who published the first thermogravimetric curves of the decomposition of NH_4Y and NH_4-mordenite. Benesi argued that the 3660 cm^{-1} band could be considered evidence for the existence of such an acid, as the relatively high frequency of the OH vibration would indicate that the OH-group is not hydrogen-bonded. In agreement with Benesi's contention, Kühl, using the AlK_β line in X-ray fluorescence spectroscopy, found no evidence of tri-coordinated Al in hydrogen zeolite Y; instead, all Al remained tetracoordinated [354, 355]. The AlO_4 tetrahedron appears to be sufficiently distorted that the ^{27}Al MAS NMR spectrum shows no evidence of aluminum, as long as the zeolite is kept meticulously dry, as demonstrated for zeolites X [356], Y [357–359], ZSM-5 [360, 361], and Rho [359, 362]. After careful hydration, the NMR signal associated with tetrahedral Al reappears [356, 361]. This observation can be explained by a ligand exchange of the proton (or sodium ion) in which water replaces the framework oxygen as the ligand, thus forming a hydronium (or hydrated sodium) ion and relieving the strain on the zeolite framework. The findings suggest that the zeolite framework bonds remain intact on dehydration although the local symmetry around the aluminum is lowered. It is convenient to formulate the hydrogen form as a transient state between the protonic and the Lewis/silanol from in Eq. 3.33, where the weakened Al–O bond is marked as a dashed line. Only the reaction of the hydrogen with a base, even a weak base such as water, restores the unstrained tetrahedral coordination.

The hydroxyl group in Eq. 3.33 has become known as the bridging OH-group. It gives rise to an IR band at 3640–3660 cm^{-1}; a band at 3540–3570 cm^{-1} is peculiar to faujasite-type zeolites [352, 363–365] and is due to OH located in the small cages [366, 367]. These acid sites are also distinguished by the different apparent activation energies [368]. Both react with ammonia; the 3540-cm^{-1} site reacts more slowly with pyridine and does not interact with olefins at room temperature [369], but reacts with cumene at temperatures above 325 °C indicating that the hydrogen is mobile and reacts with polar molecules more readily than with non-polar molecules [365]. The two sites can also be distinguished by ^{1}H MAS NMR [370–374].

The different positions of ammonium ions in zeolite Y were firmly established by DTA in an elegant piece of work reported by Chu [375]. A 96 % ammonium-exchanged zeolite Y, obtained by repeated exchange of NaY with NH_4^+ at 100°C, gave, in flowing nitrogen, a DTA curve containing four endotherms with maxima at about 150 (dehydration), 250 (NH_4^+ in small cages), 320 (NH_4^+ in supercages), and 650 °C (dehydroxylation, see 3.3.1.2). The second and third endotherms, due to decomposition of NH_4^+, were assigned by the DTA curves of samples containing ammonium ions only in the small cages or only in the supercages. The preparation of such samples was made possible by the knowledge obtained from ion exchange of zeolite Y (see 3.1.3).

For the preparation of a high-quality hydrogen zeolite, it is advisable to begin with an ammonium form with an eq. cation/Al ratio of 1.0, in order to avoid hydrolysis of framework-Al accompanying dehydration of incidentally exchanged hydronium ions. It is recommended performing the ion exchange with an ammonium salt solution with a pH close to or slightly above 7, e.g., ammonium acetate [352].

In order to prevent hydrolysis of framework-Al during the preparation of a hydrogen zeolite, all water needs to be removed from the ammonium zeolite prior to thermal decomposition of the NH_4^+ ion. Uytterhoeven et al. [352] evacuated 1–2 g of NH_4Y at 10^{-3} Pa for 2 h at ambient temperature and then heated it at 5 °C/min to the pretreatment temperature, 440 °C for the highest hydrogen content achieved. Vega and Luz [362] found that dehydroxylation has begun at 440 °C and recommend overnight drying at 110 °C and 10^{-2} Pa, followed by 7 h of heating at about 300 °C and the same pressure to obtain a fairly pure HY. Instead of evacuating the sample, Benesi [353] dried NH_4Y in flowing dry helium. In an effort to prepare a larger quantity of HY, Kerr [376] calcined NH_4Y in a thin layer (shallow bed) (Sect. 3.3.3.1). Not every zeolite can be converted to a pure hydrogen form as the thermal deammination and dehydroxylation (Sect. 3.3.1.2) reactions may overlap [377], but rapid removal of decomposition products by flushing a shallow-bed calcination with dry inert gas or by calcining in vacuo will produce HY of high purity.

The deammination reaction (Eq. 3.32) is reversible, i.e., the ammonium form can be reconstituted by sorbing ammonia on hydrogen zeolite Y, thus proving that the material obtained by calcination was the normal HY [368]. This reverse reaction can be used for the quantitative evaluation of active sites in acid catalysts by temperature-programmed desorption of ammonia [378–382]. If HY is contacted with ammonia at 500 °C, an amidozeolite, equivalent to a hydrogen zeolite, is formed by replacing OH with NH_2 [383]. This ammonia is liberated by heating at 600–800 °C, similar to dehydroxylation of HY.

Oxidative deammination of most zeolites occurs at a lower temperature than the thermal decomposition of the ammonium ion [377, 384, 385], but the water generated tends to hydrolyze framework-aluminum unless care is taken that this water is rapidly removed. In NH_4Y, however, intracrystalline oxidation of ammonium ions is not observed because the thermal deammination begins before the temperature required for the oxidation is reached [375, 377]. The maxima of the differential thermograms obtained by the oxidative decomposition of zeolitic ammonium ions shift to higher temperatures with increasing SiO_2/Al_2O_3 of the zeolites [377].

The decomposition of organic ammonium ions is considerably more complicated. Using thermogravimetric analysis at a heating rate of 10 °C/min in flowing helium, Wu et al. [386] concluded from the titration of the effluent gas and the simultaneous weight loss that methyl-substituted ammonium ions in zeolite Y release an amine in the temperature range of 250–450 °C that contains one methyl group less than the zeolitic ammonium ion. The percentage of the total nitrogen liberated in this manner, however, decreased with increasing methyl substitution, from 74% for $CH_3H_3N^+$ to 24% for $(CH_3)_4N^+$. Most of the remaining nitrogenous base was found to be released in a rapid reaction near 500 °C. The effluent gas can be analyzed by mass spectrometry [387]. Olefins generated in the decomposition react on acid sites and form carbonaceous deposits which can be burnt off at about 500 °C.

The $(C_3H_7)_4N^+$ (TPA^+) ions in ZSM-5 are decomposed in an inert atmosphere in the range of ~400–500 °C [388–393], although temperatures as high as 600 °C have been recommended by others [394]. As the SiO_2/Al_2O_3 ratio increases, the single DTG (differential thermogram) signal associated with a DTA exotherm (in nitrogen atmosphere) develops a shoulder on the low-temperature side and, in the essentially Al-free form, this shoulder becomes the only DTG feature. The two TPA^+ decomposition temperatures represent $(C_3H_7)_4N^+$ associated with Brønsted acid sites (~460 °C) and with $\equiv Si-O^-$ (~400 °C) (Fig. 3.19) [395]. Decomposition products include propane, propene, $C_3H_7NH_2$, $(C_3H_7)_2NH$, and C_6H_{12} species [388, 396]. The main decomposition scheme involves Hofmann elimination reactions [396–398]. ZSM-5 zeolite crystallized in the presence of fluoride ions contains some occluded $(C_3H_7)_4NF$ representing a third TPA species [399]. TPA^+ ions have been removed from high-silica ZSM-5 at 45 °C using oxygen or dry air radiofrequency plasma [400, 401].

The decomposition of $TPA^+O^--Si\equiv$ leaves a weakly acidic $\equiv Si-OH$ behind [402–409]. Since ZSM-5 does not have two kinds of cages as faujasite does, only one signal due to bridging OH-groups is observed in the ^1H MAS NMR spectrum of H-ZSM-5 [374, 410, 411]. H-ZSM-5 obtained from TPA-ZSM-5 shows a particularly strong ^1H NMR signal at 2 ppm [412] indicative of a high concentration of non-acidic silanol groups [413–415] and structural defects [374].

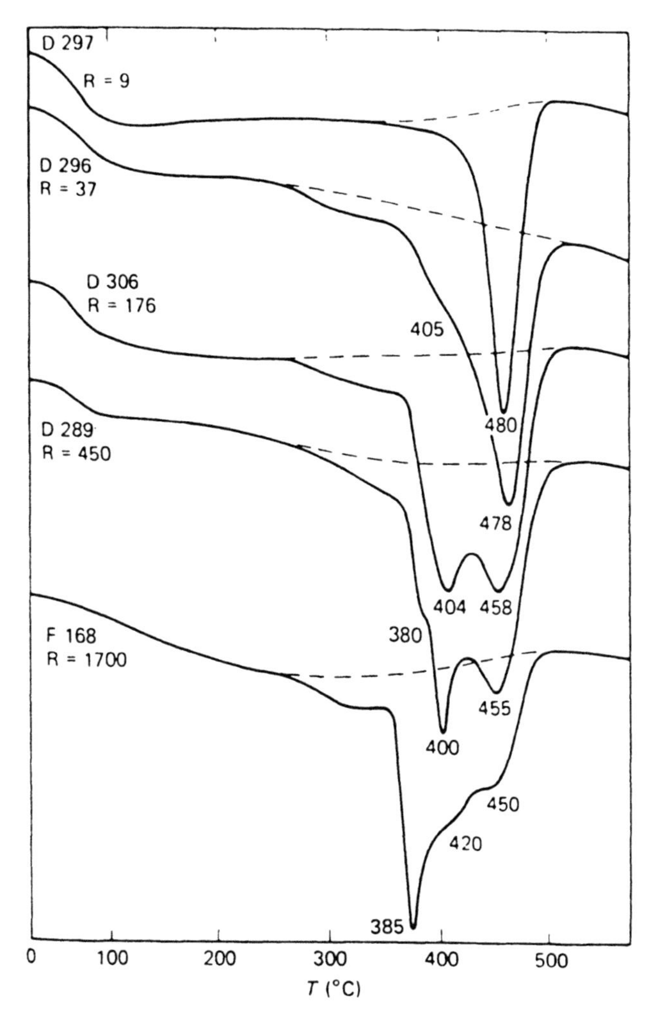

Fig. 3.19. DTA curves for ZSM-5 prepared in alkaline medium; argon flow, heating rate 10°C/min [395]

After calcination the sodium is removed by ion exchange with ammonium ion (pH 7–8) at 90°C, and the ammonium form is converted to the H-form by calcination at 450–500°C [394]. Sorption of limited quantities of water, up to 1 H_2O/framework-Al, probably hydrates the proton in zeolite Y as it does in ZSM-5 [416]. The hydronium ion thus generated can be carefully dehydrated to the H-form. An excess of water causes hydrolysis of framework-Al and leads to a collapse of the faujasite structure [376, 417]. Lower than expected acid concentrations are found in zeolites calcined at conditions permitting dehydroxylation and dealumination to occur (see Sects. 3.3.1.2 and 3.3.3). It is proposed that paired tetrahedral Al is more readily removed from the framework [417a].

3.3.1.2
Dehydroxylation

Further weight loss of HY upon heating to higher temperatures than required to produce HY [352, 353] is caused by the elimination of water corresponding to the formation of an acid anhydride and has been named "dehydroxylation". The quantity of this water agrees with the value calculated from the initial ammonium content [353, 376, 384]. The product, "dehydroxylated Y" was initially formulated as containing a tricoordinated Al and a tricoordinated, positively charged Si [352]. However, Kühl [354, 355] demonstrated that all Si remained tetracoordinated while nearly 50% of the Al was converted to octahedral instead of trigonal coordination. These results are supported by NMR studies [357, 358, 418] and IR spectroscopy [419]. Therefore, while the formula proposed by Uytterhoeven et al. [352] may be considered a formal intermediate, the final, stable product contains half of the Al in a cationic form having one charge per Al and no hydroxyl groups. Such a cationic species may be AlO^+ or a condensed aluminum oxide cluster ion coordinated to framework-oxygen [420]. The total number of OH-groups decreases during dehydroxylation, but the number of terminal silanol groups increases [421], suggesting that mesopores are formed (see 3.3.3.1). Borade et al. [422] conclude from N_{1s}XPS data that two different kinds of Lewis acid sites are present in dehydroxylated Y and interpret these as tricoordinated framework Al and non-framework octahedral aluminum, while Datka [423] observed a high concentration of Lewis acid sites, one half that of the initial OH concentration, in dehydroxylated NaHY.

Dehydroxylation is associated with an endotherm observed in the DTA profile as well as a weight loss in the TGA; the temperature at which these features occur increases with the SiO_2/Al_2O_3 ratio of the zeolite [384]. For zeolite Y at SiO_2/Al_2O_3 ratios above 4, the deammination temperature increases only slightly so that the two reactions become better separated. The maximum rate of dehydroxylation is determined by the thermal analysis data, but further gradual weight loss occurs up to 900 °C to reach the theoretical value. The reaction is accompanied by a contraction of the unit cell [384, 424], in agreement with the loss of aluminum from the framework and replacement with silicon. Carefully dehydroxylated hydrogen zeolite Y resembles "deep-bed" calcined HY [376] (see Sect. 3.3.3.1) in its stability [355, 424], but with only 50% of the total Al remaining in the framework.

Dehydroxylation of ZSM-5 at 770 K causes an intense broad band at 3250 cm^{-1} to appear in the diffuse reflectance IR spectrum, in addition to the sharp maxima at 3610 (bridging OH), 3740 (≡Si–OH), and, at higher temperature, 3680 cm^{-1} (extra-lattice AlOH species) [425]. The bands at 3610 and 3250 cm^{-1} decrease in intensity during dehydroxylation, and subsequent treatment with D_2O attenuates them. Simultaneously, the corresponding bands with approximately the same intensity ratio appear in the O–D stretching region. The broad band has been assigned to hydrogen (deuterium) bonding between a bridging OH (OD) and one of the oxygen atoms of the lattice.

Kazanski [426] examined dehydroxylated zeolites by IR spectroscopy after adsorption of molecular hydrogen and concluded that dehydroxylation of

ZSM-5 and other high-silica zeolites, but not of HY, generates two Lewis acid sites on the framework corresponding to $\equiv Si^+$ (4035 cm^{-1}) and $\equiv Al$ (4010 cm^{-1}) according to the formula proposed by Uytterhoeven at al. [352].

A different kind of dehydroxylation occurs upon calcination of certain ion-exchanged zeolites. A carefully studied example is the calcination of lanthanum-exchanged zeolite Y [427]. The lanthanum ions, initially located in the super-cages as La^{3+}, are sufficiently dehydrated at 60 °C to start migrating into the sodalite cages. Before dehydration of the lanthanum ions is complete, hydrolysis occurs in the range 200–500 °C with a reduction of the ionic charge (formation of $LaOH^{2+}$) and the generation of a proton associated with an AlO_4^- tetrahedron in the zeolite framework [428]. The reversal of this reaction at temperatures above 580 °C is a dehydroxylation [427]. Calcination of a rare earth/ammonium-exchanged zeolite Y causes little, if any, dealumination of the framework [429].

3.3.2
Extraction of Framework Aluminum with Acid

It is well known that mineral zeolites decompose when treated with strong acids [430]. Low-silica zeolites up to Si/Al of ~1.5 form a gel upon acid decomposition, whereas those of higher Si/Al ratio precipitate silica without gel formation [431]. The structures of some zeolites, especially mordenite, are resistent to acid decomposition while aluminum is slowly removed [432–434].

In the as-crystallized form, a zeolite usually contains only tetrahedrally coordinated framework aluminum. The cations, e.g., Na^+, can easily be replaced with H_3O^+ ions by mild acid treatment. This exchange establishes the equilibrium

$$[\equiv Si{-}O{-}Al^{-}\equiv]\,H_3O^+ \rightleftarrows [\equiv Si \quad \overset{\displaystyle H \atop {\displaystyle | \atop \displaystyle O^{\delta+}}}{/} \quad Al^{\delta-}\equiv] + H_2O \rightleftarrows [\equiv Si \overset{\displaystyle OH}{/} \qquad Al^{\delta-}\equiv] \;\; \text{with } H{-}^{\delta+}\underset{\displaystyle H}{O} \text{ bonded to } Al \qquad (3.34)$$

At low SiO_2/Al_2O_3 ratio, the concentration of broken SiO–Al bonds is so high that the structure collapses, as the bridging oxygen competes with the water for the proton. The water molecule is then chemisorbed on the Al Lewis acid site, thereby beginning the hydrolysis process. The reaction can be enhanced by heating and leads to the conversion of framework-Al to cationic hydroxylated aluminum.

The generation of the hydrogen form, e.g., by acid treatment or by calcining the template-containing form, is essential for the hydrolysis of framework-Al. The hydrolysis itself does not require reaction with an acid. However, the hydroxylated aluminum cations formed in the hydrolysis reaction protect remaining framework-Al from hydrolytic attack (see Sect. 3.3.3.1). These ions can be removed by ion exchange, most conveniently with an acid. The acid neutralizes the hydroxyl groups of the cationic aluminum species to form Al^{3+} ions, which are exchanged for H_3O^+, and the dealumination proceeds. Depending on the pore structure of the zeolite, it may be necessary to reduce the size of the large hydration shell of the Al^{3+} ion by conducting the reaction at elevated temperature. An attempt to dealuminate ZSM-5 by acid extraction was unsuccessful.

The first acid extraction of a zeolite was reported by Barrer and Makki [435], who treated clinoptilolite with hydrochloric acid and found that, in addition to the cations, framework aluminum was progressively removed. Without loss of crystal structure, a product without a framework charge was eventually obtained [436].

3.3.2.1
The Aluminum-Deficient Form

The crystal structure of zeolite Y is destroyed when treated with strong acids. However, when the acid is added very gradually, crystallinity can be retained up to removal of about 60% of the aluminum. The methods employed include slow dissolution of ethylenediaminetetraacetic acid (EDTA) from the thimble of a Soxhlet extractor, with NaY zeolite being stirred and refluxed as an aqueous slurry in the boiling flask [437–439], very gradual addition of hydrochloric acid to a boiling slurry of NH_4Y in a predetermined volume of 1N $(NH_4)_2H_2EDTA$ solution [440], reaction of NH_4Y with tartaric acid [441], and with an acetic acid/acetate-buffered Na_2H_2EDTA [442]. The amount of aluminum removed from the framework is proportional to the quantity of EDTA added [437] or to the hydrogen ion concentration of the acid solution; dealumination occurred only at pH of < 2.30, while complete dealumination was observed at pH of < 0.46 [443]. The thermal and hydrothermal stability of zeolite Y increased up to 50% Al removal while the unit cell dimension decreased upon calcination of the product [437, 444]. A constant sorptive capacity for cyclohexane, expressed as g/g of silica, suggests that the structure was preserved and that tetrahedral vacancies were generated by the removal of aluminum from the framework [437, 442, 445]. The generation of framework vacancies and mesopores makes some acid sites in the hexagonal prisms accessible to bulky molecules, while such sites remain inaccessible after dealumination with water vapor because of accompanying refilling of vacancies with Si (see Sect. 3.3.3.1) [445a].

The liberated Al becomes cationic, most likely Al^{3+} at the pH of acid dealumination:

$$4\left[(\equiv SiO)_3Al\ \overset{H}{O}-Si(OSi\equiv)_3\right] + 3H_2O \rightarrow (\equiv SiOH)_3\ \overset{H}{O}-Si(OSi\equiv)_3 + \left[(\equiv SiO)_3Al^{-}-O-Si(OSi\equiv)_3\right]_3 Al^{3+} \tag{3.35}$$

The "hydroxyl nest" formulated in Eq. 3.35 and postulated previously by Barrer and Makki [436] consists of four OH-groups associated with the four Si previously surrounding the framework-Al. This explanation is supported by the collapse of the structure upon further aluminum removal. New IR bands of non-acidic OH-groups observed at 3600 and 3670 cm^{-1} were attributed to silanol groups [446, 447]. The unit cell dimensions of acid-treated zeolite Y are similar to those of the original NaY, when dried at 90°C only [443], suggesting that hydroxyl nests were present. The existence of hydroxyl nests after acid extraction was confirmed by other research groups, who found that such hydroxyl nests have reduced thermal and hydrothermal stabilities [443, 448, 449], that, because a hydroxyl nest is larger than a SiO_4 tetrahedron, but smaller than an AlO_4 tetrahedron, the frequency of the IR asymmetric stretch band is affected by the presence of hydroxyl nests [449], a method proposed earlier [444]. The "nest" OH-groups are chemically similar to those in silica gel. Results indicating that the vacancies are refilled with Si were largely obtained after calcination [438, 451, 452], but very slow migration has also been observed at refluxing conditions [440].

Similar to the unextracted zeolites, samples with a low degree of dealumination are relatively stable to mild base treatment, but structurally unstable to acid, whereas highly dealuminated zeolites (Si/Al > 18) are stable to acid, but unstable to basic solutions at 80°C [453, 454].

The ratio of Brønsted/Lewis acid sites and the strength of the framework acid sites increase with the degree of acid dealumination, as determined from the IR spectra after pyridine adsorption and from the differential heats of adsorption of ammonia, respectively [455], and the acid catalytic activity increases [456, 457]. As the framework acid sites become stronger, the aluminum becomes more resistant to hydrolysis.

Commercially available hydrogen mordenite contains a considerable percentage of non-framework aluminum [347]. The highly acid-stable structure is customarily dealuminated by acid extraction [432, 458-460]. An efficient procedure for dealuminating an ammonium mordenite is thermal treatment at 700°C involving dehydroxylation followed by extraction [434, 461], or steaming at 500-600°C followed by acid extraction [462], as well as by repeated alternating steaming and acid extraction [463, 464]. However, even sodium mordenite, as crystallized, can be dealuminated with mineral acids, e.g., 0.1-11 N HCl [465-470] and 4-14 N HNO_3 [471, 472], at reflux conditions for 2-24 h. Hydrated aluminum ions are so large that they leave the pore system only with difficulty even under these conditions. In a systematic investigation of acid treatment of H-exchanged mordenite, it was found that the concentration of octahedral Al reached a maximum by washing with 0.001 N HNO_3. With increasing acid concentration up to 1 N, octahedral Al was extracted with little accompanying framework dealumination. Further increased acid concentration, up to 10 N, removed large amounts of tetrahedral framework aluminum [472a].

Oxalic acid has been recommended for dealuminizing unsteamed zeolite beta [472b].

3.3.2.2
Annealing of Tetrahedral Vacancies; High-Silica Faujasite

The nest of four silanol groups left behind after the removal of Al according to Eq. 3.35 can react in several ways depending on the reaction conditions. Water is lost upon heating under anhydrous conditions [376, 454], thus distorting the local environment, or the vacancy may be refilled with Si when calcined in the presence of water vapor [433, 445, 446].

Acid extraction tends to deplete the surface layer of aluminum more rapidly than interior portions of the crystallites, and a steaming/acid leaching process yields a more uniform composition [449, 450, 473 - 476]. The data suggest that the presence of water at high temperatures (steaming) facilitates the migration of Si to the vacancies, particularly after cationic aluminum species have been removed. This secondary synthesis generates faujasite with high SiO_2/Al_2O_3 ratios unattainable by hydrothermal crystallization, while retaining sufficient acidity for catalytic application.

The following procedure is an example of the methods described by Scherzer [477] for the preparation of a highly siliceous version of faujasite.

NH_4Y (3 wt% Na_2O) was calcined for 3 h at 760 °C under self-steaming conditions. The steamed material was ammonium-exchanged to ≤0.2 wt% Na_2O and then again calcined for 3 h at 815 °C under self-steaming conditions. The product was treated for 2-4 h with 3-5 N HCl at 90 °C. The resulting materials were reported to have very good crystallinity and contained ≤1 wt% Al_2O_3 (SiO_2/Al_2O_3 = 175-195, a_o = 24.18-24.27 Å), the intensity of the Si(0Al) signal in the ^{27}Si MAS NMR spectrum increased, and that of the Si(2Al) line decreased, indicating that framework-Al was removed by this treatment along with cationic Al [478]. In attempts to reproduce these results, others have found that the products contain large secondary pores of 1.5-1.9 nm diameter [479] and an increased concentration of silanol groups [480, 481], along with well-ordered and oriented crystalline domains throughout the crystals [480, 481]. All aluminum vacancies in the crystalline parts were found to be refilled with Si, and the aluminum atoms removed from the lattice form alumina-like fragments, as evidenced by the Al^{VI}-O and Al^{VI}-Al^{VI} distances [480]. An aluminum-rich amorphous phase (<5% SiO_2) was found to be located within the pores [482, 483].

3.3.3
Hydrothermal Treatment

3.3.3.1
The Stabilized Form

Contradictory statements in the literature concerning the stability of calcined NH_4Y [368, 484] prompted Kerr [376] to study the calcination of NH_4Y (Si/Al = 2.85) under different conditions. Identical samples were calcined at 500 °C in a muffle furnace in a wide container ("shallow bed", SB, with 0.3-cm sample depth), and in a narrow container ("deep bed", DB, with 2.9-cm sample depth), until the ammonia was completely removed, 1 and 4 h, respectively, for SB and DB samples.

The SB product had high crystallinity, but became amorphous after treatment with water and recalcination, it lost 1 mole of chemical water per Al_2O_3 upon dehydroxylation, neutralization of acid sites with NaOH gave a Na/Al of about 1,

$NH_3/Al = 0.86$ after NH_3 sorption, and the Si/Al ratio remained constant. Although the difference between Na/Al and NH_3/Al suggests reinsertion of Al with NaOH (see Sect. 3.4), the SB material had basically the properties expected from a normal HY, including the sorption capacities, the previously reported thermogravimetric results [353, 368], and the hydrothermal instability [352, 485].

In contrast, the DB sample lost aluminum when treated with aqueous NaOH [376]. It had a much lower NH_3/Al ratio of 0.24, and lost only half as much chemical water, but more gradually and at higher temperature than the SB zeolite. The sorptive capacities were somewhat lower, but the water-loaded sample retained its crystallinity upon dehydration. Furthermore, the DB material showed a lattice contraction of 1%, relative to the SB sample. The properties found for the DB sample are indicative of a zeolite that has lost aluminum from the framework and gained stabilizing cations. A product similar to the DB-calcined NH_4Y was obtained directly from the normal HY by heating HY in a closed reactor at 700°-800 °C [486] (see Sect. 3.3.3.2).

The hydrogen form can best be represented by the formula in the center of Eq. 3.33, and the left-hand formula, containing the improbable unsolvated proton, is the one that reacts with bases, while the right-hand extreme offers the Lewis acid site required for reaction of the framework-Al, e.g., with water. The strength of the Al $\cdots$ O bond, which depends on the bond length and the bond angles of the particular framework structure, determines the position of the equilibrium.

If the hydrogen form of zeolite Y reacts with water at elevated temperatures, massive hydrolysis of aluminum occurs resulting in the destruction of the framework because too many bonds are broken:

$$(\equiv SiO)_3Al\ HO{-}Si\equiv + 3\,H_2O \xrightarrow{\Delta} 4 \equiv Si{-}OH + Al(OH)_3 \tag{3.36}$$

However, if this reaction occurs more gradually, e.g., simultaneously with deammination, then the $Al(OH)_3$ assumed to be an intermediate can react with other acid sites that have not yet been hydrolyzed to generate a new cationic form of the zeolite:

$$(\equiv SiO)_4Al^- H^+ + Al(OH)_3 \rightarrow (\equiv SiO)_4Al^- \quad Al(OH)_2^+ + H_2O \tag{3.37}$$

The hydroxylated aluminum cation can react further with acid sites to yield $Al(OH)^{2+}$ and Al^{3+} ions, and Kerr found the average valence of the aluminum cations to vary between 1 and 3, e.g., 2.65 in his example [376]. While Kerr was able to remove 20% of the aluminum from a DB calcined HY (500 °C) by treatment with 0.1 N NaOH, Kühl found, by X-ray spectrometry, that 19% of the aluminum present in this DB HY was hexacoordinated [354, 355]. The ideal product from slow hydrolysis contains only Al^{3+} cations, so that 25% of the total Al is octahedral. Such a stabilized zeolite has the potential for reversibly generating the highest concentration of acid sites in the presence of water:

$$\left[(\equiv SiO)_4Al^-\right]_3 Al^{3+} + H_2O \rightleftarrows \left[(\equiv SiO)_4Al^-\right] H^+ \left[Al(OH)\right]^{2+} \tag{3.38}$$

$$[(\equiv SiO)_3Al^-]_3\, H^+ [Al(OH)]^{2+} + H_2O$$
$$\rightleftarrows [(\equiv SiO)_3Al^-]_3\, (H^+)_2 [Al(OH)_2]^+ \qquad (3.39)$$

One H_2O is formed in each of the neutralization reactions, Eq. 3.37, so that the water initially required for hydrolysis (Eq. 3.36) is recovered upon complete neutralization. Therefore, very little water is able to stabilize the zeolite, and water formed by dehydroxylation (Sect. 3.3.1.2) is adequate for such a stabilization [485, 486]. As Kerr's data show, the DB material is a stabilized form of zeolite Y. Since water is a reactant to produce this form, steaming generates the same stable zeolite modification. Too severe steaming causes the structure to collapse, too mild conditions do not allow adequate Si migration into tetrahedral vacancies, and the structure becomes unstable when the cationic Al is removed [487]. The desired conditions can be established by varying temperature and steam concentration [488].

Stablilized zeolite Y differs from an aluminum-exchanged Y mainly in the concentration of acid sites [489], probably caused by the different kinds of cations, i.e., hydroxylated Al vs. Al^{3+}. Thus, the concentration of Lewis and Brønsted acid sites is much lower in steamed zeolite Y than in AlHY, while the concentration of Si-OH-groups is increased.

The susceptibility of framework-Al to hydrolysis varies with the strength of the SiO–Al bond [488], which also determines the acid strength [490–492]. Cationic aluminum species are located in the sodalite cages (SI′ and SII′ sites); in highly dealuminated faujasite, such cations are found in SII′ sites only whereas, in mildly dealuminated samples, non-framework species are also observed in the supercages [454], suggesting that species initially located in the supercages migrate into the sodalite cages upon dehydration at higher temperature. The hydrolysis of framework-Al occurs theoretically in three stages, and FTIR evidence for partially hydrolyzed framework-Al during the early stages of dealumination has been obtained [493–495].

The removal of aluminum from the framework is accompanied by the appearance of two additional bands in the IR spectrum [446], at about 3700 and 3600 cm^{-1}. These bands and a ^1H MAS NMR signal at 2.6 ppm decrease in intensity upon mild acid extraction [373] and are due to the presence of hydroxoaluminum ions [496] or boehmite-like clusters [420, 497]. Others assign the IR band at 3610 cm^{-1}, because of a strong interaction with pyridine, to amorphous silica-alumina [498]. This amorphous material is very aluminous ($<5\%$ SiO_2) and located mainly in the supercages [482], in agreement with calorimetric data of NH_3 adsorption suggesting that an additional acid component is present within the pores of dealuminated Y [483].

In addition to the ^{27}Al NMR chemical shift of about 61 ppm from $Al(H_2O)_6^{3+}$ [499] for tetrahedral framework-Al in NaY, a signal at 0.35–3.5 ppm, corresponding to octahedrally coordinated aluminum in the zeolite channels, is observed in dealuminated zeolites [500]. A broad line at about 30 ppm [501] is assigned to pentacoordinated [502] or distorted tetrahedral [503, 504] non-framework-Al, while Rocha et al. [505] were able to resolve the ^{27}Al NMR signal at 60 ppm into two lines at 62 and 56 ppm, and assigned these peaks to tetrahedral framework and non-framework-Al, respectively. Tetrahedral non-framework-Al had previously been reported by others [506–510]. Al-oxo-hydroxo [511] and poly-nuclear aluminum species [512] are held partly responsible for the increased thermal stability [513]. As cationic aluminum species generated by dealumination migrate to the surface [474], the extent of dealumination is higher in the interior of the crystallites than near the surface [514].

The ^{29}Si MAS NMR spectrum of a sample prepared by calcining $NaNH_4Y$ at 540°C in the presence of water vapor showed evidence for the removal of aluminum from the framework and substitution with Si [470]. Hydrothermal dealumination is more rapid than silicon migration into the defects formed [454], and ^{29}Si MAS NMR spectra support the presence of hydroxyl nests [470, 515]. Complete filling of vacancies with Si was observed by Freude et al. [497].

Depending on the rate of dealumination and refilling of defects with Si, a secondary pore volume of variable size (mesopores [479, 516, 517] and micropores [518]) has been observed. The size and density of the secondary pores depend on the conditions of the hydrothermal treatment [519]. Mesopores were reported to consist of large spherical voids of 5–8 nm diameter connected by narrow restrictions. Alumina clusters were found to be present in the secondary pore system [497]. Mesopores were also observed after steaming and acid treatment of mazzite. The size and volume of these pores decrease with the Al content, and this result is shown to be similar for faujasite, offretite, mordenite, ferrierite, and beta [519a].

The DB material differs little from the so-called ultrastable Y discussed below [376] (Sect. 3.3.3.2). The unit cell parameter decreases with the degree of framework dealumination, but remains above 24.5 Å for such stabilized forms [520]. Other efforts to correlate the framework-Al content with the unit cell parameter have been made by Sohn et al. [521] and by Kerr [522]. The shift of the bridging OH-band from 3642 to 3632 cm^{-1} is evidence of an increase in the Si content of the framework [373], in agreement with the decrease of the unit cell parameter. The T–O stretching frequencies also correlate with the Al content of the zeolite unit cell [521].

Steaming of $RENH_4Y$ (RE = rare earth) causes appreciable framework dealumination, as judged from IR and ^{29}Si MAS NMR data, independent of the RE content [429], while the degree of dealumination depends on the severity of the steam treatment [523]. The unit cell shrinks upon steaming $LaNH_4Y$ at 540°C [524], and T-sites were found to be fully occupied. A 98% exchanged LaY retained the unit cell size after vacuum calcination and after steaming at 540°C. The asymmetric stretching frequency increases with the severity of steaming due to shorter T–O. Brønsted and Lewis acidity are maintained even after steaming at 820°C [525].

ZSM-20, the hexagonal counterpart of the cubic faujasite, behaves similarly to zeolite Y [526]. Dealumination of ZSM-20 by steaming was monitored by FTIR of the pyridine-treated products. The results indicate that the most rapid framework dealumination occurs in 10 min of steaming at 600°C, while simultaneously the concentration of Lewis acid sites increases. After steaming for 60 min, the band representing the pyridine adduct to Lewis acid sites splits into two components, and the higher-frequency band becomes dominant after prolonged steaming, indicating the formation of higher-condensed extra-framework species [526a, b].

3.3.3.2
The Ultrastable Form

"Ultrastable zeolite Y" (USY) is widely used as active component of cracking catalysts. It is usually supplied as the ammonium precursor and requires a final calcination to be converted into the ultrastable form. The framework of USY has a SiO_2/Al_2O_3 molar ratio above 6 and is stabilized by aluminum cations. For convenience, a USY is characterized by a unit cell parameter a_o of 24.5 Å or smaller. A variation is prepared by rare earth exchange of USY.

McDaniel and Maher [485] prepared ultrastable faujasite by repeated sequences of ammonium exchange and calcination:

$$NaY \rightarrow NH_4^+ \text{ Exchange} \rightarrow \text{Calcination at } 540\,°C$$

$$\rightarrow NH_4 \text{ Exchange} \rightarrow \text{Final Calcination at } 815\,°C$$

The product showed

- a slightly increased overall SiO_2/Al_2O_3,
- high BET surface area up to 900 °C, then slow decrease at rising temperatures,
- decrease of a_0 to < 24.5 Å (found as low as 24.35 Å),
- considerably reduced ion-exchange capacity,
- reduced sorptive capacity,
- reduced X-ray peak intensity.

A similarly stable product was obtained by Kerr [486] upon calcining HY under self-steaming conditions at 700–800 °C. About 25% of the total Al could be exchanged by treatment with 0.1 N NaOH. Similarly, Yoshida et al. [527] were able to extract about half of the non-framework aluminum by such treatment. Kühl [354, 355] found that almost 50% of the aluminum was removed from tetrahedral framework positions by dehydroxylation, so that Kerr's and Yoshida's results are in agreement. The data also indicate that dehydroxylation and ultrastabilization are related in a similar way to thermal (shallow-bed) and hydrothermal (deep-bed) treatments.

The structural rearrangement with silicon migrating into vacancies created by dealumination decreased a_o to 24.24 Å (see also [522]) and the mean T–O distance to 1.61 Å, similar to the Si–O distance [528, 529]. Al was located in the sodalite cage, coordinated to three framework oxygens. After self-steaming at 760 °C, the framework IR frequency of the T–O bond shifted from 955 ($NaNH_4Y$) to 1037 cm^{-1}.

^{29}Si NMR spectra of hydrothermally dealuminated Y show a progressive attenuation, with rising severity of steaming, of the Si(3Al), Si(2Al), and Si(1Al) signals at –90, –95, and –101 ppm, respectively, compared with the signal for

Si(0Al) at −106 to −108 ppm, which increases in intensity (see Sect. 3.3.3.1). An ultrastable Y obtained by exchanging calcined $NaNH_4Y$ with NH_4^+ ions and steaming at 815 °C showed a very intense Si(0Al) and a weak Si(1Al) line, the latter being explained mainly by the presence of $(SiO)_3Si$-OH grouping [470, 478, 510].

The extent of dealumination is limited by the degree of ammonium exchange of the starting material and depends on temperature and partial pressure of steam during the thermochemical treatment [530]. The Si, Al ordering in the zeolite framework of dealuminated Y is determined predominantly by the final Si/Al ratio attained after dealumination and is, for a given Si/Al ratio, independent of the type and external conditions of the dealumination process used [531].

Quantitative measurements of Al removal from USY faujasite during steaming at 400–500 °C were reported by Fleisch and co-workers [532]. When USY (0.46% Na) was steamed at 500 °C, the number of Al/unit cells decreased from 39 to 19 after 65 h of steaming, and the unit cell parameter, a_0, decreased from 24.55 to 24.36 Å. The intensity of the ^{27}Al NMR signal due to tetrahedrally bound Al decreased initially, as the quantity of the octahedral Al increased, but after longer steaming times no further change was observed. The observation was explained with the formation of non-framework tetrahedral Al [533]. This interpretation is supported by quadrupole nutation ^{27}Al NMR [534, 535] and ^{1}H-^{27}Al cross-polarization, resulting in the conclusion that USY contains 4-, 5-, and 6-coordinated non-framework-Al [505, 536]. A very broad signal underlying the entire spectrum may be caused by polymeric aluminum species [500]. Hydroxo-aluminum ions [446] and other octahedrally coordinated non-framework aluminum species are believed to contribute to the thermal and hydrothermal stability of hydrothermally treated zeolite Y [527] (see also [502, 503, 537–540]). Al species intermediate between NMR detectable tetrahedral and octahedral Al are resolved, in the 30-ppm region of the ^{27}Al MAS NMR spectrum by spin lattice relaxation, resulting in the conclusion that the 30-ppm peak is a composite of signals due to various extra-framework-Al species.

IR spectroscopy reveals that the acid strength of the bridging ≡Al-O(H)–Si≡ group increases with the degree of dealumination [541], i.e., with decreasing T–O bond distances [355]. Extra-framework silica-alumina species also contribute some acidity [498, 541]. An ammonia desorption peak at 570 °C prompted Mirodatos and Barthomeuf [542] to propose superacidity as an inductive effect of Lewis on Brønsted acid sites, comparable to $AlCl_3$-HCl and SbF_5-HF. Owing to the superacidity of these sites, a low concentration of such sites would be sufficient to improve the catalytic properties markedly. Beyerlein et al. [543] reported that USY samples containing low fractions of extra-framework Al exhibit enhanced activities for isooctane conversion. They conclude that strong acidity in catalysis depends on balance between tetrahedral and octahedral Al (see also [544]).

Enrichment of structural Al occurs near the surface during formation of USY. The concentration of ion-exchange sites and, presumably, acid sites is greater near the external surface of an ultrastable zeolite than in the bulk [473, 476, 545, 546].

Since removal of framework-Al by acid leaching causes formation of hydroxyl nests, acid leaching should be designed to remove extra-framework-Al without extracting aluminum from the framework, a goal achieved to some extent by slowly adding 0.1 M HCl [547] (see Sect. 3.3.4.1).

The larger the a_0 of the USY, the smaller is a_0 after acid extraction, and the lower is the crystallinity of the product indicating that the severity of steaming is the determining factor for product quality [548]. High-silica faujasite with SiO_2/Al_2O_3 = 100–200 and a_0 = 24.18–24.27 Å has been prepared in this manner [477].

More selective removal of non-framework aluminum species can be achieved by treatment with Na_2H_2EDTA or $(NH_4)_2H_2$ EDTA [437, 549, 550]. Acetylacetone is also effective in removing non-framework aluminum species [490, 551].

3.3.3.3
Dealumination of High-Silica Zeolites

The most widely used high-silica zeolite is ZSM-5, to which the following discussion will be restricted. The behavior of other high-silica zeolites is expected to be similar if the pore diameter is taken into consideration.

Since ZSM-5 can be crystallized with a wide range of Si/Al ratios, preparative dealumination appears to be unnecessary. However, low-silica ZSM-5 with SiO_2/Al_2O_3 in the range of 20–30 can be synthesized without an organic template, and it may be desirable to provide a higher SiO_2/Al_2O_3 ratio for catalytic application. Understanding the behavior of ZSM-5 under conditions of steaming and catalytic regeneration is helpful for developing a catalyst.

Steaming of the acid form causes a decrease in the concentration of tetrahedral framework aluminum while octahedral extra-framework species are generated [552]. As expected, steaming generates ≡Si–OH groups and non-framework-Al species, as demonstrated by ^1H MAS NMR [553] and ^{27}Al MAS NMR [554, 555].

The dealumination of ZSM-5 proceeds at a temperature as low as 300–350 °C, but 500–550 °C is more practical [556]. Dealumination proceeds rapidly, via dehydroxylation, when HZSM-5 is calcined in the temperature range of 700–1000 °C, and an essentially completely dealuminated framework is obtained [557]. Again, extra-framework-Al tends to migrate toward the surface of the crystals, where it accumulates dependent on the steaming severity [557]. Such surface enrichment occurred when ZSM-5 of Si/Al = 24 was steamed, while the same treatment for Si/Al = 80 caused the concentration near the surface to decrease [552]. Similar to HY, steaming of ZSM-5 causes a secondary pore volume to be generated, in which part of the aluminum removed from the framework is deposited. A broad ^{27}Al MAS NMR signal at 30 ppm is observed in the more severely steamed ZSM-5 samples, as it was in steamed HY, and is interpreted as distorted tetrahedral non-framework-Al. The ^{29}Si MAS NMR spectrum of ^{29}Si-enriched ZSM-5, shows, after calcination, ammonium exchange, and recalcination, evidence of defect centers $(\equiv SiO)_3SiO^-$ and $(\equiv SiO)_2(\equiv AlO)SiO^-$. The ^{27}Al MAS NMR spectrum includes a signal at 0 ppm (octahedral Al) and a splitting of the signal due to tetrahedral Al, suggesting the presence of Al with two different tetrahedral environments [558].

The IR spectrum of steamed HZSM-5 shows, with rising steaming temperature, a general decrease in intensity of the 3610 cm^{-1} band associated with bridging OH. After pyridine sorption, bands due to both Brønsted and Lewis acid sites are present; the ratio of B/L decreases with increasing steaming temperature [559]. The wavenumber of the asymmetric T–O–T stretching band depends linearly on the number of framework-Al per unit cell [560].

The extra-framework-Al can be extracted with acid at 100 °C. The intensity of the ^1H MAS NMR signal at 4 ppm correlates linearly with the number of tetrahedral framework-aluminum atoms per unit cell. Silicon migration into the tetrahedral vacancies generated by dealumination of the framework is indicated by the decreasing intensity of the ^1H MAS NMR signal at 1.8 ppm on steaming. Acid extraction causes this line to increase in intensity suggesting that some framework-Al is also leached by the acid treatment. However, others were unable to remove all non-framework Al by acid leaching or even alternate steaming and leaching [554, 556]. The discrepancy is apparently due to the different size of ZSM-5 crystals investigated [561].

Lago et al. [562] observed an enhanced catalytic activity in HZSM-5 after mild steaming. They explained this enhancement with the presence of aluminum pairs, one of which is non-tetrahedral and acts as a strong electron-withdrawal center for the other, tetrahedral Al, thus creating a stronger Brønsted acid site.

The dependence of the number of framework-Al left after steaming on the partial pressure of steam and the time and temperature of steaming is shown in Fig. 3.20 [563], which demonstrates that dealumination takes place, for 0.1 bar of steam and 2.5 h of steaming, in the range of 300–700 °C. A ^{29}Si MAS NMR line corresponding to Si(2Al) was not observed in activity-enhanced ZSM-5. The authors conclude that the electron-withdrawal effect can therefore not be ascribed to a partially hydrolyzed framework-Al. The presence of a ^1H MAS NMR signal at ~3.0 ppm is due to non-acidic non-framework AlOH groups, and the broad 30 ppm signal in the ^{27}Al MAS NMR spectrum is interpreted as being due to non-framework AlOOH coordinated to two framework oxygens. Others found that mild hydrothermal treatment causes the IR band at 3665 cm^{-1} to be enhanced [564] and assign this band tentatively to hydroxylated non-framework Al species [560] or to partly hydrolyzed framework-Al, still connected to the framework with fewer than four bonds [564]. A comparison of dealuminated ZSM-5 before and after acid leaching suggests an interaction between non-framework-Al and terminal OH-groups. Neither IR nor ^1H NMR shows evidence of an increase in the acid strength of the bridging ≡Si–O(H)–Al≡, due to an inductive effect of non-framework-Al, and the combined effect of framework Brønsted acid sites and non-framework Lewis acid sites on the hydrocarbon in the catalytic reaction may be responsible for the enhanced activity [563, 565]. Another IR band observed in mildly steamed ZSM-5 at 3780 cm^{-1} decreases in intensity with increasing dealumination time and gives only a weak interaction with benzene; this band is eliminated by treatment with acetylacetone [566] or $(NH_4)_2H_2EDTA$ [562] and appears to represent terminal AlOH groups, e.g. AlOOH. The 3665 cm^{-1} band is strongly influenced by benzene sorption indicat-

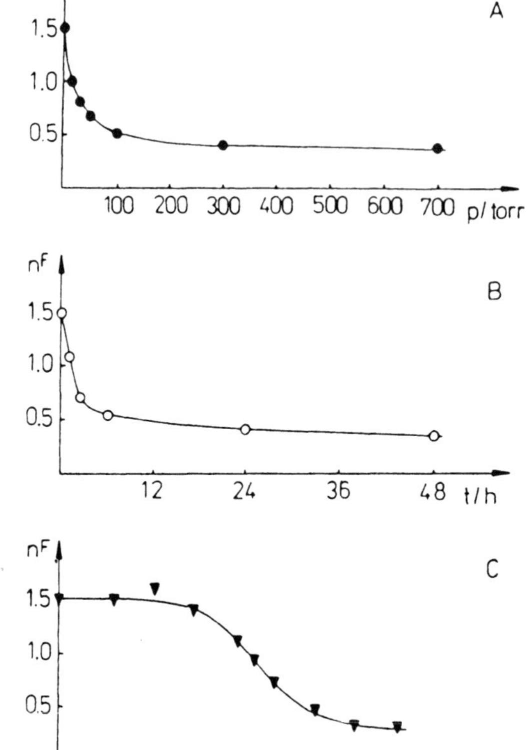

Fig. 3.20. Concentration of framework-Al atoms n^F per 24 framework tetrahedra of dealuminated HZSM-5 [563]. (**A**) Dependence of n^F on the water vapor pressure p for T = 540 °C and t = 2.5 h. (**B**) Dependence of n^F on the time on stream t for T = 540 °C and p = 0.1 bar. (**C**) Dependence of n^F on the temperature of steaming T for p = 0.1 bar and t = 2.5 h

ing a rather strong acid site, although not as strong as the bridging hydroxyl [564]. This interpretation is further supported by the IR spectra after sorption of molecular hydrogen [565, 566].

Dealumination by calcining and steaming of HZSM-5 (SiO_2/Al_2O_3 = 23.4 to 242) was monitored by XRD, ^{27}Al, ^{29}Si and ^{129}Xe NMR, FTIR, SEM, XPS, and sorption measurements, as well as chemical analysis. HZSM-5 with low Al content was found to be more resistant to dealumination by either method. Extra-framework species had low symmetry, and their presence was shown by ^{27}Al NMR after acetylacetone sorption [566a].

Similar results were obtained by Datka et al., who also made an attempt to determine the acid strength of Brønsted acid sites by measuring the intensity of the 1450 cm^{-1} band of ammonium ions after sorbing an excess of ammonia on the sample at 320 K (A_0) and after desorbing ammonia for 30 min at 660 K (A_{660}). The ratio A_{660}/A_0 was taken as the measure of acid strength of Brønsted acid sites. The maximal acid strength determined in this manner was observed in the mildly steamed (7 kPa H_2O) zeolite, correlating with the maximal catalytic activity for the cracking of n-heptane [566b].

Further information is gleaned from the comparison of aluminum-exchanged ZSM-5 and mildly steamed ZSM-5 [567]. Aluminum cations inhibit hydrolysis of framework-Al, as other cations do. Water dissociates on aluminum cations [568], generates acidic bridging hydroxyl groups and hydroxylated aluminum cations. The process is reversible.

3.3.4 Direct Replacement of Aluminum with Silicon

3.3.4.1 Reaction with Silicon Halides

Rather than just being removed from the zeolite framework, aluminum can be replaced directly with silicon. Beyer and co-workers [569–571] achieved near-perfect replacement of framework-Al with silicon by treating zeolite Y with gaseous silicon tetrachloride. The procedure described by them has to be followed very closely in order to obtain reproducible results:

- Dehydration at 620 K for 2 h in flowing nitrogen.
- Cooling to 520 K in flowing nitrogen.
- Nitrogen saturated at room temperature with $SiCl_4$.
- Temporary temperature rise (30–75 °C) in the zeolite bed upon initial contact with $SiCl_4$, temperature returns to 520 K.
- Heating at 10 °C/min to 755 K.
- Held at 755 K for 40 min.
- Supply of $SiCl_4$ is discontinued, and pure N_2 is allowed to flow at 755 K for 15 min.
- After cooling to ambient temperature, product is slurried in water, filtered and washed Cl^- free, then dried at 400 K.
- NH_4^+ exchange to obtain NH_4^+ form.
- Extracted with acid and steamed at 800 K.

The thermal stability achieved at various stages of the process is represented in Fig. 3.21 [569]. When the $SiCl_4$-treated faujasite was purged with nitrogen at temperatures above the final reaction temperature of 755 K, a severe loss of crystallinity was observed (curve a), apparently caused by the reaction of occluded $NaAlCl_4$ with the zeolite. When this complex was removed by washing with water at room temperature, the resulting product had a much enhanced thermal stability (curve b), compared with the starting material (curve d). Further improvement can be attained by extracting the washed and dried product with aqueous HCl at 100 °C and subsequent steaming at 600 °C (curve c). The latter procedure raised the SiO_2/Al_2O_3 molar ratio to 1000; the steaming served to anneal defects in the zeolite framework. The observations are interpreted as follows:

- Direct replacement of 90% of Al with Si and formation of $[AlCl_4]^-$ according to $(\equiv Si\text{-}O)_4Al^- Na^+ + SiCl_4 \rightarrow (\equiv Si\text{-}O)_4Si + Na[AlCl_4]$.

- Hydrolysis of $[AlCl_4]^-$ forms acid, which causes further dealumination: $Na[AlCl_4] + n\,H_2O \rightarrow AlCl_{3-n}(OH)_n + n\,HCl + NaCl$ (n = 1–3).
- Acid wash removes hydrolyzed Al species.
- Steaming causes silica to migrate into vacancies.
- A completely dealuminated faujasite has $a_0 = 24.35$ Å.

While the n-hexane sorption isotherm of zeolite Y dealuminated with silicon tetrachloride hardly changed from that of NaY, the water sorption capacity became very low at high SiO_2/Al_2O_3 ratio and the accompanying hydrophobicity [569] (Fig. 3.22). In contrast, hydrothermally dealuminated zeolite Y gave a n-hexane sorption isotherm indicative of a bidisperse pore system [569, 572].

Dealumination with $SiCl_4$ causes aluminum species to accumulate near the surface of the zeolite crystals [473, 573, 574]. While about 95% of the framework aluminum can be removed by treating NaY with $SiCl_4$ [575, 576], the $NaAlCl_4$ found as a product inhibits further dealumination [570]. Removing the $NaAlCl_4$ by washing and repeating the reaction with $SiCl_4$ yields a product with very low aluminum content [577]. Unlike other dealumination procedures, treatment with $SiCl_4$ in this manner produces only a small amount of hydroxyl nests [578]. In contrast to combined self-steaming and acid leaching, the reaction with $SiCl_4$ occurs progressively from the exterior of the crystals and does not generate

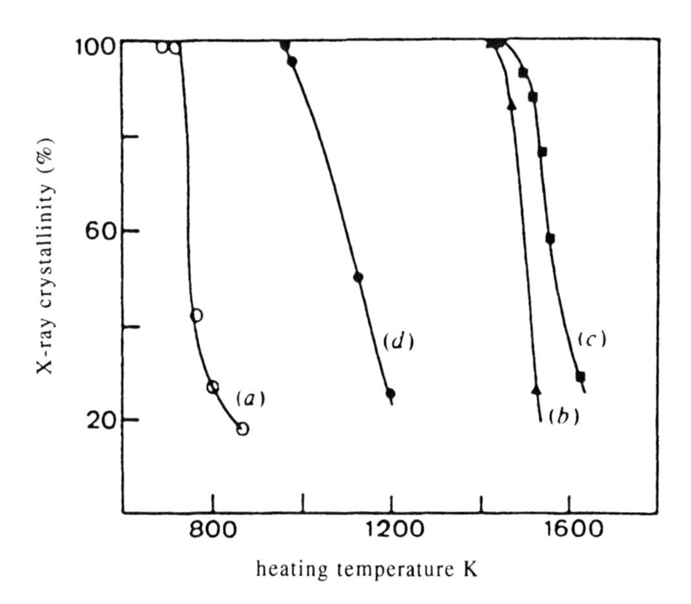

Fig. 3.21. X-ray crystallinity of NaY treated with $SiCl_4$ [569]; (**a**) heating for 1 h at the temperature indicated in N_2; (**b**) washing, drying, and heating in N_2 at the temperature indicated; (**c**) washing, drying, treatment with 1 M HCl at 373 K for 2 h, steaming at 873 K for 3 h, and further heating in N_2 (6 K/min) at the temperature indicated; (**d**) reference curve representing the thermal stability of the parent NaY in N_2

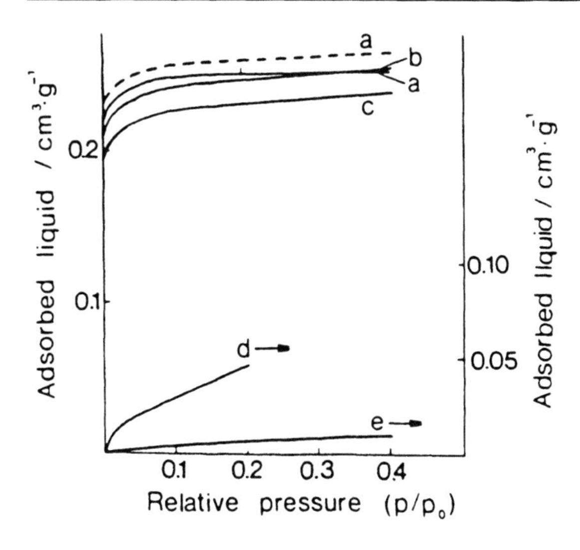

Fig. 3.22. Adsorption isotherms [569] ---- NaY; — dealuminated zeolite Y (Si/Al = 22). (**a**) n-C_6H_{14}; (**b**) n-C_4H_{10}; (**c**) C_6H_6; (**d**) NH_3; (**e**) H_2O

secondary pores [481]. Since $LiAlCl_4$ dissociates at the reaction temperature so that $AlCl_3$ volatilizes and is removed in the gas phase, a lithium-exchanged zeolite Y could be dealuminized practically completely [579], leaving only the Si NMR signal at 107.5 ppm corresponding to Si(0Al).

The dealumination reaction proceeds in at least two steps [580]:

1)

$$\equiv Si{-}\overset{M^+}{O^-}{-}Al\equiv + SiCl_4 \rightarrow \equiv Si{-}\underset{}{\overset{\displaystyle SiCl_3}{\overset{|}{O}}}\cdots Al\equiv + MCl \quad (M = Na, Li) \tag{3.40}$$

2)

$$\equiv Si{-}\overset{\displaystyle SiCl_3}{\overset{|}{O}}\cdots Al\equiv \xrightarrow[>150\,°C]{MCl} \equiv Si{-}O{-}Si\equiv + MAlCl_4 \tag{3.41}$$

In contrast to zeolite Y, zeolite X cannot be dealuminated by this method without major loss of crystallinity [569, 581, 582].

Acid extraction of NaY at pH 4.8–1.3, after $SiCl_4$ treatment, removes only non-framework aluminum species, whereas both framework – and non-framework – Al are removed at a pH of ~1.0 with a loss of structure [583]. At least three types of aluminum-independent SiOH defects and a small fraction of amorphous $Si(OSi)_4$ were found after acid extraction [584].

The ^{27}Al MAS NMR spectrum of silicon tetrachloride treated NaY [585] shows, before washing with water, some tetrahedral framework-Al, some octahedral

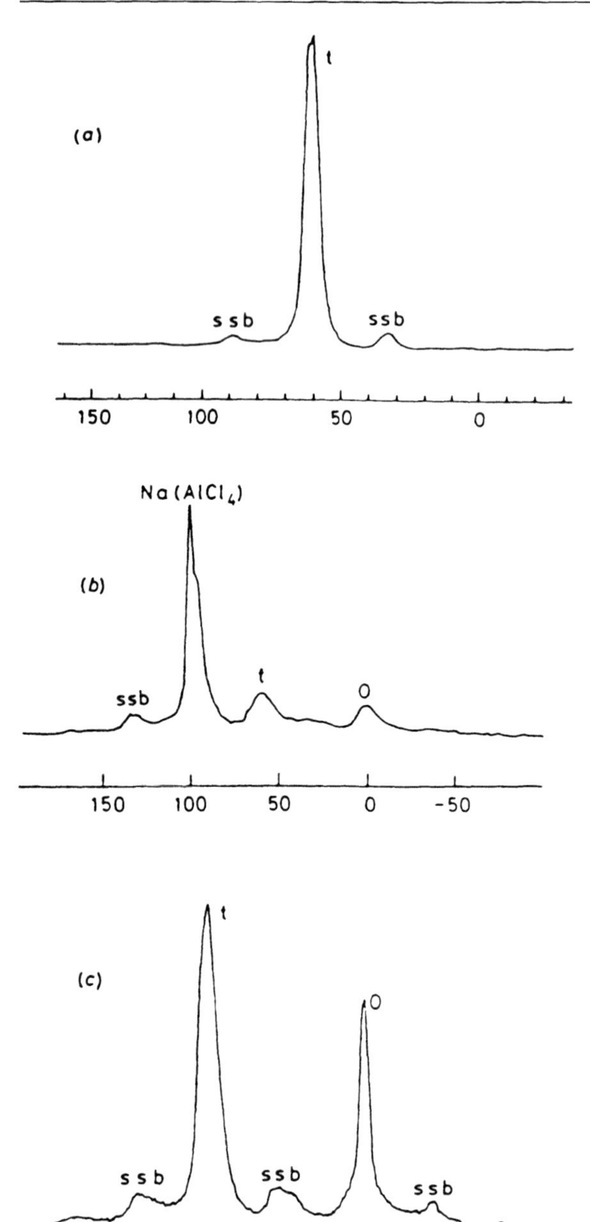

Fig. 3.23. ^{27}Al MAS NMR spectra of NaY treated with $SiCl_4$ vapor [585]. (**a**) parent zeolite; (**b**) after treatment with $SiCl_4$, but no washing; (**c**) after washing with water. t = tetrahedral, o = octahedral, ssb = spinning sideband

aluminum and an intense tetrahedral peak associated with $Na(AlCl_4)$ (Fig. 3.23). After washing with water, the latter peak disappears, and the one associated with octahedral Al is reduced somewhat in intensity as some non-framework-Al has been removed. The intense ^{29}Si MAS NMR signal at – 107.1 ppm correlates to the framework site Si(0Al), that at – 100.6 ppm is assigned to Si(1Al) [470, 510, 571]. The latter is strongly enhanced by $(\equiv SiO)_3SiOH$ groups, as revealed by ^{29}Si-^1H CP measurements, indicating the presence of Si–OH in tetrahedral vacancies [571].

The intensity of a very intense IR band at 3730 cm^-, which is enhanced with increasing degree of dealumination [586] from the water-washed to the acid-treated sample and disappears after acid extraction and steaming (Fig. 3.24), parallels the change in the NMR intensity of the Si line at – 107 ppm under CP conditions. Both features are assigned to hydroxyl nests in tetrahedral vacancies, which largely disappear upon steaming as silicon atoms migrate into the vacancies. The results confirm that not all Al removed is replaced by Si during the treatment with $SiCl_4$ and that further dealumination occurs in the washing step as a result of the hydrolysis of $AlCl_3$. Hydroxyls associated with lattice aluminum (IR bands at 3620 and 3555 cm^{-1}) or extra-framework aluminum (IR band at 3600 cm^{-1}) enhance the intensity of the ^{29}Si NMR line at – 100.6 ppm.

The properties of acid sites in HY dealuminated by treatment with $SiCl_4$ were further characterized by microcalorimetry, along with ^{29}Si NMR and FTIR spectroscopy, as well as TPAD. It is concluded that treatment with aqueous

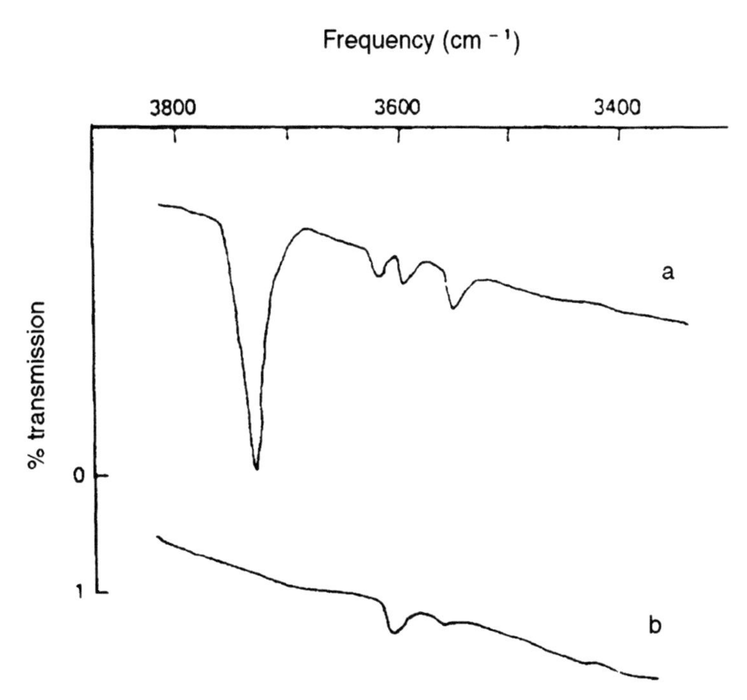

Fig. 3.24. Transmission IR spectra in the hydroxyl region [571]. (**a**) after $SiCl_4$ treatment and water wash; (**b**) after acid extraction and steaming

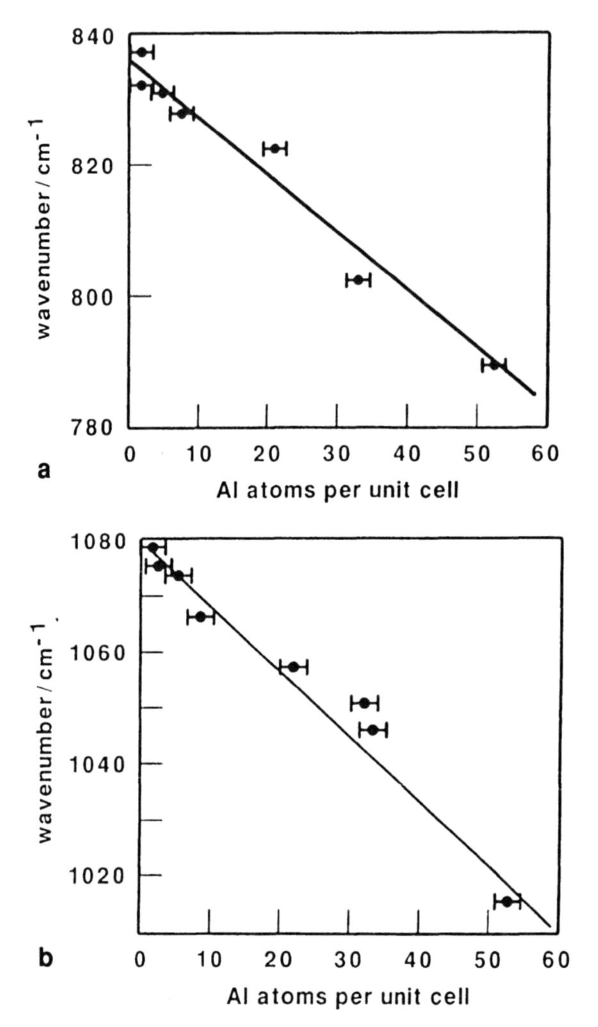

Fig. 3.25 Correlation of wavenumber and Al atoms/unit cell [585]. (**a**) external and (**b**) internal symmetric stretch frequency by IR vs. Al content of NaY treated with $SiCl_4$ vapor

NH_4OH and subsequent TPAD provides the total concentration of tetrahedral framework aluminum, whereas gaseous ammonia neutralizes only the strong Brønsted acid sites. Simultaneously, detrital silica and some cationic aluminum are removed by this treatment [586a].

Different conclusions were reached by Woolery et al. in similar work on ZSM-5 [586b].

The electron-accepting properties of cationic Al species, as provided by Al^{3+} exchange and by $SiCl_4$ treatment, with subsequent water wash (acidic solution) are much greater than of oxidic species preferentially generated in hydrothermally modified zeolites. Consequently, the latter exhibit much lower Lewis acidity than $SiCl_4$-dealuminated Y and AlHY, even though the concentration of non-framework-Al is higher [587]. Others, however, found only low levels of Lewis acidity in $SiCl_4$-dealuminized zeolite Y [578].

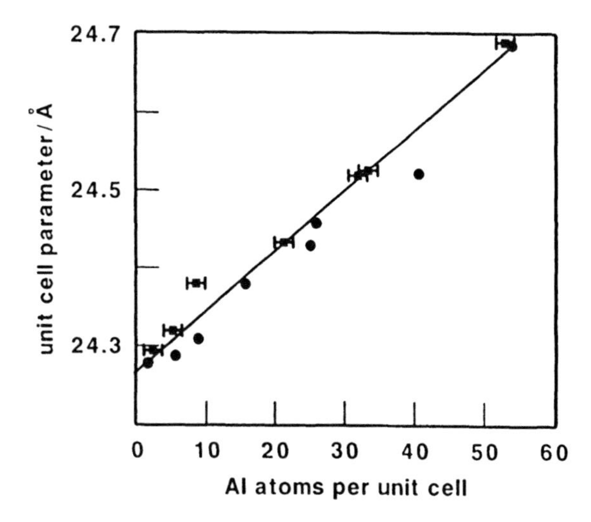

Fig. 3.26. Correlation of unit cell parameter and Al atoms/unit cell [585]. Unit cell parameter (by XRD) vs. aluminum content, (■) ^{29}Si MAS NMR; (●) EDAX

The symmetric stretch frequencies of these samples (Fig. 3.25) and the unit cell parameter (Fig. 3.26) vary linearly with the aluminum content of the dealuminated zeolite Y [585].

For more information on the reaction of faujasite with $SiCl_4$ see [588, 589].

Direct replacement of framework-Al with silicon was achieved by treatment of HZSM-5 with $SiCl_4$ at 540 °C [590], but $SiCl_4$ does not replace Al in ZSM-5 at 550 °C as completely as hydrothermal dealumination at 850 °C does [555]. Others have reported that a 12-h treatment of HZSM-5 with $SiCl_4$ at 550°C generates a product that gives a ^{29}Si NMR spectrum identical to that of a sample steamed for 4 h at 650 °C and that the ^{27}Al NMR spectrum shows that octahedral Al is present [591]. Dealumination of ZSM-5 with $SiCl_4$ above 500 °C removes surface aluminum more selectively than Al in the interior, as deduced from the surface composition determined by XPS, compared with the bulk SiO_2/Al_2O_3 ratio, but this method does not distinguish between framework - and extra-framework - aluminum [592, 593]. Treatment of mordenite with $(NH_4)_2SiF_6$ was found to limit dealumination to the pore mouth on the surface [593a].

3.3.4.2
Reaction with Hexafluorosilicates

A variation of dealumination and insertion of silicon into zeolite Y, using an aqueous solution of ammonium hexafluorosilicate, was developed by Skeels and Breck [450]:

1. The aqueous slurry of 10–25 g of NH_4Y (80–85% exchanged) per 100 ml of water is heated to 75–95°C.
2. A 1 M solution of $(NH_4)_2SiF_6$ is added at a rate of 0.005 moles of $(NH_4)_2SiF_6$ per min per mole of aluminum present in the zeolite. The amount of $(NH_4)_2SiF_6$ to be added is determined by the desired product SiO_2/Al_2O_3 ratio.

3. The concentration and addition rate provide control of the pH at about 6. Alternatively, a buffer, e.g., ammonium acetate, may be added.
4. The reaction mixture is digested for 1–3 h.

The aluminum removed and replaced by silicon is found in the filtrate as $(NH_4)_2AlF_5$ and $(NH_4)_3AlF_6$ [582, 594, 595]. It is advantageous to use the ammonium form of the zeolite for this treatment because of the low solubility of Na_3AlF_6 [582]. An excess of $(NH_4)_2SiF_6$ affects the hydrothermal stability of the product and needs to be removed by thorough washing [596, 597]. As with $SiCl_4$, dealumination of zeolite X with $(NH_4)_2SiF_6$ causes the structure to collapse. Monomeric silicic acid is an important intermediate in the reaction of a zeolite with this reagent [595].

$$SiF_6^{2-} + 4\,H_2O = Si(OH)_4 + 6\,F^- + 4\,H^+ \tag{3.42}$$

$$6\,F^- + \begin{array}{c} O\;\;O \\ \backslash\,/ \\ Al^- \\ /\,\backslash \\ O\;\;O \end{array} NH_4^+ + 4\,H^+ = \begin{array}{cc} \backslash & / \\ OH & HO \\ & \\ OH & HO \\ / & \backslash \end{array} + AlF_6^{3-} + NH_4^+ \tag{3.43}$$

$$Si(OH)_4 \;+\; \begin{array}{cc} \backslash & / \\ OH & HO \\ & \\ OH & HO \\ / & \backslash \end{array} = \begin{array}{c} O\;\;O \\ \backslash\,/ \\ Si \\ /\,\backslash \\ O\;\;O \end{array} + 4\,H_2O \tag{3.44}$$

As Al is replaced by Si, the thermal stability increases, as shown by DTA, the IR bands associated with the T–O stretch shift to higher frequencies, and the unit cell parameter decreases (Table 3.3). An enhancement of the 3710 cm^{-1} hydroxyl band indicates, especially at higher degrees of dealumination, that the products contain an increased number of Si–OH groups.

The intensities of the hydroxyl bands are considerably lower than in hydrothermally treated samples. The 3740 cm^{-1} band can be resolved into two bands at 3745 and 3738 cm^{-1}, assigned to terminal and hydroxyl nest OH-groups, respectively [598]. The band at 2680 cm^{-1} due to hydroxoaluminum species is absent. No secondary pore system is generated, but occlusion of fluoroaluminum species can cause a decrease in the pore volume [445].

In order to prevent loss of structure as a result of a greater rate of dealumination than of silicon incorporation, the conditions should be well controlled. The pH can be held constant at 6.0 by buffering with ammonium acetate [596], and the stoichiometric 25 Al/u.c. are removed within 1 h. A gradual deterioration of the structure occurs at reaction times >6 h, and 3–6 h appear to be optimum [596]. While highly crystalline products can be obtained from 60° to 95 °C, the concentration of Brønsted acid sites is at a maximum for NH_4NaY treated at 75–95 °C.

Table 3.3. Reaction of NH_4Y with Hexafluorosilicate [450]

	Starting NH_4Y (4.84)	Product NH_4Y (9.3)	Product NH_4Y (14.8)
X-ray crystallinity, %	100	106	93
unit cell, a_0, Å	24.67	24.49	24.39
framework IR:			
asymmetric stretch (cm^{-1})	1014	1044	1061
symmetric stretch (cm^{-1})	786	807	818
crystal collapse temp. (°C)	860	1037	1128
hydroxyl region IR:			
absolute absorbance, 3710 cm^{-1}	0.000	0.040	0.158
water sorption capacity, 25°C, 610 Pa wt%	32.1	29.8	28.8

Since Si insertion is slower than the removal of framework-Al, the molar ratio of $SiF_6^{=}$/framework-Al plays a significant role and should not be greater than 0.5 [596]. The effect is caused by a combination of a gradient in the dealumination from the surface to the interior and a selective deposit of silica on the surface of the crystallites [599]. The product has a unit cell parameter, a_0, of 24.50 Å, which amounts to about 50% dealumination [596]. High degrees of dealumination are difficult to achieve by treatment with $(NH_4)_2SiF_6$, but in contrast to dealumination by steaming, no mesopores are generated because of Si substitution [599a]. Repeated treatments do not increase the degree of dealumination, but intermediate calcination permits further removal of Al in a subsequent $(NH_4)_2SiF_6$ treatment, although a significant loss of crystallinity is observed. The reaction of a zeolite with $(NH_4)SiF_6$ is diffusion-controlled within the pores, and dealumination is not uniform throughout the crystallites [600–602]. Subsequent calcination at 640°C gives a product containing extra-framework-Al^{VI}, which tends to migrate toward the surface. Since catalytic application of a zeolite most commonly requires a final calcination, further framework dealumination and accumulation of extra-framework aluminum is, therefore, not avoided by this procedure [601].

Extra-framework aluminum in USY can be removed by treatment with $(NH_4)_2SiF_6$ [594], although a small quantity of framework-Al is extracted as well. Products with a_0 below 24.44 Å are difficult to obtain by this method without loss of crystallinity.

Faujasitic zeolites obtained by reaction with either $SiCl_4$ or $(NH_4)_2SiF_6$ have a less uniform distribution of framework-Al than high-silica faujasite obtained by direct synthesis in the presence of 15-crown-5 ether [603]. The dealuminated zeolites, although crystalline, are relatively inactive for cracking of n-hexane. After steaming, however, they become more active than the parent zeolite steamed to the same framework-Al content [604].

Other reagents applicable for dealumination of zeolites include phosgene [605–608], CCl_4 and $CHCl_3$ [609], NOCl [609, 610], metal halides, alkoxides and alkyls [609, 611, 612], fluorine and fluorides [613–618].

3.3.5
Removal of Other Framework Elements

The acid strength of a zeolite acid depends mainly on the type of heteroatom associated with the acid site [619], and secondarily on the topology and the SiO_2/M_2O_3 ratio. The ease with which a framework hetero-element can be removed by hydrolysis is inversely related to the acid strength, as judged by the acid catalytic activity, so that the following correlation exists:

Acid strength:		$\ll$		$<$		$<$	
	B		Fe		Ga		Al
IR band, cm^{-1}:	3725		3630		3620		3610
Ease of hydrolysis		$\gg$		$>$		$>$	

A borosilicate zeolite, eg., [B]ZSM-5 ([] designates a framework heteroatom), is a much weaker acid than [Al]ZSM-5. Although it does exist as a salt, the hydrogen form is not much more acidic than a silanol group, and the boron is tricoordinated [620–622]. Only in presence of a base can the boron be retained in or reverted to tetrahedral coordination [622].

The symmetrical BO_4 tetrahedron can be restored by hydration [623]. Addition of water to the H-form, which contains trigonal boron, establishes the equilibrium

$$Si{-}OH + B(OSi)_3 + H_2O \rightleftarrows [Si{-}O{-}B(OSi)_3]^{-1} + H_3O^+ \qquad (3.45)$$

Thus, reducing the pH shifts the equilibrium to the left and forms easily hydrolyzable trigonal boron, increasing the pH forms the salt of tetrahedral boron. Tetrahedral BO_4 gives a ^{11}B MAS NMR signal at –5 ppm (TPA^+ form) or –4.7 ppm (NH_4^+ form), while framework BO_3 shows an IR absorbance at 1380 cm^{-1} and ^{11}B MAS NMR signals at –1 and +5 ppm [622]. A hump at +15 ppm is assigned to extra-framework trigonal B. This signal disappears upon washing with NaOH at pH 11.

As a result of the low acid strength of [B]ZSM-5, the catalytic activity for most reactions is attributable to trace amounts of aluminum [624, 625]. [B]ZSM-5 is believed to catalyze double-bond shifts and surface oligomerization, but not any more demanding hydrocarbon conversions [626]. However, Lewis acid sites associated with boron in H[B]ZSM-5 were found to have greater acid strength than those in H[Al]ZSM-5 [627].

Boron is removed from the framework of an as-crystallized [B]ZSM-5 by calcination in oxygen, nitrogen, or steam. The extra-framework boron and any easily hydrolyzable trigonal framework-B can then be washed out with NaOH at pH 11, while other trigonal framework-B may revert to tetrahedral B [622]. Calcination in ammonia prevents loss of boron from the framework, but subsequent oxygen calcination causes loss of framework boron. Exchange with NaBr after ammonia calcination and prior to burning the carbon deposits in oxygen retains tetrahedral framework-B. Because of the low acid strength of the borosilicate and the inability of boron to form a stabilizing cation, hydrolysis occurs to a much greater extent than with [Al]ZSM-5, and boron can be extracted from the hydrogen form of [B]ZSM-5 by refluxing with water [628], or in a Soxhlet extractor. Removal of boron from the framework by aqueous and hydrothermal treatment causes the unit cell to expand while hydroxyl nests are formed [629]. The latter are not refilled with Si during deboronation and are well suited for post-synthesis insertion. They are stable to 400 °C and dehydrate at higher temperatures forming new Si–O–Si bonds.

The unit cell parameters of [B]ZSM-5 are smaller than for high-silica ZSM-5, because boron is smaller than silicon, and correlate with the boron content [630, 631]. Removal of B from the framework of B-containing pentasil-type zeolites by thermal and hydrothermal treatment or acid leaching causes the boron-stimulated lattice concentration to be cancelled [632]. The boric acid formed by thermal treatment at 900 °C migrates to the surface of ZSM-5 [633].

The same types of OH-groups as in aluminosilicates are found in thermally and hydrothermally treated gallosilicates, and both Brønsted and Lewis acid sites have been observed in H[Ga]beta [634].

Treatment of high-silica zeolites with aqueous alkaline or HF solution causes the crystals to dissolve, beginning at the external surface [634a, b].

3.4 Insertion into the Zeolite Framework

3.4.1 Reinsertion of Hydrolyzed Aluminum

The gradual loss of active sites in the regeneration of a catalyst resulting from dealumination makes it desirable to recover the activity by restoring the acid sites. Mild basic treatment permits the aluminum to return to the vacancy left behind, or any other vacancy, with the cation of the base supplying the cations for stabilization of the anionic sites.

The first observation of such an insertion was reported by Breck and Skeels [635], who found that titration of a slurry of calcined NH_4Y in 3.4 M NaCl required more NaOH to attain pH 11 than the filtrate of such a slurry.

Subsequent attempts to solubilize non-framework-Al with a base gave somewhat ambiguous results [520, 570, 635–642]. However, the objections were successfully overcome by further work [643, 644]. Since faujasite does not recrystallize from potassium-bearing solutions, 0.25 M KOH was used as the base [645]. Re-insertion of aluminum is evident from the ^{29}Si MAS NMR spectra, paralleling the increase of a_0. The crystallinity increases, compared with the dealuminated form, although not to a value as high as that of the parent prior to dealumination. Amorphous material, particularly in caustic-treated USY, is evident from the reduced sorptive capacities and the underlying broad hump in the ^{29}Si NMR spectrum. Both Si and Al NMR spectra indicate that aluminum is reinserted into the framework by treatment with a base [643, 644], and quadrupole nutation NMR studies demonstrate that both octahedral and non-framework tetrahedral Al are reinserted [646]. Further support is provided by the IR spectra showing shifts to lower frequencies of framework stretching bands upon realumination.

Still further evidence for reinsertion of aluminum is provided by ^1H MAS NMR [647]. $NaNH_4Y$ dealuminated by deep-bed calcination in a steam atmosphere, then treated with 0.25 M KOH at 80 °C for 24 h, ammonium exchanged at 80 °C and calcined at 400 °C overnight, gave the following signals in the ^1H MAS NMR spectrum [372]:

Line	Range, ppm	Assignment
a	1.8–2.3	non-acidic silanol groups
b	3.8–4.4	acidic bridging OH groups
c	ca. 5.2	additional acidic bridging OH in faujasite
d	6.5–7.0	residual NH_4^+ ions
e	2.6–3.6	OH groups on non-framework Al

Signals b and c were enhanced, and signal d was weak [647]. Quantitation indicates that there are 4.2 bridging OH-groups per supercage in the parent, 2.1 in the dealuminated sample, and 4.4 in the realuminated material. The SiO_2/Al_2O_3 ratios calculated from the unit cell parameters, the intensities of the Si NMR signals, and from the shifts of the IR bands also lead to the conclusion that essentially complete realumination was achieved. After realumination, the IR band at 3610 cm^{-1}, which was present in dealuminated Y, disappeared, supplying further evidence that the extra-framework-Al has been reinserted in the framework. The product contains twice as many Brønsted acid sites as the dealuminated Y precursor and approximately the same number of acid sites as the parent material.

Neutron diffraction reveals a high 95% occupancy of tetrahedral sites, in agreement with the improved crystallinity after treatment of steamed Y with highly alkaline solutions [648]. The morphology remains unchanged, indicating that any amorphous species formed are located within the zeolite crystals, e.g., in mesopores. The Al distribution after realumination, however, is different from that in the parent sample.

Steamed HY shows an increased framework-Al content in the ^{29}Si NMR spectrum after aqueous ammonium exchange [534]. Simultaneously, the concentrations of both non-framework octahedral and non-framework tetrahedral aluminum decreased, the former more than the latter. The results indicate that non-framework aluminum, especially that of octahedral coordination, was inserted into the zeolite framework while tetrahedral non-framework aluminum may be part of alumina clusters, which are less mobile and less reactive. The observation is in agreement with the explanation that octahedral aluminum cations, which are partially coordinated to framework oxygens after steaming, become mobile in the ion-exchange process. While some aluminum cations are exchanged out of the zeolite, others apparently react with silanol groups of hydroxyl nests and are reinserted into the framework.

The rate of Al reinsertion in USY increases with temperature. Silica solubilization is small with 1 N NaOH, but longer reaction times of HY at 80 °C may involve an isomorphous substitution process in this phase, i.e., after available defects have been filled [648 a].

Differences in the realumination behavior of zeolite Y dealuminated by treatment with EDTA (EDY) or $(NH_4)_2SiF_6$ (FDY) in the liquid phase, or hydrothermally (USY), become apparent on treatment with KOH, which does not permit recrystallization of zeolite Y [645]. Almost no extra-framework-Al exists in EDY samples, and realumination must, therefore, be explained by dissolution of the more severely dealuminated outer layer of the crystals, and reinsertion of the aluminum thus liberated into vacancies in the interior portions of the crystals. There are almost no structural vacancies and extra-framework-Al in FDY (Si/Al = 4.88), and no silica is dissolved by KOH treatment at concentrations up to 0.1 N. At higher concentrations, the Si/Al ratio decreases with simultaneously decreasing crystallinity and increasing Si concentration in the liquid phase, indicating removal of Si and possibly internal rearrangement of tetrahedral atoms (the intensity increase of the 3740 cm^{-1} IR band may be caused by amorphous material). In contrast to EDY and FDY, USY contains abundant non-framework aluminum, and facile reinsertion of Al into structural vacancies is observed at 0.1 N, while framework silicon is not removed at this concentration [648b].

3.4.2 Reaction with Aluminum Compounds

The catalytic activity of some high-silica zeolite structures is limited because of the small percentage of aluminum incorporated during crystallization. It may, therefore, be desirable to impart greater activity to the zeolite, which can be achieved by inserting aluminum into the framework. Several aluminum compounds were found to react in such a manner, and a considerable amount of work has been done in this field with the goal of providing improved catalysts.

3.4.2.1 Aqueous Aluminate

It may be expected that aluminum is inserted in the framework more readily when a soluble tetrahedral aluminum species is supplied. In the work described in 3.4.1, alkali hydroxide reacts with hydrolyzed cationic aluminum species to form the aluminate anion, e.g.:

$$Al(OH)_2^+ + 2\,OH^- \rightarrow Al(OH)_4^- \qquad (3.46)$$

The aluminate then reacts with the silanol groups of, e.g., a hydroxyl nest, aluminum is inserted and four water molecules are formed.

When high-silica zeolites are to be aluminated, the aluminum source, e.g., sodium aluminate, needs to be supplied externally. Aluminum is inserted into tetrahedral vacancies and/or by substituting Al for framework-Si [649]. ZSM-5 with n-hexane cracking activities of α >200 [650, 651] can easily be obtained by treating essentially inactive high-silica ZSM-5 with sodium aluminate solution in an autoclave. High-silica zeolites are more soluble in high concentrations of caustic, so that a lower molarity needs to be applied and should not be greater than 0.2 M. The ^{27}Al MAS NMR signal at 57.5 ppm of aluminated ZSM-5 is much broader than in as-crystallized ZSM-5 of the same composition indicating a wider range of Al environments [652].

Zeolite beta with SiO_2/Al_2O_3 molar ratios of ~8 can be obtained by an 8-h treatment of Hβ (SiO_2/Al_2O_3 = 38) with 3 mmol of $NaAlO_2$/g of zeolite at 70 °AC with a pH not greater than 13.0 [652 a].

3.4.2.2
Aluminum Oxide

Alumina-bound HZSM-5 (SiO_2/Al_2O_3 of 1600, 26000, 38000) catalyst, prepared by

1. wet-mulling of the as-synthesized zeolite with AlOOH,
2. extrusion,
3. drying,
4. calcination at 538 °C,
5. ammonium exchange,
6. calcination at 538 °C,

was evaluated by a series of acid-catalyzed reactions, n-hexane cracking, propene oligomerization, lube dewaxing, and conversion of methanol to hydrocarbons, along with ion exchange [653]. The n-hexane cracking activity was essentially unaffected by steaming of the zeolite in the absence of alumina, but showed considerable increase after binding and, particularly, after steaming of the bound catalyst.

The pure zeolite (SiO_2/Al_2O_3 = 38,000) had low activity for propene oligomerization, giving 16% conversion, mainly to light products. A dry physical mixture showed no activity enhancement. However, when the mixture was wet-mulled and extruded, conversion rose to 77% yielding 72% liquids, which contained 56% 165 °C + distillate range product. When tested as catalyst for lube dewaxing, the physical mixture of zeolite with alumina was inactive while the extrudate gave 15% conversion reducing the pour point from >46 to −7 °C. Methanol conversion to hydrocarbons was increased from 1.8% for the pure zeolite (SiO_2/Al_2O_3 = 26,000) to 8.0% for the extrudate. The ion-exchange capacities of HZSM-5 samples (SiO_2/Al_2O_3 of 1600 and 26000), measured by temperature-programmed ammonia desorption after ammonium exchange, were significantly higher after extrusion and increased further after steaming.

Extrusion of high-silica HZSM-5 (SiO_2/Al_2O_3 = 26,000) with alumina raised the n-hexane cracking activity from α = 0.02 to α = 1.4 [653]. Further substantial activity enhancement was achieved when the extrudate was treated with liquid water at 160–170 °C, followed by ammonium exchange and calcination. The n-hexane cracking activity was α = 40 after 18 h of this treatment, leveling off at α = 60 after 12 d, whereas steaming did not change the activity significantly (Fig. 3.27), thus indicating that the presence of a liquid phase is essential [654]. A lower level of activity was attained when zeolite crystals were treated in the same manner in the presence of alumina beads, which could be separated physically from the zeolite. The result demonstrates that, although aluminum migration is enhanced in the presence of liquid water, close contact of the zeo-

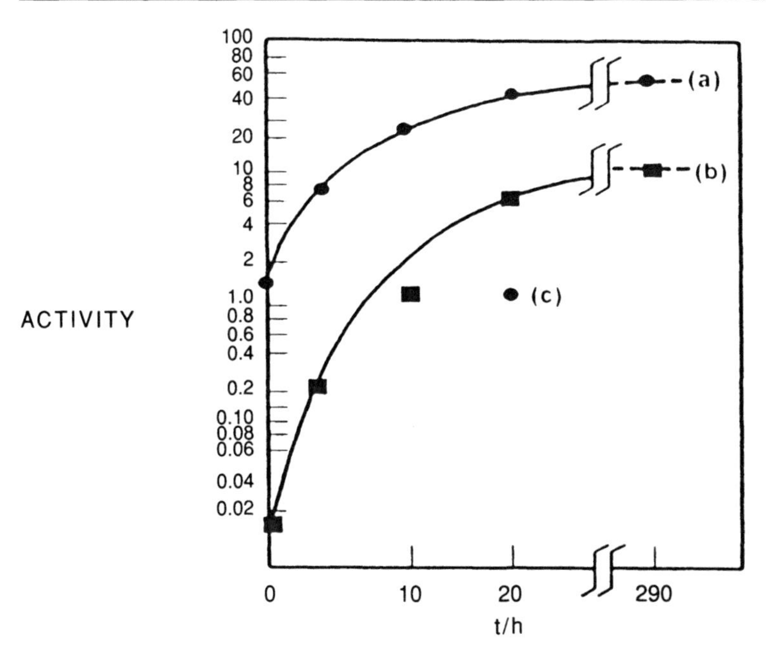

Fig. 3.27. Hydrothermal activation of high-silica ZSM-5. Effect on n-hexane cracking activity [359]. (**a**) extrudate; (**b**) ZSM-5 + alumina beads; (**c**) extrudate, dry steamed

lite with the aluminum source is advantageous. It is also further proof that the acid sites are generated in the zeolite rather than on the alumina source.

The FTIR spectra of high-silica ZSM-5 before and after hydrothermal treatment in the presence of alumina beads, and of the extrudate before and after hydrothermal treatment are shown in Fig. 3.28. The enhanced 3610 cm^{-1} band is associated with acidic OH-groups, in agreement with the ^{27}Al MAS NMR spectrum (Fig. 3.29) showing an increase in the concentration of tetrahedral Al [654]. The zeolitic Al, calculated from the latter result, is 6000–6800 ppm, in good agreement with the observed hexane cracking activity.

All four test reactions are acid catalyzed. It is concluded, therefore, that

- new acid sites are created during extrusion with alumina and subsequent calcination;
- generation of new acid sites increases the ion-exchange capacity and catalytic activity for acid-catalyzed reactions;
- the new sites are intra-crystalline because they catalyze lube dewaxing, a reaction known to depend critically on the shape-selectivity of the zeolite; and
- experiments comparing the effect of wet vs. dry binding indicate that Al is transferred through a soluble Al species rather than a solid–solid reaction.

Hydrothermal activation of a ZSM-5/alumina extrudate can be accelerated and increased by raising the pH of the aqueous phase, e.g., with sodium hydroxide, followed by ammonium exchange and calcination.

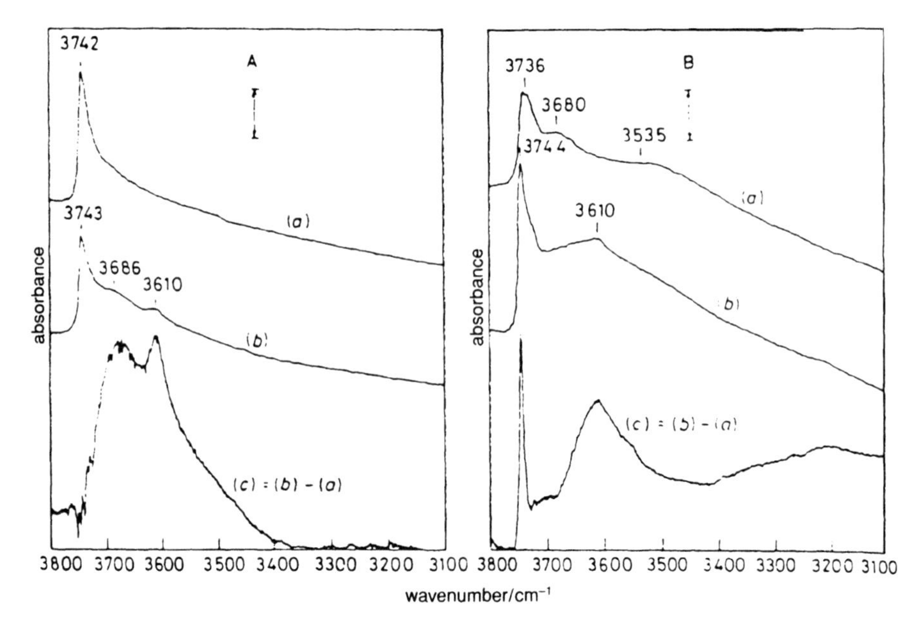

Fig. 3.28. (**A**) Fourier-transform IR spectra [359] of (**a**) high-silica ZSM-5 (SiO_2/Al_2O_3 = 26000); (**b**) after hydrothermal treatment in the presence of alumina beads; and (**c**) the difference (**b**) - (**a**). Scale bar = 0.1 for (**a**) and (**b**), 0.01 for (**c**).
(**B**) FTIR spectra of (**a**) high-silica ZSM-5 extrudate; (**b**) after hydrothermal treatment; and (**c**) the difference (**b**) - (**a**). Scale bar = 0.1 for (**a**) and (**b**), 0.04 for (**c**)

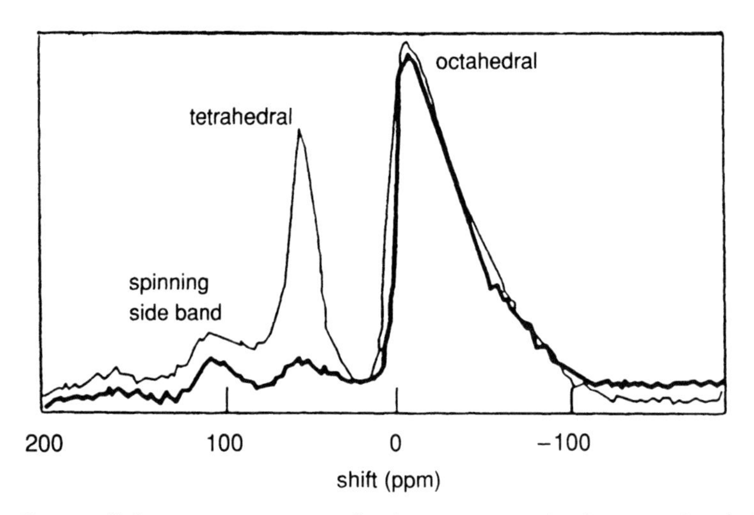

Fig. 3.29. ^{27}Al MAS NMR spectra [359]. *Top:* ZSM-5/alumina extrudate hydrothermally activated to $\alpha = 60$, calcined, and converted to NH_4-form by sorbing NH_3. *Bottom:* Alumina binder calcined at 538 °C, then treated with NH_3 gas

3.4.2.3
Aluminum Halides

A higher degree of alumination can conceivably be attained by a reaction that replaces tetrahedral Si with Al. Aluminum chloride is formed when NH_4Y or HY is treated with $SiCl_4$ at high temperature [570], and a reversal of this reaction would be of interest for enhancing the activity of high-silica zeolites.

Aluminum is inserted into the framework by a sequence of treating high-silica ZSM-5, e.g., SiO_2/Al_2O_3 of 400–50,000, with $AlCl_3$ vapor in nitrogen at temperatures in the range of 500–600 °C for 1–10 h or until no more HCl and $SiCl_4$ escape, purging with dry nitrogen, cooling, contacting with water for 0.5–17 h, washing, drying, and calcining, after slow heating, at 538 °C for 10 h [655–657]. The product was examined by ^{27}Al MAS NMR and found to contain Al_{Td} and Al_{Oh} in the ratio of 3:1. The very narrow signal of Al_{Oh} is indicative of a highly mobile species, such as hydrated Al^{3+} cation. As prolonged washing did not decrease this signal, these cations seem to compensate for the anionic charges generated in the framework. The Al-modified zeolites contained up to 2.55% Al compared to 0.15% or less prior to the $AlCl_3$ treatment, and the α-values [650, 651] increased by 70–90.

No activation was observed for $AlCl_3$-treated samples prior to hydrolysis. Samples hydrolyzed for only 30 min at room temperature had much lower activity than those hydrolyzed for 2–17 h. When the former were treated with 1 M HCl at 75 °C for 3 h and recalcined, the activity increased considerably (Table 3.4). The data show that the aluminum content does not necessarily decrease on treatment with HCl indicating that the redissolved hydrolyzed Al species insert more readily into hydroxyl nests to form acid sites.

The evidence for framework alumination is overwhelming:

- n-Hexane cracking activities are comparable to those observed for conventional ZSM-5 of 75 SiO_2/Al_2O_3.
- Constraint Index values are consistent with those obtained for conventional zeolites of the same structure, indicating that new active sites are intracrystalline.
- Increase of ion-exchange capacities from initially less than 0.01 meq/g to as high as 0.23 meq/g after alumination.
- Maximum ammonia desorption rate occurs at about 400 °C, similar to that observed for conventional ZSM-5.
- The FTIR spectrum contains an absorbance at 3610 cm^{-1}, associated with Brønsted acid sites in ZSM-5.

Table 3.4. Effect of HCl Treatment on Enhanced Acid Activity of $AlCl_3$-Treated ZSM-5 [657]

$SiO_2/Al_2/O_3$	26000		1670		2544	
	% Al	α	% Al	α	% Al	α
original	< 0.01	< 0.5	0.06	6	0.04	4 [a]
After incomplete hydrolysis	1.70	32	1.56	50	1.32	62
After 1 M HCl treatment	1.29	71	1.55	100	1.32	128

[a] Estimated.

- The ^{27}Al MAS NMR spectrum displays a sharp peak at about 55 ppm, characteristic of tetrahedral framework aluminum in ZSM-5.

The total Al incorporation, based on elemental analysis, is frequently greater than expected from the catalytic activity, indicating that only part of the total Al formed acidic sites. Anderson et al. [656] found evidence of Brønsted and Lewis acid sites in the IR spectrum of the $AlCl_3$-treated material after sorbing pyridine. They interpret their observation as a reversal of the dealumination with $SiCl_4$:

$$(SiO_2)_x + 4\,AlCl_3 \rightarrow Al^{3+}[(AlO_2)_3(SiO_2)_{x-3}] + 3\,SiCl_4 \quad (3.47)$$

Chang et al. [658], who used $AlBr_3$ as well as $AlCl_3$ as the aluminating agent, found, simultaneous with the appearance of the 3610 cm^{-1} band in the FTIR spectrum, a reduction in the intensity of the 3740 cm^{-1} absorbance, indicating that aluminum reacted with silanol groups. The $AlCl_3$-activated high-silica ZSM-5 increased methanol conversion from < 2 to $\geq 99\%$ with increased aromatics yield. The mechanism proposed by these authors involves reaction of tetrahedral lattice vacancies with $AlCl_3$ or $AlBr_3$.

This reaction occurs at temperatures as low as 350 °C while substitution of Si with Al was reported to occur at 650 °C [659]. Yamagishi et al. [660] observed later that the number of Al introduced with $AlCl_3$ at 650 °C was independent of the number of terminal silanol groups and the number of Si released, and leveled off after 1 h at 11 kPa $AlCl_3$ partial pressure. The results led to the conclusion that Si is not replaced by Al, but that Al is inserted into tetrahedral vacancies. Such hydroxyl nests, which had previously been found after dealumination of zeolite Y (see Sect. 3.3.2.1), were shown to exist in as-crystallized ZSM-5 [661, 662]. The concentration of defect sites could be estimated by exchanging the oxygen in silanol groups with ^{18}O in $C^{18}O_2$ for the exchange reaction [663]. The concentration of aluminum inserted into the framework is linearly related to the concentration of hydroxyl nests (broad IR band at 3505 cm^{-1}), with one Al replacing four silanol groups. While a high degree of alumination is observed when high-silica HZSM-5 reacts with $AlCl_3$, only minor quantities of Al are inserted into ZSM-5 of low Si/Al ratio e.g., 36, by the same procedure; ZSM-5 of Si/Al = 22 even lost aluminum when treated with $AlCl_3$ vapor [664]. Bonds formed between $AlCl_3$ and surface silanol groups are weak, and Al^{3+} ions bound in this manner are removed during acid wash [665].

Aluminum chloride is sorbed on ZSM-5 very slowly at 300 °C [666]; only 7% of the silanol groups react within a 10-h period. In contrast, dimethylaluminum chloride, having smaller critical dimensions, reacts with 68% of the silanol groups in the same period. The effect becomes particularly noticeable when large ZSM-5 crystals are used for the experiment. It is concluded that the concentration of $AlCl_3$ within the pores of ZSM-5 is low at 300 °C because of the small pore size of this zeolite compared with the critical dimensions of $AlCl_3$. The $AlCl_3$ that does diffuse into the channels preferentially reacts with hydroxyl nests to insert Al and form HCl gas. The degree of Al insertion into the framework varies with the time allowed for diffusion of $AlCl_3$ into the zeolite pores and with the length of the diffusion path, i.e., with the size of the crystallites.

As the diffusion rate increases, substitution of internal framework-Si may occur at high temperature, but needs to be demonstrated unequivocally.

Most surface OH-groups react with the comparably high concentration of $AlCl_3$ external to the zeolite. The initially formed bonds are weak, but some structural rearrangement can be expected to occur, and some additional stable Brønsted acid sites are formed in the framework, although most likely near the surface. This rearrangement, combined with further attack by $AlCl_3$, probably causes $SiCl_4$ to be generated and mesopores to be formed.

3.4.2.4 Complex Aluminum Fluoride

In contrast to $AlCl_3$ and $AlBr_3$, AlF_3 is not volatile and cannot be applied for the reaction described in 3.4.2.3. However, the compound can be solubilized by complexation with a soluble fluoride although even the fluoroaluminates are only slightly soluble, particularly the sodium form (cryolite, Na_3AlF_6). When AlF_3 is dissolved in a solution of NH_4F, the aluminum is found to be octahedral, as evidenced by ^{27}Al MAS NMR [658]. The complex formed is probably mainly $[AlF_4(H_2O)_2]^-$ in a 0.02 M solution at pH 5.9, but may contain small amounts of $[AlF_5(H_2O)]^{2-}$. When the pH is increased to 10.5 by addition of NH_4OH, the fluoride ions are gradually replaced by hydroxide, and the aluminum becomes tetrahedral. Depending on the pH and the fluoride concentration, all ions of $[AlF_{4-x}(OH)_x]^-$ may be present. The reactions of such complexes with the zeolite were carried out by the following procedure:

1. High-silica ZSM-5 ($SiO_2/Al_2O_3 = 26000$) was impregnated with an aqueous solution containing ~ 0.1% AlF_3 and dried at 130 °C.
2. The dried material was treated with 1 N NH_4OH and then calcined at 538 °C.

The products had greatly increased ion exchange capacity and n-hexane cracking activity, especially after treatment with the complex at pH 10.5, but the reduced sorptive capacity indicates that some AlF_3 remained unreacted within the pores [658]. This reaction is interpreted as a direct replacement of framework Si with Al liberating the replaced silicon as SiF_4.

3.4.2.5 Generation of Vacancies Prior to Alumination

The reaction of a high-silica zeolite with fluoroaluminate (see Sect. 3.4.2.4), and probably with hydroxoaluminate as well (see Sect. 3.4.2.1), causes tetrahedral vacancies to be generated in parallel with the insertion of aluminum into the framework [649, 658]. The concentration of tetrahedral vacancies can also be varied by adjusting the composition of the synthesis reaction mixture [663].

Boron in borosilicate zeolites can easily be hydrolyzed and removed from the framework of the hydrogen form of the zeolite, e.g., by extraction with water [667] (see Sect. 3.3.5). The tetrahedral vacancies so generated can be used to accept other elements in an insertion reaction.

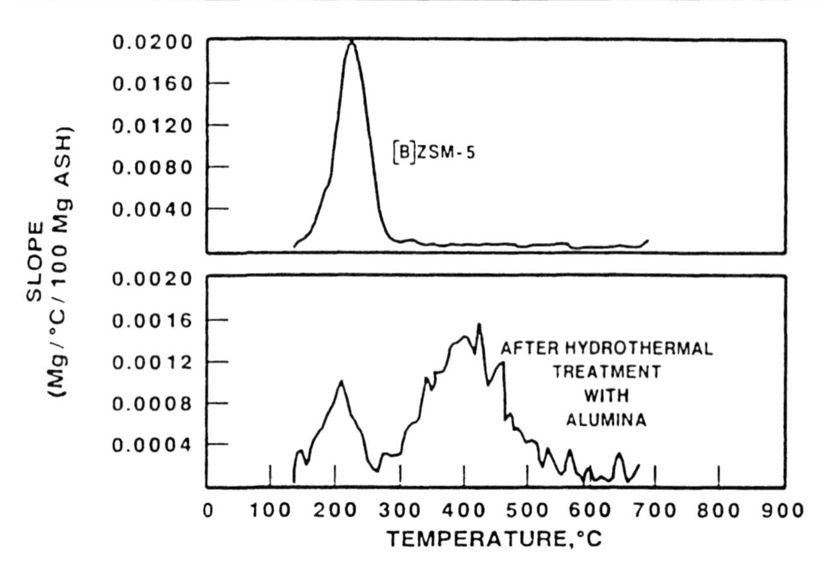

Fig. 3.30. Derivative plots of temperature-programmed ammonia desorption [654]

Because of the facile removal of boron, framework boron can directly be replaced by aluminum from an external aluminum source, e.g., alumina binder used in the extrusion of a zeolitic catalyst [668]. In a set of experiments, six samples containing from 0 to 0.95% B were contacted in the NH_4^+ and in the H^+ form with γ-alumina beads in water for 65 h at 155°C [654]. The products were separated from the alumina beads, ammonium-exchanged, washed, and calcined. In contrast to the untreated [B]ZSM-5, the α-values of the treated materials correlated with the boron content of the parent zeolite showing that framework boron facilitates aluminum insertion. Simultaneously, the temperature-programmed ammonia desorption gave a strong aluminosilicate peak at 390°C along with a greatly reduced borosilicate peak at about 200°C (Fig. 3.30). The changes in the FTIR spectrum also show the transformation of [B]ZSM-5 to [Al]ZSM-5 (Fig. 3.31). In this sample, the boron content, as measured by ^{11}B MAS NMR, was reduced to 6–8% of the initial concentration, while TPAD shows a reduction from 0.84 to 0.05 wt-% B. Simultaneously, the tetrahedral aluminum content, as measured by ^{27}Al MAS NMR, rose from 270 to 3300 ppm, and the n-hexane cracking activity increased from $\alpha = 2.3$ to $\alpha = 42$.

3.4.3
Insertion of Other Elements

The presence of internal silanol groups in ZSM-5 [669] and of hydroxyl nests [436] is well established, as is their reaction with hydroxylated aluminum species. Such tetrahedral vacancies are able to react with hydroxylated species of other elements as well, and it is tempting to predict that elements unable to be incorporated into zeolite structures by crystallization can be inserted in this

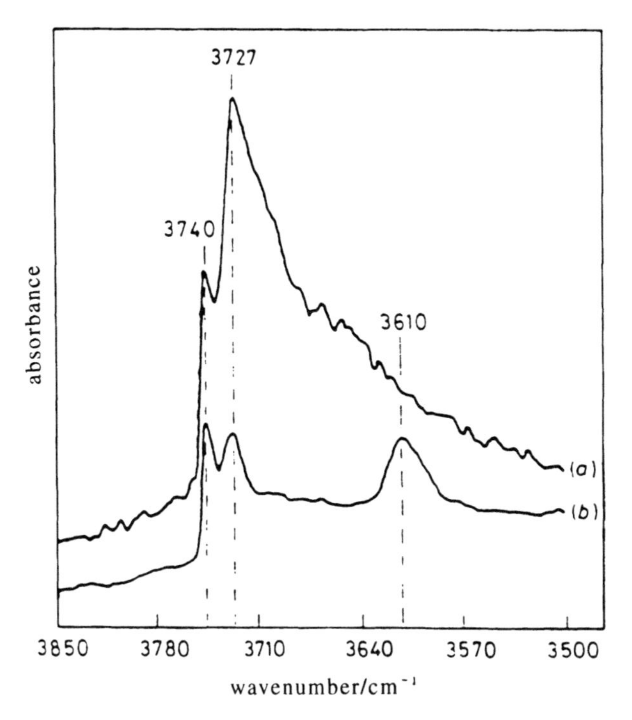

Fig. 3.31. FTIR spectra of (a) H[B]ZSM-5; (b) after hydrothermal treatment in presence of alumina [654]

manner. However, there is scant evidence that such insertion has been accomplished although some claims have been made.

3.4.3.1
Group IIIA Elements

The lower elements of this group are known to be incorporated by crystallization, and it is not surprising that they also can be inserted by secondary synthesis.

Boron is inserted in ZSM-5 by treatment with BCl_3 [670].

Boric acid introduced by impregnation reacts with SiOH groups upon calcination to yield compositions possibly containing $\equiv Si-O-B(OH)_2$, $\equiv Si-O-BO$, and $\equiv Si-O-B(O-Si\equiv)_2$ groupings. More complete insertion of boron is achieved by treatment with potassium borate at 80 °C [672]. All boron inserted in dealuminized ZSM-5 in the presence of 0.08 M KOH was tetrahedral, while only 76% of the boron introduced with 0.01 M KOH had this symmetry, the remainder was trigonal. The result indicates, not unexpectedly, that a high pH is required for complete insertion and conversion to tetrahedral coordination. The number of tetrahedral boron atoms per unit cell, 0.16–0.19, was almost constant for KOH molarities in the range of < 0.01 to 0.08, independent of whether the high-silica ZSM-5 was prepared by dealumination or direct crystallization; it increased to 0.31 and 0.36 (73% tetrahedral) when the latter was treated with 0.10 and 0.16 M

KOH, respectively. Only traces of boron were found in the solid when the treatment was carried out at ambient temperature. It appears that boron is inserted into tetrahedral vacancies at low KOH molarity, while excess boron reacts with isolated and less clustered silanol groups. Only at higher base concentration does either replacement of Si with B or rearrangement of the structure permit more boron to be incorporated in tetrahedral positions. Similar results were obtained with dealuminated forms of mordenite and zeolite Y, but not with the non-dealuminated hydrogen forms of the parent zeolites [673]. These results, along with the ^{27}Si MAS NMR data demonstrating the loss of silanol groups upon reaction with potassium borate, again indicate that boron is inserted into tetrahedral vacancies.

More recently, ZSM-5 containing framework-B was obtained when NH_4ZSM-5 was treated at 85 °C with a NH_4BF_4 solution buffered to pH 7.3 with ammonium acetate. It is important to keep all washing and ion-exchange solutions at a pH 8–9 by addition of NH_4OH. Similarly, [B]Y was prepared by treating 65% NH_4 exchanged Y at 85°C with NH_4BF_4, buffered to pH 7.1 with ammonium acetate. However, the crystallinity deteriorated gradually, as the amount of boron inserted increased. In this reaction, framework-Al is replaced with tetrahedral boron, which is so small that the imposed strain causes the structure of zeolite Y to collapse [674].

In analogy to insertion of aluminum by treatment with $AlCl_3$ [660], gallium and indium have been inserted into zeolite ZSM-5 using the respective chlorides [675]. Although the degree of substitution decreased with the size of the metal ion incorporated, the **a** and **c** unit cell parameters increased with the size of the ion thus giving rise to a larger unit cell volume. The results suggest that the metal ions were inserted into the framework. The number of acid sites generated was considerably smaller than the number of metal ions incorporated indicating that, similar to Al (Sect. 3.4.2.3), not all inserted metal ions are tetrahedrally coordinated [675]. The IR spectra after ammonium exchange and thermal conversion to the hydrogen form prove that the elements were inserted in the framework as shown below:

	Al	Ga	In
Brønsted acid (cm^{-1})	3610	3625	3650
Pyridine sorbed (cm^{-1})	1545	1545	1545

When the pyridinium ion was formed, the Brønsted acid bands disappeared. The IR band for the gallosilicate Brønsted acid had previously been established for the material synthesized by crystallization [676]. As the band of the bridging OH-group moves to higher frequency, the peak of the temperature-programmed ammonia desorption is displaced to lower temperature. IR spectra and TPAD indicate that acid sites are associated with at least some of the incorporated metal ions and that the acid strength changes in the sequence Al > Ga > In. The incorporation of indium is particularly surprising because of its large ionic radius (0.81 vs. 0.42 Å for Si).

Treatment at 100 °C of high-silica ZSM-11 with dilute sodium gallate solution for 24 h yields a product with increased unit cell size and much higher water sorption capacity. The hydrogen form gives IR bands at 3742 (surface OH) and 3622 (bridging OH) cm^{-1} [677, 678]. Similarly, Ga can be introduced into zeolite beta by treatment with sodium gallate. It seems that silica is removed from the framework by basic leaching, and gallium is then inserted into the vacancy [679].

Gallium can be inserted by treatment of zeolites Y, ZSM-20, and beta with aqueous fluorogallate in the presence of sodium acetate as buffer. This reaction replaces framework-Al with Ga. Substitution of extra-framework aluminum with gallium was observed when USY was treated in this manner, while framework-Si was not significantly attacked by treatment of USY with fluorogallate [680, 681].

3.4.3.2
Elements of Other Groups

Beside elements of group IIIA, iron and titanium had been reported previously to be incorporated by direct crystallization. Both elements have also been incorporated by post-synthesis insertion at 75–100°C using $(NH_4)_3FeF_6$ or $(NH_4)_2TiF_6$, respectively, as reagents [682]. Iron was successfully incorporated into mordenite, while the aluminum content decreased simultaneously. Several products were essentially white, others are described as beige or brown, the brown color being attributable to precipitated iron oxide hydrate or hydroxylated iron(III) cations. These samples may contain small amounts of framework iron, as inferred from the observation that few hydrogen-bonded OH-groups are generated, although the framework is dealuminated, and from the slightly elevated M^+/Al ratio in one brown sample, compared with much higher M^+/Al ratios in the white samples, where the $M^+/(Al + Fe)$ approaches unity. It also needs to be considered that some of the iron became bonded to the framework with fewer than four bonds without carrying a negative charge so that no cations are associated with such a configuration. The incompletely inserted element would be expected to be more susceptible to hydrolysis than the completely inserted ion.

Much less evidence is available for the post-crystallization insertion of titanium. Since TiO_2 and its hydrates are white, any precipitated titanium oxide is not visibly apparent, and the tetravalent Ti would not generate acid sites. The substitution of Al with Ti is inferred only from the smaller increase in the concentration of hydrogen-bonded OH-groups than predicted from the number of Al removed from the framework [682]. Again, partial insertion may be considered to explain the results. Treatment of HZSM-5 with $TiCl_4$ at 370°C caused complete loss of the acidity (IR band at 3605 cm^{-1}) and 90% of the terminal silanol groups [683].

Similarly, when NH_4Y was treated with a solution of $CrF_3 + 3NH_4F$ at 75–95°C, the Al content decreased, but the M^+/Al ratio remained essentially constant at 1.0 [684]. The reduced a_0, the higher frequency of the framework IR bands, and the increased concentration of OH-groups can be explained by dealumination alone, and the green color of the product suggests that hydroxylated chromium species are present possibly grafted (incompletely inserted) to silanol groups. The uniform distribution of Cr throughout the material shown by EDAX is in agreement with this explanation, as is the observation that ammonium oxalate solution removed about 80% of the chromium while the concentration of tetrahedral vacancies increased. The remaining 20% of the initially incorporated Cr may have been inserted completely into the framework structure, but no evidence is presented.

Others report that hydroxyl nests react with chromic acid to form grafted chromate species [685], which can then be reduced with CO to yield di- and trivalent chromium species anchored to the framework. Chromyl chloride reacts with silanol groups, as seen by a 66% decrease in the intensity of the IR band at 3740 cm^{-1}, and with Brønsted acid sites, revealed by a 30% reduction in intensity of the 3605 cm^{-1} IR band [683].

Isomorphous substitution of Be^{2+} into ZSM-5 by treatment with $(NH_4)_2BeF_4$ has been reported recently [686]. When NH_4Y is reacted with NH_4SnF_3, again the M^+/Al ratio remains essentially unchanged near unity, while Al is removed [684]. A minor decrease in the unit cell parameter and minor increases in the framework IR frequencies can be interpreted as being due to dealumination, as can the significant increase in the concentration of internal silanol groups. EDAX shows uniform distribution of tin throughout the yellow material and suggests that both hydroxylated cationic and grafted tin species may be present. Hydronium mordenite treated in the same manner yielded a white product containing 12% SnO [684]. The lack of color suggests that Sn has been inserted into the framework, although the M^+/Al ratio is only 0.77. However, since the starting material was the hydronium form, a significant number of cationic sites are occupied by H_3O^+, as shown by exchange with NaCl.

When HZSM-5 is treated with $VOCl_3$ vapor at 320–520°C, about half of the aluminum present was removed from the material, while a considerable amount of vanadium resistant to overnight contact with 1 M HCl at room temperature was incorporated. Subsequent treatment with H_2O_2 removed most of the vanadium, suggesting that it had been present as a grafted species. With increasing severity of $VOCl_3$ treatment, the IR band at 3605 cm^{-1} due to acidic bridging OH-groups associated with framework-Al gradually disappears and that at 3740 cm^{-1} assigned to isolated and surface silanol groups is reduced by about 85%. Contact with water at 320°C did not regenerate the non-acidic silanol groups, while about 60% of the acid sites were regenerated. The following reaction is proposed to occur with silanol groups:

$$n(\equiv SiOH) + VOCl_3 \rightarrow (\equiv SiO)_nV(=O)Cl_{3-n} + nHCl \tag{3.48}$$

$$(\equiv SiO)_nV(=O)Cl_{3-n} + (3\text{-}n)H_2O \rightarrow (\equiv SiO)_nV(=O)(OH)_{3-n} + (3\text{-}n)HCl \tag{3.49}$$

It is concluded that much of the vanadium is present as $(\equiv SiO)_3{\equiv}VO$ groups, and that these groups are resistant to acid hydrolysis of the Si–O–V linkages [683].

3.5 Other Modifications

3.5.1 Reactions of OH-Groups

Zeolites contain several types of silanol groups:

1. Acidic bridging hydroxyls
   a) internal: Provide shape-selective acid catalytic sites
   b) on the external surface: Non-selective sites
2. Internal silanol groups
   a) on crystallographic defects
   b) as hydroxyl nests with tetrahedral vacancies
3. External silanol groups terminate the zeolite lattice.

All zeolitic OH-groups can form hydrogen bonds with water, thus causing the zeolite to be hydrophilic. As the SiO_2/Al_2O_3 ratio of high-silica zeolites increases, the concentration of acidic OH-groups decreases, while the concentration of internal silanol groups at defect sites and tetrahedral vacancies increases, but the latter can be annealed by thermal treatment [413, 687] or steaming [413]. Surface acid sites cause non-shape-selective catalytic behavior leading to undesirable byproducts [688]. One obvious method to reduce the concentration of surface acid sites is a continuation of the lattice with aluminum-free ZSM-5, either high-silica or B-containing ZSM-5, as obtainable by seeding an Al-free reaction mixture with [Al]ZSM-5 and heating to crystallization temperature until a sufficiently thick layer has been obtained [689].

If the concentration of catalytically active sites is constant throughout the crystal, then larger crystals are more shape-selective than smaller crystallites [590] because of the smaller external surface area.

High concentrations of terminal OH-groups render the crystals of even high-silica ZSM-5 hydrophilic, which may significantly affect the selective sorption of organics from an aqueous phase if the crystals are very small. Any surface OH-groups located near the pore mouths can be modified to cause the pore apertures to become smaller (pore size engineering [691, 693]). Preferential removal of surface aluminum by acid leaching has been discussed previously (Sect. 3.3.2). In this section, modification of zeolitic OH groups with a variety of reagents will be mentioned, although more detailed discussions will be limited to reactions with Si and P compounds. Modification with fluorine compounds [694–697] and boranes [698–701] have been reported. Pyridine is used to distinguish Brønsted and Lewis acid sites by IR [702] and NMR [703] spectroscopy, quinoline and its derivatives are employed for selectively poisoning surface acid sites [704–709].

3.5.1.1
Reaction with Silanes

Hydroxyl groups of H-mordenite and NaHY react with SiH_4 in the range 30–210 °C [710] according to:

$$\equiv Si{-}OH + SiH_4 \rightarrow \equiv Si{-}O{-}SiH_3 + H_2 \tag{3.50}$$

A secondary reaction was observed in the same temperature range with zeolite Y, but not with mordenite:

$$\equiv Si{-}O{-}SiH_3 + HO{-}Si\equiv \rightarrow \equiv Si{-}O{-}SiH_2{-}O{-}Si\equiv + H_2 \tag{3.51}$$

or:

$$\equiv Si{-}O{-}SiH_3 + H_3Si{-}O{-}Si\equiv \rightarrow \equiv Si{-}O{-}SiH_2{-}SiH_2{-}O{-}Si\equiv + H_2 \tag{3.52}$$

The different behavior indicates that OH groups are located in closer vicinity in zeolite Y than in mordenite. Much less reaction was observed with NaY, suggesting that both acidic and non-acidic OH-groups are reactive. A further reaction occurs at 300–360 °C, possibly:

$$\equiv Si{-}O{-}SiH_2{-}O{-}Si\equiv + HO{-}Si\equiv \rightarrow (\equiv Si{-}O)_3{\equiv}SiH + H_2 \tag{3.53}$$

or:

$$\begin{array}{l}\equiv Si{-}O{-}SiH_2{-}SiH_2{-}O{-}Si\equiv + H_3Si{-}O{-}Si\equiv \\ \quad \rightarrow \equiv Si{-}O{-}SiH_2{-}\underset{\displaystyle \equiv Si{-}O{-}SiH_2}{\underset{|}{SiH}}{-}O{-}Si\equiv + H_2\end{array} \tag{3.54}$$

When treated with water vapor, new silanol groups are formed which are again able to react with SiH_4:

$$\equiv Si{-}O{-}SiH_3 + 3\,H_2O \rightarrow \equiv Si{-}O{-}Si(OH)_3 + 3\,H_2 \tag{3.55}$$

and

$$\equiv Si{-}O{-}SiH_2{-}O{-}Si\equiv + 2\,H_2O \rightarrow \equiv Si{-}O{-}Si(OH)_2{-}O{-}Si\equiv + 2\,H_2 \tag{3.56}$$

The sorption capacity of mordenite for N_2 decreased drastically as the percentage of added Si increased. Similar, but less dramatic, decreases in N_2 and O_2 sorption were found with silanated zeolite Y. Silanation of dealuminated H-mordenite and dealuminated Na-mordenite also influences strongly the capacity and the rate of uptake of O_2, N_2, and Ar [711]. Disilane, Si_2H_6, is chemisorbed at higher reaction temperature, preferentially at the channel entrances, thus causing pore blockage to some sorbates but no great loss in sorption capacity to others (pore size engineering [691–693].

Di- and trimethylsilane react quantitatively with the strong acid sites of HZSM-5 of Si/Al ratios of 20–50,000, while terminal silanol groups are inert [712]. Whereas $(CH_3)_2SiH_2$ reacts quantitatively with all strong acid sites at ambient temperature independent of crystallite size, sorption of $(CH_3)_3SiH$ is very rapid on high-silica HZSM-5, but slower at low Si/Al ratios, probably because of pore mouth blocking by the chemisorbed $(CH_3)_3SiH$ molecules. The newly formed Si–O–Si bonds are easily hydrolyzed. When trimethylsilane is equilibrated with HY at ambient temperature, the product shows, in addition to the 2125 cm^{-1} band of the silane, a vibration at 2050 and a weaker band at 2425 cm^{-1}, while the high-frequency hydroxyl band at 3640 cm^{-1} is reduced to a shoulder. A broad absorbance between 3500 and 3200 cm^{-1} indicates a hydroxyl bridge with the adsorbed silane. This hydroxyl bridge is believed to cause the ≡Si–H vibration at 2050 cm^{-1} by forming a trimethylsilanium ion:

$$(CH_3)_3SiH\text{-}H\text{-}O\text{-}Z \rightarrow (CH_3)_3SiOH_2^{+}\text{-}O^{-}\text{-}Z \quad (3.57)$$

The process is almost completely reversible at ambient temperature upon pumping off the silane from the zeolite. If the adsorption is carried out at 130°C, the 2050 cm^{-1} band is not seen, indicating that no free hydroxyl groups are left on the zeolite for H-bonding. This reaction is not reversible upon evacuation. Reaction at 200°C generates two products identified by their ^{29}Si MAS NMR signals at about +41 and –2 ppm. The latter is due to $(CH_3)_4Si$, while the former, representing the major product, is tentatively assigned to an undefined adduct of a silane to a zeolitic Lewis acid site. The results indicate that trimethylsilane does not form oxygen-substituted siloxanes with oxygen bridges anchored to the framework of zeolite HY [713], in contrast to the conclusions reached by von Ballmoos and Kerr [712] in their work on HZSM-5.

Zeolitic Si–OH groups of zeolite HY react above 225°C with ≡Si–Cl and ≡Si–R groups of physisorbed $(CH_3)_3SiCl$ [714] with elimination of HCl and CH_4. Such reaction generates additional Si–O–Si bridges in the supercages causing a slight contraction of the lattice. Similar bridges are formed in the reaction with $(CH_3)_2SiCl_2$, but only HCl, not CH_4, is eliminated. Trimethylchlorosilane reacts with terminal silanol groups as well as hydroxyl nests in ZSM-5 [715] and with external and internal silanol groups in zeolite Y, dealuminated with $SiCl_4$ [716]. Chloro groups of $(CH_3)_3SiCl$ are rapidly hydrolyzed by the surface hydroxyl groups, and the resulting products follow the reaction pattern of the alkoxysilanes described below [713]. When surface acid sites of HZSM-5 are blocked by reaction with alkylchlorosilane [717], 81% of the *para*-isomer was obtained in the xylene product, whereas only 24% of *p*-xylene are present at equilibrium, i.e., produced with a non-shape-selective catalyst.

Surface acid sites in HZSM-5 are passivated by reaction with a solution of di(trimethylsilyl) amine in n-hexane and decomposition of the silicon compound to deposit SiO_2 on the external surface [718]. The modification does not affect the n-hexane sorption significantly (10.9 vs. 11.3% at $p/p_o = 0.5$). Due to the loss of catalytic surface sites, the shape-selectivity of the catalyst is greatly improved, as demonstrated in the dewaxing of bright stock raffinate. The catalyst has similar activity as untreated ZSM-5, but gives higher yield of dewaxed oil with a pour point of 267 K and a higher viscosity index. Compared with large-size ZSM-5, the silane-treated catalyst yields less cracked product due to the reduction of secondary cracking because of the shorter diffusion path.

Alkoxy groups of $(CH_3)_3SiOCH_3$ and $CH_3Si(OCH_3)_3$ are readily lost by reaction with zeolitic protons to form siloxane bridges to the zeolite framework

[713]. Trimethylsilyl groups bonded to the zeolite surface are observed at a chemical shift of 10 ppm in the ^{29}Si NMR spectrum. Alkyl groups are split off from the silane at elevated temperatures, and products of condensation with the zeolite framework, $(CH_3)_2Si(OCH_3)OSi_z\equiv$ and $(CH_3)_2Si(OSi_z\equiv)_2$, are generated. The IR spectrum reveals that the high-frequency supercage hydroxyls are almost entirely consumed by the reaction with silanes, while the low-frequency sodalite cage hydroxyl groups are essentially unaffected.

Depositing tetramethyl orthosilicate (tetramethoxysilane), $Si(OCH_3)_4$, from the vapor phase onto H-mordenite reduces the size of the pore aperture, but leaves the internal acid sites unreacted [719]. While the water sorption capacity remains unchanged, the *o*-xylene sorption capacity was reduced by 92% for a sample containing 2.7 wt% deposited SiO_2. The $Si(OCH_3)_4$ molecule is larger than the pore size of H-mordenite and does, therefore, not enter the pores, but is deposited irreversibly at 320 °C by reaction with surface OH. Calcination in oxygen at 400 °C removes the carbonaceous deposits and produces SiO_2-coated H-mordenite [720]. The same procedure produces silica-coated HZSM-5 [721, 722].

Prior to depositing tetraethyl orthosilicate (TEOS), HZSM-5 is advantageously evacuated at 380 °C to 10^{-3} Pa. The zeolite is then contacted with TEOS vapor at 320 °C; the quantity of TEOS can be varied by increasing the contact time. In order to remove the organic residue of TEOS, the sample is calcined at 550 °C for 16 h. The product had improved selectivity for *p*-xylene in the conversion of methanol and ethanol to hydrocarbons, although at severely reduced yield [723]. Tetraalkyl orthosilicate has been deposited on HZSM-5 by exposing the zeolite at 180–230 °C to a vaporized blend of 50% toluene, 45% methanol, and 5% $(RO)_4Si$ [704, 724]. Compared with $(CH_3O)_4Si$ and $(C_4H_9O)_4Si$, $(C_2H_5O)_4Si$ was found to yield a superior catalyst, both in activity and selectivity. The water generated by in situ dehydration of methanol partially hydrolyzes the tetraalkyl orthosilicate deposited on the active sites thus resulting in a more uniform deposition throughout the catalyst bed. The catalyst showed high para-selectivity in ethylbenzene disproportionation and toluene alkylation with ethene. High para-selectivities were also found in the alkylation of toluene with methanol [704, 725], toluene disproportionation [704, 725], xylene isomerization [725], toluene ethylation [704], ethylbenzene disproportionation [704], and ethylation of ethylbenzene [704].

An experimentally simpler method of depositing tetraethyl orthosilicate on the surface of HZSM-5 is to suspend the zeolite in a solution of the calculated quantity of $(C_2H_5O)_4Si$ in n-hexane to achieve the desired coverage [726].

3.5.1.2
Reaction with Phosphines

Phosphines react with accessible acid sites. Trimethylphosphine was used as a probe molecule for acid sites in calcined zeolite HY prior to IR and NMR measurements [727, 728]. The compound (kinetic diameter of ~5.5 Å) is sufficiently small to enter the supercages of zeolite Y and generate $[(CH_3)_3P\text{-}H]^+$ by reaction with a Brønsted proton. The phosphonium ion gives rise to a ^{31}P MAS NMR signal at –3 ppm, relative to 85% H_3PO_4, and an IR band at 2485 cm^{-1}. Modification of ZSM-5 with trimethylphosphine may be carried out by the following procedure [729]. HZSM-5 is calcined at 600°C in vacuo (10^{-3} mbar) for 1 h. Then 9 mbar of trimethylphosphine is admitted into the reactor for 30 min, followed by evacuation at ambient temperature for 15 min. This procedure is repeated up to three times before the sample is calcined at 300 °C in air. The entire treatment may be repeated several times until no strong acid sites can be detected [730, 731]. Trimethylphosphine is adsorbed selectively on Si–O(H)–Al and terminal OH-groups in HZSM-5 [729]. The chemisorption complex on Brønsted acid sites is stable in air to high temperatures, the stability increases significantly upon calcining at 300°C, and it is impossible to oxidize the methyl groups without also losing phosphorus. Chemisorbed trimethylphosphine blocks all strong Brønsted acid sites up to reaction temperatures of 600°C. Above this temperature, it desorbed regenerating the original acid properties of HZSM-5. The increase in stability at 400°C is believed to be due to cleavage of one of the methyl groups of the adsorbed trimethylphosphine. Others [732, 733] report that trimethylphosphine oxide is formed, as concluded from NMR spectroscopy.

At ambient temperature, HY (5.0) is saturated with 5 molecules of trimethylphosphine/supercage; a sharp resonance at –69 ppm is assigned to physisorbed trimethylphosphine (2.0/s.c.), a broader signal at –3.4 ppm, proton-decoupled, represents 3.0 trimethylphosphonium ions per supercage. IR data indicate that the number of supercage protons is somewhat smaller because the 3550 cm^{-1} band is also reduced in intensity, and protonation was complete after dosing in 2.8 trimethylphosphonium ions/s.c. [734].

Similarly, dimethylphosphine yields the corresponding phosphonium ion, $[(CH_3)_2H_2]^+$. Due to the weaker base, the P NMR shift is –56 ppm, and the IR spectrum exhibits a strong band at 2289 cm^{-1} (P-H), while the intensity of the 3650 cm^{-1} band is reduced to 0. New bands due to symmetric and asymmetric stretching of P–H in zeolitic dimethylphosphonium ion appear at 2495 and 2450 cm^{-1}. The reaction is reversible by degassing [734].

Triphenylphosphine, dissolved in methylene chloride, carbon tetrachloride, or benzene, can be impregnated on HZSM-5 [735], or it can be adsorbed from the gas phase. IR spectra of adsorbed pyridine indicate interaction of the phosphine with Brønsted acid sites. The triphenylphosphine molecule is too large to penetrate the apertures of ZSM-5. However, C/P ratios determined from ESCA amount to only two phenyl groups per P and, after the surface Brønsted acid sites have reacted with triphenylphosphine, it seems that diphenylphosphine migrates into the zeolite, when the adsorption is performed via the gas phase at 400°C [736].

Oxidative calcination of zeolites containing phosphines, including chlorophenylphosphines, was reported to yield phosphoric acid derivatives of the zeolite [737]. However, this view has been challenged more recently by IR and NMR data [732, 733].

3.5.2
Reaction with Oxoacids

Certain oxoacids seem to enhance the para-selectivity of ZSM-5, in, e.g., the isomerization of xylenes, the alkylation of toluene with methanol or ethanol, the disproportionation of toluene and ethylbenzene. Unfortunately, most catalytic data published are unconvincing, the comparisons presented are frequently hard to interpret, as the catalytic reactions are carried out at conditions that give significantly different conversions. The higher selectivity is generally attained at the price of severely reduced activity. While this effect, in itself, is not necessarily unacceptable, a comparison with an unmodified ZSM-5 of the same low activity, would be most desirable. At the very least, comparisons made at the same conversion may suffice.

While a considerable number of papers have been published on zeolite modification with boric acid and other boron derivatives [698–701, 707, 735, 738–740], the following discussion is restricted to modification with oxyacids of phosphorus, which has become important commercially.

3.5.2.1
Reaction with Derivatives of Phosphorous Acid

Due to formation of hydrogen halide in the reaction of HY with PCl_3 or PBr_3, the zeolite framework is damaged, and no uniform product is obtained [698].

Trimethylphosphite reacts with zeolitic Brønsted acid sites; the reaction was speculatively formulated

```
                                              H3C– O   O– CH3
                                                    \ /
        H+                                           P
        :                                            |
   O    O    O                            O          O  O
    \  /  \ /                              \          \ /
     Al-   Si   + (CH3O)3P   →              Al-        Si  + CH3OH        (3.58)
    / \   / \                              / \        / \
   O  O  O  O                             O   O      O  O
```

Upon calcination at 500–550 °C, the P is oxidized to phosphoric acid, which is fixed on the zeolite. The para-selectivity of ZSM-5 in toluene alkylation with methanol was greatly improved at comparable conversion. The modification made it possible to produce C_2–C_4 olefins as major products from methanol (60–75% selectivity at 80–100% conversion), especially when the catalyst was calcined in the range 400–600 °C [741]. Temperature-programmed ammonia desorption [742] indicated that the number of acid sites decreased significantly and that the newly generated sites were weaker than the original ones.

The conversion of n-hexane and 1,3,5-trimethylbenzene has been utilized to probe the overall and surface activities, respectively. The P-modified HZSM-5 catalyst showed a decrease by a factor of ~3 in the conversion of n-hexane and of ~7 in that of trimethylbenzene, indicating a greater deactivation of the external surface than of the internal sites [743]. The results support the proposal that phosphorus poisons primarily acid sites at the pore mouths, based on microcalorimetry [744, 745] and XPS [745] results.

Trimethylphosphite, applied in a catalytic reactor at 360 °C in the absence of air, followed by outgassing of the ZSM-5 catalyst for 20 min, yielded a catalyst with greatly increased para-selectivity in the alkylation of toluene with methanol, although at reduced activity. IR spectroscopy showed that trimethylphosphite reacts with both surface and internal OH groups. Outgassing at 500 °C caused the bands due to methyl groups to disappear, but hydroxyl groups did not reappear suggesting that either desorption of the P species is accompanied by dehydroxylation or that only methyl groups are removed while the phosphorus remains in place [705].

Zeolite HY (5.0) saturated with trimethylphosphite by adsorption from the vapor phase was found to contain 3.6 molecules of trimethylphosphite/supercage. The ^{31}P MAS NMR and ^{13}C NMR spectra can be interpreted as due to the well-known Arbusov rearrangement of trimethylphosphite yielding dimethylhydrogenphosphite and tetracoordinated dimethylmethylphosphonate adsorbed in the zeolite [734].

A method described by Vedrine et al. [745] calls for impregnation of NH_4ZSM-5, after calcination at 230 °C in flowing nitrogen, with a trimethylphosphite solution in n-octane at 120 °C for 16 h. The catalyst is then filtered, washed with n-pentane, methylene chloride, and n-pentane, dried in air, then under vacuum at 110 °C overnight, and calcined at 540 °C for 3 h in flowing air. A similar method is described by Jentys et al. [730] and Vinek et al. [731].

The selectivity for *p*-cresol in competitive sorption of cresol isomers on HZSM-5 was greatly improved by modification with trimethylphosphite dissolved in toluene, followed by drying at 110 °C and calcination in air at 550 °C [746].

A selectivity to *p*-xylene of 84–90% was reported when ZSM-5 impregated with a solution of diphenylphosphinous acid in benzene, methylene chloride, or carbon tetrachloride, followed by calcination in air, was used as catalyst in the alkylation of toluene with methanol [735].

3.5.2.2 Reaction with Phosphoric Acid

Keading et al. [735] found unexpectedly that impregnation of HZSM-5 with aqueous solutions of phosphoric acid or diammonium phosphate, followed by drying and calcination, greatly improved the shape-selectivity of the zeolite. Up to 97% of *p*-xylene in the xylene fraction was obtained in the alkylation of toluene with methanol. It was proposed that phosphoric acid reacts with the zeolitic acid sites to firmly attach phosphorus to the zeolite via framework oxygen [741, 747]. The phosphorus was uniformly distributed over the acid sites associated with framework aluminum. The high concentration of *p*-xylene is seen as evidence that a reduction in the effective pore size occurred in the zeolite channels, in the channel intersections, and/or in the pore apertures. Such a modification would favor alkylation in the para-position and the outward diffusion of the para-isomer relative to the meta- and ortho-isomers. The residence time of the ortho- and meta-isomers within the pores would be increased, and isomerization to the more mobile para-isomer would be favored, as *p*-xylene is continuously removed from the equilibrium in the zeolite channels. Simultaneously, formation of mesitylene as an undesirable contaminant is eliminated, and formation of pseudocumene is reduced.

A particularly shape-selective catalyst was obtained by modification with both phosphate and magnesium. The alumina-bound and extruded large-size ZSM-5 (1–2 μm) was steamed for 1 h at 600 °C, impregnated with an aqueous solution of diammonium hydrogen phosphate, dried, calcined at 500 °C, then impregnated with an aqueous solution of magnesium acetate, dried, and calcined at 500 °C for 19 h. The catalyst containing 4.93% Mg and 3.48% P gave >99% *p*-diethylbenzene in the ethylbenzene product of ethylbenzene disproportionation and ethylbenzene alkylation with ethylene [748].

A complete picture of P-modified zeolite has not emerged yet. The following discussion is based on the evidence available and is not entirely consistent.

The selectivity of HZSM-5 for C_2–C_4 olefins in the conversion of methanol to olefins increases with the SiO_2/Al_2O_3 ratio, i.e., with decreasing concentration of strong Brønsted acid sites, indicating that the strong acid sites catalyze the oligomerization of the olefins, while the initial dehydration of methanol does not require such strong acidity or such a high concentration of strong acid sites. At a SiO_2/Al_2O_3 ratio of 960, the dehydration of dimethyl ether begins to decrease, but the selectivity towards low olefins is still increasing. The concentration of strong acid sites (TPAD peak near 400 °C), decreases and is eventually eliminated as HZSM-5 is impregnated with increasing quantities of phosphoric acid and calcined. A TPAD peak near 180 °C diminishes simultaneously, but is not completely removed [740, 749]. The 3620 cm^{-1} band in the IR spectrum disappears, and the bands at 3680 and 3740 cm^{-1} are reduced in intensity. Since the strong acid sites are located at the channel intersections, it is concluded that this is where the phosphorus is located. Impregnation of HZSM-5 with phosphate increases the hydrophilicity of the zeolite, which is reflected in higher water sorptive capacity, while the cyclohexane sorption is unchanged suggesting that the crystallinity is retained [750]. Hydrocarbon formation was reduced to nearly zero when the ZSM-5 contained >5% H_3PO_4. This result points to decreased acid strength rather than diffusional constraints as cause for reduced activity with increasing P content because the energy of activation for n-hexane cracking increases strongly with rising H_3PO_4 content [751]. In contrast, the activation energy for the alkylation of toluene with methanol is nearly constant [731].

While no loss in crystallinity by XRD is observed up to at least 8% P, the BET surface area decreases with increasing P content [752]. Above 5% P, platelets of unknown composition appear on the surface of the zeolite particles. Pyridine adsorbs on both external and internal acid sites of ZSM-5 after deposition of phosphoric acid. After pyridine sorption at 1.8 kPa, an IR band at 1438 cm^{-1} indicates the most abundant surface species, weakly bound pyridine. The band is removed upon outgassing at 25 °C, while the intensity of the band at 1540 cm^{-1}

increases, and a new band is observed at 1448 cm^{-1} (pyridine on Lewis acid site), which decreases with increasing P content. The authors propose a formula in which the bridging OH is replaced by $O_2P(OH)_2$, thus substituting two weak acid sites for a strong one [730], in agreement with Vedrine et al. [745]. No Al–OH species have been found after treatment with phosphoric acid. $AlPO_4$ species are intracrystalline, at least at low P loading. The platelets observed at higher P concentration may represent similar species on the external surface. IR spectroscopy indicates that, while the concentration of strong Brønsted acid sites decreases, that of weak Brønsted acid sites increases with the severity of the treatment. The conclusions are supported by NMR investigations which show that phosphoric acid enters the channels of ZSM-5. As the quantity of phosphoric acid increases, P species accumulate near the external surface, thus generating a transfer barrier to diffusing molecules. Calcination causes formation of polyphosphate, as evidenced by the appearance of strong signals at −40 and −46 ppm in the ^{31}P NMR spectrum [753]. A drastic reduction of Si–OH groups was observed simultaneously by ^{1}H NMR [754].

Seo and Ryoo [755] impregnated ZSM-5 with $(NH_4)_2HPO_4$ by the incipient wetness method, followed by calcination at 550 °C in oxygen, and found no loss in crystallinity, as determined by XRD. TPAD indicates a reduction in the concentration of acid sites up to P/Al = 0.9, in agreement with the change of ~50% of the Al from tetrahedral to octahedral coordination, as shown by ^{27}Al NMR, and with the decrease in the propene sorption capacity, while the propane sorption remains nearly constant. The ^{31}P NMR signals observed at −6 and −12 ppm are attributed to the product obtained by reaction of ~50% of the phosphorus with different Al sites (Al and P NMR results of these authors have been questioned by Lischke et al. [756]). Signals with these shifts are not known to occur in aluminum phosphates of various topologies [757]. The ratio of reacted P and Al species is about 1, and the following reaction is proposed [755] (at variance with the results of Jentys et al. [730]:

$$H_2PO_4^- + HOAl_{fr} \rightarrow {}^-H_2PO_4\overset{\displaystyle OH}{\overset{|}{Al}}_{fr} \qquad (3.59)$$

The selectivity in the methanol conversion to gasoline increased with the P loading up to 0.9 P/Al.

When ethylbenzene was alkylated with ethanol over HZSM-5 at 400 °C, the primary product was exclusively *p*-diethylbenzene (transition state selectivity). HZSM-5 modified with phosphoric acid had much higher para-selectivity than the parent zeolite, and it was concluded that the absence of strong acid sites in the modified zeolite is responsible for the suppression of *p*-diethylbenzene isomerization, whereas the reduction of the pore dimensions is of secondary importance [739]. However, poisoning of the surface acid sites with 2,4-dimethylquinoline, which is unable to enter the channels of ZSM-5, causes the same increase in para-selectivity. The relative adsorption rate of *o*-xylene, expressed as the ratio of the *m*-xylene adsorption capacity at 180 min to the *p*-xylene sorption capacity at full loading, decreases upon modification of the zeolite with phosphorus [707]. Therefore, indications are that the poisoning of surface acid sites by reaction with phosphoric acid in combination with the reduction of the acid strength of the internal acid sites, and/or the narrowing of the pores, cause the high para-selectivity.

The most extensive characterization of the products obtained by modification of HZSM-5 with phosphoric acid was reported by Lischke et al. [756]. Using a combination of TPAD, IR, ^{27}Al MAS NMR, and ^{31}P MAS NMR, these authors came to the following conclusions. Simple impregnation with H_3PO_4 causes the number of Brønsted acid sites to decrease. The Brønsted acidity can be restored completely by hot water wash, which removes the phosphoric acid. The result indicates that no dealumination occurred by impregnation. Calcining or steaming the H_3PO_4-impregnated sample causes an irreversible decrease of Brønsted acidity as a result of dealumination. Simultaneously, $AlPO_4$ species are formed. However, thermal treatment of H_3PO_4-impregnated HZSM-5 causes less dealumination than the same treatment of unimpregnated HZSM-5 indicating that phosphorus partially protects Al from being removed from the framework. In particular, they made the following observations.

- TPAD: The high-temperature peak is reduced to a shoulder upon impregnation with H_3PO_4; hot water wash restores the sites. Hydrothermal treatment reduces the concentration of strong Brønsted acid sites, but part of these sites can be recovered by hot-water wash; the degree of restoration decreases with increasing severity of the hydrothermal treatment. A peak centered in the range of 250 – 280 °C, observed after ammonium exchange, is interpreted as being due to surface-bonded species of phosphoric acid.

- IR: Impregnation with H_3PO_4 causes the 3610 cm^{-1} band to decrease in intensity and broaden. Hydrothermal treatment leaves only very broad bands in the region between 3450 and 3650 cm^{-1}. Elution of P restores some intensity at 3610 cm^{-1}. A shoulder at 3665 cm^{-1} indicates the presence of non-framework Al. Pyridine reacts with both Brønsted (1546 cm^{-1}) and Lewis (1450 cm^{-1}) acid sites causing both bands to decrease in intensity. The intensity of the 1450 cm^{-1} band increases upon elution of H_3PO_4 to values above that of the dried sample.

- ^{27}Al MAS NMR: Al_{Td} decreases upon impregnation with H_3PO_4, and weak signals appear at – 11 and + 39.5 ppm, characteristic of Al_{Oh} and Al_{Td} of aluminum phosphates, which may be present in an amorphous state. Non-framework Al_{Td} of P-free samples also gives a broad signal at ~ 40 ppm, so that superposition of Al_{Td} from different sources cannot be excluded. Washing the H_3PO_4-impregnated sample with hot water, with or without hydrothermal treatment, restores part of the lost intensity of the framework Al_{Td} signal.

- ^{31}P MAS NMR: The dealuminated, impregnated, and dried material shows signals at –5 and –13 ppm, probably due to the end and middle groups of pyrophosphate or other short-chain polyphosphates, respectively. Both lines disappear when the sample is washed with water. The remaining broad line at – 28 ppm has a shift characteristic of $AlPO_4$s, which have probably been generated by reaction of non-framework Al with H_3PO_4. Hydrothermally treated samples display a dominant ^{31}P NMR signal at ca. – 30 ppm, which is also assigned to

Al-phosphate. The relative intensity of the broad signal at -12 ppm decreases with increasing temperature of the hydrothermal treatment, while that of the -40 ppm signal increases, and an additional line at ca. -46 ppm appears after treatment at 700 °C. Both high-field lines, at -40 and -46 ppm, are assigned to highly condensed polyphosphate species. All phosphate species are partially extractable with hot water.

No evidence was found for direct modification of acid centers, as proposed by Young et al. [737] and Vinek et al. [731]. Si-O-P bonds are unlikely to occur; silicon pyrophosphate contains only octahedral Si, but no Si_{Oh} is seen in the ^{29}Si NMR spectrum.

References

1. Breck DW (1974) Zeolite Molecular Sieves, John Wiley & Sons, Inc, a) 537; b) 532; c) 534; d) 176
2. Barrer RM, Walker AJ (1964) Trans Faraday Soc 60:171
3. Townsend RP (1986) In: Murakami Y, Iijima A, Ward JW (eds) a) New Developments in Zeolite Science and Technology, Proc 7th Int Zeolite Conf, Tokyo, Japan, Stud Surf Sci Catal 28:273; b) Pure Appl Chem 58:1359
4. Gaines GL, Thomas HC (1953) J Chem Phys 21:714
5. Rees LVC (1977) Chemistry and Industry, 647
6. Dyer A, Enamy H, Townsend RP (1981) Separ Sci Technol 16(2):173
7. Barrer RM, Meier WM (1959) Trans Faraday Soc 55:130
8. Barrer RM, Rees LVC, Ward DJ (1963) Proc Roy Soc A 273:180
9. Barrer RM, Klinowski J (1972) Trans Faraday Soc 68:73
10. Sherry HS, Walton HF (1967) J Phys Chem 71:1457
11. Ames LL, Jr (1964) Amer Mineral 49:1099
12. Barri SAJ, Rees LVC (1980) J Chromatogr 201:21
13. Moore EW, Ross JW (1965) J Appl Physiolog 20:1332
14. Dyer A, Gettins RB, Brown JG (1970) J Inorg Nucl Chem 32:2389
15. Barrer RM, Townsend RP (1976) J Chem Soc, Faraday Trans 1 72:661
16. Sherry HS (1968) J Coll Interf Sci 28:288
17. Kühl GH, Sherry HS (1980) Proc 5th Int Conf Zeolites, Rees LV (ed) Heyden and Son, Ltd, p 813
18. Olson DH, Sherry HS (1968) J Phys Chem 72:4095
19. Sherry HS (1993) Zeolites 13:377
20. Drummond D, De Jonge A, Rees LVC (1983) J Phys Chem 87:1967
21. Danes F, Wolf F (1972) Z phys Chem [Leipzig] 251:329
22. Stamires DN (1962) J Phys Chem 36:3174
23. Brown LM, Sherry HS, Krambeck FJ (1971) J Phys Chem 75:3846
24. Brown LM, Sherry HS (1971) J Phys Chem 75:3855
25. Dyer H, Enamy H (1981) Zeolites 1:7
26. Dyer A, Fawcett JM (1966) J Inorg Nucl Chem 28:615
27. Dyer A, Gettins RB, Molineux A (1968) J Inorg Nucl Chem 30:2823
28. Dyer A, Gettins RB (1970) J Inorg Nucl Chem 32:319
29. Dyer A, Enamy H (1984) Zeolites 4:319
30. Fletcher P, Townsend RP (1981) J Chem Soc, Faraday Trans 2 77:955
31. Fletcher P, Townsend RP (1981) J Chem Soc, Faraday Trans 2 77:965
32. Fletcher P, Townsend RP (1981) J Chem Soc, Faraday Trans 2 77:2077
33. Townsend RP, Fletcher P, Loizidou M (1984) Proc 6th Internat Zeolite Conf, Olson and Bisio (eds) Butterworths, 110

34. Townsend RP (1986) Pure Appl Chem 58:1359
35. Franklin KR, Townsend RP (1985) J Chem Soc, Faraday Trans 1 81:1071
36. Franklin KR, Townsend RP (1985) J Chem Soc, Faraday Trans 1 81:3127
37. Franklin KR, Townsend RP (1988) Zeolites 8:367
38. Breck DW, Flanigen EM (1968) Molecular Sieves, Soc Chem Ind, London, p 47
39. Dempsey E, Kühl GH, Olson DH (1969) J Phys Chem 73:387
40. Kühl GH (1985) Zeolites 5:4
41. Loewenstein W (1942) Amer Mineral 39:92
42. Broussard L, Shoemaker DP (1960) J Am Chem Soc 82:1041
43. Olson DH (1970) J Phys Chem 74:2758
44. Baur WH (1964) Amer Mineral 49:697
45. Barrer RM, Davies JA, Rees LVC (1968) J Inorg Nucl Chem 30:3333
46. Sherry HS (1966) J Phys Chem 70:1158
47. Barrer RM, Rees LVC, Shamsuzzoha M (1966) J Inorg Nucl Chem 28:629
48. Barrer RM, Davies JA, Rees LVA (1969) J Inorg Nucl Chem 31:2599
49. Lai PP, Rees LVC (1976) J Chem Soc, Faraday Trans 1 1809
50. De Boer JJ, Maxwell IF (1974) J Phys Chem 78:2395
51. Sherry HS (1971) Molecular Sieve Zeolites-I, Gould RF (ed) Adv Chem Ser 101:350
52. Townsend RP, Franklin KR, O'Connor JF (1984) Adsorpt Sci Technol 1:269
53. O'Connor JF, Townsend RP (1985) Zeolites 5:158
54. Franklin KR, Townsend RP, Whelan SJ, Adams CJ (1986) In: Murakami Y, Iijima A, Ward JW (eds) New Developments in Zeolite Science and Technology, Proc 7th Int Zeolite Conf, Tokyo, Japan, Stud Surf Sci Catal 28:289
55. Herman RG, Bulko JB (1980) ACS Symp Ser 135:177
56. Maesen TLM, van Bekkum H, Verburg TG, Kolar ZI, Kouwenhoven HW (1991) J Chem Soc, Faraday Trans 87:787
57. Chu P, Dwyer FG (1980) J Catal 61:454
58. Barrer RM, Buser W, Grütter WF (1956) Helv Chim Acta 39:518
59. Theng BKG, Vansant E, Uytterhoeven JB (1968) Trans Faraday Soc 64:3370
60. Vansant EF, Uytterhoeven JB (1971) Molecular Sieves – I. Gould RF (ed) Adv Chem Ser 101:426
61. Wolf F, Ceacareanu D, Pilchowski K (1973) Z Phys Chem (Leipzig) 252:50
62. Kühl GH, Sherry HS (1980) Proc 5th Internat Conf Zeolites. Rees LV (ed) Heyden & Son, Ltd, p 813
63. Franklin KR, Townsend RP (1988) J Chem Soc, Faraday Trans 1 84:2755
64. Sherry HS (1980) J Phys Chem 72:4086
65. Helfferich F (1963) Ion Exchange, McGraw-Hill Book Co Inc, New York
66. Ames LL, Jr (1965) J Inorg Nucl Chem 27:885
67. Olson DH, Sherry HS (1968) J Phys Chem 72:4095
68. Sherry HS (1967) J Phys Chem 71:780
69. Maes A, Cremers A (1973) Molecular Sieves, Meier WM, Uytterhoeven JB (eds) Adv Chem Ser 121:230
70. Steinberg KH, Bremer H, Hofmann F (1974) Z anorg allg Chem 407:162
71. Fletcher P, Townsend RP (1980) Proc 5th Internat Conf Zeolites. Rees LVC (ed) Heyden & Son, Ltd, p 311
72. Lutz W, Fichtner-Schmittler H, Bülow M, Schierhorn E, Van Phat N, Sonntag E, Kosche I, Amin S, Dyer A (1990) J Chem Soc, Faraday Trans 86:1899
73. Hocevar S, Drzaj B (1980) Proc 5th Internat Conf Zeolites, LVC Rees (ed) Heyden & Son, Ltd, p 301
74. Maes A, Cremers A (1975) J Chem Soc, Faraday Trans 1 71:265
75. Gallei E, Eisenbach D, Ahmed A (1974) J Catal 33:62
76. Gal IJ, Radovanov P (1975) J Chem Soc, Faraday Trans 1 71:1671
77. Gallezot P, Imelik B (1973) J Phys Chem 77:652
78. Woolery G, Kühl G, Chester A, Bein T, Stucky G, Sayers DE (1986) J de Physique, Colloque C8, suppl no 12, 47:281

79. Barrer RM, Townsend RP (1976) J Chem Soc, Faraday Trans 1 72:2650
80. Fletcher P, Townsend RP (1983) Zeolites 3:129
81. Barrer RM, Townsend RP (1978) J Chem Soc, Faraday Trans 1 74:745
82. Fletcher P, Townsend RP (1980) J Chromatogr 201:93
83. Fletcher P, Townsend RP (1981) J Chem Soc, Faraday Trans 1 77:497
84. Kulkarni SJ, Kulkarni SB (1982) Thermochim Acta 56:93
85. Chevreau T, Cornet D, Leglise J, Roudias K (1990) Adsorp Sci Technol 6:119
86. Wang KM, Lunsford JH (1972) J Catal 34:262
87. Sherry HS (1968) J Coll Interf Sci 28:288
88. Reng HR, Chen YG (1980) Proc 5th Internat Conf Zeolites, LVC Rees (ed) Heyden & Son, Ltd, p 321
89. Sherry HS (1976) Proc 50th Internat Conf Coll Surf Sci 5:321
90. Marynen P, Maes A, Cremers A (1984) Zeolites 4:287
91. Chen SH, Chao KJ, Lee TY (1990) Ind Eng Chem 29:2020
92. Lee TY, Lu TS, Chen SH, Chao KJ (1990) Ind Eng Chem 29:2024
93. Chao KJ, Chern JY, Chen SH, Shy DS (1990) J Chem Soc, Faraday Trans 86:3167
94. Shy DS, Chen SH, Lievens J, Lin SB, Chai KJ (1991) J Chem Soc, Faraday Trans 87:2855
95. Ward JW (1969) J Catal 13:321
96. Caravajal R, Chu PJ, Lunsford JH (1990) J Catal 125:123
97. Moscou L, Lakeman M (1970) J Catal 16:173
98. Hunter FD, Scherzer J (1971) J Catal 20:246
99. Li CY, Rees LVC (1986) Zeolites 6:51
100. Jeanjean J, Aouali L, Delafosse D, Dereigne A (1991) Zeolites 11:360
101. Kokotailo GT, Lawton SL, Olson DH, Meier WM (1978) Nature 272:437
102. Zhenyi L, Wangjin Zh, Qin Y, Ganglie L, Wangrong L, Shuju W, Youshi Zh, Bingxiong L (1986) In: Murakami Y, Iijima A, Ward JW (eds) New Developments in Zeolite Science and Technology, Proc 7th Int Zeolite Conf, Stud Surf Sci Catal Tokyo, Japan, 28:415
103. Chu P, Dwyer FG (1983) Intrazeolite Chemistry, Stucky GD and Dwyer FG (eds) ACS Symp Ser 218:59
104. Matthews DP, Rees LVC (1985) Advances in Catalysis Science & Technology, Prasado Rao TSR (ed) John Wiley & Sons, p 493
105. Lin J-C, Chao K-J, Wang Y (1991) Zeolites 11:376
106. McAleer AM, Rees LVC, Nowak AK (1991) Zeolites 11:329
107. Ikeda T, Sasaki M, Yasunaga T (1984) J Coll Interf Sci 98:192
108. Chu P, Dwyer FG (1988) Zeolites 8:423
109. Sherry HS (1969) Ion exchange, Vol 2. Marinsky JA (ed) Edward Arnold Ltd, London, p 89
110. Eisenman G (1962) Biophys J 2:259
111. Kucherov AV, Slinkin AA, Kondrat'ev KhM, Bondarenko TN, Rubinshtein AM, Minachev KhM (1985) Kinet Katal 26:409; Engl Trans 353
112. Li R-S, Wen R-W, Zhang W-Y, Wei Q (1993) Zeolites 13:229
113. Fegan SG, Lowe BM (1984) J Chem Soc, Chem Commun 437
114. Chester AW, Chu YF, Dessau RM, Kerr GT, Kresge CT (1985) J Chem Soc, Chem Commun 289
115. Obermyer RT, Mulay LN, Lo C, Oskooie-Tabrizi M, Rao VUS (1982) J Appl Phys 53:268
116. Lin T-A, Schwartz LH, Butt JB (1986) J Catal 97:177
117. Gallezot P, Coudurier G, Primet M, Imelik B (1977) Molecular Sieves – II. Katzer JR (ed) ACS Symp Ser 40:144
118. Koranyi TI, van de Ven LJM, Welters WJJ, de Haan JW, de Beer VHJ, van Santen RA (1993) Catal Lett 105
119. Herron N (1986) Inorg Chem 25:4714
120. Ukisu Y, Kazusaka A, Nomura M (1991) J Mol Catal 70:165
121. Rossin JA, Davis ME (1986) J Chem Soc, Chem Commun 234
122. Weisz PB, Frilette VJ, Maatman RW, Mower EB (1962) J Catal 1:307
123. Ralek M, Jiru P, Grubner O, Beyer H (1962) Coll Czech Chem Comm 27:142

124. Kim Y, Seff K (1977) J Am Chem Soc 99:7055
125. Kim Y, Gilje JW, Seff K (1977) J Am Chem Soc 99:7057
126. Kim Y, Seff K (1978) J Am Chem Soc 100:6989
127. Jacobs PA, Uytterhoeven JB, Beyer HK (1979) J Chem Soc, Faraday Trans I 75:56
128. Gellens LR, Mortier WJ, Uytterhoeven JB (1981) Zeolites 1:11
129. Matar K, Goldfarb D (1992) J Phys Chem 96:3100
130. Naccache C, Ben Taarit Y (1978) Proc Symp Zeolites, Szeged, p 23
131. Slinkin AA, Loktev MI, Mishin IV, Plakhotnik VA, Klyachko AL, Rubinshtein AM (1979) Kinet Katal 20:181; Engl Transl 145
132. Zhang Z, Sachtler WMH (1990) J Mol Catal 67:349
133. Zhang Z, Cavalcanti FPA, Sachtler WMH (1992) Catal Lett 12:157
134. Zhang Z, Sachtler WMH (1990) J Chem Soc, Faraday Trans 86:2313
135. Beutel T, Zhang Z, Sachtler WMH, Knözinger H (1993) J Phys Chem 97:3579
136. Gandhi SN, Lei G-D, Sachtler WMH (1993) Catal Lett 17:117
137. Tauster SJ, Steger JJ (1990) J Catal 125:387
138. Larsen G, Haller GL (1992) Catal Today 15:431
139. Han W-J, Kooh AB, Hicks RF (1993) Catal Lett 18:193, 219
140. Beyer HK, Jacobs PA, Uytterhoeven JB (1979) J Chem Soc, Faraday Trans I 75:109
141. Goldwasser MR, Navas F, Perez-Zurita MJ, Cubeiro ML, Lujano E, Franco C, Gonzales Jimenez F, Jaimes E, Moronta D (1993) Appl Catal A100:85
142. Shpiro ES, Baeva GN, Sass AS, Shvets VA, Fasman AB, Kazanskii VB, Minachev KhM (1987) Kinet Katal 28:1432; Engl Transl 1236
143. Minachev KhM, Baeva GN, Shpiro ES, Antoshin GV, Fasman AB (1985) Kinet Katal 26:1265
144. Tauster SJ, Steger JJ (1988) Proc Mater Res Soc Symp 111:419
145. Alvarez WE, Resasco DE (1991) Catal Lett 8:53
146. Ostgard DJ, Kustov LM, Poeppelmeier KR, Sachtler WMH (1992) J Catal 133:342
147. Lerner BA, Carvill BT, Sachtler WMH (1992) J Mol Catal 77:99
148. Yashima T, Ushida Y, Ebisawa M, Hara N (1975) J Catal 36:320
149. Tvaruzkova Z, Bosacek V, Patzelova V (1979) React Kinet Catal Lett 11:71
150. Tvaruzkova Z, Bosacek V (1980) Coll Czech Chem Commun 45:2499
151. Naccache C, Ben Taarit Y (1973) J Chem Soc, Faraday Trans 1 69:1475
152. Atanasova VD, Shvets VA, Kazanskii VB (1977) Kinet Katal 18:1033; Engl Transl 848
153. Rubinstein AM, Slinkin AA, Loktev MI, Fedorovskaya EA, Bremer H, Vogt F (1976) Z anorg allg Chem 423:164
154. Kasai PH (1977) J Phys Chem 81:1527
155. Pearce JR, Sherwood DE, Hall MB, Lunsford JH (1980) J Phys Chem 84:3215
156. Wichterlova B, Tvaruzkova Z, Novakova J (1983) J Chem Soc, Faraday Trans 1 79:1573
157. Coughlan B, McCann WA, Carroll WM (1980) J Coll Interf Sci 74:136
158. Beran S, Jiru P, Wichterlova B (1983) J Chem Soc, Faraday Trans 1 79:1585
159. Tvaruzkova Z, Wichterlova B (1983) J Chem Soc, Faraday Trans 1 79:1591
160. Komatsu T, Miyoshi R, Namba S, Yashima T (1993) J Mol Catal 78:57
161. Özkar S, Ozin GA, Möller K, Bein T (1990) J Am Chem Soc 112:9575
162. Boulet H, Bremard C, Depecker C, Legrand P (1991) J Chem Soc, Chem Commun 1411
163. Okamoto Y, Imanaka T, Asakura K, Iwasawa Y (1991) J Phys Chem 95:3700
164. Pastore H, Ozin GA, Poe AJ (1993) J Am Chem Soc 115:1215
165. Cid R, Neira J, Godoy J, Palacios JM, Mendioroz S, Agudo AL (1993) J Catal 141:206
166. Özkar S, Ozin GA, Möller K, Bein T (1989) J Phys Chem 93:4205
167. Li Y, Armor JN (1993) Appl Catal B2:239
168. Yates DJC (1965) J Phys Chem 69:1676
169. Lawson JD, Rase HF (1970) Ind Eng Chem, Prod Res Dev 9:317
170. Abou-Kais A, Vedrine JC, Naccache C (1978) J Chem Soc, Faraday Trans 2 74:959
171. Narayana N, Kevan L (1982) J Chem Phys 76:3999
172. Schoonheydt RA (1989) J Phys Chem Solids 50:523
173. Schoonheydt RA (1993) Catal Rev - Sci Eng 35:129

174. Minachev KhM, Antoshin GV, Shpiro ES (1974) Izv Akad Nauk SSSR, Chem Ser, 23:1012; Engl Transl 972
175. Tsutsumi K, Takahashi H (1972) Bull Chem Soc Jpn 45:2332
176. Beyer H, Jacobs PA, Uytterhoeven JB (1976) J Chem Soc, Faraday Trans 1 72:674
177. Kellerman R, Texter J (1979) J Chem Phys 70:1562
178. Gellens LR, Mortier WJ, Uytterhoeven JB (1981) Zeolites 1:85
179. Handreck GP, Smith TD (1991) J Chem Soc, Faraday Trans 87:1025
180. Okamoto Y, Fukino K, Imanaka T, Teranishi S (1983) J Phys Chem 87:1983
181. Jirka I, Bosacek V (1991) Zeolites 11:77
182. Ward JW (1968) J Coll Interf Sci 28:269
183. Uytterhoeven JB, Schoonheydt R, Liengme BV, Hall WK (1969) J Catal 13:425
184. Schoonheydt RA, Vandamme LJ, Jacobs PA, Uytterhoeven JB (1976) J Catal 43:292
185. Jacobs PA, Beyer HK (1979) J Phys Chem 83:1174
186. Sass CE, Kevan L (1989) J Phys Chem 93:7856
187. Valyon J, Hall WK (1993) J Phys Chem 97:7054
188. Sarkamy J, d'Itri J, Sachtler WMH (1992) Catal Lett 16:241
189. Li Y, Hall WK (1990) J Phys Chem 94:6145
190. Iwamoto M, Maruyama K, Yamazoe, N, Seyama T (1976) J Chem Soc, Chem Commun 615
191. Sato S, Yu-u Y, Yahiro H, Mizuno N, Iwamoto M (1991) Appl Catal 70:L1
192. d'Itri J, Sachtler WMH (1992) Catal Lett 15:289
193. Petunchi JO, Hall WK (1993) Appl Catal B2:L17
194. Kharas KCC, Robota HJ, Liu DJ (1993) Appl Catal B2:225
195. Iwamoto M, Furukawa H, Mine Y, Uemura F, Mikuriya S, Kagawa S (1986) J Chem Soc, Chem Commun 1272
196. Iwamoto M, Furukawa H, Kagawa S (1986) New Developments in Zeolite Science and Technology, Proc 7th Int Zeolite Conf, Tokyo, Japan. Murakami Y, Iijima A, Ward JW (eds) Stud Surf Sci Catal 28:943
197. Iwamoto M, Yahiro H, Mine Y, Kagawa S (1989) Chem Lett 213
198. Iwamoto M, Yahiro H, Torikai Y, Yoshioka T, Mizuno N (1990) Chem Lett 1967
199. Iwamoto M, Yahiro H, Tanda K, Mizuno N, Mine Y, Kagawa S (1991) J Phys Chem 95:3727
200. Kagawa S, Ogawa H, Furukawa H, Teraoka Y (1991) Chem Lett 407
201. Sepulveda-Escribano A, Marquez-Alvarez C, Rodriguez-Ramos I, Guerrero-Ruiz A, Fierro JLG (1993) Catal Today 17:167
202. Jacobs PA, de Wilde W, Schoonheydt RA, Uytterhoeven JB (1976) J Chem Soc, Faraday Trans 1 72:1221
203. Howard J, Nicol JM (1988) Zeolites 8:142
204. Bosacek V, Drahoradova E, Jirak Z (1993) J Chem Soc, Faraday Trans 89:1833
205. Steinberg K-H, Bremer H, Hofmann F (1974) Z anorg allg Chem 407:173
206. Whyte TE Jr (1973) Catal Rev 8:117
207. Jacobs PA, Uytterhoeven JB, Beyer HK (1977) J Chem Soc, Faraday Trans 1 73:1755
208. Baba T, Akinaka N, Nomura M, Ono Y (1992) J Chem Soc, Chem Commun 339
209. Baba T, Akinaka N, Nomura M, Ono Y (1993) J Chem Soc, Faraday Trans 89:595
210. Riekert L (1969) Ber Bunsenges phys Chem 73:331
211. Jacobs PA, Linart J-P, Nijs H, Uytterhoeven JB, Beyer HK (1977) J Chem Soc, Faraday Trans 73:1745
212. Coughlan B, Keane MA (1990) J Chem Soc, Faraday Trans 86:1007
213. Herman RG, Lunsford JH, Beyer H, Jacobs PA, Uytterhoeven JB (1975) J Phys Chem 79:2388
214. Chao C-C, Lunsford JH (1972) J Chem Phys 57:2890
215. Petunchi JO, Marcelin G, Hall WK (1992) J Phys Chem 96:9967
216. Naccache CM, Ben Taarit Y (1971) J Catal 22:171
217. Beyer H, Jacobs PA, Uytterhoeven JB, Vandamme LJ (1977) Proc 6th Int Congr Catal 1:273

218. Jacobs PA, Tielen M, Linart J-P, Uytterhoeven JB, Beyer H (1976) J Chem Soc 72:2793
219. Pikus S (1991) Zeolites 11:449
220. Matsumoto H, Tanabe S (1990) J Phys Chem 94:4207
221. Barcicka A, Pikus S (1987) Zeolites 7:35
222. Moretti G, Sachtler WMH (1989) J Catal 15:205
223. Zhang Z, Xu L, Sachtler WMH (1991) J Catal 131:502
224. Gallezot P, Ben Taarit Y, Imelik B (1972) J Catal 26:295
225. Huang YY, Vansant EF (1973) J Phys Chem 77:663
226. Huang YY (1973) J Catal 30:187
227. Dai PSE, Lunsford JH (1980) Inorg Chem 19:262
228. Chao CC, Lunsford JH (1972) J Phys Chem 76:1546
229. Petunchi JO, Hall WK (1983) J Catal 80:403
230. Huang YY (1980) J Catal 61:461
231. Delgass WN, Garten RL, Boudart M (1969) J Phys Chem 73:2970
232. Petunchi JO, Hall WK (1982) J Catal 78:327
233. Fu C-M, Deeba M, Hall WK (1980) Ind Eng Chem Prod Res Dev 19:299
234. Schmidt F, Gunsser W, Adolph J (1977) Molecular Sieves - II, Katzer JR (ed) ACS Symp Ser 40:291
235. Gallezot P, Ben Taarit Y, Imelik B (1973) J Phys Chem 77:2556
236. Rabo JA, Angell CL, Kasai PH, Schomaker V (1966) Disc Faraday Soc 41:328
237. Garbowski E, Primet M, Mathieu MV (1977) Molecular Sieves - II, Katzer JR (ed) ACS Symp Ser 40:281
238. Vogt F, Bremer H, Hofmann F, Isakov JI, Minacev KM, Isakova TA (1985) Z anorg allg Chem 521:145
239. Coughlan B, Keane MA (1992) J Catal 136:170
240. Bager KH, Vogt F, Bremer H (1977) Molecular Sieves - II, Katzer JR (ed) ACS Symp Ser 40:528
241. Barth A, Hoffmann J (1985) Z anorg all Chem 521:207
242. Selenina M (1972) Z anorg allg Chem 387:179
243. Suzuki M, Tsutsumi K, Takahashi H, Saito Y (1989) Zeolites 9:98
244. Woolery G, Kühl G, Chester A, Bein T, Stucky G, Sayers DE (1986) J de Physique 47:C8-281
245. Gallezot P, Imelik B (1973) J Phys Chem 177:652
246. Coughlan B, Keane MA (1990) J Catal 123:364
247. Coughlan B, Keane MA (1991) Zeolites 11:2
248. Reman WG, Ali AH, Schuit G (1971) J Catal 20:374
249. Guilleux M-F, Kermarec M, Delafosse D (1977) J Chem Soc, Chem Commun 102
250. Jiang HJ, Tzou MS, Sachtler WMH (1988) Catal Lett 1:99
251. Bein T, McLain SJ, Corbin DR, Farlee RD, Möller K, Stucky GD, Woolery G, Sayers D (1988) J Am Chem Soc 110:1801
252. Zakharov AN (1991) Kinet Katal 32:1377; Engl Transl 1230
253. Zhang Z, Sachtler WMH, Suib SL (1989) Catal Lett 2:395
254. Lu G, Hoffer T, Guczi L (1992) Catal Lett 14:207
255. Yin Y-G, Zhang Z, Sachtler WMH (1993) J Catal 139:444
256. Kim JC, Woo SI (1991) Bull Chem Soc Jpn 64:1370
257. Zerger RP, McMahon KC, Seltzer MD, Michel RG, Suib SL (1986) J Catal 99:498
258. Nam SS, Iton LE, Suib SL, Zhang Z (1989) Chem Mater 1:529
259. Zhang Z, Zhang YD, Hines WA, Budnick JI, Sachtler WMH (1992) J Am Chem Soc 114:4843
260. Coughlan B, Narayanan S, McCann WA, Carroll WM (1977) J Catal 49:97
261. Verdonck JJ, Jacobs PA, Genet M, Poncelet G (1980) J Chem Soc, Faraday Trans 1 76:403
262. Cho SJ, Jung SM, Shul YG, Ryoo R (1992) J Phys Chem 96:9922
263. Gustafson BL, Lin M-J, Lunsford JH (1980) J Phys Chem 84:3211
264. Landmesser H, Miessner H (1991) J Phys Chem 95:10544
264a. Wu JCS, Oukaci R, Goodwin JG (1993) J Catal 142:531

265. Cisneros MD, Lunsford JH (1993) J Catal 141:191
265a. Guntow U, Rosowski F, Muhler M, Ertl G, Schlögl R (1995) Preparation of Catalysts VI. Stud Surf Sci Catal 91:217
266. Mroczek U, Steinberg K-H (1991) React Kinet Catal Lett 43:559
267. Maggiore R, Crisafulli C, Scirz S, Solarino L, Galvagno S (1992) React Kinet Catal Lett 48:367
268. Shannon RD, Vedrine JC, Naccache C, Lefebvre F (1984) J Catal 88:431
269. Naccache C, Ben Taarit Y, Boudart M (1977) Molecular Sieves - II, Katzer JR (ed) ACS Symp Ser 40:156
270. Sayari A, Morton JR, Preston KF (1988) J Chem Soc, Faraday Trans 1 84:413
271. Atanasova VD, Shvets VA, Kazanskii VB (1977) Kinet Katal 18:753; Engl Transl 628
272. Primet M, Vedrine JC, Naccache C (1978) J Mol Catal 4:411
273. Sayari A, Morton JR, Preston KF (1987) J Phys Chem 91:899
274. Sayari A, Morton JR, Preston KF, Brown JR (1988) Proc 9th Int Congr Catal 1:356
275. Corbin DR, Seidel WC, Abrams L, Herron N, Stucky GD, Tolman CA (1985) Inorg Chem 24:1800
276. Huang T-N, Schwartz J (1982) J Am Chem Soc 104:5244
277. Mantovani E, Palladino N, Zanobi A (1977/78) J Mol Catal 3:285
278. Scurrell MS, Howe RF (1980) J Mol Catal 7:535
279. Yamanis J, Yang K-C (1981) J Catal 69:498
280. Arai H, Tominaga H (1982) J Catal 75:188
281. Davis ME, Rode E, Taylor D, Hanson BE (1984) J Catal 86:67
282. Rode EJ, Davis ME, Hanson BE (1985) J Catal 96:574
283. Rode EJ, Davis ME, Hanson BE (1985) J Catal 96:563
284. Lee TJ, Gates BC (1991) Catal Lett 8:15
285. Minachev KhM, Baeva GN, Shpiro ES, Antoshin GV, Fasman AB (1985) Kinet Katal 26:1265; Engl Transl 1097
286. Nishizaka Y, Misono M (1993) Chem Lett 1295
286a. Zhang Z, Lerner B, Lei G-D, Sachtler WH (1993) J Catal 140:481
287. Bergeret G, Gallezot P, Imelik B (1981) J Phys Chem 85:411
288. Zhang Z, Sachtler WMH, Chen H (1990) Zeolites 10:784
289. Homeyer ST, Sachtler WMH (1989) J Catal 117:91
290. Homeyer ST, Sachtler WMH (1989) J Catal 118:266
291. Zhang Z, Chen H, Sheu LL, Sachtler WMH (1991) J Catal 127:213
292. Reagan WJ, Chester AW, Kerr GT (1981) J Catal 69:89
293. Zhang Z, Mestl G, Knözinger H, Sachtler WMH (1992) Appl Catal A89:155
294. Leglise J, Chambellan A, Cornet D (1991) Appl Catal 69:15
295. Homeyer ST, Sachtler WMH (1989) Appl Catal 54:189
296. Feeley O, Sachtler WMH (1990) Appl Catal 67:141
297. Feeley O, Sachtler WMH (1991) Appl Catal 75:93
298. Sheu LL, Knözinger H, Sachtler WMH (1989) J Mol Catal 57:61
299. Zhang Z, Sachtler WMH (1991) J Mol Catal 67:349
300. Zhang Z, Chen H, Sachtler WMH (1991) J Chem Soc, Faraday Trans 87:1413
301. Sheu LL, Knözinger H, Sachtler WMH (1989) J Am Chem Soc 111:8125
302. Sheu LL, Knözinger H, Sachtler WMH (1989) Catal Lett 2:129
303. Zhang Z, Wong TT, Sachtler WMH (1991) J Catal 128:13
304. Bai X, Sachtler WMH (1990) Catal Lett 4:319
305. Bai X, Sachtler WMH (1991) J Catal 129:121
306. Yin Y-G, Zhang Z, Sachtler WMH (1992) J Catal 138:721
307. Schaefer-Feeley J, Sachtler WMH (1990) Zeolites 10:738
308. Feeley JS, Sachtler WMH (1991) J Catal 131:573
309. Zhou P-L, Gates BC (1989) J Chem Soc, Chem Commun 347
310. Zhou P-L, Maloney SD, Gates BC (1991) J Catal 129:315
311. Kawi S, Gates BC (1992) J Chem Soc, Chem Commun 702
312. Kawi S, Gates BC (1991) Catal Lett 10:263

313. Kawi S, Gates BC (1991) J Chem Soc, Chem Commun 994
314. Kawi S, Chang J-R, Gates BC (1993) J Am Chem Soc 115:4830
315. Beutel T, Kawi S, Purnell SK, Knözinger H, Gates BC (1993) J Phys Chem 97:7284
316. Rabo JA, Schomaker V, Pickert PE (1965) Proc 3rd Int Congr Catal (North-Holland Publ Co, Amsterdam, and John Wiley and Sons, New York) 2:1264
317. Lewis PH (1968) J Catal 11:162
318. Mashchenko AI, Bronnikov OD, Dmitriev RV, Garanin VI, Kazanskii VB, Minachev KhM (1974) Kinet Katal 15:1603; Engl Transl 1418
319. Dalla Betta RA, Boudart M (1973) Proc 5th Int Congr Catal (North-Holland Publ Co, Amsterdam) Hightower JW (ed) 2:1329
320. Czaran E, Schnabel K-H, Selenina M (1974) Z anorg allg Chem 410:225
320a. Bastl Z, Kubelková L, Nováková J (1997) Zeolites 19:279
321. Exner D, Jaeger N, Kleine A, Schulz-Ekloff G (1988) J Chem Soc, Faraday Trans 1 84:4097
322. Exner D, Jaeger N, Schulz-Ekloff G (1980) Chem-Ing-Tech 52:734
323. Gallezot P, Alarcon Diaz A, Dalmon J-A, Renouprez AJ, Imelik B (1975) J Catal 39:334
324. Wilson WR, Hall WK (1970) J Catal 17:190
325. Gallezot P (1979) Catal Rev-Sci Eng 20:121
326. Tri TM, Massardier J, Gallezot P, Imelik B (1981) Proc 7th Int Congr Catal, Tokio 1980 (Elsevier, Amsterdam) 266
327. Gallezot P, Bergeret G (1981) J Catal 72:294
328. Tonscheidt A, Ryder PL, Jaeger NI, Schulz-Ekloff G (1993) Surface Sci 281:51
329. Dessau RM (1984) J Catal 89:520
330. Engelen CWR, Wolthuizen JP, van Hooff JHC, Zandbergen HW (1986) New Developments in Zeolite Science and Technology, Proc 7th Int Zeolite Conf, Tokyo, Japan. Murakami Y, Iijima A, Ward JW (eds) Kodansha/Elsevier, Stud Surf Sci Catal 28:709
330a. Zhang A, Nakamuro I, Fujimoto K (1997) J Catal 168:328
331. Shpiro ES, Joyner RW, Minachev KM, Pudney PDA (1991) J Catal 127:366
332. Keegan MBT, Fent AJ, Blake AB, Conyers L, Moyes RB, Wells PB, Whan DH (1991) Catal Today 9:183
333. Frilette VJ, Haag WO, Lago RM (1981) J Catal 67:218
334. Guisnet M, Giannetto P, Perot G (1983) J Chem Soc, Chem Commun 1411
335. Hirabayashi H, Yahiro H, Mizuno N, Iwamoto M (1992) Chem Lett 2235
336. Bischoff H, Jaeger NI, Schulz-Ekloff G (1990) Z phys Chem (Leipzig) 271:1093
337. Jaeger NI, Jourdan AL, Schulz, Ekloff G (1991) J Chem Soc, Faraday Trans 87:1251
338. De Mallmann A, Barthomeuf D (1990) Catal Lett 5:293
338a. Han W-J, Kooh AB, Hicks RF (1993) Catal Lett 18:193
339. Hong SB, Mielczarski E, Davis ME (1992) J Catal 134:349
340. Barrer RM, Bultitude FW, Kerr IS (1959) J Chem Soc 1521
341. Hoss H, Roy R (1960) Beitr Mineral u Petrogr 7:389
342. Taylor AM, Roy R (1965) J Chem Soc 4028
343. Peterson DL, Helfferich F, Blytas GC (1965) J Phys Chem Solids 26:835
344. Roelofsen DP, Wils ERJ, van Bekkum H (1972) J Inorg Nucl Chem 34:1437
345. Kühl GH (1973) J Catal 29:270
346. Kühl GH, Schweizer AE (1975) J Catal 38:469
347. Kühl GH (1977) Molecular Sieves-II. Katzer JR (ed) ACS Symp Ser 40:96
348. Rinne F (1896) Neues Jahrb Min 1:24
349. Rinne F (1897) ibid 1:28
350. Rinne F (1897) ibid 1:40
351. Barrer RM (1949) Nature (London) 164:113
352. Uytterhoeven JB, Christner LG, Hall WK (1965) J Phys Chem 69:2117
353. Benesi HA (1967) J Catal 8:368
354. Kühl GH (1973) Proc 3rd Internat Conf Molecular Sieves. Zurich, Switzerland, Uytterhoeven JB (ed) Leuven University Press, p 227
355. Kühl GH (1977) J Phys Chem Solids 38:1259

356. Resing HA, Rubinstein M (1978) J Coll Interf Sci 64:48
357. Bosacek V, Freude D, Fröhlich T, Pfeifer H, Schmiedel H (1982) J Coll Interf Sci 85:502
358. Bosacek V, Mastikhin VM (1987) J Phys Chem 91:260
359. Luz Z, Vega AJ (1987) J Phys Chem 91:374
360. Nagy BN, Gabelica Z, Debras G, Derouane EG, Gilson JP, Jacobs PA (1984) Zeolites 4:133
361. Kentgens APM, Scholle KFMGJ, Veeman WS (1983) J Phys Chem 87:4357
362. Vega AJ, Luz Z (1987) J Phys Chem 91:365
363. Liengme BV, Hall WK (1966) Trans Faraday Soc 62:3229
364. White JL, Jelli AN, Andre JM, Fripiat JJ (1967) Trans Faraday Soc 63:461
365. Ward JW (1968) J Catal 11:259
366. Olson DH, Dempsey E (1968) J Catal 13:221
367. Ward JW (1969) J Phys Chem 73:2086
368. Cattanach J, Wu EL, Venuto PB (1968) J Catal 11:342
369. Ward JW (1967) J Phys Chem 71:3106
370. Freude D, Hunger M, Pfeifer H, Scheler G, Hoffmann J, Schmitz W (1984) Chem Phys Lett 105:427
371. Pfeifer H, Freude D, Hunger M (1985) Zeolites 5:274
372. Freude D, Hunger M, Pfeifer H (1987) Z Phys Chem, Neue Folge 152:171
373. Lohse U, Löffler E, Hunger M, Stöckner J, Patzelova V (1987) Zeolites 7:11
374. Freude D, Hunger M, Pfeifer H, Schwieger W (1986) Chem Phys Lett 128:62
375. Chu P (1976) J Catal 43:346
376. Kerr GT (1969) J Catal 15:200
377. Beyer HK, Mihalyfi J, Kiss A, Jacobs PA (1981) J Therm Anal 20:351
378. Post JG, van Hooff JHC (1984) Zeolites 4:9
379. Hidalgo CV, Itoh H, Hattori T, Niwa M, Murakami Y (1984) J Catal 85:362
380. Niwa M, Iwamoto M, Segawa K (1986) Bull Chem Soc Jpn 59:3735
381. Hunger B, Hoffmann J, Mothsche P (1987) J Therm Anal 32:2009
382. Dima E, Rees LVC (1990) Zeolites 10:8
383. Kerr GT, Shipman GF (1968) J Phys Chem 72:3071
384. Bolton AP, Lanewala MA (1970) J Catal 18:154
385. Hopkins PD (1968) J Catal 12:325
386. Wu EL, Kühl GH, Whyte TE, Venuto PB (1971) Molecular Sieve Zeolites-I, Gould RF (ed) Adv Chem Ser 101:490
387. Jacobs PA, Uytterhoeven JB (1972) J Catal 26:175
388. Parker LM, Bibby DM, Patterson JE (1984) Zeolites 4:168
389. Choudhary VR, Pataskar SG (1986) Thermochim Acta 97:1
390. Kotasthane AN, Shiralkar VP (1986) Thermochim Acta 102:37
391. Nastro A, Ciambelli P, Crea F, Aiello R (1988) J Therm Anal 33:941
392. Nastro A, Crea F, Giordano G (1990) J Therm Anal 36:2223
393. Crea F, Giordano G, Mostowicz R, Nastro A (1990) J Therm Anal 36:2229
394. Schwieger W, Bergk KH, Alsdorf E, Fichtner-Schmittler H, Löffler E, Lohse U, Parlitz B (1990) Z phys Chem (Leipzig) 271:243
395. Soulard M, Bilger S, Kessler H, Guth JL (1987) Zeolites 7:463
396. Bilger S, Soulard M, Kessler H, Guth JL (1991) Zeolites 11:784
397. Soulard M, Bilger S, Kessler H, Guth JL (1992) Thermochim Acta 204:167
398. Soulard M, Bilger S, Kessler H, Guth JL (1991) Zeolites 11:107
399. Nowotny M, Lercher JA, Kessler H (1991) Zeolites 11:454
400. Maesen TLM, Bruinsma DSL, Kouwenhoven HW, van Bekkum H (1987) J Chem Soc, Chem Commun 1284
401. Maesen TLM, Kouwenhoven HW, van Bekkum H, Sulikowski B, Klinowski J (1990) J Chem Soc, Faraday Trans 86:3967
402. Brunner E, Ernst H, Freude D, Hunger M, Pfeifer H (1988) Innovation in Zeolite Materials Science. Grobet PJ, Mortier WJ, Vansant EF, Schulz-Ekloff G (eds) Stud Surf Sci Catal 37:155

403. Datka J, Tuznik E (1985) Zeolites 5:230
404. Dessau RM, Schmitt KD, Kerr GT, Woolery GL, Alemany LB (1987) J Catal 104:484
405. Nagy JB, Gabelica Z, Derouane EG (1982) Chem Lett 1105
406. Woolery GL, Alemany LB, Dessau RM, Chester AW (1986) Zeolites 6:14
407. Sauer J, Bleiber A (1988) Catal Today 3:485
408. Von Ballmoos R, Meier WM (1982) J Phys Chem 86:2698
409. Boxhoorn G, Kortbeek AGTG, Hays GR, Alma NCM (1984) Zeolites 4:15
410. Jacobs PA, von Ballmoos R (1982) J Phys Chem 86:3050
411. Datka J, Tuznik E (1986) J Catal 102:43
412. Hunger M, Freude D, Fröhlich T, Pfeifer H, Schwieger W (1987) Zeolites 7:108
413. Hunger M, Kärger J, Pfeifer H, Caro J, Zibrowius B, Bülow M, Mostowicz R (1987) J Chem Soc, Faraday Trans 1 83:3459
414. Hunger M, Freude D, Pfeifer H, Schwieger W (1990) Chem Phys Lett 167:21
415. Freude D, Klinowski J, Hamdan H (1988) Chem Phys Lett 149:355
416. Batamack P, Doremieux-Morin C, Fraissard J, Freude D (1991) J Phys Chem 95:3790
417. Parker LM, Bibby DM, Burns GR (1991) Zeolites 11:293
417a. Sonnemans MHW, den Heijer C, Crocker M (1993) J Phys Chem 97:440
418. Freude D, Fröhlich T, Hunger M, Pfeifer H, Scheler G (1983) Chem Phys Lett 98:263
419. Chukin GD, Kulikov AS, Sergienko SA (1985) React Kinet Catal Lett 27:287
420. Shannon RD, Gardner KH, Staley RH, Bergeret G, Gallezog P, Auroux A (1985) J Phys Chem 89:4778
421. Wichterlova B, Novakova J, Kubelkova L, Jiru P (1980) Proc 5th Internat Zeolite Conf, Naples, Italy. Rees LVC (ed) Heyden & Son, Ltd, p 373
422. Borade R, Adnot A, Kaliaguine S (1990) J Mol Catal 61:L7
423. Datka J (1981) J Chem Soc, Faraday Trans 1 77:2877
424. Marosi L (1980) Angew Chem 92:759; Angew Chem Int Ed Engl 19:743
425. Zholobenko VL, Kustov LM, Borovkov VYu, Kazansky VB (1988) Zeolites 8:175
426. Kazanski VB (1988) Catal Today 3:367
427. Lee EFT, Rees LVC (1987) Zeolites 7:545
428. Lee EFT, Rees LVC (1987) Zeolites 7:143
429. Roelofsen JW, Mathies H, de Groot RL, van Woerkom PCM, Angad Gaur H (1986) New Developments in Zeolite Science and Technology, Proc 7th Int Zeolite Conf, Tokyo, Japan. Murakami Y, Iijima A, Ward JW (eds) Stud Surf Sci Catal 28:337
430. Dana ES (1942) System of Mineralogy, 6th Ed, John Wiley & Sons, New York
431. Breck DW (1974) Zeolite Molecular Sieves, John Wiley & Sons, New York, p 503
432. Frilette VJ, Rubin MK (1965) J Catal 4:310
433. Sand LB (1968) Molecular Sieves, Soc Chem Ind (London) p 71
434. Kranich WL, Ma YH, Sand LB, Weiss AH, Zwiebel I (1971) Molecular Sieve Zeolites - I. Gould RF (ed) Adv Chem Ser 101:502
434a. Kooyman PJ, van der Waal P, van Bekkum H (1997) Zeolites 18:50
435. Barrer RM, Makki MB (1964) Can J Chem 42:1481
436. Barrer RM, Coughlan B (1968) Molecular Sieves, Soc Chem Ind (London) p 141
437. Kerr GT (1968) J Phys Chem 72:2594
438. Kerr GT (1969) J Phys Chem 73:2780
439. Kerr GT (1973) Molecular Sieves, Meier WM, Uytterhoeven JB (eds) Adv Chem Ser 121:219
440. Kerr GT, Chester AW, Olson DH (1978) Acta Phys et Chem 24:169
441. Miecznikowski A, Rzepa B (1977) Roczniki Chemii, Ann Soc Chim Pol 51:1955
442. Joshida A, Nakamoto H, Okanishi K, Tsuru T, Takahashi H (1982) Bull Chem Soc Jpn 85:581
443. Lee EFT, Rees LVC (1987) J Chem Soc, Faraday Trans 1 83:1531
444. Chandwadkar AJ, Kulkarni SB (1980) J Therm Anal 19:313
445. Akporiaye D, Chapple AP, Clark DM, Dwyer J (1986) New Developments in Zeolite Science and Technology, Proc 7th Int Zeolite Conf, Tokyo, Japan. Murakami Y, Iijima A, Ward JW (eds) Stud Surf Sci Catal 28:351

445a. Datka J, Sulikowski B, Gil B (1996) J Phys Chem 100:11242
446. Jacobs P, Uytterhoeven JB (1971) J Catal 22:193
447. Beaumont R, Pichat P, Barthomeuf D, Trambouze Y (1973) Proc 5th Internat Congr Catal 1:343
448. Bosacek V, Patzelova V, Tvaruzkova Z, Freude D, Lohse U, Schirmer W, Stach H, Thamm H (1980) J Catal 61:435
449. Zi G, Yi T, Yugin Z (1989) Appl Catal 56:83
450. Skeels GW, Breck DW (1984) Proc 6th Internat Zeolite Conf, Reno USA, Olsen DH, Bisio A (eds) Butterworths, Guildford, UK, p 87
451. Gallezot P, Beaumont R, Barthomeuf D (1974) J Phys Chem 78:1550
452. Pichat P, Beaumont R, Barthomeuf D (1974) J Chem Soc, Faraday Trans 1 70:1402
453. Aouali L, Jeanjean J, Dereigne A, Tougne P, Delafosse D (1988) Zeolites 8:517
454. Jeanjean J, Aouali L, Delafosse D, Dereigne A (1989) J Chem Soc, Faraday Trans 1 85:2771
455. Tsutsumi K, Kajiwara H, Koh HK, Takahashi H (1973) Molecular Sieves, Proc 3rd Internat Conf Mol Sieves, Zurich, Switzerland, Uytterhoeven JB (ed) p 358
456. Tsutsumi K, Kajiwara H, Takahashi H (1974) Bull Chem Soc Jpn 47:801
457. Aboul-Gheit AK (1991) Thermochim Acta 191:233
458. Eberly PE Jr, Kimberlin CN Jr, Voorhies AJ Jr (1971) J Catal 22:419
459. Karge HG, Dondur V (1990) J Phys Chem 94:765
460. Chumbhale VR, Chandwadkar AJ, Rao BS (1992) Zeolites 12:63
461. Meyers BL, Fleisch TH, Ray GJ, Miller JT, Hall JB (1988) J Catal 110:82
462. Chen NY, Smith FA (1976) Inorg Chem 15:295
463. Beyer HK, Belenykaja IM, Mishin IW, Borbely G (1984) Structure and Reactivity of Modified Zeolits, Proc Int Symp, Prague, Czechoslovakia. Jacobs PA, Jaeger NI, Jiru P, Kazansky VB, Schulz-Ekloff G (eds) Stud Surf Sci Catal 18:133
464. Hays GR, van Erp WA, Alma NCM, Couperus PA, Huis R, Wilson AE (1984) Zeolites 4:377
465. Piguzova LI, Prokefieva EN, Dubinin MM, Bursian PJ, Shavandin A (1969) Kinet Katal 10:315; Engl Trans 10:252
466. Wolf F, John H (1973) Chem Tech 25:736
467. Bodart P, Nagy JB, Debras G, Gabelica Z, Jacobs PA (1986) J Phys Chem 90:5183
468. Goovaerts F, Vansant EF, De Hulsters P, Gelan J (1989) J Chem Soc, Faraday Trans 1 85:3687
469. Stach H, Jänchen J (1992) Zeolites 12:152
470. Engelhardt G, Lohse U, Samoson A, Maegi M, Tarmak M, Lippmaa E (1982) Zeolites 2:59
471. Debras G, Nagy JB, Gabelica Z, Bodart P, Jacobs PA (1983) Chem Lett 199
472. Springuel-Huet MA, Fraissard JP (1992) Zeolites 12:841
472a. Donovan AW, O'Connor CT, Koch KR (1995) Microporous Mater 5:185
472b. Apelian MR, Fung AS, Kennedy GJ, Degnan TF (1996) J Phys Chem 100:16577
473. Dwyer J, Fitch FR, Quin G, Vickerman JC (1982) J Phys Chem 86:4574
474. Gross T, Lohse U, Engelhardt G, Richter K-H, Patzelova V (1984) Zeolites 4:25
475. Shyu JZ, Skopinski ET, Goodwin JG Jr, Sayari A (1985) Applications of Surface Science 21:297
476. Dwyer J, Fitch FR, Machado F, Qin G, Smyth SM, Vickerman JC (1981) J Chem Soc, Chem Commun 422
477. Scherzer J (1978) J Catal 54:285
478. Maxwell IE, van Erp WA, Couperus T, Hjuis R, Clague ADH (1982) J Chem Soc, Chem Commun 523
479. Lohse U, Stach H, Thamm H, Schirmer W, Isirikjan HA, Regent NI, Dubinin MM (1980) Z anorg allg Chem 460:179
480. Mauge F, Auroux A, Courcelle JC, Engelhard P, Gallezot P, Grossmangin J (1985) Catalysis by Acids and Bases, Proc Int Symp. Villeurbanne, France, 1984. Imelik B, Naccache C, Coudurier G, Ben Taarit Y, Vedrine JC (eds) Stud Surf Sci Catal 20:91

481. Lynch J, Raatz F, Delalande C (1987) Characterization of Porous Solids, Proc IUPAC Symp (COPS I), Bad Soden, Germany, Unger KK, Rouquerol J, Sing KSW, Kral H (eds) Stud Surf Sci Catal 39:547
482. Chevreau T, Chambellan A, Lavalley JC, Catherine E, Marzin M, Janin A, Hemedy JF, Khabtou S (1990) Zeolites 10:226
483. Auroux A, Ben Taarit Y (1987) Thermochim Acta 122:63
484. Rabo JA, Pickert PE, Stamires DN, Boyle JE (1960) Actes du Deuxiem Congres de Catalyse (1960) 2055
485. McDaniel CV, Maher PK (1968) Molecular Sieves, Soc Chem Ind, London, p 186
486. Kerr GT (1967) J Phys Chem 71:4155
487. Lutz W (1990) Cryst Res Technol 25:921
488. Ward JW (1972) J Catal 27:157
489. Wichterlova B, Novakova J, Kubelkova L, Jiru P (1980) Proc 5th Internat Zeolite Conf, Rees LVC (ed) Heyden & Sons, Ltd, p 373
490. Beaumont R, Barthomeuf D (1972) J Catal 27:45
491. Dempsey E (1974) J Catal 33:497
492. Kerr GT (1975) J Catal 37:186
493. Zhdanov SP, Titova TI, Kosheleva LS, Lutz W (1989) Pure Appl Chem 61:1977
494. Zhdanov SP, Kosheleva LS, Titova TI (1991) Izv Akad Nauk SSSR, Ser Khim 1299; Engl Transl 1141
495. Zhdanov SP, Kosheleva LS, Titova TI (1991) Izv Akad Nauk SSSR, Ser Khim 1303; Engl Transl 1145
496. Peri JB (1973) Proc 5th Internat Congr Catal, Hightower J (ed) Elsevier, p 329
497. Freude D, Fröhlich T, Pfeifer H, Scheler G (1983) Zeolites 3:171
498. Garralon G, Corma A, Fornes V (1989) Zeolites 9:84
499. Fyfe CA, Gobbi GC, Hartman JS, Klinowski J, Thomas JM (1982) J Phys Chem 86:1247
500. Klinowski J, Fyfe CA, Gobbi GC (1985) J Chem Soc, Faraday Trans 1 81:3003
501. Kellberg L, Linsten M, Jakobsen JH (1991) Chem Phys Lett 182:120
502. Gilson J, Edwards GC, Peters AW, Rajagopalan K, Wormsbecher RF, Roberie GT, Shatlock MP (1987) J Chem Soc, Chem Commun 91
503. Samoson A, Lippmaa E, Engelhardt G, Lohse U, Jerschkewitz H-G (1987) Chem Phys Lett 134:589
504. Freude D, Brunner E, Pfeifer H, Prager D, Jerschkewitz H-G, Lohse U, Öhlmann G (1987) Chem Phys Lett 139:325
505. Rocha J, Carr SW, Klinowski J (1991) Chem Phys Lett 187:401
506. Corbin DR, Farlee RD, Stucky GD (1984) Inorg Chem 23:2920
507. Parise JB, Corbin DR, Abrams L, Cox DE (1984) Acta Cryst C40:1493
508. Geurts FMM, Kentgens APM, Veeman WS (1985) Chem Phys Lett 120:206
509. Corma A, Fornes V, Martinez A, Sanz J (1988) ACS Symp Ser 375:17
510. Klinowski J, Thomas JM, Fyfe CA, Gobbi GC (1982) Nature 296:533
511. Bertram R, Lohse U, Gessner W (1988) Z anorg allg Chem 567:145
512. Breck DW, Skeels GW (1977) Molecular Sieves - II, ACS Symp Ser 40:271
513. Jacobs PA, Uytterhoeven JB (1973) J Chem Soc, Faraday Trans 1 69:373
514. Corma A, Fornes V, Pallota O, Cruz JM, Ayerbe A (1986) J Chem Soc, Chem Commun 333
515. Aouali L, Jeanjean J, Dereigne A, Tougne M, Delafosse D (1988) Zeolites 8:517
516. Lohse U, Mildebrath M (1981) Z anorg allg Chem 476:126
517. Dubinin MM, Isirikjan AA, Regent NI, Schirmer W, Stach H, Thamm H, Lohse U (1984) Izv. Akad Nauk SSSR, Ser Khim, 1931; Engl Transl 1758
518. Zukal A, Patzelova V, Lohse U (1986) Zeolites 6:133
519. Patzelova V, Jaeger NI (1987) Zeolites 7:240
519a. Dutartre R, de Ménorval LC, Di Renzo F, McQueen D, Fajula F, Schulz P (1996) Microporous Mater 6:311
520. Fichtner-Schmittler H, Lohse U, Engelhardt G, Patzelova V (1984) Cryst Res Tech 19:K1

521. Sohn JR, De Canio SJ, Lunsford JH, O'Donnel DJ (1986) Zeolites 6:225
522. Kerr GT (1989) Zeolites 9:350
523. Meyers BL, Fleisch TH, Marshall CL (1986) Appl Surf Sci 26:503
524. Scherzer J, Bass JL, Hunter FD (1975) J Phys Chem 79:1194
525. Scherzer J, Bass JL (1973) J Phys Chem 79:1200
526. Stöcker M, Ernst S, Karge HG, Weitkamp J (1990) Acta Chem Scan 44:519
526a. Martin A, Wolf U, Kosslick H, Tuan VA (1993) React Kinet Catal Lett 51:19
526b. Miessner H, Kosslick H, Lohse U, Parlitz B, Tuan VA (1993) J Phys Chem 97:9741
527. Yoshida A, Adachi Y, Inoue K (1991) Zeolites 11:549
528. Maher PK, Hunter FD, Scherzer J (1971) Molecular Sieve Zeolites - I, Gould RF (ed) Adv Chem Ser 101:266
529. Scherzer J, Bass JL (1973) J Catal 28:101
530. Engelhardt G, Lohse U, Patzelova V, Mägi M, Lippmaa E (1983) Zeolites 3:233
531. Engelhardt G, Lohse U, Patzelova V, Mägi M, Lippmaa E (1983) Zeolites 3:239
532. Fleisch TH, Meyers BL, Ray GJ, Hall JB, Marshall CL (1986) J Catal 99:117
533. Ray GJ, Meyers BL, Marshall CL (1986) Zeolites 7:307
534. Man PP, Klinowski J (1988) Chem Phys Lett 147:581
535. Hamdan H, Klinowski J (1989) J Chem Soc, Chem Commun 240
536. Rocha J, Klinowski J (1991) J Chem Soc, Chem Commun 1121
537. Lippmaa E, Samoson A, Mägi M (1986) J Am Chem Soc 108:1730
538. Grobet PJ, Geerts H, Martens JA, Jacobs PA (1987) J Chem Soc, Chem Comun 1688
539. Bosacek V, Freude D (1988) Innovation in Zeolite Materials Science, Grobet PJ, Mortier WJ, Vansant EF, Schulz-Ekloff G (eds) Stud Surf Sci Catal 37:231
540. Grobet PJ, Geerts H, Tielen M, Martens JA, Jacobs PA (1989) Zeolites as Catalysts, Sorbents and Detergent Builders, Karge HG, Weitkamp J (eds) Stud Surf Sci Catal 46:721
541. Janin A, Lavalley JC, Macedo A, Raatz F (1988) ACS Symp Ser 368:117
542. Mirodatos C, Barthomeuf D (1981) J Chem Soc, Chem Commun 39
543. Beyerlein RA, McVicker GB, Yacullo LN, Ziemiak JJ (1988) J Phys Chem 92:1967
544. Ashton AG, Batmanian S, Clark DM, Dwyer J, Fitch FR, Hinchcliffe A, Machado FJ (1985) Catalysis by Acids and Bases. Imelik B, Naccache C, Coudurier G, Ben Taarit Y, Vedrine JC (eds) Stud Surf Sci Catal 20:101
545. Ward MB, Lunsford JH (1984) J Catal 87:524
546. Merlen E, Lynch J, Bisiaux M, Raatz F (1990) SIA, Surf Interf Anal 16:364
547. Patzelova V, Drahoradova E, Tvaruzkova Z, Lohse U (1989) Zeolites 9:74
548. Yoshida A, Inoue K, Adachi Y (1991) Zeolites 11:223
549. Kubelkova L, Dudikova L, Bastl Z, Beyer HK, Borbely G (1987) J Chem Soc, Faraday Trans 1 83:511
550. Kubelkova L, Seidl V, Borbely G, Beyer HK (1988) J Chem Soc, Faraday Trans 1 84:1447
551. Shirinskaya LP, Komarov VS, Pryakhina NP (1975) Russian J Phys Chem, Engl Trans 49:580
552. Debras G, Gourgue A, Nagy JB, De Clippelair G (1986) Zeolites 6:241
553. Engelhardt G, Jerschkowitz H-G, Lohse U, Sarv P, Samoson A, Lippmaa E (1987) Zeolites 7:289
554. Fyfe CA, Gobbi C, Kennedy GJ (1983) Chem Lett 1551
555. Fyfe CA, Gobbi C, Kennedy GJ (1984) J Phys Chem 88:3248
556. Öhlmann G, Jerschkewitz H-G, Lischke G, Parlitz B, Richter M, Eckelt R (1988) Z Chem 28:161
557. Alsdorf E, Feist M, Gross Th, Jerschkewitz H-G, Lohse U, Schwieger W (1990) Z phys Chem (Leipzig) 271:267
558. Engelhardt G, Fahlke B, Mägi M, Lippmaa E (1985) Z phys Chem (Leipzig) 266:239
559. Topsøe N-Y, Joensen F, Derouane EG (1988) J Catal 110:404
560. Loeffler E, Peuker Ch, Jerschkewitz H-G (1988) Catal Today 3:415
561. Kornatowski J, Baur WH, Pieper G, Rozwadowski M, Schmitz W, Cichowlas A (1992) J Chem Soc, Faraday Trans 88:1339

562. Lago RM, Haag WO, Mikovsky RJ, Olson DH, Hellring SD, Schmitt KD, Kerr GT (1986) New Developments in Zeolite Science and Technology, Proc 7th Int Zeolite Conf, Tokyo, Japan. Murakami Y, Iijima A, Ward JW (eds) Stud Surf Sci Catal 28:677
563. Brunner E, Ernst H, Freude D, Hunger M, Krause CB, Prager D, Reschetilowski W, Schwieger W, Bergk K-H (1989) Zeolites 9:282
564. Löffler E, Kustov LM, Zholobenko VL, Peuker C, Lohse U, Kazansky VB, Öhlmann G (1991) Catalysis and Adsorption by Zeolites. Öhlmann G, Pfeifer H, Fricke R (eds) Stud Surf Sci Catal 65:425
565. Zholobenko VL, Kustov LM, Kazansky VB, Löffler E, Lohse U, Peuker C, Öhlmann G (1990) Zeolites 10:304
566. Löffler E, Lohse U, Peuker C, Öhlmann GG, Kustov LM, Zholobenko VL, Kazansky VB (1990) Zeolites 10:266
566a. Campbell SM, Bibby DM, Coddington JM, Howe RF, Meinwald RH (1996) J Catal 161:338
566b. Datka J, Marschmeyer S, Neubauer T, Meusinger J, Papp H, Schütze F-W, Szpyt I (1996) J Phys Chem 100:14451
567. Staudte B, Hunger M, Nimz M (1991) Zeolites 11:837
568. Plank CJ (1964) Proc 3rd Intern Congr Catal 1:727
569. Beyer HK, Belenykaja I (1980) Catalysis by Zeolites, Imelik B et al. (eds) Stud Surf Sci Catal 5:203
570. Beyer HK, Belenykaja IM, Hange F, Tielen M, Grobet PJ, Jacobs PA (1985) J Chem Soc, Faraday Trans I 81:2889
571. Grobet PJ, Jacobs PA, Beyer HK (1986) Zeolites 6:47
572. Anderson MW, Klinowski J (1986) J Chem Soc, Faraday Trans 1 82:3569
573. Andera V, Kubelkova L, Novakova J, Wichterlova B, Bednarova S (1985) Zeolites 5:67
574. Corma A, Fornes V, Martinez A, Melo F, Pallota O (1988) Innovation in Zeolite Materials Science. Grobet PJ, Mortier WJ, Vansant EF, Schulz-Ekloff (eds) Stud Surf Sci Catal 37:495
575. Mishin IV, Beyer HK, Klyachko AL, Ashavskaya GA, Nissenbaum VD, Borbely G (1987) Kinet Katal 28:706; Engl Transl 615
576. Klinowski J, Thomas JM, Fyfe CA, Gobbi GC, Hartman JS (1983) Inorg Chem 33:63
577. Klinowski J, Thomas JM, Audier M, Vasudevan S, Fyfe CA, Hartman JS (1981) J Chem Soc, Chem Commun 570
578. Anderson MW, Klinowski J (1986) Zeolites 6:455
579. Sulikowski B, Borbely G, Beyer HK, Karge HG, Mishin IW (1989) J Phys Chem 93:3240
580. Martens JA, Geerts H, Grobet PJ, Jacobs PA (1990) J Chem Soc, Chem Commun 1418
581. Sulikowski B, Klinowski J (1990) J Chem Soc, Faraday Trans 1 86:199
582. Miyake M, Komarneni S, Roy R (1987) Clay Minerals 22:367
583. Hey MJ, Nock A, Rudham R, Appleyard IP, Haines GAJ, Harris RK (1986) J Chem Soc, Faraday Trans I 82:2817
584. Ray GJ, Nerheim AG, Donohue JA (1988) Zeolites 8: 458
585. Anderson MW, Klinowski J (1986) J Chem Soc, Faraday Trans 1 82:1449
586. Kubelkova L, Seidl V, Novakova J, Bednarova S, Jiru P (1984) J Chem Soc, Faraday Trans 1 80:1367
586a. Stockenhuber M, Lercher JA (1995) Microporous Mater 3:457
586b. Woolery GL, Kuehl GH, Timken HC, Chester AW, Vartuli JC (1987) Zeolites 19:288
587. Kubelkova L, Beran S, Malecka A, Mastikhin V (1989) Zeolites 9:12
588. Martens JA, Grobet PJ, Jacobs PA (1991) Preparation of Catalysts V, Stud Surf Sci Catal 63:355
589. Goyvaerts D, Martens JA, Grobet PJ, Jacobs PA (1991) ibid 63:381
590. Thomas JM, Klinowski J, Anderson MW (1983) Chem Lett 1555
591. Jacobs PA, Tielen M, Nagy JB, Debras G, Derouane EG, Gabelica Z (1984) Proc 6th Internat Zeolite Conf, Reno, USA, July 10–15, 1983. Olson D, Bisio A (eds) Butterworth & Co, Ltd, p 783
592. Namba S, Inaka A, Yashima T (1984) Chem Lett 817

593. Namba S, Inaka A, Yashima T (1986) Zeolites 6:107
593a. Silva JM, Ribeiro MF, Ramôa Ribeiro F, Benazzi E, Gnep NS, Guisnet M (1996) Zeolites 16:275
594. Corma A, Fornes V, Rey F (1990) Appl Catal 59:267
595. He Y, Li C, Min E (1989) Zeolites: Facts, Figures, Future. PA Jacobs, RA van Santen (eds) Stud Surf Sci Catal 49A:189
596. Garralon G, Fornes V, Corma A (1988) Zeolites 8:268
597. Wang QL, Giannetto G, Guisnet M (1990) Zeolites 10:301
598. Dwyer J, Dewing J, Thompson NE, O'Malley PJ, Karim K (1989) J Chem Soc Chem Commun 843
599. Wang QL, Torrealba M, Giannetto G, Perot G, Cahoreau M, Caisso J (1990) Zeolites 10:703
599a. Matharu AP, Gladden LF, Carr SW (1995) Stud Surf Sci Catal 94:147
600. Ness JN, Joyner DJ, Chapple AP (1989) Zeolites 9:250
601. Cruz JM, Corma A, Fornes V (1989) Appl Catal 50:287
602. Chauvin B, Boulet M, Massiani P, Fajula F, Figuera SF, Des Courieres T (1990) J Catal 126:532
603. Dwyer J, Karim K, Smith WJ, Thompson NE, Harris RK, Apperley DC (1991) J Phys Chem 95:8826
604. Lonyi F, Lunsford JH (1992) J Catal 136:566
605. Fejes P, Kiricsi I, Hannus I, Kiss A, Schöbel GY (1980) React Kinet Catal Lett 14:481
606. Fejes P, Hannus I, Kiricsi I (1982) React Kinet Catal Lett 19:239
607. Fejes P, Kiricsi I, Hannus I (1982) Acta Phys Chem 28:173
608. Fejes P, Hannus I, Kiricsi I (1984) Zeolites 4:73
609. Fejes P, Kiricsi I, Hannus I, Schöbel GY (1985) Zeolites, Drzaj B, Hocevar S, Pejovnik S (eds) Stud Surf Sci Catal 24:263
610. Fejes P, Schöbel G, Kiricsi I, Hannus I (1985) Acta Phys Chem 31:119
611. Fejes P, Kiricsi I, Hannus I (1982) Metal Microstructures in Zeolites: Prep-Prop-Appl. Jacobs PA, Jaeger NI, Jiru P, Schulz-Ekloff G (eds) Stud Surf Sci Catal 12:159
612. Garwood WE, Chen NY, Bailar JC Jr (1976) Inorg Chem 15:1044
613. Lok BM, Izod TPJ (1982) Zeolites 2:66
614. Lok BM, Gortsema FP, Messina CA, Rastelli N, Izod TPJ (1983) Intrazeolite Chemistry. Stucky GD, Dwyer FG (eds) ACS Symp Ser 218:41
615. Penchev V, Sariev I, Zhelyazkova M (1981) Kinet Catal 23:732; Engl Transl 23:562
616. Kowalak S (1985) React Kinet Catal Lett 27:441
617. Becker KA; Kowalak S (1987) J Chem Soc, Faraday Trans 1 83:535
618. Becker KA, Fabianska K, Kowalak S (1985) Acta Phys Chem 31:63
619. Chu CT-W, Chang CD (1985) J Phys Chem 89:1569
620. Kessler H, Chezeau JM, Guth JL, Strub H, Coudurier G (1987) Zeolites 7:360
621. Datka J, Piwowarska Z (1989) J Chem Soc, Faraday Trans 1 85:47
622. De Ruiter R, Kentgens APM, Grootendorst J, Jansen JC, van Bekkum H (1993) Zeolites 13:128
623. Scholle KFMGJ, Veeman WS (1985) Zeolites 5:118
624. Chu CT-W, Kuehl GH, Lago RM, Chang CD (1985) J Catal 93:451
625. Coudurier G, Vedrine JC (1986) Pure Appl Chem 58:1389
626. Cornaro U, Wojciechowski BW (1989) J Catal 120:182
627. Datka J, Cichocki A, Piwowarska Z (1991) Catalysis and Adsorption by Zeolites. Oehlmann G et al. (eds) Stud Surf Sci Catal 65:681
628. Brunner E, Freude D, Hunger M, Pfeifer H, Reschetilowski W, Unger B (1988) Chem Phys Lett 148:226
629. Cichocki A, Lasocha W, Michalik M, Sawlowicz Z, Bus M (1990) Zeolites 10:583
630. Meyers BL, Ely SR, Kutz NA, Kaduk JA, van den Bossche E (1985) J Catal 91:352
631. Cichocki a, Parasiewicz-Kaczmarska J, Michalik M, Bus M (1990) Zeolites 10:577
632. Unger B, Wendtland K-P, Toufar H, Schwieger W, Bergk K-H, Brunner E, Reschetilowski W (1991) J Chem Soc, Faraday Trans 87:3099

633. Simon MW, Nam SS, Xu W, Suib SL, Edwards JC, O'Young C-L (1992) J Phys Chem 96:6381
634. Hegde SG, Abdullah RA, Bhat RN, Ratnasamy P (1992) Zeolites 12:951
634a. Iwasaki A, Sano T (1997) Zeolites 19:41
634b. Le Van Mao R, Le ST, Ohayon D, Caillibot F, Gelebart L, Denes G (1997) Zeolites 19:270
635. Breck DW, Skeels GW (1980) Proc 5th Int Conf Zeolites. Rees LV (ed) Heyden & Son Ltd p 335
636. Kerr GT (1982) J Catal 77:307
637. Engelhardt G, Lohse U (1984) J Catal 88:513
638. Liu X, Klinowski J, Thomas JM (1986) J Chem Soc, Chem Commun 582
639. Bezman RD (1987) JI Chem Soc, Chem Commun 1562
640. Lutz W, Lohse U, Fahlke B (1988) Cryst Res Technol 23:925
641. Sohn JR, De Canio SJ, Fritz PO, Lunsford JH (1986) J Phys Chem 90:4847
642. Mortier W (1982) Compilation of Extra-Framework Sites in Zeolites. Butterworths, Guildford
643. Hamdan H, Klinowski J (1989) Zeolite Synthesis. Occelli ML, Robson HE (eds) ACS Symp Ser 398:448
644. Hamdan H, Sulikowski B, Klinowski J (1989) J Phys Chem 93:350
645. Barrer RM (1982) Hydrothermal Chemistry of Zeolites, Academic Press, London
646. Hamdan H, Klinowski J (1989) Zeolite Synthesis. Occelli ML, Robson HE (eds) ACS Symp Ser 398:465
647. Klinowski J, Hamdan H, Corma A, Fornes V, Hunger M, Freude D (1989) Catal Lett 3:263
648. Barrie PJ, Gladden LF, Klinowski J (1991) J Chem Soc, Chem Commun 592
648a. Calsavara V, Falabella Sousa-Aguiar E, Fernandes Machado NRC (1996) Zeolites 17:340
648b. Liu D-S, Bao S-L, Xu Q-H (1997) Zeolites 18:162
649. Sulikowski B, Rakoczy J, Hamdan H, Klinowski J (1987) J Chem Soc, Chem Commun 1542
650. Weisz PB, Miale JN (1965) J Catal 4:527
651. Miale JN, Chen NY, Weisz PB (1966) J Catal 6:278
652. Zhang Z, Liu X, Xu Y, Xu R (1991) Zeolites 11:232
652a. Yang C, Xu Q (1997) J Chem Soc, Faraday Trans 93:1675
653. Shibabi DS, Garwood WE, Chu P, Miale JN, Lago RM, Chu CT-W, Chang CD (1985) J Catal 93:471
654. Changa CD, Hellring SD, Miale JN, Schmitt KD, Brigandi PW, Wu EL (1985) J Chem Soc, Faraday Trans 1 81:2215
655. Jacobs PA, Tielen M, Nagy JB, Debras G, Derouane EG, Gabelica Z (1984) Proc 6th Intern Zeolite Conf, Reno, USA, July 10–15, 1983. Olson DH, Bisio A (eds) Butterworth & Co, Ltd, p 783
656. Anderson MW, Klinowski J, Liu X (1984) J Chem Soc, Chem Commun 1596
657. Dessau RM, Kerr GT (1984) Zeolites 4:315
658. Chang CD, Chu CT-W, Miale JN, Bridger RF, Calvert RB (1984) J Am Chem Soc 106:8143
659. Yashima T, Yamagishi K, Namba S, Nakata S, Asaoka S (1988) Innovation in zeolite Materials Science, Grobet PJ, Mortier WJ, Vansant EF, Schulz-Ekloff G (eds) Stud Surf Sci Catal 37:175
660. Yamagishi K, Namba S, Yashima T (1990) J Catal 121:47
661. Kraushaar B, van de Ven LJM, de Haan JW, van Hooff JHC (1988) Innovation in Zeolite Materials Science. Grobet PJ, Mortier WJ, Vansant EF, Schulz-Ekloff G (eds) Stud Surf Sci Catal 37:167
662. Kraushaar B, de Haan JW, Van Hooff JHC (1988) J Catal 109:470
663. Yamagishi K, Namba S, Yashima T (1991) J Phys Chem 95:872
664. Yamagishi K, Namba S, Yashima T (1988) Acid-Base Catalysis, Proc Int Symp on Acid-Base Catalysis. Tanabe K, Hattori H, Yamagishi T, Tanaka T (eds) Kodansha-VCH, p 297
665. Handreck GP, Smith TD (1990) Zeolites 10:746
666. De Ruiter R, Jansen JC, van Bekkum H (1992) Molecular Sieves: Synthesis of Microporous Materials, Occelli ML, Robson H (eds) 1:167

667. Brunner E, Freude D, Hunger M, Pfeifer H, Reschetilowski W, Unger B (1988) Chem Phys Lett 148:226
668. Chu CT-W, Kuehl GH, Lago RM, Chang CD (1985) J Catal 93:451
669. Chester AW, Chu YF, Dessau RM, Kerr GT, Kresge CT (1985) J Chem Soc, Chem Commun 289
670. Derouane EG, Baltusis L, Dessau RM, Schmitt KD (1985) Catalysis by Acids and Bases, Imelik B et al. (eds) Stud surf Sci Catal 20:135
671. Sayed MB (1987) J Chem Soc, Faraday Trans 1, 83:1751
672. Sulikowski B, Klinowski J (1989) Zeolite Synthesis. Occelli ML, Robson HE (eds) ACS Symp Ser 398:393
673. Gaffney TR, Pierantozzi R, Seger MR (1989) Zeolite Synthesis, Occelli ML, Robson HE (eds) ACS Symp Ser 398:374
674. Han S, Schmitt KD, Schramm SE, Reischmann PT, Shihabi DS, Chang CD (1994) J Chem Phys 98:4118
675. Yamagishi K, Namba S, Yashima T (1991) Bull Chem Soc Jpn 64:949
676. Chu CT-W, Chang CD (1985) J Phys Chem 89:1569
677. Liu X, Thomas JM (1985) J Chem Soc, Chem Commun 1544
678. Thomas JM, Liu X (1986) J Phys Chem 90:4843
679. Liu X, Lin J, Liu X, Thomas JM (1992) Zeolites 12:936
680. Dwyer J, Karim K (1991) J Chem Soc, Chem Commun 905
681. Karim K, Dwyer J, Rawlence DJ, Tariq M, Nabhan A (1992) J Mater Chem 2:1161
682. Skeels GW, Flanigan EM (1989) Zeolite Synthesis, Occelli ML, Robson HE (eds) ACS Symp Ser 398:420
683. Whittington BI, Anderson JR (1991) J Phys Chem 95:3306
684. Skeels GW, Flanigen EM (1989) Zeolites: Facts, Figures, Future. Jacobs PA, van Santen RA (eds) Stud Surf Sci Catal 49A:331
685. Spoto G, Bordiga S, Garrone E, Ghiotti, G, Zecchina A, Petrini G, Leofanti G (1992) J Mol Catal 74:175
686. Han S, Schmitt KD, Shihabi DS, Chang CD (1993) J Chem Soc, Chem Commun 1287
687. Engelhardt G, Fahlke B, Mägi M, Lippmaa E (1985) Z phys Chem (Leipzig) 266:239
688. Fraenkel D (1990) Ind Eng Chem Res 29:1814
689. Lee CS, Park TJ, Lee WY (1993) Appl Catal A96:151
690. Chen NY, Kaeding WW, Dwyer FG (1979) J Am Chem Soc 101:6783
691. Yan Y, Verbiest J, Vansant EF, Philippaerts J, De Hulsters P (1990) Zeolites 10:137
692. Yan Y, Vansant EF (1990) J Phys Chem 94:2582
693. Vansant EF (1990) Pore Size Engineering in Zeolites. John Wiley & Sons
694. Kowalak S (1985) React Kinet Catal Lett 27:441
695. Becker KA, Kowalak S (1987) J Chem Soc, Faraday Trans 1 83:535
696. Becker KA, Kowalak S, Kozlowski M (1985) React Kinet Catal Lett 29:1
697. Becker KA, Fabianska K, Kowalak S (1985) Acta Phys Chem 31:63
698. Geismar G, Westphal U (1982) Z anorg allg Chem 487:207
699. Peeters G, Thys A, Vansant EF, De Bievre P (1984) Proc 6th Int zeolite Conf, Reno 1983. Olson D, Bisio A (eds) Butterworths, p 651
700. Vansant EF, Peeters G, Thijs A, Verhaert I (1985) Zeolites, Synthesis, Structure, Technology and Application. Drzaj B, Hocevar S, Pejovnik S (eds) Stud Surf Sci Catal 24:329
701. Vansant EF (1988) Innovation in Zeolite Materials Science. Grobet PJ, Mortier WJ, Vansant EF, Schulz-Ekloff (eds) Stud Surf Sci Catal 37:143
702. Ward JW (1976) Zeolite Chemistry and Catalysis. Rabo JA (ed) ACS Monograph 171:118
703. Haw JF, Chuang I-S, Hawkins BL, Maciel GE (1983) J Am Chem Soc 105:7206
704. Wang I, Ay C-L, Lee B-J, Chen M-H (1989) Appl Catal 54:257
705. Nunan J, Cronin J, Cunningham J (1984) J Catal 87:77
706. Namba S, Nakanishi S, Yashima T (1984) J Catal 88:505
707. Kim J-H, Namba S, Yashima T (1989) Zeolites as Catalysts. Sorbents and Detergent Builders. Karge HG, Weitkamp J (eds) Stud Surf Sci Catal 46:71

708. Gaffney AM, Jones CA, Sofranco JA, Tsiao C, Dybowski C (1993) Selctivity in Catalysis. ACS Symp Ser 517:316
709. Goldsein MS, Morgan TR (1970) J Catal 16:232
710. Barrer RM, Jenkins RG, Peeters G (1977) Molecular Sieves - II. Katzer JR (ed) ACS Symp Ser 40:258
711. Barrer RM, Trombe J-C (1978) J Chem Soc, Faraday Trans 1 74:2798
712. Von Ballmoos R, Kerr GT (1985) Zeolites - Synthesis, Structure, Technology and Application. Drzaj B, Hocevar S, Pejovnik S (eds) Stud Surf Sci Catal 24:307
713. Bein T, Carver RF, Farlee RD, Stucky GD (1988) J Am Chem Soc 110:4546
714. Geismar G, Westphal U (1982) Z anorg allg Chem 484:131
715. Kraushaar B, van der Ven LJM, de Haan JW, van Hooff JHC (1988) Innovation in Zeolite Materials Science, Grobet PJ et al. (eds) Stud Surf Sci Catal 37:167
716. Ray GJ, Nerheim AG, Donohue JA, Hriljac JA, Cheetham AK (1992) Coll Surf 63:77
717. Chen NY et al. (1981) In: Kaeding WW, Chu C, Young LB, Weinstein B, Butter SA. J Catal 67:159
718. Sivasanker S, Reddy KM (1989) Catal Lett 3:49
719. Niwa M, Itoh H, Kato S, Hattori T, Murakami Y (1982) J Chem Soc, Chem Commun 819
720. Niwa M, Itoh H, Kato S, Hattori T, Murakami Y (1984) J Chem Soc, Faraday Trans 1 80:3135
721. Niwa M, Kato M, Hattori T, Murakami Y (1986) J Phys Chem 90:6233
722. Niwa M, Kawashima Y (1985) J Chem Soc, Faraday Trans 1 81:2757
723. Kva LD, Ponomareva OA, Sinitsyna OA, Moskovskaya IF, Khusid BL, Chukin GD, Parenago OO, Lunina EV (1989) Kinet Katal 30:1461; Engl Transl 1272
724. Wang I, Ay C-L, Lee B-J, Chen M-H (1988) Proc 9th Int Congr Catal 1:324
725. Hibino T, Niwa M, Murakami Y (1991) J Catal 128:551
726. Wichterlova B, Cejka J (1992) Catal Lett 16:421
727. Rothwell WP, Shen WX, Lunsford JH (1984) J Am Chem Soc 106:2452
728. Lunsford JH, Rothwell WP, Shen W (1985) J Am Chem Soc 107:1540
729. Rumplmayr G, Lercher JA (1990) Zeolites 10:283
730. Jentys A, Rumplmayr G, Lercher JA (1989) Appl Catal 53:299
731. Vinek H, Rumplmayr G, Lercher JA (1989) J Catal 115:291
732. Lunsford JH, Tutunjian PN, Chu P, Yeh EB, Zalewski DJ (1989) J Phys Chem 93:2590
733. Zalewski DJ, Chu P, Tutunjian PN, Lunsford JH (1989) Langmuir 5:1026
734. Bein T, Chase DB, Farlee RD, Stucky GD (1986) New Developments in Zeolite Science and Technology. Proc 7th Int Zeolite Conf, Tokyo, Japan. Murakami Y, Iijima A, Ward JW (eds) Stud Surf Sci Catal 28:311
735. Kaeding WW, Chu C, Young LB, Weinstein B, Butter SA (1981) J Catal 67:159
736. Rahman A, Lemay G, Adnot A, Kaliaguine S (1988) J Catal 112:453
737. Young LB, Butter SA, Kaeding WW (1982) J Catal 76:418
738. Kaeding WW, Chu C, Young LB, Butter SA (1981) J Catal 69:392
739. Kim J-H, Namba S, Yashima T (1988) Bull Chem Soc Jpn 61:1051
740. Balkrishnan I, Rao BS, Hegde SG, Kotasthane AN, Kulkarni SB, Ratnaswamy P (1982) J Mol Catal 17:261
741. Kaeding WW, Butter SA (1980) J Catal 61:155
742. Kerr GT, Chester AW (1971) Thermochim Acta 3:113
743. Gilson JP, Derouane EG (1984) J Catal 88:538
744. Auroux A, Vedrine JC, Gravelle PC (1982) Adsorption at the gas-Solid and Liquid-Solid Interface, proc Int Symp, Ecully (Lyon). Imelik B et al. (eds) Stud Surf Sci Catal 10:305
745. Vedrine JC, Auroux A, Dejaifve P, Ducarme V, Hoser H, Zhou S (1982) J Catal 73:147
746. Namba S, Kanai Y, Shoji H, Yashima T (1984) Zeolites 4:77
747. Csicsery SM (1969) J Org Chem 34:3338
748. Kaeding WW (1985) J Catal 95:512
749. Jentys A, Rumplmayr G, Vinek H, Lercher JA (1987) Proc 6th Int Symp Heterog Catal, Sofia, Pt 2, 264

750. Derewinski M, Haber J, Ptaszynski J, Shiralkar VP, Dzwigaj S (1984) Structure and Reactivity of Modified Zeolites. Jacobs PA et al. (eds) Stud Surf Sci Catal 18:209
751. Lercher JA, Rumplmayr G, Noller H (1985) Acta Phys Chem 31:71
752. Lercher JA, Rumplmayr G (1986) Appl Catal 25:215
753. Caro J, Bülow M, Derewinski M, Hunger M, Kärger J, Kürschner U, Pfeifer H, Storek W, Zibrowius B (1989) Recent Advances in Zeolite Science. Klinowski J, Barrie PJ (eds) Stud Surf Sci Catal 52:295
754. Caro J, Bülow M, Derewinski M, Haber J, Hunger M, Kärger J, Pfeifer H, Storek W, Zibrowius B (1990) J Catal 124:367
755. Seo G, Ryoo R (1990) J Catal 124:224
756. Lischke G, Eckelt R, Jerschkewitz H-G, Parlitz B, Schreier E, Storek W, Zibrowius B, Öhlmann G (1991) J Catal 132:229
757. Kühl GH, Schmitt KD (1990) Zeolites 10:2

Characterization of Zeolites – Infrared and Nuclear Magnetic Resonance Spectroscopy and X-Ray Diffraction

Hellmut G. Karge, Michael Hunger, and Hermann K. Beyer

List of Abbreviations

2D	two-dimensional
Al^{f}	framework aluminum
Al^{nf}	non-framework aluminum
$AlPO_4$	aluminophosphate
CAW	calcium tungstate aluminum sodalite
COSY	correlation spectroscopy
CP	cross-polarization
CRAMPS	combined rotational and multiple-pulse spectroscopy
DOR	double oriented rotation
DRIFT	diffuse reflectance infrared spectroscopy
DRS	diffuse reflectance spectroscopy
D4R	double four ring
D6R	double six ring
ESR	electron spin resonance
F1	frequency domain 1 of a two-dimensional spectrum
F2	frequency domain 2 of a two-dimensional spectrum
FD	fixed delay
FID	free induction decay
FT	Fourier transform
HF	high frequency
INADEQUATE	incredible natural abundance double quantum transfer experiment
IR	infrared
LF	low frequency
MAS	magic angle spinning
MeOH	metal hydroxyl groups
NMR	nuclear magnetic resonance
NOESY	nuclear overhauser enhancement and exchange spectroscopy
PAS	photoacoustic infrared spectroscopy
PFG	pulsed field gradient
Q^4	tetrahedrally coordinated
REDOR	rotational-echo double resonance
rf	radio frequency
SAPO	silicoaluminophosphate

SEDOR	spin-echo double resonance
SI, SII...	site I, site II...
SiOH	silanol groups
SiOHAl	bridging hydroxyl groups
SP	single-pulse
TEDOR	transferred-echo double resonance
TMP	trimethylphosphine
TMS	tetramethylsilane
VT	variable temperature
WAHUHA	multiple-pulse sequence denoted by Waugh, Huber, and Haeberlen
XRD	X-ray diffraction

Introduction

A number of spectroscopic techniques are advantageously used to characterize zeolites and/or zeolite-adsorbate systems. The general feature of all of these techniques is the following one. A sample, which is built up from quantum mechanical systems (e.g., electrons, nuclei, atoms, molecules), is exposed to radiation. The quantum mechanical systems absorb (or emit) certain radiation quanta according to certain quantum mechanical transitions. These transitions occur between discrete energy levels which are described by the stationary Schroedinger equation. This equation is an example of a so-called eigenvalue equation, where an operator is applied to a function reproducing this function multiplied with a parameter, i.e., the eigenvalue:

$$\hat{H}\Phi = E \cdot \Phi \tag{4.1}$$

In the Schroedinger Eq. 4.1, $\hat{H}$ stands for the Hamilton or energy operator, pertinent to the particular problem; Φ is the eigenfunction describing the quantum mechanical system, and E is the energy eigenvalue. Usually, a set of parameters E_i satisfies Eq. 4.1. Not all of the mathematically possible transitions between the various energy levels are "allowed"; rather, they are subjected to selection rules.

In the case of electromagnetic radiation, the energy quanta are given by $h \cdot \nu$, where h is Planck's constant and ν the frequency of the radiation. Only those quanta may be absorbed or emitted which exactly fit the energy difference $\Delta E = E_l - E_k$ for a permitted transition between two energy levels k and l. The particular frequency for which this interaction between matter and radiation occurs is called resonance frequency ν_r. Thus, in case of resonance,

$$\Delta E = h \cdot \nu_r \tag{4.2}$$

Resonance frequencies, ν_r, are measured in appropriate apparatuses, i.e., spectrometers. Measurements of ν_r and the amount of radiation absorbed at ν_r provide information about the nature and/or status of the sample investigated. The various types of spectroscopy are distinguished by the energy range or

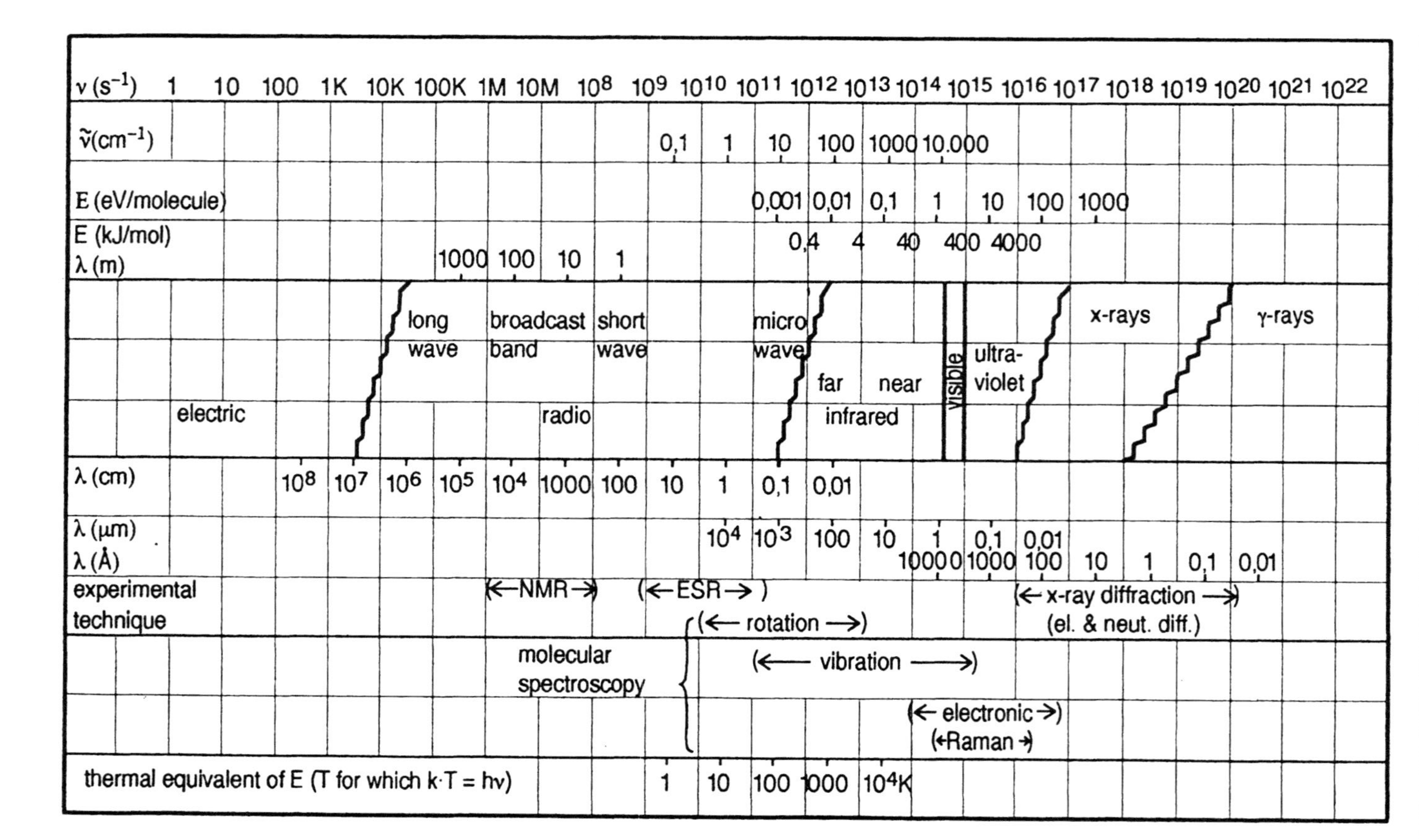

Fig. 4.1. Relationship between frequency, wavelength or energy of electromagnetic radiation and spectroscopic techniques

frequency range to which the observed transitions belong. The correlation between the type of spectroscopy and the energy or frequency range is illustrated in Fig. 4.1.

Various spectroscopic techniques have been and are being increasingly employed for investigations of solid surfaces and their interactions with adsorbates, particularly in view of heterogeneous catalysis. There are a number of excellent review articles, books or book chapters devoted to the application of spectroscopy in this field. A selection of these are included in the list of references even though they are not predominantly related to zeolite characterization [1–4].

4.1 IR-Spectroscopy

4.1.1 Introduction

For a long time, IR spectroscopy had been the mainly used spectroscopic technique to characterize zeolites and zeolite/adsorbate systems. It was only with the advent of solid-state NMR spectroscopy (see Sect. 4.2) that a spectroscopic method became available in zeolite research that was a similarly powerful tool. Even though IR and NMR spectroscopy in some cases of application in zeolite characterization may appear to be competitive, in general they are complementary in a very valuable sense.

IR studies on zeolites and zeolite/adsorbate systems reported in the literature before 1975 have been extensively reviewed by Ward [5]. More recently, an excellent and very concise overview has been published by Foerster [6].

The main areas of application of IR spectroscopy have been and still are (i) investigation of framework properties, (ii) study of sites of the zeolite lattice, being relevant, e.g., for adsorption or catalysis, (iii) characterization of zeolite/adsorbate systems, and (iv) measurements related to the motion of guest molecules in the pores and cavities of the zeolites. These most important cases, where IR spectroscopy is employed for zeolite characterization, will be dealt with in what follows, after a brief description of the theoretical background and the experimental aspects.

4.1.2 Theoretical Background

The infrared section of the electromagnetic radiation covers the wavelength range from about 9×10^2 nm (0.9 µm) to 5×10^5 nm (500 µm) or ca. 12000 cm^{-1} to about 20 cm^{-1}. The units reciprocal centimeters or wavenumbers are usually preferred in IR spectroscopy, with $\tilde{\nu}$ $(cm^{-1}) = 10^4/\lambda$ (µm). The range from 12000 cm^{-1} to 4000 cm^{-1} is called near infrared, that from 4000 cm^{-1} to 400 cm^{-1} middle infrared (most frequently used), and the range below 400 cm^{-1} is the far infrared region. In order to interact with electromagnetic radiation in the

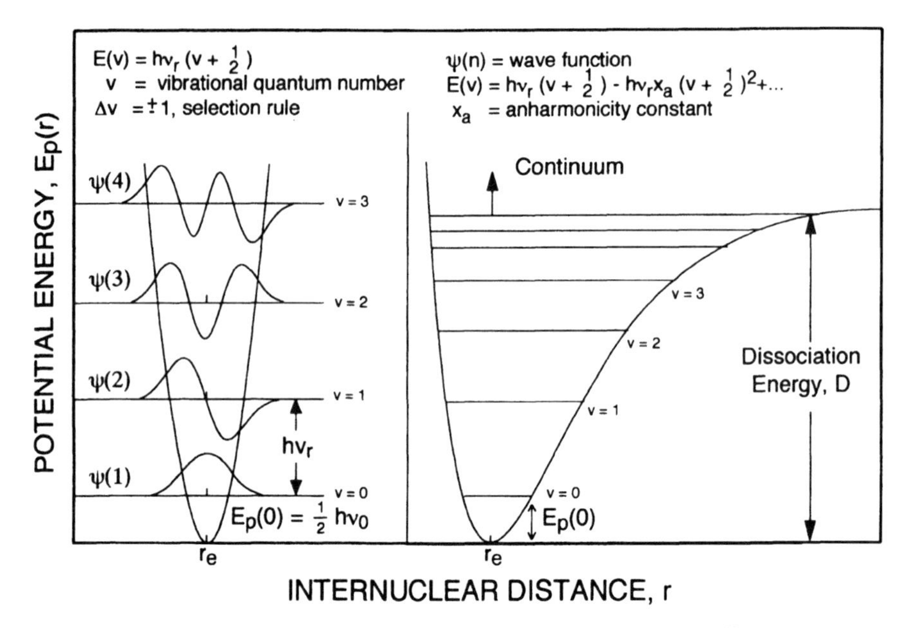

Fig. 4.2. Potential energies of a harmonic (*left*) and anharmonic (*right*) oscillator

infrared range, the species exposed to it must have a permanent or induced dipole. Then the radiation can be absorbed and excite certain states, e.g., states of translation, rotation and/or vibration. In principle, the resonance frequencies (or wavenumbers, $\tilde{v}_r = v_r/c_0$, where c_0 represents the velocity of light) can be derived from Eq. 4.1 and 4.2. However, application of Eq. 4.1 would require an appropriate model to evaluate the potential and kinetic energies to be used for solving the stationary Schroedinger equation. In a most simple case such a model may be the rigid rotor or the harmonic oscillator. This would lead to an energy scheme as depicted schematically in Fig. 4.2 (*left*). More refined models would be the non-rigid rotor, the anharmonic oscillator (Fig. 4.2 (*right*)), etc. (for a detailed discussion see [7–9]). On application of IR spectroscopy in zeolite research, however, the spectra are usually not interpreted by comparison with theoretically derived resonance frequencies but on the base of empirical data, for instance via comparison with known spectra of solids, liquids or gases. Very valuable and widely used is, in that context, the concept of group frequencies (typical of particular groups such as $\tilde{v}$ – OH, =CO, ≡CH etc.) and their shifts (compare, e.g., [9]). Only in a few cases were frequencies computed and related to experimentally obtained data (*vide infra*).

Correct interpretation of the signals (IR bands), which appear upon resonance with the radiation, leads to an *identification* of the absorbing entities, i.e., information about their *nature*. However, in most cases one also wishes to have information about the *number* of species involved. Such *quantitative* analysis is based on the intensities of the bands, viz. the maximum or integrated absorbance. This is illustrated in Fig. 4.3, where I_0 represents the incident energy

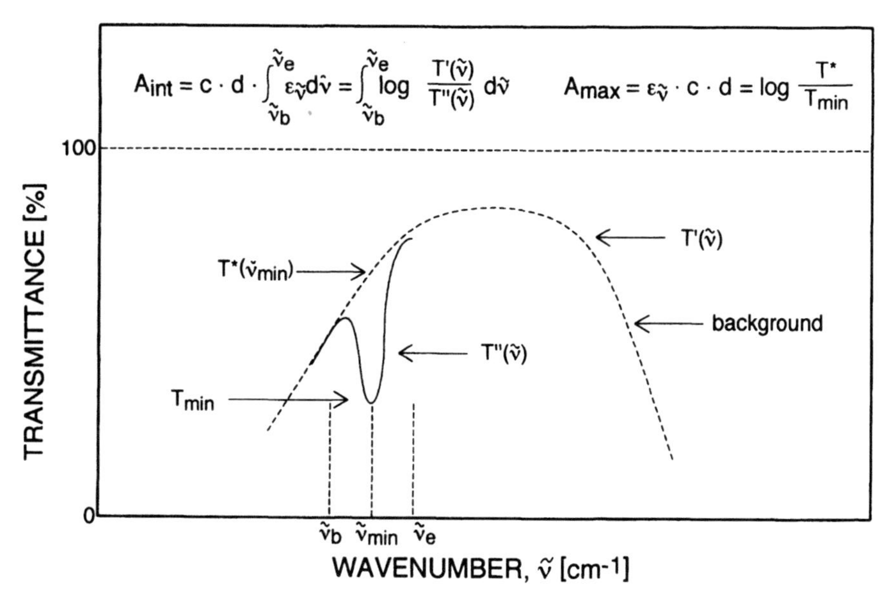

Fig. 4.3. Quantitative evaluation of IR spectra-integrated and maximum absorbance

of radiation, I the transmitted energy, and T the transmittance, $T = I/I_0$. The absorbance is defined as:

$$A = -\log T = \log I_0/I$$

For quantitative evaluation of the concentration, c, of the species involved one may use the Lambert-Bouguer law as shown in Eq. 4.3.

$$A = \varepsilon_{\tilde{\nu}} \cdot c \cdot d \tag{4.3}$$

where $\varepsilon_{\tilde{\nu}}$ is the extinction coefficient and d the thickness of the sample. Then, the integrated absorbance as a quantitative measure of the IR absorbing species is as depicted in Eq. 4.4.

$$A_{int} = c \cdot d \int_{\tilde{\nu}_b}^{\tilde{\nu}_e} \varepsilon_{\tilde{\nu}}\, d\tilde{\nu} = \int_{\tilde{\nu}_b}^{\tilde{\nu}_e} \log \frac{T'(\tilde{\nu})}{T''(\tilde{\nu})}\, d\tilde{\nu} \tag{4.4}$$

Eq. 4.4 is strictly valid only for infinitely diluted systems. Thus, one must be aware of the fact that $\varepsilon_{\tilde{\nu}}$ may depend on the concentration (coverage). In IR studies on zeolites, $\varepsilon_{\tilde{\nu}}$ is usually unknown; experimental determination of $\varepsilon_{\tilde{\nu}}$ is troublesome and has been carried out only in a few cases (compare [10] and *vide infra*). Very often the absorbances are, therefore, given in arbitrary units, and only relative concentrations or numbers of species are evaluated. Frequently, the bands are not integrated, and the maximum absorbance, $A_{max} = \log (T^*/T_{min})$, is used as a measure of the relative concentration.

The integrated absorbance is proportional to the so-called oscillator strength, f, and this in turn is proportional to the square of the transition moment:

$$f \propto \left[\int_{-\infty}^{+\infty} \Psi_m \hat{\mu} \Psi_n \right]^2 \tag{4.5}$$

where Ψ_m and Ψ_n are the wave functions of the ground and excited states, respectively, and $\hat{\mu}$ stands for the operator of the dipole moment. Consideration of the transition moments provides the selection rules. A transition is "allowed" only if the corresponding integral is different from zero. This is only then the case if the integrand remains unchanged upon application of the symmetry operations which are allowed by the symmetry of the absorbing species.

4.1.3
Experimental Techniques

4.1.3.1
Transmission IR Spectroscopy

IR spectrometers measure the energy of the radiation transmitted through a sample as a function of the frequency (wavenumber or wavelength). One distinguishes between dispersive and non-dispersive (Fourier transform) infrared spectrometers. The older design of dispersive instruments employs prisms or gratings as monochromators in order to select IR radiation of a certain frequency (or frequency interval) from the originally continuous radiation generated by the source (e.g., a globar or a Nernst glower). The dispersive spectrometers are generally double-beam instruments, where one beam (I_o) does not transmit the sample and is used for comparison. Due to a continuous rotation of the monochromator, which is coupled with the movement of the recorder device, radiation of successively changed frequency is directed to the exit slit, transmits the sample and reaches the detector (e.g., a thermocouple or a Golay detector). According to the interactions of the radiation with the molecules of the sample at the resonance frequencies the energy I received by the detector is, in comparison with that of the second beam I_o, lower at these frequencies. A plot of the transmittance vs. the frequencies (or wavenumbers) provides the conventional IR spectrum with typical bands of decreased transmittance at certain frequencies (compare, e.g., Fig. 4.3).

The concept of a Fourier transform IR spectrometer is completely different. Here, the whole spectrum is obtained via a process of interference. The interference is provided by an "interferometer" device, e.g., a Michelson interferometer (see Fig. 4.4).

The polychromatic light, made parallel by a collimator, is directed to the sample S. One part of the light transmitted through the sample is reflected by the beam splitter and thrown to the fixed mirror, M1. The other part reaches the moving mirror, M2. Both partial beams are returned to point C where they

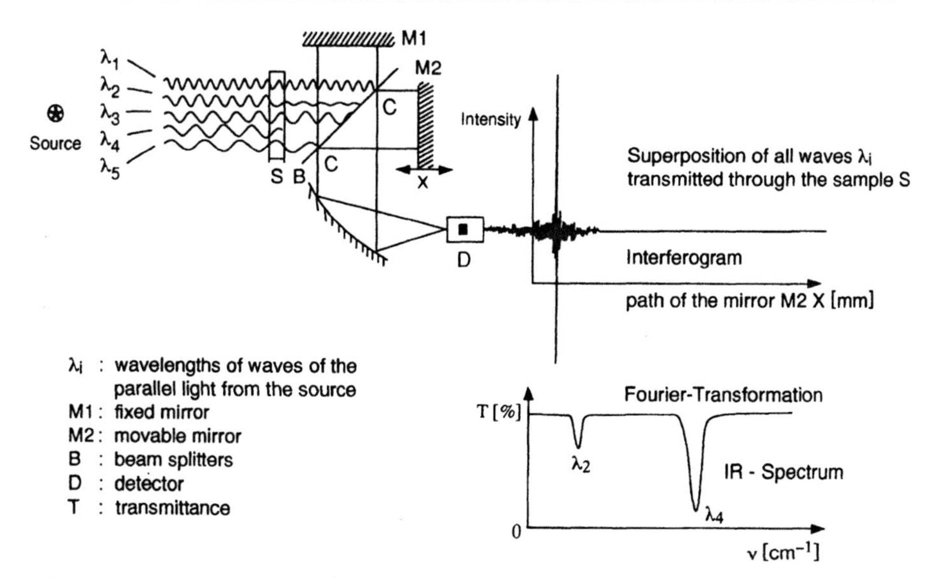

Fig. 4.4. Scheme of the principle of a Fourier transform infrared spectrometer (FTIR) [11]

interfer. After interference, half of the radiation is re-directed to the source and is lost, the other half is directed to the detector. If, for a particular wavelength λ_i (frequency ν_i) of the polychromatic light, the total pathway difference created by the movement of mirror M2 is just λ_i, then the corresponding waves will be enhanced by interference. If the total pathway difference is for instance $\nu_i/2$, the interference for that particular wave is destructive, the wave is extinct. Thus, after reflection at point C, the radiation of wavelength λ_i will create at the detector D a sinusoidal fluctuation of the energy. This holds for any wave of wavelength λ_i (frequency ν_i) being included in the polychromatic light and transmitted through the sample. All the partial energy fluxes corresponding to the various λ_i's are superimposed to give the total energy arriving at the detector, thus generating a complex interferogram (Fig. 4.4). The features of the interferogram depend, *inter alia*, on the energies (amplitude) of the waves with λ_i transmitted through the sample and the speed of the moving mirror M2. As a function of the pathway difference **x** the total energy flux of the interferogram is given by Eq. 4.6:

$$f(x) = I(x) - \frac{I(0)}{2} = 2 \int_{\tilde{\nu}=0}^{\infty} C(\tilde{\nu}) \cos(2\pi\tilde{\nu})\, d\tilde{\nu} \tag{4.6}$$

The mathematical operation called Fourier transformation, which is carried out by the computer of the FTIR spectrometer, generates therefrom the energy as a function of the wavenumber ν:

$$C(\tilde{\nu}) = 2 \int_{x=0}^{x=\infty} \left[I(x) - \frac{I(0)}{2} \right] \cos(2\pi\tilde{\nu}x)\, dx. \tag{4.7}$$

This may be approximated by a summation,

$$C(\tilde{\nu}) = \sum_{n=0}^{N} [2I(nq) - I(0)] \cos(2\pi\tilde{\nu}nq)q \tag{4.8}$$

with $n \times q = x$, where q is the difference between two consecutive values of the path difference at which the interferogram is sampled and is called the sampling period.

The summation carried out according to Eq. 4.8 should provide the IR spectrum in its conventional form.

Precise computation has to account for the fact that (i) the sampling period q is finite instead of infinitesimal and (ii) that the integral is truncated (infinity is replaced by $N \times q$). Appropriate convolution and weighting functions are used for corrections of the sampling and truncating errors [12].

Advantages of FTIR spectrometers over dispersive instruments are that they provide (i) a large signal to noise (S/N) ratio (Fellgett advantage); (ii) greater light throughput; (iii) better resolution; and (iv) very rapid generation of the spectra.

4.1.3.2 Diffuse Reflectance IR (Fourier Transform) Spectroscopy (DRIFT)

For a long time, IR spectroscopic studies on zeolites were almost exclusively carried out in the transmission mode. During more recent years diffuse reflectance spectroscopy (DRS), which used to play an important role in UV-visible spectroscopy on zeolites, has also been employed in IR studies. Pioneering work in this area was done by Kazansky and co-workers (compare, e. g., [13–15]). DRS is particularly helpful in exploring the near IR region, where transmission techniques essentially fail due to the severe scattering of the zeolite samples. However, this region is of great interest because, for example, of the occurrence of overtone and combination modes of the hydroxyls which might be studied with or without the use of probe molecules.

Two important requirements for successful diffuse reflectance measurements must be fulfilled: (i) the sample must be sufficiently thick and large to avoid light loss by forward scattering and/or scattering at its edges; (ii) scattering must be predominant compared with absorption so that illumination is essentially diffuse.

The theory of diffuse reflectance spectroscopy is well developed [16–17]. The features of DRS measurements are described by two parameters, K and S, related to absorption and scattering, respectively. The ratio of these parameters is the Schuster-Kubelka-Munk remission function:

$$F(R_\infty) = K/S = (1 - R_\infty)^2/(2R_\infty) \tag{4.9}$$

where R_∞ represents the reflectance of a very thick sample with zero background reflectance [17]. If K is a function of the wavelength (wavenumber), and S is not, $F(R_\infty)$ reflects the dependence of the energy of absorption on the wave-

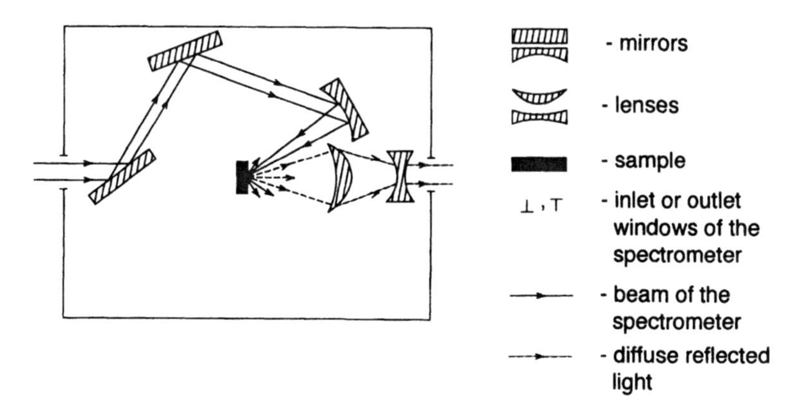

Fig. 4.5. Scheme of a unit for diffuse reflectance infrared spectroscopy (DRS)

length (wavenumber), and a plot of F (R_∞) should correspond to the transmission spectrum of the material under investigation, according to Eq. 4.10:

$$\log F(R_\infty) = \log K(\lambda) - \log S(\text{constant}). \qquad (4.10)$$

In the IR region, however, essentially no *quantitative* measurements of the diffuse reflectance are reported. This is mainly due to the fact that no general standard (as in UV-visible) is available. Nevertheless, DRS is particularly sensitive in the overtone and combination band section in near IR of zeolites and zeolite/adsorbate systems, and the information provided by DRS about zeolites and zeolite/adsorbate systems is most valuable (*vide infra*).

Reflectance accessories for IR spectrometers are commercially available. They can, however, also be relatively easily designed in a laboratory using commercially produced mirrors or lenses. Fourier transform IR spectroscopy can also be used in the reflectance mode (DRIFT). An example of a laboratory designed DRIFT accessory is shown in Fig. 4.5 (compare [18]).

4.1.3.3
Photoacoustic IR Spectroscopy (PAS)

Photoacoustic spectroscopy, not only for the IR region but also for the UV and visible range, was developed and particularly recommended for investigations with opaque samples. The experimental set-up uses a powerful source, the radiation of which is modulated, e.g., by a chopper, and monochromatized. Then the periodically chopped monochromatized radiation is directed into the photoacoustic cell (see Fig. 4.6), which is filled with a non-adsorbate gas. There the light generates, via periodic heating of the boundary layer of the filler gas, an acoustic wave with the same frequency as that used for the chopper.

The radiation penetrates the sample to a certain depth, this depth being a function of the optical absorption at a particular frequency. The radiation

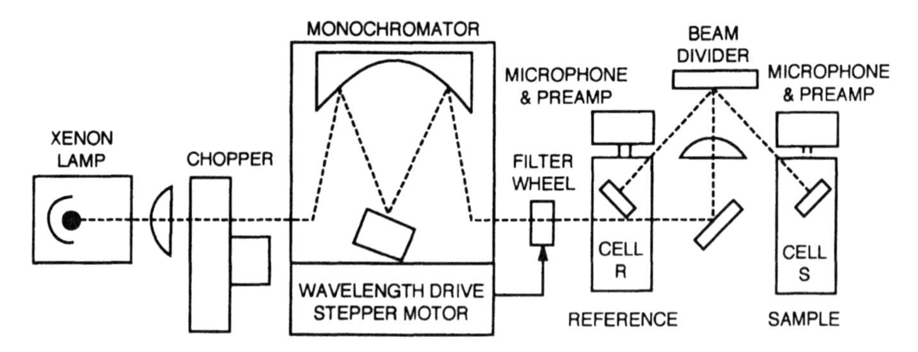

Fig. 4.6. Scheme of a photoacoustic spectrometer [19]

absorbed generates a corresponding amount of heat. This heat, in turn, causes an increase in the temperature in that volume element where the absorption occurred. This effect is partly counterbalanced by the flow of the heat to the sample surface. A temperature profile is built up, and the surface temperature will depend (i) on the depth where the heat is produced, i. e., on the optical absorption length, and (ii) on the efficiency of heat transfer, i. e., on thermal diffusivity. The surface temperature determines, in a sealed cell, the pressure of a surrounding gas phase. Now, if the frequency of the radiation illuminating the sample is varied, e. g., by a monochromator, the surface temperature will also change because of the variation in wavelength-dependent light absorbance and heat generation. Consequently, the gas pressure will also vary, and this can be indicated via the membrane of a microphone giving rise to an acoustic or, after appropriate conversion, an electrical signal. The intensity of the sound is a function of the wavelength-dependent absorption. Thus, the plot of the sound intensity vs. the wavenumber provides the photoacoustic spectrum in analogy to the IR absorption spectrum of the sample. A possible design of a photoacoustic spectrometer is described by Blank and Wakefield [19] and partially reproduced in Fig. 4.6.

Application of photoacoustic IR spectroscopy is sometimes recommended for optically opaque samples with the thermal diffusion lengths being smaller than the inverse of the optical absorption coefficients.

The theoretical treatment by Rosencwaig and Gersho [20], however, applies for non-scattering solids, and it seems that the theoretical base for PAS of systems which involve light scattering is still not satisfactorily developed. Photoacoustic IR spectroscopy may also take advantage of the Fourier transform of interferograms. Examples of applications of Photoacoustic Fourier Transform IR Spectroscopy (FTIR-PAS) to zeolites are available [21–23].

4.1.3.4
Cells for Studying Zeolites and Zeolite Adsorbate Systems by IR Spectroscopy

It is essential to have suitable cells at one's disposal in order to investigate, after appropriate pre-treatment, the pure zeolite samples and/or the samples

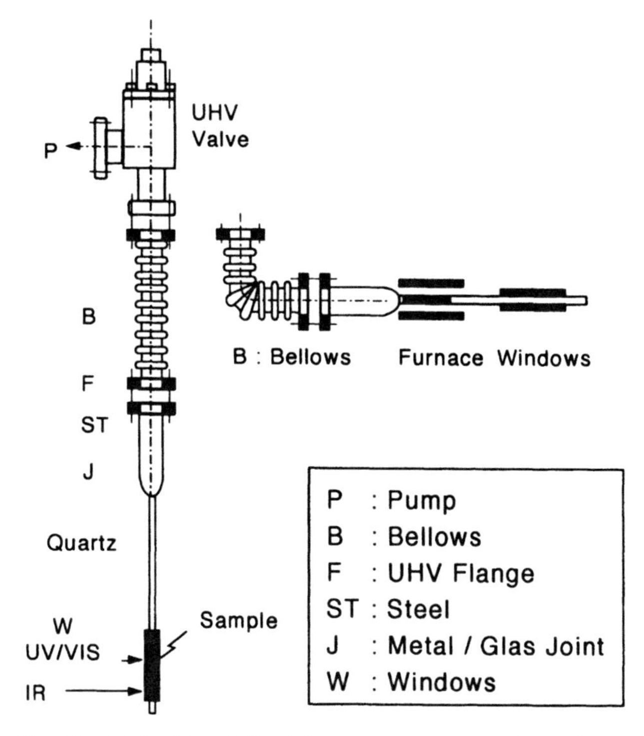

Fig. 4.7. Cell for IR and UV-visible measurements under static conditions

after contact with adsorbates or reactants. In many cases it will be desirable to carry out *in situ* IR spectroscopy, e.g., at a certain temperature and pressure in the case of (static) adsorption or reaction. In the latter case a flow system may be used, where the IR cell operates as a microflow reactor. Flow reactor cells may be advantageously connected to an analytical instrument such as a mass spectrometer or gas chromatograph. A number of useful cell designs have been described in the literature (see, e.g., [24–32]). Figures 4.7–4.9 present, as some examples, three types of cells which have been successfully employed.

The cell depicted in Fig. 4.7 is a relatively simple one. It may be connected by an all-metal valve to the high-vacuum and gas-dosing system and, after closing the valve, safely disconnected, if necessary. The sample is pre-treated at higher temperatures in the upper part in order to avoid heat-damage to the sealed windows. After treatment, the metal (gold or tantalum) frame containing the sample wafer is slid down into the compartment. With appropriately chosen windows this design enables measurements to be carried out with the very same sample in UV-visible, near infrared and middle infrared or even ESR (see [33]) using different spectrometers. Of course, if the range of interest is constrained to UV-visible and near infrared, the whole cell may be made of quartz glass only. Such a cell is also suitable for DRIFT measurements.

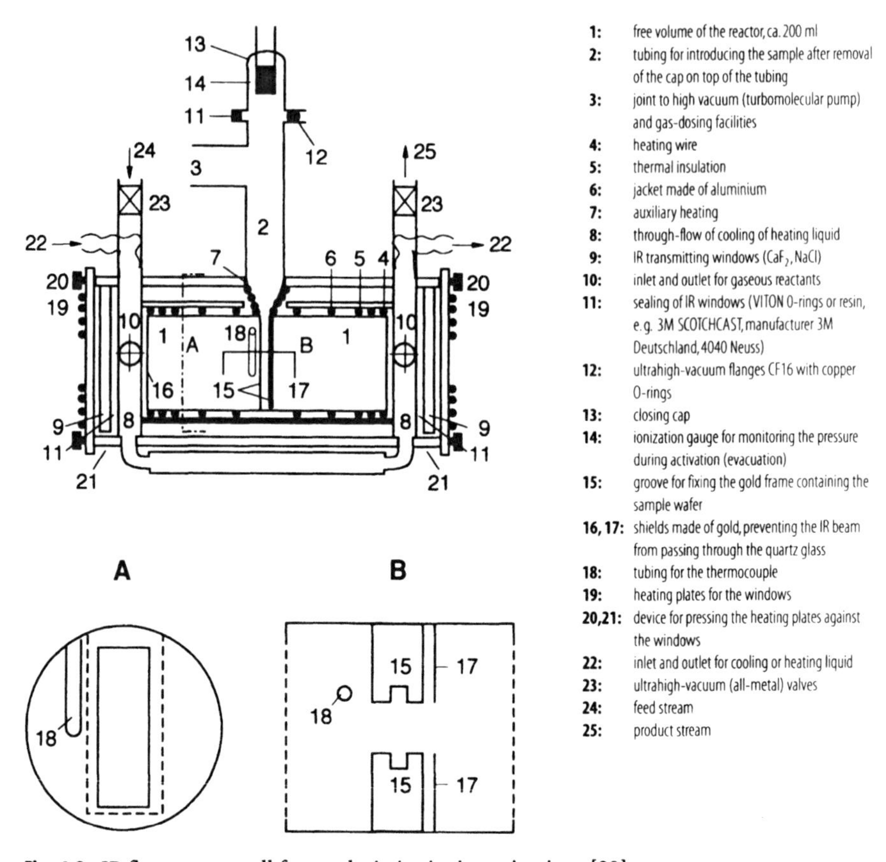

Fig. 4.8. IR flow reactor cell for catalytic *in situ* investigations [30]

Figure 4.8 shows a cell, manufactured from quartz glass which has proved its worth as a flow reactor cell. The sample is introduced into the middle of the cell through the upper conflat flange prior to connecting the cell with an ultra-high vacuum system. The design allows *in situ* measurements at high temperatures, the temperature gradient between the position of the sample and the (thermostated) windows being still acceptable. Because of the position of the sample in the middle of the cell, it is desirable, even though not indispensible, to have the focus of the beam close to the middle of the spectrometer compartment. For simultaneous monitoring of the spectra of a catalyst sample and the conversion of the reactant flowing through the cell, the outlet is connected with a gas chromatograph or mass spectrometer.

The cell drafted in Fig. 4.9 has the advantage of a very short optical path. The lower compartment can be heated up to 525 K (depending on the seal of the windows). For treatment at higher temperatures, the sample, the holder of which is connected to a rod with glass-sealed iron, is moved into the upper part of the stainless steel tube, and the rod is held in position by a strong magnet. The

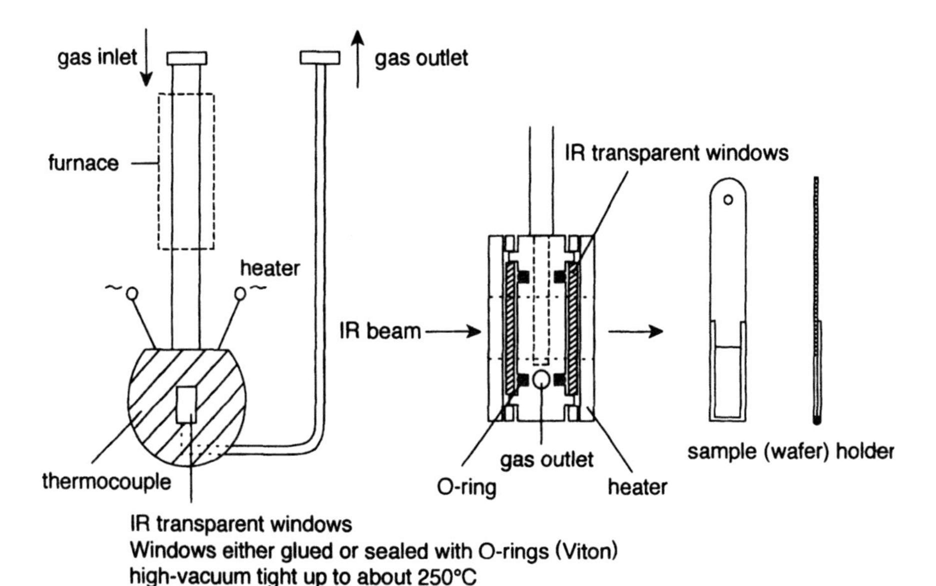

Fig. 4.9. IR flow reactor cell for *in situ* measurements of diffusion and reactions

cell can be used for *in situ* investigations in a flow of adsorbates or reactants, e.g., for studying the sorption kinetics of a zeolite/hydrocarbon system (*vide infra*).

IR cells for experiments at low temperatures have also been described in the literature (compare, e.g., [34]). Similarly, cells particularly suitable for IR microscopy and for DRIFT have been designed [35].

4.1.4 Study of Framework Vibrations of Zeolites

In a classic paper, Flanigen et al. [36] studied systematically the relationship between IR spectral features of zeolites and structural properties thus extending earlier IR investigations of silica and non-zeolitic aluminosilicate frameworks [37–38] to zeolites. The bands observed in mid infrared were classified into two main categories, viz. bands due to internal vibrations of the $TO_{4/2}$ tetrahedra (T = Si or Al) and external vibrations of tetrahedra linkages, e.g., in double rings (as in A-, X-, Y-type) or pore openings (as, e.g., in mordenite). This is illustrated in Fig. 4.10 (adopted from [36]), where the internal tetrahedra vibrations are indicated by solid lines, whereas the external linkage vibrations are indicated by broken ones.

The most prominent bands of the first category occur in the ranges from 1250 to 950, from 790 to 650 and from 500 to 420 cm^{-1}, which are tentatively assigned to the asymmetric stretching ($\leftarrow$ OT $\rightarrow$ $\leftarrow$ O) mode, the symmetric stretching

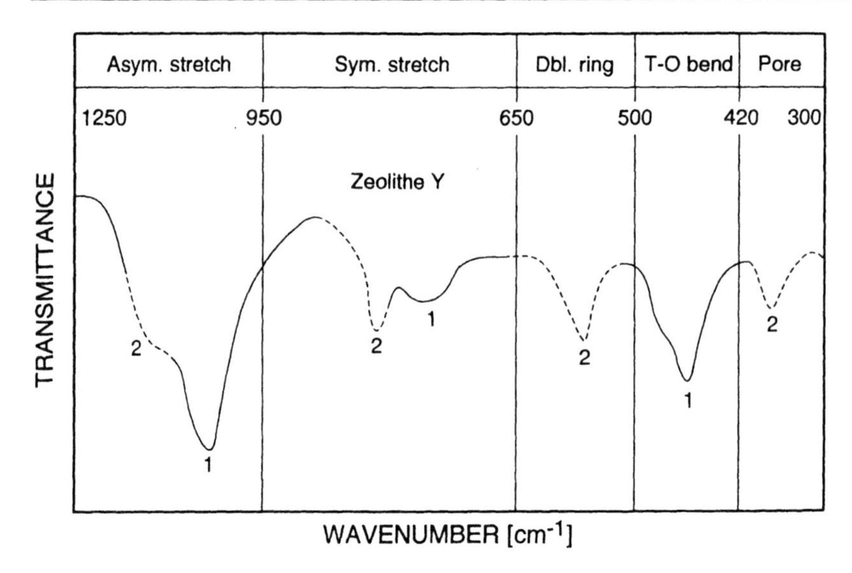

Fig. 4.10. Lattice vibrations of zeolite Y (from [36])

mode (← OTO →) and the T–O bending mode of the $TO_{4/2}$ tetrahedra, respectively. Similarly, bands around 650–500 cm^{-1} and 420–300 cm^{-1} were ascribed to external linkage vibrations, viz. vibrations of double four-membered rings (D4R) or double six-membered rings (D6R), and pore opening vibrations, respectively. These proposed assignments as listed in Table 4.1 were substantiated by data obtained for a large number of zeolites with a great variety in structure and composition.

It turned out by comparison of spectral and structural features that the internal vibrations are largely structure insensitive, whereas the position of the bands due to vibrations of external linkages are often pronouncedly structure sensitive. This was confirmed, *inter alia*, by the observation that the bands of the first category remained almost unaffected when the structure of, e.g., a Y-type zeolite was successively destroyed by thermal treatment. In contrast, the bands of the external linkage vibrations were caused to disappear upon the same treatment.

An essentially linear relationship, even though exhibiting some scatter due to some structure-related features of the IR band, was obtained if the wave-

Table 4.1. Assignments of Zeolite Lattice Vibrations [36]

Internal Tetrahedra	Vibrations	External Linkages	Vibrations
Asym. stretch	1250–950	Double ring	650–500
Sym. stretch	720–650	Pore opening	420–300
T–O bend	500–420	Sym. stretch	820–750
		Asym. stretch	1150–1050 sh

sh: shoulder.

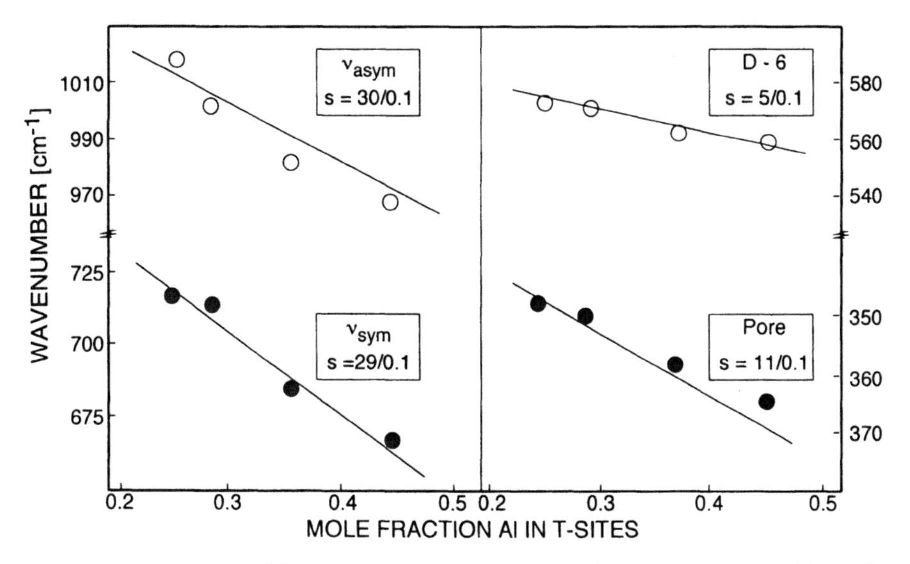

Fig. 4.11. Wavenumbers of lattice vibrations as a function of the atom fraction Al/(Si + Al) (from [36])

numbers of the asymmetric stretch was plotted vs. the atom fraction of Al in T sites for all the zeolites studied. Very good correlations between ν and Al/(Si + Al) were observed if ν(asym), ν(sym), ν(DR) or ν ("pore opening") were plotted for homologous series of only one structure type such as faujasite, i.e., X and Y with different Si/Al ratios (Fig. 4.11). In fact, such relations were later on frequently used to determine the Al/(Si + Al) ratio in zeolite frameworks (compare, e.g., [39,40]). The T–O bending band did not vary significantly, in particular not with the same structure and different Si/Al ratios.

Flanigen et al. [36] were also able to show that the position of the framework IR bands respond to dehydration and rehydration as soon as this is connected with a migration of multivalent cations accompanied by a change in the framework distortion. This is, for example, the case with Ca^{2+} and La^{3+} in faujasite-type zeolites, where the dehydration causes a migration of the cations from inside the sodalite cage to a site close to the center of the D6R. This, in turn, distorts the framework and changes the symmetry of the D6R which is reflected in the shift of, e.g., the corresponding IR band from 570 to 635 cm^{-1}.

Theoretical attempts in more recent years have resulted in the proposal of a modification of the approach by Flanigen et al. and a re-assignment of the bands observed in the mid infrared of zeolite frameworks. Thus, Geidel [41] and Geidel et al. [42] carried out a normal coordinate analysis of the framework vibrations of NaX with the help of subunit cluster modeling, increasing the cluster size from single tetrahedra over ditetrahedra, four- and six-rings and double six-rings to a $SiO_4\,(AlO_3)_4$ cluster. They arrived at the conclusion that the Flanigen concept of strictly separated external and internal tetrahedral vibrations must be modified: Since the zeolite framework vibrations are strongly coupled, each mode is supposed to exhibit simultaneously the character of internal tetrahedral and bridging vibrations. Moreover, the coupling between adjacent tetrahedra was shown to be more significant than that within a tetrahedron.

Computation of the normal modes of deformation vibrations δ(OTO) yielded a sequence of frequencies in the range 550–140 cm^{-1} due to the occurrence of mixed forms. Furthermore, the calculations suggest bands originating from deformations of the type δ(TOT) and torsional vibrations to appear in the range below 200 cm^{-1}. Pore opening modes are interpreted in a somewhat different view compared to that of Flanigen [36]. Geidel [41] and Geidel et al. [42] concluded from their calculations that the so-called pore opening mode (420–300 cm^{-1}) is not due to a breathing vibration of the pore mouth ring, rather it is related to a complex motion which in total includes an opening (rupture) of the rings (4-ring, 6-ring) of the structure.

The vibration around 650–500 cm^{-1} was, in agreement with the other authors, ascribed to the vibration of one ring against the other in a double ring subunit (D6R) which, however, is accompanied by an opening of the four-rings of the structure.

Frequencies in the range 700–600 cm^{-1} were assigned to symmetric stretching vibrations, similar to the proposal by Flanigen et al. However, these vibrations were not interpreted as internal displacements inside the $TO_{4/2}$ tetrahedra as they were by those authors. They were rather described as motions of bridging SiO and AlO groups for which the notation $\nu_{b,s}$ (SiO, AlO) was chosen. Similarly, the high-frequency bands in the range 1200–950 cm^{-1} were attributed to asymmetric stretching modes of bridging SiO, AlO oscillators, $\nu_{b,as}$ (SiO, AlO). Bands due to coupled asymmetric vibrations of bridging AlO oscillators $\nu_{b,as}$ (AlO) appear at lower frequencies, i.e., around 750 cm^{-1}. At even lower frequencies, bands are observed which result from (symmetric) displacements of all bonds of the Al tetrahedra, indicated by $\nu_{T,s}$ (AlO).

Table 4.2. Approximate Assignments of Observed Lattice Vibrational Frequencies of NaX Zeolite [43] Based on Calculations [41,42]

Observed bands (cm^{-1})		Calculated (cm^{-1})	Assignment
IR	Raman		
1078 sh	1082 m	1095–1065	$\nu_{b,as}$ (SiO,AlO)
975 vs	985 m	936–901	
752 m	800 w	812–791	$\nu_{b,as}$ (AlO)
700 m		744–692	$\nu_{b,s}$ (SiO,AlO)
612 vw		609–604	$\nu_{T,s}$ (AlO)
562 m		569	double ring mode
	510 vs	507–494	δ_{bbs} (OTO)
461 s			δ (OTO)
408 w			
364 m	375 m	365–260	pore opening mode

sh: shoulder, w: weak, vw: very weak, m: medium, s: strong, vs: very strong.
b: displacement of bridging tetrahedron; T: displacement of total tetrahedron.

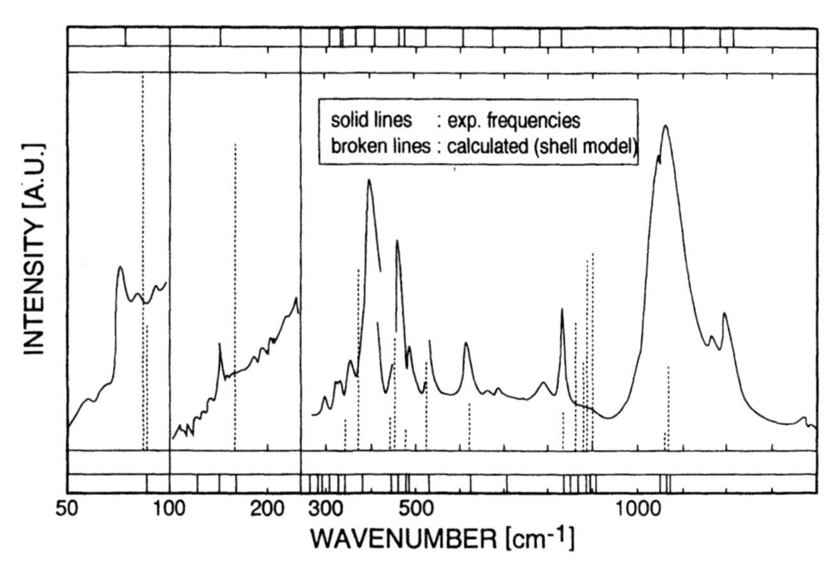

Fig. 4.12. Calculated and experimental IR spectra of lattice modes of faujasite (from [44])

On the basis of their normal coordinate analyses, Geidel et al. [42] proposed assignments of the IR bands experimentally found by Miecznikowski and Hanuza [43] for a NaX zeolite (Si/Al = 1.12). These assignments are reproduced in Table 4.2. The assignments for Raman active lattice vibrations are supported by measurements carried out by the same authors [42].

While the calculations by Geidel [41] were based on the consideration of small clusters, van Santen et al. [44] presented results of infinite lattice calculations providing data on the O–Si–O bending vibrations (below 500 cm^{-1}) and the Si–O stretch region (800 to 1100 cm^{-1}). A rigid ion and a shell model (Bethe lattice model) were used, the latter giving results in closer agreement with experimental spectra. An example is shown in Fig. 4.12: (i) The computations reproduce the intensity-free "quartz frequency gap" between 500 and 700 cm^{-1}; (ii) the agreement between the calculated and measured vibration modes originating from O–Si–O bending (below 500 cm^{-1}) is rather good; (iii) the calculated asymmetric Si–O stretch (900 cm^{-1}) is significantly lower than the experimentally observed one (1080 cm^{-1}), whereas the data for the lower stretching mode (about 800 cm^{-1}) coincide.

4.1.5
IR Investigation of Acidic and Basic Sites in Zeolites

Acidic sites, such as acidic OH groups (Brønsted acid centers), true Lewis sites (aluminum-containing extra-framework species) and cations as well as basic sites (such as basic oxygen atoms or alkaline metal clusters) are encountered in zeolites and are of paramount importance in acid-base catalysis on zeolites. By IR spectroscopy, only the Brønsted acid sites may be investigated with and without probe molecules, whereas acidic Lewis sites, cations and basic sites can be identified and quantitatively determined only with the help of probes. Probe molecules frequently employed for acidic sites are pyridine, substituted pyridine, ammonia, amines, carbon monoxide, methane, hydrogen and more recently [45–47] fluoro-/chloroethane and ethene for acidic centers and carbon dioxide or pyrrole for basic sites.

4.1.5.1
Brønsted Acid Sites (Acidic Hydroxyls)

4.1.5.1.1
Fundamental Stretching Bands

In a classic paper, Uytterhoeven et al. [48] systematically studied the development of acidic hydroxyls (now usually called "bridging hydroxyls", see Scheme 4.1 in Sect. 4.1.5.1.3) upon deammoniation of the ammonium form of a faujasite-type Y-zeolite, i.e., NH_4-Y. Two prominent bands appeared in the region of OH stretching vibrations at 3640 and 3550 cm^{-1} (Fig. 4.13). Deammoniation was essentially completed around 625 K. Above 625 K, further heat-treatment caused these two bands to be weakened which indicated dehydroxylation of the zeolite (see below). The fact that two well-separated bands were observed was first ascribed to interaction of a fraction of the acidic hydroxyls, i.e., of suitably adjacent ones, via hydrogen bonding [49–50]. Later on, Jacobs and Uytterhoeven [51] pointed out that the duplicity of OH stretching modes originates from different crystallographic positions of the oxygens involved. The high-frequency

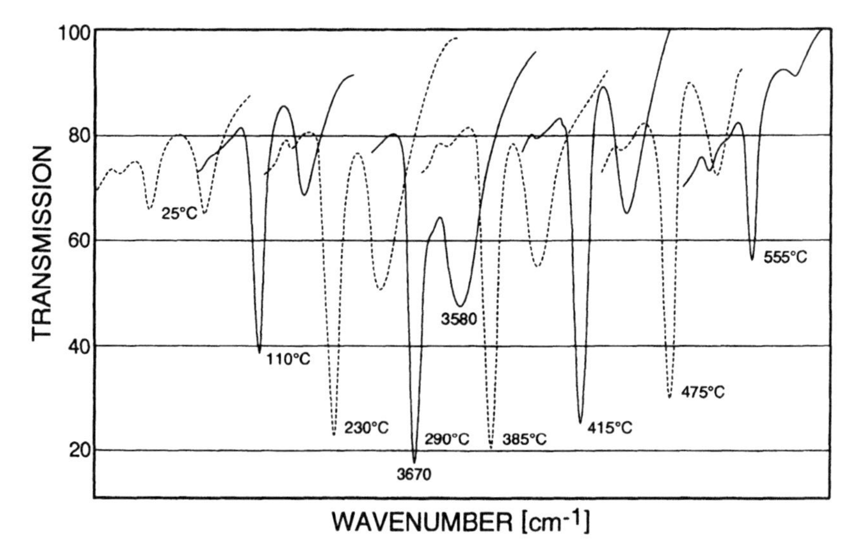

Fig. 4.13. Development of the low-frequency (3580 cm^{-1}) and high-frequency (3670 cm^{-1}) bands on deammoniation of zeolite NH_4-Y in high vacuum at increasing temperatures, dehydroxylation above 400 °C [48]

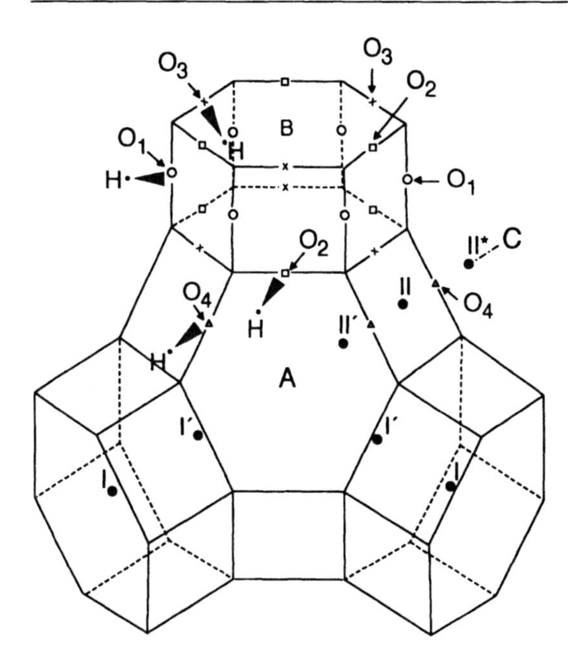

Fig. 4.14. Fragment of the faujasite-type structure indicating the four crystallographically different oxygens (O_1, O_2, O_3, O_4) and corresponding types of OH groups (O_1H, O_2H, O_3H, O_4H)

band (usually indicated as the HF band) was ascribed to OH oscillators formed upon attachment of protons to O_1-oxygens (see Fig. 4.14). These OH groups point into the large cavity. By contrast, the LF–OH band at 3550 cm^{-1} was attributed to OH groups in which O_2, O_3, O_4 oxygens of the framework are involved and which are directed into the sodalite cages. This assignment was re-considered by, e.g., Czjzek et al. [52], who showed that the HF band was mainly caused by O_1H and the LF band by O_3H; the concentration of O_2H is distinctly lower whereas O_4 is essentially not populated by protons.

More than one type of stretching bands resulting from acidic OH groups was also found in the case of hydrogen forms of, e.g., erionite [53], ZSM-20 [54], SAPO-37 [55] and, in analogy to the case of faujasite-type zeolites, explained by a distinct and sufficiently pronounced difference in the crystallographic positions of the framework oxygens involved. In other cases such as mordenite [56, 57], clinoptilolite [58, 59], H-ZSM-5 (compare [60] and [61]), only one band due to acidic hydroxyls was observed. In several IR studies hydroxyl groups attached to (framework or non-framework) aluminum, both acidic and non-acidic were also identified [62]. The data reported until 1976 have been reviewed in articles by Ward [5] and Foerster [6] and in a book by Jacobs [63].

As early as 1978 a particular proposal had been advanced by Barthomeuf [64, 65] to correlate the band position (wavenumber, frequency) of the acidic OH groups with other chemical or physicochemical properties of zeolites (compare also [66], especially p. 59). Figure 4.15 shows a plot of a great number of measurements of the OH band position which were obtained with a variety of zeolites as a function of the Si/Al ratio of the respective materials. Similarly, a decrease of the wavenumber with increasing degree of exchange of alkaline metal

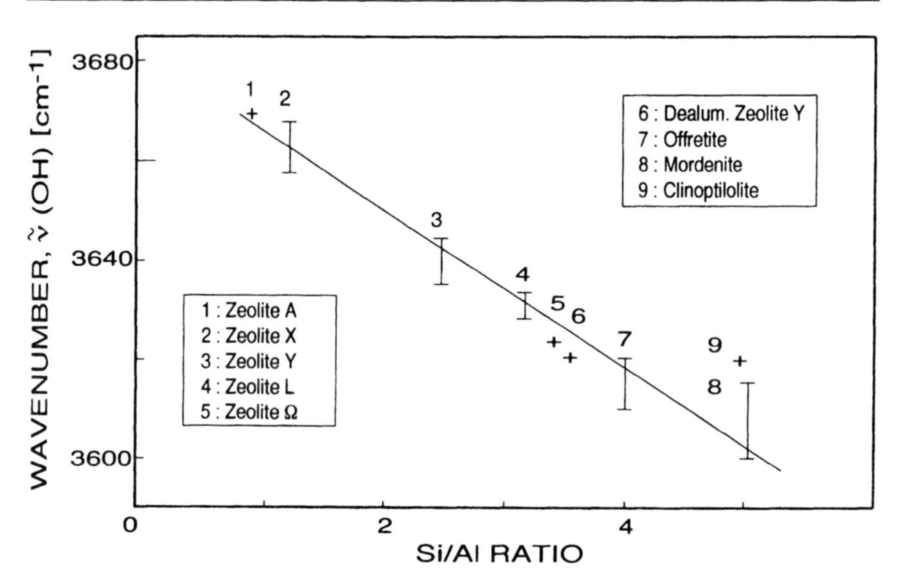

Fig. 4.15. Wavenumbers of stretching vibrations of acidic OH groups, ν(OH), of hydrogen forms of various zeolites (from [60])

cations for ammonium (hydrogen after deammoniation) in a series of faujasites was observed [66, 67]. In a homologous series of mordenites exchanged with alkaline-earth metal cations a corresponding shift of the wavenumber of the OH band to lower frequencies was measured if the coulomb potential of the introduced cation increased. Thus, $\tilde{\nu}$(OH) decreased in the sequence Ba,Na-MOR > Sr,Na-MOR > Ca,Na-MOR > Mg,Na-MOR > Be,Na-MOR [68, 69].

In fact, the relationship between $\tilde{\nu}$(OH) and Si/Al seems to suggest a correlation between the acidic strength of the OH groups and the wavenumber, since it was frequently claimed that, for instance, H,Na-X (Si/Al $\simeq$ 1.25) has Brønsted acid sites of lower strength (lower activity in acid-catalyzed hydrocarbon reactions) than H,Na-Y (Si/Al $\simeq$ 2.5). H,Na-Y in turn was assumed to exhibit lower strength and activity of OH groups than mordenite (Si/Al $\simeq$ 5). Moreover, an increase of acidity strength and activity was reported for an increase of dealumination of the framework of H,Na-Y [70-74]. In this context, the strength of the acidity of the respective zeolites has usually been characterized by titration with bases, temperature-programmed desorption of ammonia or pyridine and/or microcalorimetric measurements of the differential heats of adsorption of such probe molecules (see [70-72, 74]). Looking at a relationship as demonstrated in Fig. 4.15 it may be concluded that low values of $\tilde{\nu}$ correspond to high acid strength and higher values of $\tilde{\nu}$ to weak acid strength. In fact, this approach has been frequently used in attempts to rationalize experimental results [21, 75].

However, since in some cases of different zeolites with similar (weak) acid strengths a variety of $\tilde{\nu}$ (OH)'s were observed, Barthomeuf [64, 76] concluded that the acid strength cannot be related simply to the wavenumber of the band of acidic hydroxyls alone. Rather, this author related the acidity strength to the charge density of the framework which also depends on the aluminum content. A decrease in the aluminum content (rise in Si/Al) decreases the charge density of the anion framework of the zeolite. Thus, the hydroxyl groups are subjected to less intense interaction with the framework, which should decrease the force constant of the OH oscillator, shift the wavenumbers to lower values and facilitate the deprotonation, i.e., enhance the acidity strength. It was suggested that the aluminum content [Al/(Al + Si)] affects the activity coefficient which might be defined for zeolites in a similar way as for solutions. A small aluminum content (low charge density) would correspond to a higher activity coefficient. The activity coefficient would increase, e.g., in the sequence, X, Y, mordenite, i.e., opposite to $\tilde{\nu}$(OH). The activity coefficient was, in the case of faujasite-type zeolites, identified with the efficiency coefficent as defined earlier on the basis of titration experiments [70, 76]. The main difficulty in the application of the concept of activity coefficients in the case of zeolites lies in the problem of defining an appropriate reference and standard state.

Mortier [77], Jacobs et al. [78] and Jacobs [79] have correlated the wavenumbers of IR bands originating from acidic hydroxyls in zeolites, with the Sanderson electronegativity [80] and the acidity strengths. Since the intermediate Sanderson electronegativity, S_{int}, and the partial charges on the atoms of the zeolite structure, e.g., on the hydrogen of the hydroxyl groups, are linearly related, S_{int} will directly correlate with the acidity strength of Brønsted acid sites. Higher values of S_{int} correspond to higher hydrogen charges and, therefore, to greater acid strengths [77]. Since the earlier studies [64, 65] used data from various authors with unknown accuracy of the measurements and without consideration of residual cation content, for this re-investigation a complete series of hydrogen zeolites of known chemical composition was prepared. Any dealumination (formation of cationic Al-containing extra-framework species, see below) was avoided by careful preparation and pretreatment. A plot of $\tilde{\nu}$(OH) vs. the intermediate Sanderson electronegativity yielded a linear relationship including H-ZSM-5 (MFI) and H-ZSM-11 [79, 81].

The results summarized in Fig. 4.16 show that there is no discontinuity in the physicochemical properties as a function of the Al content. By contrast, a plot of $\tilde{\nu}$(OH) vs. Al/(Al + Si) would show a break in the straight line at about Si/Al = 6.4 close to the ratio Si/Al = 7 expected on the basis of earlier approaches [76, 79].

The main difference between the plot of $\tilde{\nu}$ or $\tilde{\nu}^2$ vs. Al/(Al + Si) or a pseudo-electric field parameter E [82] on the one hand and vs. S_{int} on the other is that only the intermediate Sanderson electronegativity accounts for differences in chemical composition, whereas Al/(Al + Si) and E do not (compare [79]). Thus, the non-linearity in plots of the latter types, which had already been encountered earlier, was ascribed to small differences in the chemical composition, particularly in the nature and number of residual cations, of the zeolites. Consequently, the correlation between the wavenumber of the stretching mode of acidic hydroxyls and the acidity strength of these OH groups does not any longer hold as soon as inhomogeneities or unknown changes in the chemical composition of the zeolites occur. In view of these results it was concluded that, in particular with siliceous zeolites, the wavenumber $\tilde{\nu}$(OH) is not an adequate measure of acidity strength [79].

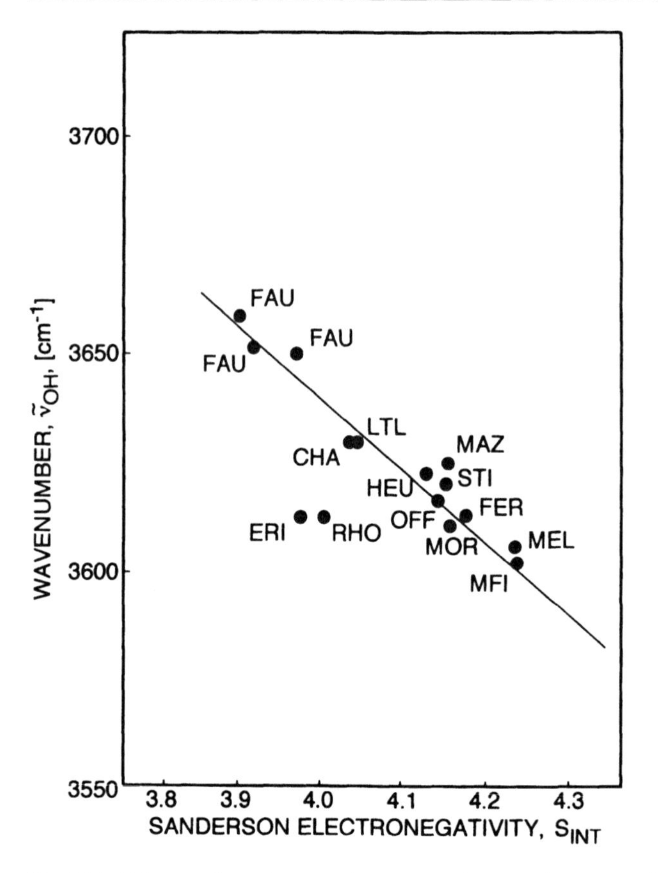

Fig. 4.16. OH stretching frequencies vs. Sanderson electronegativity of various zeolites (from [74])

Even though the concepts of both Barthomeuf and Mortier et al. imply long-range interactions and describe collective phenomena, it is important to consider also the local configuration of the OH groups. Mortier therefore pointed out that in a strict sense the intermediate electronegativity will be quantitatively correlated to physicochemical properties such as $\tilde{\nu}$(OH) and acidity strength only within a homologous series of zeolite samples. Both aspects, i.e., the long range effect as well as that of local configuration, is also involved in the concept advanced by Rabo and Gajda [83, 84]. These authors pointed out that the specific zeolitic acidity results from the ability of the crystalline structure to respond to a proton attack via relaxation of the whole zeolite lattice. In this way major local distortions of the structure (bond lengths, bond angles) in the environment of the OH groups are avoided. This gives rise to a tendency of equivalization of the Al–O and Si–O bonds and an increased interaction of the oxygen with the adjacent aluminum which results in a weakening of the O–H bond and enhancement of acid strength. Because of the lack of long-range interaction, an amorphous material such as silica/alumina cannot re-adjust like a zeolite lattice.

In the view of Rabo and Gajda this may be the basic reason for the differences in the properties of hydroxyls in zeolites and silica/alumina, in particular with respect to acidity. This concept remains to be tested by comparative IR (and ^1H MAS NMR) spectroscopic investigations.

At some variance with Rabo and Gajda [83], studies by van Santen [85] and van Santen et al. [86] on the strength of the OH bond in the bridging Si(OH)Al configuration stress much more the local effects. Their considerations of changes in partial charges on the atoms involved, in bond angles and bond lengths are, however, based on rather small clusters. It is still a matter of debate whether this approach is adequate to the problem, and it remains an open question as to whether long-range or short-range influences are predominant in molecular sieve chemistry and, in particular, in respect to rationalizing IR spectroscopic observations.

There are only a few contributions providing theoretical calculations of IR stretching frequencies of OH groups in zeolites. For example, Sauer [87] computed $\tilde{\nu}$(OH) values and deprotonation energies using methods of quantum chemical cluster calculations which yield the harmonic force field including the coupling terms. From that the vibrational frequencies and normal modes were derived. The calculated frequencies for the stretching mode of the terminal SiOH groups and for their in-plane deformation vibration were 4140 cm^{-1} and 960 cm^{-1}, respectively, whereas the observed frequencies were 3745 cm^{-1} and 795 to 835 cm^{-1}. Similarly, the calculations yielded for the stretching mode, the in-plane and the out-of-plane deformation vibration of the bridging hydroxyls, respectively, the wavenumbers 4000 to 4090 cm^{-1}, 1180 to 1200 cm^{-1} and 420 to 440 cm^{-1} which have to be compared with the observed values of 3550 to 3660 cm^{-1}, 970 to 1055 cm^{-1} and 360 to 380 cm^{-1}. Thus, as frequently encountered, the calculated absolute frequencies turn out to be too high. The differences, however, are similar. For instance, the differences between the $\tilde{\nu}$(OH) frequencies calculated for terminal and bridging OH groups are between (4140 – 4090 =) 50 cm^{-1} and (4140 – 4000 =) 140 cm^{-1}. The corresponding values for the observed frequencies are (3745 – 3660 =) 85 cm^{-1} and (3745 – 3550 =) 195 cm^{-1}.

Improvements were achieved by a combination of long-range electrostatic two-body potentials and short-range pair and three-body potentials [88]. The OH stretching frequencies obtained for O_1–H, O_2–H, O_3–H and O_4–H were now 3772, 3702, 3736 and 3751 cm^{-1}. More recent progress in quantum chemical calculations are provided [89 – 92]. For example, for O_1–H, O_2–H, O_3–H and O_4–H, the following frequencies have now been calculated: 3626, 3506, 3569 and 3772 cm^{-1}, respectively, which should be compared with the observed band positions at 3623 (HF, O_1–H) and 3550 cm^{-1} (LF, O_3–H). Theoretical treatment is now able to take into account different zeolite structures [90 – 92] and even isomorphous substitution of T-atoms such as Ti in titanium silicalite [92]. In summary, there is rapid promising progress in the application of theoretical techniques in order to study vibrations of zeolite lattices and surface species.

4.1.5.1.2
Hydroxyls – Overtone and Combination Bands

The IR region above 4000 cm^{-1} includes the overtone vibrations of the hydroxyls as well as their combinations with the hydroxyl bending modes and lattice vibrations. This range is, for instance, very interesting with respect to the construction of the potential curves of the oscillators, such as Si–OH. Also, in many cases, it provides additional information about the surface species under study. However, the IR range above 4000 cm^{-1} is extremely difficult to investigate in the transmission mode because of the low intensity of the respective signals which is usually 3 to 5% of that of the fundamental vibration modes. In this context it should be noted that the absolute transmission of a zeolite wafer used for transmission IR spectroscopy in the fundamental stretching region is sometimes only 1% or even lower. Therefore, it was an important step forward when Kazansky and co-workers, in a series of pioneering papers [13–15, 93–94], introduced diffuse reflectance IR spectroscopy into zeolite research. IR reflectance spectroscopy takes advantage of the fact that the many reflections of the IR light experienced when interacting with the particles of the zeolite powder are equivalent to an extended optical pathway. The use of zeolite powder instead of self-supporting wafers as for transmission spectroscopy is, in some cases, another advantage. A drawback of diffuse reflectance spectroscopy is the poorer resolution; the difficulty in obtaining quantitative data has already been mentioned.

Kazansky et al. [93] investigated, *inter alia*, a series of zeolites including very siliceous ones. The spectra obtained in the range from 3400 to 7400 cm^{-1} are displayed in Fig. 4.17. For comparison, the diffuse reflectance spectrum of amorphous silica was also studied. It showed bands at 3745 cm^{-1} (fundamental stretch of silanol groups, ν [SiOH]), at 4540 cm^{-1} (combination of stretching and bending, $\nu + \delta$ [SiOH]) and at 7325 cm^{-1} (overtone, $\nu(0-2)$ [SiOH]). With the help of the silica spectrum and the well-established assignment of the fundamental stretching bands of H-ZSM-5 [60–62] and hydrogen mordenite [56–58], Kazansky et al. [93] were able to assign the bands of, e.g., H-ZSM-5 (Fig. 4.17) as follows: 3610 $cm^{-1} \simeq$ fundamental stretch of the bridging acidic hydroxyls, ν[Si(OH)Al]; 3745 $cm^{-1} \simeq$ fundamental stretch of silanol groups of the crystallite (surface, silica contaminations or defects), ν[Si–OH]; 4540 $cm^{-1} \simeq$ $\nu + \delta$ [Si-OH]; 4660 $cm^{-1} \simeq \nu + \delta$ [Si(OH)Al]; 7065 $cm^{-1} \simeq \nu(0-2)$ [Si(OH)Al]; 7325 $cm^{-1} \simeq \nu(0-2)$ [Si–OH]. On heat-treatment of the H-ZSM-5 sample, a new band appeared in the fundamental stretch region, viz. at 3680 cm^{-1}, which was ascribed to OH groups attached to extra-framework Al-containing species, Al–OH, formed upon dealumination.

Another interesting example out of the increasing number of diffuse reflectance studies of zeolite OH groups is that reported by Staudte et al. [95]. These authors investigated the combination and overtone vibration region 4000 to 5600 cm^{-1} and were able to assign the frequencies observed at 5130, 4655, 4545 and 4380 cm^{-1} to the combination vibrations $\nu(OH) + \delta(OH)$ of adsorbed water, acidic bridging OH groups, silanol Si–OH groups and extra-framework Al–OH, respectively (see Fig. 4.18). From these data the wavenumbers of the deformation mode, $\delta(OH)$, can be derived which, in the case of zeolites, are not

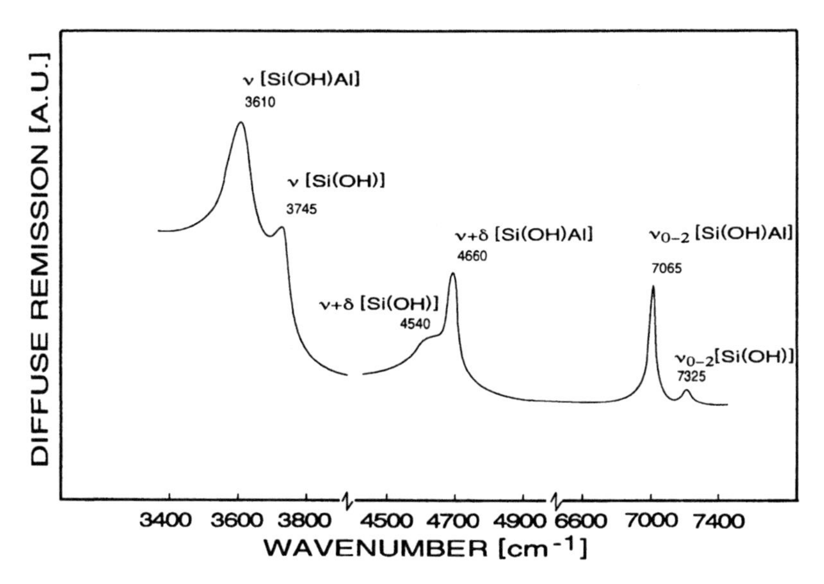

Fig. 4.17. Diffuse reflectance IR spectra of H-ZSM-5 (from [93])

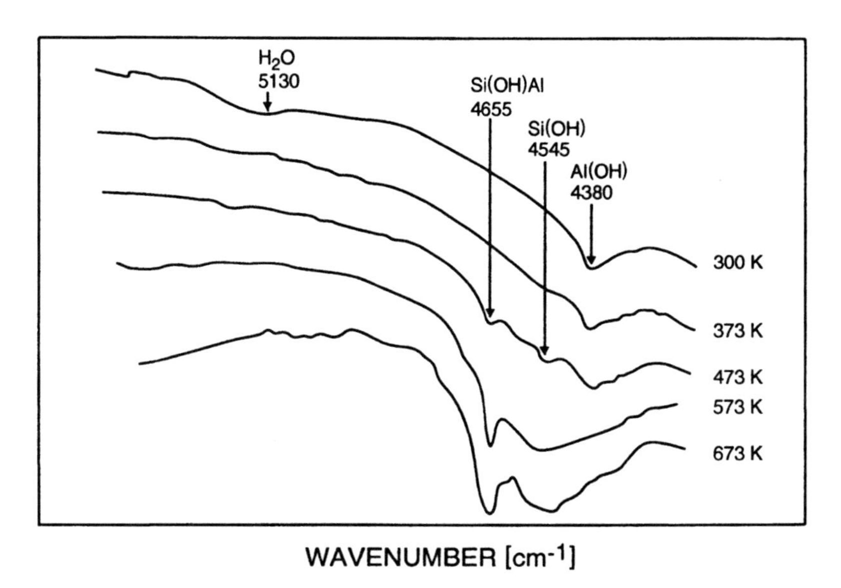

Fig. 4.18. Near infrared spectra with combination bands, ν(OH) + δ(OH), of Al-ZSM-5 for increasing activation temperatures [95]

Table 4.3. Assignments of Combination Bands Measured by DRIFT of H-ZSM-5 and Derived Lattice Vibration Modes (in cm^{-1}) [96]

Measured frequencies	Assigned to combination vibrations	Derived assignments	Computed [97]
3614	ν(OH) of Si–OH–Al	–	
4060	$\nu(OH) + \delta(OTO)$	δ(OTO):446	470
4340	$\nu(OH) + \nu_s(TO)$	ν_s(TO):726	765
4661	$\nu(OH) + \delta(OH)$	δ(OH):1047	
4760	$\nu(OH) + \nu_{as}(TO)$	ν_{as}(TO):1146	1130
4800	$\nu(OH) + \delta(OTO) + \nu_s(TO)$	$\delta(OTO) + \nu_s(TO)$:1186	1196
5700	$\nu(OH) + 2\delta(OH)$	δ(OH):1043	

available via direct measurements because the bands are obscured by the strong lattice vibrations in the region 800–1000 cm^{-1}.

In a more recent study, Beck et al. [96] were able to evaluate, inter alia, by careful analyses of their diffuse reflectance Fourier transform (DRIFT) spectra, the Si–OH and Si(OH)Al bending modes which appeared above 4000 cm^{-1} in combination with ν(OH) and lattice vibrations. The proposed assignments are collected in Table 4.3 (see also [97]).

4.1.5.1.3
Hydroxyls – Investigation by IR and Probe Molecules

The presence of acidic hydroxyls (Brønsted acid sites) on the external and internal surface of zeolite crystallites may be proven not only by the direct observation of the respective OH bands in IR but also by features which appear in the IR spectrum of zeolites containing acidic OH groups upon interaction with suitable probe molecules. These probes may be strong, medium or weak bases such as ammonia, pyridine or benzene, but even light paraffins and hydrogen have been employed. However, the most prominent probe molecule used in IR spectroscopy of zeolites is still pyridine. Pyridine adsorbed on or sorbed into zeolite structures gives rise to very sharp bands which selectively indicate, in a well-established specific way, Brønsted acid centers, so-called true Lewis sites (see Scheme 1) and cations. Moreover, pyridine is relatively thermally stable and will start to decompose or be reduced only above about 675 K (see [98]). Therefore, it can be used to study acidic sites at elevated temperatures being relevant with respect to catalysis. A drawback of this probe molecule is its relatively large size which, in the case of narrow pores or pore mouths or surface barriers etc., might prevent the probe molecules from having access to the sites.

Application of pyridine as a probe for IR studies goes back to early work on other adsorbents such as silica and alumina [99, 100]. Liengme and Hall [101] adopted this method to investigate the Brønsted acidity and its conversion to Lewis acidity in faujasite type zeolites. Figure 4.19 shows, as an example, the IR bands of pyridine adsorbed on hydrogen Y-zeolite (faujasite structure), hydrogen mordenite and sodium mordenite [102]. The signals at 1540, 1490,

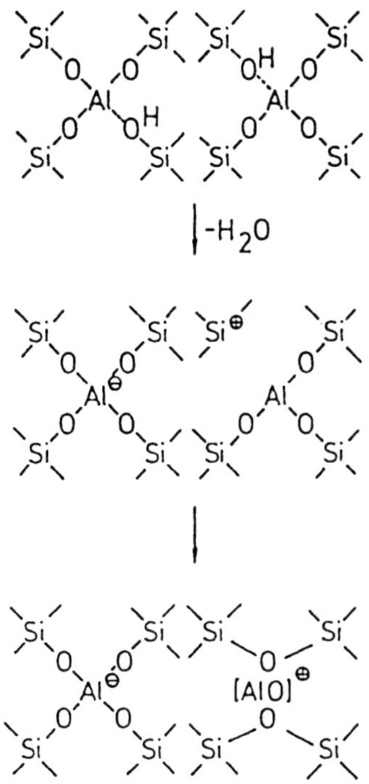

Scheme 4.1 Schematic representation of the dehydroxylation of a hydrogen zeolite [48, 123, 124]

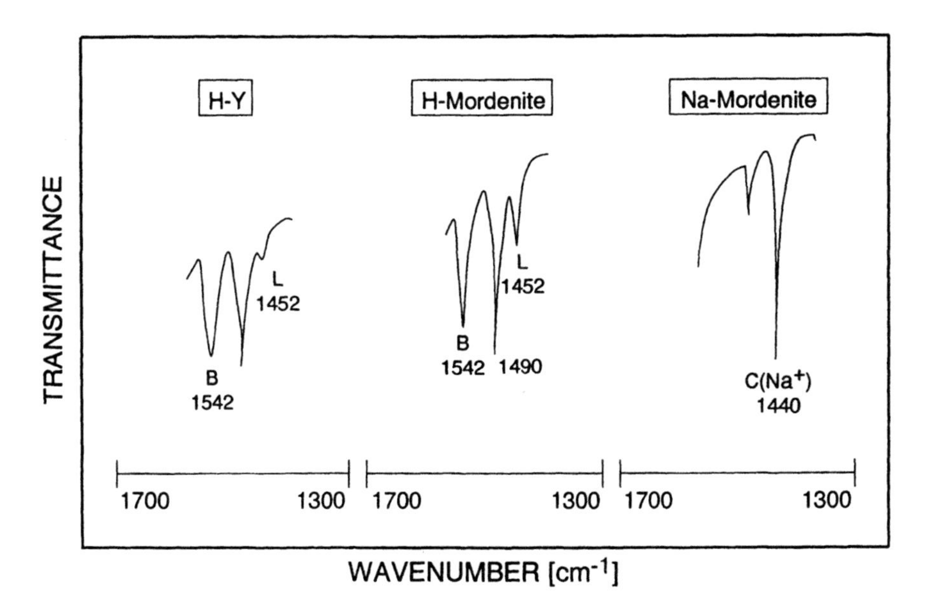

Fig. 4.19. IR bands of pyridine adsorbed on Brønsted acid sites, true Lewis sites and cations in H-FAU(Y), H-MOR and Na-MOR

1450 and 1440 cm^{-1} are ascribed, respectively, to (i) pyridinium ions (PyH^+), i.e., pyridine indicative of Brønsted acid sites (B-sites), (ii) unspecifically adsorbed pyridine, (iii) pyridine attached to true Lewis acid sites (L-sites) and (iv) pyridine attached to cations. All of these bands are due to the ν_{19b} ring deformation mode of pyridine [103–105] which, however, is affected differently by the various interactions with the different adsorption sites. The band at 1490 cm^{-1} is non-specific in that it may originate from pyridine interaction with either of the sites mentioned or from hydrogen bonding. Other bands, sometimes preferred for investigation of sites indicated by pyridine, appear at 1620 and 1632 cm^{-1} (mode 8a, 8b of the pyridine ring vibration, see [56]), the first resulting from both B- and L-sites, the latter one being specific for B-sites. The complete spectra for pyridine adsorbed on hydrogen-faujasite (H-Y) and hydrogen-mordenite (H-MOR) are, for instance, provided in [5, 101] and [56], respectively, where the assignments are also discussed in more detail.

The position of the bands typical of pyridine attached to cations depends on the Coulomb potential, i.e., the frequency increases with increasing charge and decreasing radius of the respective cation (*vide infra*; compare Fig. 4.24 and also [5, 68, 106]). A large collection of spectra and relevant bands observed in IR both for hydroxyls and pyridine attached to the various sites is included in [5].

Since the pyridine bands are rather intense and sharp, they are frequently used for quantitative evaluation. Also, extinction coefficients (integrated absorbances) were derived in a few cases [10, 50, 107–109]. Thus, Hughes and White obtained for the integrated absorbance of the PyB band observed upon pyridine adsorption on H,Na-Y with an exchange degree of 90% an integrated absorbance of 3.1 cm/μmol. Bielánski and Datka [109] arrived at a somewhat lower value, i.e., 1.07 cm/μmol. The integrated absorbances derived by Uytterhoeven et al. [107] for the acidic hydroxyls of H-Y upon pyridine adsorption provided a good agreement between IR and ^{1}H MAS NMR data if the total of the acidic OH groups (corresponding to the HF and LF band) was calculated (compare [110] and Sect. 4.2). In any case, it turns out that extinction coefficients determined with the use of pyridine and similar probes provide more reliable results than direct measurements of the intensities of OH groups as a function of their density. The extinction coefficients of, e.g., pyridine are less dependent on coverage and temperature. More recently, Khabtou et al. published an extended quantitative IR study on distinct acidic hydroxyl groups in modified Y-type zeolites and re-evaluated their extinction coefficients ε(HF), ε(LF) as well as the integrated absorbance of the pyridinium ion band at 1542 cm^{-1} [ε(HPy^+)]. The results were at some variance with data reported earlier (see [111] and references cited therein).

By contrast, the IR bands observed upon adsorption of ammonia, which is another frequently used thermally stable probe molecule, are relatively broad. However, the band around 1445 cm^{-1} being indicative of NH_4^+ may also be used for quantitative measurements yielding absolute numbers of acidic sites or the density of acidic hydroxyls in arbitrary units [112].

The advantage of NH_3 is its small size so that it can be employed in cases where the pyridine technique fails because of the bulkiness of this probe molecule. However, some care must be taken with ammonia. As a strong base it also adsorbs on rather weakly acidic sites which are irrelevant for acid-based properties and performance of the investigated acidic zeolite. As in the case of pyridine it might be necessary to define a standard procedure to include the removal of ammonia from weak sites. Moreover, in some cases, NH_3 seems to form, at higher temperatures, $-NH_2$ groups on the surface of bridging-hydroxyl-containing zeolites [112, 113]. In the case of hydrogen-mordenite the corresponding band appeared at 1550 cm^{-1}; it disappeared on heating at about 525 K.

Other strongly basic molecules which have been employed for probing acidic hydroxyls in connection with IR spectroscopy are, e.g., piperidine [50], methyl- and ethylamine [114] and substituted pyridines. Bulky organic bases such as substituted pyridines, for instance 2,6-di-*tert*-butylpyridine, were successfully used to selectively determine the acidic hydroxyls on the external surface of zeolites excluding those of the interior surface because the molecule could not penetrate into the structure due to steric hindrance [115]. Such experiments have shown that, in contrast to earlier assumptions [48], the external surface of zeolite crystallites does not only bear lattice-terminating silanol groups but also acidic bridging hydroxyls. The latter sometimes play a role in secondary isomerization of products which have first catalytically formed inside the zeolite structure with a product distribution prescribed by geometric constraints, i.e., in a shape-selective manner.

Utilization of strong bases raises the question as to whether the results of experiments with such probes will provide a relevant description of the acidic properties of the zeolites investigated since the action of the first adsorbed probe molecules might alter the acidity of the remaining sites. This should be particularly true if long range effects are operative (*vide supra*). Moreover, the acidic properties of a zeolite investigated with the help of strongly basic probe molecules is frequently correlated with the performance (activity, selectivity, time-on-stream behavior) of the same zeolite in catalyzing the conversion of hydrocarbons. However, hydrocarbons are rather weak bases. It was therefore suggested that as more suitable probes hydrocarbons or hydrogen, which is frequently involved in hydrocarbon reactions, be used.

As an example, benzene as a weak base is proposed as a probe for the acidic strength of bridging hydroxyls [116, 117]. The shift of the OH-stretching bands originating from Brønsted acid centers to lower frequencies is a convenient, although somewhat rough, measure of acidity strength. Figure 4.20 shows the results for a series of acidic zeolites and the correlation between the shift, $\delta\tilde{\nu}(OH)$, and the Sanderson eletronegativity, S, [79]. The shifted OH bands are relatively broad and certainly cover a broad distribution of sites with varying acidity strength similar to the curves of TPD spectra. In fact, Datka et al. [117] have pointed out that such bands originating from the interaction of hydroxyls with benzene reveal a considerable heterogeneity of the internal zeolite surface with respect to its acidity.

Kazansky and co-workers [13–15, 93–94] employed light hydrocarbons and hydrogen as probe molecules for the investigation of both Brønsted and Lewis acidity by diffuse reflectance spectroscopy. The molecule H_2 is a homonuclear diatomic species without a dipole. Therefore, the transition from the vibrational ground state to the first excited vibrational state is, according to the selection

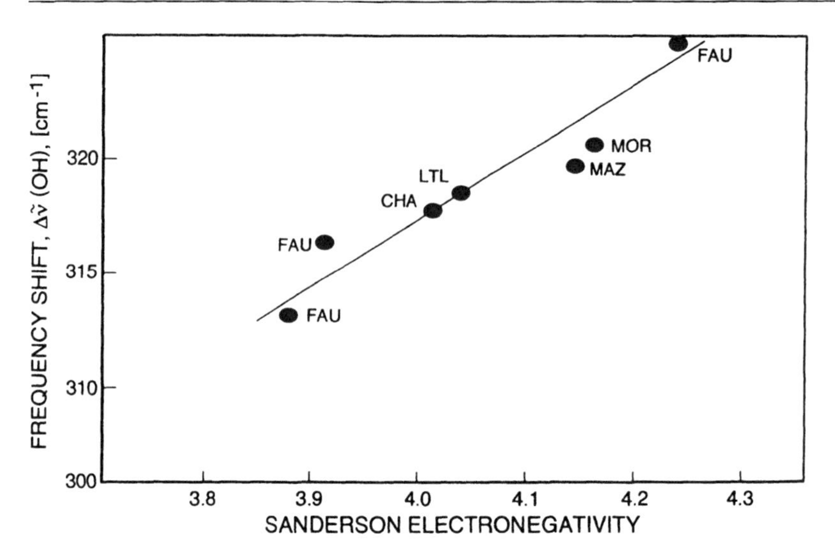

Fig. 4.20. Frequency shift of the high-frequency band of hydrogen zeolites upon benzene adsorption (from [116])

rules (*vide supra*), forbidden and the fundamental stretching mode is normally IR inactive. However, when at 77 K, H_2 is adsorbed on certain adsorption sites of the zeolite, the hydrogen molecule is disturbed and typical bands appear in the near infrared region (*vide infra*). Kazansky and co-workers [15] investigated the interaction of H_2 with acidic zeolites such as H-Y, H-MOR, H-ZSM-5 and bivalent cation-containing mordenites. A set of typical spectra is shown in Figure 4.21 for the example of H-ZSM-5 heat-treated at increasing temperatures. Spectrum *a* obtained after pretreatment at 770 K, characterizes the pure Brønsted acid form of H-ZSM-5, i.e., the zeolite prior to any dehydroxylation. The band at about 4125 cm^{-1} is also obtained after adsorption of H_2 on silica; therefore it is ascribed to the interaction of H_2 with the silanol groups (Si–OH) of H-ZSM-5, whereas the second band at 4105 cm^{-1} is supposed to be due to a disturbance of H_2 adsorbed on the acidic bridging OH groups of the Si(OH)Al configuration. The changes which occur on treatment at higher temperatures will be discussed below.

As another probe molecule for acidic sites in zeolites including Brønsted acid centers, carbon monoxide has been used. The potential and applicability of this probe in a great variety of cases was extensively reviewed by Knoezinger [118]. Usually, CO is adsorbed at 77 K and low pressures (see, e.g., [119–122]). On interaction with hydroxyls, the charge donation from an orbital on the carbon atom into an antibonding orbital of the OH bond weakens the latter and gives rise to a shift, $\Delta\tilde{\nu}$(OH), to lower wavenumbers. As an example, Fig. 4.22 demonstrates how with increasing doses of CO (spectra 2 to 6) the intensity of the HF band of a H,Na-Y sample (degree of exchange 20%) decreases and the intensity of the shifted band around 3400 cm^{-1} increases. While the number of adsorbed CO molecules may be used as a measure of the density of the Brønsted

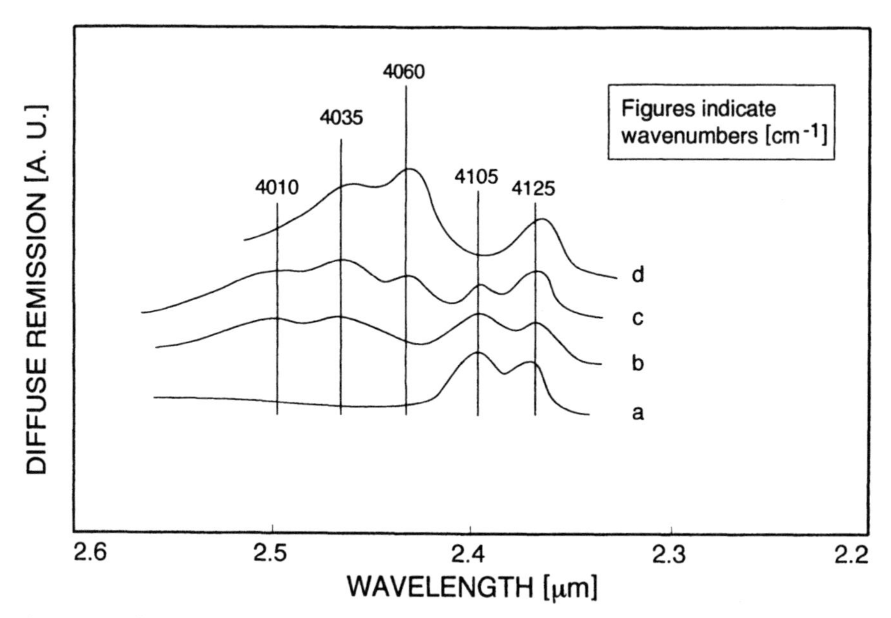

Fig. 4.21. Diffuse reflectance spectra of hydrogen adsorbed on H-ZSM-5 at 77 K; **a, b, c** after heat-pretreatment in high vacuum at 770, 970 and 1220 K, respectively; **d**, after "deep-bed treatment" at 770 K; abscissa linear in wavelength; important bands indicated by wavenumbers in cm^{-1}; (from [15])

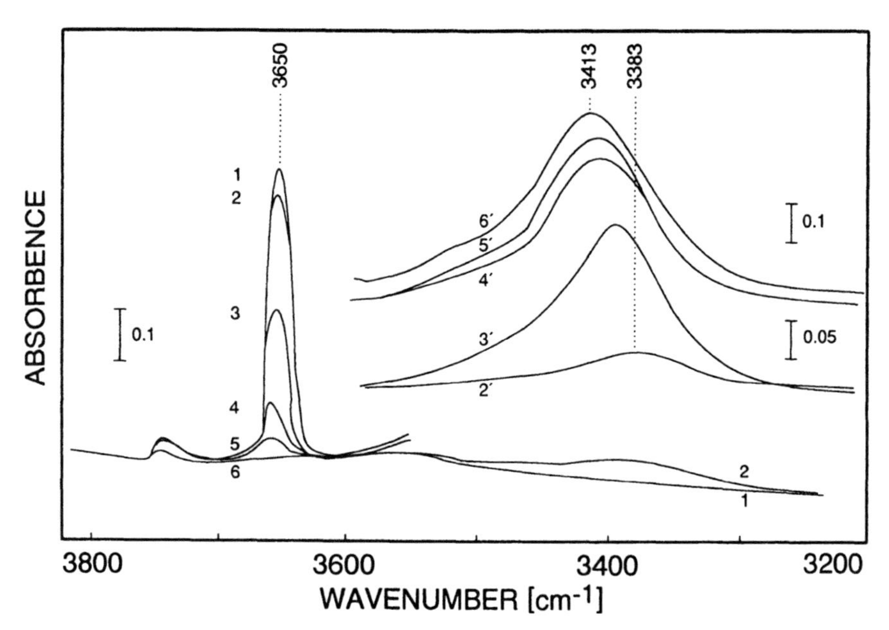

Fig. 4.22. IR spectra of OH groups of $H_{20}Na_{80}$-Y before (*1*) and after (*2–6*) adsorption of increasing amounts of CO (from [120])

acid sites [120], the value of $\Delta\tilde{\nu}(OH)$ is indicative of their strength. Thus, Kubelkova et al. [121] obtained, on the basis of the measured shifts (the wavenumbers of which are given in brackets), the following ranking with respect to acidic strength:

Si(OH)Al [240–340] > P–OH [175–195] > Al(OH) [140–195] > SiOH [90].

4.1.5.2
Lewis Acid Sites – True Lewis Sites

Lewis acid sites in zeolites may occur as threefold-coordinated aluminum or silicon (see p. 225, Scheme 1, upper part) and/or extra-framework aluminum-containing species such as AlO^+ (see p. 225, Scheme 1, lower part). Lewis sites form upon dehydroxylation and dealumination of zeolites. In fact, Kuehl [123, 124] reported that X-ray fluorescence spectroscopy did not provide any evidence for the existence of $\equiv$Al or $\equiv Si^+$ in the structure of dehydroxylated H-Y. This prompted Kuehl [124] to extend the dehydroxylation scheme developed by Szymanski et al. [125] and Uytterhoeven et al. [48] in the way indicated in Scheme 1. The concept of "true Lewis sites" as extra-framework Al-containing species was further supported by the work of Jacobs and Beyer [126]. However, it is still a matter of debate as to whether Lewis sites in zeolites may also exist in the form of only threefold coordinated Al and/or Si, in particular in view of more recent work by Kazanski and co-workers (see below).

As already mentioned, utilization of probe molecules is a prerequisite to detect Lewis acid sites by IR. With pyridine adsorbed onto true Lewis sites, a prominent band appears at 1450–1452 cm^{-1} as shown in Fig. 4.19. Sometimes more than one band around 1450 cm^{-1} is observed, e.g., at 1465 and 1452 cm^{-1} [26, 127–128], which could be due to the occurrence of different types of Lewis sites. However, this is still not clarified and requires further investigation.

Lewis sites are also detectable in IR by adsorption of carbon monoxide [120] and hydrogen [14] at low temperatures. Carbon monoxide attached to Lewis sites gives rise to a band at around 2190 cm^{-1} [118]. The disturbing effect of Lewis sites on hydrogen molecules adsorbed on such centers gives rise to bands at 4010, 4035 and 4060 cm^{-1} (Fig. 4.21). They were assigned by Kazansky et al. [14] to H_2 on $\equiv$Al, $\equiv Si^+$ and extra-framework Al-containing species such as AlO^+, respectively. These assignments were based on the following observations.

(i) Upon dehydroxylation of H-ZSM-5 at lower temperatures (e.g., 970 K), where aluminum removal from the framework was supposed to be unlikely, only two bands appear, viz. those at 4010 and 4035 cm^{-1}.
(ii) When dehydroxylation was performed at higher temperatures and under more severe conditions ("deep-bed calcination"), the third IR band at 4060 cm^{-1} developed. It was ascribed to H_2 interacting with species like AlO^+ since it was assumed that more severe treatment of H-ZSM-5 resulted in removal of Al from the framework according to the lower part of Scheme 1, i.e., due to the mechanism proposed by Kuehl [123, 124].
(iii) In agreement with argument (ii), the increase of the band at 4060 cm^{-1} was accompanied by a weakening of the band at 4010 cm^{-1} ($\equiv$Al).

(iv) When, in a complementary experiment, silica was severely heat-treated and subsequently contacted with H_2 at 77 K, a band at 4035 cm^{-1} was also detected. It was ascribed to $\equiv Si^+$ formed from strongly disturbed siloxane bridges.

It has been suggested that the dehydroxylation yielding $\equiv Si^+$ and $\equiv Al$ according to the mechanism by Uytterhoeven et al. [48] is the predominant one with H-ZSM-5 and, similary, H-MOR. In contrast, zeolites less resistant to dealumination upon heat-treatment, such as H-Y, are more likely to follow Kuehl's mechanism and should form only small amounts of "framework Lewis sites", i.e., $\equiv Al$ and $\equiv Si^+$. Upon very strong dehydroxylation of H-Y an additional IR band due to interaction with H_2 appeared at 4105 cm^{-1} and was assigned to extra-framework Al^{3+} [14].

Thus, the probe molecule H_2 reveals a greater variety of Lewis sites occurring in zeolites than pyridine does. In fact, this needs further clarification since one would expect that the ring vibration (ν_{19}) of pyridine also responds to differences in the adsorption sites such as presented by $\equiv Al$ and $\equiv Si^+$. When attached to various cations, changes in the position of the band typical of coordinatively bonded pyridine are indeed observed (see next section). However, pyridine adsorption on dehydroxylated H-ZSM-5, H-MOR, H-Y, Al_2O_3 etc. usually results in the formation of the same band at about 1450 cm^{-1}. Furthermore, Dwyer [129] has pointed out that a band around 4030 cm^{-1}, as was assigned by Kazansky et al. [14] to $\equiv Si^+$, appears upon H_2 interaction with η-Al_2O_3 pre-treated at 870 K as well and could, therefore, also be indicative of threefold coordinated $\equiv Al$.

4.1.5.3
Lewis Acid Sites – Cations

Vibrations of cations like alkaline, alkaline earth or rare earth cations etc., compensating the negative charge of the anionic framework of zeolites, can be observed in the far infrared region from 400 to 20 cm^{-1}. Brodskii et al. [130, 131] derived a relationship between the frequency, ν, the mass, m, and the radius r of the vibrating cation, viz.

$$\nu = C_B \cdot m^{-1/2} \cdot r^{-3/2} \qquad (4.11)$$

where C_B is the so-called Brodskii constant. As an example, the spectrum due to vibrations of Na^+ is reproduced from [131] in Fig. 4.23. The assignments proposed by Brodskii and Zhdanov [131] were as follows: bands at $\nu \approx 66$ cm^{-1} were supposed to be due to vibrations of Na^+ at site III (in front of four-rings), signals in the range from about 90 to 110 cm^{-1} were ascribed to vibrations of Na^+ in the hexagonal prisms (site I) and, finally, those between 155 and 200 cm^{-1} to Na^+ vibrations in front of six-rings (sites II, II', I'). More recently, Ozin et al. [132, 133] have extensively studied cation vibrations in zeolites, both experimentally and theoretically. Foerster [6], and Esemann and Foerster [134] showed that FTIR spectroscopy is a powerful tool for obtaining information about the siting of extra-framework cations in zeolite structures. Computer modelling successfully

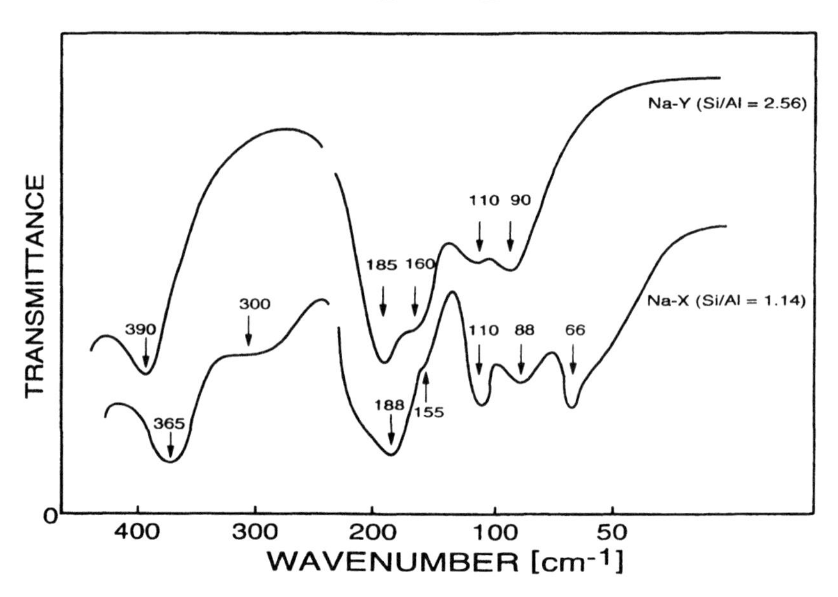

Fig. 4.23. Far infrared spectra of Na-X and Na-Y (from [131])

reproduced the particular features in the far infrared spectra of alkaline metal-containing Y-type zeolites [135]. Similar to observations of the changes of lattice vibrations upon interaction of adsorbates with an adsorbent zeolite, the vibrations of cations are also affected by guest molecules [136] resulting in typical band shifts. Investigation of the cation vibrations provides information about the location of cation sites, the occupancies of these sites and the motion of cations in the zeolite structure (compare [6] and references cited therein).

In the IR region more frequently used for zeolite characterization, i.e., from 4000 to 1000 cm^{-1}, the various cations which may occur in zeolites are conveniently identified and analyzed employing probe molecules such as pyridine or carbon monoxide. The bands typical of pyridine coordinatively bonded to the various cations in different zeolite structures and well suited for diagnostic purposes have been extensively studied and collected by Ward [5, 106]. This author already realized the close relationship between, e.g., the position of the band, which originates from the ν_{19b} mode of pyridine attached to a cation and the Coulomb field of this adsorbent site. For the sake of illustration, this relationship is demonstrated in Fig. 4.24 for a series of alkaline-metal cations in Y-type zeolites. Similar relationships were found for alkaline earth metal cations with pyridine or water as a probe [5, 68, 106, 137], whereas with transition metal cations no simple relationship between the pyridine band position and the physical properties of the respective cations was observed [138].

The other frequently used probe molecule, i.e., CO, is particularly sensitive to the oxidation state of the cation to which it is coordinated. Thus, it is advantageously employed for the characterization of zeolites loaded with transition metal cations and/or atoms. For instance, Sheu et al. [139] demonstrated via CO

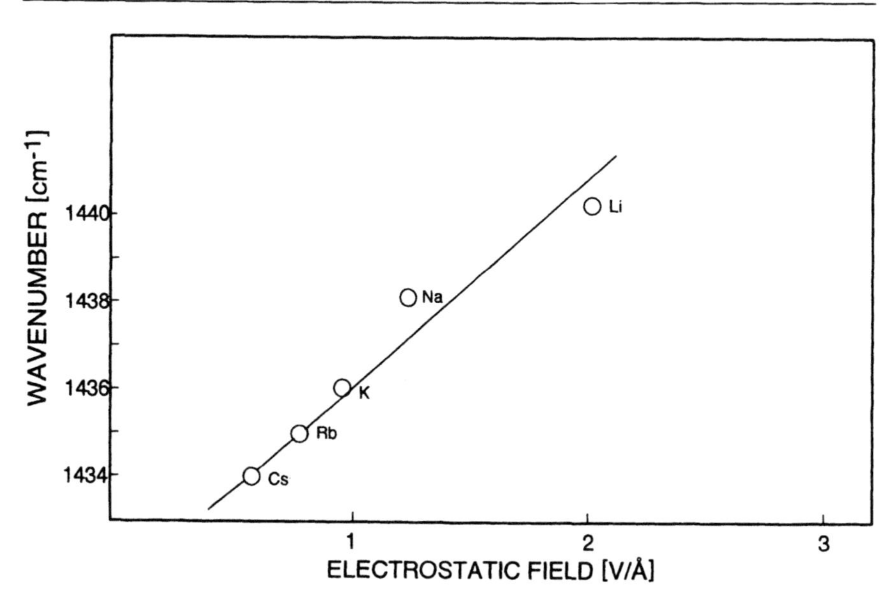

Fig. 4.24. Frequency of the IR band of pyridine coordinatively bound to alkali metal cations in M^+-Y (from [106])

interaction with reduced Pd/Na-Y zeolite the formation of $Pd_{13}(CO)_x$ clusters entrapped in the supercages of the faujasite-type zeolite. The frequencies of the bands due to CO attached to Pd^{3+}, Pd^{2+}, Pd^+ or electron-deficient Pd_x^0 decreased in this sequence, viz. from 2218/2207 over 2142/2176 to 2118 cm^{-1} [140]. There are many other interesting cases of application of CO for identification and analysis of the state and properties of cations in zeolites [5]. However, in some cases, a rather complex picture emerges because of the multiplicity of possible CO coordination, since this probe molecule may be linearly, doubly (in a bridged form) or even triply coordinated or geometrically arranged in a "butterfly" structure [6, 9, 122, 140].

4.1.6 Basic Sites (Basic Hydroxyls, Basic Oxygens)

In the past years there has been an increasing interest in basicity of zeolites because of the role of basic sites in (i) adsorption/separation and (ii) base-catalyzed reactions. Examples for (i) are the separation of C_8 aromatics (ethylbenzene, xylenes; see [141, 142]) and for (ii) reactions such as chain alkylation of toluene by methanol [143, 144], condensation of benzaldehyde by derivatives of malonic esters [145] and processes involving H_2S [146–148]. Possible basic sites are basic hydroxyls (Brønsted base sites), basic oxygens (Lewis base sites) and metal clusters such as Na_4^+ [149]. The most important basic sites are the framework oxygens. Their basicity depends on the fractional negative charge they bear and, therefore, may be related to the intermediate Sanderson electronegativity in an analogous way as was discussed with respect to acidity (*vide infra*). Barthomeuf, who gave an instructive review on various aspects of zeolite basicity (see [142] and, also, [150–151]) pointed out that significant basicity is to be expected for $S_{int} \leq 3.5$ and an oxygen charge of about $\leq -0.35\ e^-$. However, similar to the case of acidity, not only the composition of the zeolite accounted for by S_{int} but also structural parameters (bond lengths, bond angles, Al distribution) have to be considered.

As in the case of Lewis acid sites, the basic sites are not detected by IR without the help of probe molecules. Unfortunately, there are only a few suitable ones. Carbon dioxide has been employed for investigation of basicity but its use suffers from the drawback that CO_2 may form different carbonates when interacting with zeolites (see [5], pp. 197–200 and [9] pp. 189–190). So far, application of the amphoteric molecule pyrrole as a probe for basic sites in zeolites seems to provide relatively relevant results [142, 150]. In some cases, benzene was used for probing basic oxygen sites in the 12-membered ring windows of faujasite-type zeolites, where this molecule can weakly interact through its hydrogens. It is suggested that the IR shift of the band of the CH deformation vibration provides a measure of the basicity strength [152]. In a similar way the shift of the NH vibration of pyrrole is used to characterize the basicity of, e.g., alkaline-metal-containing X, Y, MOR, beta and ZSM-5. Figure 4.25 displays an example of a spectrum of pyrrole adsorbed on K-LTL [152]. In Table 4.4 [142] the shifts are listed of the NH vibration of pyrrole when this probe molecule was attached to basic sites in a series of alkaline-metal forms of zeolites. The table provides the calculated charges on the framework oxygens. The parallelism of basicity strength indicated by pyrrole and the partial charges on the basic sites is obvious.

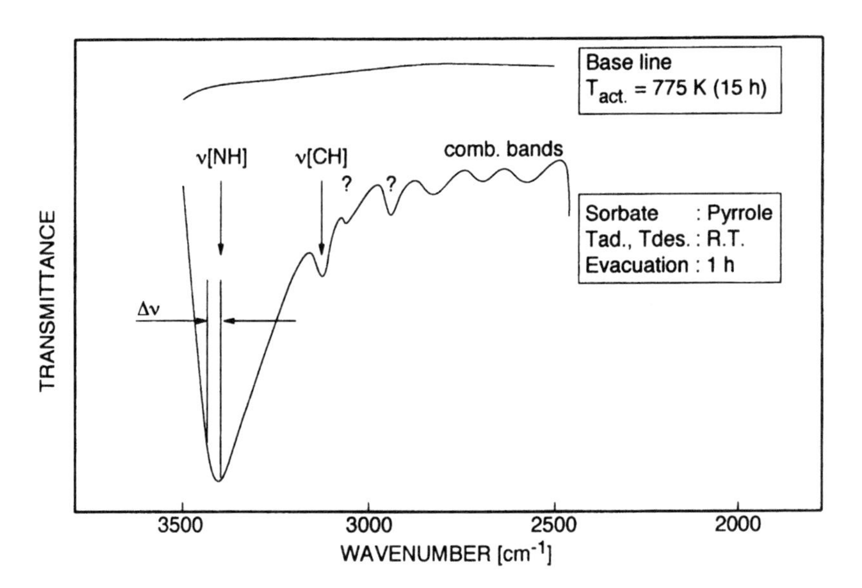

Fig. 4.25. Adsorption of pyrrole on zeolite K-LTL (from [150])

Table 4.4. Shift of ν(NH) Vibration of Pyrrole Adsorbed on Zeolites and Calculated Average Charge q_O on Oxygen [142]

Zeolite	$\Delta\nu$(NH)[a]	q_O[b]
CsX	240	-0.461
NaX	180	-0.413
KY	70	-0.383
NaY	30-40	-0.352
KL	30	-0.356
Na-MOR	30	-0.278
Na-beta	30	-0.240
Cs-ZSM-5	0	-0.236
Na-ZSM-5	0	-0.225

[a] Shift of ν(NH) referenced to ν(NH) of the liquid.
[b] Charge on oxygen calculated from Sanderson electronegativity.

4.1.7
Zeolite-Adsorbate Systems

Besides the use of adsorbate molecules as probes for active sites such as Brønsted and Lewis acid or basic centers (see Sects. 4.1.5.1.3, 4.1.5.2, 4.1.5.3 and 4.1.6) and their study by IR spectroscopy, there is an overwhelming number of IR investigations concerned with adsorption of molecules in zeolites. Since these studies are, in general, not directly aimed at an IR spectroscopic characterization of zeolites, they will only be briefly reviewed here. In most cases the interaction of molecules with zeolites was IR spectroscopically investigated in view of adsorption, separation and catalysis. Thus, a great variety of hydrocarbons (paraffins, olefins, aromatics and "coke"), oxygenates like alcohols, aldehydes, ketones, organic acids and bases, as well as inorganic compounds (H_2O, CO_2, SO_2, H_2S) sorbed onto or into zeolite structures have been studied.

Significant progress in IR studies of adsorbates in zeolites was achieved by Lercher and co-workers who demonstrated that very important subtle features in the zeolite/adsorbate systems could be observed when rather small adsorbate pressures were applied [153, 154]. Thus they discussed the formation of hydronium and oxonium species when water and alcohols were adsorbed onto H-ZSM-5. As an example, Fig. 4.26 displays a set of spectra obtained by Lercher et al. [153] upon adsorption of increasing amounts of water. Between about 10^{-3} and 10^{-2} mbar (equivalent to about three H_2O molecules per strong acidic OH group) hydroxonium ions were assumed to cause bands at 2885 and 2463 cm^{-1}. At higher coverages strongly hydrogen-bonded water species seemed to predominate. Later on, the authors [153] reconsidered and revised the above-indicated assignments [154], in particular in view of more recent theoretical results [89].

Coke is a particular type of adsorbate which may be located inside the porous structure and/or on the external surface of zeolites. Because of the important role which coke plays in deactivation of zeolite catalysts many IR investigations have been devoted to elucidating the formation and nature of such carbon-

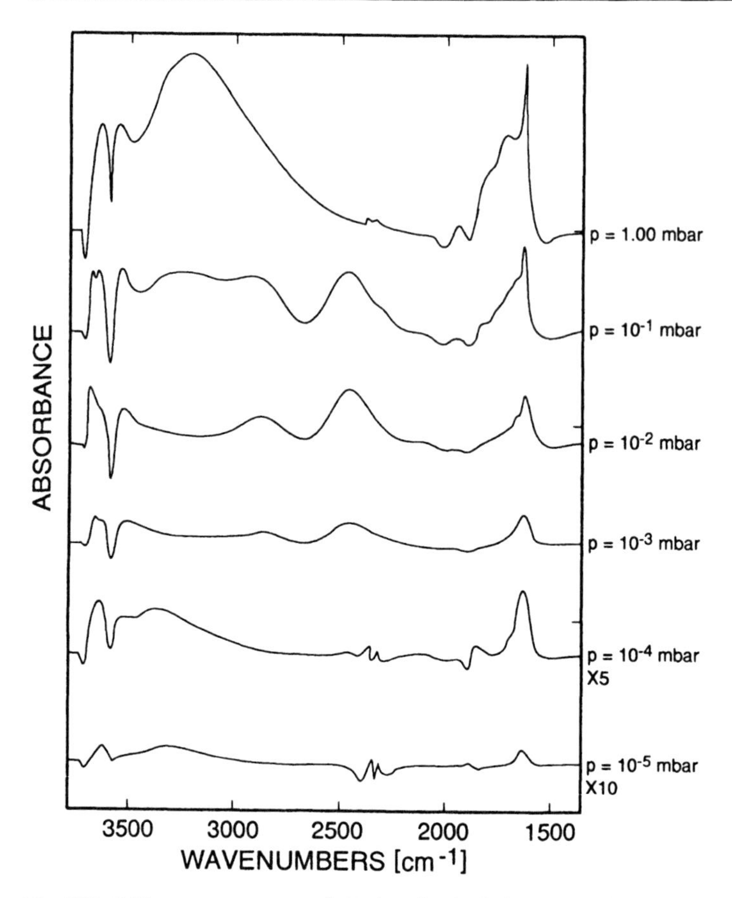

Fig. 4.26. Difference spectra of H_2O adsorbed in increasing amounts on H-ZSM-5 (from [153])

aceous deposits and to characterize the zeolite catalysts with respect to their propensity to coking (see, e.g., [155] and references cited therein).

It turns out that "coke" is almost never a uniform adsorbate but a very complex mixture of various carbonaceous species, the formation of which as well as their nature is strongly dependent on the properties of the zeolite catalyst and the reactants employed, on the reaction temperature, on time on stream, etc.

The IR study of very small and simple molecules (such as N_2, H_2 and O_2) entrapped in zeolites and their interaction with the host has also attracted considerable interest. The use of H_2 as a probe for acidic Brønsted and Lewis sites has already been discussed. However, the interaction of small molecules (CO, CH_4, N_2O, N_2, O_2 mainly studied in A-type zeolites, see [156–160]) with the zeolite framework has also been successfully studied by IR in order to characterize the main electric field and the activation energy of the motion experienced by the molecule inside the cavities.

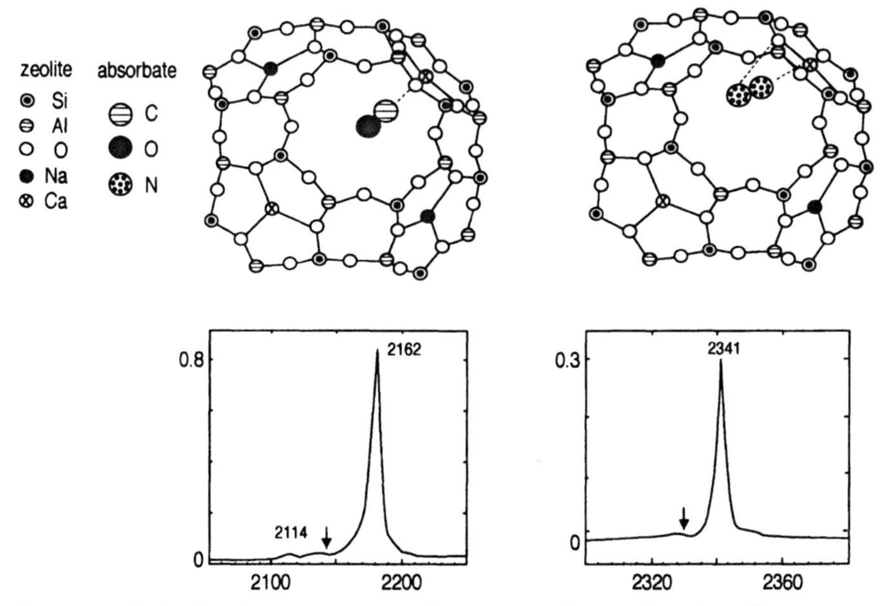

Fig. 4.27. Calculated sorbate geometries and IR spectra of CO and N_2 in zeolite Na_4Ca_4A; gas phase wavenumbers indicated by arrows (from [161])

Theoretical work was carried out on the state of adsorption of molecules encaged in zeolites, in particular in a series of studies by Foerster and co-workers. Thus, employing an electrostatic interaction model, Boese and Foerster [161] were able to calculate the sorption geometries of, e.g., CO and N_2 in zeolite Na_4Ca_4A as shown in Fig. 4.27 together with the corresponding IR spectra. Computed band shifts as well as the calculated field strengths and heats of adsorption were in fairly good agreement with experimentally derived data. A similar theoretical treatment as well as molecular orbital calculations were applied to the systems CO_2/Na_4Ca_4A and N_2O/Na_4Ca_4A yielding reasonable results for the geometries of the adsorption complex, the field strength, heats of adsorption and charge distributions [162]. Semiempirical molecular orbital calculations were carried out on Ca- and Ni-containing A zeolites [163], providing valuable information about the charge transfer due to adsorption and showing a rather good agreement between calculated and measured wavenumbers of the fundamental CO stretch, viz. $\nu(CO)_{calc.} = 2167$ and $2150\ cm^{-1}$ for CaA-CO and NiA-CO, respectively, in comparison with 2182 and $2190\ cm^{-1}$ measured for the same systems. In addition, normal coordinate analyses have been carried out for adsorption complexes where larger molecules are involved, e.g., for propene and butene adsorbed in variously exchanged zeolites A [164, 165] and propene in Na-Y, Ca-Y and Mg-Y. Again, these calculations led to information about the changes in frequencies (force constants) and the geometries of the adsorption complexes.

4.1.8
Motion, Diffusion and Reaction of Guest Molecules in Zeolites

Similar to many IR investigations of adsorbates (see Sect. 4.1.7), that of diffusion and reaction of molecules inside the zeolite structure is generally not, in a strict sense, a technique of zeolite characterization, even though diffusion and reaction are frequently closely related to the properties of the respective zeolite systems. Therefore, the possibility of IR studies on diffusion and/or reaction in zeolites will only be briefly considered.

The mobility of small molecules has been studied by IR in several contributions from Cohen de Lara et al. (see, e.g., [160, 166] and related references in [6]) and Foerster et al. [167, 168]. Cohen de Lara and Kahn [166] analyzed the band shape of molecules like CO_2 or N_2O at temperatures in the range from 180 to 450 K as well as the intensity and shape of induced bands of N_2, O_2 and CH_4. These parameters were related to the potential distribution and the average electric field inside the cavities affecting the mobility of the respective molecule. Activation energies for diffusion and diffusivities for H_2 and CH_4 at 77 K were estimated.

Foerster et al. [168, 169] derived from their IR data on CO and D_2, which were encaged inside Na-A and NaCa-A zeolites, that translational excitations of the adsorbed guest molecules occur resulting in jumps into adjacent potential wells.

In a paper by Moeller et al. [170] observations in the far infrared of translational motions of water in the cavities of a Ca,Na-A zeolite were reported.

IR measurements on diffusion in zeolites closely related to conventional uptake experiments were reported by Karge and Klose [171] who investigated the sorption of pyridine into mordenites and Y-type zeolites. Recently, this technique was considerably improved employing fast Fourier transform spectroscopy and an experimental set-up which allowed *in situ* measurements of sorption and desorption under conditions close to those of catalytic runs in a microflow reactor at elevated temperatures. Single-component diffusion data obtained by this novel method were in good agreement with well-established results in the literature. However, the FTIR technique also permitted the measurement of diffusivities under the conditions of co-diffusion and counter-diffusion in two-component systems such as benzene/ethylbenzene [172, 173] or benzene/*p*-xylene [174].

A very recent development in diffusion studies makes use of the IR microscopy technique [175] which renders possible the investigation of fast processes in single zeolite crystallites.

A number of experiments have been described in the literature where reactions catalyzed by zeolites in a flow reactor cell were monitored simultaneously by IR and gas chromatography or mass spectrometry [153, 154, 176–178]. This enables the simultaneous observation and correlation of concentrations of reactants and/or coke deposition on the one hand and activity/selectivity on the other.

Finally, *in situ* IR spectroscopy has been successfully employed in many investigations of H/D exchange [179, 180] and solid-state reactions, e.g., in the solid-state ion exchange of cations into hydrogen or ammonium forms of zeolites [134, 181–183]. In these cases, the removal of the bands of acidic OH groups upon interaction with D_2 or reaction of the zeolite with metal compounds such as halides or oxides was usually monitored.

4.2
NMR Spectroscopy

4.2.1
Introduction

In the 1960s Nuclear Magnetic Resonance (NMR) spectroscopy started to play an important role in zeolite characterization. The application of NMR spectroscopy to zeolites and adsorbate systems in zeolites suffered from the considerable solid-state line-broadening in contrast to NMR experiments with liquids and solutions. The reason for the line-broadening in the case of zeolite powder samples, as well as experimental techniques for overcoming or at least diminishing these difficulties due to solid-state interactions, will be briefly discussed in Sect. 4.2.2. With sophisticated experimental techniques such as Magic Angle Spinning (MAS) and Double Oriented Rotation (DOR) which significantly reduce the problem of line-broadening, and Cross-Polarization (CP) which increases the sensitivity, solid-state NMR spectroscopy has become an indispensible tool for zeolite characterization. The most important isotopes used for multi-nuclear NMR spectroscopy of zeolites are ^{29}Si, ^{27}Al, ^{31}P, ^{11}B and ^{17}O incorporated in framework positons, ^{1}H in hydroxyl groups, ^{23}Na and ^{133}Cs on non-framework sites, and ^{129}Xe, ^{31}P, ^{13}C and ^{15}N in probe atoms and probe molecules.

The main fields of characterization are: (i) evaluation of the environment of silicium framework atoms, (ii) n_{Si}/n_{Al} ratio of the zeolite framework, (iii) silicium and aluminum ordering and evaluation of structural problems, (iv) discrimination between framework aluminum (Al^{f}) and non-framework aluminum (Al^{nf}), (v) dealumination and aluminum re-insertion, incorporation (isomorphous substitution) of metals into the zeolite frameworks, (vi) the formation and acidity of hydroxyl groups, (vii) adsorption and entrapping of organics such as templates, reactants, coke etc., (viii) cation exchange and localization, and (ix) use of probes such as pyridine, carbon monoxide and xenon atoms. This enumeration is by no means exhaustive, and more possibilities of utilizing NMR spectroscopy for zeolite characterization are expected to emerge.

A number of very valuable books and review articles covering the field of solid-state NMR spectroscopy on zeolites have already been published [184–191]. Furthermore, wide-line NMR studies and PFG (Pulsed Field Gradient) NMR have provided valuable information about, e.g., mobilities and diffusivities of adsorbate molecules in zeolites. Important contributions to this field of zeolite research came, in particular, from the groups of Pfeifer and Kaerger [192] and Boddenberg [193]. In addition, the recently developed multiple-quantum MAS NMR (MQMAS) has to be mentioned which finds an increasing application in the study of ^{17}O, ^{23}Na and ^{27}Al nuclei of zeolites [194–196].

The following section presents a brief review of the basic principles which enable the experimentalist to evaluate wide-line spectra, to obtain NMR spectra of medium resolution, and, finally, to approach the goal of high-resolution solid-state NMR spectroscopy in zeolite powder investigations. In view of the very large and steadily increasing number of papers on NMR spectroscopic characterization of zeolites, a selection of the most important applications will be illustrated by examples adopted from the literature in Sect. 4.2.4.

4.2.2
Theoretical Background

4.2.2.1
Zeeman Interaction and Relaxation Effects

The phenomenon fundamental to NMR spectroscopy is the behavior of atomic nuclei with nulcear spins $I \geq 1/2$ in an external magnetic field, B_0, which is in turn related to the permanent magnetic dipolar moment μ [197–201]

$$\mu = \gamma_I \hbar I \tag{4.12}$$

with the magnetogyric ratio, γ_I. $\hbar$ is Planck's constant h divided by 2π (see Fig. 4.28a). In the classical description, the magnetic dipolar moments, μ, precess with the Larmor frequency, ν_0, around the z-axis of the laboratory frame (i.e., a fixed x, y, z-frame) which corresponds to the direction of the external magnetic field, B_0. ν_0 is given by

$$\nu_0 = -\gamma_I B_0/2\pi \tag{4.13}$$

In the quantum-mechanical picture, the external magnetic field, B_0, leads to a quantization of the nuclear spins in the z-direction which is described by the magnetic spin quantum number, m. For nuclear spins $I = 1/2$ the magnetic spin quantum numbers amount to $m = +1/2, -1/2$. Due to the nuclear spin quantiza-

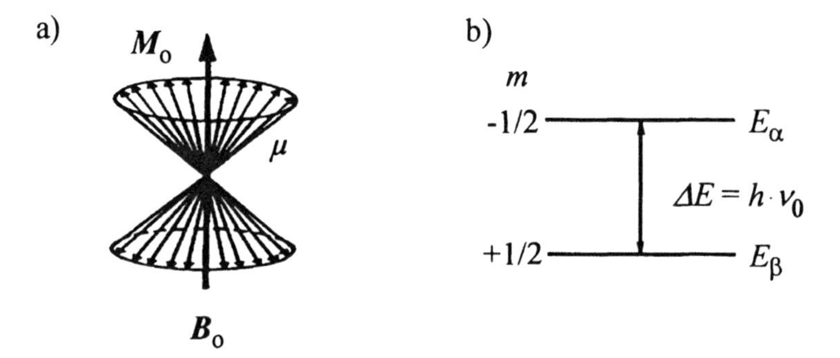

Fig. 4.28. The macroscopic magnetization, M_0, due to the precession of the magnetic dipolar moments, μ (**a**) and the Zeeman interaction of nuclei with spin $I = 1/2$ in an external magnetic field, B_0 (**b**)

tion, the energies of spins $I = 1/2$ are split into two levels (Zeeman interaction), a high energy level E_α and a low energy level E_β. The energy difference between E_α and E_β is given by $\Delta E = h\nu_0$ and correponds to the transition frequency of the nuclear spins (see Fig. 4.28b). In equilibrium, the numbers of spins with the energies E_α and E_β can be derived from Boltzmann's equation.

In an external magnetic field, B_0, an ensemble of N spins I will attain a magnetization, M_0, given by Curie's equation [197–199]:

$$M_0 = \frac{N\gamma_I^2 \hbar^2 I(I+1)}{3kT} \cdot B_0 \tag{4.14}$$

where k is Boltzmann's constant and T the absolute temperature. Generally, in a fixed x,y,z-frame the magnetization has the components M_x, M_y, and M_z. In equilibrium, M_z corresponds to M_0 while M_x and M_y are zero.

The principle of Fourier Transform (FT) NMR spectroscopy is the observation of the so-called Free Induction Decay (FID) after application of a radio frequency (rf) pulse with the magnetic field, $B_1(t) = [B_1 \cos \omega t, \pm B_1 \sin \omega t, 0]$, to the resonating spin ensemble. It is convenient to describe the effect of an rf-pulse in a frame x′, y′, z′ which rotates with the Larmor frequency, ν_0, around the direction z′ = z of the external magnetic field. In the x′, y′, z′-frame, the application of an rf-pulse results in a rotation of the magnetization from the z′-direction into the x′,y′-plane. Since the frequency of this rotation is given by $\nu_{rf} = -\gamma_I B_1(t)/2\pi$, the rotation angle, ζ, can be calculated by $\zeta = -\gamma_I B_1(t)\, t_p$, where t_p is the pulse length.

The time dependence of the magnetization, $M(t) = [M_x(t), M_y(t), M_z(t)]$, affected by the external magnetic field and the application of rf-pulses is described by Bloch's equation [197–199]

$$\mathrm{d}M(t)/dt = \gamma_I [M(t) \times B(t)] - \frac{M_x(t)e_x - M_y(t)e_y}{T_2} - \frac{M_z(t) + M_0}{T_1} e_Z \tag{4.15}$$

where $1/T_2$ and $1/T_1$ are the transversal and longitudinal relaxation rate, respectively. T_1 is the spin-lattice relaxation time which is responsible for the reconstruction of the equilibrium magnetization, M_0, and $B(t)$ is

$$B(t) = [B_1 \cos \omega t, \pm B_1 \sin \omega t, B_0]\,. \tag{4.16}$$

In the so-called time domain, the NMR signal is described as a function $G(t)$. This function $G(t)$ corresponds to the decay (FID) of the magnetization components $M_x(t)$ and $M_y(t)$ due to their dephasing. The spectrum, $S(\nu)$, in the frequency domain is obtained from $G(t)$ by Fourier transformation [197, 198]

$$S(\nu) = \int_{-\infty}^{\infty} G(t) \exp\{-\mathrm{i}\, 2\pi \nu t\}\, dt\,. \tag{4.17}$$

where ν denotes the frequency. It is advantageous to describe the NMR signal of solids (rigid spins) by the so-called second moment, M_2. The second moment with respect to the center of signal gravity, ν_{cg}, is defined by [197]

$$M_2 = 4\pi^2 \int_0^{-\infty} (\nu - \nu_{cg})^2 f(\nu)\, d\nu. \tag{4.18}$$

with the normalized line shape function, $f(\nu)$. The full width at half maximum of an NMR signal, i.e., the static linewidth, $\Delta\nu_{1/2}$, amounts to [188, 197]

$$\Delta\nu_{1/2} = \frac{1}{\pi} \sqrt{2 \ln 2}\, \sqrt{M_2} \tag{4.19}$$

while the transversal relaxation time T_2^{rigid} of a rigid spin ensemble is given by $\sqrt{2/M_2}$.

If the thermal motion of a spin system can be described by a single correlation time, τ_c, the free induction decay, $G(t)$, is given by [202]

$$G(t) = \exp\{-M_2(\tau_c t + \tau_c^2[\exp\{-t/\tau_c\} - 1])\}. \tag{4.20}$$

At high temperatures ($\tau_c \ll T_2^{rigid}$) the thermal motion leads to a deceleration in the observed decay

$$G(t)^{mobile} = \exp\{-M_2 \cdot \tau_c \cdot t\} = \exp\{-\frac{2\tau_c}{(T_2^{rigid})^2} \cdot t\} \tag{4.21}$$

which corresponds to a decrease in the width, $\Delta\nu_{1/2}$, of the Lorentzian line in the frequency domain. The signal of a rigid spin system ($\tau_c \gg T_2^{rigid}$) is decribed in the frequency domain by a Gaussian line with the second moment, M_2, which corresponds to a decay in the time domain of [202]

$$G(t)^{rigid} = \exp\{-\frac{M_2 \cdot t^2}{2}\} = \exp\{-(t/T_2^{rigid})^2\}. \tag{4.22}$$

4.2.2.2
Solid-State Interactions

Solid-state interactions of nuclear spins are described by the total Hamiltonian [197–199]

$$\hat{H} = \hat{H}_{CSA} + \hat{H}_{DD} + \hat{H}_Q \tag{4.23}$$

The terms in Eq. 4.23 are the Hamiltonians of the spatially *anisotropic chemical shift*, $\hat{H}_{CSA}$, of the *dipole-dipole interaction*, $\hat{H}_{DD}$, and the *quadrupole inter-*

action, $\hat{H}_Q$. All these interactions are responsible for the line-broadening of solid-state NMR spectra, and their analysis allows the determination of parameters which yield insight into the chemical nature and the local structure of the investigated atoms.

The *anisotropic chemical shift* originates from the anisotropic shielding of the external magnetic field by the electron cloud around the resonating spins. The shielding effect is described by the shielding tensor with principal values σ_{11}, σ_{22}, and σ_{33} or, in the same manner, by the chemical shift tensor with principal values δ_{11}, δ_{22}, and δ_{33} and $\delta_{ii} \propto -\sigma_{ii}$. The values σ_{ii} are denoted such that $\sigma_{11} \leq \sigma_{22} \leq \sigma_{33}$. In solutions and liquids, the anisotropic shielding effect is eliminated by the random isotropic reorientation of the species containing the nuclei and, therefore, the chemical shift is averaged to the isotropic value

$$\delta = (\delta_{11} + \delta_{22} + \delta_{33})/3 . \tag{4.24}$$

The experimentally observed resonance position, ν, is referenced to that of a standard, ν_s, e.g., tetramethylsilane (TMS) in ^1H, ^{13}C and ^{29}Si NMR spectroscopy. Therefore, the chemical shift follows as

$$\delta = (\nu - \nu_s)/\nu_o . \tag{4.25}$$

The Hamiltonian of the anisotropic chemical shift is given by [184]

$$\hat{H}_{CSA} = \hbar \gamma_I B_o \{\sigma + \frac{\Delta\sigma}{2} [(3 \cos^2\theta - 1) + \eta_{CSA} \sin^2\theta \cdot \cos 2\phi]\} \cdot \hat{I}_z . \tag{4.26}$$

where θ and ϕ are the angles between the principal axes of the chemical shift tensor and those of the laboratory frame and $\hat{I}_z$ is the spin angular momentum operator of the *I*-spins. According to Eq. 4.26 each value of θ and ϕ yields a signal at a chemical shift in the range of δ_{11} and δ_{33}. Therefore, the sum of all possible values θ and ϕ in a powder sample leads to the so-called powder pattern. The broadening and the shape of this pattern depend on the anisotropy, $\Delta\sigma$, and the asymmetry parameter, η_{CSA}, of the chemical shift tensor which are defined as:

$$\Delta\sigma = \frac{2\sigma_{33} - \sigma_{22} - \sigma_{11}}{3} \tag{4.27}$$

and

$$\eta_{CSA} = \frac{\sigma_{11} - \sigma_{22}}{\Delta\sigma} . \tag{4.28}$$

In addition to the shielding effect of the electron cloud, there is a coupling between bonding electrons of different atoms and of bonding electrons with nuclei having a spin $I = 1/2$ which results in an *indirect nuclear coupling*, also denoted *scalar coupling* or *J-coupling* [197, 198]. Depending on the orientation of the nuclear spin of the atom X ($m = +1/2$ or $m = -1/2$), the *J*-coupling leads to a resonance shift of the nuclear spin of the atom A to lower or higher magnetic fields. The *J*-coupling is restricted to directly bonded atoms A and X or to atoms A and X which are bonded via a maximum of two other atoms. However, since the resonance shifts due to the *J*-coupling are in the order of maximal 100 Hz, this interaction only has a small influence on the lineshapes of solid-state NMR spectra and is, therefore, not discussed in detail in the present section.

The *magnetic dipole–dipole interaction* between the magnetic moments of the resonating nuclei with magnetic moments of nuclei in their environment is described by the second term in Eq. 4.23. For spins $I = 1/2$, this interaction provides the main contribution to the line-broadening of the NMR signals. Dipole–dipole interactions are not affected by the external magnetic field, B_o, but are strongly dependent on the distance r_{kl} of the interacting magnetic moments μ_k and μ_l. Therefore, the homonuclear dipole–dipole interaction of diluted systems, e.g., of systems with nuclei of low natural abundance such as ^{13}C or with low concentrations due to chemical composition, is usually small. The Hamiltonian, e.g., of the heteronuclear dipole–dipole interaction between *I*-*S*-spin-pairs, is given by [198, 201]

$$\hat{\mathbf{H}}_{DD}^{IS} = -\frac{\mu_o \hbar^2}{4\pi} \cdot \frac{\gamma_I \gamma_S}{2r^3} \cdot (3\cos^2\theta - 1) \cdot \hat{\mathbf{I}}_z \cdot \hat{\mathbf{S}}_z \tag{4.29}$$

where μ_o is the permeability of the vacuum, θ is the angle between the direction of the external magnetic field and the distance vector r, and $\hat{\mathbf{I}}_z$ and $\hat{\mathbf{S}}_z$ are the spin angular momentum operators of the *I*- and *S*-spin, respectively. The second moments M_2^{II} and M_2^{IS}, due to homonuclear and heteronuclear dipole–dipole interaction, respectively, are defined by [188]

$$M_2^{II} = \frac{3}{5}\left(\frac{\mu_o}{4\pi}\right)^2 \gamma_I^4 \hbar^2 I(I+1) \frac{1}{N_I} \sum_{i \neq j}^{N_I} r_{ij}^{-6} \tag{4.30}$$

and

$$M_2^{IS} = \frac{4}{15}\left(\frac{\mu_o}{4\pi}\right)^2 \gamma_I^2 \gamma_S^2 \hbar^2 S(S+1) \frac{1}{N_I} \sum_{j=1}^{N_I} \sum_{k=1}^{N_S} r_{jk}^{-6} . \tag{4.31}$$

Analysis of the solid-state NMR line-broadening and application of Eqs. 4.18, 4.30 and 4.31 allow the determination of internuclear distances of the resonating *I*-spins to spins in the surroundings (cf. Sect. 4.2.4.6).

For nuclei with nuclear spins $I > 1/2$ the last term of Eq. 4.23 describes the *quadrupole interaction* of the electric quadrupole moment, eQ, and the electric field gradient, $eq = V_{33}$, at the sites of the resonating nuclei. In zeolites these nuclei are, e.g., ^{27}Al ($I = 5/2$), ^{23}Na ($I = 3/2$), ^{17}O ($I = 5/2$), ^{11}B ($I = 3/2$), etc. The electric field gradient (EFG) originates from the charge distribution of the surroundings of the resonating nuclei which has in most cases non-spherical symmetry. The Hamiltonian of the quadrupole interaction is given by [198, 201]

$$\hat{H}_Q = \frac{e^2qQ}{8I(2I-1)}\{(3\cos^2\theta - 1) + \eta_Q \sin^2\theta \cos 2\phi\} \cdot [3\,\hat{I}_z^2 - \hat{I}\,(\hat{I}+1)] \qquad (4.32)$$

where η_Q is the asymmetry parameter

$$\eta_Q = \frac{V_{11} - V_{22}}{V_{33}}. \qquad (4.33)$$

The principal values V_{11}, V_{22}, and V_{33} of the electric field gradient tensor are denoted such that the inequality $|V_{11}| \leq |V_{22}| \leq |V_{33}|$ is fulfilled. The strength of the quadrupole interaction is described by the quadrupole coupling constant

$$C_Q = \frac{e^2qQ}{h} \qquad (4.34)$$

which is related to the quadrupole frequency, ν_Q, given by $\nu_Q = 3C_Q/(2I(2I-1))$.

Quadrupole interactions usually result in a considerable change in the spin energy levels described by pertubation terms of first and second order (see [201] and references cited therein). The spatial term of the first order quadrupole interaction corresponds to the second Legendre polynomial, $P_2(\cos\theta) = 1/2 \cdot (3\cos^2\theta - 1)$, while the spatial term of the second order quadrupole interaction is a function of $P_2(\cos\theta)$ and the fourth Legendre polynomial, $P_4(\cos\theta) = 1/8 \cdot (35\cos^4\theta - 30\cos^2\theta + 3)$. The main contribution to the spectroscopically observed line-broadening of the central transition is due to the second order quadrupole interaction. This second order line-broadening decreases with increasing external magnetic fields. An additional effect of the second order quadrupole interaction is the field-dependent quadrupolar shift, ν_{qs} [201]

$$\nu_{qs} = \nu_{cg} - \nu = -\frac{1}{30} \cdot \frac{\nu_Q^2}{\nu_o} \cdot \left[I(I+1) - \frac{3}{4}\right] \cdot \left(1 + \frac{\eta_Q^2}{3}\right) \qquad (4.35)$$

where ν is the resonance position of the NMR signal without quadrupolar interaction, also denoted the isotropic chemical shift. The investigation of the second order quadrupole shift at different magnetic fields allows the determination of the quadrupole frequency, ν_Q, of the resonating nuclei.

4.2.3 Experimental Techniques

4.2.3.1 Methods of High-Resolution Solid-State NMR

The principle of modern solid-state NMR experiments consists of the acquisition of the Free Induction Decay (FID) after exciting the spin ensemble by application of a high-power radio frequency (rf) pulse with a carrier frequency of ν_o

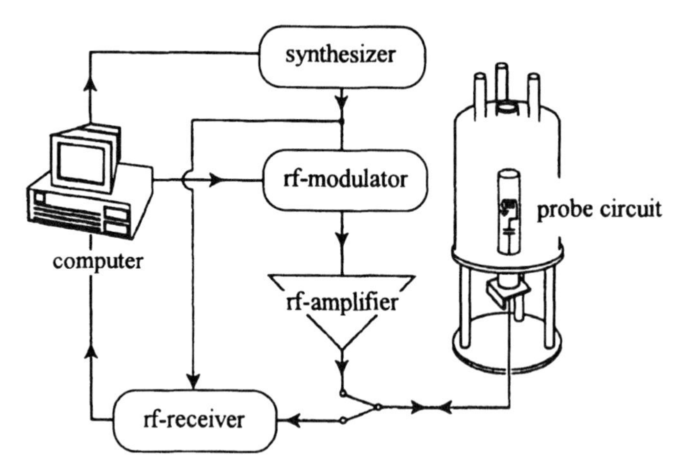

Fig. 4.29. Components of a modern NMR spectrometer for pulse excitation and Fourier transform technique

and subsequent Fourier transformation [198, 200]. The components of a Fourier Transform (FT) NMR spectrometer are depicted in Fig. 4.29. The sample is situated in the coil of a circuit which is placed at the center of a superconducting magnet. The radio frequency signal is produced by a synthesizer, is modulated to form rf-pulses, and is amplified to about 1 kW. The weak free induction decay, which is recorded immediately after the excitation, is phase-sensitively rectified using a reference signal from the synthesizer. The Fourier transformation, the control of the pulse sequence, and of the spectrometer equipment are carried out by a computer. Commercially available instruments for high-resolution solid-state NMR spectroscopy use external magnetic fields of up to 19.2 T corresponding to a Larmor frequency of the ^1H isotope of $\nu_0 = 800$ MHz.

Often, line-broadening by solid-state interactions prevents a separation of NMR signals of nuclei in various chemical surroundings. On the other hand, the analysis of these interactions yields interesting data on the chemical nature and local structure of the atoms under investigation. The dipole–dipole interaction depends on the magnitude of the nuclear magnetic moments. Because of the large magnetic moment of protons and their high natural abundance, there is a strong dipolar line-broadening in proton-containing samples. Therefore, it is neccessary to remove the heteronuclear dipole–dipole interactions such as ^1H–^{13}C, ^1H–^{29}Si, or ^1H–^{31}P by application of a high-power proton decoupling-pulse after the excitation of the resonating ^{13}C, ^{29}Si, or ^{31}P nuclei, i.e., during the detection of its free induction decay.

Magic Angle Spinning (MAS) is an experimental technique of high-resolution solid-state NMR which is based on the behavior of the spatial terms in Eqs. 4.29 and 4.32 [203, 204]. The principle of the MAS technique is the rapid sample rotation around an axis on an angle of $\theta_m = 54.7°$ to the external magnetic field. The line-narrowing is due to the behavior of the second Legendre polynomial $(3\cos^2\theta - 1)$ which is zero at the so-called *magic angle* θ_m. If an NMR signal is broadened by inhomogeneous interactions such as, e.g., the heteronuclear magnetic dipole–dipole interaction and the anisotropic chemical shift, MAS leads to signals consisting of a narrow central line at the center of gravity, ν_{cg}, and spinning sidebands at the frequencies

$$\nu = \nu_{cg} + k \cdot \nu_{rot} \tag{4.36}$$

where $k = \pm 1, \pm 2, \ldots$ denotes the order of the spinning sidebands and ν_{rot} is the spinning rate. In the present chapter spinning sidebands are denoted in the figures by asterisks.

A suitable method of averaging strong homonuclear dipole–dipole interactions is the application of multiple-pulse sequences. In the so-called WAHUHA (Waugh–Huber–Haeberlen) [205, 206] sequence, four $\pi/2$-pulses with alternating phases – $[\tau, (\pi/2)_x, 2\tau, (\pi/2)_{-x}, \tau, (\pi/2)_y, 2\tau, (\pi/2)_{-y}]$ – are applied on the resonating spins. By application of these $\pi/2$-pulses the resonating spins are subsequently aligned along the x-, y-, and z-axes which corresponds to a rotation around the magic angle axis. The FID is built up over a large number of cycles by taking one data point per cycle in one of the 2τ-windows. It has, however, to be stated that most of the multiple-pulse sequences lead only to a suppression of the homonuclear dipole–dipole interaction. The combination of multiple-pulse sequences with magic angle spinning is denoted CRAMPS (combined rotation and multiple-pulse sequence) [199].

As mentioned above, the spatial term of second order quadrupole interactions is given by the second and the fourth Legendre polynomials. Therefore, narrowing of the quadrupole interaction requires a simultaneous sample rotation around two axes. The corresponding device consists of an outer and an inner rotor (see Fig. 4.30). The angle θ_1 between the external magnetic field and the rotational axis of the outer rotor corresponds to the magic angle θ_m (*vide supra*), and the angle θ_2 between the rotational axes of the inner and that of the

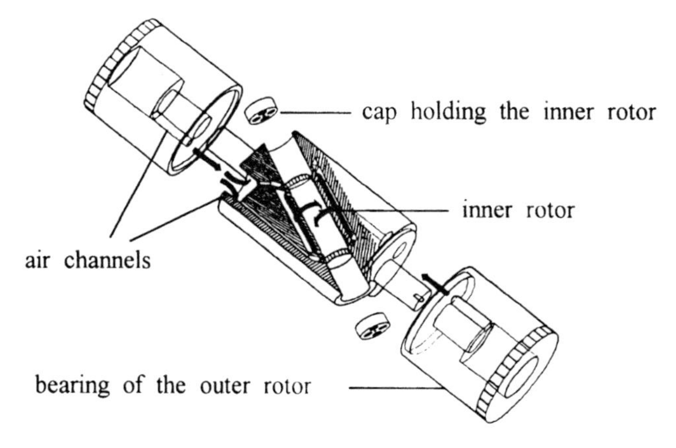

Fig. 4.30. Schematic drawing of a DOR rotor consisting of a larger outer rotor and a small inner rotor [207]

outer rotor amounts to 30.56°. The advantage of doubly oriented rotation in comparison with the magic angle spinning technique is the complete suppression of all second order quadrupolar line-broadenings. The corresponding experimental technique is denoted Double Rotation (DOR) [207, 208]. Until now, the rotation rate of the outer DOR rotor is limited to about 1500 Hz and that of the inner rotor to ca. 8 kHz.

4.2.3.2
Cross-Polarization and Other Selected Pulse Techniques

An important aspect of solid-state NMR spectroscopy of zeolites is the sensitivity. Cross-Polarization (CP) is an effective technique to significantly increase sensitivity which is, therefore, advantageously employed in the case of dilute spins in the neighborhood of dipolar coupled abundant spins [209, 210]. The principle of this technique is demonstrated in Fig. 4.31. The experiment starts with a $\pi/2$-pulse in the channel of the abundant S-spins. During a subsequent contact time, τ, magnetization (also denoted spin polarization) is transferred from the abundant S-spins to the dilute I-spins. The magnetic fields, B_{1S} and B_{1I}, of the contact pulses, which are applied simultaneously to the S- and the I-spins, have to fulfill the so-called Hartmann–Hahn condition:

$$\alpha_S \cdot \gamma_S \cdot B_{1S} = \alpha_I \cdot \gamma_I \cdot B_{1I} \tag{4.37}$$

where α_S and α_I are equal to 1 for $S = I = 1/2$. Otherwise, α_I is equal to $[I(I+1) - m(m-1)]^{1/2}$ provided that the high-power rf-pulse applied to the S-spins induces transitions between the levels with the magnetic spin quantum numbers m and m-1. The cross-polarization experiment leads to an enhancement of the signal intensity for the dilute I-spins if the abundant S-spins exhibit a higher magnetogyric ratio and/or a higher concentration than the I-spins. The maxi-

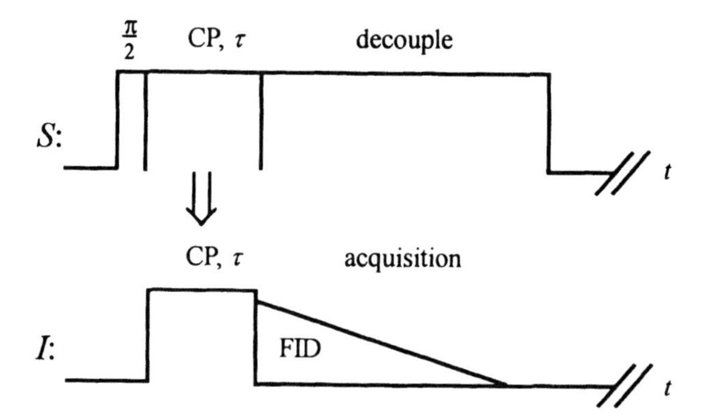

Fig. 4.31. Scheme of a cross-polarization experiment, starting with the excitation of the abundant S-spins ($\pi/2$-pulse) and subsequent magnetization transfer from the abundant S-spins to the dilute I-spins by application of a CP-pulse

mum signal enhancement achievable by cross-polarization (I_{CP}) for spin $I = 1/2$ nuclei in comparison to single-pulse excitation (I_{SP}) amounts to:

$$\frac{I_{CP}}{I_{SP}} = \frac{\gamma_S}{\gamma_I} \cdot \frac{1}{1+\varepsilon} \tag{4.38}$$

with

$$\varepsilon = \frac{N_I}{N_S} \cdot \left(\frac{\alpha_I}{\alpha_S}\right)^2 \tag{4.39}$$

where N_I and N_S denote the numbers of *I*- and *S*-spins, respectively, in the sample.

An interesting method which allows the determination of internuclear distances, r_{IS}, within isolated *I*-*S*-spin-pairs is the Spin-Echo Double Resonance (SEDOR) experiment [211]. SEDOR is based on a [$\pi/2$, τ, π, τ, echo] sequence applied to the resonating *I*-spins. A π-pulse applied to the non-resonating *S*-spins during the pulse delay of the *I*-spins leads to a damping of the echo. This echo damping depends on the strength of the magnetic dipole–dipole interaction within the *I*-*S*-spin-pair and, therefore, on the distance r_{IS} (see Eq. 4.29). An important advantage of SEDOR is the selective determination of internuclear distances within heteronuclear spin-pairs separated from other solid-state interactions. However, since this technique is carried out under static conditions (without MAS), only mean distances r_{IS} of different *I*-*S*-spin-pairs are derived. Combinations of the heteronuclear spin-echo technique with the application of magic angle spinning are referred to as Rotational-Echo Double Resonance (REDOR) and Transferred-Echo Double Resonance (TEDOR) [211].

4.2.3.3
Two-Dimensional NMR Spectroscopy

Considerable progress has been achieved by the introduction of two-dimensional (2D) NMR techniques into solid-state NMR spectroscopy and their application to zeolites [200, 212]. Typically, a 2D experiment starts with the preparation of the spin system (preparation period) by direct excitation or cross-polarization. In a subsequent evolution period t_1, the spin ensemble evolves under the influence of the Hamiltonian of a selected interaction which may be the indirect *J*-coupling, the dipole–dipole interaction, the quadrupole interaction, etc. The evolution of the spins under the influence of these interactions is mapped out point-by-point by incrementing t_1 and by monitoring the free induction decays, $G(t_2)$, during the detection period t_2. Hence, information about the selected interaction is obtained indirectly, by observing its influence on a set of free induction decays which leads to an array $G(t_1, t_2)$ of free induction decays. The maximum duration of the evolution time is limited effectively by the spin-spin relaxation time, T_2, of the nuclei. The 2D Fourier transformation of the array $G(t_1, t_2)$ yields the 2D dimensional spectrum $F(\nu_1, \nu_2)$. In this 2D spectrum, the selected interaction, mapped out by incrementing t_1, is separated along the F1-dimension while the F2-projection corresponds to the spectrum in the 1D experiment. A coupling between different spins may cause a magnetization transfer from spins resonating at the frequency ν_a in the time period t_1 to spins resonating at the frequency ν_b in the time period t_2. In the 2D spectrum $F(\nu_1, \nu_2)$ this coupling leads to a cross peak at $\nu_1 = \nu_a$ and $\nu_2 = \nu_b$ with $\nu_a \neq \nu_b$. If there is no coupling, a signal appears at $\nu_a = \nu_b$ which lies on the diagonal line of the spectrum (see Sect. 4.2.4.1 and Fig. 4.36). The so-called Correlation Spectroscopy (COSY) experiment which uses the *J*-coupling enables an investigation of the connectivities within a zeolite framework. The cross-peaks of 2D COSY spectra indicate a coupling between at least two kinds of nuclei at different chemical shifts. The signals of these nuclei are situated at points off the diagonal line which are crossed by perpendicular and horizontal lines going through the cross peaks (see Fig. 4.36). Other 2D NMR experiments considered in Sect. 4.2.4 are useful, e.g., to gain insight into dipole-dipole couplings and the strength of quadrupolar interactions.

4.2.4
Applications

4.2.4.1
^{29}Si MAS NMR Spectroscopy of SiO_4 Tetrahedra in the Zeolite Framework

The basic units of aluminosilicate-type zeolites are TO_4 tetrahedra with silicon or aluminum atoms at the central T-position. Therefore, silicon atoms incorporated in these zeolite frameworks are tetrahedrally coordinated (Q^4), resulting in five different silicon environments denoted as Si(nAl) units, where n corresponds to the number of aluminum atoms in the second coordination sphere. Each type of Si(nAl) unit (n = 0, 1, 2, 3, or 4) yields ^{29}Si MAS NMR signals in a well defined range of chemical shifts. These ranges are summarized for the various silicon units in Fig. 4.32. The spectra of zeolites consisting of silicon atoms on crystallographically equivalent T-sites, e.g., zeolite X and Y, are a function of the framework composition only (see top of Fig. 4.32). Therefore, the framework n_{Si}/n_{Al} ratio of these materials may be calculated directly from the ^{29}Si MAS NMR intensities using the formula [184]

$$n_{Si}/n_{Al} = \sum_{n=0}^{4} I_{Si(nAl)} / \sum_{n=0}^{4} 0.25 \cdot n \cdot I_{Si(nAl)} \tag{4.40}$$

where $I_{Si(nAl)}$ is the intensity of the Si(nAl) line, i.e., of the NMR signal caused by silicon atoms with a number n of aluminum atoms in the second coordination sphere. Moreover, by a comparison of the ^{29}Si MAS NMR derived framework n_{Si}/n_{Al} ratio with the bulk composition determined by chemical analysis, the amount of non-framework aluminum can be calculated.

However, in each zeolite, hydroxyl groups exist, bound to silicon atoms (Q^3, Q^2) located at the outer surface of zeolite particles or at framework defects. It is important to note that the chemical shifts of Si(1Al) units (ca. -95 to -105 ppm) are in the same range as those for Si(3Si,OH) units (ca. -100 to -103 ppm) [213].

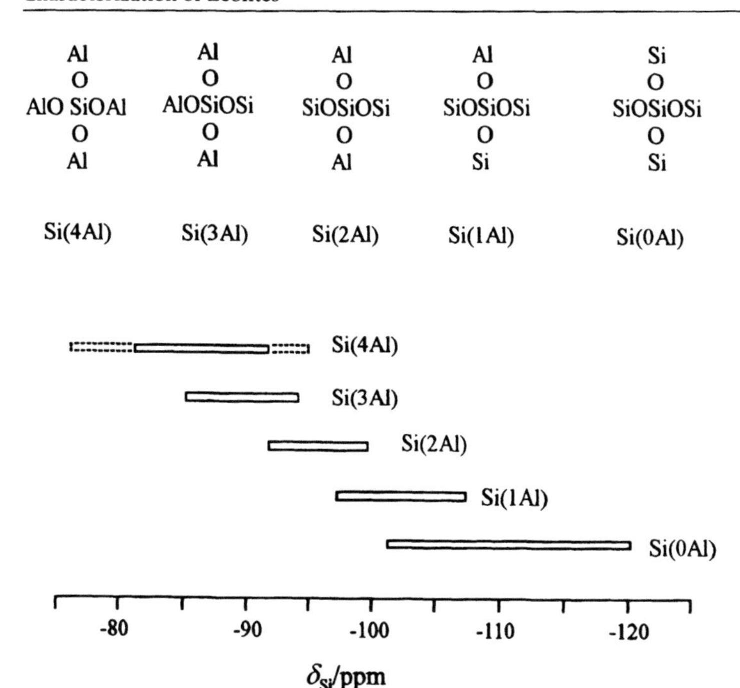

Fig. 4.32. ^{29}Si chemical shifts of Si(nAl) units in zeolite frameworks. The *dotted lines* for Si(4Al) designate the chemical shift range observed for 1:1 aluminosilicate sodalites with different cage fillings [184, 213]

In the same manner as described for ^{29}Si MAS NMR spectroscopy on aluminosilicate-type zeolites, the chemical composition of zincosilicates and of the gallium analogue of ZSM-5 (Ga-MFI) can be determined. The ^{29}Si MAS NMR spectra of zeolites VPI-7 and VPI-9, examined by Camblor and Davis [214], show signals of Si(1Zn) and Si(2Zn) units at resonance positions between −88.5 ppm and −95.6 ppm and between −77.9 ppm and −81.0 ppm, respectively.

The order of silicon and aluminum atoms at the T-positions in zeolite frameworks can be derived by comparing the relative Si(nAl) populations obtained from ^{29}Si MAS NMR spectroscopy with generated model populations. The models are derived using different distribution patterns of the silicon and aluminum atoms on the T-sites and certain self-repeating units of the zeolite framework. This method has been used to investigate the Si–Al-order in zeolites X and Y and in several other structure types [184, 215].

Most zeolite structures are characterized by crystallographically inequivalent T-sites in chemically equivalent environments. Silicon atoms on these T-sites may have different chemical shifts causing a split of the ^{29}Si MAS NMR lines of certain Si(nAl) units. For the investigation of crystallographically inequivalent T-sites it is advantageous to use high-silica zeolites which consist of Si(0Al) units only. If the ammonium form of a zeolite is calcined in the presence of water vapor at temperatures between 873 and 1073 K a dealumination and ultrastabil-

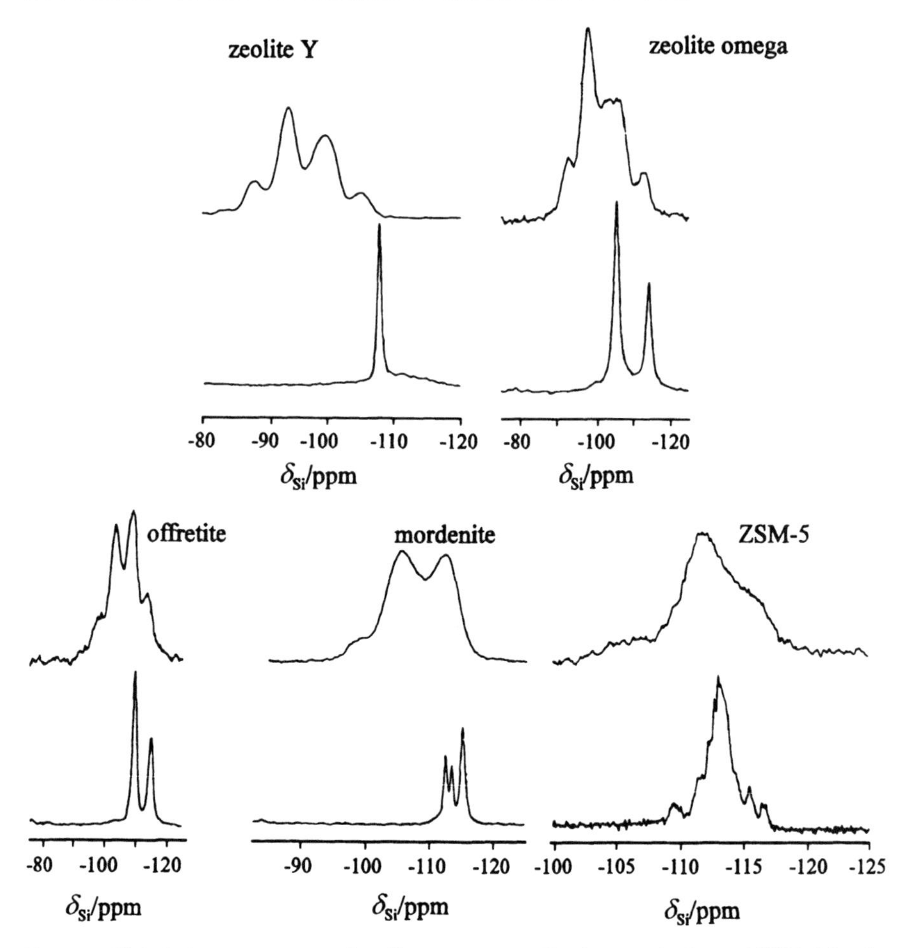

Fig. 4.33. ^{29}Si MAS NMR spectra of zeolites Y, omega, offretite, mordenite and ZSM-5 (*top*), and of their dealuminated forms (*bottom*) [216]

ization of the framework take place which decrease the ^{29}Si MAS NMR line width dramatically [216, 217]. Figure 4.33 shows the spectra of several zeolites before (top) and after (bottom) dealumination and ultrastabilization. In agreement with the structure type derived by X-ray diffraction (see [218]), the spectrum of dealuminated zeolite Y consists of one Si(0Al) line only. In contrast, all the other ^{29}Si MAS NMR spectra show Si (0Al) signals of at least two crystallographically inequivalent T-sites. The intensity ratios of these lines, e.g., of 2:1 for zeolites omega and offretite, correspond to the population ratios of these T-sites. ^{29}Si chemical shift data of Si(nAl) units in some selected zeolites are given in Table 4.5.

Different bonding geometries of the SiO_4 tetrahedra on crystallographically inequivalent T-sites affect the ^{29}Si MAS NMR shifts of the framework silicon atoms. Empirical correlations and theoretical considerations show, that the ^{29}Si

Table 4.5. ^{29}Si Chemical Shifts of Si(nAl) Units (in ppm, referenced to TMS) of Some Selected Zeolites

Zeolite	n_{Si}/n_{Al}	Site	Si(4Al)	Si(3Al)	Si(2Al)	Si(1Al)	Si(0Al)	Ref.
A	1.0	T	-89.6					[219]
	∞	T					-112.9	[220]
Y	2.5	T	-83.8	-89.2	-94.5	-100.0	-105.5	[219]
	∞	T					-107.8	[221]
Ω	3.1	T1		-89.1	-93.7	-98.8	-103.4	[222]
		T2	-89.1	-93.7	-98.8	-107.0	-112.0	
	∞	T1					-106.0	[216]
		T2					-114.4	
offretite	2.8	T1		-93.5	-97.5	-101.9	-106.9	[184]
		T2		-97.5	-101.9	-106.9	-112.5	
	∞	T1					-109.7	[216]
		T2					-115.2	
mordenite	5.0	T1 to T4			-100.1	-105.7	-112.1	[223]
	∞	T1					-112.2	[216]
		T4					-113.1	
		T2 + T3					-115.0	
ZSM-5	20	T1 to T12				-106.0	-112.0	[224]

NMR shifts, δ_{Si}, of Si(nAl) units are linearly related to the average value of the four Si–O–T bond angles, $\bar{\alpha}$, at the central silicon atom. With linear regression analysis, quantitative relationships between the values of δ_{Si} and $\bar{\alpha}$, $\overline{\sec\alpha}$, $\overline{\sin(\alpha/2)}$ and $\overline{\cos\alpha/(\cos\alpha-1)}$ have been found [184, 216, 225–228]. For Si(4Al) units in sodalite, cancrinite, thomsonite and zeolites A, X, Y, ABW, Engelhardt et al. [228] obtained a correlation coefficient of 0.988 for the following equation

$$\delta_{Si}/\text{ppm} = -5.230 - 0.570\,\bar{\alpha} \tag{4.41}$$

The correlation between δ_{Si} and $\overline{\cos\alpha/(\cos\alpha-1)}$ was theoretically derived and led to the following experimentally confirmed relationship [225]

$$\delta_{Si}/\text{ppm} = -223.9 \cdot \overline{\cos\alpha/(\cos\alpha - 1)} + 5n - 7.2 \tag{4.42}$$

where n denotes the number of aluminum atoms in the second coordination sphere. Comparing recent data of ^{29}Si MAS NMR and XRD studies on unloaded zeolite ZSM-5, Fyfe et al. [229] obtained a correlation coefficient of 0.97 in Eq. 4.41. Chemical shifts where calculated from the Si–O–Si bond angles derived from XRD and using the gradient of the curve defined by Eq. 4.42. This method allows the examination of structure models which are correct if the experimental ^{29}Si MAS NMR spectrum can be described by the X-ray data. Figure 4.34 shows, e.g., the experimentally (top) and theoretically (bottom) derived ^{29}Si

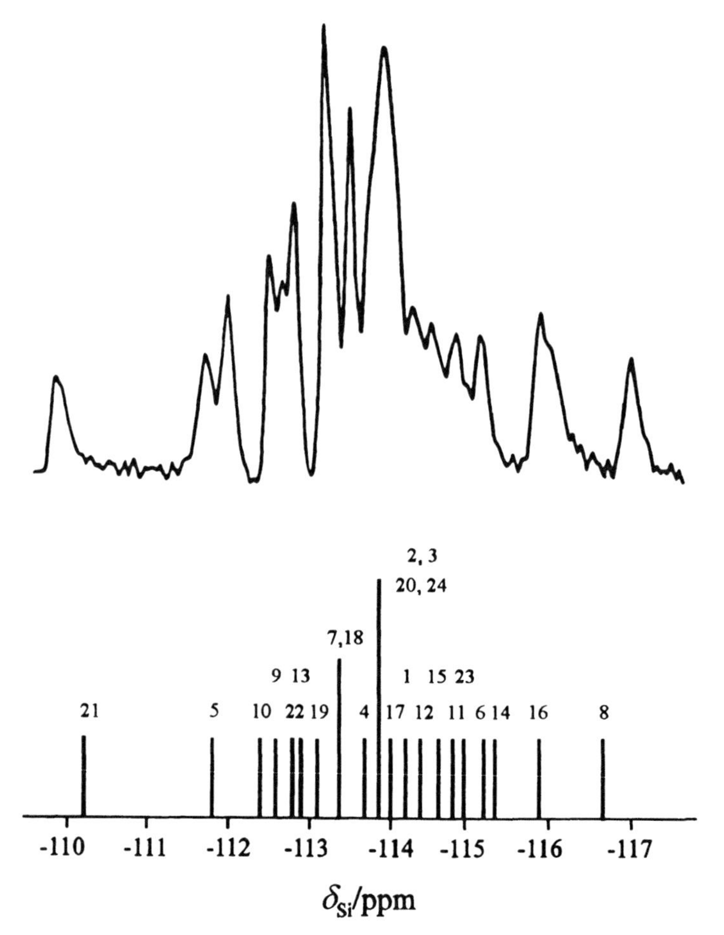

Fig. 4.34. Experimental (*top*) and calculated (*bottom*) ^{29}Si MAS NMR signals of silicon atoms at crystallographically inequivalent T-positions in the framework of monoclinic high-silica zeolite ZSM-5 [230]

NMR spectra of a monoclinic high-silica zeolite ZSM-5 [230, 231]. The numbers at the top of the calculated spectrum assign the crystallographically inequivalent T-positions. As shown, there is good agreement in the line positions and intensities which demonstrates the usefulness of this method for the verification of structure models derived from XRD.

As shown by Fyfe et al. [232, 233], adsorption of organic molecules on zeolites also induces changes in the zeolite framework and, therefore, in the ^{29}Si MAS NMR spectra which are characteristic of the adsorbate. Adsorption of *p*-xylene on zeolite ZSM-5 induces a change from the monoclinic crystal structure with 24 crystallographically inequivalent T-sites to the orthorhombic crystal structure with 12 crystallographically inequivalent T-sites. Heating the unloaded high-silica zeolite ZSM-5 also causes transition from the monoclinic to the orthorhombic structure. Tuel and Ben Taarit [234] used the temperature dependence of the monoclinic/orthorhombic transition of the titanosilicate TS-1 to determine the Ti content. TS-1 is a zeolite with the same structure-type MFI as zeolite ZSM-5 in which a fraction of the silicon atoms is replaced by Ti atoms. The authors found that the transition temperature, observed by the change in the shape of the ^{29}Si MAS NMR spectrum (appearance of a shoulder at ca. – 116 ppm), shows an almost linear decrease as a function of increasing number of Ti atoms per unit cell.

Changes in the local geometry of the zeolite framework allow the investigation of the cation migration from the supercages into the small cavities of zeolite Y by ^{29}Si MAS NMR spectroscopy [235, 236]. Framework strains originating from multivalent cations located at position SI′ in sodalite cages give rise to resonance shifts of SiO_4 tetrahedra located in six-membered oxygen rings close to these cations. Figure 4.35 shows the effect of lanthanum migration on the ^{29}Si MAS NMR spectra of zeolites 30LaNaY and 60LaNaY (lanthanum exchange degrees of 30% and 60%, respectively) before (suffix a) and after (suffix b) calcination at 433 K [236]. The as-exchanged samples yield ^{29}Si MAS NMR spectra with lines at chemical shifts of – 89.2, – 94.5, – 100.5 and – 105.0 ppm for Si(3Al), Si(2Al), Si(1Al) and Si(0Al) units, respectively. Calcination at 433 K results in an intensity decrease of the low-field lines and the appearance of a new peak at – 97.2 ppm. The decomposition (bottom) indicates that in the spectra of zeolites 30LaNaY and 60LaNaY about 19% and 30%, respectively, of the Si(nAl) lines were shifted by about 3 ppm to a higher magnetic field (denoted n = 3′, 2′, and 1′). Applying Eq. 4.41, this shift corresponds to a variation in the mean Si–O–T (T = Si or Al) bond angles of about 5°.

Fyfe et al. [237 – 243] investigated 3D connectivities between crystallographically inequivalent T-sites in zeolites ZSM-5, ZSM-12, ZSM-22, ZSM-23, ZSM-39, and DD3R by 2D COSY (correlation spectroscopy) [244] and Incredible Natural Abundance Double Quantum Transfer Experiment (INADEQUATE) [244] experiments. The homonuclear ^{29}Si COSY MAS NMR experiments yielded 2D spectra consisting of ^{29}Si MAS NMR signals in both F1- and F2-dimensions and enabled the study of the connectivities of the zeolite framework (see also Sect. 4.2.3.3). The COSY experiment uses the J-coupling between the nuclear spins in a molecular structure. The pulse sequence, shown in Fig. 4.36a, consists of: (a) the preparation of the resonating spins by cross-polarization, (b) the evolution of the spin system during the time period t_1, (c) a $\pi/2$-pulse, which induces the magnetization transfer between the coupled spins [244], and (d) the acquisition of the free induction decays in the time period t_2. The array $G(t_1, t_2)$ was recorded with 128 increments in t_1 and 64 scans for each t_1. In Fig. 4.36c and 4.36d the stacked plot and the contour plot of the 2D spectrum $F(\nu_1, \nu_2)$ of zeolite ZSM-39 are shown, respectively. Cross-peaks originating from silicon atoms on sites T1 and T2 (cross-peak T1T2) and from silicon atoms on sites T2 and T3 (cross peak T2T3) clearly indicate connectivities within these two pairs of silicon atoms which agree with the structure model of zeolite ZSM-39 depicted in Fig. 4.36b.

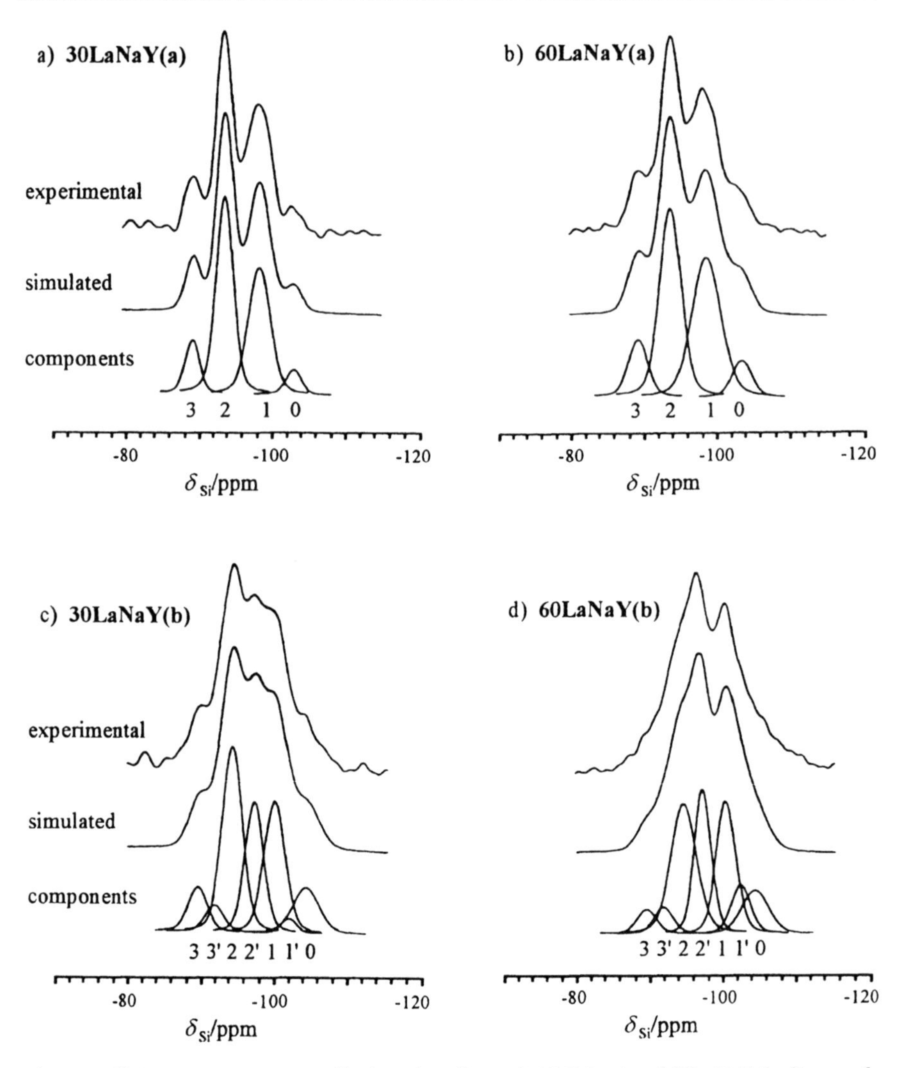

Fig. 4.35. ^{29}Si MAS NMR spectra of hydrated zeolites 30LaNaY (**a, c**) and 60LaNaY (**b, d**), recorded before (**a, b**) and after (**c, d**) thermal treatment at 433 K and rehydration. The figures denote the signals of the undisturbed ($n = 1, 2, 3$) and of the strained ($n = 1', 2', 3'$) Si(nAl) units [236]

4.2.4.2
^{27}Al NMR Spectroscopy of Framework and Non-Framework Aluminum in Zeolites

Comparing the NMR spectra of aluminum and silicon atoms incorporated in zeolite frameworks, the ^{27}Al NMR spectra are much simpler than their ^{29}Si NMR counterparts (see [184] and [213]). According to Loewenstein's rule, the formation of Al–O–Al bondings is forbidden and only Al(4Si) units exist in the framework of aluminosilicates. Therefore, the ^{27}Al NMR spectra of hydrated zeolites

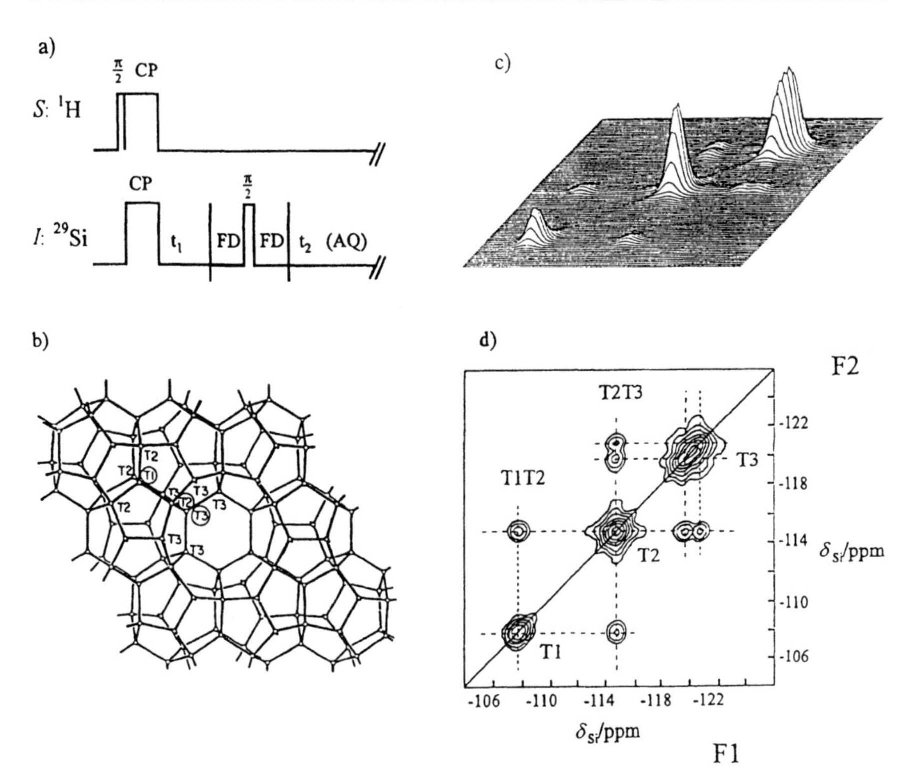

Fig. 4.36. Schematic representation of the modified COSY experiment (**a**), of the zeolite ZSM-39 framework (**b**) and the stacked plot (**c**) and the contour plot (**d**) of the 2D ^{29}Si COSY MAS NMR spectrum of ZSM-39 recorded at 373 K with 128 experiments, 64 scans in each experiment, and a fixed delay of 5 ms. The total experimental time was about 23 h [239].

consist, in general, of only one line of framework aluminum atoms (Al^f) in a relatively small range of chemical shifts between 55 ppm and 68 ppm [referenced to 0.1 M aqueous $Al(NO_3)_3$ solution]. In hydrated zeolites there are only small deviations from the tetrahedral symmetry of the AlO_4 units which result in a weak quadrupolar line broadening. Non-framework aluminum (Al^{nf}) species of hydrated zeolites, originating from dealumination by thermal treatment or acid leaching, are octahedrally coordinated (AlO_6) and cause a ^{27}Al NMR signal at about 0 ppm. If non-framework aluminum exists as polymeric aluminum oxide or oxide hydrates in zeolite cages or channels, strong quadrupolar line broadening may be observed owing to distortions of the octahedral symmetry of AlO_6 units. A broad signal at 30–50 ppm may be caused by a non-framework aluminum species in a disturbed tetrahedral coordination or a penta-coordinated state [245]. For the signals of tetrahedrally coordinated framework aluminum atoms no definite relationship was found between the ^{27}Al chemical shift and the n_{Si}/n_{Al} ratio or the Si–Al-order scheme of the zeolite framework. However, a linear relationship between the ^{27}Al NMR line positions and the mean Al–O–Si bond angles has been established [246]. An important prerequisite of a suitable correlation is the careful correction of the experimentally derived shift data for the quadrupolar shift contribution. By ^{27}Al and ^{29}Si MAS NMR spectroscopy on lithium and sodium halide aluminosilicate sodalites with high-speed spinning, Jacobsen et al. [247] found the following linear relationship between the ^{27}Al and ^{29}Si chemical shifts

$$\delta_{Al}/\text{ppm} = 1.03 \cdot \delta_{Si}/\text{ppm} + 151.94 \qquad (4.43)$$

For chemical shift data covering a range of 12 ppm a correlation coefficient of 0.9998 was calculated. This result suggests that accurate ^{27}Al chemical shifts may be used for predicting mean Al-O-Si bond angles and ^{29}Si chemical shift values of Si(0Al) units in high-silica zeolites (for the latter see Table 4.5).

Dealuminated zeolites Y, mordenite and ZSM-5 have been extensively investigated by multi-nuclear solid-state NMR spectroscopy [184, 248–257]. While ^{29}Si MAS NMR spectroscopy can show that aluminum is removed from the zeolite framework, ^{27}Al NMR spectroscopy yields insight into the formation of non-framework aluminum species (Al^{nf}). In order to elucidate the nature of such non-framework aluminum species, ^{27}Al MAS NMR and ^{1}H-^{27}Al CP/MAS NMR investigations were carried out on a series of hydrated zeolites with increasing degree of dealumination [245, 256]. As shown in Fig. 4.37 the decrease of the MAS NMR signal of tetrahedrally coordinated framework aluminum at about 60 ppm is accompanied by the appearance of signals at about 0 and 30 ppm. Furthermore, Fig. 4.37 shows that for the spectra recorded with the cross-polarization technique (right-hand side) the intensities of the signals at 0 and 30 ppm increase inversely to that of the signal at 60 ppm. Therefore, the authors suggested that ^{1}H-^{27}Al CP/MAS NMR experiments monitor preferentially non-framework aluminum atoms. In addition, this behavior indicates that the 30 ppm line is an independent signal, possibly due to penta-coordinated aluminum atoms [245]. The possibility that the 60 ppm line is caused by more than one component, was checked by ^{27}Al nutation NMR accompanied by fast MAS (ν_{rot} ca. 13 kHz). F1-cross-sections of these 2D spectra yielded two components at 56 and 62 ppm. The first one originated from four-coordinated Al^{nf} and the second one from tetrahedral framework aluminum [245].

While the standard ^{27}Al MAS NMR technique reduces the second-order quadrupolar line-broadening by a factor of about 3 only, application of the DOR (double oriented rotation) technique results in a complete averaging of this solid-state interaction. Line-narrowing of about two orders of magnitude may be achieved by DOR providing well separated lines for structurally inequivalent quadrupolar nuclei. The narrow DOR central lines are generally accompanied by a pattern of DOR sidebands separated by the spinning frequency of the outer rotor. However, the central lines may be identified from spectra recorded at different spinning rates. Adopting the method published by Samoson and Lippmaa [257], rotor-synchronized single-pulse excitation supresses the odd-numbered DOR sidebands. A prerequisite of a suitable line separation is the absence of a distribution of chemical shifts or quadrupolar parameters, caused, e.g., by a distribution of geometric parameters in the local structure of resonating nuclei.

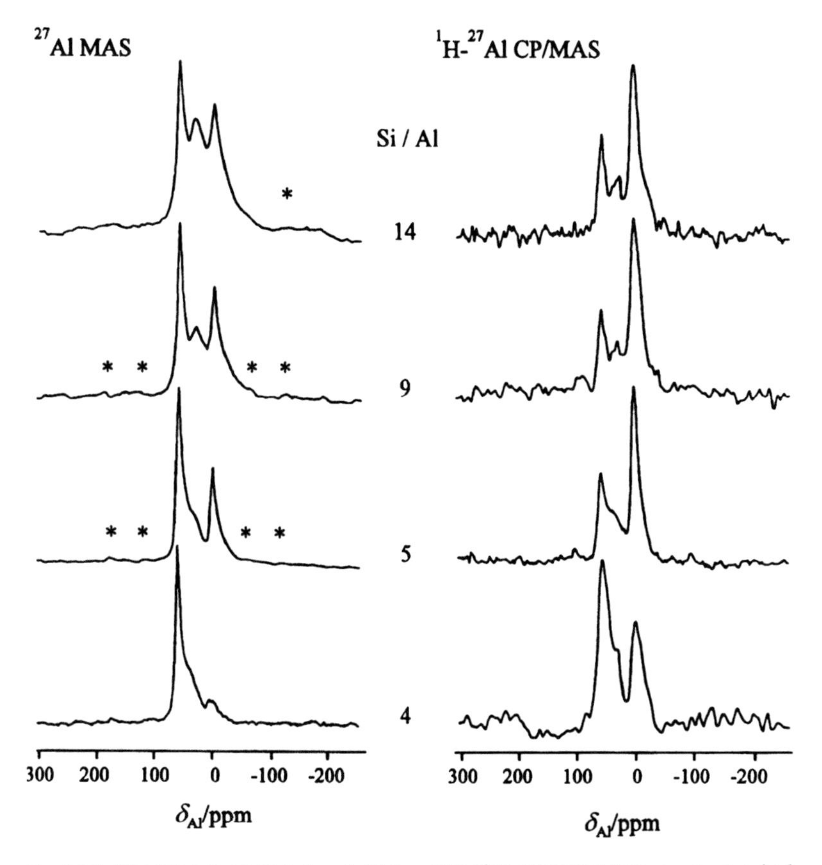

Fig. 4.37. ^{27}Al MAS NMR (*left-hand side*) and ^{1}H-^{27}Al CP/MAS NMR spectra (*right-hand side*) of increasingly dealuminated (from *bottom* to *top*) zeolite HY, recorded at a resonance frequency of 104.3 MHz and with a MAS rate of ca. 13 kHz (*: MAS sidebands) [245]

Dramatic line-narrowing was observed in ^{27}Al DOR NMR investigations of calcium tungstate aluminate sodalite (CAW) and sodium aluminosilicate sodalite [258]. The frameworks of these compounds are built exclusively of AlO_4 tetrahedra and alternating arrangements of AlO_4 and SiO_4 tetrahedra, respectively. The ^{27}Al DOR NMR spectra of CAW measured at 9.4 and 11.7 T and with spinning rates of the outer DOR rotor in the range between 800 and 1150 Hz showed seven sharp central lines corresponding to seven crystallographically inequivalent aluminum sites of the CAW framework [258].

Crystalline aluminophosphates ($AlPO_4$) are materials with a high degree of framework order. The framework of aluminophosphate-type zeolites is built of AlO_4 and PO_4 tetrahedra in an alternating arrangement. The ^{27}Al DOR NMR spectra of these materials consist of narrow lines due to aluminum atoms on crystallographically inequivalent T-sites [259–263] or caused by framework aluminum atoms interacting with probe molecules [264–267]. At the top of

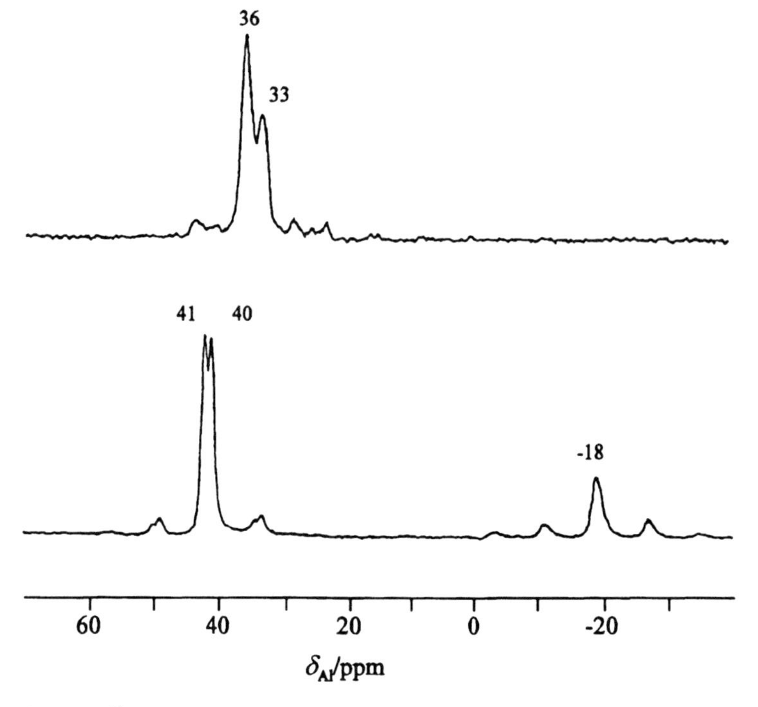

Fig. 4.38. ^{27}Al DOR NMR spectra of dehydrated (*top*) and hydrated (*bottom*) aluminophosphate VPI-5, recorded at 104.2 MHz with a rotation frequency of the outer DOR rotor of ν_{rot} = 800 Hz and rotor-synchronized pulses [260]

Fig. 4.38, the ^{27}Al DOR NMR spectrum of dehydrated VPI-5 is depicted showing two peaks at 36 and 33 ppm with an intensity ratio of 2:1. These signals were attributed to different tetrahedrally coordinated framework aluminum species [260]. Based on the experimentally observed intensity ratio the line at 36 ppm was assigned to aluminum sites in the six-membered oxygen rings and the line at 33 ppm to aluminum sites in double four-membered rings (see at the top of Fig. 4.39) [259]. Addition of water molecules dramatically alters the local bonding of the framework aluminum atoms of VPI-5, as reflected by the spectrum at the bottom of Fig. 4.38. The lines at 36 and 33 ppm have disappeared and two partially resolved signals associated with tetrahedrally coordinated framework aluminum atoms are present at 41 and 40 ppm. In addition, a broader upfield line at -18 ppm in the range ascribed to octahedrally coordinated aluminum sites can be observed. The integrated intensities of these three lines occur in a 1:1:1 ratio. Spectra recorded in different magnetic fields show different second order quadrupolar shifts (Eq. 4.35). This was used to determine isotropic chemical shifts of 43.6, 41.6 and – 10.4 ppm and quadrupole coupling constants of 2.0 – 2.3 MHz for the first line and of 1.0 – 1.2 MHz for the second line [268]. A strong argument for the assignment of the peaks at 43.6 and 41.6 ppm is based on the relationship between the strength of the quadrupole interaction

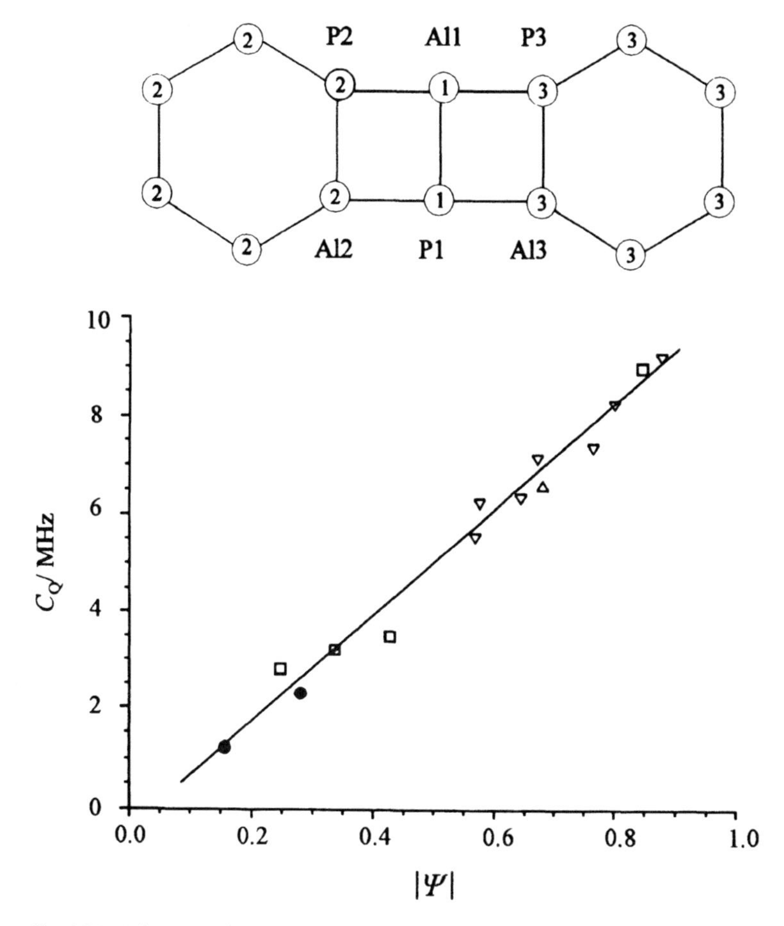

Fig. 4.39. Schematic drawing (*top*) of the VPI-5 structure and the plot (*bottom*) of the quadrupole coupling constants, C_Q, of ^{27}Al nuclei as a function of the *shear strain parameter* $|\Psi|$ of AlO_4 tetrahedra in aluminate sodalites (△, ▽), feldspars (□), and the aluminophosphate VPI-5 (●) [269, 270]

of the framework aluminum atom and the local distortion of the AlO_4 tetrahedron. This distortion can be described quantitatively by the *shear strain parameter* [269]

$$|\Psi| = \sum_{i=1}^{6} |\tan(\theta_i - \theta_0) \qquad (4.44)$$

where the sum runs over the six individual O–Al–O bond angles θ_i. The angle θ_o is the ideal bond angle of 109.5°. The geometries of AlO_4 tetrahedra given in [269] and [270] yield a linear interdependence of $|\Psi|$ and C_Q with a correlation coefficient of 0.923 and a slope of 0.106 (see Fig. 4.39). From the crystal structure data of VPI-5 [271] it follows that $|\Psi| = 0.157$ for Al2 sites and $|\Psi| = 0.280$ for Al3 sites. According to the correlation between $|\Psi|$ and C_Q, shown in Fig. 4.39, the quadrupole coupling constant of aluminum atoms at Al3 sites must be significantly larger than that of aluminum atoms at Al2 sites. Hence, the low-field ^{27}Al DOR NMR signal at 43.6 ppm is due to Al3 sites (C_Q ca. 2 MHz) and the signal appearing at 41.6 ppm has to be attributed to Al2 sites (C_Q ca. 1 MHz) [270].

4.2.4.3
^{31}P MAS NMR Spectroscopy of PO_4 Tetrahedra in Aluminophosphate-, Silicoaluminophosphate-, and Gallophosphate-Type Zeolites

^{31}P MAS NMR spectroscopy is a suitable method for obtaining data on PO_4 tetrahedra in aluminophosphate-, silicoaluminophosphate-, and gallophosphate-type zeolites. First ^{31}P NMR investigations on $AlPO_4$-5, $AlPO_4$-11, $AlPO_4$-17 and $AlPO_4$-31 were carried out by Blackwell and Patton [272]. Typically, the signals of PO_4 units in aluminophosphates cover a chemical shift range between –20 and –33 ppm (referenced to 85% H_3PO_4) [272–278]. As a result of ^{31}P MAS NMR studies on crystalline $AlPO_4$ polymorphs and aluminophosphate-type zeolites, Mueller et al. [279] found a linear relationship between the resonance positions due to PO_4 tetrahedra and their mean P–O–Al angles. The gradient was determined to be –0.61 ppm/deg with a correlation coefficient of 0.995. According to this relationship an increase in the mean P–O–Al angles causes a high-field shift of the PO_4 signal.

A number of studies have shown that the incorporation of metal atoms into the aluminophosphate-type zeolites affects the ^{31}P resonances of PO_4 tetrahedra [280–290]. Investigating magnesium aluminophosphate MgAPO-20, Barrie and Klinowski [280] observed ^{31}P MAS NMR signals at chemical shifts of –14.0, –21.1, –28.0, and –34.9 ppm due to P(1Al, 3Mg), P(2Al, 2Mg), P(3Al, 1Mg), and P(4Al) units. The relatively small chemical shift of the phosphorus atoms surrounded by four aluminum atoms [P(4Al)] was explained by the high P–O–Al bond angle of 160 °C (*vide supra*). The concentrations of framework atoms in MeAPOs can be calculated from the ^{31}P MAS NMR intensities in a manner similar to that used for the calculation of n_{Si}/n_{Al} ratios from ^{29}Si MAS NMR spectra. With P(nAl) instead of Si(nAl) and using the integrated intensities $I_{P(nAl)}$ of the P(nAl) signals, Eq. 4.40 yields the framework n_P/n_{Al} ratio. The expression for the n_P/n_{Me} ratio, and (since 50% of the T-sites are occupied by phosphorus) the fraction [*Me*] of T-sites occupied by the metal atom, Me, can be calculated by Eq. 4.45 [280]

$$[\mathrm{Me}] = \sum_{n=0}^{4} (4-n) I_{P(n\mathrm{Al})} / 8 \sum_{n=0}^{4} I_{P(n\mathrm{Al})} \tag{4.45}$$

The aluminophosphate VPI-5 is quite unusual and is, therefore, considered here in the context of its structure and structural changes. As shown at the top of Fig. 4.39 the structure of VPI-5 consists of two phosphorus sites contributing to six-membered oxygen rings and one phosphorus site belonging to two adjacent

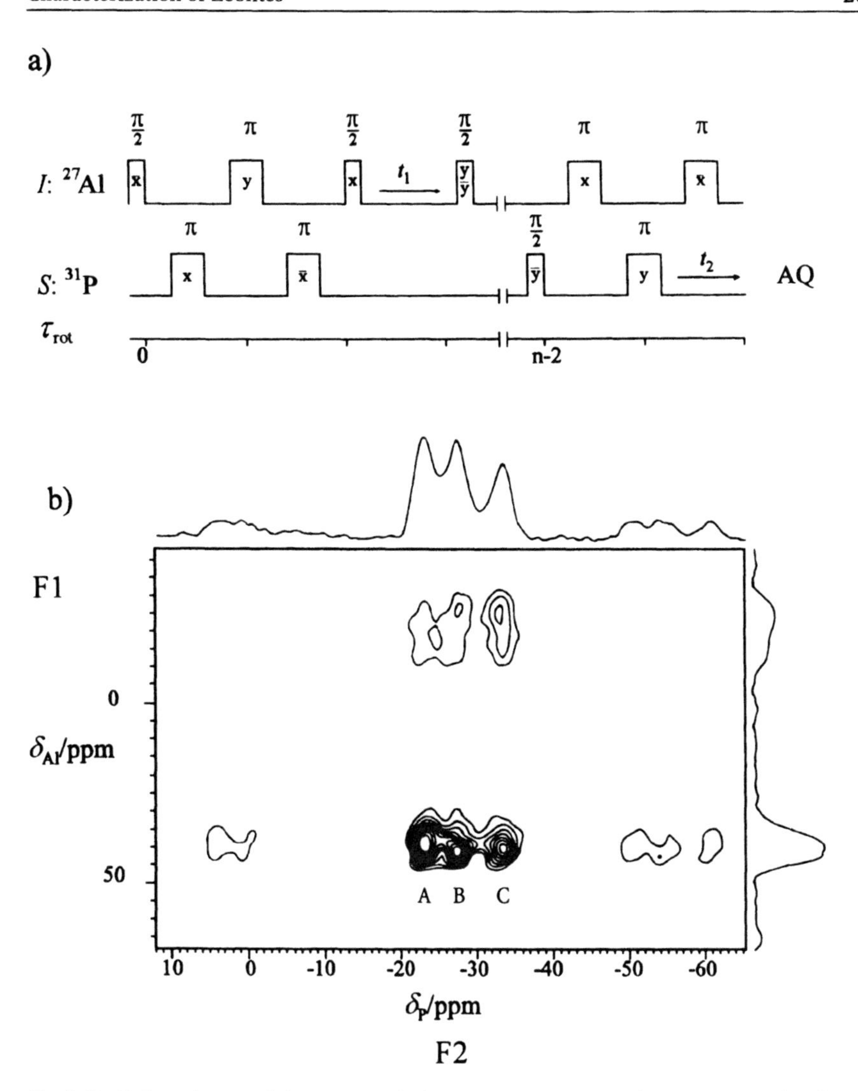

Fig. 4.40. Pulse scheme of the 2D correlation experiment (**a**) and the contour plot (**b**) and projections (F1: ^{27}Al, F2: ^{31}P) of the ^{27}Al-^{31}P correlation spectrum of the aluminophosphate VPI-5 [281]

four-membered rings. Indeed, the ^{31}P MAS NMR spectrum of dehydrated VPI-5 consists of two lines at about −26 and −31 ppm with an intensity ratio of close to 2:1 [282–284]. In contrast, the spectrum of hydrated VPI-5 reveals three lines of equal intensities at chemical shifts of about −23, −27, and −33 ppm [281, 285–287]. A number of papers have been published [285–290] dealing with the assignment of these ^{31}P MAS NMR signals to distinct phosphorus sites in VPI-5. In this context, van Eck and Veeman [281] applied a heteronuclear correlation pulse sequence which is depicted in Fig. 4.40a. This 2D experiment allows the correlation of NMR signals via heteronuclear dipolar coupling of the nuclei. The preparation period of the 2D experiment corresponds to a REDOR (rotational-echo double resonance) pulse sequence (see Sect. 4.2.3.2). The REDOR sequence transfers magnetization from the I- to the S-spins. During the evolution period, t_1, the spin system evolves under the Hamiltonian of the S- and I-spin chemical shifts and the heteronuclear dipolar coupling. Two $\pi/2$-pulses at the beginning and the end of the evolution period rotate the magnetization along the z-axis and back into the x-y-plane, respectively. Before the detection of the S-spin magnetization in t_2, a second REDOR pulse sequence is applied to the S-spins which produces observable S-spin magnetization. A more extensive treatment of the pulse sequence is available [281]. At the bottom of Fig. 4.40 the ^{31}P–^{27}Al correlation spectrum of dehydrated VPI-5 recorded at resonance frequencies of 121.4 MHz for ^{31}P and 78.2 MHz for ^{27}Al and at the MAS frequency of 3.3 kHz is shown [281]. At the magnetic field of B_0 = 7.0 T the ^{27}Al NMR signals of aluminum atoms at Al3 and Al2 sites have a chemical shift of 38 and 40 ppm, respectively [281]. The cross-peaks in the 2D spectrum indicate that each phosphorus atom is coupled with both tetrahedrally and octahedrally coordinated aluminum atoms. However, the intensities of the cross-peaks differ which indicates differences in the magnetization transfer due to different neighbors (compare top of Fig. 4.39). From the crystal structure, it can be derived that the three different phosphorus sites have the following aluminum atoms in their surroundings: P1(2 · Al1, Al2, Al3), P2(Al1, 2 · Al2, Al3), and P3(Al1, Al2, 2 · Al3). The strong intensity of cross-peak A indicates a preferred coupling of phosphorus nuclei at the ^{31}P NMR shift of –23 ppm with aluminum nuclei at the ^{27}Al NMR shift at 38 ppm (Al3). In the same manner, the strong intensity of cross-peak B indicates a preferred coupling of nuclei at the ^{31}P NMR shift of –27 ppm with nuclei at the ^{27}Al NMR shift of 40 ppm (Al2). Hence, the ^{31}P NMR signals at –23 and –27 ppm are due to phosphorus atoms at P3(Al1, Al2, 2 · Al3) and P2(Al1, 2 · Al2, Al3) sites, respectively. The ^{31}P NMR signal at –33 ppm is attributed via cross peak C to the phosphorus atoms at P1(2 · Al1, Al2, Al3) which show a preferred coupling with the octahedrally coordinated aluminum atoms at Al1. The same signal assignment was deduced by Fyfe et al. [291, 292] from 2D TEDOR (transferred-echo double resonance) experiments.

Extensive studies of the thermal stability of VPI-5 and its structural transformation to $AlPO_4$-8 have been carried out [282, 286, 287, 293, 294]. Depending on the treatment conditions (vacuum, heating rate, final temperature) the structure of VPI-5 can be preserved up to a dehydration temperature of 673 K, as proved by ^{31}P MAS NMR spectroscopy. At slightly elevated temperatures (343 to 423 K), VPI-5 undergoes a reversible dehydration and rehydration which results in a merging of the two low-field ^{31}P MAS NMR lines. The splitting occurs again after cooling down to ambient temperature [287, 293, 294]. Under more drastic conditions (unsealed samples, fast heating rates, high temperature) VPI-5 transforms to $AlPO_4$-8 which can be monitored by ^{31}P and ^{27}Al MAS NMR spectroscopy (Fig. 4.41). The ^{31}P MAS NMR spectrum of hydrated $AlPO_4$-8 at the bottom of Fig. 4.41 consists of three signals at about –21, –25, and –30 ppm with an intensity ratio of 1:2:6. Simultaneously, the center of gravity of the ^{27}Al MAS NMR signal is shifted from about 42 (VPI-5) to about 37 ppm ($AlPO_4$-8).

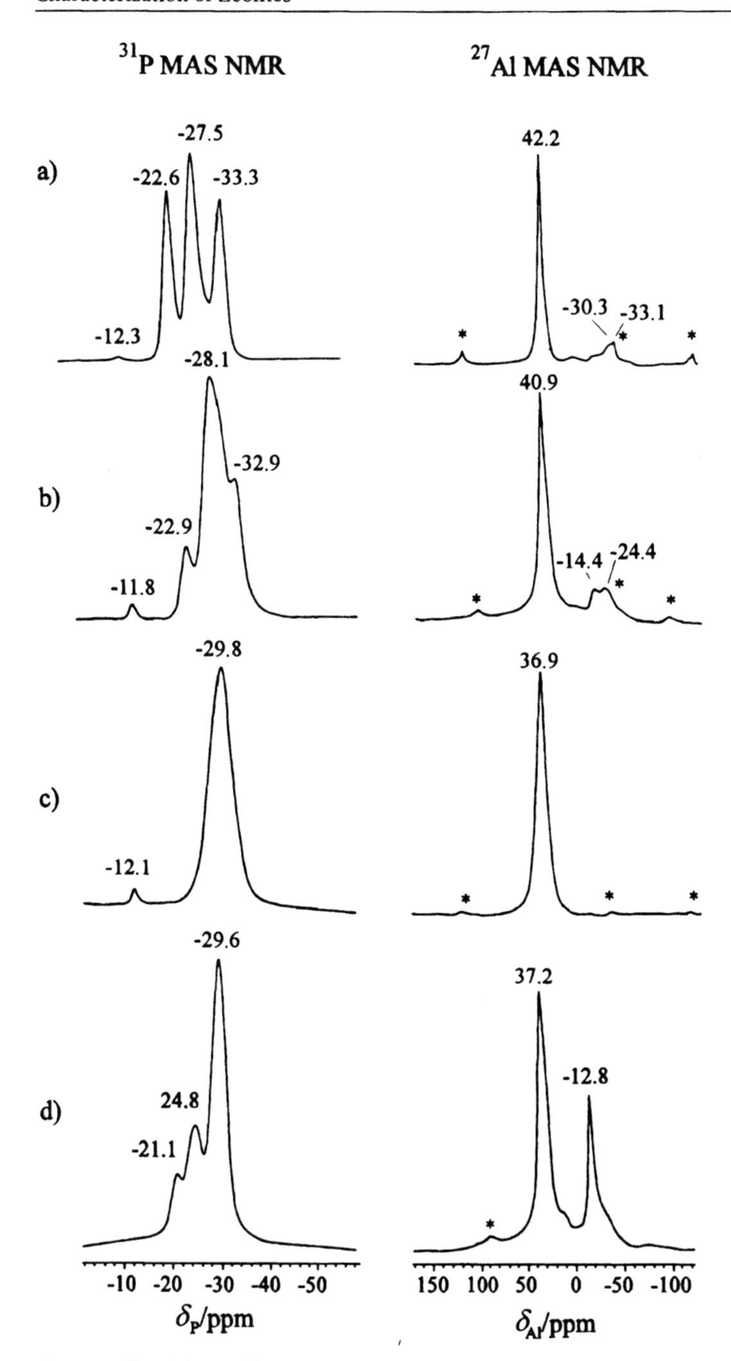

Fig. 4.41. ^{31}P (*left*) and ^{27}Al MAS NMR (*right*) spectra of the aluminophosphate VPI-5 dried at 333 K overnight (**a**), VPI-5 evacuated at 327 K overnight and calcined at 523 K overnight (**b**), $AlPO_4$-8 formed from VPI-5 by calcination at 673 K overnight (**c**), and hydrated $AlPO_4$-8 (**d**) [282]. The asterisks denote MAS sidebands

4.2.4.4
^{11}B MAS NMR Spectroscopy of Boron-Modified Zeolites

In zeolites, boron may occur in tetrahedral BO_4 or trigonal BO_3 units. At external magnetic fields of $B_0 \geq 7$ T and with the application of MAS, BO_4 tetrahedra cause a narrow ^{11}B MAS NMR signal with a symmetric line shape. As observed for framework aluminum atoms ($I = 5/2$) in hydrated aluminosilicate-type zeolites, tetrahedrally coordinated boron atoms ($I = 3/2$) in hydrated boron-containing zeolites exist in a symmetric surrounding as well with a negligible electric field gradient. Owing to its strong quadrupolar interaction, the trigonal boron (BO_3) causes a broad quadrupolar pattern [295–298]. For hydrated boron-substituted ZSM-5, Scholle and Veeman [296] derived a superposition of the narrow BO_4 signal and the broad BO_3 pattern. Typically, the narrow BO_4 line has a chemical shift of about –3 ppm (referenced to $BF_3 \cdot OEt_2$) and a line width of about 1 ppm in fully hydrated samples. At the resonance frequency of 96.2 MHz the singularities of the BO_3 pattern have a splitting of about 20 ppm [297]. In the spectra of dehydrated boron-containing zeolites the broad BO_3 pattern predominates and is superimposed by the narrow signal of residual BO_4 units. Rehydration causes a partial reconstruction of the original spectrum consisting of a narrow BO_4 line [297, 298].

Axon and Klinowski [299] utilized 2D nutation MAS NMR spectroscopy to investigate boron atoms in zeolite ZSM-5 synthesized via the fluoride method. In this 2D experiment a high-power radio frequency pulse is applied to the resonating spins in order to measure the nutation of the quadrupole nuclei which is sampled with increments in the time t_1. The frequency ν_1 of the signal in the F1-dimension depends on the quadrupole frequency, ν_Q, of the resonating nuclei and is independent of the chemical shift of the corresponding signal in the F2-dimension. Three different cases can be distinguished: (1) ν_Q is small in relation to the applied radio frequency, ν_{rf}, i.e., $\nu_Q \ll \nu_{rf}$, and the signal appears at $\nu_1 = 1\nu_{rf}$ in the F1-dimension; (2) $\nu_Q \gg \nu_{rf}$ leading to a nutation frequency of $\nu_1 = (I + 1/2) \cdot \nu_{rf}$, i.e., for ^{11}B nuclei with $I = 3/2$, a signal follows at $\nu_1 = 2\nu_{rf}$ in the F1-dimension; (3) $\nu_Q \approx \nu_{rf}$ causes a broad pattern in the range between $\nu_1 = 1\nu_{rf}$ and $\nu_1 = 2\nu_{rf}$. Using this 2D technique, Axon and Klinowski [299] found two signals of boron atoms in boron-modified zeolite ZSM-5 at $\nu_1 = 1\nu_{rf}$ due to tetrahedral boron ($\nu_Q \ll \nu_{rf}$) and at $\nu_1 = 2\nu_{rf}$ caused by trigonal boron ($\nu_Q \gg \nu_{rf}$). Hence, in agreement with the quadrupolar line broadening observed in the 1D ^{11}B MAS NMR spectra, tetrahedral and trigonal boron atoms in boron-modified zeolite ZSM-5 are characterized by a negligibly weak and a strong quadrupole interaction, respectively.

4.2.4.5
Solid-State ^{17}O NMR Spectroscopy of the Zeolite Framework

^{17}O NMR studies on zeolites have always been performed on isotopically enriched samples because of the low natural abundance of the isotope ^{17}O. Timken et al. [300, 301] carried out static ^{17}O NMR spectroscopic studies on zeolites NaA and NaY, on dealuminated Y-type zeolites and on gallosilicates and aluminophosphates. The spectra of aluminosilicate-type zeolites consist of broad quadrupolar patterns corresponding to a quadrupole coupling constant, C_Q, of about 4.6–5.2 MHz and 3.1–3.2 MHz due to oxygen atoms in Si–O–Si bridges and Si–O–Al bridges, respectively. The asymmetry parameters, η_Q, are 0.1 for Si–O–Si and 0.2 for Si–O–Al arrangements. These values were verified by MAS and variable-angle spinning experiments at different magnetic field strengths. The isotropic ^{17}O chemical shifts, δ_{iso}, amount to 44–57 ppm (referenced to H_2O) for Si–O–Si and to 31–45 ppm for Si–O–Al [300]. ^{17}O NMR investigations on aluminophosphates and gallosilicates yielded the following quadrupole parameters: $C_Q = 5.6-6.5$ ppm, $\eta_Q \approx 0$ and $\delta = 61-67$ ppm for oxygen atoms in Al–O–P bridges and $C_Q = 4.0-4.8$ ppm, $\eta_Q \approx 0.3$ and $\delta = 28-29$ ppm for oxygen atoms in Si–O–Ga bridges [302].

Hence, ^{17}O NMR investigations on zeolites of different chemical compositions have demonstrated that the spectra of these materials consist of quadrupolar patterns characteristic of oxygen atoms in distinct T–O–T bridges (T = Si, Al, P, Ga). The presence of oxygens in more than one type of T–O–T bridges, e.g., in Si–O–Al and Si–O–Si bridges in zeolite Y, causes a superposition of quadrupolar patterns with intensities according to their relative populations in the zeolite framework. The observed order of quadrupole coupling constants of oxygen atoms in different T–O–T bridges is: Al–O–P > Si–O–Si > Si–O–Ga > Si–O–Al, and that of isotropic chemical shifts: Al–O–P > Si–O–Si > Si–O–Al > Si–O–Ga.

4.2.4.6
^1H MAS NMR Spectroscopy of Acidic and Non-Acidic Hydroxyl Groups in Zeolites

The first ^1H MAS NMR investigations of dehydrated zeolites were carried out in 1982 in an external magnetic field with a flux density of $B_o = 1.4$ T yielding residual line widths of ca. 1 kHz [303]. The application of external magnetic fields of $B_o \geq 7$ T and CRAMPS [304, 305] (see Sect. 4.2.3.1) or the MAS technique with sample spinning rates of $\nu_{rot} \geq 10$ kHz [306] result in well-resolved ^1H MAS NMR spectra. Prerequisites for such investigations are a careful dehydration of the zeolite samples and either the application of gas-tight MAS rotors filled in a glove box or the preparation of symmetric glass ampules suitable for spinning in MAS rotors [303, 305, 307]. ^1H NMR positions of hydroxyl protons in zeolites cover a range from 0 to 7 ppm. The lowest chemical shifts were observed for metal OH groups (MeOH), e.g., hydroxyl groups at Mg^{2+} cations located on position SII in the supercages of zeolite Y (Fig. 4.42) which appear at about 0 ppm (Fig. 4.43a) [308, 309]. Silanol groups (SiOH) on framework defects cause signals in the shift range between 1.3 and 2.2 ppm (Fig. 4.43 and Table 4.6) [310, 311]. The proton NMR shifts of bridging OH groups in large zeolite cavities and channels cover a range between 3.8 and 4.3 ppm. The exact resonance position depends on the framework n_{Si}/n_{Al} ratio [311, 312], and, therefore, on the mean electronegativity of the zeolite framework [313]. In Fig. 4.44 the chemical shifts, δ_H, of SiOHAl groups located in large cages or channels of zeolites, i.e. of bridging OH groups which are not involved in hydrogen bonds (entry 6 of Table 4.7 and [311, 313, 314]), are depicted as a function of the mean Sanderson electronegativity [315], S^m, of zeolite frameworks (column 5 of Table 4.7 and

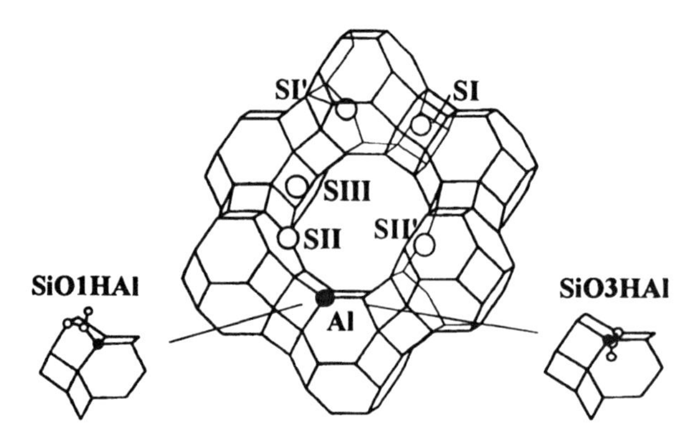

Fig. 4.42. Faujasite structure with bridging OH groups pointing into the supercage (SiO1HAl) and the sodalite cages (SiO3HAl) and cation positions in the supercage (SII, SIII), the sodalite cages (SI′, SII′) and the hexagonal prisms (SI)

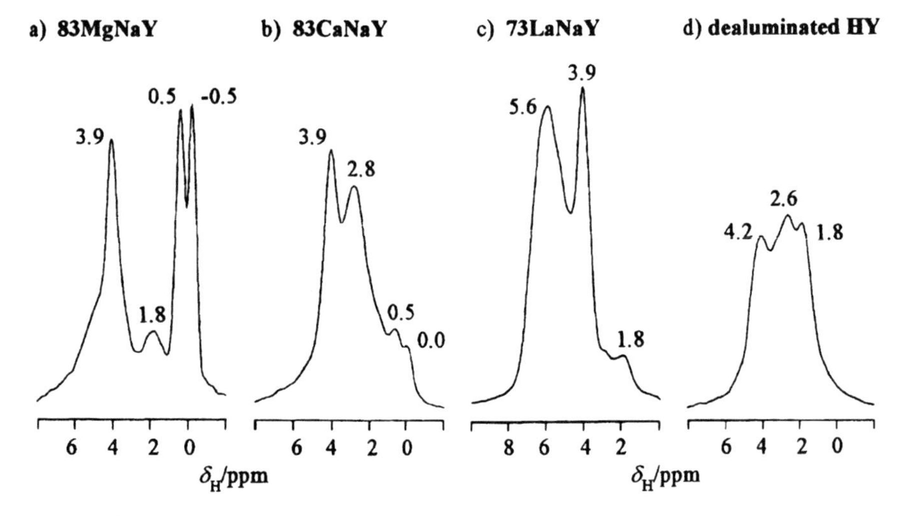

Fig. 4.43. ^{1}H MAS NMR spectra of zeolites 83MgNaY (**a**) [308], 83CaNaY (**b**) [308] and 73LaNaY (**c**) [306] calcined at 433 K and of dealuminated zeolite HY (**d**) [310] calcined at 673 K

[313]). The values of S^m are derived from the geometric averaging of the Sanderson electronegativities S_i of the atoms i ($S_{Al} = 2.22$, $S_O = 5.21$, $S_{Si} = 2.84$, and $S_H = 3.55$) related to the number of these atoms per unit cell. Figure 4.44 shows an increase of the chemical shift, δ_H, with increasing mean electronegativity of the zeolite framework. The significant deviation of the chemical shift of bridging OH groups in zeolite 25 HNaA is noteworthy. This deviation may be caused by the location of these bridging OH groups in eight-membered oxygen rings.

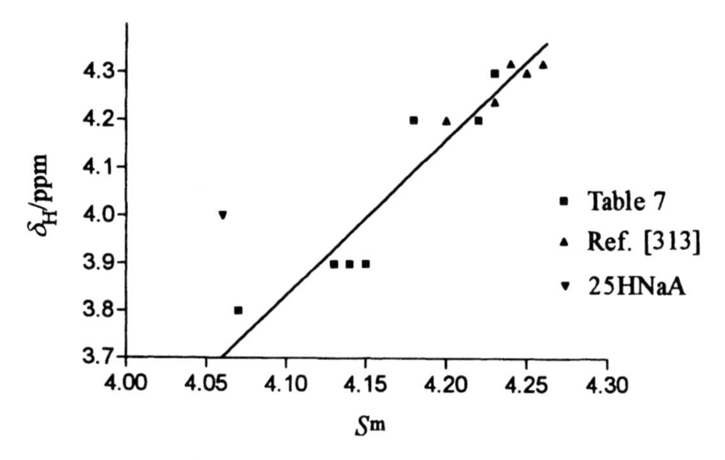

Fig. 4.44. Proton NMR shifts, δ_H, of bridging OH groups in the supercages and channels of aluminosilicate type zeolites as a function of the mean Sanderson electronegativity, S^m, given in columns 5 and 6 of Table 4.7 and in [313]

Table 4.6. ^1H NMR Shifts (referenced to TMS), δ_H, and Assignments of Hydroxyl Groups in Dehydrated Zeolites

^1H NMR shift δ_H/ppm	Abbreviation	Type of hydroxyl groups	Ref.
-0.5 to 0.5	MeOH	metal OH groups in large cages or at the outer surface of zeolite particles	[308, 316]
1.3 to 2.2	SiOH	silanol groups at the outer surface or at lattice defects	[305, 311, 314]
2.6 to 3.6	AlOH	OH groups at non-framework aluminum species, hydrogen bonded	[310, 317-319]
2.8 to 6.2	CaOH, AlOH, LaOH	cation OH groups located in sodalite cages of zeolite Y and in channels of ZSM-5, hydrogen bonded	[308, 319, 320]
3.8 to 4.3	SiOHAl, SiO1HAl*	bridging OH groups in large cages and channels of zeolites	[311, 314, 321, 322]
4.6 to 5.2	SiOHAl, SiO3HAl*	bridging OH groups in small cages of zeolites, electrostatic interaction	[311, 314]
6.1 to 7.0	SiOHAl	disturbed bridging OH groups in zeolite HZSM-5, Hbeta, and HMCM-22, electrostatic interaction	[313, 322, 323-325]

* Assignment of bridging OH groups in faujasite-type zeolites.

In the ^1H MAS NMR spectra of dehydrated zeolite HY an additional line appears at about 4.8 ppm. This signal is due to bridging OH groups pointing to the centers of six-membered oxygen rings of hexagonal prisms (columns 3 and 6 of Table 4.7). The low-field shift of this ^1H MAS NMR signal in comparison to the resonance positions of the bridging OH groups at about 4 ppm was explained by electrostatic interactions with neighboring framework oxygens [305, 311, 326]. A quantitative relationship between the O–O distance of hydrogen bonds (d_{OH-O}) and the resonance shift, $\Delta\delta_H$, of MeOH groups in inorganic solids was empirically derived by Yesinowski and Eckert [327] as follows:

$$\Delta\delta_H/\text{ppm} = 79.05 - 0.255\, d_{OH-O}/\text{pm}. \tag{4.46}$$

CaOH groups located in sodalite cages of dehydrated zeolite CaNaY show a low-field resonance shift, $\Delta\delta_H$, from the resonance position of 0 ppm (undisturbed MeOH groups) to about 2.8 ppm (Fig. 4.43b) [308]. Assuming the formation of hydrogen bonds, the resonance shift of $\Delta\delta_H = 2.8$ ppm corresponds to an O–O distance of $d_{OH-O} = 0.299$ nm (Eq. 4.46). XRD investigations of dehydrated zeolite CaY have shown that OH groups bonded at calcium cations, which are located at position SI′ in sodalite cages (Fig. 4.42), exhibit an O–O distance to the next nearest framework oxygen atom of $d_{OH-O} = 0.297$ nm [328]. This value is in good agreement with the O–O distance derived from ^1H MAS NMR. In zeolite LaNaY a low-field shift of the proton NMR signal of LaOH groups of about $\Delta\delta_H = 5.6$ ppm was found (Fig. 4.43c) [306]. Hence, depending on the location of non-acidic metal OH groups their ^1H MAS NMR signals cover a chemical shift range of up to 6 ppm. In the spectra of dealuminated zeolites HY (Fig. 4.43d) [310], H-mordenite [311] and HZSM-5 [317, 318], signals of hydroxyl protons at non-framework aluminum species appear at chemical shifts in the range between 2.6 and 3.6 ppm. The comparison of these resonance positions with those of AlOH groups on the outer surface of aluminum oxides [316] indicates a low-field shift of the resonance position by about 3 ppm. In fact, this is a hint that hydroxyl protons bound to non-framework aluminum species form hydrogen bonds to neighboring oxygen atoms.

In Table 4.6 a summary of ^1H NMR shifts of hydroxyl protons in dehydrated zeolites is given. Using the empirical relation [329]

$$\delta_H/\text{ppm} = 42.2 - 0.01\, \nu_{OH}/\text{cm}^{-1}, \tag{4.47}$$

proton NMR shifts, δ_H, of OH groups in zeolites can be estimated from the wavenumbers of their IR stretching vibrations, ν_{OH}.

Zeolites exchanged with multivalent cations possess strong intracrystalline electric fields. The strengths of these electric fields are strictly proportional to the ratio of cation charge to cation radius. According to the mechanism of Hirschler [330] and Plank [331] dissociation of water molecules upon adsorption on multivalent cations results in the formation of cation and bridging OH groups. ^1H MAS NMR spectroscopy has been applied to investigate quantitatively the formation of hydroxyl protons due to water adsorption on dehydrated zeolites MgNaY and CaNaY [308]. After adsorption, the samples were thermally treated at temperatures between 473 and 673 K. The ratio of the numbers of metal OH groups and bridging OH groups, determined by ^{1}H MAS NMR intensities, was found to be about 1:1 which corresponds to the hydrolysis of bivalent cations in zeolite Y yielding one cation OH group and one bridging OH group per dissociated water molecule

$$2(\equiv Si-O-Al^{-}\equiv) + Me(H_2O)_n^{2+} \rightleftarrows \equiv Si-O-Al^{-}\equiv$$
$$+ Me(OH)^{+} + \equiv Si-OH-Al\equiv + (n-1)H_2O.$$

After thermal treatment at temperatures between 473 and 673 K a simultaneous decrease in the ^{1}H MAS NMR intensities of cation OH and bridging OH groups was observed which indicates the reversibility of this mechanism. For zeolite LaNaY a maximum concentration of bridging OH groups occurs after dehydration at 450 K [320]. The hydrolysis of aluminum cations was studied on zeolite ZSM-5 exchanged with $Al(NO_3)_3$ [319]. Upon dehydration of zeolite AlZSM-5 at 473 K and subsequent water adsorption a strong increase in the signal of AlOH groups at the chemical shift of 2.8 ppm was found. As observed for zeolites MgNaY and CaNaY, dehydration of zeolite AlZSM-5 at temperatures between 473 K and 673 K causes a simultaneous decrease of the concentrations of AlOH and bridging OH groups.

An often applied method of preparing bridging OH groups in zeolites is the exchange of alkaline cations by ammonium ions with subsequent thermal treatment. Calcination of ammonium-exchanged zeolites at temperatures of $T > 573$ K results in a simultaneous deammoniation and protonation of the zeolite framework. For each desorbed ammonia molecule one hydroxyl proton remains at an oxygen bridge. Applying IR spectroscopy and neutron diffraction Ward [332], Ward and Hansford [333], Jirak et al. [334] and Czjzek et al. [52] investigated the occupation of the proton sites Si-O1-Al and Si-O3-Al in dehydrated zeolites HNaY. For a zeolite HNaY prepared by exchange of 95% of sodium cations by NH_4^+ ions and subsequent calcination, populations of 28.6 OH/u.c. for SiO1HAl groups in supercages and of 24.5 OH/u.c. for SiO3HAl groups in sodalite cages were determined [52]. Figure 4.45 shows the ^{1}H MAS NMR spectra of dehydrated zeolites HNaY ($n_{Si}/n_{Al} = 2.56$) with exchange degrees of 36% (36HNaY), 52% (52HNaY), 71% (71HNaY) and 88% (88HNaY). Decompositions of these spectra were carried out using lines of SiOH groups at 1.8 ppm, of SiO1HAl groups at 3.9 ppm and of SiO3HAl groups at 4.8 ppm. The concentrations of SiO1HAl, SiO2HAl and SiO3HAl groups as a function of the sodium exchange degree are depicted in Fig. 4.46. The data for zeolite HNaY with an exchange degree of more than 99% are taken from [335]. Considering Fig. 4.46, exchange degrees of up to 25% of sodium cations cause an exclusive formation of SiO1HAl groups in the supercages of zeolite Y. For exchange degrees larger than 25% there is a linear increase in the SiO3HAl group population.

Recent NMR studies on dehydrated zeolites HZSM-5 and Hbeta [313, 317, 312-324] yielded hints for a new type of OH groups in zeolites. Low-field shifted signals with a maximum chemical shift of 7.0 ppm were found in addition to the lines of bridging OH groups at the chemical shift of ca. 4 ppm. These signals

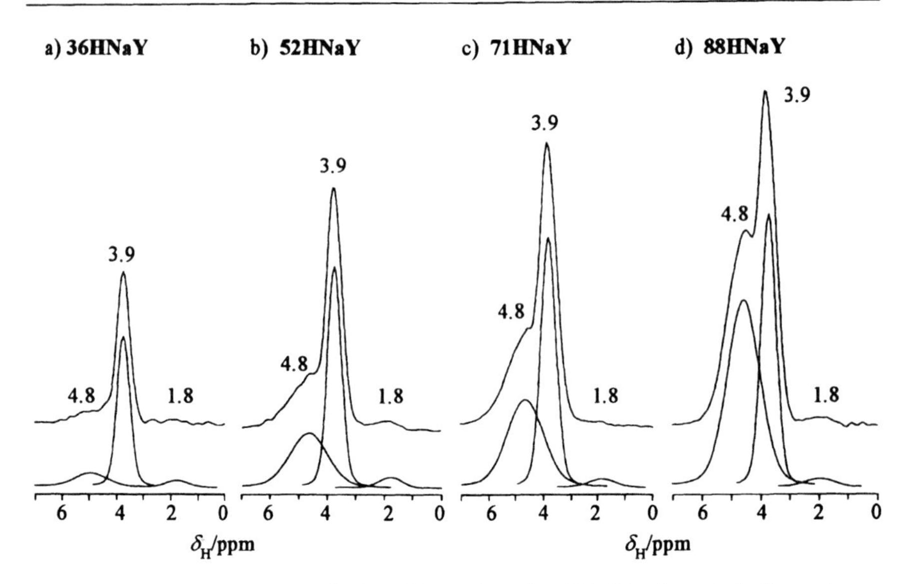

Fig. 4.45. ^{1}H MAS NMR spectra of calcined (673 K) zeolites HNaY ($n_{Si}/n_{Al} = 2.6$) after exchange of 36% (**a**), 52% (**b**), 71% (**c**) and 88% (**d**) of sodium cations by ammonium ions [306]. The spectra were recorded at the resonance frequency of 400.13 MHz and with a MAS frequency of ν_{rot} = 12 kHz. At the bottom the components are plotted

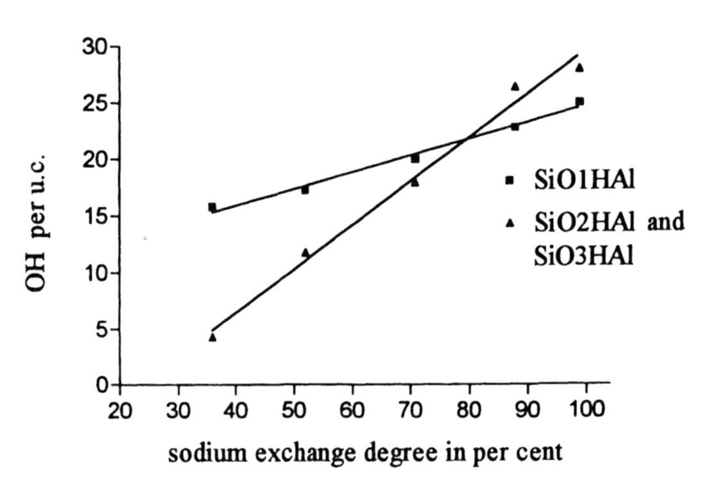

Fig. 4.46. Concentrations of bridging OH groups in the supercages (signal at 3.9 ppm) and the sodalite cages (signal at 4.8 ppm) of dehydrated zeolites HNaY as a function of the degree of exchange of sodium cations. The values of zeolites 36HNaY, 52HNaY, 71HNaY and 88HNaY were determined by decomposition of the ^{1}H MAS NMR spectra depicted in Fig. 4.45. The values for zeolite 99HNaY were taken from [335]

were assigned to hydroxyl protons located in -Al-(OH)-Si-O-Si-(OH)-Al- arrangements [313], to bridging hydroxyl protons hydrogen-bonded to neighboring framework oxygens [322, 323], and to mobile hydroxyl protons in dipole-dipole interaction with framework aluminum atoms [324]. All these models are based on hydroxyl protons located in the neighborhood of framework aluminum atoms. To investigate the acidity of these unknown hydroxyl groups, carbon monoxide was adsorbed on zeolite HZMS-5 [323]. The authors found that at a temperature of 123 K, the hydroxyl protons of the unknown signal at ca. 7 ppm and the bridging hydroxyl groups at ca. 4 ppm interact in the same manner with carbon monoxide indicating an equal acid strength of these OH groups.

A tuning of the zeolite acidity can be achieved by dealumination and/or realumination of the framework. ^{1}H MAS NMR studies of realuminated and rehydroxylated zeolite HNaY were carried out on hydrothermally treated samples [336, 337]. These samples were dealuminated at 800 K, subsequently treated with an aqueous solution of KOH, and ammonium-exchanged and calcined at $T = 673$ K. ^{1}H MAS NMR spectroscopy of the samples after calcination yielded a doubling in the number of bridging OH groups in comparison to the dealuminated parent zeolite. This behavior was explained by reinsertion of a corresponding number of non-framework aluminum atoms into the zeolite framework. In agreement with the variation in ^{1}H MAS NMR signals of bridging OH groups, the ^{29}Si MAS NMR spectra of the KOH-treated samples show a decrease in the framework n_{Si}/n_{Al} ratio from 5.1 to 2.9. In addition, a redistribution of the Si(nAl)-lines in the ^{29}Si MAS NMR spectrum was found which corresponds to a redistribution of the framework aluminum atoms in comparison to those in the dealuminated sample [336]. Similar results were obtained for hydrothermally dealuminated zeolite HZSM-5 treated with an aqueous solution of NaOH [338].

The local structures of bridging OH groups in zeolites have been investigated in a number of quantum-chemical approaches [339–341]. Schroeder et al. [340] published a paper with local structures of bridging OH groups in zeolite HY with H–Al distances of 0.239 nm and 0.233 nm for SiO1HAl and SiO3HAl groups [339, 340], respectively. Stevenson [342] measured the second moment M_2^{HAl} (Eq. 4.31) of the broad line ^{1}H NMR signal of bridging OH groups in dehydrated zeolite HY and calculated the H–Al distance of r_{HAl} to be 0.238 nm. This value was verified by Freude et al. [343, 344] using the same method determining $r_{HAl} = 0.248$ nm for bridging OH groups in dehydrated zeolite HZSM-5 ($n_{Si}/n_{Al} = 15$). Kenaston et al. [345] applied SEDOR (spin-echo double resonance) for the determination of the H–Al distance of bridging OH groups in dehydrated HZSM-5 ($n_{Si}/n_{Al} = 30$). The authors detected the spin-echoes of ^{27}Al nuclei which are damped by application of ^{1}H NMR π-pulses during the ^{27}Al NMR pulse delays. Analyzing the SEDOR decay the authors derived an H–Al distance of $r_{HAl} = 0.243$ nm.

In a recent approach, the second moment of dipolar H–Al interaction, M_2^{HAl}, and the anisotropy of the chemical shift, $\Delta\sigma_{OH}$, of hydroxyl protons in bridging OH groups were derived from an analysis of the ^{1}H MAS NMR side-band patterns [314, 346, 347]. An important advantage of this method is the selective analysis of the ^{1}H MAS NMR signals of hydroxyl protons with different isotropic chemical shifts. Hence, the NMR parameters for bridging OH groups located in

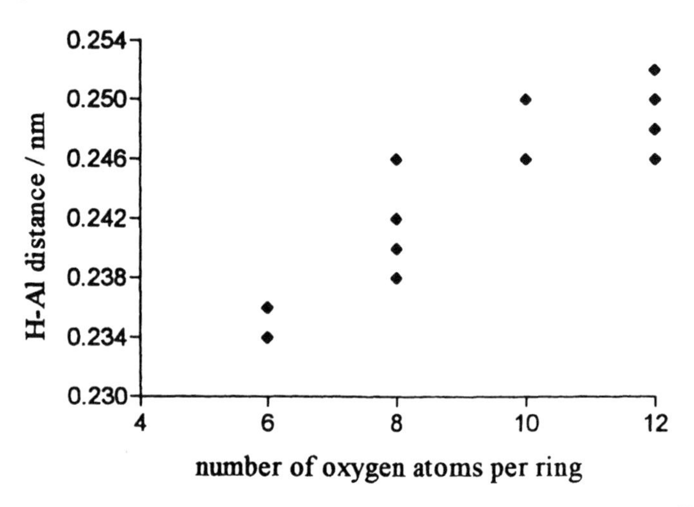

Fig. 4.47. Plot of the experimentally determined H–Al distances, r_{H-Al}, of bridging OH groups as a function of the number of oxygen atoms per ring (entries 3 and 7, Table 4.7) in which the hydroxyl protons are located

6- and 12-membered oxygen rings, which occur in zeolites HY, SAPO-5, and SAPO-37, can be determined selectively [314, 347]. Table 4.7 gives a summary of the experimentally determined H–Al distances. The experimentally determined chemical shift anisotropies, $\Delta\sigma_{OH}$, cover a range from 13.5 to 17.5 ppm [314]. In Fig. 4.47 the H–Al distances in the local structure of bridging OH groups are depicted over the number of oxygen atoms in the rings in which the hydroxyl protons are located (columns 3 and 7 of Table 4.7). The variation in these values suggests that the H–Al distances of bridging OH groups depend on the size of the local structure element. It is interesting to note that the difference in the H–Al distances of SiO1HAl (r_{HAl} = 0.250 ± 0.002 nm) and SiO3HAl (r_{HAl} = 0.235 ± 0.002 nm) groups in zeolites HY and SAPO-37 agrees qualitatively with that derived from quantum-chemical calculations (r_{HAl} = 0.239 and r_{HAl} = 0.233 nm, respectively) [339, 340]. The same relationship follows from the T–O1–T and T–O3–T bond angles of bridging OH groups in zeolites HY and SAPO-37 determined by neutron diffraction which cover a range of 135.7° to 137.9° for SiO1HAl and 139.8° to 144.5° for SiO3HAl groups [52, 348]. Koch et al. [349, 350] investigated the influence of carbon monoxide adsorption on the H–Al distance of bridging OH groups in zeolite HZSM-5 by MAS NMR side-band analysis. For unloaded and loaded (about one CO/SiOHAl) zeolite HZSM-5, the same H–Al distance was determined. However, after adsorption of carbon monoxide, the chemical anisotropy increases significantly from $\Delta\sigma_{OH}$ = 14 ± 2 to $\Delta\sigma_{OH}$ = 22 ± 2 ppm due to the formation of a linear adsorbate complex. In addition, the equilibrium distance of r_{HC} = 0.20 ± 0.02 nm between the hydroxyl proton and the carbon atom of the probe molecule was investigated at a temperature of 123 K. This value agrees well with the H–C distances derived from quantum-chemical calculations [351, 352].

Table 4.7. Structural Data [218], Molar Ratios of T-atoms [314], Mean Sanderson Electronegativities S^m [315] of Aluminosilicate Frameworks, Chemical Shifts (referenced to TMS) [314], δ_H, of Bridging OH Groups and H–Al Distances [314], r_{HAl}, in the Local Structure of Bridging OH Groups Derived from an Analysis of the ^{1}H MAS NMR Side-Band Patterns

Zeolite	Structure type*	Oxygen rings*	n_{Si}/n_{Al} [a] $n_{Si}/n_{Al+Si+P}$ [b]	S^m	δ_H/ppm	r_{HAl}/nm**
25HNaA	LTA	8-membered	1.0[a]	4.06	4.0	0.238
50HNaX	FAU	12-membered	1.4[a]	4.07	3.8	0.246
52HNaY	FAU	12-membered	2.6[a]	4.13	3.9	0.250
90HNaY	FAU	12-membered	3.5[a]	4.14	3.9	0.250
		6-membered			4.8	0.236
37HKERI	ERI	8-membered	2.9[a]	4.15	3.9	0.246
HM	MOR	12-membered	7.1[a]	4.18	4.2	0.248
HZSM-5/1	MFI	10-membered	15.0[a]	4.22	4.2	0.250
HZSM-5/2	MFI	10-membered	26.0[a]	4.23	4.3	0.246
SAPO-5	AFI	12-membered	0.036[b]		3.8	0.252
		6-membered			4.8	0.236
SAPO-17	ERI	8-membered	0.021[b]		3.7	0.242
		8-membered			4.0	0.240
SAPO-34	CHA	8-membered	0.114[b]		3.8	0.242
SAPO-37	FAU	12-membered	0.121[b]		3.8	0.250
		6-membered			4.3	0.234

* see [218].
** accuracy of ± 0.002 nm.
[a] n_{Si}/n_{Al}.
[b] $n_{Si}/n_{Al+Si+P}$.

4.2.4.7
Solid-State ^{23}Na NMR Spectroscopy of Sodium Cations in Hydrated and Dehydrated Zeolites

In most as-synthesized zeolites the negative framework charges are compensated by sodium cations. The application of ^{23}Na NMR spectroscopy allows insight into the cation migration in the process of zeolite modification. ^{23}Na has a natural abundance of 100%, a nuclear spin of $I = 3/2$, and a moderate quadrupole moment. In a number of approaches [236, 353–357] ammonium-, calcium- and lanthanum-exchanged zeolites Y were investigated in the hydrated state before and after the first thermal treatment. The ^{23}Na MAS NMR spectra of these samples consist of two lines at –9 and –13 ppm, relative to solid NaCl (Fig. 4.48). While the first line is caused by the signal of hydrated sodium cations in the supercages, the second one was assigned to hydrated sodium cations located in the sodalite cages. Measurements carried out at different magnetic fields have shown that the signal at –13 ppm is a superposition of two components due to sodium cations on position SI in the hexagonal prisms and hydrated sodium cations in the sodalite cages [236]. In Fig. 4.48 the ^{23}Na MAS NMR spectra of a series of hydrated lanthanum-exchanged zeolites Y are depicted, on the left-hand side before (suffix a) and on the right-hand side after the first thermal

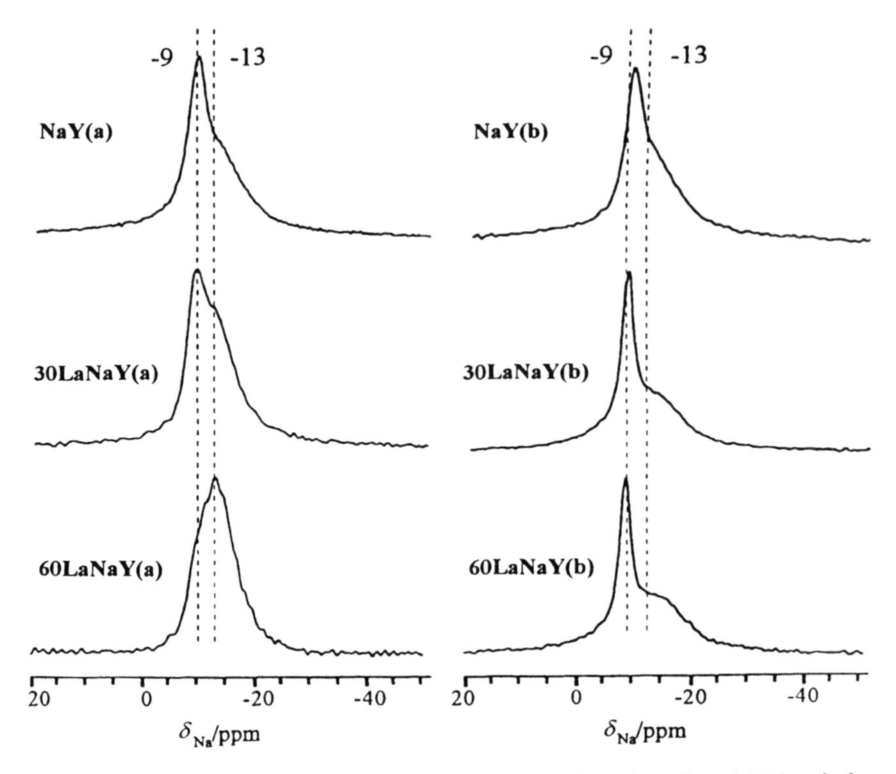

Fig. 4.48. ^{23}Na MAS NMR spectra of hydrated zeolites Y in the sodium form (NaY) and after exchange of 30% and 60% of the sodium cations by lanthanum (30LaNaY and 60LaNaY, respectively), recorded before (*left-hand side*, suffix a) and after (*right-hand side*, suffix b) thermal treatment at 433 K and rehydration [236]

treatment at 433 K and rehydration (suffix b). The notations 30LaNaY and 60LaNaY apply to zeolites after an exchange of 30% and 60%, respectively, of sodium cations by trivalent lanthanum cations. As shown on the left-hand side of Fig. 4.48 there is a strong decrease in the relative ^{23}Na MAS NMR intensity of the –9 ppm signal of zeolites 30LaNaYa and 60 LaNaYa in relation to the parent zeolite NaY. This behavior is indicative of an initial exchange of hydrated sodium cations in the large cavities. Hydrated lanthanum cations with a diameter of about 0.8 nm are too bulky to diffuse into the sodalite cages through the six-ring windows and, therefore, cannot exchange sodium cations located in these cavities. The thermal treatment (433 K) and rehydration of zeolites LaNaY (right-hand side of Fig. 4.48) results in a significant decrease in the ^{23}Na MAS NMR intensity at –13 ppm while the signal at –9 ppm increases. This fact proves that the calcination-induced migration of lanthanum cations to positions in the sodalite cages is accompanied by a counter-migration of sodium cations from the sodalite cages into the supercages. A quantitative discussion of the ^{23}Na MAS NMR spectroscopic data of hydrated lanthanum-exchanged zeolites Y is given in

[236]. The effects observed in the ^{23}Na MAS NMR spectra of the ammonium- and calcium-exchanged series [353, 354] are very similar to those described for zeolite LaNaY [236, 353].

For cations of dehydrated zeolites, which are strongly coordinated with framework oxygens, the site-exchange can be neglected on the time scale of NMR spectroscopy. In this case, the NMR line shapes of sodium cations are determined only by the strength of quadrupole interactions. The observed quadrupole coupling constant, C_Q, depends on the electric field gradients which, as a first approximation, are caused by a superposition of the electrostatic fields of the negatively charged framework oxygens. Simply, C_Q is large for sodium cations which are non-spherically coordinated to oxygen atoms, e.g., for sodium cations close to a six-membered oxygen ring as on position SII in zeolite Y, and is small for sodium cations in a spherical oxygen surrounding, e.g., octahedrally coordinated sodium cations on position SI in zeolite Y (see Fig. 4.42). Tijink et al. [358] investigated dehydrated zeolite NaA by static ^{23}Na NMR and determined the quadrupole parameters of $C_Q = 3.2$ MHz and $\eta_Q = 0.9$ for sodium cations located on four- and eight-membered oxygen rings and of $C_Q = 5.8$ MHz and $\eta_Q = 0$ for sodium cations located close to the centers of six-membered oxygen rings. In this case the assignment was found by the asymmetry parameter, η_Q, which should be small for sodium cations in the symmetric surrounding of six-membered oxygen rings. However, more sophisticated approaches are often necessary to assign ^{23}Na NMR signals of sodium cations in dehydrated zeolites. One way is the theoretical estimation of quadrupole parameters using a point-charge model [359, 360–362]. Table 4.8 gives a summary of quadrupole parameters of sodium cations located on distinct positions in dehydrated zeolite NaY, calculated with oxygen charges between -0.8 and -1.2 e and the coordinates of framework oxygens and sodium cations, determined by X-ray diffraction. The characteristic values of quadrupole coupling constants calculated for sodium cations of different framework positions are useful for a rough interpretation of experimental ^{23}Na NMR spectra. However, this assignment of NMR signals should be supported by detailed studies on samples exchanged with various cations and/or by a comparison of population numbers determined by XRD and solid-state NMR [359, 363]. The ^{23}Na MAS NMR spectra of dehydrated zeolites Y in the sodium form (NaY) and after exchange of 52% and

Table 4.8. Quadrupole Coupling Constants, C_Q, Asymmetry Parameters, η_Q, and Isotropic Chemical Shifts (referenced to solid NaCl), δ, of Sodium Cations in Dehydrated Zeolite NaY Calculated by a Point-Charge Model (see text) and Experimentally Determined by Solid-State ^{23}Na MAS NMR Spectroscopy [359, 361, 362]

Zeolite /Sites	Calculated: C_Q/MHz	Experimental: C_Q/MHz	Experimental: η_Q	Experimental: δ/ppm
NaY:				
SI	0.1–0.5	ca. 1.0	0	−12
SI′	4.6–5.6	4.8	0	− 4
SII	3.4–4.2	3.9	0	−12

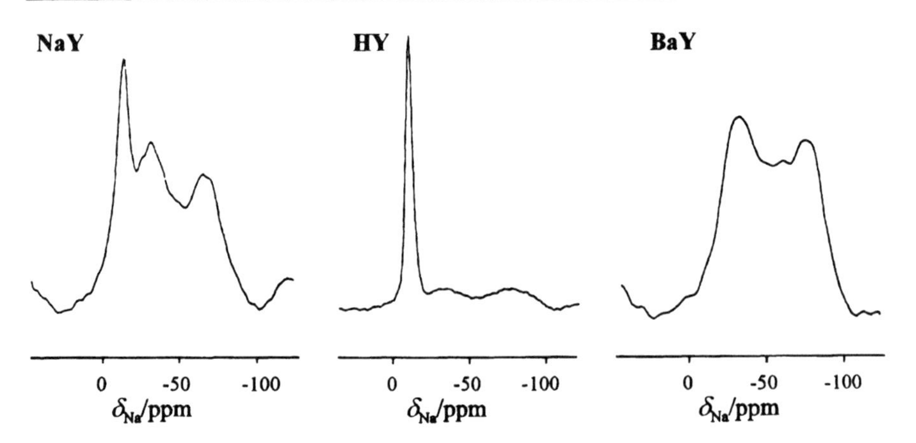

Fig. 4.49. ^{23}Na MAS NMR spectra of dehydrated (673 K) zeolites Y in the sodium form (NaY) and after exchange of 52% and 83% of the sodium cations by hydroxyl protons (HY) and by barium cations (BaY), respectively, recorded at the resonance frequency of 105.8 MHz with an MAS rate of ca. 12 kHz [359]

83% of sodium cations by hydroxyl protons (HY) and barium cations (BaY), respectively, are depicted in Fig. 4.49. The low-field range of the spectra of dehydrated zeolites NaY and HY (left and middle) shows an isotropic line at about −13 ppm while the high-field range of all spectra consists of a pattern with two singularities at about −32 and −63 ppm [359]. The replacement of 52% of sodium cations by hydroxyl protons causes a strong decrease in the relative intensity of the high-field pattern which is characteristic of cations initially located on positions SII and SI′ [364]. On the other hand, in the spectrum of the dehydrated zeolite BaY, the low-field line has completely disappeared which shows that this signal originates from sodium cations on position SI [364]. By application of 2D ^{23}Na nutation MAS NMR spectroscopy, signals of sodium cations characterized by different quadrupole frequencies can be separated in the F1-dimension. In the 2D spectrum depicted at the top of Fig. 4.50 the high-field pattern of dehydrated zeolite NaY appears at $\nu_1 = 2\ \nu_{rf}$ in the F1-dimension corresponding to a quadrupole frequency of $\nu_q \gg \nu_{rf}$ [359]. In contrast, the low-field line shows small intensities in the range between $\nu_1 = 1\nu_{rf}$ and $\eta_1 = 2\nu_{rf}$. Hence, this signal is caused by sodium cations with a quadrupole frequency of $\nu_q \approx \nu_{rf}$ (ν_{rf} = 125 kHz) [359]. As expected for dehydrated zeolite NaEMT, which has a negligible sodium population on position SI (low-field line) [359], the spectrum shown at the bottom of Fig. 4.50 consists of the high-field pattern at $\nu_1 = 2\nu_{rf}$, caused by sodium cations at SII and SI′.

The observed center of gravity, δ_{cg}, of the NMR signal of quadrupole nuclei is shifted from the isotropic chemical shift value, δ, by the second order quadrupolar shift, δ_{qs}, which, according to Eq. 4.35, is a function of the external magnetic field and the quadrupole parameters ν_Q and η_Q. The value of δ can be calculated from the line positions of NMR spectra measured at different Larmor frequencies [359, 365]. Figure 4.51 shows the ^{23}Na DOR NMR spectra

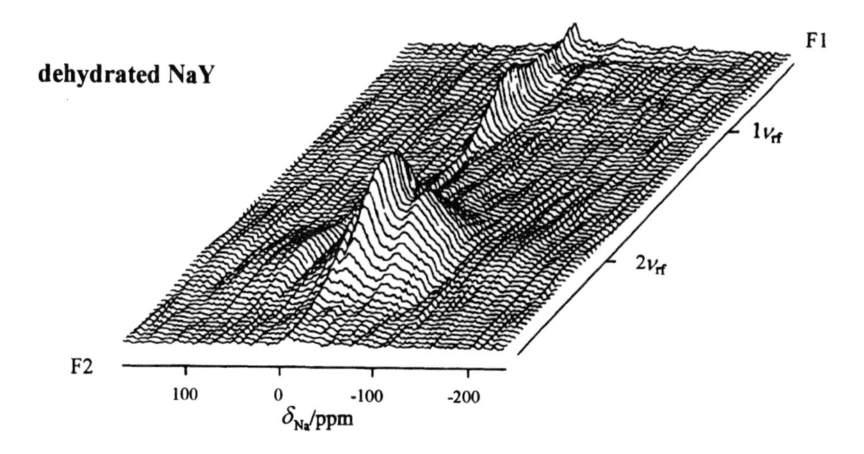

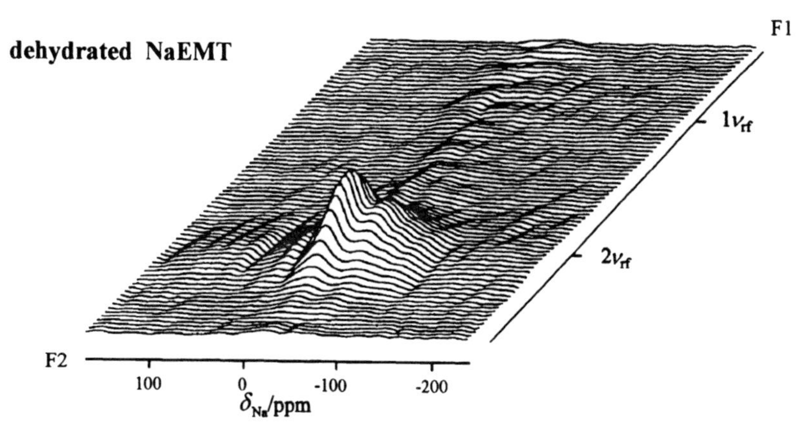

Fig. 4.50. Two-dimensional ^{23}Na nutation MAS NMR spectra of dehydrated (673 K) zeolites NaY (*top*) and NaEMT (*bottom*); the spectra were recorded at the resonance frequency of 105.8 MHz and with a radio frequency field of $\nu_{rf} = 125$ kHz [359]

of dehydrated (673 K) zeolite NaY recorded at resonance frequencies of $\nu_{o,1} = 79.4$ MHz and $\nu_{o,2} = 105.8$ MHz, with rotor-synchronized pulse excitation and at different rotation frequencies of the outer DOR rotor [359]. At least two DOR central lines, marked by arrows, can be observed. The low-field signal is slightly shifted from -15 ppm at $\nu_{o,1}$ to -13 ppm at $\nu_{o,2}$ which corresponds to an isotropic chemical shift of -10.5 ppm (Eq. 4.35). The high-field signal shows a shift from -78 ppm at $\nu_{o,1}$ to -48 ppm at $\nu_{o,2}$ which yields an isotropic shift of -9 ppm (Eq. 4.35). In the same way, quadrupole coupling constants of ca. 1 MHz for the low-field line and of ca. 4.2 MHz for the high-field line were calculated (see also [359, 366]). These parameters agree very well with the theoretically derived values (Table 4.8) and with the results of the 2D nutation MAS NMR spectroscopy (Fig. 4.50).

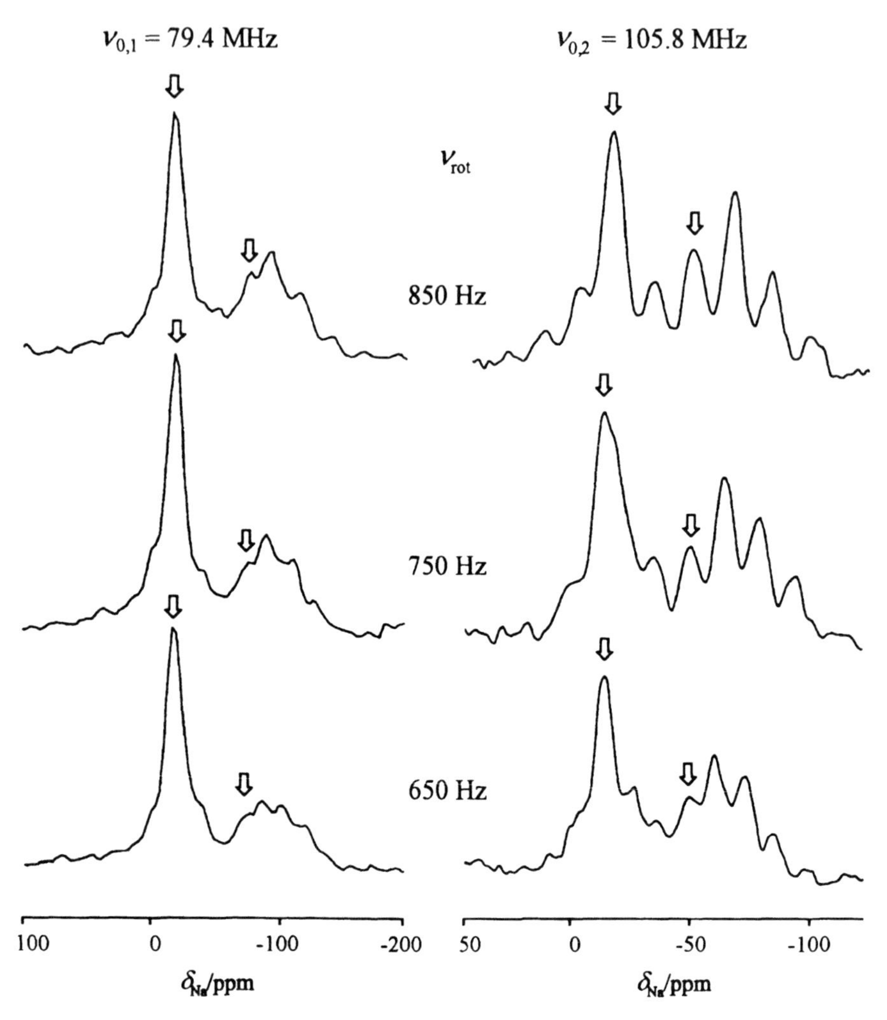

Fig. 4.51. ^{23}Na DOR NMR spectra of dehydrated (673 K) zeolite NaY recorded at resonance frequencies of 79.4 MHz (*left-hand side*) and 105.8 MHz (*right-hand side*) with rotor-synchronized pulse excitation and different rotation frequencies of the outer DOR rotor [359]. The arrows indicate the DOR central lines

^{23}Na MAS NMR spectra of dehydrated zeolite NaY recorded at resonance frequencies of 105.8 and 198.4 MHz (see Fig. 4.52) show that more than two signals are necessary to fit the pattern. Using the quadrupole parameters derived by the point-charge model (Table 4.8), a suitable fit of the experimental MAS NMR spectra was achieved with a narrow low-field line (C_Q of ca. 1.0 MHz) due to sodium cations located on positions SI and of two high-field patterns with quadrupole coupling constants of 3.9 and 4.8 MHz due to sodium cations at SII and SI′, respectively (see Fig. 4.42). In general, selective excitation of the

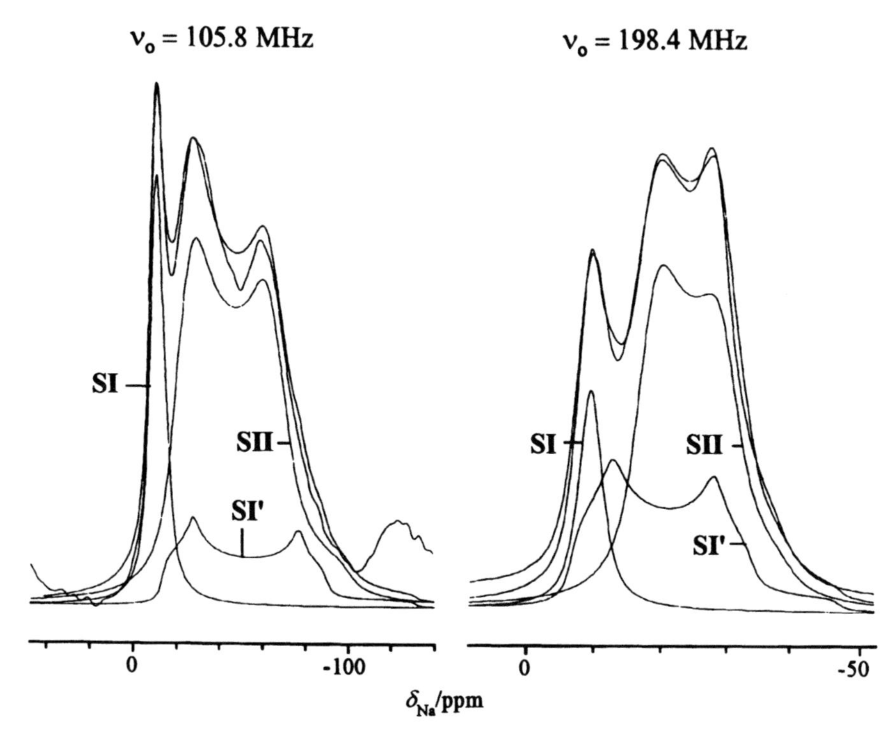

Fig. 4.52. ^{23}Na MAS NMR spectra of dehydrated (673 K) zeolite NaY recorded at resonance frequencies of 105.8 and 198.4 MHz and with an MAS rate of 12 kHz [361]

$\{+1/2 \rightleftarrows -1/2\}$ transition is observed for nuclei which are involved in strong quadrupole interactions while non-selective excitation of all transitions, i.e., central transition and satellite transitions, occurs for nuclei with vanishing quadrupole frequencies. For sodium nuclei with a spin $I = 3/2$ the central transition contains only 40% of the total intensity observed for non-selective excitation [201, 367]. In addition, depending on the MAS rate, ν_{rot}, the Larmor frequency, ν_0, and the quadrupole parameters C_Q and η_Q, a certain portion of the MAS NMR intensity is distributed in MAS side bands. Using a method similar to that proposed by Massiot et al. [367], correction factors for the experimentally derived intensities of the central transitions were determined. In this way, the sodium populations of cation positions in dehydrated zeolite NaY with the framework n_{Si}/n_{Al} ratio of 2.6 and 53.3 Na^+/u.c., measured by ^{23}Na MAS NMR, are in reasonable agreement with the values derived by XRD (Table 4.9). Verhulst et al. [370] determined sodium populations of cation sites in dehydrated zeolite NaY by simulation of ^{23}Na DOR NMR spectra. However, strong overlap of DOR sidebands complicates this approach.

Table 4.9. Sodium Populations of Cation Sites in Dehydrated (673 K) Zeolite NaY ($n_{Si}/n_{Al} = 2.5$) Determined by ^{23}Na MAS NMR Spectroscopy (Fig. 4.52) [362] and XRD (NaY: $n_{Si}/n_{Al} = 2.37$ [368, 369])

Zeolite/Sites	^{23}Na MAS NMR: Na^+/u.c.	X-ray diffraction: Na^+/u.c.
NaY:		
SI	7	8
SI′	18	13
SII	29	30

4.2.4.8
^{133}Cs MAS NMR Spectroscopy of Cesium Cations in Hydrated and Dehydrated Mordenites and Faujasites

Cesium exchange of zeolites is an initial step in the preparation of basic catalysts [371, 372]. A suitable way to characterize these materials is the application of ^{133}Cs MAS NMR spectroscopy. The isotope ^{133}Cs has a natural abundance of 100%, a nuclear spin of $I = 7/2$, and a weak quadrupole moment. Starting with the work of Chu et al. [373], static ^{133}Cs NMR and ^{133}Cs MAS NMR spectroscopy was applied to monitor the local structures of cesium cations in hydrated and dehydrated mordenite. The crystallographic sites of cations in mordenite are labeled SII, SIV, and SVI and have relative cesium occupancies of 3.78:1.86:1.75 [374]. Cesium cations on SII and SIV are located close to the center of an eight-membered oxygen ring, while cesium cations on SVI are placed close to a six-membered oxygen ring. The quadrupole coupling constant was determined from the split of the singularities of the first satellite transitions ($\{+3/2 \rightleftarrows +1/2\}$ and $\{-1/2 \rightleftarrows -3/2\}$) [375] and, additionally, by simulating MAS NMR line shapes. In the ^{133}Cs MAS NMR spectrum of the fully hydrated mordenite a single line at the chemical shift of −64 ppm, relative to saturated CsCl solution, with a quadrupole coupling constant of 210 kHz and an asymmetry parameter of zero was found. After dehydration, the ^{133}Cs MAS NMR spectrum consisted of three components at isotropic shifts of −157, −186, and −24 ppm, due to cesium cations located on positions SII, SIV and SVI, respectively. The quadrupole parameters were found to be $C_Q = 3.1$ MHz and $\eta_Q = 0.6$ for cesium cations at SII and SIV and $C_Q = 3.2$ MHz and $\eta_Q = 0.7$ for cesium at SVI. The NMR intensity ratio of cesium cations on positions SII, SIV, and SVI of 2:1:1 agrees very well with the values obtained by X-ray diffraction.

As mentioned above, there are two reasons for the differences in the solid-state ^{133}Cs NMR parameters and the line shapes of hydrated and dehydrated cations: (i) the formation of hydration shells and (ii) the high mobility of hydrated cations. The site-exchange of cations in hydrated zeolites can be significantly slowed down by cooling the samples. This results in line broadening, but the cations retain water molecules in their coordination shell. The resulting low-temperature spectra can be narrowed by use of variable-temperature (VT) MAS spectroscopy. Ahn et al. and Tokuhiro et al. [376–378] utilized this technique to

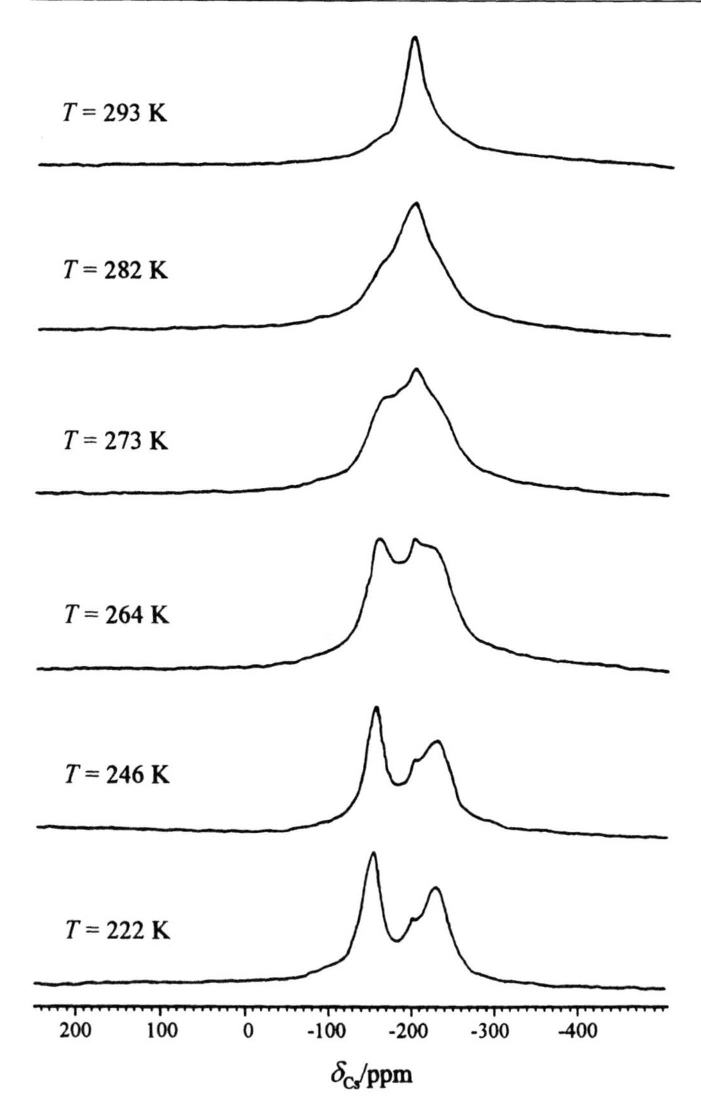

Fig. 4.53. ^{133}Cs MAS NMR spectra of hydrated CsLiA zeolite (3.8 Cs^+/u.c.). The spectra were recorded at temperatures between 222 and 293 K, at the resonance frequency of 39.8 MHz and with an MAS rate of 2.5 kHz [377]

study hydrated zeolites CsNaA and CsLiA (3.8 Cs^+/u.c.). They recorded ^{133}Cs MAS NMR spectra at 293 K consisting of only one signal at around −200 ppm caused by a rapid cesium exchange between six-ring and eight-ring sites in the large α-cages. The temperature decrease down to 222 K led to a split of this signal into three lines (Fig. 4.53). Two lines were readily assigned to cesium cations occupying sites in eight-membered (−226.8 ppm) and six-membered oxygen rings (−151.7 ppm) of the α-cages. The third component at the chemical shift of

-200.3 ppm was attributed to hydrated cesium cations located on sites in the large cages near four-membered oxygen rings interacting to a moderate degree with the framework. The simulation of the experimental line shapes using a modified Bloch equation gave a lifetime of $\tau = 0.12$ ms for hydrated cesium cations at sites in the eight-ring windows. The activation energy for the site-exchange process of hydrated cesium cations was determined to be 44 kJ/mol. Based on crystallographic studies, Subramanian and Seff [379] proposed that cesium cations in dehydrated zeolite CsNaA start occupying positions close to the centers of six-ring windows inside the sodalite cages before beginning to fill positions in the α-cages. The chemical shifts of cesium cations located on equivalent positions inside and outside the sodalite cages are expected to be similar. However, considering the diameter of the six-ring windows, the site-exchange of hydrated cesium cations between sites inside and outside the sodalite cages is too slow to cause one averaged signal at 293 K (top of Fig. 4.53). Therefore, Ahn and Iton [378] proposed that cesium cations enter the sodalite cages first upon dehydration.

Applying solid-state NMR spectroscopy and X-ray powder diffraction, Koller et al. [380] investigated a homologous series of cesium-exchanged zeolites Y. Rietveld refinement of the structure of 72% cesium-exchanged zeolite CsNaY, dehydrated in vacuum at 623 K, revealed cesium cations at positions SI in the center of hexagonal prisms, at SI′ close to the center of six-membered oxygen rings in sodalite cages, at SII close to the center of six-ring windows in the supercages, and at SIII close to four-membered oxygen rings in the supercages. Figure 4.54 shows the ^{133}Cs MAS NMR spectrum of this sample recorded at a resonance frequency of 52.4 MHz and with an MAS rate of 6 kHz. The spectrum is com-

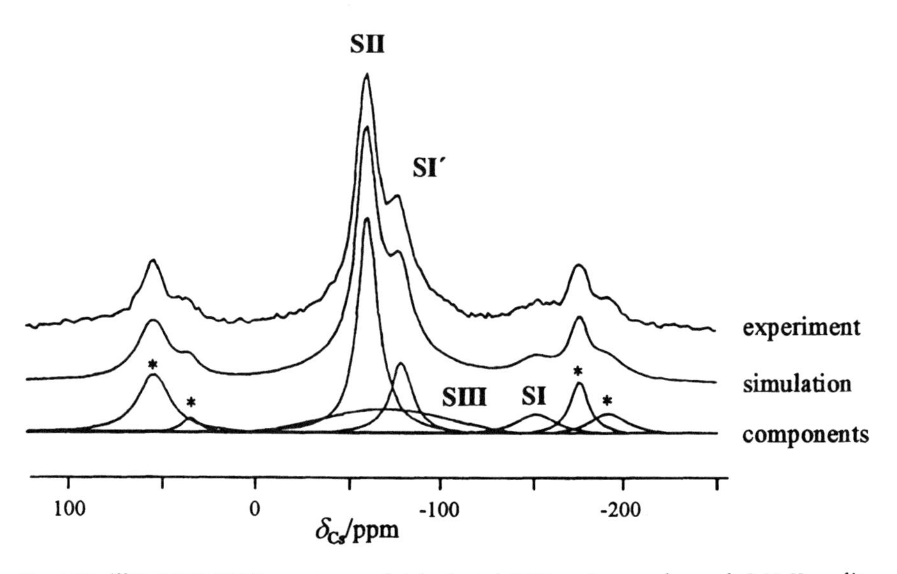

Fig. 4.54. ^{133}Cs MAS NMR spectrum of dehydrated 72% cesium-exchanged CsNaY zeolite (*top*), simulation (*middle*), components (*bottom*), and spinning sidebands (*) [380]

posed of four overlapping signals consisting of two narrow MAS central lines at -62 and -79 ppm, a broad signal at -71 ppm and a weak line at about -153 ppm. ^{133}Cs MAS NMR investigations carried out at resonance frequencies of 52.4 and 26.2 MHz indicated a negligible quadrupole interaction of cesium cations [380]. Therefore, the strong spinning sidebands of the signals at -62 and -79 ppm are exclusively caused by chemical shift anisotropy. This makes the NMR properties of cesium cations in zeolite Y more reminiscent of a nucleus with spin $I = 1/2$ than of a quadrupolar one. However, the large chemical shift range makes the line positions in the NMR spectra very sensitive to the coordination environment of cesium cations in zeolite cavities. The line at -62 ppm has the highest intensity and was, therefore, attributed to cesium cations located on position SII with the largest population. The signal at -79 ppm with a much smaller intensity, but a line width and MAS sideband pattern comparable to the first one, was assigned to cesium cations located on position SI'. Both positions SII and SI' have equivalent site symmetries and, therefore, similar NMR parameters. Because of the larger shielding of cesium cations by the higher number of oxygen atoms in the hexagonal prism, the high-field shift at -153 ppm was attributed to cesium cations located on position SI and the remaining broad signal at about -71 ppm to cesium cations at SIII. By the NMR intensities of these signals (including the spinning sidebands), cation populations of 23.6 Cs^+/u.c. on SII, 8.3 Cs^+/u.c. on SI', 5.5 Cs^+/u.c. on SIII and 2.4 Cs^+/u.c. on SI were determined. These values agree very well with the values derived from XRD, viz. 26.4 Cs^+/u.c. on SII, 4.5 Cs^+/u.c. on SI', 5.5 Cs^+/u.c. on SIII and 0.9 to 1.2 Cs^+/u.c. on SI. Considering the free diameter of six-ring windows of 0.22 to 0.27 nm compared with the ionic diameter of cesium cations of 0.33 to 0.38 nm, migration of these cations into the sodalite cages and the hexagonal prisms seems to be improbable. However, as proven by solid-state NMR and XRD [380] cesium cations do migrate during thermal treatment (623 K) from the supercages into the sodalite cages, perhaps due to the flexibility of the zeolite framework.

4.2.4.9
^{129}Xe NMR Investigations of the Zeolitic Pore Architecture

Besides structure determination by X-ray diffraction, NMR studies of adsorbed xenon atoms provide a suitable approach to the characterization of the zeolitic pore architecture. The isotope ^{129}Xe has the nuclear spin $I = 1/2$, a natural abundance of 26.4% and, adsorbed in zeolites, a T_1 relaxation time in the range from 10 ms to a few seconds, which make this nucleus suitable for NMR spectroscopy.

Assuming a fast exchange of xenon atoms adsorbed in the pores and cavities of zeolites, the ^{129}Xe NMR shift can be described by the sum of several additive terms [381, 382]:

$$\delta = \delta_0 + \delta_S + \delta_E + \delta_M + \delta_{Xe-Xe} \cdot \rho_{Xe} \tag{4.48}$$

where δ_0 is the reference, δ_S is due to collisions between xenon atoms and the walls of the zeolite channels and cavities, δ_E takes into account the electric field effect caused by exchangeable cations, and δ_M is a term to account for the presence of paramagnetic species. The $\delta_{Xe-Xe} \cdot \rho_{Xe}$ term arises from xenon-xenon collisions (two-body collisions) and disappears upon extrapolating the xenon density to zero. For aluminosilicate-type zeolites in the H^+-form or exchanged with alkali metal cations, and for AlPO's, the δ_E und δ_M terms can also be neglected. In a number of ^{129}Xe NMR investigations the empirical relationship given in Eq. 4.48 was utilized to characterize the void space and non-framework species in zeolite particles. For a review on ^{129}Xe NMR spectroscopy of xenon-loaded zeolites see [381] and [382].

Demarquay and Fraissard [383] developed a model which allows the correlation of the δ_S term in Eq. 4.48 with the mean free path, $\bar{l}$, of a single xenon atom within the zeolite channels and cavities. According to the model proposed by Johnson and Griffiths [384], the ^{129}Xe NMR shift δ_s depends on the fractional time which the xenon atoms spend on the cavity walls. The plot of the ^{129}Xe NMR shifts δ_s of xenon atoms adsorbed on various zeolites as a function of the surface area to volume ratios yields a linear relationship. Derouane and Nagy [385] considered the effect of the surface curvature of zeolite pores and, therefore, of the adsorption energy on the chemical shifts δ_s. They developed a model for physisorption of xenon atoms in zeolites which treats the pore walls as an isotropic dielectric continuum. The authors obtained a reasonable linear relationship between the chemical shift values δ_s and the above-mentioned curvature term which is a function of the pore radius. Hence, there are several approaches leading to the estimation of the dimensions of the void spaces in zeolite particles making ^{129}Xe NMR spectroscopy a useful tool.

In zeolites Y, xenon atoms (diameter of ca. 0.26 nm) can only have access to the supercages since they are too large to enter the six-ring windows of the sodalite cages (diameter of ca. 0.22 nm). Typically, the relationship between the ^{129}Xe NMR shifts δ and the xenon coverage for zeolites X and Y with n_{Si}/n_{Al} ratios ranging from 1.2 to 50 and exchanged with univalent cations such as Na^+, Li^+ and H^+ is approximately linear [386]. This is due to the $\delta_{Xe-Xe} \cdot \rho_{Xe}$ term in Eq. 4.48 arising from the two-body xenon–xenon collisions and indicates that for these zeolites the δ_E and δ_M terms can be neglected. In addition, the experimentally determined shift values are independent of the framework composition. In Fig. 4.55 the ^{129}Xe NMR chemical shifts of xenon atoms adsorbed on zeolites with different structure types are depicted [381]. All curves show an increase in the total chemical shifts, δ, with increasing xenon coverage $[Xe]$ which is, at least for low coverages, approximately linear. Extrapolation of the data to $[Xe] = 0$ enables the determination of the chemical shift δ_S ranging from about 60 ppm for zeolite Y to 113 ppm for zeolite ZSM-5. Table 4.10 gives a summary of the structural data of selected zeolites and the δ_S term of xenon atoms adsorbed on these samples. The data given in columns 2 and 3 of Table 4.10 demonstrate clearly the relationship between the δ_S and the channel sizes. For xenon atoms adsorbed in the small side-pockets of mordenite the shift value δ_S reaches up to 250 ppm. The smallest ^{129}Xe NMR shift of 60 ppm was observed for xenon atoms adsorbed in the supercages of zeolite Y. Chemical shift values of xenon atoms adsorbed on aluminophosphates and silicoaluminophosphates are also given [382] and [387–389].

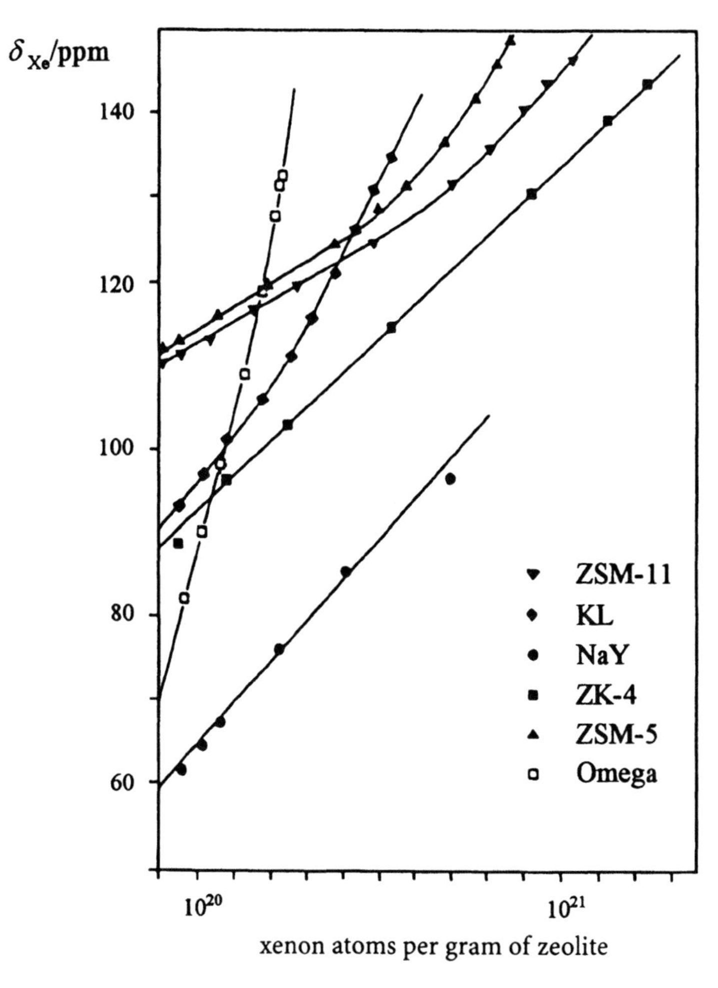

Fig. 4.55. ^{129}Xe NMR chemical shifts as a function of the number of xenon atoms per mass of powder material adsorbed on various zeolites [381]

For zeolites exchanged with divalent cations, the ^{129}Xe NMR shift of adsorbed xenon atoms shows a parabola-like dependence on the loading. Assuming an exchange rate between xenon atoms in the gas phase and strongly adsorbed xenon atoms much faster than the ^{129}Xe NMR relaxation rate and using an approach similar to that for the Langmuir isotherm, Cheung et al. [390] found the following dependence of the chemical shift $\delta(T,\rho)$:

$$\delta(T,\rho) = (n_s \cdot \delta_s + \zeta_g(T) \cdot n_s^2)/\rho - 2n_s \cdot \zeta_g(T) + \zeta_g(T) \cdot \rho \qquad (4.49)$$

where δ_s and $\zeta_g(T)$ are the chemical shift of xenon atoms in the adsorbed phase and the temperature-dependent chemical shift gradient in the gaseous phase,

Table 4.10. ^{129}Xe NMR Shifts (referenced to isolated xenon atoms), δ_S, of Xenon Atoms Adsorbed in Selected Zeolites and Their Structural Data (∅: diameters of cages or pores) [381]

Zeolite	δ_S/ppm	Cages and pore sizes
X and Y	60	spherical supercages with ∅ ≈ 1.3 nm
omega	73	unidimensional 12-ring pores with openings of 0.74 nm
A and ZK-4	87	spherical cages with ∅ ≈ 1.14 nm, six 8-ring openings with ∅ ≈ 0.4–0.5 nm
L	90	unidimensional 12-ring pores with openings of ∅ ≈ 0.71 nm, maximum ∅ ≈ 0.9 nm
ZSM-11	110	tridimensional interconnecting 10-ring pores, 0.51 nm × 0.55 nm
ferrierite	110	pseudo-spherical cages with ∅ ≈ 0.7 nm, two 8-ring openings, 0.35 × 0.48 nm
	165	bidimensional interconnecting 10-ring pores, 0.42 nm × 0.55 nm
ZSM-5	113	tridimensional interconnecting 10-ring pores, 0.51 mm × 0.55 nm and 0.53 nm × 0.56 nm
rho	114	tridimensional interconnecting pores forming spherical cages with ∅ ≈ 1.0 nm,
	230	prisms, 8-ring, 0.36 nm
mordenite	115	unidimensional 12-ring pores, 0.65 nm × 0.70 nm,
	250	8-ring side-pockets, 0.26 nm × 0.57 nm

respectively. n_s is given by N_s/V with the number N_s of strong adsorption sites and the free volume V of the cavities, and ρ is the total density of xenon atoms. The $1/\rho$ dependence of the first term in Eq. 4.49 results in an initial decrease in the chemical shift $\delta(T,\rho)$ with increasing coverage before the third term, linear in ρ, begins to dominate. Equation 4.49 describes the behavior of the ^{129}Xe NMR shift of xenon atoms in the gas phase exchanging with strongly adsorbed xenon atoms and, hence, of xenon atoms adsorbed on zeolites in the presence of divalent cations.

Figure 4.56 shows experimentally determined ^{129}Xe NMR shift values observed for xenon atoms adsorbed on calcium- and magnesium-exchanged zeolites Y. The number before the cation gives the percentage of exchange degree. At low levels of divalent cations (< 55% exchange) there is a linear relationship between the chemical shifts and the coverage. In these samples the divalent cations are located in the sodalite cages and, therefore, are not accessible for xenon atoms. However, at higher levels of exchange, the divalent cations occupy cation positions in the supercages where they interact with adsorbed xenon atoms. Therefore, xenon atoms adsorbed on these samples show a parabola-like dependence of their chemical shift which agrees with the behavior described by Eq. 4.49.

In a number of studies xenon adsorption on zeolites exchanged with divalent cations like Cd^{2+} and Zn^{2+} [391], trivalent cations like Ce^{3+} and La^{3+} [392–394] and loaded with metal particles (Ni, Rh, Pd, Ir, Pt) [381, 395–400] have been

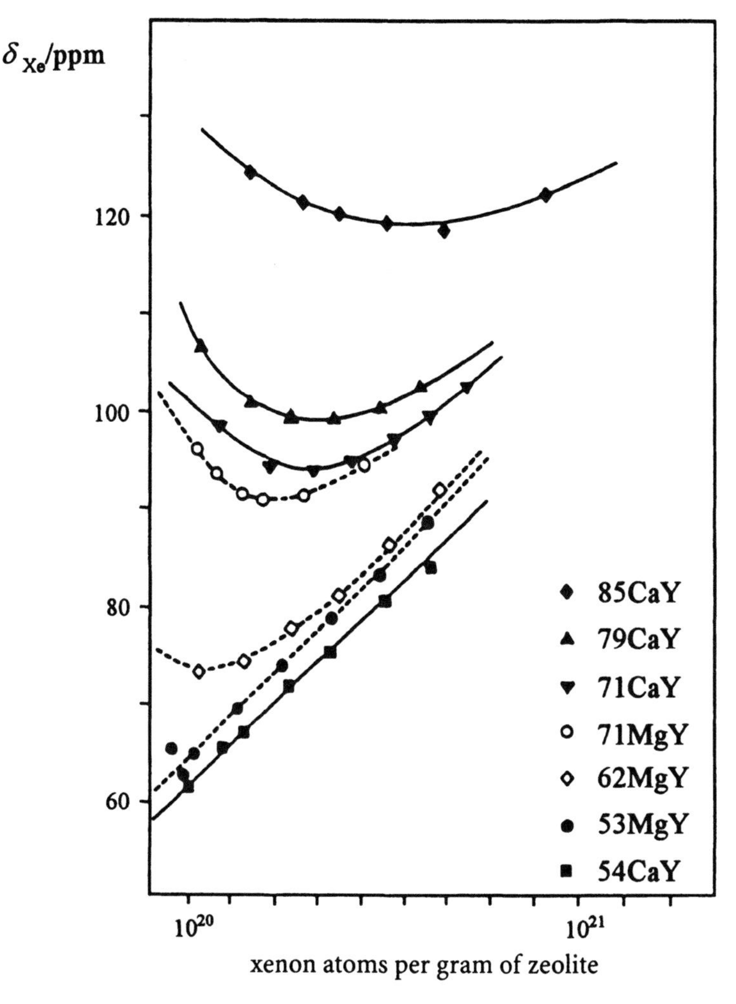

Fig. 4.56. ^{129}Xe NMR chemical shifts as a function of the number of xenon atoms per mass of powder material adsorbed on zeolites Y exchanged with various cations [381]

investigated. The chemical shifts of xenon atoms adsorbed on these zeolites are, at least for low coverages, always greater than that of xenon adsorbed on zeolites NaX and NaY. The sensitivity of ^{129}Xe NMR spectroscopy for the location of cations in zeolites was utilized by Scharpf et al. [395] to study the behavior of nickel in zeolite Y. On reduction with hydrogen the nickel migrates out of the sample as metallic nickel. For zeolites NiNaY with less than 14% nickel, Bansal and Dybowski [397] found that nickel predominantly occupies positions in the sodalite cages. With increasing Ni^{2+} content the authors observed a paramagnetic shift arising from the δ_M term in Eq. 4.48. Zeolites exchanged with Co^{2+} were studied by Bonardet et al. [401]. Shoemaker and Apple [396] used

^{129}Xe NMR spectroscopy to follow the migration of ruthenium in zeolite RuNaY during oxidation and reduction. According to Gedeon et al. [402], xenon atoms adsorbed on dehydrated zeolites CuY are characterized by chemical shift curves which are displaced to lower ^{129}Xe NMR shift values running almost parallel to each other and to the NaY curve. To date, only dehydrated silver- and copper-exchanged zeolites X and Y [391, 403–407] have shown negative or, at least, lower ^{129}Xe NMR shift values than those of xenon atoms adsorbed on zeolites NaX or NaY. In CuY zeolites this negative ^{129}Xe NMR shift is due to a specific interaction of xenon atoms with the copper ions via a $3d^{10}-5d^{0}$ electron donation mechanism from Cu^{+} to Xe [391, 402, 406]. The absence of parabola-like chemical shift curves indicates that in these zeolites no Xe-Cu^{2+} interactions occur, i. e., that Cu^{2+} cations are located mainly in the sodalite cages [402].

Fraissard and Ito [381] noted that ^{129}Xe NMR spectroscopy is also a suitable method for investigating the crystallinity and the presence of intergrowths and of non-framework species. The ^{129}Xe NMR intensity is not only proportional to the quantity of xenon atoms adsorbed per cage, but also to the number of cavities. Consequently, if a sample consists of a mixture of zeolites with different structure types or exchanged with cations of different sizes, the ^{129}Xe NMR spectrum is a direct measure of the composition of the sample. Springuel-Huet et al. [408] demonstrated this using a heterogeneous mixture of zeolites CaA and NaY. However, the size of the zeolite particles has to be considered in this kind of investigation since it determines the diffusion of xenon atoms between different aggregates [409, 410].

Furthermore, ^{129}Xe NMR spectroscopy can be applied to investigate the blockage of certain pores. For example, in zeolites with identical and unidimensional channels, the chemical shifts δ_S may differ. Different gradients of the chemical shift curves indicate that the effect of Xe–Xe collisions in various parts of the samples is different. This result corresponds to partial obstruction of channels, probably due to stacking defects, intergrowths or incomplete elimination of template molecules. Smith et al. [411, 412] investigated the flexibility of the zeolite rho framework. The authors found that, at a temperature of 160 K, xenon is strongly adsorbed in the double eight-rings of zeolite rho (see Table 4.10). Above 195 K, xenon is in rapid exchange between these sites and positions in the α-cages. According to the authors this behavior can be explained by a structural change which destabilizes the adsorption sites in double eight-rings relative to positions in the α-cages.

Bonardet et al. [413, 414] and Barrage et al. [415] applied ^{129}Xe NMR spectroscopy to monitor the formation of coke inside the zeolite particles. The authors demonstrated that non-framework aluminum species are involved in the coke formation. For coke deposits up to 10 wt%, the supercages are filled with coke and the remaining internal void space consists of narrow channels. Beyond 10 wt%, the coke deposits on the external surface of the zeolite particles as well, which gives rise to microcavities between the crystallites [415]. Coking and regeneration of catalysts applied in fluid catalytic cracking have been studied by ^{129}Xe NMR spectroscopy at different stages of regeneration [414].

4.2.4.10
Investigations of Brønsted and Lewis Acid Sites by Probe Molecules

Adsorption of probe molecules on dehydrated zeolites allows the characterization of the Brønsted and Lewis acidity in terms of acid–base interactions. A classical probe molecule used in ^1H MAS NMR spectroscopy to distinguish between acidic and non-acidic hydroxyl groups and to investigate the accessibility of OH groups is deuterated pyridine [306, 307, 311, 416]. As C_5ND_5 shows only weak H–D exchange with acidic hydroxyl protons, only weak lines originating from ring protons (6.9 ppm) contribute to the ^1H MAS NMR spectrum of dehydrated zeolites. In the ^1H MAS NMR spectra of organic acids dissolved in pyridine, lines in the chemical shift range from 12 to 20 ppm were observed which indicate the formation of pyridinium ions (PyH^+) [417]. Adsorption of pyridine on zeolite HZSM-5 with a large number of defect SiOH groups produces a low-field shift of the silanol signal from about 2 to 10 ppm, caused by the formation of hydrogen bonds to the adsorbed pyridine molecules [416]. This behavior corresponds to a low acid strength of SiOH groups. The adsorption of

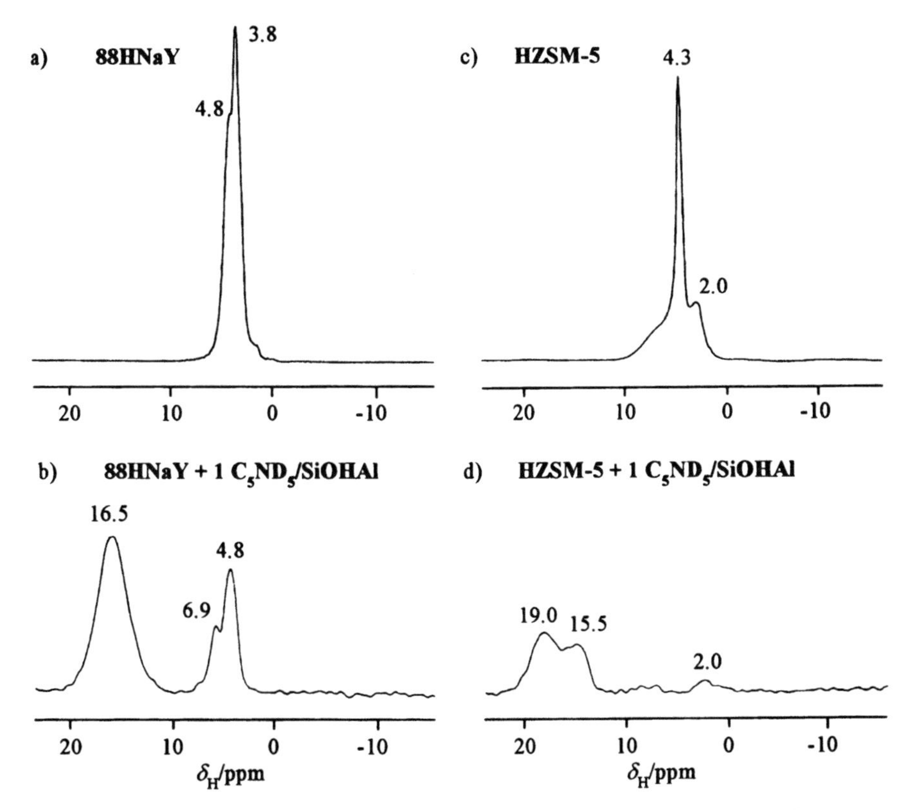

Fig. 4.57. ^1H MAS NMR spectra of dehydrated (673 K) zeolites 88HNaY (**a, b**) and HZSM-5 (**c, d**), unloaded (**a, c**) and loaded (**b, d**) with one molecule of deuterated pyridine per bridging OH group. The spectra were recorded at 400.13 MHz and with a MAS frequency of 8 kHz [306]

pyridine on acidic zeolites, e.g., HNaY [306, 311] and HZSM-5 [311, 416], gives lines at chemical shifts in the range between 15 and 20 ppm (Fig. 4.57). This signal results from a proton transfer to the probe molecules forming pyridinium ions. At the same time, the lines of the acidic and accessible bridging OH groups which are involved in this proton transfer disappear. Another way to detect the formation of PyH^+ in zeolites is the application of ^{15}N CP/MAS NMR spectroscopy [418, 419]. The corresponding signal has a chemical shift of about 110 ppm relative to pure pyridine. Pyridine at Lewis sites and physisorbed pyridine give rise to signals at about 60 and 10 ppm, respectively [419]. However, due to the low natural abundance and low sensitivity of the isotope ^{15}N the chemical application of cross-polarization is necessary. The chemical exchange of pyridine molecules at different adsorption sites with different cross-polarization efficiencies renders a quantitative evaluation difficult [210].

In a number of studies on zeolite acidity, the strong base trimethylphosphine (TMP) was used as a probe molecule [420–422]. Since the isotope ^{31}P occurs in a natural abundance of 100%, this method does not necessarily require the application of the CP technique. Lunsford et al. [420] studied TMP adsorbed on zeolite HY by ^{31}P MAS NMR spectroscopy (Fig. 4.58). After adsorption of TMP the samples under study were degassed at 353 K for 30 min to avoid signals of physisorbed probe molecules. For zeolite HY calcined at 673 K (Fig. 4.58a), the spectrum is dominated by a signal at about –3 ppm (referenced to 85% H_3PO_4) due to $(CH_3)_3P\text{–}H^+$ complexes arising from chemisorption of TMP at Brønsted acid sites. The spectra of the sample calcined at 773 K (Fig. 4.58b) and above (Figs. 4.58c and 4.58d) show additional resonances in the region from about

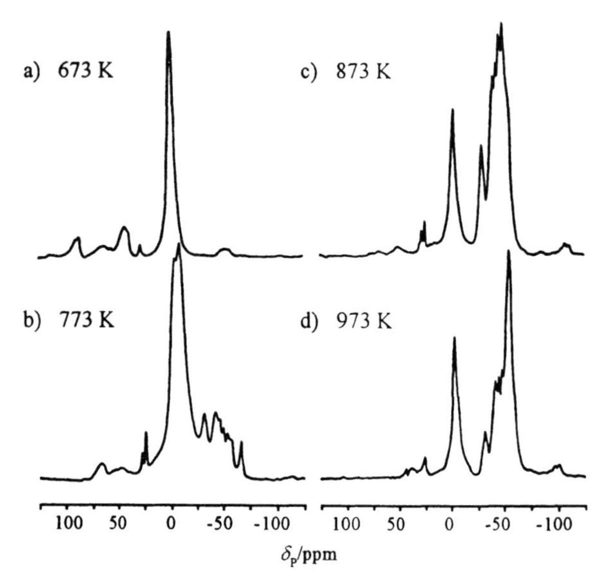

Fig. 4.58. ^{31}P MAS NMR spectra of trimethylphosphine adsorbed on zeolite HY which was calcined at temperatures of 673 K (**a**), 773 K (**b**), 873 K (**c**), and 973 K (**d**). The spectra were recorded at 80.9 MHz with a recycle delay of 10 s and an MAS frequency of ca. 4 kHz [420]

−32 to −58 ppm due to chemical interaction of TMP with Lewis acid sites. These investigations have shown that TMP is a sensitive NMR probe molecule for acidity measurements. However, TMP is a rather large molecule and the determination of acid sites in zeolites is therefore restricted by the maximum adsorption capacity of the pore system as well as steric hindrances [421]. In another approach, Biaglow et al. [423] carried out ^{13}C MAS NMR measurements of C2 labelled acetone adsorbed on zeolites HZSM-5, HZSM-12, HZSM-22, HY, H-mordenite, SAPO-5, and MeAPO-5. To investigate the slow-exchange range in the presence of adsorbed molecules, the measurements were carried out at 125 K. The adsorption of one acetone molecule per bridging OH group leads to a low-field shift of the resonance position of C2 atoms relative to pure acetone between 10.1 ppm for SAPO-5 and 18.7 ppm for HZSM-22. Therefore, the authors suggest that the ^{13}C NMR shift of carbonyl carbon atoms in acetone may be used as a measure of the strength of Brønsted acid sites in zeolites.

Variable-temperature ^{1}H MAS NMR studies of zeolite HZSM-5 loaded with weak bases interacting with zeolitic hydroxyl groups via hydrogen bonds have been reported by White et al. [424], Haw et al. [425], and Brunner et al. [323, 426]. The probe molecules used were acetylene, ethylene, carbon monoxide, benzene, ethane and nitrogen. After adsorption of three molecules of acetylene per bridging OH groups in zeolite HZSM-5, a low-field shift of the ^{1}H resonance position of the bridging OH groups from 4.3 to 7.3 ppm was observed at 298 K (top of Fig. 4.59a). This corresponds to an observed low-field shift of the resonance position by $\Delta\delta_{obs} = 3.0$ ppm. After adsorption of about one molecule of acetonitrile, ethylene, carbon monoxide, and ethane per bridging OH group, White et

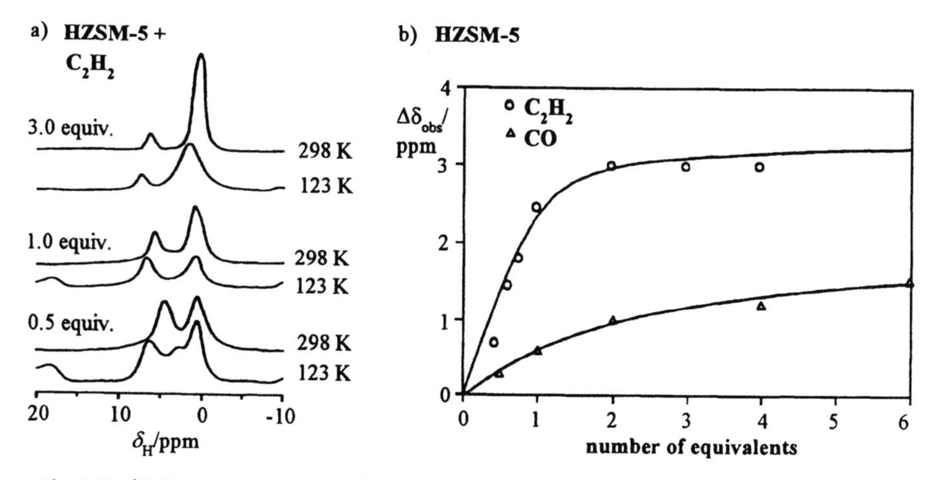

Fig. 4.59. ^{1}H MAS NMR spectra of dehydrated (673 K) zeolite HZSM-5 ($n_{Si}/n_{Al} = 38$) loaded with various numbers of acetylene molecules per bridging OH groups (1 equiv. = 1 acetylene/SiOHAl). The spectra were recorded at the resonance frequency of 299.7 MHz, with an MAS rate of ca. 6 kHz, and at temperatures of 298 and 123 K (**a**). The curves on the right-hand side (**b**) show the observed low-field shifts, $\Delta\delta_{obs}$, of the ^{1}H resonance positions of hydroxyl protons in zeolite HZSM-5 loaded with one equivalent of acetylene and carbon monoxide at 298 K [424]

al. [424] and Haw et al. [425] found resonance shifts of the ^{1}H MAS NMR signals of these hydroxyl protons in zeolite HZSM-5 of 7.2, 2.0, 1.5, and 0.3 ppm, respectively. At 123 K the adsorbate-induced low-field shift amounts to 2.7 ppm for ethylene, 1.8 ppm for carbon monoxide and 0.6 ppm for ethane [425]. Hence, the adsorbate-induced resonance shift is extremely sensitive to the type of probe molecule used. The observed loading dependencies of the resonance shift of acetylene and carbon monoxide at 298 K are plotted in Fig. 4.59b. The curves suggest an exchange, averaging the signals of free bridging OH groups with those of bridging OH groups associated via hydrogen bonds with the probe molecules. Adsorbing less than one probe molecule per bridging OH group, the ^{1}H MAS NMR lines of free and bonded hydroxyl protons are clearly resolved at 123 K (bottom of Fig. 4.59a). Koch [350] studied the adsorbate-induced low-field shift of the ^{1}H MAS NMR signal of bridging OH groups as a function of the temperature. For zeolite HZSM-5 loaded with one molecule of carbon monoxide per bridging OH group, a low-field shift of the resonance position by $\Delta\delta_{obs} = 2.0$ ppm was observed at 123 K. In this case, the adsorbate-induced wavenumber shift in the IR stretching vibration of bridging OH groups amounts to 300 cm^{-1} [350]. From the linear correlation between the IR wavenumbers and ^{1}H NMR shifts of hydrogen-bonded hydroxyl protons [329], a corresponding low-field NMR shift of 2.8 ppm was estimated [350]. This indicates that even at 123 K the influence of thermal motion on the ^{1}H resonance position of the adsorbate complex is not completely removed.

Using water molecules to study the strength of Brønsted acid sites in zeolites has several advantages. Water molecules are small enough to gain access to OH groups in the supercages and pores as well as those in small cages. In addition, water has a convenient base strength and hardness resulting in an equilibrium between hydrogen-bonded (I) and protonated complexes (II). Complex II corresponds to the formation of hydroxonium ions (H_3O^+). Using quantum chemical methods, Sauer et al. calculated the geometries of the complexes I and II [427] and the affinity of water molecules to sorption sites in zeolites [428, 429]: 791 kJ/mol for H_2O at Al^{3+}, 117 kJ/mol at Na^+, 58 kJ/mol at SiOHAl groups, 20 kJ/mol for bulk water, and 16 kJ/mol for H_2O at SiOH groups. According to these values, the hydration of zeolites starts with the adsorption of water molecules at Lewis acid sites and cationic species, followed by adsorption at bridging OH groups. However, in H^+-form zeolites not containing non-framework aluminum, bridging OH groups are favored sorption sites. In a number of studies [430–435] wide-line ^{1}H NMR at 4 K was successfully applied to determine the numbers of the above-mentioned complexes I and II in the frozen state. For water adsorbed on H^+-form zeolites, the ^{1}H NMR signal depends on the following species: (a) free hydroxyl protons of bridging OH groups (n_{OH}); (b) free water molecules (n_{H_2O}); (c) water molecules physisorbed at bridging OH groups via hydrogen bonds (n_{OH-OH_2} complexes I); and (d) hydroxonium ions ($n_{H_3O^+}$ complexes II). Deconvolution and simulation of the wide-line ^{1}H NMR spectrum of zeolite HY ($n_{Si}/n_{Al} = 2.4$) loaded with one water molecule per bridging OH group yielded negligibly small numbers n_{OH} and n_{H_2O} [432, 433]. The numbers $n_{H_3O^+}$ and n_{OH-OH_2} were determined to 9.7 and 38.6 u.c.$^{-1}$, corresponding to an equilibrium constant of $K = n_{H_3O^+}/n_{OH-OH_2} = 0.25$. Batamack et al. [433]

proposed a scale of acid strength of Brønsted acid sites in solids given by the ratio of $n_{H_3O^+}$ and n_{OHac} (number of initially observed acidic OH groups) which they called the ionization coefficient. Adsorbing one water molecule per OH group, the value of the ionization coefficient is zero for silanol groups, 0.2 for bridging OH groups in zeolite HY, 0.4 for hydroxyl protons in the superacid $H_2Sb_4O_{11} \cdot 2H_2O$, and 1.0 for sulfonic acid groups in Nafion-H (perfluorinated resinsulfonic acid).

4.3 Application of Powder X-Ray Diffractometry in Zeolite Research

4.3.1 Introduction

With improvements and computerization of the single-crystal X-ray diffractometric (XRD) technique in the last decades, large progress has been made in the structure determination of crystalline materials. However, zeolite research has not much benefited from this progress since zeolite crystals of good quality and large size needed for single-crystal XRD are available only in exceptional cases. Thus, most of the structural information on zeolites was derived from powder XRD and neutron diffraction, generally applied in combination with other techniques such as MAS NMR spectroscopy, adsorption methods etc.

XRD provides information about the Bragg angles, Θ, of the reflection of an incident monochromatic X-ray beam on crystal planes, according to Bragg's equation

$$n\lambda = 2d \sin \Theta \tag{4.50}$$

where n represents the order of reflection, λ the wavelength of the incident X-ray beam and d the spacing between reflecting crystal planes. Also, information is provided about the intensities of the reflected X-rays. An XRD pattern provides, in principle, all data necessary to solve crystal structures. However, powder patterns of complex structures with low symmetry might consist of a great number of Bragg reflections with either severe peak overlap or of very low intensity. Thus, in most cases it would be practically impossible to obtain information on the precise position and intensity of enough individual reflections.

Two computer-based analytical procedures are generally used to obtain structural information from powder XRD patterns. The so-called profile-fitting method uses integrated intensities of individual Bragg reflections which are treated in the same way as single-crystal data to solve or refine structures. It is evident that this procedure can only be applied if no severe peak overlap occurs in the pattern. In contrast to the profile-fitting method, the second procedure [1], first reported by Rietveld for neutron diffraction data [436], is not based on integrated intensities of individual reflections, but regards each intensity measured in small (e.g., 0.02) 2Θ intervals as an individual observation.

[1] This procedure is generally known as Rietveld method but sometimes (e.g., in [437] also named "pattern-fitting structure-refinement method".

The problem of peak overlapping does not arise and the maximum of information is obtained from the XRD pattern. It should be emphasized that both procedures are structure refinement and not structure solution methods in the strict sense of the word. Therefore, a reasonably good structural starting model is an imperative prerequisite for the successful application of them both.

Powder XRD is not only the current method in structural investigations of zeolites, but is also widely used for the identification, quantitative determination and characterization of zeolitic phases. Since a typical XRD pattern is obtained for each crystalline phase, the pattern feature may be regarded as a "fingerprint" of the respective material. Thus, zeolite structures can be easily identified by comparison of the d-spacings or 2Θ positions of the typical Bragg reflections with those given for known zeolites in straightforward compilations (e.g., [438, 439]). Even in mixtures with other zeolites or crystalline materials the identification of a zeolitic phase generally does not encounter difficulties. In zeolite synthesis, powder XRD is an indispensable analytical method which allows a quick identification of the prepared crystalline phases and monitoring of the progress of the crystallization processes.

The present section deals with

- factors influencing the integrated intensity of XRD peaks,
- structural refinement methods based on powder XRD data,
- the utilization of powder data for the quantitative determination of aluminum in zeolitic frameworks,
- the so-called "X-ray crystallinity" of zeolitic materials,
- the estimation of the crystallite size from diffraction peak broadening.

4.3.2 Parameters Affecting the Intensity of Bragg Reflections

The objective envisaged in this section is to sum up, without going too far into theoretical details, the experimental, physical and crystallographic parameters which determine the intensity of XRD reflections. Knowledge of these relationships is indispensable for the understanding of structure refining methods and the assessment of possibilities and limits of XRD methods applied for the characterization of zeolites.

The intensity I of an XRD reflection from the plane (hkl) of a crystal occurring at the corresponding Bragg angle Θ can be described by

$$I = k \cdot Lp \cdot P \cdot A \cdot F^2 \tag{4.51}$$

However, reflection also occurs at angles deviating slightly from the Bragg angle, though the X-ray beam reflected exactly at the Bragg angle is greatest in intensity. This results in intensity vs. 2Θ curves which are similar to Gaussians but, generally, they can be described by convolution of more complex functions. The *integrated intensity* I of the reflection is the energy of the sum of the reflected beams when the reflecting plane rotates through the Bragg angle. It is proportional to the area under the diffractogram peak after subtraction of the background.

The proportionality factor k involves some constant numerical factors and instrumental parameters (e.g., intensity and spectral distribution of the incident beam, slit sizes, geometrical features of the cameras) which can easily be kept constant when diffractograms of different materials are registered with the same diffractometer.

The effects on the integrated intensity, which are associated with the use of unpolarized X-ray beams, are taken into account by the polarization factor $f_P = 0.5(1 + \cos^2 2\Theta)$. Certain trigonometrical factors which influence the intensity of the reflected beam are combined in the Lorentz factor $f_L = 1/(4 \sin^2\Theta \cos \Theta)$.

For practical use f_P and f_L, both only dependent on Θ, are combined to the so-called Lorentz-polarization factor Lp, and the constant numerical factor 1/8 is included in k, which gives $Lp = (1 + \cos^2 2\Theta)/(\sin^2\Theta \cos \Theta)$.

In powder techniques, beams reflected from several planes generally superimpose. For example, in cubic crystals a general plane such as (731) has the same spacing as (173), (317) and the other 3 permutations of this index triple. Moreover, planes with different index triples but identical values for $h^2 + k^2 + l^2$, such as the pairs (551) and (711), reflect in the cubic system at the same angle Θ. Accordingly, the intensity of such diffraction peaks is higher than if it resulted from a single plane. The intensity increase due to reflections from more than one plane at the same Bragg angle is taken into account by an integer, the so-called multiplicity factor P. The multiplicity factors of the different hkl reflections depend on the crystal symmetry, and they are listed for the different space groups in [440].

It is obvious that the parameters Lp and P may be regarded as constants when, in XRD patterns of zeolite samples, the intensities of reflections appearing at the same 2Θ value are compared. Thus, these factors are irrelevant to the application of XRD for the determination of standard-related crystallinity (see Sect. 4.3.5).

The absorption of the incident and reflected X-ray beams in the specimen itself is taken into account by the absorption factor A. It is a function of

- the wavelength λ,
- the linear absorption coefficient μ of the specimen material,
- the morphology of the crystallites,
- the specimen geometry.

In the case of flat specimens as applied in the diffractometric technique the absorption is independent of the Bragg angle Θ. Moreover, the effect of the crystal morphology on A is not very pronounced and may be generally neglected, the more so as a particular zeolite exhibits mostly, even after the usual modifications, similar crystal morphology.

The absorption coefficient μ involved in the absorption factor A essentially depends, at a given wavelength λ, on the chemical nature (and physical state) of the investigated material. Thus, changes in the composition of the zeolitic phase by ion exchange, isomorphic substitution of the framework, dehydration etc. necessarily result in intensity changes. Absorption decreases the intensities of all reflected beams by the same factor which has two important consequences:

- the intensity decrease is not accompanied by peak broadening,

- the relative intensities (related to the intensity of one, generally the most intense diffraction peak) of reflections at different 2Θ angles are independent of the absorption phenomenon.

Thus, the relative intensities, i. e., the intensity distribution in diffractometric patterns, are not affected when the absorption coefficient is changed due to variation of the chemical composition of the zeolitic material. The absolute values of integrated intensities, however, significantly depend on compositional changes of the crystal lattice, especially if constituents are involved which contribute by highly different degrees to the coefficient μ as, for example, in the case of Na^+-Cs^+ ion exchange. In general, the larger the nuclear charge number of the constituents of the particular material, the higher the degree of absorption.

It is a rather difficult task to take into account all effects on the reflection intensity associated with the absorption of X-rays in zeolites of the same framework topology but different chemical composition. This causes problems first of all when the intensity of X-ray diffraction peaks are used as a measure of the crystallinity (see Sect. 4.3.5).

The integrated intensity I_{hkl} of an X-ray reflection typically depends on the so-called structure factor F_{hkl} which enters into the intensity expression in Eq. 4.51 as the square $F_{hkl}{}^2$. This quantity can be described by the equation

$$F_{hkl} = [(\Sigma f_n p_n e^{-2M} A_n)^2 + (\Sigma f_n p_n e^{-2M} B_n)^2]^{1/2} \qquad (4.52)$$

where

- n is the number of scattering atoms in the asymmetric unit,
- f_n is the scattering factor of the n-th atom of the asymmetric unit occupying the lattice position (x_n, y_n, z_n),
- A_n and B_n are the so-called geometrical structure factors which are a complex function of (h,k,l) and the lattice positions (x_n, y_n, z_n) of the scattering atoms in the asymmetric unit; equations for the calculation of the geometrical structure factors A_n and B_n are given for the different space groups in [441],
- p_n is the number of the n-th atoms of the asymmetric unit in the crystallographic unit cell, the so-called site occupancy, and
- M is the so-called Debye-Waller temperature factor of the n-th atom of the asymmetric unit occupying the lattice position (x_n, y_n, z_n).

The atomic scattering factor f_{n0} of the n-th atom of the asymmetric unit equals, for $\Theta = 0$, the atomic number of the element, i. e., the number of electrons in the shells of the atom. The scattering efficiency decreases with increasing sin Θ and decreasing wavelength λ. For convenience, f_n data are tabulated (e. g., in [442]) as a function of $(\sin \Theta)/\lambda$.

The atoms in a crystal undergo thermal vibrations at all finite temperatures above absolute zero so that they are located rather in more or less thick platelike regions than exactly on parallel crystal planes as presumed for the deduction of the fundamental laws of X-ray scattering. Temperature increase results in a gradual decrease of the diffraction peak intensity, but does not affect the width of the diffraction profile.

The temperature factor M deduced from theoretical considerations is a function of λ and Θ, commonly expressed as

$$M = b(\sin^2 \Theta)/\lambda^2. \quad (4.53)$$

which appears in the structure factor (Eq. 4.52) as e^{-2M}.

The straightforward application of the temperature factor is highly problematic because b is a complicated function of numerous variables. It is evident that the thermal motion of an atom and, hence, the displacement from its lattice point cannot be independent of atoms in neighboring sites. Consequently, b depends on the composition as well as on the structure of the crystal, and each atom in the asymmetric unit has its own temperature factor. Fortunately, neglecting the variation in b has generally no serious consequences, and it may be regarded as a constant, the value of which may be about 4. The effect of this simplification is that the calculated intensities decrease in a less pronounced manner in the high-Θ region than the observed integrated intensities. In refinement procedures, the temperature factor, if not available from, e.g., single-crystal studies, may be refined using starting values between 3 and 5 for b.

Equation 4.52 reveals that the scattering at all atoms of the asymmetric unit contributes to the intensity of all XRD reflections but not in a simple additive way. It must be kept in mind that the values of the geometrical structure factors may be quite different and also partly negative for different atom positions and Bragg angles. Further, the scattering coefficients mainly depend on the electron density at the particular lattice point, i.e., on the chemical nature of the scattering particle and the occupancy of the corresponding lattice sites and, hence, on the chemical composition of the zeolite crystal. Since the structure factor enters in Eq. 4.51 as the square, negative contributions to the structure factor may cause an increase and positive ones a decrease in the diffraction peak intensity. This is the case if the absolute numerical value of the structure factor is increased by addition of a negative quantity and decreased by a positive contribution.

Up to now only the influence of specimen-intrinsic and instrumental parameters on the diffraction peak intensity have been regarded, and it was tacitly presumed that the crystals in the specimen are randomly oriented. Preferred orientation inevitably occurring to a certain degree during sample preparation, especially in the case of crystals with layer or needle morphology, is in fact the phenomenon which most seriously intervenes in the measurement and interpretation of diffraction peak intensities. Rotation of the sample holder during the registration of the diffractogram may mitigate this effect, but it is less efficient in the case of a layer morphology. Grinding samples to particle sizes generally smaller than 10 µm, spray drying and other appropriate sample mounting techniques as well as dilution with a second non-interfering phase can essentially reduce the problems associated with preferred crystal orientation.

4.3.3 Calculation of Structure Factors

In this section the calculation of structure factors will be demonstrated for a hypothetical dehydrated, occluded NaCl-containing Y zeolite of the chemical composition $Na_{56}[Al_{56}Si_{138}O_{384}] \cdot 8NaCl$.

The following assumptions are made:

1. The origin is fixed, as normally, in the center of the hexagonal prism (site SI) with the point symmetry 3 m.
2. Chloride ions are located at site U in the center of the sodalite cages, one in each of the eight sodalite units in the unit cell. (It follows from the geometry of the faujasite structure that site U is on the space diagonal at a distance of 1/8 from the origin.)
3. Each chloride ion is tetrahedrally surrounded by four sodium ions located in the sodalite cages at the six-rings communicating with the hexagonal prisms, i.e., at sites SI′.
4. The remaining 32 sodium ions are located in the large cavities (8/unit cell) at the six-rings communicating with the sodalite cages, i.e., at site SII.
5. Atomic positions and temperature factors of framework and extra-framework atoms are considered to be identical with those reported for natural faujasite [443] and dehydrated NaY [444], respectively.
6. The cubic lattice constant of the hypothetical zeolite and, hence, the 2Θ- and d-values of its reflections are given by the framework Si/Al ratio using the Breck-Flanigen relationship (Eq. 4.59 in Sect. 4.3.6).

The data consistent with these assumptions and used for the calculations of the structure factors and relative intensities of ten selected reflections are listed in Table 4.11 and in the first 3 columns in Table 4.13.

For the space group of faujasite, F3dm, the geometrical structure factors B_n equal zero. The geometrical structure factors A_n are given by the following equation [445]:

$$A_n = 8\cos^2(2\pi U)\cos^2(2\pi W\, G) \tag{4.54}$$

where

$$\begin{aligned}
G = {} & \cos 2\pi(hx + ky + lz) + \cos 2\pi(hx + ky - lz - U) \\
& + \cos 2\pi(hx - ky + lz - V) + \cos 2\pi(-hx + ky + lz - W) \\
& + \cos 2\pi(hy + kz + lx) + \cos 2\pi(hy + kz - lx - U) \\
& + \cos 2\pi(hy - kz + lx - V) + \cos 2\pi(-hy + kz + lx - W) \\
& + \cos 2\pi(hz + kx + ly) + \cos 2\pi(hz + kx - ly - U) \\
& + \cos 2\pi(hz - kx + ly - V) + \cos 2\pi(-hz + kx + ly - W) \\
& + \cos 2\pi(hy + kx + lz) + \cos 2\pi(hy + kx - lz - U) \\
& + \cos 2\pi(hy - kx + lz - V) + \cos 2\pi(-hy + kx + lz - W) \\
& + \cos 2\pi(hz + ky + lx) + \cos 2\pi(hz + ky - lx - U) \\
& + \cos 2\pi(hz - ky + lx - V) + \cos 2\pi(-hz + ky + lx - W) \\
& + \cos 2\pi(hx + kz + ly) + \cos 2\pi(hx + kz - ly - U) \\
& + \cos 2\pi(hx - kz + ly - V) + \cos 2\pi(-hx + kz + ly - W)
\end{aligned}$$

and $U = (h + k)/4$, $V = (l + h)/4$, $W = (k + l)/4$.

This equation was used to calculate for ten selected reflections the geometrical structure factors A_n of the lattice positions defined in Table 4.11. The individual values are listed in Table 4.12.

The structure factors F_{hkl} presented in Table 4.13 were obtained using Eq. 4.52 which is, in the case of faujasite, reduced to

Table 4.11. Lattice Positions (x, y, z), Temperature Factors (b) and Population Parameters of Framework and Extra-Framework Atoms in a Hypothetical Sodium Y Zeolite Containing Occluded Sodium Chloride

Site		Atomic positions			Temp. factor	Site occupancy	
designation	number per u.c.	x	y	z	b in Å^2	element	factor
T	192	0.1254	0.9466	0.0363	1.2	Si Al	0.708 0.292
O1	96	0.1742	0.1742	0.9680	2.8	O	1.00
O2	96	0.1733	0.1733	0.3232	2.5	O	1.00
O3	96	0.2527	0.2527	0.1439	2.5	O	1.00
O4	96	0.1053	0.8947	0.0000	2.8	O	1.00
SI	16	0.000	0.000	0.000		–	–
SI′	32	0.055	0.055	0.055	2.5	Na	1.00
U	8	0.125	0.125	0.125	3.0	Cl	1.00
SII	32	0.233	0.233	0.233	2.5	Na	1.00

Table 4.12. Geometrical Structure Factors for Lattice Positions in the Faujasite Structure

hkl	T	O1	O2	O3	O4	SI	SI'	U	SII
111	-52.2	-83.8	-31.1	-56.9	-26.2	-96.0	-111.1	-135.8	-30.5
220	23.6	14.1	61.7	4.7	45.1	0.0	-78.0	-192.0	-8.6
311	7.7	17.7	12.6	-5.2	5.0	96.0	-3.8	-135.8	-10.9
222	6.4	42.2	-49.8	-57.0	14.5	192.0	87.8	0.0	-179.2
400	2.6	10.6	-61.1	55.4	-33.6	192.0	36.0	-192.0	174.7
331	-29.8	-13.6	-49.8	-62.5	7.5	96.0	61.8	135.8	-46.8
620	25.5	60.5	26.0	12.1	-41.7	0.0	-107.2	192.0	-24.3
533	-55.9	25.8	-55.2	-45.2	-8.2	96.0	-98.1	-135.8	-0.5
551	24.0	-10.0	29.1	-11.2	44.9	-96.0	-80.3	-135.8	-93.9
642	13.2	-44.1	6.2	11.7	-23.7	0.0	-20.1	-192.0	-22.3

$$F_{hkl} = [(\Sigma f_n p_n e^{-2M} A_n)^2]^{1/2} \tag{4.55}$$

since B_n equals zero for the space group Fd3m. The scattering factors f_n were obtained by interpolation from the values of the respective elements tabulated as a function of $(\sin \Theta)/\lambda$ [442], and p_n, M_n and A_n were taken from Tables 4.11 and 4.12, respectively, taking into account the relationship between b_n and M_n (see Eq. 4.53). To show the role of extra-framework components, contributions of framework and extra-framework species to the structure factors were also separately calculated and presented.

The geometrical structure factors in Table 4.12 clearly show that atoms located at different sites contribute to highly different degrees to the structure factor calculated according to Eq. 4.52.

Scattering at framework atoms of zeolites of the same framework type contributes practically to the same degree to the structure factors F_{hkl} since the atomic positions and, hence, the geometrical structure factors do not signifi-

Table 4.13. Relative Intensities and Structure Factors of Selected Reflections of NaY

hkl	2Θ	δ[Å]	Relative intensity	Structure factor	Contribution to structure factor by framework	32 Na at SI′	8 Cl at U	32 Na at SII
111	6.20	14.25	3.65	- 1842	- 1491	- 201	- 95	- 55
220	10.14	8.73	0.26	513	796	- 137	- 131	- 15
311	11.89	7.44	0.02	97	214	- 7	- 91	- 19
222	12.43	7.12	0.04	- 266	- 109	151	0	- 307
400	14.36	6.17	0.01	153	- 75	60	- 126	293
331	15.65	5.66	0.60	- 692	- 804	102	88	- 77
620	22.79	3.90	0.13	387	475	- 163	112	- 37
533	23.64	3.76	1.00	- 1115	- 889	- 147	- 78	- 1
551	25.78	3.46	0.01	93	423	- 117	- 75	- 137
642	27.04	3.30	0.29	- 457	- 293	- 29	- 104	- 32

cantly differ and the crystallographic framework sites are generally fully occupied, i.e. the occupancy is also identical. Even differences in the framework Si/Al ratios are not of great importance, since the scattering factors of Si and Al as neighboring elements in the periodic system are not very different. In contrast, extra-framework species located at crystallographically defined lattice positions, especially lattice cations and occluded compounds (e.g., templates), may contribute to highly different degrees to the structure factors depending on their location in the unit cell and their chemical nature and concentration.

4.3.4 Powder-Data Structure Refinement

The common feature of structure refinement methods is that all parameters affecting the diffraction pattern are varied in a least-squares procedure until the calculated pattern matches the observed one as closely as possible. In the previous sections it has been demonstrated how intrinsic properties and the chemical nature of the scattering matter affect structure factors and intensities of individual reflections. Further, an experimental X-ray diffraction pattern is determined also by effects associated with

- aberrations (displacement, transparency) and properties (crystallite size and microstrain) of the specimen,
- instrumental geometrical-optical features,
- spectral characteristics of the X-ray beam (e.g., K_{α} doublet).

The refinement of so many parameters is not straightforward and is even not *a priori* possible in all cases.

Nowadays numerous sophisticated computer programs for powder-data structure refinements are offered by professional crystallographers, scientific software companies and X-ray diffractometer manufacturers. Usually, the zeolite researcher will use one of them based either on the "profile-fitting method" or, preferentially, the Rietveld method. However, one should be aware that the basic requirements for any powder-data refinement techniques are

- accurate determination of the positions of the Bragg reflections in the powder pattern which involves correction of systematic errors arising from zero-shift (of the 2Θ scale), peak asymmetry and sample displacement,
- precise intensity data measured in small intervals of 2Θ in the step-scan mode and accurate determination of the background intensity,
- elimination or, at least, minimization of preferred crystallite orientation or incomplete randomness,
- a starting structure model which is a reasonable approximation of the actual crystal structure under investigation.

Thus, crucial points are both adequate sample preparation and the formation of conceptional ideas on possible structures.

4.3.4.1
Profile-Fitting Method

The profile-fitting procedure has been described in detail by Will et al. [446]. It is a two-step procedure in which first the instrument function of the experimental diffractometer set-up is determined from diffraction profiles of carefully selected powder specimens not giving rise to peak-broadening. The instrument function represents the effects of the intensity distribution of the K_α spectral lines and all the geometrical and instrumental factors (e.g., slit widths) which contribute to the shape of a diffraction peak. Thus, the instrument function must be separately measured for each combination of experimental conditions contributing to the line shape. In the second step the integrated intensities of the material under investigation are calculated. The experimentally determined peak intensity is regarded as the convolution of Lorentzian curves describing the instrumental function and the true diffraction effects associated with the specimen. 21 parameters have to be adjusted by a least-squares procedure until the calculated profile best matches the observed diffraction peak.

The structure factors obtained from the thus-obtained integrated intensities of individual Bragg reflections are used to refine structural and/or compositional parameters by least-squares procedures in the way generally followed in single-crystal data analysis. This method works well for high-symmetry structures with no or minimal peak overlap in the XRD patterns, provided the suggested starting model is reasonably close to the real crystal structure of the material being analyzed.

There are also some other, less recent methods which, essentially, are based on the same method of calculation. They differ from the profile-fitting method in the manner in which the contribution of instrumental and spectral factors to the integrated peak intensities are taken into account and are known as "trial-and-error" methods. The profile-fitting method and the related "trial-and-error" procedures have been successfully used to determine the crystal structure of some zeolites, e.g., Theta-1 [447] and, as a straight classical example, ZSM-5 [448]. However, in most cases these methods were applied for the determination of lattice positions of cations and their occupancy by different cations.

4.3.4.2
Rietveld Method

In the last decade the Rietveld method has been comprehensively reviewed (cf., e.g., [449–453]). The commercially available computer programs are also delivered with more or less detailed descriptions of theoretical and practical aspects concerning this method. Therefore, the present review is restricted to the fundamental features of this method and to a quite general description of the procedure.

In the Rietveld method, the intensity measured for each 2Θ step is considered as an individual piece of information consisting of contributions of the background and Bragg reflections at that step. During the refinement process based on a least-squares procedure, all parameters determining the contributions of background and Bragg reflections are varied systematically until all the individual data calculated for a presumed structure model best match the observed pattern. The quantity minimized by the least-squares method is:

$$R = \sum w_i (I_{i0} - I_{ic}) \tag{4.56}$$

where I_{i0} and I_{ic} are the observed and calculated intensities, respectively, at step i, and w_i is the weight attributed to step i. Since the whole diffractometric pattern, subdivided in very small 2Θ intervals, is involved in the Rietveld refinement, problems associated with peak overlapping do not arise and a maximum of structural information is obtained.

Background contribution may be composed of detector noise, thermal-diffuse and incoherent scattering, fluorescence and scatter from air, slits, sample holder and disordered or amorphous components in the sample. In some cases the background can be determined by extrapolation between selected pattern points, where the intensity is due only to background effects and definitely unaffected by Bragg reflections. However, background variables are usually refined together with other parameters influencing the position, intensity and shape of diffraction peaks as

- structure factors (atomic lattice positions, occupancies, temperature factors),
- scale factor and unit cell parameters,
- instrumental geometrical-optical parameters (e.g., axial divergence of the X-ray beam),
- specimen-intrinsic parameters (e.g., transparency, microstrain, structure disorder),
- specimen aberrations (e.g., specimen displacement),
- crystallite size.

The peak shape is a complex convolution of effects related to several instrumental or specimen-intrinsic parameters and, therefore, depends on the distribution of the spectral lines (since generally not strictly monochromatic radiation is used), the diffractometer geometry and the sample (crystallite size and shape, non-uniform strains). In its original form developed for the evaluation of neutron diffraction patterns [436], the Rietveld method was based on Gaussian-shaped diffraction peak profiles. However, because of the complexity of the effects in X-ray diffractometry, the shapes of Bragg peaks measured with that technique deviate considerably from the Gaussian distribution which, in general, describes only instrumental effects well enough. Sample-related contributions (e.g., line-broadening) can be better described by a Lorentzian function. Thus, the Voigt function, a convolution of Gaussian and Lorentzian distributions, has proved to be appropriate to describe accurately the profiles of X-ray diffraction peaks. Because of its simpler mathematical structure, most of the Rietveld programs use at present the pseudo-Voigt function which also allows the refinement of the parameters of separate half-width functions for the Lorentzian and Gaussian components of the diffraction profile. The right choice of the most appropriate profile-shape function out of the equations included in a given program is an important condition for obtaining reasonable refinements.

Usually, CuK_α radiation is used in X-ray diffractometry. The two components of the X-ray doublet, $K_{\alpha 1}$ and $K_{\alpha 2}$, make the modeling of the profiles more complicated due to the small differences in their wavelength λ which result in two separate beams reflected from the same crystal plane at different Bragg angles. This effect may cause, especially in the range of larger 2Θ values, pronounced distortion of the peak shape by partial overlapping of the two profiles. In recent Rietveld refinement programs, this problem is overcome by calculating both $K_{\alpha 1}$ and $K_{\alpha 2}$ profiles and assuming a constant intensity ratio between the intensity of the two components of the doublet, typically $K_{\alpha 1}/K_{\alpha 2} = 2$.

Most of the commonly used profile-shape functions including pseudo-Voigt are symmetrical with the maximum at the Bragg angle. However, some instrumental (e.g., axial divergence of the X-ray beam) and specimen-intrinsic (e.g., structure disorder) effects cause a pronounced asymmetry of the peaks, especially in the low-angle range. Therefore, it may be advisible to omit in Rietveld refinements data measured at very low angles. More recent Rietveld programs include a semi-empirical correction term as variable which takes the asymmetric profile deformation into account.

In 1986, Baerlocher [454] reported on another method for modeling of powder X-ray diffraction profiles based on the so-called "learned" peak shape function. Here, a non-analytical function is used in the Rietveld program containing symmetric and asymmetric parts which have to be determined ("learned") in numerical form from a single non-overlapping peak in the pattern. A prerequisite for the applicability of this method is the existence of at least one completely resolved reflection which, generally, is fulfilled in zeolite patterns.

Among the criteria used in Rietveld refinements, the R_{wp} value is the most meaningful as far as mathematics is concerned. The "goodness of fit" S also proved to be very useful. These numerical criteria are defined as

$$R_{wp} = [\,\sum w_i (I_{i0} - I_{ic})^2 / \sum w_i I_{i0}^{\,2}\,]^{1/2} \tag{4.57}$$

and

$$S = R_{wp}/R_{exp} \tag{4.58}$$

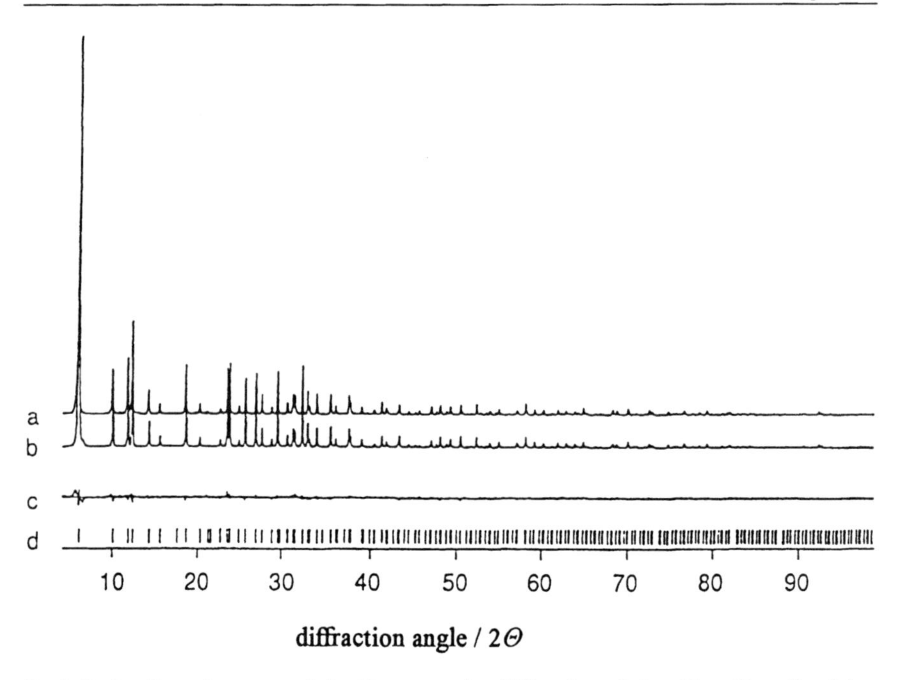

Fig. 4.60. Profile refinement of the X-ray powder diffraction of $Cs_{39.5}Na_{15.4}Y$ zeolite dehydrated in vacuum at 623 K: (**a**) experimental, (**b**) calculated and (**c**) difference profiles; (**d**) positions of the Bragg reflections [380]

where R_{exp} = (R-expected) = $[(n - P)/\sum w_i I_{i0}]^{1/2}$; I_{i0} and I_{ic} are the observed and calculated intensities, respectively, at step i; w_i is the weight attributed to the step i; n is the number of steps and P is the number of parameters refined. Values of 0.1 and 1.3 or less for R_{wp} and S, respectively, represent a quite satisfactory fit.

It is common practice to illustrate the goodness of fit by superimposing the experimentally determined pattern with that calculated with the refined parameters and showing the difference plot (Fig. 4.60). This presentation sometimes reveals deviations which are due not to insufficient refinement arising from a poor starting model or not adequately performed measurements but to extraneous diffraction peaks which may belong to a second crystalline phase in the specimen or arise from the material of improperly placed sample holders. Such extraneous reflections may considerably increase the R_{wp} and S values and will result in false conclusions, if no graphical refinement criterion is applied simultaneously.

4.3.4.3
Application of the Rietveld Method in Zeolite Structure Analysis

In the last decade the Rietveld method has been successfully applied to solve a great number of structural problems concerning, e.g., the topology of zeolite frameworks, lattice coordinates and occupancies of cation sites, and the position of templates or guest molecules and clusters introduced into the structure. An extended review of these publications would go beyond the scope and frame of this book. In this section the usage of Rietveld refinements is outlined presenting only one example which well illustrates the typical way of how to proceed in the application of this method.

Koller et al. [380] used the Rietveld method to refine positions and site occupancies of sodium and cesium cations in dehydrated Y zeolites containing, as determined by wet chemical analysis, 39.6 Cs^+ and 15.4 Na^+ per unit cell. The XRD pattern of the dehydrated samples sealed in a 0.2 mm thin-walled glass capillary were measured using a powder diffractometer with $CuK_{\alpha 1}$ radiation and a small linear position-sensitive detector in the 2Θ interval from 4°–100° in steps of 0.02°.

The refinement procedure was started in the known space group of faujasite, Fd3m, with the coordinates given in [455] for the Si(Al)-atom and the four oxygen atoms of the asymmetric unit. Si and Al were treated as a mixed atom T with an average scattering calculated with regard to the actual Si/Al ratio. After adjustment of the profile parameters and relaxation of the framework in the first refinement step, a difference Fourier synthesis was carried out which revealed electron density maxima along the space diagonal at (0 0 0) in the center of the hexagonal prism (SI), at (0.09 0.09 0.09) in the sodalite cages (SI′) and at (0.26 0.26 0.26) in the large cages (SII) [380]. Supported by results of ^{133}Cs and ^{23}Na MAS NMR spectroscopy the model to be refined was extended assuming cesium at SI′ and SII and sodium at SI and again refined. A value of 0.289 was obtained for R_{wp} (see Eq. 4.57). Then, a further difference Fourier synthesis was carried out. Some scattering matter found in the Fourier maps at position (0.43 0.125 0.125) in the supercage (SIII) was assumed to also represent cesium which was taken into account in the model. After a third Rietveld refinement, which resulted in R_{wp} = 0.204, a subsequent Fourier synthesis revealed electron density at (0.2 0.2 0.2) in the sodalite cages (SII′). This scattering matter was assigned to sodium cations because of the short distances between this site and the framework oxygen atoms. Finally, this repeatedly improved model was again refined in a final Rietveld cycle in which a total of 30 atomic, temperature, population and profile parameters were treated simultaneously. The final R_{wp} value was found to be 0.095 which represents a reliable fit. The structural parameters obtained by the described refinement procedure for the studied NaCs-Y zeolite are summarized in Table 4.14. The degree of refinement becomes obvious if the X-ray diffraction pattern calculated from the refined parameters is compared with the experimental one (Fig. 4.60). The difference pattern is very close to a straight line to be expected for a complete fit. Nevertheless, a final difference Fourier synthesis showed some electron density maxima still at the positions (0.2600 0.2642 0.2642), (0.1200 0.1245 0.1245) and (0.0200 0.0234 0.02307). These remaining maxima are due to the fact that from 39.6 Cs^+ and 15.4 Na^+ ions found by chemical analysis only 37.3–37.6 Cs^+ and 12.0–13.6 Na^+ could be located in the refined structure (see Table 4.14). Naturally, further sound modifications of the model and subsequent refinement cycles may result in, but not necessarily lead to, an even better approach.

Table 4.14. Lattice Positions (x, y, z) and Population Parameters in Dehydrated NaCs-Y Zeolite [380]

Atom	x	y	z	Sites/u.c.	Site occupancy
T(Si, Al)	0.12381(8)	0.94655(9)	0.03532(9)	192	1
O1	0.1070	$-x$	0	96	1
O2	0.2551	x	0.1416	96	1
O3	0.1782	x	0.9665	96	1
O4	0.1745	x	0.3186	96	1
Cs1	0.26331(5)	x	x	32 (SII)	0.820(2)
Cs2	0.0942(3)	x	x	32 (SI′)	0.139(2)
Cs3	0.4314(5)	1/8	1/8	48 (SIII)	0.115(2)
Cs4	0	0	0	16 (SI)	0.055–0.075
Na1	0	0	0	16 (SI)	0.20–0.30
Na1	0.1925(9)	x	x	32 (SII′)	0.28(1)

The Rietveld method only gives reliable structural information if the refinement is based on, besides fundamental structure data available in the literature, a model derived from crystallographically and chemically sound assumptions. Therefore, the input of information on the chemical nature and crystallographic location of scattering matter obtained by different techniques is generally of decisive significance. Solid-state NMR spectroscopic techniques proved to be in many cases especially useful to contribute, as in the study reviewed above, details to structure models, and Fourier synthesis based on the difference diffraction pattern is in general the method of choice to improve the model. Nevertheless, no generally applicable procedures exist of how to apply the Rietveld method, and anyway, the outcome of a sole refinement will in most cases not be satisfactory.

4.3.5 Crystallinity Determination

X-ray diffractometry is widely used to determine the degree of the so-called "X-ray crystallinity" of zeolites and zeolite-like materials. The intensities of one or more arbitrarily selected X-ray reflections of the studied zeolite are simply related to those of the respective peaks in the XRD pattern of a standard material containing the zeolite in a known or presumed amount, and this intensity ratio is generally believed to reflect the crystallinity, i.e., the amount of the crystalline zeolitic phase in the studied material. There is no problem with such a simple application of the XRD method provided the zeolitic phase in the materials to be compared is identical not only in the framework topology but also with respect to both chemical composition and occupancy of extra-framework sites. However, this fundamental restriction connected with the basic principles of X-ray diffraction and crystallography outlined in Sect. 4.3.2 has frequently been neglected. In the literature numerous examples can be easily found in which changes in the intensities of X-ray reflections upon ion exchange processes, chemical modifications (e.g., dealumination), heat treatment, steaming etc. were interpreted exclusively as differences in the crystallinity without taking into account the effect of lattice position and chemical nature of extra-framework species (e.g., exchangeable lattice cations) on the peak intensities and intensity distribution in XRD patterns of zeolites. However, it is not intended to refer in this review to concrete publications in which XRD was applied in a non-acceptable, over-simplified way to crystallinity determination.

It is evident that, for example, the crystallization during the synthesis procedure can be followed by an increase in the intensity of selected XRD pattern peaks since the chemical composition of and the site occupancy in the crystalline phase will not essentially change during the process.

Also, in the case of silica varieties of zeolites (e.g., silicalite) it is, naturally, allowed to correlate peak intensities with the crystallinity since exclusively framework atoms occupying the same lattice sites contribute to the structure factors. The XRD method can also be used to determine the crystallinity of high-silica zeolite samples (e.g., ZSM-5) since the scattering factors of silicon and aluminum are not very different and, due to the low lattice cation content, the contribution of scattering matter in extra-framework lattice positions to the structure factors is generally small compared to that of the framework.

In contrast, isomorphous incorporation of trivalent and/or tetravalent framework elements (e.g., B, Ga, Fe, Ge, Ti) in amounts corresponding to about 2% of the T-atoms affects the peak intensities to an intolerable degree, not only by their different contributions to the structure factors due to different scattering factors but also by changes of the absorption factor A (*vide supra*). At such high substitution degrees the crystallinity of zeolites on the one hand and their substituted varieties on the other cannot any longer be directly correlated to reflection intensities.

When applying the XRD method for the quantitative determination of a zeolitic phase, one must be aware of the effects of scattering extra-framework matter on the intensity of Bragg reflections. In order to observe the limits of the method, first of all the nature of lattice cations and cation site positions and occupancy have to be taken into account. The experimental error due to differences in the concentration and nature of cations can be estimated from the geometrical structure factors (see Table 4.12) and the site populations (Table 4.11), and it can be minimized if intensities of that reflection are compared, for which the contribution of extra-framework matter to the structure factor is smallest.

It is evident that the effects dealt with in Sect. 4.3.2 may result in dramatic changes of the intensity distribution including extinction of reflections or appearance of new ones in XRD patterns of zeolites with low Si/Al framework ratios, but they may be less significant in the case of highly siliceous template-free zeolites. In any case, an intensity decrease of reflections in the XRD patterns of zeolites cannot be exclusively and *a priori* regarded as a loss in crystallinity.

4.3.6
Determination of Framework Aluminum from X-Ray Data

The increase of tetrahedrally coordinated framework aluminum atoms in the lattice of aluminosilicates results in an increase in the unit cell size, due to the difference between the Al–O and Si–O distances amounting to 1.73 and 1.61 Å, respectively, and to resulting alterations of the T–O–T angles. Breck and Flanigen [456] were the first to recognize, in 1968, the utilization of this effect for the quantitative determination of framework aluminum in zeolites. They empirically derived, from the chemical composition and the lattice constant a_0 of well-crystallized faujasite-type X and Y zeolites with Si/Al ratios varying from 1 to 3, the linear equation

$$N_{al} = 115.2(a_0 - 24.191) \tag{4.59}$$

where N_{al} is the number of framework aluminum atoms per unit cell, i.e., per 192 T-atoms, and a_0 the lattice constant in A. The linear relationship provides the slope of 115.2 in $Å^{-1}$ and the intercept of 24.191 in Å.

Later, as methods became known to prepare highly siliceous and even aluminum-free silica faujasite varieties, this equation was found to be valid for faujasite-type zeolites with molar Si/Al ratios up to infinity [457, 458].

In numerous papers the application of Eq. 4.59 was found to be very useful. Especially in former times, when MAS NMR spectrometers were not yet generally available, utilization of Eq. 4.59 proved to be, in combination with wet chemical analysis of the bulk, an essential method to differentiate between framework and extra-framework aluminum in faujasite-type zeolites, first of all in ultrastabilized varieties used as cracking catalysts [459].

In order to improve the validity of Eq. 4.59 attempts were made to derive more precise values of the numerical coefficients by regression analysis of experimental data. Modified empirical equations reported in the literature are:

$$N_{al} = 107.1(a_0 - 24.238) \quad [460] \tag{4.60}$$

$$N_{al} = 112.4(a_0 - 24.233) \quad [461] \tag{4.61}$$

$$N_{al} = 115.2(a_0 - 24.225) \quad [462] \tag{4.62}$$

As to the units of the numerical coefficients see Eq. 4.59.

Since the numerical coefficients are obtained by regression analysis of plots of a_0 vs. N_{al} of zeolites characterized by XRD on the one hand and by chemical analysis in combination with methods allowing distinction between the different aluminum species (e.g., ^{29}Si MAS NMR spectroscopy) on the other, it is evident that the accuracy of the determination of both the cell constant and the framework aluminum content are the crucial points of this method.

Unit cell parameters of high accuracy can be determined by Rietveld refinement of the experimental pattern (see Sect. 4.3.4.2). If this is the only goal, and at least approximate structural parameters of the zeolite are known from the literature, the refinement procedure quickly approaches the best matches. However, it is absolutely necessary to include in the refinement all crystalline components present in the specimen to be investigated.

In contrast, the calculation of unit cell parameters based on the d-spacing of individual diffraction peaks is not disturbed by reflections of different crystalline phases in a particular sample provided that reflections belonging to the same phase are selected. Nowadays all crystallographic computer program packages offer appropriate evaluation facilities for optimal cell constant determinations from individual reflections including corrections based on least-square refinement procedures for 2Θ zeropoint shift, peak asymmetry and sample displacement. Nevertheless, d-spacings of as many reflections as possible and reasonable should be used for the calculation of the lattice constant.

The accuracy of the cubic lattice constant of well-crystallized faujasites determined with thoroughly adjusted commercial powder X-ray diffractometers may be estimated, for routine analyses, to be about ± 0.005 Å which corresponds to an experimental error of about ± 0.6 Al atoms per unit cell, independent of the absolute aluminum content. Without doubt, chemical analysis allows the determination of the bulk aluminum content with greater accuracy, especially at low aluminum contents. The framework aluminum content can be calculated from the framework Si/Al ratio determined by ^{29}Si MAS NMR spectroscopy and the aluminum content of the bulk, or directly obtained from the equivalent maximum ion-exchange capacity. However, both methods are associated with greater experimental errors. Thus, the inaccuracy of the numerical values of the coefficients in the Eqs. 4.59 to 4.62 and, hence, of the method described in this section is first of all due to the error made in the distinction between framework and extra-framework aluminum. It is known, that Y zeolite partially dealuminated with $SiCl_4$ [462] or by ultrastabilization and extraction methods [459] contains considerable amounts of extra-framework aluminum species. Thus, coefficients obtained from experimental data of dealuminated Y zeolite samples in the range $N_{al} < 50$, except if $N_{al} = 0$, may be less reliable.

However, the unit cell size of zeolites does not only depend on the framework Si/Al ratio but also on the chemical nature and amount of lattice cations and, of course, to a minor degree, on crystal strain and hydration state. For example, data reported by Rubio et al. [463] and later reproduced by Beyer et al. [462] and Fritz et al. [464] show that the unit cell dimension of fully hydrated sodium Y zeolite (about 24.66 Å) increases by 0.08–0.13 Å on complete dehydration. The lattice parameter was also found to change with the sodium content in NH_4,Na-Y and H,Na-Y zeolites. Gradual exchange of sodium ions by protons in $Na_{27}H_{29}$-Y zeolite to a residual sodium content of 2 ions per unit cell results in a steady decrease of the lattice constant of the dehydrated zeolites from 24.75 to 24.67 Å [465]. However, lattice parameters of obviously hydrated NH_4, Na-Y zeolites were found to increase with decreasing sodium content [464, 466]. More or less pronounced changes in the lattice constant upon ion exchange are reported in a great number of papers. Thus, such effects may affect the accuracy of framework Si/Al determinations to a higher degree than experimental errors of the lattice constant determination. They should be taken into consideration if high accuracy is required.

A further attempt to derive numerically modified forms of Eqs. 4.59 to 4.62 has been made by Jorik [467] who reported a semi-empirical equation derived from an idealized structural faujasite model in which deviations of the real structure from its idealized geometrical form are taken into account by introduction of the correction factor α:

$$N_{al} = 101.202(a_0/\alpha - 24.2115). \quad (4.63)$$

As to the units of the numerical coefficients see Eq. 4.59.

The correction factor α, in its physical sense, is a measure of the framework deformation induced by the combined effects of nature, amount and distribution of lattice cations and adsorbed material. It is 0.9957 and 0.9966 for as-synthesized sodium X and Y zeolite, respectively [467]. Unfortunately, no numerical values of the correction factor are given for other ion-exchanged and/or dealuminated faujasite-type zeolites.

It stands to reason that the validity of Eqs. 4.59 to 4.62 is restricted to zeolites with faujasite structure while the effect of lattice contraction upon isomorphous substitution of framework aluminum by silicon is a general one. Nevertheless, only a few attempts have been made to establish quantitative relationships of this kind for other zeolites as well. For mordenites, the question of whether a linear relationship between lattice parameters and the unit cell composition exists seems to be controversial. Olsson and Rollmann [468] determined unit cell dimensions of mordenites dealuminated by acid leaching (Si/Al ratios between 5 and 50) and found the lattice contraction on aluminum removal to be anisotropic and nonlinear. In contrast, Senderov et al. [469] reported linear relationships between the number of aluminum atoms in the unit cell and the lattice dimensions of Na-mordenites synthesized with Si/Al ratios between 5 and 10. However, both findings may not be contradictory. Olsson and Rollmann [468] postulate that the aluminum atoms are distributed in three basic types of siting within the mordenite framework, where they affect the lattice constants to different degrees. Upon acid leaching the aluminum atoms are believed to be successively eliminated from the three sitings which should, theoretically, result in three linear intervals of the lattice dimension vs. aluminum content curves. Indeed, the curves reported [469] may be regarded within the limits of deviation as linear in the Si/Al range of 5–10. Mishin et al. [470] also found for Na- and NH_4-mordenites a linear, anisotropic dependence in the same Si/Al range. The most pronounced changes were observed for the lattice constant b which could best be described, for the ammonium form, by

$$b = 20.258 + 0.021\,N_{al} \quad (4.64)$$

where the numerical coefficients are given in Å.

However, for hydrogen-mordenites (Si/Al ratios: 6.5–191), dealuminated by both high-temperature treatment of NH_4-mordenites and acid leaching, a significantly smaller intercept and a greater slope were found:

$$b = 20.209 + 0.035\,N_{al}. \quad (4.65)$$

Thus, it is difficult to decide whether both equations represent only an approach to more or less linear segments of a nonlinear curve or whether they reflect the influence of lattice cations differing in concentration and chemical nature. It may be, however, that these equations can be applied for the approximate determination of framework aluminum in mordenites of the respective cation form and in the respective aluminum concentration interval.

It is evident that, for the determination of framework aluminum in high-silica zeolites (e.g., ZSM-5 or beta), the applicability of the methods reviewed in this section is first of all limited by the accuracy of the lattice constant determination which may cause differences of about ±0.4 Al per 100 T-atoms. This would result, for example, in a relative error of about 20% for zeolites with Si/Al = 50.

A simplified method proposed by Bibby et al. [471] for the determination of the framework aluminum content in ZSM-5 is based on the difference, Δ, between the Bragg angles of the two peaks at about $2\Theta = 45$ and $45.5°$ (using CuK_α radiation). They found empirically a linear relationship between Δ and the framework aluminum content:

$$\text{percentage } Al_2O_3 = 16.5 - 30.8\,\Delta\,. \tag{4.66}$$

Assuming a standard deviation of 0.003° for Δ, the Al_2O_3 content of ZSM-5 can be determined to about ± 0.1% corresponding to about 0.1 Al per unit cell. It is reported that the coefficients in Eq. 4.66 will vary for different cation forms. For example, Δ values of 0.417°, 0.433° and 0.431° were found for the hydrogen, ammonium and lanthanum form, respectively, of the same ZSM-5 zeolite.

Since the Miller indices of the two selected peaks are 10,0,0 and 0,10,0, the spacing depends on only one of the orthorhombic cell parameters, a or b. Thus, the difference Δ of the spacings reflects the anisotropy of the lattice expansion in the a and b direction with increasing framework aluminum content. The effect is especially pronounced due to the high Miller index (10) of the selected reflections.

A similar method for mordenites is based on the 2Θ distance Δ_1 between the (402) peak of mordenite and the (311) reflection of silicon added as internal standard and Δ_2 between the (004) reflection of mordenite and the (511) of silicon [472]. These distances can be related to the aluminum content by the following linear equations:

$$\Delta_1 = 2.828 - 0.0613\,N_{al} \tag{4.67}$$

$$\Delta_2 = 1.812 - 0.114\,N_{al}\,. \tag{4.68}$$

The error of N_{al} is estimated to be about ± 0.25 Al per unit cell.

Very early on it was observed that dealumination not only affects the exact position of reflections in an XRD pattern but also the intensity of at least some of them. Therefore, attempts were made to correlate such intensity changes to the framework aluminum content of zeolites in order to utilize empirically deduced relationships for the determination of Si/Al ratios [472, 473]. It was, for example, reported that in the range up to 8 Al per unit cell the framework aluminum content of NH_4-mordenites can be approximately described by

$$N_{al} = 14.955 - 12.411\,R + 2.714\,R^2 \tag{4.69}$$

where R is the intensity ratio of the (200) and (150) reflections [472].

However, the changes in diffraction peak position and intensity are basically different phenomena. Framework aluminum affects directly the distances between lattice planes in aluminosilicate crystals and, hence, the spacing of XRD reflections. In contrast, the intensity of XRD reflections depends only to a negligible degree on the framework Si/Al ratio since Al and Si contribute, because of their similar scattering factors, to nearly the same degree to the structure factors and, consequently, to the peak intensities. The observed effects are mainly due to the contributions of lattice cations and molecules (water) coordinatively bound to them. The amount of these extra-framework species depends, of course, on the framework aluminum content but not in a straightforward manner. Moreover, structure factors depend not only on the concentration but also, often decisively, on the nature and lattice positions of the cations. The peak intensities may also be substantially influenced by the dehydration degree. This is not only due to additional X-ray scattering on water molecules coordinatively bound to cations but also to the population at different lattice cation positions which varies with the dehydration degree. In addition, in the case of fibrous or layered morphology, crystal orientation may occur which is associated with a considerable intensity increase of reflections on planes parallel to the distinguished faces. Thus, when applying the method described in [472, 473] one must be aware of its extensive restrictions.

4.3.7 Determination of the Crystallite Size

The experimentally observed breadth B of a diffraction line is caused by

- the instrumental broadening b due to instrumental parameters and, eventually, to peak overlapping associated with the application of a non-strictly monochromatic incident X-ray beam, e.g., the CuK_α doublet,
- the pure diffraction breadth β depending on the crystallite size and lattice strains.

Scherrer [474] first deduced the equation which describes the relationship between the diffraction breadth β of a reflection and the mean crystallite thickness

$$D = K\lambda/(\beta \cos \Theta) \tag{4.70}$$

wherein K is a constant the numerical value of which is approximately unity, provided that Θ is measured in radians. K depends on the particular definition of D and β, the indices hkl of the reflecting crystal planes and the shape of the crystallites. β is generally defined as the angular width at half-maximum intensity, $\beta_{1/2}$, but sometimes also as integral breadth. D may be considered as D_{hkl}, the mean thickness of the crystallites in the direction perpendicular to the reflecting plane or defined, for spherical morphology, as the cube root of the volume of the crystallite, $V^{1/3}$, since the diameter D of a sphere is related to its volume V by

$$V^{1/3} = 0.806\,D\,. \tag{4.71}$$

It is obvious that the accuracy of crystallite size determinations from the diffraction breadth according to the Scherrer equation (Eq. 4.70) is limited by uncertainties in the determination of both K and β. In numerous studies it has been shown that K depends only slightly on the crystallite shape and amounts to about 0.9, provided D is defined as D_{hkl} and β as $\beta_{1/2}$. Equation 4.70 may then be written as

$$D_{hkl} = (0.89\,\lambda)/(\beta_{1/2\,hkl} \cdot \cos\Theta)\,. \tag{4.72}$$

The concept based on spherical morphology is generally a rather good approach if the crystallite size is of the same order of magnitude in all dimensions, i.e., if there is no pronounced needle or layer morphology. In these cases Eq. 4.70 changes to

$$0.806\,D = V^{1/3} = (1.107\,\lambda)/(\beta_{1/2\,hkl} \cdot \cos\Theta)\,. \tag{4.73}$$

It is an experimental fact that the contribution of the diffraction breadth β to the observed breadth B becomes negligibly small when the crystallite size exceeds 200–300 nm. Hence, postulating additivity of β and the peak breath b observed for materials with large crystallite size, i.e.,

$$\beta = B - b \tag{4.74}$$

the diffraction breadth should be obtained by subtraction of the half-maximum width of peaks at similar 2Θ positions in patterns obtained from the sample and a standard material consisting of crystals definitely greater than 300 nm. However, the assumption of strict additivity implicitely involved in Eq. 4.74 has been found to not be generally justified. Assuming that the shape of the diffraction peak can be described by the Gaussian function, Warren and Biscoe [475] derived the equation

$$\beta^2 = B^2 - b^2 \quad \text{or} \quad \beta^2 = B_m^2 - B_{st}^2\,, \tag{4.75}$$

where B_m and B_{st} are the peak breadths of a particular reflection in the patterns of the sample and the standard material, respectively, measured at half their maximum intensity in radians.

Other approaches to the subject of instrumental broadening correction have also been reported (see, e.g., in [476]). They are based on more exact considerations and may be somewhat more accurate; however, their application in practice is rather difficult.

In zeolite research the correction of the instrumental diffraction peak broadening according to Eqs. 4.74 or 4.75 and application of Eqs. 4.72 or 4.73 will, generally, meet all requirements, especially if the investigation is not aimed at determination of the absolute particle size but at its relative changes.

To obtain Bragg reflections, a minimum crystallite size corresponding to a unit cell periodicity of at least 6–8 is required. It follows that in the case of, for example, faujasite with a cubic unit cell parameter of about 2.45, the detection limit of the method described in this section corresponds to a crystallite size of about 15–20 nm. The applicability of the method ranges, as already mentioned, up to 200–300 nm. The crystallite size of zeolites is generally beyond this upper limit, however, the method covers a size interval important in regard to crystal growth and crystallization mechanism.

This method may also be applied to determine the size of crystalline particles of separate phases formed on the outer surface of zeolite crystals by agglomeration (recrystallization) of products of intracrystalline chemical reactions. For example, bifunctional metal/acid catalysts are commonly prepared by reduction of metal lattice cations previously introduced into zeolites by ion exchange, using, e.g., hydrogen as reducing agent. It is evident that the reviewed method offers excellent possibilities for the determination of the actual size of the formed metal particles and to study how this parameter, which is important in catalysis, depends on experimental conditions such as reduction temperature etc.

In Sect. 4.3.4.2 it was pointed out that in recent Rietveld programs the crystal size is also considered and refined as a parameter which, in a particular size interval, significantly affects the diffraction peak profile. Thus, the Rietveld procedure presents itself as an alternative method to the method for crystallite size determination based on $\beta_{1/2}$ values of individual diffraction peaks.

References

1. Delgass WN, Haller GL, Kellermann R, Lunsford JH (1979) Spectroscopy in Heterogeneous Catalysis. Academic Press, New York
2. Fierro JLG (1990) Stud Surf Sci Catal 57A: 1 ibid B: 1
3. Boehm HP, Knoezinger H (1983) In: Anderson JR, Boudart M (eds) Catalysis – Science and Technology, vol 4. Springer, Berlin Heidelberg New York, p 49
4. Kane PF, Larrabee GB (eds) (1976) Characterization of solid surfaces. Plenum Press, New York
5. Ward JW (1976) In: Rabo JA (ed) Zeolite chemistry and catalysis. Am Chem Soc, Mon 171, Washington, DC, p 118
6. Foerster H (1992) In: Davies JE (ed) Spectroscopic and computational studies of supramolecular systems. Kluver Academic Publishers, Dordrecht, The Netherlands, p 29
7. Herzberg GH (1950) Molecular spectra and molecular structure, vol I and II, 2nd ed. Van Nostrand Reinhold Comp, New York
8. Barrow GM (1962) Introduction to molecular spectroscopy. McGraw-Hill Book Comp Inc, New York
9. Hair ML (1967) Infrared spectroscopy in surface chemistry. Marcel Dekker Inc, New York
10. Kiricsi I, Tasi G, Foerster H, Fejes P (1987) Acta Phys Chem Szeged 33:69
11. Geppert G (1964) Experimentelle Methoden der Molekuelspektroskopie. Akademie-Verlag, Berlin, p 51
12. Martin AE (1980) In: Durig JR (ed) Vibrational Spectra and Structure, Elsevier, Amsterdam, pp 27, 34
13. Kustov LM, Borovkov VY, Kazansky VB (1981) J Catal 72:149
14. Kazansky VB, Borovkov VY, Kustov LM (1984) Proc 8th Int Congr Catal, vol 3. Berlin (West), Germany, Verlag Chemie, Weinheim, p 3
15. Kazansky VB (1991) Stud Surf Sci Catal 65:117
16. Kubelka P (1948) J Opt Soc Am 38:448
17. Kellermann R (1979) In: Delgass WN, Haller GL, Kellermann R, Lunsford JH (eds) Spectroscopy in Heterogeneous Catalysis. Academic Press, New York, p 86 (and references therein)

18. Kazansky VB, Borovkov VY, Karge HG (1997) J Chem Soc Faraday Trans 93:1843
19. Blank RE, Wakefield T (1979) Anal Chem 51:50
20. Rosencwaig A, Gersho A (1976) J Appl Phys 47:64
21. Phillippaerts J, Vasant EF, Yan YA (1989) Stud Surf Sci Catal 46:555
22. Chu YF, Keweshan CF, Vasant EF (1981) J Catal 72:749
23. Post MFM, Huizinga T, Emeis CA, Nanne JM, Stork WHJ (1981) J Catal 72:365
24. Kiselev AB, Lygin BU (1962) Uspechi Chim 31:351
25. Knoezinger H, Stolz H, Buehl H, Clement G, Meyer W (1970) Chem Ing Tech 42:548
26. Karge HG (1980) Z Phys Chem NF 122:103
27. Karge HG, Ziółek M, Łaniecki M (1987) Zeolites 7:197
28. Gallei E, Schadow E (1974) Rev Sci Instrum 45:1504
29. Karge HG, Klose K (1973) Z Phys Chem NF 83:92
30. Karge HG, Abke W, Boldingh EP, Laniecki M (1984) Proc 9th Iberoamerican Symp Catal, Lisbon, Portugal, Jorge Fernandes, Lisbon, p 582
31. Karge HG, Niessen W (1991) Catal Today 8:451
32. Pastore HO, Ozin GA, Poe AJ (1993) J Am Chem Soc 115:1215
33. Karge HG, Trevizan de Suarez S, Dalla Lana IG (1984) J Phys Chem 88:1782
34. Foerster H, Meyn V, Schuldt M (1978) Rev Sci Instrum 49:74
35. Narbeshuber T, Lercher JA (1992) Preprints of the 4th German Workshop on Zeolite Chemistry, Mainz, Germany, Johannes-Gutenberg-University, Mainz, p 137
36. Flanigen EM, Szymanski HA, Khatami H (1971) Adv Chem Ser 101:201
37. Dutz H (1969) Ber Deutsch Keram Ges 46:75 (and references therein)
38. White WB, Roy R (1964) Am Mineral 49:1670
39. Beyer HK, Belenykaja IM, Hange F (1985) J Chem Soc Faraday Trans I 81:2889
40. Kubelkova L, Seidl V, Borbély G, Beyer HK (1988) J Chem Soc Faraday Trans I 84:1447
41. Geidel E (1989) PhD thesis, University of Leipzig, Germany
42. Geidel E, Boehlig H, Peuker Ch, Pilz W (1991) Stud Surf Sci Catal 65:511
43. Miecznikowski A, Hanuza J (1985) Zeolites 5:188
44. van Santen RA, van Beest BWH, de Man AJM (1990) NATO ASI Ser 221:201
45. Sachsenroeder H, Brunner E, Koch M, Pfeifer H, Staudte B (1996) Microporous Mater 6:341
46. Brunner E, Kaerger J, Koch M, Pfeifer H, Sachsenroeder H, Staudte B (1997) Stud Surf Sci Catal 105:463
47. Sachsenroeder H (1997) PhD thesis, University of Leipzig, Germany
48. Uytterhoeven JB, Christner LC, Hall WK (1965) J Phys Chem 69:2117
49. Angell CL, Schaffer (1965) J Phys Chem 69:3463
50. Hughes TR, White HM (1967) J Phys Chem 71:2112
51. Jacobs PA, Uytterhoeven JB (1973) J Chem Soc Faraday Trans I 69:359
52. Czjzek M, Jobic H, Fitch AN, Vogt T (1992) J Phys Chem 96:1535
53. Roessner F, Steinberg KH, Rechenburg S (1987) Zeolites 7:488
54. Stoecker M, Ernst S, Karge HG, Weitkamp J (1990) Acta Chemica Scandinavia 44:519
55. Dzwigaj S, Briend M, Shikholeslami A, Peltre MJ, Barthomeuf D (1990) Zeolites 10:157
56. Karge HG (1971) Z Phys Chem NF 76:133
57. Karge HG (1973) Z Phys Chem NF 83:100
58. Karge HG (1975) Z Phys Chem NF 95:241
59. Detrekoy EJ, Jacobs PA, Kalló D, Uytterhoeven JB (1974) J Catal 32:442
60. Hatada K, Ono Y, Ushiki Y (1979) Z Phys Chem NF 117:37
61. Jacobs PA, v. Ballmoos R (1982) J Phys Chem 86:3050
62. Loeffler E, Lohse U, Peuker C, Oehlmann G, Kustov LM, Zholobenko VL, Kazansky VB (1990) Zeolites 10:266
63. Jacobs PA (1977) Carboniogenic activity of zeolites. Elsevier, Amsterdam
64. Barthomeuf D (1978) Proc Symp Zeolites, Szeged, Hungary, Acta Universitatis Szegediensis, Acta Physica et Chemica, Nova Ser 24:71
65. Barthomeuf D (1979) J Phys Chem 83:249
66. Schoonheydt RA, Uytterhoeven JB (1970) J Catal 19:55

67. Jacobs PA, Uytterhoeven JB (1973) J Chem Soc Faraday Trans I 69:373
68. Karge HG (1977) ACS Symp Ser 40:584
69. Karge HG, Ladebeck J, Sarbak Z, Hatada K (1982) Zeolites 2:94
70. Beaumont R, Barthomeuf D (1972) J Catal 26:218
71. Beaumont R, Pichat P, Barthomeuf D, Trambouze Y (1972) Proc 5th Int Congr Catal, Miami Beach, Florida, US, North Holland Publ Co Inc, New York, p 343
72. Neuber M, Dondur V, Karge HG, Pacheco L, Ernst S, Weitkamp J (1988) Stud Surf Sci Catal 37: 461
73. Barthomeuf D (1987) Mater Chem Phys 17:49
74. Stach H, Jaenchen J, Jerschkewitz HG, Lohse U, Parlitz B, Hunger M (1992) J Phys Chem 96:8480
75. Kazansky VB (1978) Proc 4th National Symp Catal, Indian Inst Catal, Bombay, India, p 14
76. Barthomeuf D (1980) Stud Surf Sci Catal 5:55
77. Mortier WJ (1978) J Catal 55:138
78. Jacobs PA, Mortier WJ, Uytterhoeven JB (1978) J Inorg. Nucl Chem 40:1919
79. Jacobs PA (1982) Catal Rev Sci Eng 24:415
80. Sanderson RT (1976) Chemical bonds and bond energy, 2nd ed. Academic Press, New York, p 75
81. Jacobs PA, Mortier WJ (1982) Zeolites 2:226
82. Abbas SH, Al-Dawood TK, Dwyer J, Fitch FR, Georgopoulos A, Machado FJ, Smyth SM (1980) Stud Surf Sci Catal 5:127
83. Rabo JA, Gajda GJ (1990) NATO ASI Ser 221:273
84. Rabo JA, Gajda GJ (1989-90) Catal Rev Sci Eng 31:385
85. van Santen R, van Beest BWH, de Man AJM (1990) NATO ASI Ser 221:201
86. van Santen R, de Man AJM, Kramer GJ (1992) NATO ASI Ser 352:493
87. Sauer J (1989) Stud Surf Sci Catal 52:73
88. Schroeder KP, Bleiber A, Sauer J (1991) Proc 3rd German Workshop on Zeolite Chemistry, Berlin, Germany, paper O1
89. Sauer J (1994) Stud Surf Sci Catal 84:2039
90. Schroeder KP, Sauer J (1996) J Phys Chem 100:11043
91. Sauer J (1997) Symp on Advances and Applications of Computational Chemistry Modelling to Heterogeneous Catalysis, Am Chem Soc, San Francisco, ACS Ser 42, Washington, DC, p 53
92. de Man AJM, Sauer J (1996) J Phys Chem 100:5025
93. Kazansky VB, Kustov LM, Borovkov VY (1983) Zeolites 3:77
94. Kazansky VB (1988) Catal Today 3:367
95. Staudte B, Hunger M, Nimz M (1991) Zeolites 11:837
96. Beck K, Pfeifer H, Staudte B (1993) Microporous Mater 2:1
97. Jacobs WPJH, Jobic H, van Wolput JHMC, van Santen RA (1992) Zeolites 12:315
98. Defosse C, Cannesson P, Delmon B (1976) J Phys Chem 80:1028
99. Richardson RL, Benson SW (1957) J Phys Chem 61:405
100. Parry EP (1963) J Catal 2:371
101. Liengme BV, Hall WK (1966) Trans Faraday Soc 62:3229
102. Karge HG, Weitkamp J (1986) Chem Ing Tech 58:946
103. Turkevich J, Stevenson PC (1943) J Chem Phys 11:328
104. Kline CH, Turkevich J (1944) J Chem Phys 12:300
105. Basila MR, Kantner TR, Rhee KH (1964) J Phys Chem 68:3197
106. Ward JW (1968) J Coll Interf Sci 28:269
107. Uytterhoeven JB, Jacobs PA, Makay K, Schoonheydt R (1968) J Phys Chem 72:1768
108. Kubelkova L, Beran S, Malecka A, Mastikhin VM (1989) Zeolites 9:12
109. Bielánski A, Datka J (1974) Bull Acad Polon Ser Sci Chim 27:341
110. Karge HG (1991) Stud Surf Sci Catal 65:133
111. Khabtou S, Chevreau T, Lavalley JC (1994) Microporous Mater 3:133
112. Karge HG, Schweckendiek J (1983) Proc 5th Int Symp Heterogeneous Catalysis, Varna, Bulgaria, Publ. House of the Bulgarian Academy of Sciences, Sofia, p 429

113. Schweckendiek J (1982) PhD thesis, University of Bremen, Germany
114. Jacobs PA, Theng BKG, Uytterhoeven JB (1972) J Catal 26:191
115. Karge HG, Koesters H, Wada Y (1984) Proc 6th Int Zeolite Conf, Reno, Nevada, USA, Butterworths, Guildford, Surrey, UK, p 308
116. Jacobs PA, Martens JA, Weitkamp J, Beyer HK (1982) Faraday Disc Chem Soc 72:353
117. Datka J, Boczar M, Rymarowicz P (1988) J Catal 114:368
118. Knoezinger H (1989) Proc Int Symp Acid-Base-Catal, Sapporo, Japan, Kodansha, Tokyo, p 147
119. Paukshtis EA, Soltanov RI, Yurchenko EN (1982) React Kinet Catal Lett 19:105
120. Kustov LM, Kazansky VB, Beran S, Kubelkova L, Jiru P (1987) J Phys Chem 91:5247
121. Kubelkova L, Beran S, Lercher JA (1989) Zeolites 9:539
122. Zecchina A, Spoto G, Bordiga S, Padovan M, Leofanti G, Petrini G (1991) Stud Surf Sci Catal 65:671
123. Kuehl GH (1973) Proc 3rd Int Conf on Molecular Sieves, Zuerich, Switzerland, University Leuven Press, p 227
124. Kuehl GH (1977) J Phys Chem Solids 38:1259
125. Szymanski HA, Stamires DN, Lynch GR (1960) J Optical Society of America 50:1323
126. Jacobs PA, Beyer HK (1979) J Phys Chem 83:1174
127. Cannings FR (1968) J Phys Chem 72:4691
128. Ghosh AK, Curthoys G (1983) J Chem Soc Faraday Trans I 79:805
129. Dwyer J (1988) Stud Surf Sci Catal 37:333
130. Brodskii IA, Zhdanov SP, Stanevich AE (1971) Opt Spektrosk 30:58
131. Brodskii IA, Zhdanov SP (1980) Proc 5th Int Zeolite Conf, Naples, Italy, Heyden, London, p 234
132. Ozin GA, Baker MD, Godber J, Gil GJ (1989) J Phys Chem 93:2899
133. Baker MD, Ozin GA, Godber J (1985) Catal Rev Sci Eng 27:591
134. Esemann H, Foerster H (1995) Z Phys Chem 189:263
135. Krause K, Geidel E, Kindler J, Foerster H, Smirnow KS (1996) Vibrational Spectroscopy 12:45
136. Peuker C, Möller K, Kunath D (1984) J Mol Structure 114:215
137. Ward JW (1968) J Phys Chem 72:4211
138. Ward JW (1971) J Catal 22:237
139. Sheu LL, Knoezinger H, Sachtler WMH (1989) Catal Lett 2:129
140. Sheu LL, Knoezinger H, Sachtler WMH (1989) J Mol Catal 57:61
141. Ruthven DM (1984) Principles of adsorption and adsorption processes. John Wiley & Sons, New York, p 332
142. Barthomeuf D (1991) Stud Surf Sci Catal 65:157
143. Itoh H, Hattori T, Suzuki K, Murakami Y (1983) J Catal 79:21
144. Giordano N, Pino L, Cavallaro S, Vitarelli P, Rao BS (1987) Zeolites 7:131
145. Corma A, Fornes V, Martin-Aranda RM, Garcia H, Primo J (1990) Appl Catal 59:237
146. Karge HG, Rasko J (1978) J Coll Interf Sci 64:522
147. Ziółek M, Szuba D, Leksowski R (1988) Stud Surf Sci Catal 37:427
148. Ziółek M, Karge HG, Niessen W (1990) Zeolites 10:662
149. Martens LR, Vermeiren WJ, Huybrechts DR, Grobet PJ, Jacobs PA (1988) Proc 9th Int Congr Catal, vol 1, Calgary, Alberta, Canada, Chemical Institute of Canada, Ottawa, p 420
150. Barthomeuf D, de Mallmann A (1988) Stud Surf Sci Catal 37:365
151. Barthomeuf D (1984) J Phys Chem 88:42
152. de Mallmann A, Barthomeuf D (1988) Zeolites 8:292
153. Jentys A, Warecka G, Derewinski M, Lercher JA (1989) J Phys Chem 93:4837
154. Lercher JA, private communication
155. Karge HG (1991) Stud Surf Sci Catal 58:531
156. Yates DJC (1969) Catal Rev 2:113
157. Cohen de Lara E (1980) Proc 5th Int Zeolite Conf, Naples, Italy, Heyden, London, p 414
158. Foerster H, Frede W, Peters G (1989) Stud Surf Sci Catal 46:545

159. Boese H, Foerster H, Frede W, Schumann M (1984) Proc 6th Int Zeolite Conf, Reno, Nevada, USA, Butterworths, Guildford, UK, p 201
160. Cohen de Lara E, Kahn R (1990) NATO ASI Ser 221:169
161. Boese H, Foerster H (1990) J Mol Structure 218:393
162. Grodzicki M, Zakharieva-Pencheva O, Foerster H (1988) J Mol Structure 175:195
163. Grodzicki M, Foerster H, Pfiffer R, Zakharieva-Pencheva O (1988) Catal Today 3:75
164. Foerster H, Zakharieva-Pencheva O (1988) J Mol Structure 175:189
165. Foerster H, Zakharieva-Pencheva O (1990) J Mol Structure 218:399
166. Cohen de Lara E, Kahn R (1984) Proc 6th Int Zeolite Conf, Reno, Nevada, USA, Butterworths, Guildford, Surrey, UK, p 172
167. Foerster H, Schuldt M (1978) J Mol Structure 47:339
168. Foerster H, Frede W, Schuldt M (1980) J Mol Structure 61:75
169. Foerster H, Schuldt M (1980) J Mol Structure 61:361
170. Moeller K, Kunath D, Spangenberg HJ (1971) Spectrochimica Acta 27:195
171. Karge HG, Klose K (1975) Ber Bunsenges Phys Chem 79:454
172. Karge HG, Niessen W (1991) Catal Today 8:451
173. Niessen W, Karge HG (1991) Stud Surf Sci Catal 60:213
174. Niessen W, Karge HG (1993) Microporous Mater 1:1
175. Hermann M, Niessen W, Karge HG (1995) Stud Surf Sci Catal 94:131
176. Karge HG and Boldingh EP (1988) Catal Today 3:379
177. Karge HG, Łaniecki M, Ziółek M, Onyestyak G, Kiss A, Kleinschmit P, Siray M (1989) Stud Surf Sci Catal 49:1327
178. Mirth G, Cejka J, Lercher JA (1993) J Catal 139:24
179. Roland U, Winkler H, Bauch H, Steinberg KH (1991) J Chem Soc Faraday Trans 87:3921
180. Salzer R, Dressler J, Steinberg KH, Roland U, Winkler H (1989) Proc 2nd Int Conf Spillover, University of Leipzig, Germany, p 70
181. Karge HG (1992) NATO ASI Ser 352:273
182. Karge HG, Beyer HK (1991) Stud Surf Sci Catal 69:43
183. Karge HG (1996) Stud Surf Sci Catal 105:1901
184. Engelhardt G, Michel D (1987) High-resolution solid-state NMR of silicates and zeolites. Wiley, Chichester
185. Klinowski J (1984) Progress in NMR Spectroscopy 16:237
186. Klinowski J (1991) Chem Rev 91:1459
187. Stoecker M (1994) Stud Surf Sci Catal 85:429
188. Pfeifer H (1994) NMR Basic Principles and Progress 31:32
189. Pfeifer H, Ernst H (1994) Annual Reports on NMR Spectroscopy 28:91
190. Haw JF (1994) In: Bell AT, Pines A (eds) NMR techniques in catalysis. Marcel Dekker Inc, New York, p 139
191. Barrie PJ, Klinowski J (1992) Progress in NMR Spectroscopy 24:91
192. Kaerger J, Pfeifer H (1987) Zeolites 7:79
193. Boddenberg B, Burmeister R (1988) Zeolites 8:480 and 8:488
194. Fernandez C, Amoureux JP, Chezeau JM, Delmotte L, Kessler H (1996) Microporous Materials 6:331
195. Rocha J, Lourenco JP, Ribeiro MF, Fernandez C, Amoureux JP (1997) Zeolites 19:156
196. Amoureux JP, Bauer F, Ernst H, Fernandez C, Freude D, Michel D, Pingel UT (1998) Chem Phys Lett 285:10
197. Abragam A (1994) The principles of nuclear magnetism. Clarendon Press, Oxford
198. Slichter CP (1996) Principles of magnetic resonance. Springer, Berlin
199. Schmidt-Rohr K, Spiess HW (1994) Multidimensional solid-state NMR and polymers. Academic Press, London, New York
200. Haarer D, Spiess HW (1995) Spektroskopie amorpher und kristalliner Festkörper, Steinkopff Verlag, Darmstadt
201. Freude D, Haase J (1993) NMR, Basic Principles and Progress 29:1
202. Anderson PW, Weiss PR (1953) Rev Mod Phys 25:269
203. Andrew ER, Bradbury A, Eades RG (1958) Nature (London) 182:1659

204. Lowe IJ (1959) Phys Rev Lett 2:285
205. Waugh JS, Huber LM, Haeberlen V (1968) Phys Rev Lett 20:180
206. Mehring M (1980) Principles of high resolution NMR of solids. Springer, Berlin
207. Samoson A, Lippmaa E, Pines A (1988) Mol Phys 65:1013
208. Wu Y, Sun BQ, Pines A, Samoson A, Lippmaa E (1990) J Magn Reson 89:297
209. Goldman M (1970) Spin temperature and nuclear magnetic resonance in solids. Oxford University Press, London
210. Michel D, Engelke F (1994) Basic Principles of NMR Spectroscopy 34:27
211. Hing AW, Vega S, Schaefer J (1992) J Magn Reson 96:205
212. Freeman R (1988) Nuclear magnetic resonance. Wiley, New York
213. Engelhardt G, Lohse U, Samoson A, Maegi M, Tarmak M, Lippmaa E (1982) Zeolites 2:59
214. Camblor MA, Davis ME (1994) J Phys Chem 98:13151
215. Melchior MT, Vaughan DEW, Pictroski CF (1995) J Phys Chem 99:6128
216. Thomas JM, Klinowski J, Ramdas S, Hunter BK, Tennakoon DTB (1983) Chem Phys Lett 102:158
217. Klinowski J (1989) Coll Surf 36:133
218. Meier WM, Olson DH, Baerlocher C (1996) Atlas of zeolite structure types, 4th ed. Elsevier, Amsterdam
219. Lippmaa E, Maegi M, Samoson A, Engelhardt G, Grimmer AR (1980) J Am Chem Soc 102:4889
220. Fyfe CA, Kennedy GJ, Kokotailo GT, DeSchutter CT (1984) J Chem Soc Chem Commun 1093
221. Engelhardt G, Lohse U, Lippmaa E, Tarmak M, Maegi M (1981) Z Anorg Allg Chem 482:49
222. Jarman RH, Jacobsen AJ, Melchior MT (1984) J Phys Chem 88:5748
223. Hays GR, van Erp WV, Alma NCM. Couperus PA, Huis R, Wilson AE (1984) Zeolites 4:377
224. Engelhardt G, Fahlke B, Maegi M, Lippmaa E (1985) Z Phys Chem (Leipzig) 266:239
225. Radeglia R, Engelhardt G (1985) Chem Phys Lett 114:28
226. Smith JV, Blachwell CS (1983) Nature 303:223
227. Newsam JM (1987) J Phys Chem 91:1259
228. Engelhardt G, Luger S, Buhl JC, Felsche J (1989) Zeolites 9:182
229. Fyfe CA, Feng Y, Grondey H (1993) Microporous Mater 1:393
230. Engelhardt G, van Koningsveld H (1990) Zeolites 10:650
231. Engelhardt G, Rageglia R, Lohse U, Samoson A, Lippmaa E (1985) Z Chem 25:253
232. Fyfe CA, Kennedy GJ, DeSchutter CT, Kokotailo GT (1984) J Chem Soc Chem Commun 541
233. Fyfe CA, Strobl H, Kokotailo GT, Kennedy GJ (1988) J Am Chem Soc 110:3373
234. Tuel A, Ben Taarit Y (1992) J Chem Soc Chem Commun 1578
235. Chao KJ, Chern JY (1989) J Phys Chem 93:1401
236. Hunger M, Engelhardt G, Weitkamp J (1995) Microporous Mater 3:497
237. Fyfe CA, Gies H, Feng Y (1989) J Chem Soc Chem Commun 1240
238. Fyfe CA, Gies H, Feng Y (1989) J Am Chem Soc 111:7702
239. Fyfe CA, Feng Y, Gies H, Grondey H, Kokotailo GT (1990) J Am Chem Soc 112:3264
240. Fyfe CA, Grondey H, Feng Y, Kokotailo GT (1990) Chem Phys Lett 173:211
241. Fyfe CA, Grondey H, Feng Y, Kokotailo GT (1990) J Am Chem Soc 112:8812
242. Fyfe CA, Gies H, Feng Y, Grondey H (1990) Zeolites 10:278
243. Fyfe CA, Grondey H, Feng Y, Kokotailo GT, Ernst S, Weitkamp J (1992) Zeolites 12:50
244. Ernst RR, Bodenhausen G, Wokau A (1990) Principles of nuclear magnetic resonance in one and two dimensions. Clarendon Press, Oxford
245. Rocha J, Carr SW, Klinowski J (1991) Chem Phys Lett 187:401
246. Lippmaa E, Samoson A, Maegi M (1986) J Am Chem Soc 108:1730
247. Jacobsen HS, Norby P, Bildsøe, Jakobsen HJ (1989) Zeolites 9:491
248. Samoson A, Lippmaa E (1983) Phys Rev B28:6565
249. Samoson A, Lippmaa E (1983) Chem Phys Lett 100:205
250. Man PP (1986) J Magn Reson 76:78

251. Man PP, Klinowski J (1988) J Magn Reson 77:148
252. Klinowski J, Thomas JM, Fyfe CA, Gobbi GC (1982) Nature (London) 296:533
253. Maxwell IE, van Erp WA, Hays GR, Couperus T, Huis R, Clague DH (1982) J Chem Soc Chem Commun 523
254. Engelhardt G, Lohse U, Samoson A, Maegi M, Tarmak M, Lippmaa E (1982) Zeolites 2:59
255. Freude D, Froehlich T, Hunger M, Pfeifer H, Scheler G (1983) Chem Phys Lett 98:263
256. Rocha J, Klinowski J (1991) J Chem Soc Chem Commun 1121
257. Samoson A, Lippmaa E (1989) J Magn Reson 84:410
258. Engelhardt G, Koller H, Sieger P, Depmeier W, Samoson A (1992) Solid State Nucl Magn Reson 1:127
259. Wu Y, Chmelka BF, Pines A, Davis ME, Grobet PJ, Jacobs PA (1991) Nature (London) 346:550
260. Chmelka BF, Wu Y, Jelinek R, Davis ME, Pines A (1991) Stud Surf Sci Catal 69:435
261. Rocha J, Liu X, Klinowski J (1991) Chem Phys Lett 182:531
262. Grobet PJ, Samoson A, Gerts H, Martens JA, Jacobs PA (1991) J Phys Chem 95:9620
263. Jelinek R, Chmelka BF, Wu Y, Davis ME, Ulan JG, Gronsky R, Pines A (1992) Catal Lett 15:65
264. Peeters MPJ, de Haan JW, van de Ven LJM, van Hooff JHC (1992) J Chem Soc Chem Commun 1560
265. Peeters MPJ, van de Ven LJM, de Haan JW, van Hooff JHC (1993) J Phys Chem 97:8254
266. Peeters MPJ, de Haan JW, van de Ven LJM, van Hooff JHC (1993) J Phys Chem 97:5363
267. Jaenchen J, Peeters MPJ, de Haan JW, van de Ven LJM, van Hooff JHC (1993) J Phys Chem 97:12042
268. Grobet PJ, Samoson A, Gerts H, Martens JA, Jacobs PA (1991) J Phys Chem 95:9620
269. Ghose S, Tsang T (1973) Am Mineral 58:748
270. Engelhardt G, Veeman W (1993) J Chem Soc Chem Commun 622
271. McCusker LB, Baerlocher Ch (1991) Zeolites 11:308
272. Blackwell CS, Patton RL (1984) J Phys Chem 88:6135
273. Freude D, Ernst H, Hunger M, Pfeifer H (1988) Chem Phys Lett 143:477
274. Jahn E, Mueller D, Becker K (1990) Zeolites 10:151
275. Zibrowius B, Lohse U, Richter-Mendau J (1991) J Chem Soc Faraday Trans 87:1433
276. Zibrowius B, Loeffler E, Hunger M (1992) Zeolites 12:167
277. Zibrowius B, Lohse U (1992) Solid State Nucl Magn Reson 1:137
278. Borade RB, Clearfield A (1994) J Mol Catal 88:249
279. Mueller D, Jahn E, Ladwid G (1984) Chem Phys Lett 109:332
280. Barrie PJ, Klinowski J (1989) J Phys Chem 93:5972
281. van Eck ERH, Veeman WS (1993) J Am Chem Soc 115:1168
282. Stoecker M, Akporiaye D, Lillerud KP (1991) Appl Catal 69:L7
283. Davis ME, Montes C, Hathaway PE, Garces J, Crowder C (1989) Stud Surf Sci Catal 69:199
284. Davis ME, Montes C, Hathaway PE, Arhancet JP, Hasha DL, Garces JM (1989) J Am Chem Soc 111:3919
285. Grobet PJ, Martens JA, Balakrishnan I, Mertens M, Jacobs PA (1989) Appl Catal 56:L21
286. Maistriau L, Gabelica Z, Derouane EG, Vogt ETC, van Oene J (1991) Zeolites 11:583
287. Akporiaye D, Stoecker M (1992) Zeolites 12:351
288. Derouane EG, Maistriau L, Gabelica Z, Tuel AB, Nagy J, v. Ballmoos R (1989) Appl Catal 51:L 13
289. Perez JO, Chu PJ, Clearfield A (1991) J Phys Chem 95:9994
290. Kolodziejski W, He H, Klinowski J (1992) J Chem Phys Lett 191:117
291. Fyfe CA, Grondey H, Mueller KT, Wong-Moon KC, Markus T (1992) J Am Chem Soc 114:5876
292. Fyfe CA, Mueller KT, Grondey H, Wong-Moon KC (1993) J Phys Chem 97:13484
293. Rocha J, Kolodziejski W, He H, Klinowski J (1992) J Am Chem Soc 114:4884
294. He H, Kolodziejski W, Klinowski J (1992) Chem Phys Lett 200:83
295. Gabelica Z, Nagy JB, Bodart P, Debras G (1984) Chem Lett 1059
296. Scholle KFMGJ, Veeman WS (1985) Zeolites 5:118

297. Brunner E, Freude D, Hunger M, Pfeifer H, Reschetilowski W, Unger B (1988) Chem Phys Lett 148:226
298. Kessler H, Chezeau JM, Guth JL, Strub H, Coudurier G (1987) Zeolites 7:360
299. Axon SA, Klinowski J (1994) J Phys Chem 98:1929
300. Timken HKC, Turner GL, Gilson JP, Welsh LB, Oldfield E (1986) J Am Chem Soc 108:7231
301. Timken HKC, Janes N, Turner GL, Lambert SL, Welsh LB, Oldfield E (1988) J Magn Reson 76:106
302. Kyung H, Timken C, Janes N, Turner GL, Lambert SL, Welsh LB, Oldfield E (1986) J Am Chem Soc 108:7236
303. Freude D, Hunger M, Pfeifer H (1982) Chem Phys Lett 91:307
304. Freude D, Hunger M, Pfeifer H, Scheler G, Hoffmann J, Schmitz W (1984) Chem Phys Lett 105:427
305. Pfeifer H, Freude D, Hunger M (1985) Zeolites 5:274
306. Hunger M (1996) Solid State Nucl Magn Reson 6:1
307. Hunger M (1997) Catal Rev Sci Eng 39:345
308. Hunger M, Freude D, Pfeifer H, Prager D, Reschetilowski W (1989) Chem Phys Lett 163:221
309. Mastikhin VM, Mudrakovsky IL, Shmachkova VP, Kotsarenko NS (1987) Chem Phys Lett 139:93
310. Lohse U, Loeffler E, Hunger M, Stoeckner J, Patzelova V (1987) Zeolites 7:11
311. Freude D, Hunger M, Pfeifer H (1987) Z Phys Chem (NF) 152:429
312. Freude D, Hunger M, Pfeifer H, Schwieger W (1986) Chem Phys Lett 128:62
313. Hunger M (1991) Thesis (habilitation), University of Leipzig, Germany
314. Hunger M, Anderson MW, Ojo A, Pfeifer H (1993) Microporous Mater 1:17
315. Sanderson RT (1976) Chemical bonds and bond energy. Academic Press, New York
316. Mastikhin VM, Mudrakovsky IL, Nosov AV (1991) Progress in NMR Spectroscopy 23:259
317. Engelhardt G, Jerschkewitz HG, Lohse U, Sarv P, Samoson A, Lippmaa E (1987) Zeolites 7:289
318. Brunner E, Ernst H, Freude D, Hunger M, Krause CB, Prager D, Reschetilowski W, Schwieger W, Bergk KH (1989) Zeolites 9:282
319. Staudte B, Hunger M, Nimz M (1991) Zeolites 11:837
320. Schlunk R (1994) Thesis (diploma), University of Stuttgart, Germany
321. Hunger M, Horvath T, Engelhardt G, Karge HG (1995) Stud Surf Sci Catal 94:756
322. Beck LW, White JL, Haw JF (1994) J Am Chem Soc 116:9657
323. Brunner E, Beck K, Heeribout L, Karge HG (1995) Microporous Mater 3:395
324. Freude D (1995) Chem Phys Lett 235:69
325. Beck LW, Haw JF (1995) J Phys Chem 99:1076
326. Makarova MA, Ojo AF, Karim K, Hunger M, Dwyer J (1994) J Phys Chem 98:3619
327. Yesinowski JP, Eckert H (1987) J Am Chem Soc 109:6274
328. Costenoble ML, Mortier WJ, Uytterhoeven JB (1978) J Chem Soc Faraday Trans I, 174:466, 477
329. Brunner E, Karge HG, Pfeifer H (1992) Z Phys Chem 176:173
330. Hirschler AE (1963) J Catal 2:428
331. Plank CJ (1963) Proc 3rd Int Congr Catal, vol 1, North-Holland, Amsterdam, p 727
332. Ward JW (1969) J Catal 9:225
333. Ward JW, Hansford RC (1969) J Catal 13:364
334. Jirak Z, Vratislav S, Bosacek V (1980) J Phys Chem Sol 41:1089
335. Brunner E (1993) Microporous Mater 1:431
336. Klinowski J, Hamdan H, Corma A, Fornes V, Hunger M, Freude D (1989) Catal Lett 3:263
337. Klinowski J (1993) Anal Chim Acta 283:929
338. Reschetilowski W, Einicke WD, Jusek M, Schoellner R, Freude D, Hunger M, Klinowski J (1989) Appl Catal 56:L15
339. Sauer J (1994) Stud Surf Sci Catal 84:2039
340. Schroeder KP, Sauer J, Leslie M, Catlow CRA, Thomas JM (1992) Chem Phys Lett 188:320
341. Curtiss LA, Brand H, Nicholas JB, Iton LE (1991) Chem Phys Lett 184:215

342. Stevenson RL (1971) J Catal 21:113
343. Freude D, Klinowski J, Hamdan H (1988) Chem Phys Lett 149:355
344. Freude D, Klinowski J (1988) J Chem Soc Chem Commun 1411
345. Kenaston NP, Bell AT, Reimer JA (1994) J Phys Chem 98:894
346. Fenzke D, Hunger M, Pfeifer H (1991) J Magn Reson 95:477
347. Hunger M, Freude D, Fenzke D, Pfeifer H (1992) Chem Phys Lett 191:391
348. Bull LM, Cheetham AK, Hopkins PD, Powell BM (1993) J Chem Soc Chem Commun 1196
349. Koch M, Brunner E, Fenzke D, Pfeifer H, Staudte B (1994) Stud Surf Sci Catal 84:709
350. Koch M (1996) PhD thesis, University of Leipzig, Germany
351. Beran S (1983) J Phys Chem 87:55
352. Geerlings P, Tariel N, Botrel A, Lissilour R, Mortier WJ (1984) J Phys Chem 88:5752
353. Welsh LB, Lambert SL (1988) In: Flank WH, Whyte TE (eds) Perspectives in molecular sieve science. ACS Symp Ser 368, Washington DC, p 33
354. Welsh LB, Lambert SL (1989) In: Bradley SA, Gattuso MJ, Bertolacini RJ (eds) Characterization and catalyst development – an interactive approach. ACS Symp Ser 411, Washington DC, p 262
355. Beyer HK, Pál-Borbély G, Karge HG (1993) Microporous Mater 1:67
356. Challoner R, Harris RK (1991) Zeolites 11:265
357. Hunger M, Engelhardt G, Weitkamp J (1994) Stud Surf Sci Catal 84:725
358. Tijink GA, Janssen R, Veeman WS (1987) J Am Chem Soc 109:7301
359. Hunger M, Engelhardt G, Koller H, Weitkamp J (1993) Solid State Nucl Magn Reson 2:111
360. Koller H, Engelhardt G, Kentgens APM, Sauer J (1994) J Phys Chem 98:1544
361. Engelhardt G, Hunger M, Koller H, Weitkamp J (1994) Stud Surf Sci Catal 84:421
362. Feuerstein M, Hunger M, Engelhardt G (1996) Solid State Nucl Magn Reson 7:95
363. Jelinek R (1992) J Am Chem Soc 114:4907
364. Mortier WJ (1983) Compilation of extra-framework sites in zeolites. Butterworth, Guildford, p 26
365. Mueller D (1982) Ann Phys (Leipzig) 39:451
366. Hunger M, Sarv P, Samoson A (1997) Solid State Nucl Magn Reson 9:115
367. Massiot D, Bessada C, Coutures JP, Taulelle F (1990) J Magn Reson 90:231
368. Mortier WJ, van den Bossche E, Uytterhoeven JB (1984) Zeolites 4:41
369. Lievens JL, Mortier WJ, Chao KJ (1988) J Phys Chem Sol 53:1163
370. Verhulst HAM, Welters WJJ, Vorbeck G, van de Ven LJM, de Beer VHJ, van Santen RA, de Haan JW (1994) J Chem Soc Chem Commun 639
371. Kim JC, Li HX, Chen CY, Davis ME (1994) Microporous Mater 2:413
372. Hathaway PE, Davis ME (1989) J Catal 119:497
373. Chu PJ, Gerstein BC, Nunan J, Klier K (1987) J Phys Chem 91:3588
374. Schlenker JL, Pluth JJ, Smith JV (1979) Mater Res Bull 14:751
375. Torgeson DR, Barnes RG (1974) J Chem Phys 62:3968
376. Ahn MK, Iton LE (1989) J Phys Chem 93:4924
377. Tokuhiro T, Mattingly M, Iton LE, Ahn MK (1989) J Phys Chem 93:5584
378. Ahn MK, Iton LE (1991) J Phys Chem 95:4496
379. Subramanian V, Seff K (1980) J Phys Chem 84:2928
380. Koller H, Burger B, Schneider AM, Engelhardt G, Weitkamp J (1995) Microporous Mater 5:219
381. Fraissard J, Ito T (1988) Zeolites 8:350
382. Barrie PJ, Klinowski J (1992) Progress in NMR Spectroscopy 24:91
383. Demarquay J, Fraissard J (1987) Chem Phys Lett 136:314
384. Johnson DW, Griffiths L (1987) Zeolites 7:484
385. Derouane EG, Nagy JB (1987) Chem Phys Lett 137:341
386. Ito T, Fraissard J (1982) J Chem Phys 76:5225
387. Davis ME, Saldarriaga C, Montes C, Hanson BE (1988) J Phys Chem 92:2557
388. Chen QJ, Fraissard J, Cauffriez, Guth JL (1991) Zeolites 11:534
389. Annen MJ, Davis ME, Hanson BE (1990) Catal Lett 6:331
390. Cheung TTP, Fu CM, Wharry S (1988) J Phys Chem 92:5170

391. Gedeon A, Fraissard J (1994) Chem Phys Lett 219:440
392. Kim JG, Kompany T, Ryoo R, Ito T, Fraissard J (1994) Zeolites 14:427
393. Chen QJ, Ito T, Fraissard J (1991) Zeolites 11:239
394. Shy DS, Chen SH, Lievens J, Liu SB, Chao KJ (1991) J Chem Soc Faraday Trans 87:2855
395. Scharpf EW, Crecely RW, Gates BC, Dybowski C (1986) J Phys Chem 90:9
396. Shoemaker R, Apple T (1987) J Phys Chem 91:4024
397. Bansal N, Dybowski C (1988) J Phys Chem 92:2333
398. Ryoo R, Cho SJ, Pak C, Kim JG, Ihm SK, Lee JY (1992) J Am Chem Soc 114:76
399. Ihee H, Becue T, Ryoo R, Potvin C, Manoli JM, Djega-Mariadassou G (1994) Stud Surf Sci Catal 84:765
400. Gedeon A, Bonardet JL, Ito T, Fraissard J (1989) J Phys Chem 93:2563
401. Bonardet JL, Gedeon A, Fraissard J (1995) Stud Surf Sci Catal 94:139
402. Gedeon A, Bonardet JL, Lepetit C, Fraissard J (1995) Solid State Nucl Magn Reson 5:201
403. Grosse R, Burmeister R, Boddenberg B, Gedeon A, Fraissard J (1991) J Phys Chem 95:2443
404. Gedeon A, Burmeister R, Grosse R, Boddenberg B, Fraissard J (1991) Chem Phys Lett 79:191
405. Grosse R, Gedeon A, Watermann J, Fraissard J, Boddenberg B (1992) Zeolites 12:909
406. Gedeon A, Bonardet JL, Fraissard J (1993) J Phys Chem 97:4254
407. Hartmann M, Boddenberg B (1994) Microporous Mater 2:127
408. Springuel-Huet MA, Ito T, Fraissard J (1984) Stud Surf Sci Catal 18:13
409. Tway C, Apple T (1992) J Catal 133:42
410. Chen QJ, Fraissard J (1992) J Phys Chem 96:1814
411. Smith ML, Corbin DR, Abrams L, Dybowski C (1993) J Phys Chem 97:7793
412. Smith ML, Corbin DR, Dybowski C (1993) J Phys Chem 97:9045
413. Bonardet JL, Barrage MC, Fraissard J (1993) Proc 9th Int Zeolite Conf, Butterworth-Heinemann, Stoneham, vol 2, p 475
414. Bonardet JL, Barrage MC, Fraissard J (1993) Am Chem Soc – Preprints 38:628
415. Barrage MC, Bonardet JL, Fraissard J (1990) Catal Lett 5:143
416. Hunger M, Freude D, Froehlich T, Pfeifer H, Schwieger W (1987) Zeolites 7:108
417. Ikenoue T, Yoshida N (1988) J Phys Chem 92:4883
418. Maciel GE, Haw JF, Chuang I-S, Hawkins BL, Early TA, McKay DR, Petrakis L (1983) J Am Chem Soc 105:5529
419. Ripmeester JA (1983) J Am Chem Soc 105:2925
420. Lunsford JH, Rothwell WP, Shen W (1985) J Am Chem Soc 107:1540
421. Lunsford JH, Tutunjian PN, Chu P-J, Yeh EB, Zalewski DJ (1989) J Phys Chem 93:2590
422. Baltusis L, Frye JS, Maciel GE (1987) J Am Chem Soc 109:40
423. Biaglow AI, Gorte RJ, Kokotailo GT, White D (1994) J Catal 148:779
424. White JL, Beck LW, Haw JF (1992) J Am Chem Soc 114:6182
425. Haw JF, Hall MB, Alvarado-Swaisgood AE, Munson EJ, Lin Z, Beck LW, Howard T (1994) J Am Chem Soc 116:7308
426. Brunner E (1997) Catal Today 38:361
427. Sauer J, Horn H, Haeser M, Ahlrichs R (1990) Chem Phys Lett 173:26
428. Sauer J (1985) J Acta Chim Phys (Szeged) 31:19
429. Sauer J, Hobza P (1984) Theor Chim Acta (Berlin) 65:279
430. Doremieux-Morin C (1976) J Magn Reson 21:419
431. Doremieux-Morin C (1979) J Magn Reson 33:505
432. Batamack P, Doremieux-Morin C, Vincent R, Fraissard J (1991) Chem Phys Lett 180:545
433. Batamack P, Doremieux-Morin C, Vincent R, Fraissard J (1993) J Phys Chem 97:9779
434. Batamack P, Doremieux-Morin C, Vincent R, Fraissard J (1991) Microporous Mater 2:515
435. Batamack P, Doremieux-Morin C, Fraissard J, Freude D (1991) J Phys Chem 95:3790
436. Rietveld HM (1967) Acta Cryst 22:151
437. Young RA, Mackie PE, von Dreele RB (1977) J Appl Cryst 10:262
438. Treacy MMJ, Higgins JB, von Ballmoos R (1996) Collection of simulated XRD powder patterns for zeolites, 3rd ed. Elsevier, Amsterdam

439. Powder Diffraction File, JCPDS International Centre for Diffraction Data, Swarthmore
440. Henry NFM, Londsdale K (ed), (1952) International tables for X-ray crystallography, vol 1, Symmetry Group, The Kynoch Press, Birmingham, p 31
441. Shmueli U (ed), (1993) International tables for X-ray crystallography, vol B, Reciprocal space. Kluwer Academic Publishers, Dordrecht, p 128
442. Wilson AJC (ed), (1992) International tables for X-ray crystallography, vol C, Mathemathical, physical and chemical tables. Kluwer Academic Publishers, Dordrecht, p 476
443. Baur WH (1964) Am Miner 49:697
444. Gallezot P, Beaumont R, Barthomeuf D (1974) J Phys Chem 78:1550
445. Shmueli U (ed), (1993) International tables for X-ray crystallography, vol B, Reciprocal space. Kluwer Academic Publishers, Dordrecht, p 152
446. Will G, Parrish W, Huang TC (1983) J Appl Cryst 16:611
447. Highcock RM, Smith GW, Wood D (1985) Acta Cryst C 41:1391
448. Olson DH, Kokotailo GT, Lawton SL, Meier WM (1981) J Phys Chem 85:2238
449. Young RA (1995) The Rietveld method. Oxford University Press, Oxford
450. Bish DL, Post JE (ed), (1989) The Modern X-ray powder diffraction. Reviews in Mineralogy, vol 20, Mineralogical Society of America, Washington
451. McCusker LB (1994) Stud Surf Sci Catal 84:341
452. Baerlocher C, McCusker LB (1994) Stud Surf Sci Catal 85:391
453. Proffen T, Hradil K (1993) Z Kristallogr Suppl 7:155
454. Baerlocher C (1986) Zeolites 6:325
455. Mortier WJ, van den Bossche E, Uytterhoeven JB (1984) Zeolites 4:41
456. Breck DW, Flanigen EM (1968) Zeolite molecular sieves. Society of Chemical Industry, London, p 47
457. Scherzer J (1978) J Catal 54:285
458. Beyer HK, Belenykaja IM (1980) Stud Surf Sci Catal 5:222
459. Scherzer J (1989) Catal Rev Sci Eng 31:215
460. Sohn JR, DeCanio SJ, Lunsford JH, O'Connell DJ (1986) Zeolites 6:225
461. Fichtner-Schmittler H, Lohse U, Engelhardt G, Patzelová V (1984) Cryst Res Technol 19:K1
462. Beyer HK, Belenykaja IM, Hange F, Tielen M, Grobet J, Jacobs PA (1985) J Chem Soc Faraday Trans I 81:2889
463. Rubio JA, Soria J, Cano FH (1980) J Coll Interf Sci 73:312
464. Fritz PO, Lunsford JH, Fu C-M (1988) Zeolites 8:205
465. Gallezot P, Imelik B (1970) Compt Rend Acad Sc Paris, Ser B 271:912
466. McDaniel CV, Maher PK (1976) In: Rabo JA (ed) Zeolite chemistry and catalysis, ACS, Washington, DC, p 285
467. Jorik V (1993) Zeolites 13:187
468. Olsson RW, Rollmann LD (1977) Inorg Chem 16:651
469. Senderov EE, Zubkov AM, Lipkind BA, Dadashev FB, Voroveva IG, Lebedkova AV (1981) Dokl Akad Nauk SSSR 256:884
470. Mishin IV, Beyer HK, Karge HG (1993) Kinet Katal 34:155
471. Bibby DM, Aldridge LP, Milestone NB (1981) J Catal 72:373
472. Mishin IV, Beyer HK (1993) Kinet Katal 34:307
473. Habashi K, Fukushima T, Igawa K (1986) Zeolites 6:30
474. Scherrer P (1918) Göttinger Nachrichten 2:98
475. Warren BE, Biscoe J (1938) J Am Ceram Soc 21:49
476. Klug HP, Alexander LE (1978) X-ray diffraction procedures for polycrystalline and amorphous materials. John Wiley and Sons, New York

CHAPTER 5

Shape-Selective Catalysis in Zeolites

Jens Weitkamp, Stefan Ernst, and Lothar Puppe

5.1
Scope

Among the primary tasks of catalysis is the control of the selectivity in chemical reactions. In heterogeneously catalyzed reactions in zeolites and zeolite-related microporous solids, this can, *inter alia*, be achieved by exploiting the phenomenon of shape-selective catalysis. In a very simplified manner, shape-selective catalysis can be described as the combination of catalysis with the molecular sieve effect. Shape selectivity effects can occur, if the sizes and shapes of reactants, of products, of transition states or of reaction intermediates are similar to the dimensions of the pores and cavities of the zeolite.

The first examples of shape-selective catalysis in zeolites were reported more than 35 years ago by Weisz and Frilette from the Mobil laboratories [1,2]. Since those days, research in the field of shape-selective catalysis has expanded both at universities and in industrial laboratories in an impressive manner. This has resulted in numerous examples of shape selectivity effects in the course of catalytic conversions over zeolitic catalysts, which are nowadays no longer restricted to acid-catalyzed reactions as described in the early days. Rather, there are also examples for shape-selective catalysis on metals, on redox active sites, on basic guests in zeolites, and even for shape-selective photochemistry in a zeolitic environment. Moreover, from the intensive research done in the field of shape-selective catalysis during the last decades, several large-scale commercial applications in the refining and the petrochemical industries have emerged, demonstrating the usefulness and importance of the phenomenon of shape-selective catalysis.

In the past, numerous review papers on shape-selective catalysis, its principles and its commercial applications have been published [e.g., 3–19]. In view of the vast amount of work that has been done in this field in recent years, it is obvious that a review on shape-selective catalysis cannot be exhaustive. Rather, the discussion has to be restricted to selected topics. In the present contribution, emphasis will be placed on a thorough discussion of the classical and of more recent principles of shape-selective catalysis and on how these principles can be used to rationalize the observed catalytic behavior in selected examples. In addition, it is the aim of the present contribution to discuss the available methods for tailoring the shape-selective properties of zeolite catalysts and for their characterization by catalytic methods. Finally, an attempt will be undertaken to review more recent research directions and challenges in the field of shape-selective catalysis.

5.2 Introduction

Essentially, zeolites have two properties which make them particularly suitable as starting materials for the preparation of catalysts, viz. (i) they are cation exchangers, hence it is possible to introduce a large variety of cations with different catalytic properties into their intracrystalline pore system, which in turn offers the opportunity to create different catalytic properties, e.g., in acid- or metal-catalyzed reactions and (ii) zeolites are crystalline porous materials with pore dimensions in the same order as the dimensions of simple molecules (e.g., from ca. 0.3 nm to ca. 1.5 nm). Hence they possess molecular sieving properties. In the case of shape-selective catalysis in zeolites, the combination of both properties, viz. pores of molecular dimensions and catalytically active sites inside these pores, is exploited to control the selectivity of catalytically conducted reactions. Obviously, it is not only the size of the pores, but also the dimensions of the reacting and diffusing molecules which play a decisive role. Therefore, a "shape-selective zeolite" per se cannot exist. Rather, the occurrence or not of shape-selectivity effects always depends on the combination of the specific zeolite pore sizes and the molecular dimensions of the reacting and diffusing species. Hence the selection of a proper zeolite for the shape-selective conduction of a given reaction must be based on both the sizes of the intracrystalline cavities and the dimensions of the reacting molecules. In the following, the methods for describing molecular dimensions and the pore sizes of porous solids will be briefly discussed.

5.2.1 Molecular Dimensions

In principle, the sizes of molecules can be determined by calculating their equilibrium diameters from the known molecular shape, bond distances, bond angles and van der Waals radii. For a comparison with zeolite pore dimensions, however, the use of the kinetic diameter seems to be much more popular [20]. The kinetic diameter for spherical, non-polar molecules can be obtained from the Lennard-Jones potential which describes the potential energy of interaction between two molecules in dependence of their distance from each other, viz.

$$\Phi(r) = 4\,\varepsilon\,[(\sigma/r)^6 - (\sigma/r)^{12}]$$

In this equation, σ and ε denote the characteristic diameter ("collision diameter") and the characteristic energy of interaction (the maximum energy of attraction between two molecules), respectively. Numerical values for σ and ε can be derived from the second virial coefficient, from viscosity data or from the critical parameters of the pure substances [21, 22]. A schematic graphical representation of the Lennard-Jones potential is shown in Fig. 5.1. At large

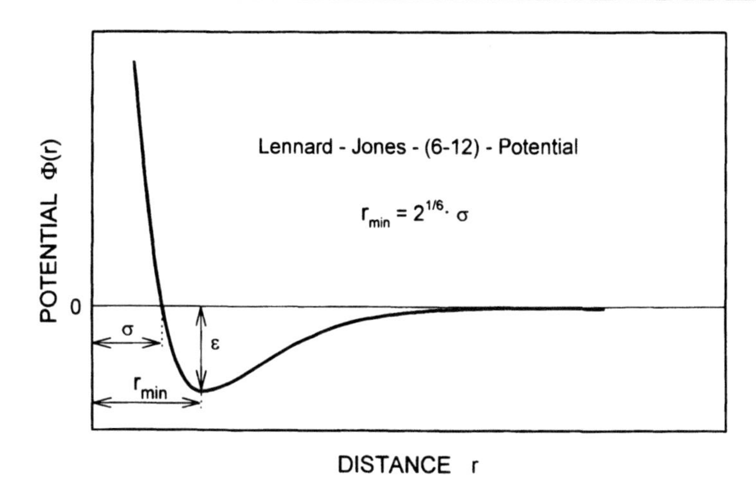

Fig. 5.1. Schematic representation of the Lennard-Jones-(6-12)-potential (σ and ε denote the characteristic diameter and the characteristic energy of interaction, respectively; r_{min} is the minimum distance between two approaching molecules, where the maximum energy of interaction occurs)

separations, the attractive component [proportional to $(1/r)^6$] is dominant and describes the dipole-dipole interaction. For small distances, the repulsive forces [proportional to $(1/r)^{12}$] predominate. The kinetic or collision diameter σ is the smallest distance between two molecules. The maximum energy of attraction (ε) occurs at the minimum distance r_{min} with $r_{min} = 2^{1/6} \times \sigma$ (cf. Fig. 5.1). While the use of the kinetic diameter is often appropriate for truly spherical molecules, it is not very well suited for more complex molecules, e.g., n-alkanes. In these cases, the use of the minimum cross-sectional diameter is recommended [20].

5.2.2 Porous Solids: Crystallographic and Effective Pore Diameter

Zeolites and zeolite-like materials are unique among porous solids due to their crystalline structure and their strictly regular pore and cage dimensions. Moreover, the size of the pores and intracrystalline cavities is in the same order of magnitude as the one of common organic molecules. Figure 5.2 visualizes these specific properties of zeolites which make them in many cases superior over other (amorphous) solids. It can be seen from Fig. 5.2 that zeolites, due to their crystalline structure, possess discrete pore sizes, in contrast to amorphous porous solids like silica gel and activated charcoal which have relatively broad pore size distributions and pore widths approximately one to two orders of magnitude larger than those of zeolites. Due to this fact, shape-selective catalysis cannot be expected to occur in these latter types of non-crystalline porous solids. It can also be seen from Fig. 5.2 that there would be a large gap in the available pore sizes, if just zeolites and the classical porous materials were

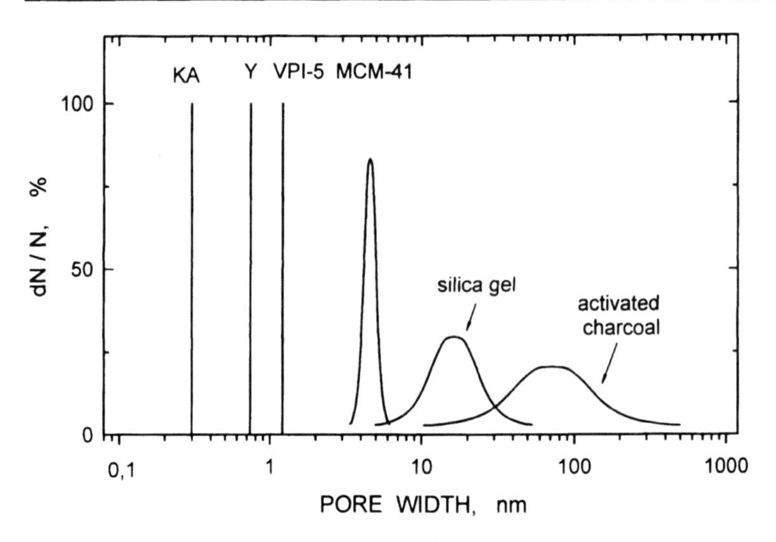

Fig. 5.2. Pore width distributions for selected porous solids

known. This gap has fortunately been closed in recent years by the discovery of ordered mesoporous solids like MCM-41, MCM-48 and the like (e.g., [23, 24]). These materials, too, possess very narrow pore size distributions.

The pore dimensions of zeolites are determined by the number of (SiO_4 or AlO_4) tetrahedra in the ring which circumscribe the pore. If the tetrahedra are arranged in an ideally circular and planar manner, the maximum possible free diameter results. However, there is only a relatively limited number of zeolite structures possessing this ideal arrangement of T-atoms (e.g., zeolites A, X and Y). Rather, in most cases, the shapes of the pore openings deviate from the ideal one, e.g., they are oval or they even possess peculiar shapes like the tear-drop pore opening in the MTT (i.e., ZSM-23) framework topology [25], or the tetrahedra are puckered, i.e., not all of them are arranged in the same plane.

The channel dimensions of zeolites and zeolite-like microporous solids are usually calculated from their known crystallographic structures (atomic coordinates) based on an oxygen radius of 0.135 nm [25]. The so-called crystallographic pore dimensions obtained in this manner are of limited value only if a decision has to be made whether or not a molecule of a given size could enter the pores. In some cases they could even lead to misleading results. One well known example is the adsorption of benzene or cyclohexane in the pores of zeolite ZSM-5: its crystallographic pore sizes are 0.53 × 0.56 nm for the almost circular pores and 0.51 × 0.55 nm for the pores with a somewhat elliptical shape [25]. From a comparison of these data with the kinetic diameters of benzene (0.585 nm) and cyclohexane (0.60 nm) obtained from the Lennard-Jones relationship [20], one is tempted to conclude that neither of these two molecules will be able to enter the pores of zeolite ZSM-5. However, it is an experimental fact that these molecules are readily adsorbed in zeolite ZSM-5, even at room temperature [e.g., 26]. This apparent contradiction is usually explained by the fact that neither the crystal structure nor the molecules are really rigid. Rather, the pore sizes enlarge with increasing temperature, and also molecular vibration of molecules allows them to wriggle through somewhat narrower pores than expected [8]. Moreover, even the cleavage of bonds followed by their reconstruction after the diffusing species have passed the pore mouth have been invoked as a possible mechanism to explain the diffusion of larger molecules through narrow pores [27].

5.2.3 Molecular Sieving

Molecular sieving is the selective adsorption of molecules, whose dimensions are below a certain critical size, into the intracrystalline void system of a molecular sieve. An industrially relevant example is illustrated in Fig. 5.3: A gaseous mixture of n-hexane and 3-methylpentane was continuously passed over a fixed bed of calcium-exchanged zeolite A (or zeolite 5A) with an effective pore diameter of ca. 0.5 nm. This is large enough to allow for the diffusion of the n-alkane molecules through the eight-membered ring windows, but it is too small for the uptake of the branched alkane. Hence, 3-methylpentane breaks through at the adsorber outlet directly after the onset of the experiment, while the smaller n-hexane is adsorbed by the zeolite until its adsorption capacity is exhausted. Molecular sieving of normal and branched alkanes is commercially exploited for different purposes in a variety of adsorption processes, e.g., to separate the (low-octane) unbranched C_5/C_6 alkanes from their (high-octane) branched isomers [28]. Other examples of molecular sieving are the separation of *para*-xylene/*meta*-xylene mixtures with zeolite HZSM-5 as adsorbent [29, 30],

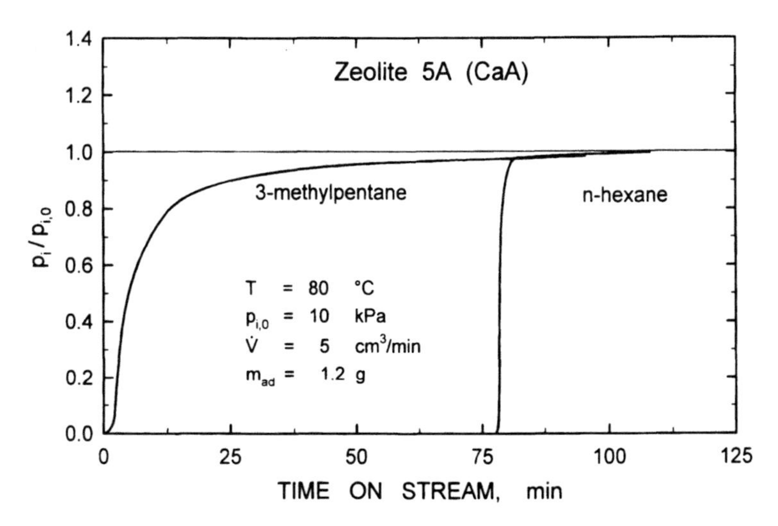

Fig. 5.3. Breakthrough curves for the adsorption of a n-hexane/3-methylpentane mixture over zeolite CaA (5A) in a fixed-bed flow-type reactor

and the separation of 1- and 2-methylnaphthalene, again with an HZSM-5-type zeolite as adsorbent [31].

It has been demonstrated several times that it is not only the kinetic diameter of a molecule that determines whether it is adsorbed in a given zeolite or not, but also the shape of the molecule in relation to the shape of the pore openings. This has been pointed out by Wu et al. [32], who studied the adsorption of hydrocarbons over selected high-silica zeolites. They were able to show that although cyclohexane and 2,2-dimethylbutane have almost the same kinetic diameter, the former is adsorbed much faster (higher diffusion coefficients) than the latter in the medium pore zeolites ZSM-5, ZSM-11 and ZSM-48. This result was interpreted in terms of the differences in the shapes of the adsorbing molecules: while the critical cross-section of the 2,2-dimethylbutane molecule is almost circular, the one of cyclohexane appears to be elliptical. The authors further argued that, under adsorption conditions, the pore openings of the investigated zeolites have a more or less elliptical shape, hence the faster uptake of cyclohexane. Similar conclusions for the shape of the pore cross-sections in zeolites ZSM-5 and ZSM-23 under catalytic conditions were drawn by Bendoraitis et al. from catalytic data obtained during dewaxing of distillates [33]. They also concluded that pore sizes determined under catalytic conditions are always larger than computed crystallographic pore sizes, as also observed in adsorption measurements. As a further example for the influence of the shape of the adsorbed molecules, data reported by den Exter et al. should be mentioned [34]. These authors studied the adsorption of several small molecules on aluminum-free decadodecasil 3R, which has larger intracrystalline cavities, accessible through eight-membered ring windows only. In essence, they found that this molecular sieve readily adsorbs gases like CO_2, N_2O and NH_3. On the other hand, alkanes and alkenes with three or four carbon atoms show only a retarded adsorption at 296 K, while 1,3-butadiene is easily adsorbed under these conditions. This peculiar behavior has been attributed to the more elliptical cross-section of the diolefin due to the presence of sp^2-hybridized carbon atoms. 1,3-Butadiene obviously has a better "fit" to the elliptical windows in the DDR structure than the saturated hydrocarbons with their circular cross-section.

The effect of molecular sieving is frequently exploited for the characterization of the pore width of microporous materials. Indeed, adsorption of a set of components with different molecular dimensions (pore gauging) is a widely accepted procedure. In numerous patents dealing with the synthesis of novel materials, characterization by adsorption is described. For small pore and medium pore zeolites (crystallographic pore widths between 0.4 and 0.6 nm), adsorption of vapors of water (kinetic diameter: 0.26 nm), n-hexane (0.43 nm), benzene (0.585 nm) and cyclohexane (0.60 nm) is often employed. For characterizing the range from medium to puckered 12-membered ring materials (crystallographic pore widths between 0.5 and 0.7 nm), a suitable choice can often be made from normal and branched alkanes in the carbon number range from C_5 to C_9 and/or benzene and its alkyl derivatives. This has been demonstrated by Wu et al. [32] for an in-depth characterization of the pore widths of zeolites ZSM-5, ZSM-11, ZSM-12, ZSM-23 and ZSM-48. For large pore (12-membered ring) materials, tertiary alkylamines with alkyl groups of varying bulkiness or with perfluorinated alkyl groups have been used successfully [35]. Super-large pore materials can be readily distinguished from large pore (12-membered ring) zeolites by adsorption experiments with 1,3,5-triisopropylbenzene. Indeed, the recognition of VPI-5 as a super-large pore molecular sieve by Davis et al. [36] was, to a large extent, based on such adsorption experiments. The triisopropylbenzene molecules are too bulky (ca. 0.85 nm) to enter the pore system of faujasite (0.75 nm) with reasonable uptake rates, but they do have easy access to the channels of VPI-5 (ca. 1.2 nm). More recently, Lobo et al. used the adsorption of 1,3,5-triisopropylbenzene to identify the novel high-silica zeolite UTD-1 as a very large pore material [37]. In agreement with its crystallographic structure (linear, non-interconnected channels with pore openings of ca. 0.75 × 1.0 nm, [38]) the new 14-membered ring material adsorbs ca. 0.11 cm^3/g of the probe molecule.

5.3 Catalysis and Selectivity

Various older definitions of the term "catalysis", though they continue to be used in part of the scientific literature, are nowadays considered to be less appropriate. A scientifically sound definition of a catalyst describes it as a substance which transforms reactants into products via an uninterrupted and repeated cycle of elementary steps in which the catalyst participates while being regenerated in its original form at the end of each cycle during its life [39, 40]. In the following, the incentives for applying a catalyst and the tools to describe selectivity are discussed before the term shape selectivity is considered.

5.3.1 Incentives for Applying a Catalyst

The incentives for using a catalyst for a desired chemical reaction can be manifold. There are, however, some frequently encountered cases which will be briefly discussed in the following.

Catalysis is often very useful if an exothermic reaction is to be performed. In this case, high temperature would be advantageous for the reaction kinetics, but in many instances high temperature shifts the thermodynamic equilibrium in the undesired direction. Lowering the reaction temperature, on the other hand, would improve thermodynamics in accordance with the principle of Le Châtelier, but at the same time this would reduce the rate of the chemical reaction. In this situation, using a catalyst which efficiently accelerates the reaction helps to achieve both, viz. favorable thermodynamics and reasonably high rates at relatively low temperatures. One well known example for such a reaction is the oxidation of SO_2 to SO_3 in the course of sulfuric acid production, which is industrially performed on vanadium pentoxide catalysts.

If, on the other hand, an endothermic reaction is to be performed, a high reaction temperature would be advantageous for both the position of the thermodynamic equilibrium and reaction rates. However, under certain circumstances engineering aspects (such as temperature stability of the materials used for construction of the reactor) lead to the necessity to reduce the reaction temperature. Under these conditions, using a catalyst helps to conduct the desired reaction at lower temperature (where the position of thermodynamic equilibrium is worse but still acceptable) with reasonable rates.

Probably the most frequently encountered motivation for using a catalyst is to control the selectivity of a chemical reaction: Among the variety of possible reaction pathways the chosen reactant(s) may undergo, a catalyst may selectively accelerate the one which is particularly desired. A prominent example of selectivity control using a catalyst is the conversion of synthesis gas ($CO + H_2$): besides the reaction conditions (temperature, pressure, composition of the feed) the nature of the catalyst fundamentally determines which products are formed preferentially. For example, methane is obtained on catalysts containing nickel, methanol is formed over copper/zinc catalysts, and higher hydrocarbons are favored on Fischer-Tropsch-type catalysts containing, e.g., cobalt.

5.3.2
Intrinsic, Grain and Reactor Selectivity

For the vast majority of industrially relevant chemical reactions, high selectivity to the desired products (beside high activity of the catalyst) is the key issue. The selectivity S_B for the formation of a product B out of the reactant A is usually defined as

$$S_B = \dot{n}_{B,out}/(\dot{n}_{A,in} - \dot{n}_{A,out})$$

with $\dot{n}$ representing the molar fluxes of the components at the reactor inlet and outlet, respectively. As pointed out by Riekert [41], however, the situation is more complex for a reactor containing pellets of a solid porous catalyst (e.g., a fixed bed reactor, cf. Fig. 5.4). Different kinds of selectivity pertaining to different control volumes have to be discerned here: The intrinsic selectivity describes the selectivity of the active site, e.g., of the Brønsted acid OH-groups in the intracrystalline voids of a zeolite, or of small platinum clusters located in the pores (Fig. 5.4A). The selectivity of the porous grain, consisting, for example, of zeolite crystallites embedded in the matrix of a catalyst pellet (cf. Fig. 5.4B), can be rather different from the intrinsic selectivity: diffusion through the zeolite pores as well as mass transfer through the pores of the catalyst pellet modify the macroscopically observable selectivities, viz. a finite residence time of the reacting substrate in the catalyst due to mass transfer resistances will have a similar influence on the product distribution as an increase of the residence time in the reactor. Finally, the reactor selectivity (Fig. 5.4C) depends not only on the intrinsic selectivity of the catalytically active site and the mass transfer through the zeolite crystallites and/or the catalyst grain, but also on the reactor configuration, i.e., the residence time of the fluid in the reactor, the conditions of heat exchange with the surroundings etc. Moreover, Riekert showed [41] in which way conclusions on the intrinsic selectivity can be drawn from the measurable reactor selectivities using mathematical treatments. This, however,

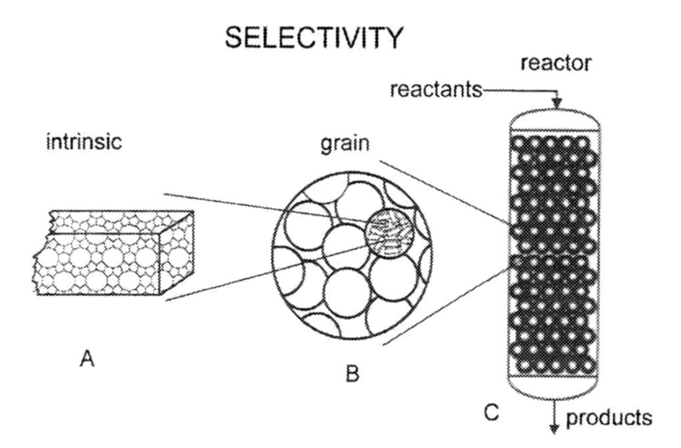

Fig. 5.4. Control volumes for intrinsic, grain and reactor selectivity [41]

requires information on the mass transfer coefficients in the catalyst pellet under reaction conditions, i.e., data which is usually difficult to acquire in the case of zeolite-based catalysts.

5.3.3 Shape-Selective Catalysis

The phenomenon of shape-selective catalysis has been explained in the simplest manner by the combination of the molecular sieve effect and a catalytic reaction. In many instances, however, this is an oversimplified and unsatisfactory definition. It has also been stated that shape-selective catalysis may occur in microporous solids, if their pore width and the dimensions of the reactant molecules, transition states, intermediates and/or product molecules which take part in the reaction, are similar. We propose the following definition of shape-selective catalysis: it encompasses all effects in which the selectivity of a heterogeneously catalyzed reaction depends, in an unambiguous manner, on the pore width or pore architecture of the microporous or mesoporous catalyst.

Today, zeolites and zeolite-like microporous solids are almost exclusively used as catalysts for shape-selective conversions. In principle, however, any other porous solid having pores of molecular dimensions can be the basis for a shape-selective catalyst. Examples are artificially expanded ("pillared") clays (e.g., [42]), modified carbon molecular sieves (e.g., [43]), or certain crystalline cyanometallates (e.g., [44]). Whatever the chemical nature of the catalyst, it is essential for shape-selective catalysis to occur that the active sites are completely (or at least to a very large extent) located inside the pores of the catalyst, rather than at the outer surface of the catalyst particles. In the following, we will discuss the possible contribution of the external surface of zeolites to their catalytic activity.

5.4
Internal vs. External Surface of Zeolites

In order to maximize shape selectivity effects in the intracrystalline voids of microporous catalysts, it is often desirable (or even indispensable) to minimize the contribution of the (non-shape-selective) external surface of the zeolite crystallites to the overall catalytic activity. This can, in principle, be achieved by minimizing the number of catalytically active sites on the external surface of the crystallites. Assuming an even distribution of the active sites over the volume of the zeolite, this means that the specific external surface area has to be as small as possible. Besides simple geometric calculations based on the size and shape of the zeolite crystallites, some experimental techniques have been developed in an attempt to quantify the role of the external surface. In the following, these methods will be briefly described and discussed.

5.4.1
Effect of the Crystallite Size

The effect of the crystallite size on the specific external surface area $A_{ext.}$ (expressed in m^2/g) can be principally derived from the geometry (size and shape) and the known density of the crystallites. Thus, for the case of zeolite ZSM-5 (density: 1.79 g/cm^3) and assuming spherical particles with a diameter of 1 μm, an external surface area of 3.35 m^2/g is calculated. Compared to the usually reported values for the total specific surface area of zeolite ZSM-5 between 300 and 400 m^2/g (e.g., [45, 46]), the external surface area is indeed low, viz. ca. 1% of the total surface area. If, however, the crystallite diameter is further reduced to 0.1 μm, the external surface area amounts to ca. 10% of the total specific surface area. Moreover, it should be stressed that it has been assumed for the above calculation that the surface is completely plane and that the crystallites are regularly shaped and of equal size. In practice, however, there is always a particle size distribution and the surface of the crystallites possesses a certain roughness. For very small crystallites, it may not even be possible to detect such a surface roughness due to the limited resolution of the scanning electron microscope. As a consequence, the true contribution of the external surface to the total surface of zeolite crystallites is almost always underestimated.

Gilson and Derouane [47] derived the following correlation which estimates the relationship between the percentage of external tetrahedral sites and the mean crystallite size of pentasil zeolites (viz. ZSM-5- and ZSM-11-type materials) assuming that no concentration gradient for aluminum occurs in the lattice:

$$n_{T,ext.}/n_{T,total} = 181 \cdot 1/(2R),$$

with $n_{T,ext.}/n_{T,total}$ in % and the mean crystallite radius R in nm. This correlation is visualized in Fig. 5.5 together with a very similar correlation for zeolites of the faujasite structure, which has been derived by Farcasiu and Degnan [48] based on the known crystallite density and again under the assumption of a homo-

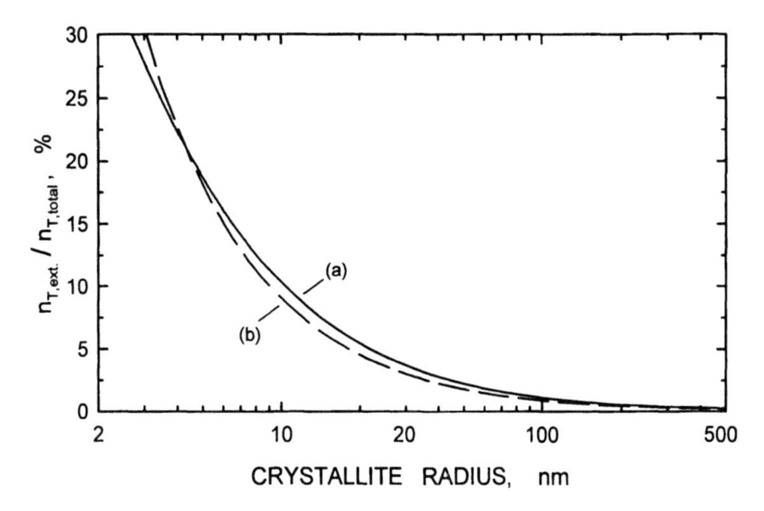

Fig. 5.5. Relative amount of tetrahedral framework atoms located at the external surface of zeolites with the FAU (curve "**a**") and the MFI (curve "**b**") framework topology in dependence of the crystallite size (numerical values calculated according to the equations given in the text)

geneous distribution of aluminum atoms among the tetrahedral sites. This correlation (with $n_{T,ext.}/n_{T,total}$ in % and R in nm) is as follows:

$$n_{T,ext.}/n_{T,total} = 1/(1 + 8.68 \cdot 10^{-1}\,R)$$

It can be seen from Fig. 5.5 that the calculated fraction of T-atoms on the external surface is very similar for both equations. Even though these equations are based on simple assumptions, they safely indicate that the external surface of zeolites can contribute significantly to catalysis under unfavorable circumstances.

5.4.2 Experimental Techniques

There are at least three different experimental approaches for probing the external surface and/or catalytically active sites on the external surface of zeolites. They are based on adsorption [45, 49–54], ion exchange [46, 55] and catalytic reactions of suitable probe molecules [47, 56–60].

One type of adsorption measurements for determining the external surface of zeolite crystallites is based on the so-called t-plot method [61–63] or its variant, the α_S- method [62, 64]. Both methods not only rely on the adsorption of nitrogen at 77 K on the porous materials under investigation, but also on non-porous reference materials of known surface area. It has been reported that the t-plot technique is not very reliable for samples with a small external surface area [51].

Another technique, which is based on adsorption measurements, relies on plugging the zeolite micropores with a strongly held adsorbate before determining the remaining surface area [45, 49–51]. The method involves filling the micropores with large molecules (e.g., benzene [45] or n-nonane [50]) which are not removed by outgassing the adsorbent at room temperature, followed by the adsorption of nitrogen at 77 K using the standard BET techniques. It is assumed that the micropores are blocked by the adsorbed hydrocarbon molecules and that nitrogen adsorption occurs exclusively on the external surface of the zeolite crystallites. Crucial points in using this method are to ensure that complete pore filling has occurred and that only negligible amounts of hydrocarbon are desorbed during the outgassing step.

A third adsorption-based procedure for determining the external surface area of zeolites was used by Suzuki et al. [49]. In this method, adsorption *kinetics* of relatively large molecules (e.g., neopentane for the case of zeolite ZSM-5) are followed. In this case, it is assumed that the rate-controlling step of adsorption is the diffusion of the probe molecules into the micropores. Hence, the external surface can be determined from the amount of adsorptive which is rapidly adsorbed after the onset of the experiment. For the application of this method, it is essential that the adsorption is monitored with high accuracy during the very early stages of the experiment. Moreover, the choice of the best suited probe molecule depends on the structure (in particular on the size of the pore openings) of the zeolite under investigation in order to ascertain diffusion-controlled uptake.

There have also been attempts to estimate the amount of acid sites on the external surface of zeolites by using bulky basic probe molecules which are not able to diffuse inside the pores. For example, Karge et al. used 2,6-di-*tert*-butylpyridine to characterize the external acidity of mordenite- and clinoptilolite-type zeolites [52]. These authors found that the activity of the investigated catalysts for the dehydration of cyclohexanol correlated almost linearly with the absorbance of the IR-band characteristic for di-*tert*-butylpyridinium ions. Other bulky bases which have been used as probe molecules, especially for characterizing the acid sites on the external surface of zeolite ZSM-5, are 4-methylquinoline [53] and 2,6-dimethylpyridine [54].

Ion exchange of bulky cations in aqueous solution has been explored as a means to determine the number of cation exchange sites and hence the possible number of acid sites on the external surface of zeolites, in particular of zeolite ZSM-5 [46, 55]. Using the cationic dye methylene blue in an aqueous ion exchange with zeolite ZSM-5 in the Na^+-form, Handreck and Smith [55] observed a release of sodium ions from the external surface of the zeolite which they exploited to estimate the number of exchange sites per unit area of the external surface of the zeolite. Depending on the zeolite sample, these authors found that ca. 15 to 25% of the total number of exchange sites are located on the external surface of the crystallites. This is an unexpectedly large contribution of the external surface, which can be attributed to either the highly intergrown nature of most of the zeolite samples used in their study, to a compositional zoning with aluminum (and the cation exchange sites associated herewith) preferentially located in an outer shell of the zeolite crystallites, or both.

Crystal violet, which is also a cationic dye, has been used by Noack and Spindler to probe the external surface of ZSM-5-type zeolites having medium to low aluminum contents (viz. n_{Si}/n_{Al} = 100 to ca. 1000) [46]. These authors did not observe a correlation between the aluminum content of the zeolite and the amount of crystal violet which is adsorbed from aqueous solution. However, they were able to establish calibration curves between the amount of dye taken up by the zeolite and its external surface area, which was determined independently by a combination of micropore filling with n-butane and nitrogen adsorption on the preloaded sample. Once the calibration curves have been established, a relatively simple experiment suffices to determine the size of the external surface. Unfortunately, however, this method is not applicable to aluminum-rich ZSM-5-type zeolites (n_{Si}/n_{Al} at or below 100) since in these cases interference occurs between adsorption of the dye at the external surface and its ion exchange with zeolitic cations which are located at the external crystallite surface [46].

In addition to these adsorption and ion-exchange techniques, catalytic tests have been proposed for estimating the activity of the external surface of ZSM-5-type zeolite catalysts [47, 56–60]. Namba et al. [56] suggested the conversion of 1,3,5-triisopropylbenzene as a probe molecule which is too bulky to diffuse through the ten-membered ring windows of zeolite ZSM-5. Alternatively, the (separate) conversion of n-hexane and 1,3,5-trimethylbenzene at temperatures around 450 °C has been suggested by Gilson and Derouane [47]. From a comparison of specific catalytic activities for the conversion of n-hexane and 1,3,5-trimethylbenzene, these authors have drawn conclusions on the relative roles played by the acid sites located in the intracrystalline environment and at the external surface, respectively. However, bearing in mind that diffusion coefficients for 1,3,5-trimethylbenzene in zeolite ZSM-5 (measured at 315 °C) in the order of 10^{-12} cm^2/s [65] have been reported, a certain influence of acidic sites located in the channels of zeolite ZSM-5 on the conversion of even the relatively bulky 1,3,5-trimethylbenzene could also be invoked.

More recently, Röger et al. [57] proposed the transformation of 1,2,4-trimethylbenzene as a new reaction to monitor the catalytic contributions of the internal and of the external surface sites of HZSM-5 catalysts using one single probe molecule. From the isomerization of this probe molecule to 1,2,3-trimethylbenzene and 1,3,5-trimethylbenzene the authors were able to draw conclusions on the activity of the external surface of the zeolite crystallites, while its disproportionation to 1,2,4,5-tetramethylbenzene and 1,2,3,5-tetramethylbenzene indicated changes in the overall (i.e., external and internal) activity and shape selectivity. These authors, moreover, could show that their conclusions correlate well with those observed for the conversion of 1,3,5-triisopropylbenzene (to probe the activity of the external surface) and n-hexane (monitoring the total activity). They were also able to probe the influence of a silicalite-1 coating on aluminum-containing HZSM-5 on the shape selective properties [57].

Other probe molecules which have been considered useful for a characterization of the external surface of zeolite ZSM-5 or even zeolites with larger pore entrances (i.e., zeolites beta and mordenite) are 1,3,5-tri-*tert*-butylbenzene [58], 2-methoxynaphthalene [59] and allyl-3,5-di-*tert*-butylphenylether [60]. In the latter case, an attempt was made to independently characterize the contribution of Brønsted acid sites and noble metal clusters located at the external surface of zeolite crystallites.

5.5 Examples for Shape-Selective Reactions and Models for Rationalizing the Observed Effects

5.5.1 Early Observations

Shape-selective catalysis in zeolites was discovered by Weisz and co-workers in the Mobil laboratories. In their publications which appeared starting in 1960 [1, 2, 66, 67] several different examples were described. Among them is the selective acid-catalyzed dehydration of butanol-(1) from a mixture with the bulkier 2-methylpropanol-(1) on the small-pore zeolite CaA (cf. Fig. 5.6). In a complementary experiment with the large-pore zeolite CaX as catalyst it was found that, in the absence of shape selectivity, the alcohol with the branched carbon skeleton is more reactive than the one with the unbranched chain. The very slow dehydration of 2-methylpropanol-(1) on zeolite CaA can most probably be attributed to the external surface of the crystallites. Another example described by Weisz et al. is the almost complete discrimination in the catalytic combustion of linear alkanes (n-butane) or alkenes (2-butene) without any noticeable reaction of their branched counterparts on Pt-containing zeolite CaA [67].

While the examples for shape selectivity effects reported by Weisz and co-workers could be interpreted as the result of a hindered diffusion of certain reactants or products through the zeolite pores, Csicsery demonstrated that

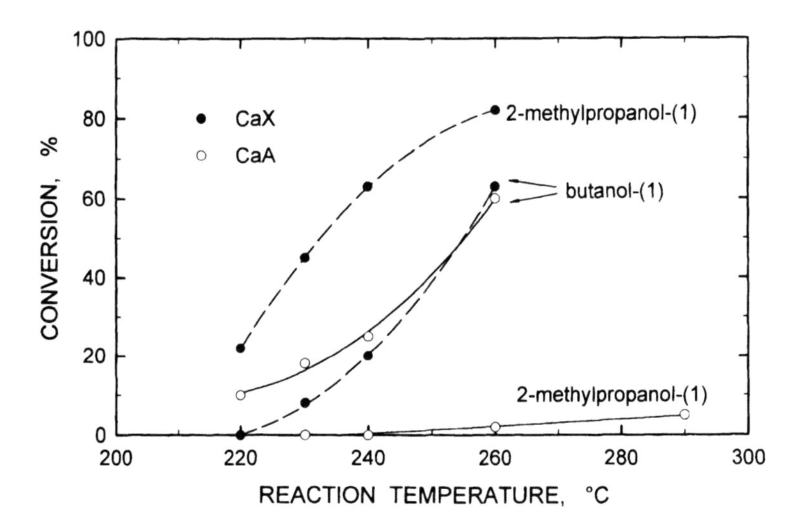

Fig. 5.6. Shape-selective conversion of alcohols over zeolite CaA [2]

there must be at least one other type of shape selectivity [68, 69]: He studied the acid-catalyzed disproportionation of 1-ethyl-2-methylbenzene (to toluene and diethylmethylbenzenes or to ethylbenzene and ethyldimethylbenzenes). Among the trialkylbenzenes, the 1,3,5-isomers predominate if thermodynamic equilibrium is established, and this is indeed observed experimentally if the reaction is conducted over catalysts with sufficiently large pores (i.e., amorphous SiO_2-Al_2O_3 or zeolite HY). In zeolite H-mordenite, however, which possesses a more restricted pore system, the same 1,3,5-trialkylbenzenes are formed only very slowly. Csicsery was able to show that this is not due to a hindered diffusion of these species in the pores of mordenite, but rather, that the intermediates and/or transition states required for the direct formation of the 1,3,5-substituted isomers are too bulky to be formed in the intracrystalline voids of zeolite mordenite [68, 69].

5.5.2
The Classical Concept After Weisz and Csicsery

From the early observations of shape-selective catalysis in zeolites, a concept has been developed to rationalize the observed effects by discerning between three types of shape selectivity, viz. reactant, product and restricted transition state shape selectivity [3, 4, 8]. This concept will be described in more detail in the following sections.

5.5.2.1
Mass Transfer Effects: Reactant and Product Shape Selectivity

Both reactant and product shape selectivity have their origins in mass transfer effects, viz. the hindered diffusion of reactants and products, respectively, in the pores of the zeolite catalyst. As an example for reactant shape selectivity the competitive cracking of n-octane and 2,2,4-trimethylpentane is depicted in Fig. 5.7. This situation can indeed be understood as molecular sieving combined with catalytic conversion: one type of molecules (2,2,4-trimethylpentane) out of the mixture of reactants is too bulky to enter the pores of the zeolite. Therefore, these molecules are hindered from reaching the catalytically active (Brønsted acid) sites in the pores of the zeolite. They can only be converted at catalytic sites located on the external surface of the zeolite crystallites, or they can leave the reactor without being converted. By contrast, the slim species (n-octane) do have access to the pores and the catalytic sites in the intracrystalline voids of the zeolite, where they are readily converted. The net effect which can be measured at the reactor outlet is the selective conversion of the slim reactant molecules. Preferential cracking of unbranched over branched aliphatics is or has been exploited in commercial processes like selectoforming [71], M-forming [72] and various dewaxing processes [73, 74] on erionite-, ZSM-5- or mordenite-type zeolites. Another example is the addition of zeolite HZSM-5 in the fluid catalytic cracking process (where the catalyst is an acidic form of zeolite Y) to enhance the octane number of the gasoline produced by a shape selective consecutive cracking of unbranched and mildly branched aliphatics [75].

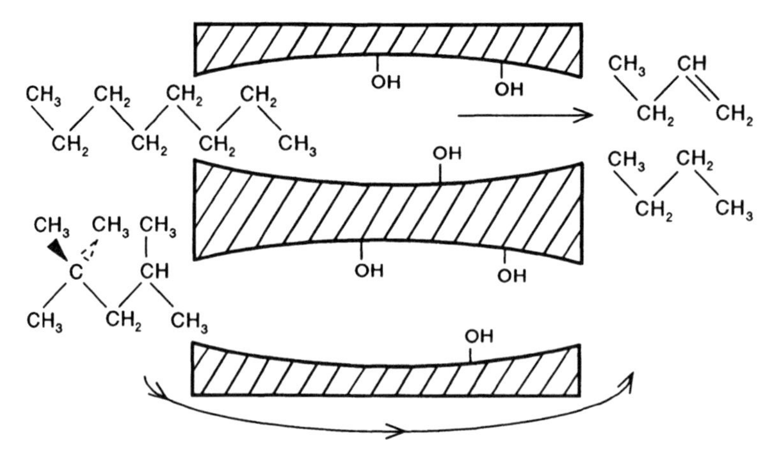

Fig. 5.7. Selective cracking of n-octane in the presence of 2,2,4-trimethylpentane as an example for reactant shape selectivity (adapted from [70])

As an example for product shape selectivity, the acid-catalyzed ethylation of toluene has been chosen in Fig. 5.8. Product shape selectivity may be looked upon as the reverse of reactant shape selectivity: Both reactants are small enough to enter the zeolite pores, but of the potential products (*ortho*-, *meta*- and *para*-ethyltoluene), only one species (the *para*-isomer) is small enough to escape from the pore system. The two bulkier ethyltoluene isomers, even though they may form in relatively spacious intracrystalline cages or at channel intersections, are unable to escape from the pores and do not occur in the reactor effluent. Ultimately, these entrapped products may be catalytically transformed into smaller molecules (e.g., isomerized into *para*-ethyltoluene) which are able

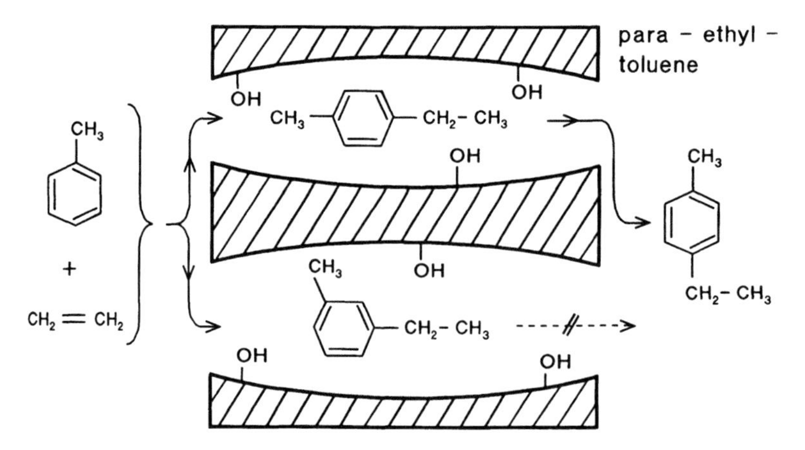

Fig. 5.8. Selective formation of *para*-ethyltoluene in the alkylation of toluene with ethylene as an example for product shape selectivity (adapted from [70])

to escape from the pores or into coke which is deposited inside the pores. Through shape-selective alkylation of toluene with ethylene an ethyltoluene fraction can be obtained containing up to 99% of *para*-ethyltoluene [76]. This particular isomer is attractive as an intermediate in the manufacture of poly-*para*-methylstyrene, which is superior to conventional polystyrene in certain applications [77].

Reactant and product shape selectivity have the same principal origin, viz. hindered diffusion of bulky reactant or product molecules in the pores of the zeolite. The complete exclusion of bulky reactant molecules as well as the complete encapsulation of bulky product molecules are, of course, limiting cases. In many instances, the hindered diffusion just results in a rate of disappearance of a reactant or a rate of formation of a product which is lower than it would have been in the absence of mass transfer limitations, but which is not zero. In all these cases, the observed rates of reaction limited by diffusion effects, and hence the observed selectivities, will be dependent, under otherwise constant conditions, on the length of the intracrystalline diffusion path, i.e., the crystallite size of the zeolite catalyst.

5.5.2.2
Intrinsic Chemical Effects: Restricted Transition State Shape Selectivity

Restricted transition state shape selectivity was first identified by Csicsery [69]. A typical example for this type of shape selectivity is depicted in Fig. 5.9. Under the influence of an acid site, *meta*-xylene can undergo isomerization into *para*-xylene (and *ortho*-xylene, which is omitted in Fig. 5.9 for clarity) and transalkylation into toluene and one of the trimethylbenzene isomers. It is evident that transalkylation is a bimolecular reaction and as such it necessarily proceeds via bulkier transition states and intermediates than the monomolecular isomeriza-

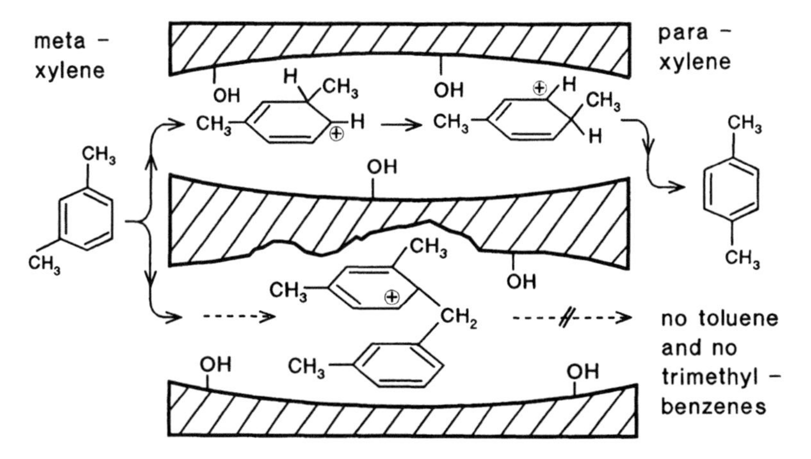

Fig. 5.9. Suppression of trimethylbenzene formation during *meta*-xylene isomerization as an example of restricted transition state shape selectivity (adapted from [70])

tion. In a zeolite with the appropriate pore width, there will be just enough space for the accommodation of the transition states and intermediates for the monomolecular reaction, but no room for the formation of the bulky transition states and intermediates of the bimolecular reaction. The net effect will be a complete suppression of the latter reaction.

The principles of restricted transition state shape selectivity are, among others, commercially exploited on a very large scale in the production of *para*-xylene. Here, the pore geometry of the catalyst (zeolite HZSM-5) strongly suppresses the formation of trimethylbenzenes (and toluene) via transalkylation [78].

While reactant and product shape selectivity are essentially based on mass transfer effects, restricted transition state selectivity is due to intrinsic chemical effects which emerge from the limited space around the intracrystalline active sites. Hence, the resulting selectivities will not depend on the crystallite size.

5.5.2.3
Discrimination Between Mass Transfer and Intrinsic Chemical Effects

For a given shape selectivity effect, a clear-cut discrimination between the two possible origins (mass transfer effects vs. intrinsic chemical effects) may be difficult. One method for distinguishing between both effects consists of a separate determination of the effective diffusion coefficients for all reactants or all products in the zeolite under consideration. Apart from the fact, however, that there is, up to now, no reasonably simple, yet reliable technique available for measuring such diffusion coefficients, this approach is necessarily limited to subcatalytic temperatures.

In another approach, the length of the intracrystalline diffusion paths is varied in catalytic experiments by using zeolite catalysts of the same structure, chemical composition and intrinsic activity, but significantly different crystallite size. A prerequisite for the application of this technique is, of course, that synthesis procedures for the required zeolite samples are available. To date, such sophisticated syntheses are restricted to a few types of molecular sieves, the most prominent example being zeolite ZSM-5.

Haag et al. [79] investigated the reasons for the preferential catalytic cracking of linear alkanes (e.g., n-hexane) over that of its methyl-branched isomers (e.g., 3-methylpentane) in acid medium pore zeolites, such as HZSM-5. One could be tempted to assume that this is an effect of reactant shape selectivity with a progressively hindered diffusion of the bulkier 3-methylpentane, as the pores are getting narrower. However, using two different HZSM-5 samples of equal concentration of acid sites but strongly different crystallite sizes (0.05 versus 2.7 μm), Haag et al. convincingly demonstrated that neither the measurable rate of cracking of n-hexane nor that of 3-methylpentane depends on the length of the intracrystalline diffusion paths. From this it was inferred that the selectivity effects encountered in the competitive cracking of n-hexane and 3-methylpentane over zeolite HZSM-5 have to be interpreted in terms of intrinsic chemical effects (i.e., restricted transition state shape selectivity) rather than by mass transport effects. Haag et al. suggested that, in the rate controlling step of acid-catalyzed alkane cracking, i.e., in the hydride transfer between a cracked carbenium ion (e.g., a secondary propyl cation) and a feed molecule, significantly more space is required for the transition state, if the feed alkane is branched, the net effect being a significant inhibition of cracking of 3-methylpentane. The principle findings of Haag et al. were later confirmed by other groups [80, 81]. Voogd and van Bekkum extended the cracking experiments to HZSM-5 samples with considerably larger crystallite sizes [81]. In this way, they were able to show that catalytic cracking of both n-hexane and 3-methylpentane in zeolite HZSM-5 is free from diffusion limitations as long as the crystallite size is below 40 μm.

5.5.3
Other Concepts

Beside the classical concept of Weisz and Csicsery, a variety of refined or novel shape selectivity concepts have been advanced. Some of these concepts have, in the meantime, been falsified, while others continue to be a matter of debate.

5.5.3.1
The Cage or Window Effect

Usually, reactivities, diffusivities and most other properties of the members of a homologous series vary monotonously with the chain length. A deviation from this rule was reported by Gorring for the diffusion coefficients of n-alkanes with 2 to 14 carbon atoms in K^+-exchanged zeolite T [82]. In particular, the diffusion coefficients for n-alkanes with carbon numbers around 8 were unexpectedly low. For such a non-monotonous dependence of the carbon number the terms "window effect" [82] or "cage effect" [83] have been coined.

Zeolite T is an intergrowth of the structurally related zeolites offretite and erionite. Its diffusion behavior is governed by the void structure of the small-pore zeolite erionite. The free space in an erionite cage can be approximated by a cylinder with a length of ca. 1.3 nm and a diameter of ca. 0.6 nm. It is reasonable to assume that n-alkanes diffusing through the pore system of erionite will adopt an almost linear molecular shape. In this case, the diameter of n-alkanes amounts to roughly 0.5 nm, while their length increases with increasing carbon number. For n-octane it is ca. 1.3 nm, which is equal to the length of the erionite cage. From this coincidence, Gorring deduced a particular intense interaction between the adsorbate and the zeolite crystal, and in this way he interpreted the slow diffusion of n-octane in zeolite T. It has, moreover, been reported that, depending on the kind of experiment, the hindered diffusion of n-alkanes with carbon numbers around C_8 can be observed either as reactant or product shape selectivity: Chen and Garwood studied hydrocracking of mixtures consisting of n-alkanes up to C_{16} over a bifunctional erionite-type catalyst at temperatures around 450 °C [84]. The rate constants for the disappearance of the n-alkanes exhibited a minimum around C_8H_{18}. Chen et al. [83], on the other hand, observed a peculiar distribution of the cracked products in catalytic cracking of n-$C_{23}H_{48}$ over an acidic form of erionite at 340°C (cf. Fig. 5.10): C_8 hydrocarbons were almost absent in the cracked products, and this was again

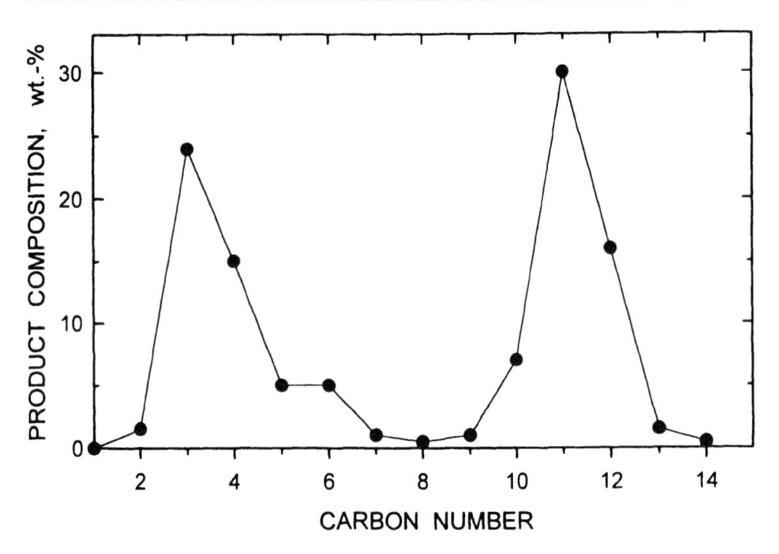

Fig. 5.10. Distribution of the cracked products from n-$C_{23}H_{48}$ in zeolite H-erionite at 340 °C [83]

tentatively interpreted by the authors in terms of a hindered diffusion of product hydrocarbons with carbon numbers around eight.

The cage effect has attracted considerable attention among zeolite researchers shortly after the pertinent publications. Nevertheless, the system n-alkanes/zeolite T (or erionite) remained an isolated example for a non-monotonous dependence of diffusion coefficients on the chain length. Recently, the soundness of Gorring's early diffusion measurements was even questioned: Neither Cavalcante et al. [85] nor Magalhães et al. [86] succeeded in reproducing the non-monotonous behavior of the diffusion coefficients of homologous n-alkanes in erionite/offretite intergrowths. The possible shortcomings of Gorring's experimental procedures are discussed in [86].

5.5.3.2
Molecular Traffic Control

The concept of "molecular traffic control" was first proposed by Derouane and Gabelica [87] for zeolites with intersecting channels of different sizes. Examples are zeolite ZSM-5, ferrierite, mordenite or the more recently discovered zeolite SSZ-26. It is envisaged that, in such materials, the smaller channels are accessible only for the smaller molecules formed in a catalytic reaction, while the larger channels are accessible for both the larger and the smaller molecules. Based on such considerations a new type of shape selectivity has been envisaged [87]. Derouane and Gabelica used this concept to account for the unexpected absence of counterdiffusional effects in the conversion of methanol to hydrocarbons in zeolite HZSM-5. They suggested that the smaller molecules (i.e., methanol) enter the (somewhat smaller) sinusoidal channels, while the bulkier hydrocarbon products leave the intracrystalline voids of the catalyst via the straight elliptical channels.

The concept of molecular traffic control has been debated controversially in the literature [88–91]. One argument against this model is the stochastic nature of diffusion. Why should the diffusion of methanol molecules be restricted to the narrower pores? Moreover, as pointed out by Lowe et al. [88], water is formed during methanol conversion to hydrocarbons, and one would expect the water molecules to also diffuse through the somewhat smaller channels. Finally, the whole concept necessarily remains speculative as long as there is no experimental technique available which allows a safe measurement of which molecules diffuse in the different types of channels.

5.5.3.3
Shape Selectivity at the External Surface: The Nest Effect

Fraenkel et al. postulated [92, 93] that under certain circumstances shape selectivity occurs not only inside the pores of zeolites (especially of zeolite ZSM-5) but also at the external surface. These authors presumed that two kinds of acid sites exist in zeolite HZSM-5, which differ in their accessibility for larger molecules and the available space for catalytic transformations: one located inside the intracrystalline channel system (probably at the intersections of the two different pore systems); the other one presumably in so-called half-cavities at the external surface of the ZSM-5 crystallites. These half-cavities are claimed to exist at the (001)-plane of ZSM-5, where no sinusoidal channels emerge from the straight channels.

The model of Fraenkel et al. is based on the observation that in the alkylation of naphthalene with methanol on HZSM-5 pronounced selectivities for 2,6- and 2,7-dimethylnaphthalene are encountered which are much higher than those observed on zeolites H-mordenite or HY. Fraenkel et al. ruled out that acid sites at the internal surface of HZSM-5 are responsible for the shape selectivity effects, simply because naphthalene and its alkyl derivatives were considered to be too bulky to be accommodated inside ten-membered ring pores.

The concept advanced by Fraenkel et al. was later generalized by Derouane et al. [94, 95] who coined the term "nest effect". It is assumed that optimum van der Waals interactions exist between certain molecules adsorbed in holes or "nests" at the external surface, due to the curvature of the walls. It is further assumed that these curvatures can affect the selectivity if the adsorbed molecules undergo catalytic reactions.

The whole concept of shape selectivity occurring at the external surface of zeolite crystallites was later severely questioned [96, 97]: From the results of more detailed and thorough experiments on the catalytic alkylation of 2-methylnaphthalene with methanol on various zeolites (including ten-membered ring zeolites with a structure which rules out the existence of half-cavities at their external surface), Neuber and Weitkamp concluded [96, 97] that all experimental findings concerning the catalytic conversion of naphthalene and its derivatives in zeolite HZSM-5 can be readily accounted for by conventional shape selectivity effects *inside* the pore systems, especially at the usually applied reaction temperatures of ca. 300 to 400 °C. In other words, the bulky molecules with a naphthalene skeleton do have access to the pore system under reaction conditions, yet the observed rates of reaction are often severely influenced by hindered diffusion. In summary, the concept of shape selectivity at the external surface has been advanced without a true need, since the experimental facts from which it was derived can be smoothly rationalized in terms of conventional mechanisms for shape-selective catalysis.

The concept of shape-selective catalysis at the external surface of zeolite crystallites was recently revisited by Martens et al. [98, 99] in connection with an entirely different catalytic reaction, viz. the isomerization of long-chain n-alkanes over bifunctional medium-pore zeolites: If n-heptadecane is converted over a bifunctional form of zeolite ZSM-22, monobranched isomers with a single methyl group (i.e., 2-, 3-, 4-, 5-, and 6-methylhexadecane) are formed exclusively at low conversions. Among these, 2-methylhexadecane, i.e., the isomer with the methyl branching at the very end of the carbon chain, predominates. The absence of isomers with longer side chains (i.e., ethylpentadecanes or propyltetradecanes) was explained in terms of the restricted space in the narrow pores of zeolite ZSM-22. The pronounced preponderance of methyl branching in the 2-position is in sharp contrast to what is observed on large-pore zeolites (e.g., zeolite Y): On bifunctional forms of such catalysts, methyl isomers having the branching at or near the center of the carbon chain are preferentially formed, and isomers with ethyl or propyl branchings occur in the product even at very low conversions. To account for the unique selectivities observed on zeolite ZSM-22, Martens et al. invoked acid-catalyzed reactions in the pore mouth, close to the external surface [98]. In their more recent work [99], the authors went one step further: In their view, 2-methylhexadecane formed in the first branching step arranges on the surface of the zeolite crystallites in such a manner that the methyl group is pinned in one pore mouth, while the rest of the molecule is stretched across an external (001) crystal face. A second branching is then introduced in a position where a neighboring pore mouth with an active site can be reached. Martens et al. found that out of the theoretically possible 219 dibranched heptadecane isomers, only five are formed at higher conversions over zeolite ZSM-22, viz. 2,7-, 2,8-, 2,9-, 2,10- and 2,11-dimethylpentadecane. The distance between the two methyl branchings in these five isomers correlates well with the distance between two adjacent pore entrances in the (001)-plane of zeolite ZSM-22. The authors therefore concluded that, at least for the example reported in their study, shape-selective catalysis at the external surface of zeolite crystallites may occur, during which the geometric arrangement of the pore mouths acts as kind of a template for the positions of the methyl branchings to be formed.

It will be interesting to see, whether and how the concept of pore mouth catalysis will develop in the future. In particular, the following questions remain to be answered: Can the concept be transferred to other reactants and/or other zeolite structures? Which role do active sites at the external surface of the crystallites play and what is the influence of the crystallite size on the activity and selectivity of the zeolite?

5.5.3.4
Tip-on Adsorption of Molecules Diffusing Inside the Pore System

Gallezot et al. [100] investigated the selective hydrogenation of cinnamaldehyde in the liquid phase at 60°C under a hydrogen pressure of 40 bar. The first hydrogenation step can lead to either 3-phenylpropanal or cinnamyl alcohol (cf. Fig. 5.11), depending on whether hydrogen adds to the carbon–carbon double bond or the carbonyl group. Ultimately, 3-phenylpropanol is formed. Two samples of Pt-containing Y-type zeolites were prepared via ion exchange of NaY with $Pt(NH_3)_4^{2+}$ and subsequent decomposition of the ammine complex under different, but well defined conditions. The Pt loadings were 11 and 14 wt%, respectively. Transmission electron microscopy revealed that, in the first sample, the noble metal was homogeneously distributed inside the zeolite with a particle diameter of 1 to 2 nm. In the second sample, the platinum had an average diameter of 5 nm, and these metal particles were preferentially located in a shell near the external surface.

At first glance, it is difficult to imagine how Pt particles of 5 nm in diameter could be located inside the faujasite crystallites, because the maximum space which is available is that in the supercages, which are ca. 1.3 nm in diameter [101, 102]. However, Schulz-Ekloff, Jaeger and co-workers have shown that even such large Pt crystallites can be located inside the zeolite matrix. They proposed that the growth mechanism of the metal particles involves atomic rearrangements of the zeolite lattice in analogy to the processes occurring during steam stabilization [102].

The results of the catalytic experiments of Gallezot et al. are summarized in Fig. 5.11. It can be seen that hydrogenation on platinum in zeolite Y gives much better selectivities for the desired cinnamyl alcohol than on a non-zeolitic reference catalyst like platinum on charcoal. This was interpreted by a *special*

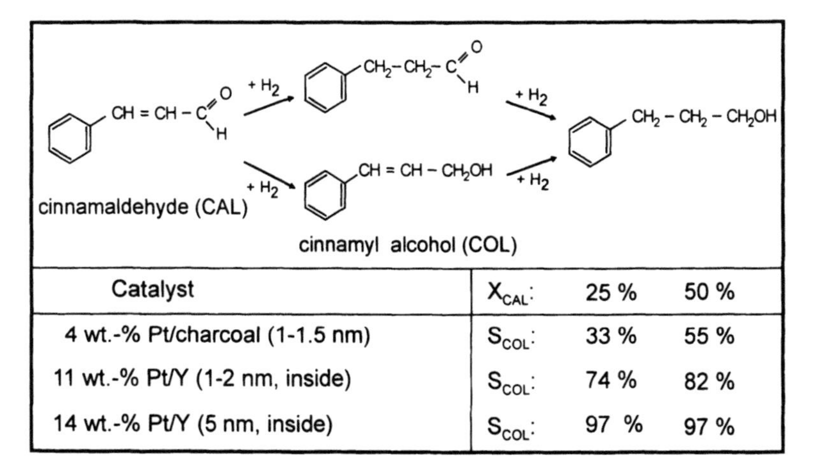

Catalyst	X_{CAL}:	25 %	50 %
4 wt.-% Pt/charcoal (1-1.5 nm)	S_{COL}:	33 %	55 %
11 wt.-% Pt/Y (1-2 nm, inside)	S_{COL}:	74 %	82 %
14 wt.-% Pt/Y (5 nm, inside)	S_{COL}:	97 %	97 %

Fig. 5.11. Reaction pathways and selectivities in the conversion of cinnamaldehyde over selected hydrogenation catalysts [100]

shape selectivity effect. Inside the faujasite pores, the cinnamaldehyde molecule can adsorb on the platinum clusters sitting in the supercages only in an "end-on" (or "tip-on") mode, i.e., via its carbonyl group, through the 12-membered ring window. In contrast, the adsorption and activation of the carbon–carbon double bond in the center of the molecule is considered to be strongly hindered, but only as long as the supercage is completely filled with platinum; a small cluster of six or less metal atoms would leave enough space for cinnamaldehyde molecules to enter the supercage and adsorb laterally, i.e., via the carbon–carbon bond, in such a way that hydrogenation would lead to 3-phenylpropanal. This is how the authors explain that on Pt/Y with clusters of only 1 to 2 nm in size, the undesired hydrogenation of the carbon–carbon double bond occurs to a certain extent. More recently, tip-on adsorption was also demonstrated to occur on platinum aggregates in the pores of zeolite beta [103]. Again, the selectivity to the desired cinnamyl alcohol was higher over the catalyst with larger platinum particles (ca. 2 to 5 nm in diameter) than over the one containing relatively small Pt-clusters (below 1 nm). Thus the concept of "tip-on" adsorption was nicely confirmed.

5.5.3.5
Secondary Shape Selectivity/Inverse Shape Selectivity

Primary (or conventional) shape selectivity results directly from the spatial constraints experienced by guest molecules or transition states inside the micropores of the zeolite host. If there are mixtures of reactants or products, the following effects of competitive adsorption are usually superimposed to primary shape selectivity: (i) molecules with higher molecular weight are preferentially adsorbed on the catalytic sites and (ii) more polar molecules are preferentially adsorbed if the zeolite surface is polar. The superimposition of such competitive adsorption effects is customery in shape-selective catalysis, and the net result is therefore still classified as primary shape selectivity. There are, however, some experimental findings which cannot be rationalized in terms of conventional shape selectivity: Namba et al. [104] found that catalytic cracking of n-octane over zeolite HZSM-5 at 400 °C is retarded by the addition of 2,2-dimethylbutane or 2,3-dimethyl-butane, but not by 3-methylpentane. This effect cannot be accounted for by competitive adsorption since 2,2- and 2,3-dimethylbutane are neither larger nor more polar than n-octane. However, it has been explained by assuming that 2,2- or 2,3-dimethylbutane hinder the diffusion of n-octane inside zeolite HZSM-5. Hence, the term secondary shape selectivity was coined for the effect that one compound affects the reactivity of another one in a manner which does not occur in large-pore catalysts. Another example for secondary shape selectivity was reported by Santilli and Zones [105]. These authors studied cracking of n-hexane, n-hexadecane and mixtures of both over the small pore zeolite SSZ-16 (AFX framework topology) at 332 °C. Whereas for the pure hydrocarbon feedstocks the reactivity increased from C_6 to C_{16}, n-hexadecane remained almost unconverted, if it was passed in a 1:1 mixture over the catalyst, while the conversion of n-hexane remained unchanged or even increased slightly.

In conventional shape-selective catalysis, the diffusion and/or formation of those molecules with the smaller cross-section is favored. However, it has been found that zeolite SSZ-24 (AFI-topology) and molecular sieves of the $AlPO_4$-5 family exhibit a preference for the adsorption (from a mixture of C_6-alkanes) or formation (from hydrocracking of n-hexadecane) of the bulkier isomers (especially dimethylbutanes and methylpentanes) [106, 107]. To describe such findings, the term "inverse shape selectivity" has been coined. The preference for the sterically more demanding species has neither been found for molecular sieves with smaller pore diameters (e.g., zeolite ZSM-5), nor in the absence of shape selectivity (e.g., zeolite Y). The experimental results have been rationalized by computer calculations on the basis of increased attractive forces between the branched isomers and the walls of zeolites with pore widths in the range from 0.65 to 0.74 nm [107].

5.6 Tailoring the Shape-Selective Properties of Zeolite Catalysts

Maximizing the selectivities (and yields) of the desired products is among the most important tasks of catalysis. Shape-selective catalysis in zeolites offers a unique opportunity of controlling reaction selectivities via the pore width and architecture. Several methods for tailoring the shape-selective properties of molecular sieve catalysts have been employed so far, and they have been reviewed in detail by Vansant [108]. In brief, they may be divided into two groups, viz. (i) methods applied during the hydrothermal synthesis of the molecular sieve and (ii) modification steps which can be applied after the synthesis of the crystalline material (post-synthesis modification).

5.6.1 Variation of the Zeolite Type and Isomorphous Substitution

One possibility of enhancing shape selectivity effects in a given catalytic application is to vary the pore size and pore architecture by using different zeolite structures. The most recent edition of the "Atlas of Zeolite Structure Types" (Ref. [25] and current updates via the Internet) teaches that at least 121 different structure types of zeolites and zeolite-like materials are known, in addition to the microporous materials whose structures have not yet been determined. In practice, the selection of the zeolites to be evaluated can be based on their known crystallographic structures, on information about their effective pore width from adsorption measurements or on results obtained in catalytic reactions including special catalytic test reactions for probing the pore width (cf. Sect. 5.7). Recently, the assortment of zeolites with most promising structures for novel catalytic applications has been significantly expanded [109]. Among these are zeolites NU-87 [110] and MCM-22 [111] with spacious intracrystalline cavities which are only accessible via ten-membered ring windows. These two materials promise to be particularly attractive catalysts for conversions in which the desired products are relatively slim (and can, hence, diffuse through the

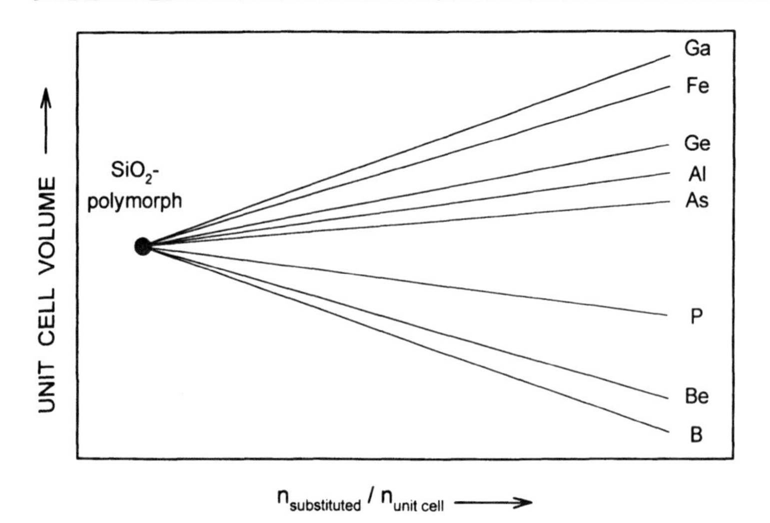

Fig. 5.12. Schematic representation of the changes in the unit cell volume of silicates with the MFI framework topology upon progressive incorporation of heteroelements; $n_{substituted}/n_{unit\ cell}$ denotes the number of heteroatoms incorporated per unit cell of the MFI structure (adapted from [116])

10-membered ring pore openings), but in which the transition states required for the formation of these desired products are relatively bulky. Further examples for interesting new structures are SSZ-26, the first synthetic zeolite possessing intersecting 10- and 12-membered ring pores [112] and UTD-1, the first silica-based microporous material with 14-membered ring channels [113].

Isomorphous substitution of silicon and/or aluminum by other elements is another possibility of tuning the pore width. The aluminophosphates and their derivatives represent the best known family of crystalline microporous materials besides the aluminosilicates. For one and the same framework topology, a large variety of different compositions can often be synthesized which differ not only in the nature and density of active sites but also in a very subtle manner in their pore dimensions [114, 115]. For MFI-type materials, Tielen et al. [116] demonstrated how isomorphous substitution can change the unit cell volume and hence the pore sizes. Figure 5.12 shows in a qualitative manner the changes in the unit cell volume upon progressively incorporating into the framework the heteroelements indicated. These results suggest that both a gradual increase and a gradual decrease in the pore size can be achieved. Tielen et al. also explored the effect of isomorphous substitution on the shape-selective properties of MFI-type materials in hydrocracking of n-decane [116]. They found, for example, that the yield of isopentane formed via hydrocracking (for comparable yields of hydrocracked products) is considerably lower on beryllosilicate MFI than on aluminosilicate MFI. Since, according to Tielen et al. [116], the yield of isopentane can be taken as a measure of the intracrystalline space available for catalytic reactions, this result is indicative of a reduced pore width in the Be-containing material, which is in general agreement with the trends sketched in Fig. 5.12.

Great care should, however, be applied when discussing the effect of isomorphous substitution on the pore width, since the nature of the active sites (in particular the strength of acid sites) changes simultaneously, and it is almost always difficult to separate the influence of both factors.

5.6.2
Variation of the Crystallite Size and Compositional Zoning

If a shape selectivity effect is based on mass transfer effects, it can be amplified (or diminished) by increasing (or reducing) the size of the zeolite crystallites. One illustrative example of the influence of the crystallite size on product selectivities is depicted in Fig. 5.13. Olson and Haag [78] studied the influence of the crystallite size of zeolite HZSM-5 on the selectivity for *para*-xylene formed from toluene disproportionation over three different catalysts with similar intrinsic activities. They observed a significant increase in *para*-xylene selectivity with increasing crystallite size. Thus, product shape selectivity, favoring the diffusion of the smallest xylene isomer, is strongly enhanced. Similar effects were observed in the cases of toluene alkylation with methanol [e.g., 117, 118] or with ethylene [e.g., 119]. However, the effects of an increased selectivity for the *para*-substituted products cannot only be attributed to a lengthening of the diffusion path, but also to a decreasing external surface with increasing crystallite size. Hence, secondary (non-shape-selective) reactions of the product molecules leaving the crystallites are reduced. A reduction in the contribution of active sites at the external surface cannot only be achieved by selecting large crystallites, but also by coating the crystallites with an aluminum-free shell having the same structure as the core. This is possible during hydrothermal synthesis by first initiating crystallization of the desired (aluminum-containing) zeolite and

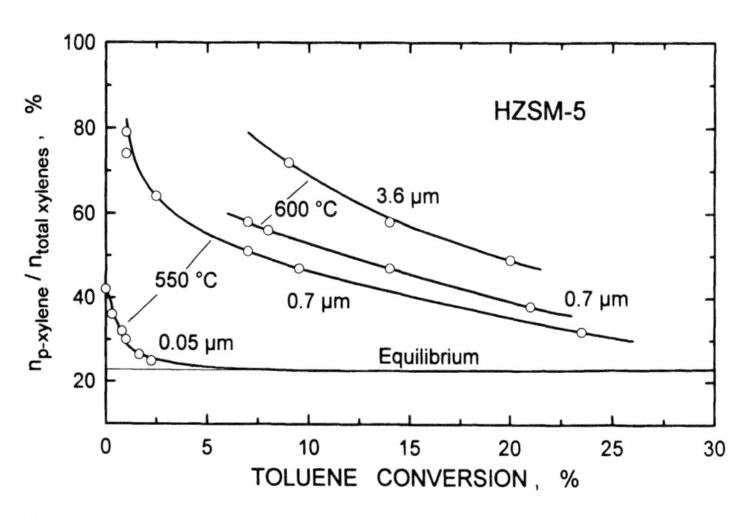

Fig. 5.13. The influence of the crystallite size and the reaction temperature on the selectivity to *para*-xylene in the disproportionation of toluene over HZSM-5-type catalysts [78]

then altering the crystallization medium to eliminate the aluminum therein, as demonstrated by Rollmann [120]. He added as a complex-forming ligand sodium gluconate to a gel for the synthesis of zeolite ZSM-5 after a crystallinity of 50% had been reached. Alternatively, preformed and purified ZSM-5-type crystallites can be added to an aluminum-free synthesis gel containing tetrapropylammonium cations in order to induce the crystallization of a silicalite-1 layer on top of the ZSM-5 crystallites.

Similarly, the reverse is also possible, as has been demonstrated by Angevine et al. [121] for zeolite ZSM-12 as an example. They started from an aluminum-free gel for the synthesis of zeolite ZSM-12 and added an aluminum source to the synthesis mixture after the product was 20 to 80% crystalline. In this way, it was possible to prepare a material with aluminum, and hence acid sites, enriched in an outer shell of the crystallites.

Besides such an "artificially" induced zoning of aluminum in zeolite crystallites, it has been repeatedly found that aluminum zoning in zeolite ZSM-5 can occur depending on the reaction medium and the synthesis conditions [e.g., 122–124]. The early work has been summarized and critically evaluated by Jacobs and Martens [125]. Althoff et al. have recently shown how the spatial distribution of aluminum in ZSM-5 crystallites can be systematically controlled [126].

It is obvious that both a systematic variation of the crystallite size and a tailored compositional zoning in zeolite crystallites require highly sophisticated synthesis techniques. Several reliable methods have been published which allow the synthesis of ZSM-5 zeolites of different crystallite dimensions, especially large crystallites [e.g., 127, 128]. For other structures, however, information on how the crystallite size can be varied and how aluminum zoning can be created, is very scarce.

5.6.3
Ion Exchange and Pore Size Engineering

Ion exchange with cations of different size is a well known method for influencing the effective pore width of zeolites. An example is the change in the pore size of zeolite NaA (0.4 nm) upon cation exchange with potassium (0.3 nm) or calcium (0.5 nm) cations [20]. Likewise, the effective pore width of H-mordenite can be varied by post-synthetic ion exchange with large metal cations [129–132]. The density and strength of the catalytically active sites may, however, also be varied by this method [132].

Pore size engineering in zeolites has been extensively reviewed by Vansant [108]. It comprises all kinds of measures to modify the external and/or internal surface of zeolite crystallites. In most cases, the zeolites are subjected to treatment with chemically reacting species. Depending on their size and the pore width of the zeolite, they may penetrate into the pore system or modify the external surface only [133]. For example, loading with boric acid, trimethylborate, phosphoric acid or trimethylphosphine followed by calcination may lead to a narrowing of the pore dimensions. This effect can be exploited to increase the *para*-xylene selectivity in toluene disproportionation to more than 90% [134]. Likewise, treatment with silane vapors or silicon alkoxides (chemical vapor

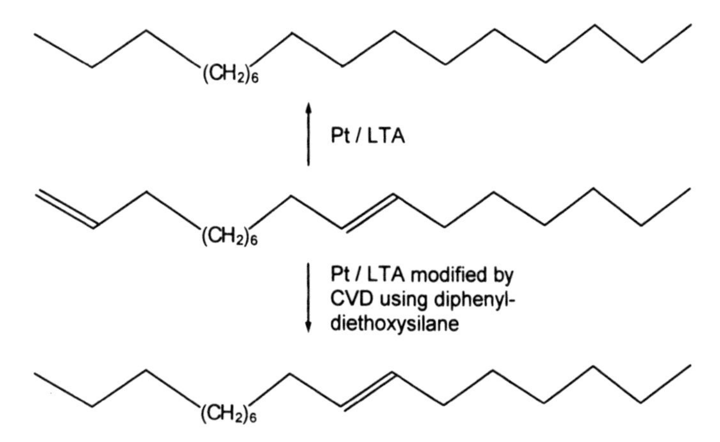

Fig. 5.14. Selective hydrogenation of 1,11-octadecadiene on zeolite Pt/LTA modified by chemical vapor deposition [138]

deposition, CVD) followed by calcination can be used to introduce SiO_2 deposits [e.g., 134–138]. By a variation of the temperature, the partial pressure of the silane and the duration of the treatment, the location and amount of deposited material can be controlled [108]. If the deposits are located at the surface of the zeolite crystallites, they may narrow the pore entrances and thereby influence the diffusion of reactants or products into or from the intracrystalline voids.

An example of how a regioselective hydrogenation catalyst can be prepared by pore size engineering is shown in Fig. 5.14. Kuno et al. [138] modified a platinum-loaded A-type zeolite by repeated (5-fold) treatment with vapors of diphenyldiethoxysilane followed by hydrolysis at 300 °C. The resulting catalyst was tested in the hydrogenation of a diolefin with one terminal and one internal double bond, viz. 1,11-octadecadiene. It was found that over the modified catalyst only the terminal double bond is hydrogenated, while over an unmodified reference catalyst completely saturated products are obtained [138].

5.6.4 Selective Poisoning of the External Surface

Several methods have been reported in the literature to deactivate or poison the catalytically active sites at the external surface of zeolite crystallites. One technique relies on the selective dealumination of the external surface using acids or complexing agents [139–142]. Namba et al. [56, 140] and Anderson et al. [141] made use of a treatment with $SiCl_4$ vapors at elevated temperatures (ca. 450 to 650 °C) to remove framework aluminum (i.e., acid sites) at the external surface of ZSM-5 crystallites. According to the method of Apelian et al. [139], surface dealumination of organic-containing zeolites can be achieved by contacting the zeolite with dicarboxylic acids, especially oxalic acid in the liquid phase at temperatures below 100 °C. This treatment resulted in a reduction of surface acidity (as determined from the catalytic conversion of 1,3,5-tri-*tert.*-butylbenzene) without a significant reduction in the overall activity. The latter method seems to be superior to $SiCl_4$ dealumination since it can, in principle, be applied to a wide range of pore sizes, provided they are filled with organic matter, while the $SiCl_4$ method is restricted to those materials which do not admit the dealuminating agent to the intracrystalline pore system. Moreover, dealumination in the liquid phase and at temperatures below 100 °C seems to be a more straightforward and easier to perform technique for surface dealumination.

Another approach for deactivating the external surface of zeolites is the selective poisoning of acidic sites by the adsorption of bases which are too bulky to penetrate the micropores [142–145]. Examples for such bases are 4-methylquinoline [142], 2,6-di-*tert.*-butylpyridine [58, 143], *ortho*-phenanthroline [144] or certain organophosphorus compounds like tributylphosphite [145] and triphenylphosphine [58]. Moreover, reagents may be applied which, after adsorption and calcination, irreversibly react with the external surface of zeolite crystallites. An example is the reaction of triphenylchlorosilane with acidic OH groups at the external surface of zeolites [146]. This molecule is too bulky to penetrate even the large 12-membered ring pore openings of faujasite-type zeolites. Hence a selective poisoning of the external surface is possible.

5.7 Catalytic Test Reactions for Probing the Effective Pore Width of Microporous Materials

The use of catalytic test reactions is a well known tool for characterizing the effective pore width of molecular sieve catalysts under catalytically relevant conditions [147–153]. The principal experimental approach is to measure the selectivities of a given catalytic reaction over a variety of zeolites with known structures. Once the method has been calibrated in this way, the selectivities of the same reaction observed on a zeolite with unknown structure give valuable information about its effective pore width or its pore architecture. Several catalytic test reactions for the characterization of the effective pore widths of zeolites have been proposed so far. The requirements which should be fulfilled by such a test reaction have been summarized previously [153]. In essence, these requirements encompass that the selectivity of the test reaction should be dependent in a pronounced and well defined manner on the pore width of the zeolite. Moreover, the shape selectivity effect should not be influenced by the crystallite size of the zeolitic materials, i.e., it should be based on restricted transition state shape selectivity rather than on mass transfer effects. Furthermore, the underlying reaction mechanism should be understood in considerable detail, and the selectivities should, as far as possible, be independent of conversion and reaction temperature. Finally, the catalytic test reaction should be easy to perform, and the possibility of expressing its result in terms of a quantitative criterion or index is most desirable [153]. In the following, the most prominent test reactions of this type will be described and their usefulness discussed. Test reactions occurring on monofunctional (acidic) forms of zeolites will be dealt with first, followed by test reactions requiring bifunctional forms (acidic zeolite modified by a hydrogenation/dehydrogenation component, e.g., a noble metal).

5.7.1
Test Reactions for Acidic Molecular Sieves

5.7.1.1
Competitive Cracking of n-Hexane and 3-Methylpentane (the Constraint Index, CI)

The method of characterizing the effective pore width of zeolites by catalytic tests was first employed by researchers at Mobil. They introduced the Constraint Index [154] which is based on the competitive cracking of an equimolar mixture of n-hexane and 3-methylpentane on the acidic form of the zeolite. As long as the catalyst pores are sufficiently spacious, branched alkanes are cracked at higher rates than their unbranched isomers. The opposite holds for medium-pore zeolites like ZSM-5. The Constraint Index (CI) is defined as

$$CI = k_{n\text{-}Hx}/k_{3\text{-}M\text{-}Pn}$$

where $k_{n\text{-}Hx}$ and $k_{3\text{-}M\text{-}Pn}$ are the first order rate constants for the cracking of n-hexane and 3-methylpentane, respectively. Using the performance equation for an integral fixed-bed reactor [155] this equation can be rewritten as

$$CI = \log(1 - X_{n\text{-}Hx})/\log(1 - X_{3\text{-}M\text{-}Pn})$$

where X denotes the conversion. The Constraint Index has to be measured under specified conditions (reaction temperature between 290 and 510°C; LHSV between 0.1 and 1 h^{-1}; 10 vol.-% of each reactant in helium as carrier gas, mass of catalyst around 1 g; overall conversion 10 to 60%). According to [154], the Constraint Index allows a classification into large-pore (12-membered ring), medium-pore (10-membered ring) and small-pore (8-membered ring) molecular sieves:

$CI < 1$: large-pore materials;
$1 \leq CI \leq 12$: medium-pore materials;
$12 > CI$: small-pore materials.

Constraint Indices taken from (mainly the patent) literature are summarized in Table 5.1. By and large, a correct classification of 8-, 10- and 12-membered ring materials can be achieved. Exceptions are zeolites ZSM-12 and, in a sense, zeolite MCM-22. ZSM-12 possesses 12-membered ring channels; however, on the basis of its Constraint Index it would be classified as a material with 10-membered ring pores. Zeolite MCM-22 has two different intracrystalline pore systems, one consisting of a two-dimensional (i.e., interconnected) system of 10-membered channels, the other one consisting of large intracrystalline cages which are accessible by 10-membered ring windows only [111]. In this case, a Constraint Index of 1.5 can be looked upon as an averaged value characterizing the mean space available in the 10-membered ring pores and in the large cavities.

Table 5.1. Constraint Indices of Selected Zeolites[a]

Zeolite	Structure type	CI	Classification after CI	Ring size
Erionite	ERI	38	narrow pores	8-ring
ZSM-23	MTT	9.1		10-ring
ZSM-22	TON	7.3 (7.4)		10-ring
ZSM-5	MFI	6–8.3 (4.6)		10-ring
			medium pores	
ZSM-11	MEL	5–8.7		10-ring
ZSM-50	EUO	2.1		10-ring
MCM-22	MWW	1.5		10-ring + cage
ZSM-12	MTW	2.3		12-ring
mordenite	MOR	0.5 (1.0)		12-ring
			large pores	
beta	BEA	0.6–2.0		12-ring
X or Y	FAU	0.4 (0.2)		12-ring

[a] Data from the patent literature [139, 156], values in parantheses from the open literature [151, 157].

The main advantages and disadvantages of the Constraint Index have been discussed in detail recently [153]. In brief, the reaction suffers from a relatively fast catalyst deactivation for large pore zeolites, difficulties in the correct assigment in the transient region between medium- and large-pore zeolites and a (sometimes pronounced) temperature dependence for medium-pore zeolites. The reason for the latter has been investigated in much detail by Haag et al. [158, 159]. These authors were able to show that two basically different cracking mechanisms may occur in medium-pore zeolites, viz. the classical chain-type mechanism involving tri-coordinated alkyl carbenium ions and a non-classical mechanism via penta-coordinated carbonium ions, their relative importance depending on the reaction temperature. While the former mechanism proceeds via a bimolecular transition state and is sensitive to the space available around the active sites, the latter requires only a monomolecular transition state which is much less demanding in space. Haag et al. [158, 159] demonstrated that the contribution of the monomolecular cracking mechanism increases with increasing reaction temperature at the expense of the classical bimolecular pathway, hence a decrease of CI with increasing temperature. As to the nature of the shape selectivity effects, Haag et al. [79] have shown, in a very impressive study, that the Constraint Index is based on restricted transition state shape selectivity rather than on product shape selectivity.

Historically, the Constraint Index was the first test reaction introduced for probing the shape-selective properties of zeolite catalysts with unknown structure. In spite of its shortcomings (*vide supra*), it has been widely used, and ample CI data are available in the literature.

5.7.1.2
Isomerization and Disproportionation of meta-Xylene

On catalysts with Brønsted acid sites, *meta*-xylene can undergo isomerization (to *para*- and *ortho*-xylene) or disproportionation to 1,2,3-, 1,2,4- or 1,3,5-trimethylbenzene (and toluene) (cf. Fig. 5.15). While it is obvious that disproportionation is necessarily a reaction involving a bimolecular transition state, there is some ambiguity as to whether the acid-catalyzed isomerization of xylenes proceeds via a monomolecular or a bimolecular pathway [160, 161].

The use of *meta*-xylene conversion for the characterization of the effective pore width of zeolites was first proposed by Gnep et al. [162]. These authors identified three selectivity criteria which may furnish valuable information on the effective pore width: (i) the relative rates of formation of *ortho*- and *para*-xylene, (ii) the ratio of rates of disproportionation and isomerization and (iii) the distribution of the trimethylbenzene isomers (cf. Fig. 5.15) formed in the disproportionation reaction.

Criterion (i) is based on the finding that, in the absence of shape selectivity, *ortho*- and *para*-xylene are formed at virtually the same rate. With decreasing pore width, however, the formation of *para*-xylene is increasingly favored over the formation of the bulkier *ortho*-substituted isomer. Criterion (ii) is a quantitative expression of the observation that with decreasing pore width, *meta*-xylene isomerization is more and more favored over its disproportionation, viz. the bimolecular transition state required for disproportionation is increasingly hindered [162]. Similar results were published slightly later by Olson and Haag [78] who studied the conversion of pure *meta*-xylene and a mixture of *ortho*- and *meta*-xylene over four acid zeolites with different pore widths. These authors demonstrated that a linear relationship results if the ratio of rate constants of disproportionation and isomerization is plotted vs. the dimensions of the largest zeolite cavities. One of the conclusions of this study was that the suppression of disproportionation is due to restricted transition state shape selectivity rather than to mass transfer effects.

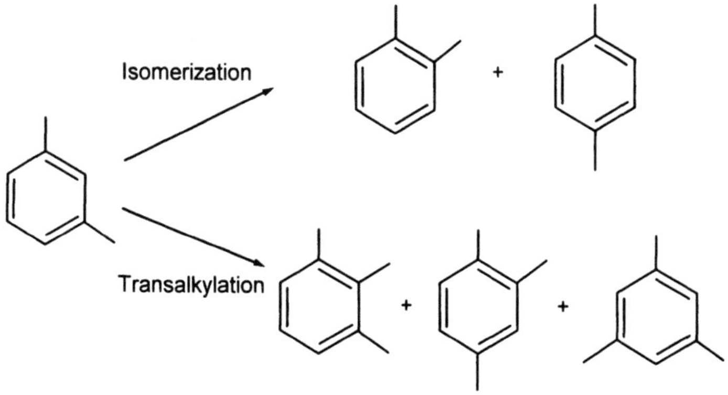

Fig. 5.15. Principal reaction pathways of *meta*-xylene over acidic catalysts

Criterion (iii) was deduced from the experimental finding that the disproportionation of *meta*-xylene in the spacious cages of zeolite Y and the much narrower channels of mordenite resulted in significantly different distributions of the trimethylbenzene isomers: While in the product obtained over H-mordenite about 95% of the trimethylbenzene fraction consisted of the 1,2,4-isomer, 24% of the 1,3,5-isomer as well as 74% of the 1,2,4-isomer are produced over zeolite HY (on both zeolites, only negligible amounts of 1,2,3-trimethylbenzene were formed, and this was attributed to its low concentration in thermodynamic equilibrium). The hindered formation of 1,3,5-trimethylbenzene from *meta*-xylene in zeolite mordenite was interpreted [162] as an example for transition state shape selectivity.

meta-Xylene conversion as a catalytic test was later adopted by quite a number of other groups [147, 163–170], and the criteria proposed initially by Gnep et al. were refined [163] and applied to a broader structural variety of molecular sieves [163–170]. A critical evaluation of published data [153] reveals that criterion (i) is useful for a discrimination between 10- and 12-membered ring zeolites, while it does not allow the ranking of zeolites with respect to their pore width in the individual classes of medium- and large-pore materials. Moreover, it has been repeatedly found that the *para/ortho* selectivity does not only depend on the pore width of a given structure, but also (especially for large pore zeolites) on its n_{Si}/n_{Al} ratio [171–174]. However, there seems to be valuable information on the pore architecture in the distribution of the trimethylbenzenes formed from *meta*-xylene in large-pore zeolites, as has been pointed out by Martens et al. [163]. With decreasing space available around the acid sites, systematically higher selectivities to the least bulkiest of the three trimethylbenzenes (1,2,4-trimethylbenzene) at the expense of the bulkier 1,2,3- and 1,3,5-trimethylbenzenes are observed.

On the whole, *meta*-xylene conversion has found widespread use, probably because it is easy to handle, e.g., the number of products which have to be considered is small. Criteria (i) and (ii) initially proposed by Gnep et al. [162], i.e., the relative rates of formation of *ortho*- and *para*-xylene and the relative rates of isomerization vs. disproportionation may be useful for a coarse discrimination between 10- and 12-membered ring molecular sieves. Their applicability and usefulness for a subtler differentiation of the pore widths within the group of large pore zeolites is, however, doubtful [153]. Criterion (iii) may be useful to distinguish large-pore zeolites [163].

5.7.1.3
Reactions of Other Alkyl Aromatics

A variety of other alkyl aromatics have been proposed as probe molecules for estimating the effective pore width of zeolites. Based on his earlier studies [68, 69], Csicsery has more recently [175] proposed isomerization and transalkylation of 1-ethyl-2-methylbenzene as a suitable test reaction. He furthermore suggested using this test for probing the state of the hydrogenation metal in bifunctional zeolite catalysts [176]. A wealth of information can be derived from Csicsery's catalytic test, the interpretation of which has been discussed in much detail [175].

Disproportionation of ethylbenzene has been shown to be a valuable tool for collecting information on the number of Brønsted acid sites in zeolites [132, 177, 178] as well as on their effective pore width [178, 179]. From comparative catalytic experiments with a variety of 10- and 12-membered ring zeolites [178, 179], the following criteria were established: 12-membered ring zeolites exhibit an induction period, whilst 10-membered ring zeolites do not; (ii) after the induction period there is very little or no deactivation in 12-membered ring zeolites, whereas in 10-membered ring zeolites, there is considerable deactivation from the very beginning of a catalytic experiment; (iii) on 12-membered ring zeolites, the distribution of the diethylbenzene isomers is ca. 5% *ortho*-, 62% *meta*- and 33% *para*-isomer; on 10-membered ring zeolites, the *ortho*-isomer is often completely absent or it appears as a very minor component (only 1 to 2% of the diethylbenzenes), often exclusively at the onset of the catalytic experiment. The combined application of the above criteria allowed for a safe discrimination between medium- and large-pore zeolite catalysts. So far, however, it has not been refined to such an extent as to allow for a ranking of 12- (or 10-) membered ring zeolites according to their effective pore width.

More recently, Kim et al. [180] have proposed a test reaction particularly suitable for probing the effective pore width of large pore zeolites. They chose a very bulky reactant, viz. *meta*-diisopropylbenzene, which is not able to enter the channels of medium pore zeolites, and alkylated it with propene. Under reaction conditions, both isomerization of *meta*-diisopropylbenzene and its alkylation to the three isomeric triisopropylbenzenes occurred. From the selectivities to the individual products, conclusions could be drawn on the effective pore width of the investigated materials which are in general agreement with the results of another catalytic test (for the characterization of bifunctional zeolites), viz. the determination of the Spaciousness Index (cf. Sect. 5.7.2.2). Hence, the conversion of *meta*-diisopropylbenzene seems to be a promising test reaction for large pore acid zeolites, and one that certainly deserves further investigation.

The alkylation of biphenyl with propene has recently been suggested as a test reaction for probing the pore size of pillared clays, large pore zeolites and related microporous materials [42]. The major selectivity criterion discussed so far is the content of *ortho*-isopropylbiphenyl in the monoalkylated product fraction. The reaction could have a potential for characterizing large and even superlarge pore molecular sieves, such as UTD-1 [113], but more systematic work on the influence of the pore size and architecture of the catalyst on the distribution of the mono-, di- and, desirably, trialkylated products is needed.

5.7.2
Test Reactions for Bifunctional Molecular Sieve Catalysts

Since the early 1980s, the potential of catalytic reactions occurring on bifunctional forms of zeolites (i.e., the Brønsted acid form modified by a small amount of a hydrogenation/dehydrogenation component, typically platinum or palladium) for characterizing the pore size and pore architecture of zeolites has been explored. In this context, particular attention was placed on the isomerization and hydrocracking of n-decane [146, 181–186] and to hydrocracking of naphthenes, in particular butylcyclohexane [187, 188]. These reactions have in common that they are conducted under hydrogen, which is activated by the metal component of the catalyst. Thus, coke formation and the concomitant catalyst deactivation are completely absent or very slow. As an advantageous consequence, neither the conversion nor the selectivities vary with time-on-stream. In addition, the risk of narrowing the pores by the deposition of carbonaceous residues is eliminated.

The mechanisms of hydrocarbon conversion over bifunctional zeolite catalysts with respect to their use as test reactions have been extensively discussed several times [e.g., 146, 150, 153]. In brief, the saturated hydrocarbon reactant is first dehydrogenated on the noble metal to the corresponding olefins which, in turn, are protonated at Brønsted acid sites. The resulting carbenium ions, while adsorbed at acid sites, undergo skeletal rearrangements and β-scissions, and all these complex steps are nowadays known in considerable detail. Finally, the product carbocations are desorbed from the acid sites as alkenes which are hydrogenated on the noble metal.

Essentially two test reactions of this type have turned out to be particularly useful, viz. isomerization and hydrocracking of n-decane [146, 181] (and, occasionally, n-alkanes with even longer chains [189–191]) and hydrocracking of butylcyclohexane [187, 188] (or other naphthenes with ten carbon atoms). These two test reactions will be discussed in the following sections.

5.7.2.1
Isomerization and Hydrocracking of Long-Chain n-Alkanes (the Refined or Modified Constraint Index, CI*)

If an n-alkane is converted on a bifunctional catalyst in the absence of shape selectivity (i.e., using a zeolite with very large pores), the feed molecule is first isomerized to a mixture of all possible alkanes with one branching. Starting from, e.g., n-decane, the monobranched isomers are: 2-, 3-, 4- and 5-methylnonane, 3- and 4-ethyloctane and 4-propylheptane. In consecutive reactions, iso-alkanes with two branchings (dimethyloctanes, ethylmethylheptanes) are formed. In subsequent steps, these can undergo hydrocracking reactions, either directly or via tribranched intermediates. In this way, a characteristic pattern of isomerized and hydrocracked products is generated. If a medium-pore zeolite is used instead of a large-pore zeolite, pronounced differences in the product distributions, e.g., from n-decane, are observed [192, 193]: Ethyl- and propyl-branched isomers are no longer formed, and the distribution of the four methylnonanes differs substantially from the one observed with Y-type zeolites. While in the absence of shape selectivity and at low conversions, 2-methylnonane forms about half as fast as 3- or 4-methylnonane, a drastically different pattern is observed in medium-pore zeolites (e.g., ZSM-5). On such catalysts, 2-methylnonane is the preferred isomer. Based on the observation that, with decreasing pore width of the zeolite, the amount of 2-methylnonane formed from n-decane at low conversions increases relative to the other methylnonanes, the Refined or Modified Constraint Index (CI*) was defined [146]. It is calculated as the yield

Definition:

$$CI^* \equiv \frac{Y_{2\text{-methylnonane}}}{Y_{5\text{-methylnonane}}} \quad (\text{at } X_{n\text{-decane}} \approx 5\%)$$

in isomerization of n-decane.

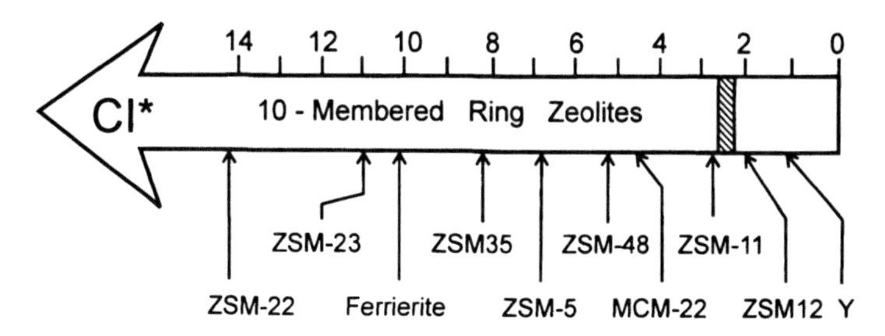

Fig. 5.16. Refined or Modified Constraint Indices for various zeolites (data from [149])

ratio of 2-methylnonane and 5-methylnonane formed at an n-decane conversion of 5%. It should be noted that the reaction mechanisms, on which Mobil's original Constraint Index (cf. Sect. 5.7.1.1) and the Modified Constraint Index are based, are entirely different. The only feature both indices have in common is that they increase with decreasing pore size of the zeolite.

Modified Constraint Indices for a variety of different zeolite structures are summarized in Fig. 5.16. It can be seen that the CI* values for ten-membered ring zeolites extend over a broad range, viz. from ca. 2.3 to 15, hence this is the range where the Modified Constraint Index is particularly useful. On the other hand, a very narrow range, namely from ca. 1 to 2.3, is available only for 12-membered ring zeolites.

The Modified Constraint Index exploits the selectivity occurring in the first step of the n-alkane reaction network. Numerous additional shape-selectivity effects are encountered in the consecutive reactions, i.e., the formation of dibranched isomers and hydrocracking. After a careful inspection of all these effects, Martens et al. [146, 181] defined a total of eight quantitative criteria. For a thorough discussion of these criteria and their usefulness for characterizing the pore width of microporous materials, the reader is referred to the pertinent literature [146, 153, 181].

In certain cases, e.g., if zeolites with very large pore sizes or void dimensions are to be characterized, it can be of advantage to use a probe molecule with more than ten carbon atoms. For example, an attempt was undertaken [189] to detect differences between the pore systems of zeolite Y and zeolite ZSM-20, which is an intergrowth of zeolites Y and EMT. For this purpose, n-tetradecane was selected as feed hydrocarbon with the rationale that there are three isomers with a propyl and one with a butyl side chain among the monobranched isomers and that such bulky isomers may be suitable for detecting subtle differences in the pore systems. Indeed, such differences were found: 5-butyldecane was lacking in the product obtained on Pd/HZSM-20 at low conversion ($X_{n\text{-tetradecane}} \approx 2\%$), while it did appear at the same conversion on zeolite Pt/CaY. Moreover, the three isomeric propylundecanes appeared in significantly different distributions on both bifunctional catalysts.

More recently, Martens et al. [190] went one step further and suggested the use of n-heptadecane as probe molecule for the characterization of large pore zeolites [191]. In this case, however, the information is acquired from the selectivity of hydrocracking, rather than from feed isomers of different bulkiness.

5.7.2.2
Hydrocracking of Butylcyclohexane (the Spaciousness Index)

It has been found that hydrocracking of C_{10}-naphthenes over bifunctional catalysts and in the absence of spatial constraints is surprisingly selective: methylcyclopentane and iso-butane are almost exclusively formed [194]. This can be well understood in terms of modern carbocation chemistry [194]. It can be shown that the energetically favored reaction pathways leading to iso-butane and methylcyclopentane require bulky intermediates [194]. It has been predicted on the basis of carbocation chemistry that, as the pores become narrower, hydrocracking of C_{10}-naphthenes proceeds much less selectively and a larger number of hydrocracked products is formed [195], and this has indeed been found experimentally [150, 187, 188, 195]. A thorough investigation of all features of shape-selective hydrocracking of C_{10}-naphthenes (preferentially butylcyclohexane, pentylcyclopentane is, from a chemical viewpoint, equally well suited, though usually not readily available) revealed that the yield ratio of iso-butane and n-butane in the hydrocracked products is a most valuable indicator for the effective pore width of the zeolite. Since this ratio increases with increasing pore width of the zeolite, it was named the Spaciousness Index (SI) [187, 188]. It has been found that the Spaciousness Index is, within a relatively broad range, virtually independent of the reaction temperature and, hence, the butylcyclohexane conversion. In addition, the Spaciousness Index relies on restricted transition state shape selectivity, rather than on mass transport effects, as has been concluded from the results of hydrocracking butylcyclohexane over ZSM-12 zeolites of different crystallite sizes (0.5 µm vs. 11 × 1.5 µm, [188]). All these features, along with the very easy analysis for iso- and n-butane, add to the usefulness of the Spaciousness Index in practice.

Figure 5.17 shows the Spaciousness Indices of selected zeolites. It is evident that SI is not appropriate for ranking the 10-membered ring zeolites, they all have Spaciousness Indices around or below 1. On the other hand, the Spaciousness Indices for 12-membered ring molecular sieves cover a very wide range from ca. 3 (ZSM-12) to more than 20 (faujasite). SI is very suitable in this range of pore widths. For super-large pore materials (i.e., those having larger than 12-membered ring pore openings, like the molecular sieve UTD-1 [113]) the Spaciousness Index will probably not be very sensitive, as may be forecasted from the fact that amorphous SiO_2-Al_2O_3 possesses the same SI as zeolite Y. The potential of other test reactions, such as isomerization and hydrocracking of

Definition:

$$SI \equiv \frac{Y_{i\text{-butane}}}{Y_{n\text{-butane}}}$$

in hydrocracking of butylcyclohexane or pentylcyclopentane.

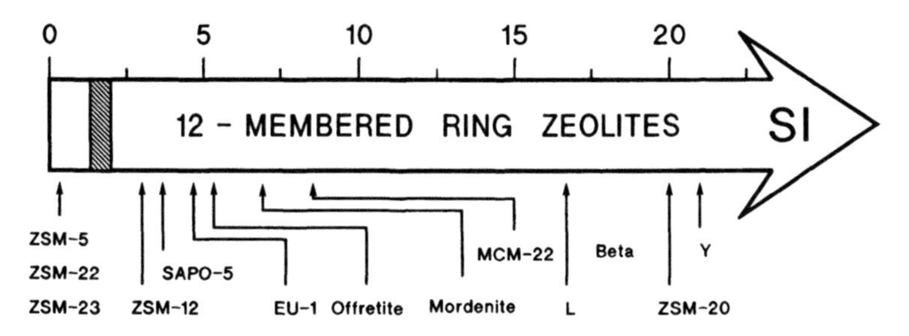

Fig. 5.17. Spaciousness Indices for various zeolites (data from [188])

n-heptadecane, alkylation and isomerization of *meta*-diisopropylbenzene or alkylation of biphenyl with propene, remains to be explored.

5.8 More Recent Directions and Challenges in Shape-Selective Catalysis

Shape-selective catalysis in zeolites and zeolite-related molecular sieves continues to be a rapidly expanding field. The use of shape-selective catalysis in the manufacture of organic intermediates and fine chemicals has been excellently summarized in two more recent articles by Venuto [19, 196]. A comprehensive overview of (in part shape-selective) oxidation reactions of organic compounds over transition metal containing redox molecular sieves has been given more recently by Arends et al. [197]. There are many more potential applications of shape-selective catalysis. In the following, selected topics will be briefly addressed.

5.8.1 Trend Towards Bulkier Molecules

In the early days, almost all shape selectivity effects were related to the conversion of aliphatic compounds or mononuclear aromatics in ten-membered ring zeolites, in particular zeolite HZSM-5 [4]. More recently, however, more and more attempts have been made for the shape-selective production of larger molecules, e.g., derivatives of substituted binuclear aromatics, viz. 2,6-dialkylnaphthalenes [e.g., 97, 198–201] and 4,4′-diisopropylbiphenyl [e.g., 42, 200–206], which are valuable intermediates for the manufacture of high-temperature resistant polyesters or liquid crystals. For example, zeolite mordenite, whose pore system has been tailored by dealumination [198, 201, 203] and pore size

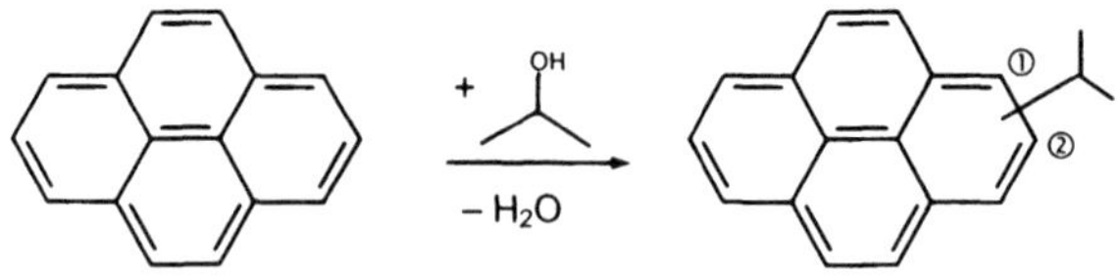

	X_{Py}, %	$S_{1\text{-}IPP}$, %	$S_{2\text{-}IPP}$, %	S_{pIPP}, %	m_{coke}/m_Z, %
HY	16.7	30.4	35.5	34.1	2.5
HMCM-41	22.5	14.3	85.7	0	20.4
HMCM-48	33.9	12.8	74.9	12.3	15.4

Fig. 5.18. Shape selective isopropylation over mesoporous catalysts; S_{pIPP} denotes the selectivity for poly-isopropylpyrenes and m_Z the mass of the catalyst, Ref. [209]

engineering [204, 206], has been successfully used for the production of 4,4′-diisopropylbiphenyl. In addition, zeolites with other pore structures have been explored as well as catalysts for the shape-selective conversion of polynuclear aromatics, e.g., zeolites beta or ZSM-12 [97, 199].

In the last few years, the shape-selective conversion of even trinuclear aromatics has been reported. Examples are the direct synthesis of anthraquinone from phthalic anhydride and benzene in the shape selective environment offered by the pores of zeolite beta [207] and the isomerization of symmetrical octahydrophenanthrene to symmetrical octahydro-anthracene over zeolite H-mordenite [208].

Most interestingly, even shape selectivity in mesoporous molecular sieves has been claimed very recently [209]. Pu et al. studied the isopropylation of pyrene with isopropanol over acidic MCM-41- and MCM-48-type materials and, for comparison, over zeolite HY. Pertinent results of their study are summarized in Fig. 5.18. As evidence for the occurrence of shape selectivity effects, the following experimental observations are quoted: (i) The mesoporous catalysts showed a significantly higher activity than the large-pore zeolite, which was interpreted as an indication for the pores in MCM-41 and MCM-48 being accessible for pyrene. (ii) MCM-48, possessing a three-dimensional pore system, was more active than MCM-41 with its one-dimensional pores. It was argued that the reaction over zeolite Y occurs at its external surface since 1- and 2-isopropylpyrene are formed unselectively in almost equal amounts, and large quantities of poly-isopropylpyrenes are formed. Moreover, there is only very little coke formation on zeolite Y which suggests that the large polynuclear pyrene molecules, which could undoubtedly act as coke precursors, do not have access to the intracrystalline void volume of the zeolite. As an indication for the (shape-selective) conversion of pyrene inside the mesopores of MCM-41 and MCM-48, the large amount of coke formed after one hour time-on-stream accompanied by a large surplus of the less bulky 2-isopropylpyrene (as compared to 1-isopropylpyrene) and a strongly suppressed formation of polyisopropylpyrenes has been quoted. Even though many questions concerning shape-selective catalysis in mesoporous materials are still not answered, this is an attractive field of contemporary research in heterogeneous catalysis. Once the feasibility has been proven unambiguously, the door will be open for a whole bunch of desirable reactions.

5.8.2
Shape-Selective Catalysis on Transition Metals in Zeolites

Catalytic reactions over small transition metal clusters located in the intracrystalline environment of molecular sieves may proceed in a shape-selective manner. This has been demonstrated by Dessau [210, 211], among others. He showed that over platinum clusters located inside the pores of zeolite ZSM-5, linear olefins (e.g., hexene-1) are preferentially hydrogenated over branched olefins (e.g., 4,4-dimethylhexene-1), and this finding was exploited as a diagnostic test for the location of the noble metal. *para*-Xylene could also be selectively oxidized in the presence of the bulkier *ortho*-xylene over copper-loaded ZSM-5 [210]. Other examples for shape-selective, metal-catalyzed reactions are the conversion of prehnitene over platinum clusters located in the pores of silicalite-1 [212] and the hydrogenation [213, 214] or hydroformylation of olefins [215] on intracrystalline rhodium clusters.

More recently, Weitkamp et al. [216] have described the preparation of small noble-metal clusters (i.e., of palladium, platinum and rhodium) in different small-pore molecular sieves, such as ZSM-58, Rho, ZK-5 or SAPO-42, via solid-state ion exchange. These materials proved to be highly selective in the hydrogenation of hexene-1 in the presence of 2,4,4-trimethylpentene-1.

For the future, we expect many more examples of shape-selective catalysis on small metal particles inside microporous materials.

5.8.3
Stereoselective Catalysis in Zeolites

One of the ultimate goals in zeolite science is certainly stereoselective catalysis [14, 17]. Recent examples have been reported for the liquid-phase alkylation of α-chiral benzaldehydes by butyllithium, which results in the formation of an increased proportion of the so-called Cram product in the diastereomeric mixture [217], and for the dimerization of styrene followed by subsequent cyclization of the open dimers to the indane dimers [218]. In the latter case, a significant degree of diastereoselectivity was observed with large-pore zeolites, such as faujasite or beta.

Enantioselective catalysis would require zeolitic materials with chiral channels or cavities. Since the preparation of such materials has been so far unsuccessful, the accommodation of chiral guests inside the pores which either act themselves as catalytically active sites or induce chirality onto an adjacent site is required. Enantioselective catalysis on zeolite-supported systems has been demonstrated with chiral complexes anchored in ultrastabilized zeolite Y [219-221] or platinum-loaded zeolites modified by cinchonidine [222].

Very recently, enantioselective epoxidation of alkenes has been achieved on manganese complexes with a chiral salen-derived ligand [e.g., *trans*-(*R*,*R*)-1,2-bis(salicylidene-amino)cyclohexane] by two different groups [223, 224]. The complexes were assembled from smaller species inside the large cavities of zeolite Y via the ship-in-a-bottle method. In both cases, high enantiomeric excess values were found in the epoxidation of aromatic alkenes with NaClO. Such catalytic systems are at the same time interesting examples for zeolitic host/guest compounds. It remains to be seen how the sizes of the zeolite pores and channels and the size of the immobilized species limit the size of the reactants which can be converted.

5.8.4
Host/Guest Chemistry in Zeolites

Zeolites and zeolite-related microporous solids are ideal inorganic hosts for a large variety of guest species. Host/guest chemistry and catalysis in zeolites have previously been reviewed [14]. Here, only a few selected examples will be highlighted. For example, host/guest photochemistry in zeolites has been shown to allow a photochemical reaction to be influenced by the pore geometry of the zeolitic host. A large variety of photochemical reactions in zeolites has been studied so far, the most prominent ones being Norrish type I and type II reactions of ketones and photochlorination or photolysis of carbonyl compounds.

Recently, a modest enantioselectivity was achieved during the photolysis of two ketones capable of undergoing the Norrish-Yang reaction of, e.g., 1-adamantyl-*p*-carbomethoxyacetophenone (cf. Fig. 5.19) [225, 226]. This was possible by employing a strategy which involved the use of chirally modified zeolites as the reaction medium. As chiral modifier, ephedrine was used in a concentration of ca. one molecule per supercage of zeolite NaY. The observed values for the enantiomeric excess of the cyclobutanols formed during the reaction were 5% for the *trans* and 35% for the *cis* configuration. Although these ee-values are still low, this novel approach offers interesting perspectives for chiral induction in photochemical reactions.

A rapidly expanding field of host/guest chemistry in zeolites is the immobilization of transition metal complexes in the intracrystalline voids of microporous hosts (ship-in-a-bottle catalysts). Although their preparation and properties have been reviewed in a number of recent articles [e.g., 227–230], it is difficult to keep track of the fast development. Whereas in the 1980s, only faujasites were used as hosts, this basis has recently been extended to other structures (e.g., VPI-5, EMT, MCM-22). Also, many more ligand systems are nowadays under investigation, viz. phthalocyanines, salene and some of its derivatives, bipyridines etc. The incentive for preparing such ship-in-a-bottle catalysts can be twofold: (i) zeolite-encapsulated catalytically active transition metal complexes offer some specific advantages over their counterparts in homogeneous

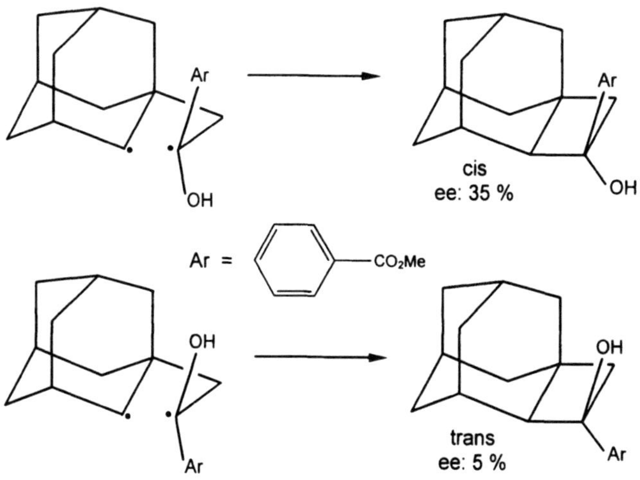

Fig. 5.19. Enantioselective photochemical Norrish–Yang reaction of 1-adamantyl-*p*-carbomethoxyacetophenone over zeolite NaY loaded with ca. one ephedrine molecule per supercage Ref. [226]

solution, viz. easy catalyst separation from the reaction medium and the possibility of using a large variety of different solvents and reaction conditions [227, 230]. Moreover, the activity and selectivity of the immobilized complexes could be modified by constraints superimposed by the restricted intracrystalline environment of the zeolite. In addition, a higher catalytic activity and stability of the encapsulated complexes are usually observed upon immobilization [230]. (ii) Zeolite-encapsulated complexes have been suggested as model compounds for enzyme mimicking [227]. In this context, the term "zeozymes" has been coined in order to describe a catalytic system in which the zeolite replaces the protein mantle of the enzyme, while the entrapped metal complex (e.g., a metal-phthalocyanine) mimics the active site of the enzyme (e.g., an iron-porphyrin). An efficient mimic of cytochrome P-450 was recently successfully prepared by Parton et al. [231]. It comprises the encapsulation of iron-phthalocyanine complexes in the supercages of zeolite Y and embedding this inclusion compound in a polydimethylsiloxane membrane. This system oxidizes alkanes at room temperature at rates comparable to the true enzyme. It will be interesting to watch how this concept can be generalized or transferred to other types of encaged complexes and host materials with other structures.

Acknowledgements. J.W. and S.E. gratefully acknowledge financial support from Deutsche Forschungsgemeinschaft, Fonds der Chemischen Industrie and Max-Buchner-Forschungsstiftung.

References

1. Weisz PB, Frilette VJ (1960) J Phys Chem 64:382
2. Weisz PB, Frilette VJ, Maatman RW, Mower EB (1962) J Catal 1:307
3. Csicsery SM (1976) Shape-selective catalysis. In: Rabo JA (ed) Zeolite chemistry and catalysis. ACS monograph, vol 171, American Chemical Society, Washington DC, p 680
4. Weisz PB (1980) Pure Appl Chem 52:2091
5. Chang CD, Lang WH, Bell WK (1981) In: Moser WR (ed) Catalysis of organic reactions, Chemical Industries, vol 4, Marcel Dekker, New York, p 73
6. Derouane EG (1982) In: Whittingham MS, Jacobsen AJ (eds) Intercalation chemistry, Academic Press, New York, p 101
7. Derouane EG (1984) In: Kaliaguine S, Mahay, A (ed) Catalysis on the energy scene, Studies in surface science and catalysis, vol 19, Elsevier, Amsterdam, p 1
8. Csicsery SM (1984) Zeolites 4:202
9. Haag WO, Olson DH, Weisz PB (1984) In: H. Grünewald (ed) Chemistry for the future, Pergamon, Oxford, p 327
10. Csicsery SM (1985) Chemistry in Britain 21:473
11. Csicsery SM (1986) Pure Appl Chem 58:841
12. Chen NY, Garwood WE (1986) Catal Rev Sci Eng 28:185
13. Weitkamp J, Ernst S, Dauns H, Gallei E (1986) Chem Ing Tech 58:623
14. Weitkamp J (1993) In: von Ballmoos R, Higgins JB, Treacy MMJ (eds) Proceedings from the ninth international zeolite conference, vol 1. Butterworth-Heinemann, Boston, p 13
15. Chen NY, Degnan TF Jr, Smith CM (1994) Molecular transport and reaction in zeolites. VCH, Weinheim
16. Csicsery SM (1995) In: Beyer HK, Karge HG, Kiricsi I, Nagy JB (eds) Catalysis by microporous materials, Studies in surface science and catalysis, vol 94. Elsevier, Amsterdam, p 1
17. Weitkamp J, Weiss U, Ernst S (1995) In: Beyer HK, Karge HG, Kiricsi I, Nagy JB (eds) Catalysis by microporous materials, Studies in surface science and catalysis, vol 94. Elsevier, Amsterdam, p 363
18. Chen NY, Garwood WE, Dwyer FG (1996) Shape selective catalysis in industrial applications, 2nd edn. Marcel Dekker, New York
19. Venuto PB (1996) In: Chon H, Ihm S-K, Uh YS (eds) Progress in zeolite microporous materials, Studies in surface science and catalysis, vol 105, part B. Elsevier, Amsterdam, p 811
20. Breck DW (1984) Zeolite molecular sieves - structure, chemistry and use, 2nd edn. Robert E. Krieger, Malabar, Florida, pp 633-645
21. Bird BB, Stewart WE, Lightfoot EN (1960) Transport phenomena. John Wiley, New York, pp 19-26
22. Reid RC, Prausnitz JM, Sherwood TK (1977) The properties of gases and liquids. McGraw-Hill, New York, pp 23-24, 678-679
23. Beck JS, Vartuli JC, Leonowicz ME, Kresge CT, Schmitt KD, Chu CTW, Olson DH, Sheppard EW, McCullen SB, Higgins JB, Schlenker JL (1992) J Am Chem Soc 114:10834
24. Casci JL (1994) In: Jansen JC, Stöcker M, Karge HG, Weitkamp J (eds) Advanced zeolite science and applications, Studies in surface science and catalysis, vol 85. Elsevier: Amsterdam, p 329
25. Meier WM, Olson DH, Baerlocher C (1996) Atlas of zeolite structure types, 4th edn. Elsevier, London
26. Dwyer FG, Chu P (1985) US Patent 4 526 879 assigned to Mobil Oil Corp
27. Rabo JA (1976) In: Rabo JA (ed) Zeolite chemistry and catalysis. American Chemical Society, Washington DC, p 332
28. Ruthven DM (1988) Chem Eng Progr 84 (No 2):42
29. Yan TY (1988) Ind Eng Chem Res 27:1665
30. Boger T, Fritz F, Ascher R, Ernst S, Weitkamp J, Eigenberger G (1997) Chem-Ing-Tech 69:475

31. Weitkamp J, Schwark M, Ernst S (1989) Chem-Ing-Tech 61:887
32. Wu EL, Landolt GR, Chester AW (1986) In: Murakami Y, Iijima A, Ward JW (eds) New developments in zeolite science and technology. Kodansha, Tokyo and Elsevier, Amsterdam, p 547
33. Bendoraitis JG, Chester AW, Dwyer FG, Garwood WE (1986) In: Murakami Y, Iijima A, Ward JW (eds) New developments in zeolite science and technology. Kodansha, Tokyo and Elsevier, Amsterdam, p 669
34. den Exter MJ, Jansen, JC, van Bekkum H (1994) In: Weitkamp J, Karge HG, Pfeifer H, Hölderich W (eds) Zeolite and related microporous materials: state of the art 1994, Studies in surface science and catalysis, Vol 84, Part B. Elsevier, Amsterdam, p 1159
35. Breck DW (1984) Zeolite molecular sieves - structure, chemistry and use, 2nd edn. Robert E. Krieger, Malabar, Florida, p 596
36. Davis ME, Saldarriaga C, Montes C, Garces J, Crowder C (1988) Nature 331:698
37. Lobo RF, Tsapatsis M, Freyhardt CC, Khodabandeh S, Wagner P, Chen C-Y, Balkus KJ Jr, Zones SI, Davis ME (1997) J Am Chem Soc 119:8474
38. Freyhardt CC, Tsapatsis M, Lobo RF, Balkus KJ Jr, Davis ME (1996) Nature 381:295
39. Boudart M (1992) In: Thomas JM, Zamaraev KI (eds) Perspectives in catalysis. Blackwell, Oxford, p 183
40. Boudart M (1997) In: Ertl G, Knözinger H, Weitkamp J (eds) Handbook of heterogeneous catalysis, vol 1. VCH, Weinheim, p 1
41. Riekert L (1985) Appl Catal 15:89
42. Butruille J-R, Pinnavaia TJ (1992) Catal Lett 12:197
43. Garces JM, Vrieland GE, Bates SI, Scheidt FM (1985) In: Imelik B, Naccache C, Ben Taarit Y, Vedrine JC, Coudurier G, Praliaud H (eds) Catalysis by zeolites. Elsevier, Amsterdam, p 67
44. Boxhoorn G, Moolhuysen J, Coolegem JGF, van Santen R (1985) J Chem Soc Chem Commun 1305
45. Inomata M, Yamada M, Okada S, Niwa M, Murakami Y (1986) J Catal 100:264
46. Noack M, Spindler H (1990) Z Chem 30:188
47. Gilson JP, Derouane EG (1984) J Catal 88:538
48. Farcasiu M, Degnan TF (1988) Ind Eng Chem Res 27:45
49. Suzuki I, Oki S, Namba S (1986) 100:219
50. Sato H, Sakamoto A, Hirose K, Chikaishi K (1989) Chem Lett 1695
51. Sayari A, Crusson E, Kaliaguine S, Brown JR (1991) Langmuir 7:314
52. Karge HG, Kösters H, Wada Y (1984) In: Olson D, Bisio A (eds) Proceedings of the sixth international zeolite conference, Butterworths, London, p 308
53. Weber RW, Fletcher JCQ, Möller KP, O'Connor CT (1996) Microporous Materials 7:15
54. Melson S, Schüth F (1997) J Catal 170:46
55. Handreck PG, Smith TD (1989) J Chem Soc, Faraday Trans 1, 85:645
56. Namba S, Inaka A, Yashima T (1986) Zeolites 6:107
57. Röger HP, Möller KP, O'Connor CT (1997) Microporous Materials 8:151
58. Chen CSH, Schramm SE (1996) Microporous Materials 7:125
59. Harvey G, Binder G, Prins R (1995) In: Beyer HK, Karge HG, Kiricsi I, Nagy JB (eds) Catalysis by microporous materials, Studies in surface science and catalysis, vol 93. Elsevier, Amsterdam, p 397
60. Creyghton, EJ, Elings JA, Downing RS, Sheldon RA, van Bekkum H (1996) Microporous Materials 5:299
61. Lippens BC, de Boer JH (1965) J Catal 4:319
62. Sing KSW, Rouquerol J (1997) In: Ertl G, Knözinger H, Weitkamp J (eds) Handbook of heterogeneous catalysis, vol 2. VCH, Weinheim, p 427
63. Bellat J-P, Pilverdier E, Simonot-Grange M-H, Jullian S (1997) Microporous Materials 9:213
64. Sing KSW (1970) In: Everett DH, Ottewills RH (eds) Surface area determination, Butterworths, London, p 25
65. Olson DH, Kerr GT, Lawton SL, Meier WM (1981) J Phys Chem 85:2238

66. Chen NY, Weisz PB (1967) In: Weisz PB, Hall WK (eds) Kinetics and catalysis, Chem Eng Progr Symp Ser 63 (no 73), AIChE, New York, p 86
67. Weisz PB (1965) Erdöl Kohle Erdgas, Petrochem 18:527
68. Csicsery SM (1970) J Catal 19:394
69. Csicsery SM (1971) J Catal 23:124
70. Weitkamp J, Ernst S (1994) Catal Today 19:107
71. Chen NY, Maziuk Z, Schwartz AB, Weisz PB (1968) Oil Gas, J 66 (no 47):154
72. Chen NY, Garwood WE, Heck RH (1987) Ind Eng Chem Res 26:706
73. Bennett RN, Elkes GJ, Wanless GJ (1975) Oil Gas, J 73 (no 1):69
74. Donelly SP, Green JR (1980) Oil Gas, J 78 (no 43):77
75. Yanik SJ, Demmel EJ, Humphries AP, Campagna RJ (1985) Oil Gas, J 83 (no 19):108
76. Kaeding WW, Young LB, Chu CC (1984) J Catal 89:267
77. Kaeding WW, Young LB, Prapas AG (1982) Chemtech 12:556
78. Olson DH, Haag WO (1984) In: Whyte Jr TE, Dalla Betta RA, Derouane EG, Baker RTK (eds) Catalytic materials: relationship between structure and reactivity, ACS Symp Ser, vol 248, American Chemical Society, Washington DC, p 275
79. Haag WO, Lago RM, Weisz PB (1982) Faraday Discuss Chem Soc 72:317
80. Hölderich W, Riekert L (1986) Chem-Ing-Tech 58:412
81. Voogd P, van Bekkum H (1990) Appl Catal 59:311
82. Gorring RL (1973) J Catal 31:13
83. Chen NY, Lucki SJ, Mower EB (1969) J Catal 13:329
84. Chen NY, Garwood WE (1973) In: Meier WM, Uytterhoeven JB (eds) Molecular sieves. American Chemical Society, Washington DC, p 575
85. Cavalcante Jr CL, Eic M, Ruthven DM, Occelli ML (1995) Zeolites 15:293
86. Magalhães FD, Laurence RL, Conner WC (1996) AIChE J 42:68
87. Derouane EG, Gabelica Z (1980) J Catal 65:486
88. Lowe BM, Whan DA, Spencer MS (1981) J Catal 70:237
89. Derouane EG, Gabelica Z, Jacobs PA (1981) J Catal 70:238
90. Pope GC (1981) J Catal 72:174
91. Derouane EG (1981) J Catal 72:177
92. Fraenkel D, Cherniavsky M, Levy M (1984) In: Proceedings from the 8th international congress on catalysis, vol 4, VCH, Weinheim, p 545
93. Fraenkel D, Cherniavsky M, Levy M (1986) J Catal 101:273
94. Derouane EG (1986) J Catal 100:541
95. Derouane EG, Andre J-M, Lucas AA (1988) J Catal 110:58
96. Neuber M, Weitkamp J (1989) In: Jansen JC, Moscou L, Post MFM (eds) Zeolites for the nineties; Recent research reports, 8th International zeolite conference, Amsterdam, p 425
97. Weitkamp J, Neuber M (1991) In: Inui T, Namba S, Tatsumi T (eds) Chemistry of microporous crystals, Studies in surface science and catalysis, vol 60. Kodansha, Tokyo, and Elsevier, Amsterdam, p 291
98. Martens JA, Parton R, Uytterhoeven L, Jacobs PA, Froment GF (1991) Appl Catal 76:95
99. Martens JA, Souverijns W, Verrelst W, Parton R, Froment GF, Jacobs PA (1995) Angew Chem Int Ed Engl 34:2528
100. Gallezot P, Giroir-Fendler A, Richard D (1990) Catal Letters 5:169
101. Schulz-Ekloff G, Wright D, Grunze M (1982) Zeolites 2:70
102. Jaeger NI, Ryder P, Schulz-Ekloff G (1982) In: Jacobs PA, Jaeger NI, Jiru P, Kazansky VB, Schulz-Ekloff G (eds) Structure and reactivity of modified zeolites, Studies in surface science and catalysis, vol 18, Elsevier, Amsterdam, p 299
103. Gallezot P, Blanc B, Barthomeuf D, Païs da Silva MI (1994) In: Weitkamp J, Karge HG, Pfeifer H, Hölderich W (eds) Zeolites and related microporous materials: State of the art 1994, Studies in surface science and catalysis, vol 84, part B. Elsevier, Amsterdam, p 1433
104. Namba S, Sato K, Fujita K, Kim JH, Yashima T (1986) In: Murakami Y, Iijima A, Ward JW (eds) New developments in zeolite science and technology, Kodansha, Tokyo, and Elsevier, Amsterdam, p 661

105. Santilli DS, Zones SI (1990) Catal Letters 7:383
106. Van Nordstrand RA, Santilli DS, Zones SI (1992) In: Occelli M, Robson HE (eds) Synthesis of microporous materials, vol 1. Van Nostrand Reinhold, New York, p 373
107. Santilli DS, Harris TV, Zones SI (1993) Microporous Materials 1:329
108. Vansant EF (1990) Pore size engineering in zeolites. John Wiley, Chichester
109. Ernst S (1998) In: Weitkamp J, Karge HG (eds) Molecular sieves-science & technology, vol 1. Springer, Heidelberg, p 65
110. Shannon MD, Casci JL, Cox PA, Andrews SJ (1991) Nature 353:417
111. Leonowicz ME, Lawton JA, Lawton SL, Rubin MK (1994) Science 264:1910
112. Lobo RF, Pan M, Chan I, Medrud RC, Zones SI, Crozier PA, Davis ME (1993) Science 262:1543
113. Freyhardt CC, Tsapatsis M, Lobo RF, Balkus KJ Jr, Davis ME (1996) Nature 381:295
114. Flanigen EM, Patton RL, Wilson ST (1988) In: Grobet PJ, Mortier WJ, Vansant EF, Schulz-Ekloff G (eds) Innovation in zeolite materials science, Studies in surface science and catalysis, vol 37. Elsevier, Amsterdam, p 13
115. Rabo JA, Pellet RJ, Coughlin PK, Shamshoum ES (1989) In: Karge HG, Weitkamp J (eds) Zeolites as catalysts, sorbents and detergent builders-applications and innovations. Elsevier, Amsterdam, p 1
116. Tielen M, Geelen M, Jacobs PA (1985) Acta Phys Chem 31:1
117. Shiralkar VP, Joshi PN, Eapen MJ, Rao BS (1991) Zeolites 11:511
118. Beschmann K, Riekert L, Müller U (1994) J Catal 145:243
119. Bhat YS, Das J, Rao KV, Halgeri AB (1996) J Catal 159:368
120. Rollmann LD (1980) US Patent 4 203 869 assigned to Mobil Oil Corp
121. Angevine PJ, Kühl GH, Mizrahi S (1985) US Patent 4521297 assigned to Mobil Oil Corp
122. Suib SL, Stucky GD, Blattner RJ (1980) J Catal 65:174
123. von Ballmoos R, Meier WM (1981) Nature 289:782
124. Chao KJ, Chern, JY (1988) Zeolite 8:82
125. Jacobs PA, Martens JA (1987) Synthesis of high-silica aluminosilicate zeolites, Studies in surface science and catalysis, vol 33. Elsevier, Amsterdam, p 91
126. Althoff R, Schulz-Dobrick B, Schüth F, Unger K (1993) Microporous Materials 1:207
127. Jansen, JC, Engelen CWR, van Bekkum H (1989) In: Occelli ML, Robson HE (eds) Zeolite synthesis, ASC Symp Ser, vol 398, American Chemical Society, Washington DC, p 257
128. Müller U, Brenner A, Reich A, Unger KK(1989) In: Occelli ML, Robson HE (eds) Zeolite synthesis, ACS Symp Ser, vol 398, American Chemical Society, Washington DC, p 346
129. Namba S, Iwase O, Takahashi N, Yashima, T, Hara N (1979) J Catal 56:445
130. Cartraud P, Cointot, A, Dufour M, Gnep, NS, Guisnet M, Joly G, Tejada J (1986) Appl Catal 21:85
131. Vinek H, Derewinski M, Mirth G, Lercher JA (1991) Appl Catal 68:277
132. Karge HG, Ladebeck J, Sarbak Z, Hatada K (1982) Zeolites 2:94
133. Ramôa Ribeiro F (1993) Catal Letters 22:107
134. Kaeding WW, Chu C, Young LB, Weinstein B, Butter SA (1981) J Catal 67:159
135. Niwa M, Senoh N, Hibino T, Nakatsuka Y, Murakami Y (1994) In: Hattori T, Yashima T (eds) Zeolites and microporous crystals, Studies in surface science and catalysis, vol 83. Kodansha, Tokyo, and Elsevier, Amsterdam, p 155
136. Niwa M, Yamazaki K, Murakami Y (1994) Ind Eng Chem Res 33:371
137. Choplin A (1994) J Molec Catal 86:501
138. Kuno H, Shibagaki M, Takahashi K, Honda I, Matsushita H (1993) In: Guczi L, Solymosi F, Tétényi (eds) New frontiers in catalysis, Elsevier, Amsterdam, p 1727
139. Apelian MR, Degnan TF, Fung AS (1993) US Patent 5 234 872 assigned to Mobil Oil Corp
140. Namba S, Inaka A, Yashima T (1984) Chem Letters 817
141. Anderson JR, Chang Y-F, Hughes AE (1989) Catal Letters 2:279
142. Anderson JR, Foger K, Mole T, Rajadhyaksha RA, Sanders JV (1993) Zeolites 13:518
143. Chen CSH, Bridger RF (1996) J Catal 161:871
144. Rollmann, LD (1991) In: Bartholomew CH, Butt JB (eds) Catalyst deactivation 1991, Studies in surface science and catalysis, vol 68. Elsevier, Amsterdam, p 791

145. Matsuda T, Kikuchi E (1993) In: Hattori T, Yashima T (eds) Zeolites and microporous crystals, Studies in surface science and catalysis, vol 83. Elsevier, Amsterdam, p 295
146. Martens JA, Tielen M, Jacobs PA, Weitkamp J (1984) Zeolites 4:98
147. Dewing J (1984) J Molec Catal 27:25
148. Ribeiro FR, Lemos F, Perot G, Guisnet M (1986) In: Setton R (ed) Chemical reactions in organic and inorganic constrained systems, D. Reidel, Dordrecht, p 141
149. Jacobs PA, Martens JA (1986) Pure Appl Chem 58:1329
150. Weitkamp J, Ernst S (1988) In: Ward JW (ed) Catalysis 1987, Studies in surface science and catalysis, vol 38. Elsevier, Amsterdam, p 367
151. Ernst S, Kumar R, Neuber M, Weitkamp J (1988) In: Unger KK, Rouquerol J, Sing KSW, Kral H (eds) Characterization of porous solids, Studies in surface science and catalysis, vol 39. Elsevier, Amsterdam, p 531
152. Weitkamp J, Ernst S (1988) Catal Today 3:451
153. Weitkamp J, Ernst S (1994) Catal Today 19:107
154. Frilette VJ, Haag WO, Lago RM (1981) J Catal 67:218
155. Levenspiel O (1972) Chemical reaction engineering, 2nd ed. John Wiley, New York, p 484
156. Morrison RA (1987) US Patent 4 686 316 assigned to Mobil Oil Corp
157. Ernst S, Weitkamp J, Martens JA, Jacobs PA (1989) Appl Catal 48:137
158. Haag WO, Dessau RM (1984) In: Proc. 8th Intern Congr Catal, vol II, Verlag Chemie, Weinheim, p. 305
159. Haag WO, Dessau RM, Lago RM (1991) In: Inui T, Namba S, Tatsumi T (eds) Chemistry of microporous crystals, Studies in surface science and catalysis, vol 60. Elsevier, Amsterdam, p. 255
160. Csicsery SM (1969) J Org Chem 34:3338
161. Corma A, Sastre E (1991) J Catal 129:177
162. Gnep NS, Tejada J, Guisnet M (1982) Bull Soc Chim Fr I:5
163. Martens JA, Perez-Pariente J, Sastre E, Corma A, Jacobs PA (1988) Appl Catal 45:85
164. Joensen F, Blom N, Tapp NJ, Derouane EG, Fernandez C (1989) In: Jacobs PA, van Santen RA (eds) Zeolites: Facts, figures, future, Studies in surface science and catalysis, vol 49, part B. Elsevier, Amsterdam, p 1131
165. Richter M, Fiebig, Jerschkewitz H-G, Lischke G, Öhlmann G (1989) Zeolites 9:238
166. Kumar R, Rao GN, Ratnasamy P (1989) In: Jacobs PA, van Santen RA (eds) Zeolites: Facts, figures, future, Studies in surface science and catalysis, vol 49, part B. Elsevier, Amsterdam, p 1141
167. Rao GN, Kumar R, Ratnasamy P (1989) Appl Catal 49:307
168. Kumar R, Ratnasamy P (1989) J Catal 116:440
169. Kumar R, Ratnasamy P (1989) J Catal 118:68
170. Ratnasamy P, Bhat RN, Pokhriyal SK, Hegde SG, Kumar R (1989) J Catal 119:65
171. Corma A, Fornés V, Pérez-Pariente J, Sastre E, Martens JA, Jacobs PA (1988) In: Flank WH, Whyte TE Jr (eds) Perspectives in molecular sieves science, ACS Symp Ser, vol 368. American Chemical Society, Washington DC, p 555
172. Pérez-Pariente J, Sastre E, Fornés V, Martens JA, Jacobs, PA, Corma A (1991) Appl Catal 69:125
173. Corma A, Llopis F, Monton JB (1993) In: Guczi L, Solymosi F, Tétényi (eds) New frontiers in catalysis, Elsevier, Amsterdam, p 1145
174. Morin S, Gnep NS, Guisnet M (1996) J Catal 159:296
175. Csicsery SM (1987) J Catal 108:433
176. Csicsery SM (1988) J Catal 110:348
177. Karge HG, Hatada K, Zhang Y, Fiedorow R (1983) Zeolites 3:13
178. Karge HG, Wada, Y, Weitkamp J, Ernst S, Girrbach U, Beyer HK (1984) In: Kaliaguine S, Mahay A (eds) Catalysis on the energy scene, Studies in surface science and catalysis, vol 19. Elsevier, Amsterdam, p 101
179. Weitkamp J, Ernst S, Jacobs PA, Karge HG (1986) Erdöl Erdgas Kohle, Petrochem 39:13

180. Kim M-H, Chen C-Y, Davis ME (1993) In: Davis ME, Suib SL (eds) Selectivity in catalysis, ACS Symp Ser, vol 517. American Chemical Society, Washington DC, p 222
181. Martens JA, Jacobs PA (1986) Zeolites 6:334
182. Schulz H, Weitkamp J (1972) Ind Eng Chem Prod Res Dev 11:46
183. Weitkamp J (1975) In: Ward JW, Qader SA (eds) Hydrocracking and hydrotreating, ACS Symp Ser, vol 20. American Chemical Society, Washington DC, p 1
184. Weitkamp J (1978) Erdöl Kohle Erdgas, Petrochem 31:13
185. Steijns M, Froment G, Jacobs PA, Uytterhoeven J, Weitkamp J (1981) Ind Eng Chem Prod Res Dev 20:654
186. Weitkamp J (1982) Ind Eng Chem Prod Res Dev 21:550
187. Weitkamp J, Ernst S, Kumar R (1986) Appl Catal 27:207
188. Weitkamp J, Ernst S, Chen CY (1989) In: Jacobs PA, van Santen RA (eds) Zeolites:facts, figures, future, Studies in surface science and catalysis, vol 49, part B. Elsevier, Amsterdam, p 1115
189. Weitkamp J, Ernst S, Cortés-Corbéran V, Kokotailo GT (1986) In: Preprints of poster papers, 7th International zeolite conference, Aug, 17-20, Tokyo, p 239
190. Martens JA, Vanbutsele G, Jacobs PA (1993) In: von Ballmoos R, Higgins JB, Treacy MMJ (eds) Proceedings from the ninth international zeolite conference, vol 2. Butterworth-Heinemann, Stoneham, p 355
191. Feijen EJP, Martens JA, Jacobs PA (1996) In: Hightower JW, Delgass WN, Iglesia E, Bell AT (eds) 11th International congress on catalysis, Studies in surface science and catalysis, vol 101. Elsevier, Amsterdam, p 721
192. Jacobs PA, Martens JA, Weitkamp J, Beyer HK (1982) Farad Discuss 72:353
193. Weitkamp J, Jacobs PA, Martens JA (1983) Appl Catal 8:123
194. Weitkamp J, Ernst S, Karge HG (1984) Erdöl Kohle Erdgas, Petrochem 37:457
195. Ernst S, Weitkamp J (1985) Acta Phys Chem 31:457
196. Venuto PB (1994) Microporous Materials 2:297
197. Arends IWCE, Sheldon RA, Wallau M, Schuchardt U (1997) Angew Chem 109:1190
198. Horseley JA, Fellmann JD, Derouane EG, Freeman CM (1994) J Catal 147:231
199. Pu S-B, Inui T (1996) Appl Catal A General 146:305
200. Matsuda T, Kikuchi E (1993) Res Chem Intermed 19:319
201. Sugi Y, Toba M (1994) Catal Today 19:187
202. Lee GS, Maj JJ, Rocke SC, Garces JM (1989) Catal Letters 2:243
203. Tu X, Matsumoto M, Matsuzaki T, Hanaoka T, Kubota Y, Kim J-H, Sugi Y (1993) Catal Letters 21:71
204. Sugi Y, Matsuzaki T, Hanaoka T, Kubota Y, Kim J-H, Tu X, Matsumoto M (1994) Catal Letters 27:315
205. Sugi Y, Matsuzaki T, Hanaoka T, Kubota Y, Kim J-H, Tu X, Matsumoto M (1994) Catal Letters 26:181
206. Takeuchi G, Shimoura Y, Hara T (1996) Appl Catal A General 137:87
207. Kikhtyanin GV, Ione K, Snytnikova GP, Malysheva LV, Toktarev AV, Paukshtis EA, Spichtinger R, Schüth F, Unger KK (1994) In: Weitkamp J, Karge HG, Pfeifer H, Hölderich W (eds) Zeolites and related microporous materials: state of the art 1994, Studies in surface science and catalysis, vol 84, part C. Elsevier, Amsterdam, p 1905
208. Song C, Moffat K (1994) Microporous Materials 2:459
209. Pu SB, Kim JB, Seno M, Inui T (1997) Microporous Materials 10:25
210. Dessau RM (1982) J Catal 77:304
211. Dessau RM (1984) J Catal 89:520
212. Martens JA, Jacobs PA (1984) In: Jacobs PA, Jaeger NI, Jiru P, Kazansky VB, Schulz-Ekloff G (eds) Structure and reactivity of modified zeolites, Studies in surface science and catalysis, vol 18. Elsevier, Amsterdam, p 189
213. Yamaguchi I, Joh T, Takahashi S (1986) J Chem Soc Chem Commun 1412
214. Corbin DR, Seidel WC, Abrams L, Herron N, Stucky GD, Tolman CA (1985) Inorg Chem 24:1800
215. Davis RJ, Rossin JA, Davis ME (1986) J Catal 98:477

216. Weitkamp J, Ernst S, Bock T, Kiss A, Kleinschmit P (1995) In: Beyer HK, Karge HG, Kiricsi I, Nagy JB (eds) Catalysis by microporous materials, Studies in surface science and catalysis, vol 94. Elsevier, Amsterdam, p 278
217. Mahrwald R, Lohse U, Girnus I, Caro J (1994) Zeolites 14:486
218. Benito A, Corma A, Garcia H, Primo J (1994) Appl Catal 116:127
219. Corma A, Iglesias M, del Pino C, Sanchez F (1991) J Chem Soc Chem Commun 1253
220. Corma A, Iglesias M, del Pino C, Sanchez F (1993) In: Guczi L, Solymosi F, Tétényi (eds) New frontiers in catalysis. Elsevier, Amsterdam, p 2294
221. Carmona A, Corma A, Iglesias M, Sanchez F (1996) Inorganica Chimica Acta 244:79
222. Reschetilowski W, Böhmer U, Wiehl J (1984) In: Weitkamp J, Karge HG, Pfeifer H, Hölderich W (eds) Zeolites and related microporous materials: state of the art 1994, Studies in surface science and catalysis, vol 84, part C. Elsevier, Amsterdam, p 2021
223. Ogunwumi SB, Bein T (1997) J Chem Soc Chem Commun 901
224. Sabater M, Corma A, Domenech A, Fornés V, Garcia H (1997) J Chem Soc Chem Commun: 1285
225 Leibovitch M, Olovsson G, Sundararabu G, Ramamurthy V, Scheffer JR, Trotter J (1996) J Am Chem Soc 118:1219
226. Sundararabu G, Leibovitch M, Corbin DR, Scheffer JR, Ramamurthy V (1996) J Chem Soc Chem Commun 2159
227. Parton RF, De Vos D, Jacobs PA (1992) In: Derouane EG, Lemos F, Naccache C, Ramoa Ribeiro F (eds) Zeolite microporous solids: synthesis, structure and reactivity. Kluwer Academic Publishers, Dordrecht, p. 555
228. De Vos DE, Thibault-Starzyk F, Knops-Gerrits PP, Parton R, Jacobs PA (1994) Macromolecular Symposia 80:157
229. Romanovsky B (1994) Macromolecular Symposia 80:185
230. Schulz-Ekloff G, Ernst S (1997) In: Ertl G, Knözinger H, Weitkamp J (eds) Handbook of heterogeneous catalysis, vol 1. VCH, Weinheim, p 374
231. Parton RF, Vankelecom IFJ, Casselman MJA, Bezoukhanova CP, Uytterhoeven JB, Jacobs PA (1994) Nature 370:54

CHAPTER 6

Zeolite Effects in Organic Catalysis

Patrick Espeel, Rudy Parton, Helge Toufar, Johan Martens, Wolfgang Hölderich, and Pierre Jacobs

6.1
Reported Catalytic Technology with Zeolites

New catalytic technology developed during the 1980s in Japan [1, 2], USA [3], China [4] and Europe [5] has been reviewed recently. Table 6.1 contains the reported catalytic technology extracted from this work which makes use of zeolite molecular sieves. It is clear that most of the technological developments have been made in the area of the refining of oil fractions (Table 6.1.I). Strong acidity in a shape-selective environment in which the zeolite pore and cage walls exert stereoselective constraints on the transition states and/or on the movement of products and reagents, together with intense hydrogen transfer reactions, are at the basis of this specific behavior (*vide infra*).

Generation of ultra-pure olefinic feedstocks for particular applications constitutes another set of new processes (Table 6.1.II). Generation of (alkyl)benzenes from light alkanes, and their interconversion via isomerization, disproportionation and alkylation (Table 6.1.III) are again typical zeolite-catalyzed processes which result from shape-selective catalysis in the proton-decorated micropores of certain zeolites (see below and in other chapters).

In Tables 6.1.IV and 6.1.V technology for the synthesis of substrates which contain heteroatoms has been collected. In the former case, the organic chemistry is again governed by steric discrimination of the zeolite micropores among the different product isomers, while in the latter case new catalytic chemistry is involved. Finally, zeolite-based technology in the area of environmental catalysis is emerging (Table 6.1.VI).

It is striking that from the very large number of zeolite topologies which are known today [6], only a limited number are being used in these applications. For the traditional zeolites (such as Y, erionite, ZSM-5, mordenite, L) catalyst properties are changed by the framework Si/Al ratio, the Si/Al profile across the crystals, the pore size and structure and the way of pore interconnection, the isomorphous substitution of framework Si by other atoms (such as B or Ti) [5]. Brønsted acid catalysis is the dominating catalytic act among this established technology, although bifunctional reactions (Brønsted and metal sites) are at the basis of several processes. Occasionally, metal and redox catalysis is involved as well.

Table 6.1. Catalytic Technology Using Zeolite Catalysts Developed in the 1980s

Process	Catalyst	Ref.
I.		
hydrocracking of distillates	Ni/W on zeolite	[3]
hydrocracking	zeolite	[3]
distillate dewaxing	ZSM-5	[3]
lube dewaxing	ZSM-type medium pore	[3]
high octane gasoline	stabilized, partially dealuminated Y	[3]
	ZSM-5	[3]
deep cracking of vacuum gas oil	ultrastable Y and pentasil	[4]
resid cracking	ultrastable Y	[1]
resid hydrocracking	iron on zeolite	[1]
shape-selective hydrocracking	Pt/erionite	[5]
light naphtha reforming	Pt/zeolite L	[5]
C_5/C_6 paraffin isomerization	metal on zeolite	[3]
gasoline from methane (oxychlorination)	iron chloride and ZSM-5	[5]
gasoline from synthesis gas	zeolite	[5]
methanol to gasoline	ZSM-5	[3]
II.		
methanol to olefins (C_3, C_4)	modified ZSM-5	[3]
(C_2, C_3)	SAPO-34	[3]
butene-1 catalytic purfication (isobutene/butene codimerization)	high-silica mordenite	[2]
MTBE cracking for high purity isobutene	boron-pentasil zeolite	[5]
III.		
LPG aromatization	zeolite with promoter	[3]
aromatics from C_3/C_4 paraffins	Ga-modified ZSM-5	[5]
aromatization of hexanes	Pt/L zeolite	[3]
disproportionation of BTX aromatics	zeolite	[3]
xylene isomerization	ZSM-5	[3]
selective toluene disproportionation	ZSM-5	[3]
ethylbenzene from benzene and ethene	high-silica pentasil	[4]
IV.		
dimethylamine from methanol/ammonia	steamed ion-exchanged mordenite	[1]
tert-butylamine from isobutene/ammonia	probably a zeolitic catalyst	[5]
ethanolamines from ethene oxide and ammonia	Al-Si zeolites	[5]
isomerization of *m*-dichlorobenzene *m*-chlorotoluene dichlorotuluene	high-silica H-zeolite	[2]
adamantane from cyclopentadiene	Pt on zeolite Y	[2]
cyclohexanol from cyclohexene	novel zeolite	[1]
phenylacetone from 2-phenylpropanal	zeolite	[5]
neo-acids from olefins, CO and water[a]	Cu/Co-modified zeolite	[5]

Table 6.1 (continued)

V.		
hydroquinone from phenol and hydrogen peroxide	Ti-silicalite-1	[5]
propene oxide from propene and hydrogen peroxide[a]	Ti-silicalite-1	[5]
cyclohexanonoxime from cyclohexanone, ammonia and hydrogen peroxide[b]	Ti-silicalite-1	[5]
VI.		
NO_x reduction with ammonia	zeolite	[3]
selective reduction of NO_x by ammonia, oxidation of SO_2	zeolite	[5]
CO removal from air	Pd-Pt zeolite	[2]

[a] In development.
[b] Demonstration plant.

6.2 Generalities on Catalytic Organic Chemistry with Zeolites

The topic of zeolite catalysis of organic reactions has been extremely well reviewed. A multiplicity of reactions has been combined with an impressive number of zeolite topologies modified in many different ways. The large number of excellent reviews focusses often on the reaction type catalyzed, the topics treated reflecting almost the contents of a standard handling on organic chemistry [7–23]. The classification of organic reactions catalyzed by zeolites has been given in every detail by Hölderich et al. [10, 11] and recently by Venuto [9]. For the sake of completeness a survey of the reaction types is given in Table 6.2. It is clear that this compilation of reactions covers the traditional content of handbooks on organic chemistry. It follows that acid zeolites are preferentially used and that data on basic and metal catalysis are only scarcely present. Oxidation with peroxides seems to be an exploding area as far as research is concerned. Reference in the present work in most cases is made to reviews. The work is therefore only exhaustive as far as reviews of topics and general areas are concerned.

From Table 6.2 it also follows that, in principle, all acid organic chemistry in mineral acids can be performed with zeolites. The specific action of zeolites is often an increase in (i) conversion (rate) resulting from the high zeolite Brønsted acidity, (ii) selectivity (yield) as a result of intrazeolitic chemistry and (iii) time stability as a result of suppression of coke or debris-forming side reactions.

The specific zeolite effects are often categorized in terms of pore size and configuration, size and morphology of the zeolite crystals, type at T-atom site, and type of charge compensating cation [18, 19, 21]. Perot and Guisnet [22] enumerate the following advantages of zeolites which are identical to those of more conventional solid catalysts: (i) decreased or absence of corrosion; (ii) no

Table 6.2. Survey of the Different Categories of Organic Reactions [10] and Typical Zeolite Effects, Advantages and/or Difficulties to Overcome

Substitution reactions with arenes

alkylation of arenes*: *often enhanced para-selectivity; heterogeneous Brønsted-type Friedel–Crafts catalysis, often environmentally beneficial and energy saving*
side chain alkylation: *basic catalysis on alkali-metal zeolites; solid base catalysts*
alkylation of heteroarenes*: *replacement of homogeneous Lewis acids by solid Brønsted acids*
acylation of arenes*: *stability against acylating agent (in case of organic chloride) or its decomposition product (in case of anhydride) required; reactions often not catalytic; fast deactivation*
halogenation and nitration of arenes*: *a possible alternative to homogeneous processes, provided the zeolite is stable in reaction conditions against hydrogen halides*

Substitution reactions in aliphatic compounds

ether & ester formation*: *no direct energy saving or environmental incentive; longer life-times might stimulate industrial application*
thiols from alkanols and H_2S*: *zeolites seem to show better in-time stability over traditional homogeneous and heterogeneous catalysts*
amines from alkanols and NH_3*: *no significant reduction in reaction pressure can be achieved, but great energy saving up to 40% in the case of dimethylamine production, high product shape selectivity*
isomerization of arenes*: *deviations from thermodynamic equilibrium among the product isomers is often possible in favor of the geometrically smaller isomer by zeolite shape selectivity; tuning of pore size and geometry to product size and shape is required*

Isomerization of aliphatics

double-bond isomerization of alkenes*: *difficult to obtain deviations from equilibrium by shape selectivity; accompanied by oligomerization in some cases*
skeletal rearrangement of alkanes & alkenes*: *used often in refinery operated technology; shape selectivity is crucial*
skeletal rearrangement of carbonyl containing molecules*: *specific zeolites overcome many drawbacks encountered with traditional technology; shape selectivity is at the basis of the performance*
pinacolone rearrangement*: *heterogenization of classic organic chemistry in mineral acids*
Wagner–Meerwein rearrangement*: *better yields generally obtained than with alumina catalysts*
epoxide rearrangement*: *better activity, selectivity and time-stability than homogeneous or non-zeolitic solid acids*
cyclic acetal rearrangement*: *superior to conventional catalysts based on selectivity and activity*
Beckmann rearrangement*: *only careful catalyst design enhances sufficiently activity and selectivity; service time (catalyst life time) still too short*

Additions and eliminations

hydration/dehydration (of alkenes/alkanols)*: *surprisingly few data are available for hydration; dehydration gives often equilibrium distribution of olefins; the methanol to olefins and aromatics is zeolite (ZSM-5/ZSM-11) specific; high selectivity and service time*
dehydration of aldehydes*: *manipulation of the acid strength via, e.g., isomorphous substitution is a prerequisite for selectivity*
addition/elimination of alkanols*: *MTBE synthesis remains a challenge in this respect*
addition/elimination of acids*: *esterification of olefins again remains a challenge for catalyst design; not much systematic data are available*
addition of S-compounds*: *better yields and stability than with conventional acid catalysts*
carbonylation/decarbonylation: *competition with the established homogeneous processes does not seem possible at the present stage; catalysis with metal clusters in zeolites*

Table 6.2 (continued)

Additions and eliminations

addition to epoxides*: *according to the scarce data, competition with established chemistry is not yet possible*
addition of ammonia*: *very high selectivity, high pressure; in the case of tert-butylamine environmentally friendly*

Reactions with hydrogen

hydrogenation/dehydrogenation: *metal catalysis; only well studied in the frame of bifunctional catalysis; hydrogenation on shape-selective carriers remains a challenge*
dehydrocyclization: *arene formation from alkanes on bifunctional shape-selective zeolites (see technology, Table 6.1)*
hydroformylation: *metal catalysis; removal of traces of catalyst acidity to avoid side reactions and metal leaching from the catalyst into solution remain hurdles to overcome; it is not clear how the activity of classic systems compares to that of zeolites*

Oxidation reactions

with dioxygen: *only Wacker chemistry has been successfully transplanted from homogeneous corrosive and polluting conditions to a transition ion-exchanged zeolite*
with peroxides: *this area made a recent explosive growth with the advent of Ti- and V-substituted zeolites*

Condensation reactions

aldol condensation*: *few data on base-catalyzed reactions; intracrystalline catalysis (pore size) determines product selectivity*
synthesis of aromatic N-heterocycles*: *the condensation of several small molecules (formaldehyde, aldehydes, ketones) in a cage or pore is very zeolite specific chemistry*
synthesis of isocyanates and nitriles: *only few data available*
O/N replacements*: *possible with acidic or basic zeolites; the only zeolite specific effect seems to be suppression of side reactions via shape selectivity*

* Acid-catalyzed reaction.

waste or disposal problems; (iii) easy engineering of continuous fixed-bed processes; (iv) high thermostability. The advantages of zeolites over other solids are according to these authors their great acid strength and practically unlimited applicability to different types of catalysis. As main weakness of such catalytic materials are recognized:

(i) their sensitivity to deactivation by irreversible adsorption or pore blockage by heavy side products;
(ii) the often encountered incompatibility in size of molecules from the fine chemicals area and the cages of the micropores.

As specific zeolite effects are categorized:

(i) the possible adjustment of acidic and basic properties, by variation of the overall acid strength, the site density, by the coexistence of acidic and basic sites, as well as by association of redox and basic sites; at the basis of this effect is the extremely large variability in terms of chemical composition of the materials;

(ii) the availability of a large structural variability and accessible surface area which is reflected in selectivity effects in parallel as well as consecutive reactions;
(iii) the existence of external surface shape selectivity; in this respect the alkylation of naphthalene and the Beckmann rearrangement of cyclohexanone oxime will be treated below;
(iv) the occurrence of adsorption and concentration effects; this is inter alia reflected in the higher selectivity to bimolecular reactions.

It is evident that for the design of an active, selective and stable catalyst, these zeolite effects have to be adapted and tuned. At the present stage of knowledge this procedure still has to be repeated for every specific substrate. Only in certain areas have general rules started to emerge which try to establish *structure-activity-selectivity relationships*. Their quantification is the next challenge.

6.3 Established Generalities on Shape Selectivity with Zeolites

Although the fundamentals on shape selectivity in zeolites were originally derived for A, erionite and mordenite-type zeolites [24–30], it is evident that for reacting molecules and micropores of comparable size, shape selectivity is possible with all zeolites and substrate molecules. Indeed, the perfect fitting of akylbenzenes into the pores of ZSM-5-type zeolites constitutes a perfect example (see Table 6.1) for validation of these earlier established principles. Weitkamp and Ernst [31] recently proposed the following definition for shape-selective catalysis: "*the term shape-selective catalysis encompasses all effects in which the selectivity of a reaction depends, in an unambiguous manner, on the pore width or pore architecture of the microporous solid*". As Brønsted acid sites as well as metal clusters, basic or redox sites may occupy the intracrystalline voids of the zeolite, shape-selective phenomena can be encountered in acid, base, metal, bifunctional and redox catalysis.

Out of a mixture of molecules only the smaller ones have access to the intracrystalline zeolitic space. This phenomenon is now generally described as *reactant selectivity*. It is evident that the selectivity is determined by the size of the pore mouths. When out of the mixture of products formed in the zeolite voids, only the smaller ones with the proper dimensions are able to diffuse out and to appear as reaction products, *product selectivity* is at the origin of the occurrence of a thermodynamically non-equilibrated product mixture. Structures which contain cages that are only accessible through smaller windows are required to achieve this kind of shape selectivity. This selectivity is an example of shifting chemical equilibria as a selective product removal [17]. *Transition state shape selectivity* occurs when, for geometric reasons, the formation of certain transition states is inhibited. As at the external surface of zeolite crystals the same active sites may be present as in the intracrystalline space, the latter may "equilibrate" again the products from the selective sites or even catalyze side reactions. It should, however, be realized that the interrupted framework at the external surface also contains pore mouths or half-open cages, which reflect the architecture of the intracrystalline pores. They can be at the origin of special effects of shape selectivity.

It remains difficult to distinguish between product diffusion and transition state shape selectivity, as it remains difficult to discriminate between mass transfer effects and intrinsic chemical effects, as until now no reliable method has existed to estimate diffusion coefficients in reaction conditions [31]. Use of *in situ* measurement via MAS NMR of the product composition of "a working zeolite cage" [32], as well as molecular graphics modelling, may allow shape selective catalysis to be given a more quantitative basis.

These classic types of molecular shape selectivity with zeolites are well documented and have often been reviewed [27, 33–36]. Examples are, however, very often with non-functional organic substrates. Alkanes and alkyl aromatics accommodate themselves easily in the zeolitic voids, their catalytic conversion (via bifunctional or acid catalysis, respectively) being very susceptible to the inorganic microenvironment. Such molecules are therefore used to probe the architecture of the intracrystalline environment based mainly on transition state shape selectivity in the case of alkanes, and a combination of product shape selectivity and transition state shape selectivity in the case of alkyl aromatics. Recently, Guisnet [30] and later Weitkamp and Ernst [31] have reviewed this area thoroughly. A very recent example illustrating the power of methods using test molecules to characterize zeolite pore architecture is available. Before the crystallographic structure of zeolite MCM-22 became available, three groups were independently probing the void structure of this zeolite with the help of catalytic test reactions [37–39]. From a single criterion test (the butylcyclohexane bifunctional conversion giving the *spaciousness index*) it was already clear that the zeolite had to be classified between the 10- and 12-membered ring (-MR) structures [37]. A combination of xylene isomerization and the decane test (a multi-criterion test) allowed the statement that the structure contained "10-membered rings and larger cavities, which would give to MCM-22 zeolite the typical features of 10-MR and 12-MR" [37]. In the hands of the original inventors [38], the "decane test reaction clearly shows that the pore architecture of MCM-22 is more complex than that of a medium pore zeolite". Comparison of several sets of product selectivities, from the n-decane conversion reaction on this unknown zeolite, with the corresponding values for known structures, advances evidence for a channel system with 10- and 12-MR rings and with monodimensionality. "Protons are present in both sets of rings, the stronger ones being located in the smaller rings". The crystallographic structure of ZSM-22 is compared with the structural picture derived from catalytic test reactions in Fig. 6.1.

In a given solid catalyst, competitive sorption between feed molecules of different molecular weight or polarity, or between reagents and products, will determine activity, selectivity and poisoning. The adsorption coefficients in Langmuir–Hinshelwood (LH) kinetics quantify these effects. In microporous materials such *competitive sorption effects* are the result of interactions between substrate and micropore wall. The rules that govern competitive sorption in microporous environments devoid of steric constraints are:

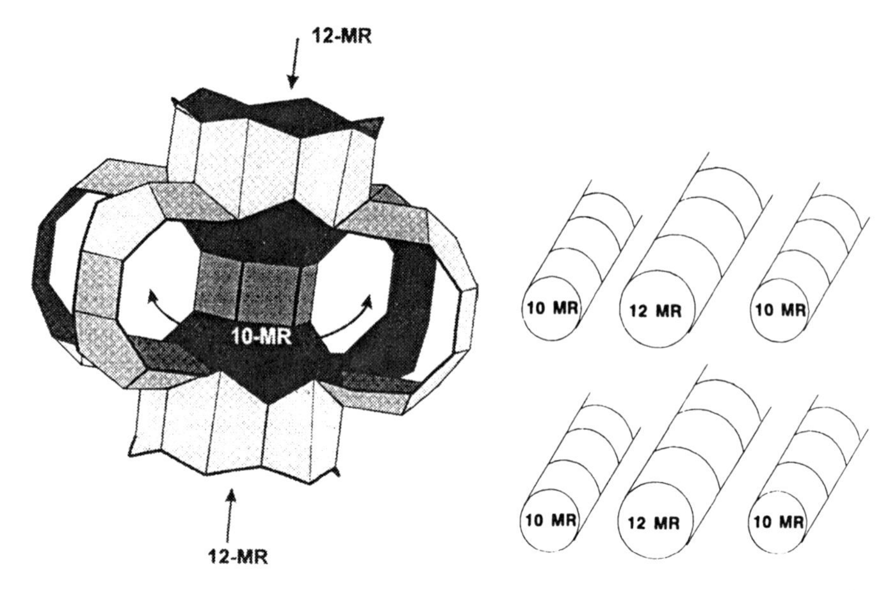

Fig. 6.1. Representation of the structure of zeolite MCM-22, determined by straight crystallographic methods [40] and by catalytic test reactions [39]

(i) the highest boiling molecule most strongly adsorbs in a given microporous environment;
(ii) molecules with higher polarity preferentially sorb in more polar zeolites and *vice versa*. An illustrative example will be given in the section on Friedel-Crafts chemistry.

It is well established in bifunctional catalysis in the vapor phase that competitive sorption in large pore zeolites dominate the reactivity of individual compounds in feed mixtures. Dauns and Weitkamp [41] reported that with Pt/LaY in an equimolar mixture of n-decane and n-dodecane, the presence of the latter molecule caused a strong activity reduction, while that of the higher boiling compound remained unaffected by the presence of lower boiling molecules. In this way, most 12-MR zeolites are able to discriminate between molecules belonging to the same homologous series but differing only by one or two carbon numbers [42]. In terms of LH kinetics, this points to the existence of largely different adsorption constants for molecules of comparable size and shape.

Santilly and Zones recently proposed a distinction between *primary* and *secondary shape selectivity* [43]. Primary shape selectivity is the result of an interaction between a molecule, or a transition state and a microporous environment. It encompasses product, reactant and transition state shape selectivity. Secondary effects occur when this primary selectivity is perturbed. The phenomena are schematically illustrated in Fig. 6.2.

The following examples illustrate the presence and absence of secondary shape selectivity [43]. The cracking of n-octane on H-ZSM-5 is not affected by the presence of 3-methylheptane or 2,2,4-trimethylpentane, while it is by 2,2-dimethylbutane (Fig. 6.2). As all three molecules are lower boiling than n-octane, competitive adsorption cannot be involved. 2,2,4-Trimethylpentane for obvious steric reasons is excluded from the ZSM-5 pores (reactant shape selectivity) and can only interact weakly with the external surface, which explains its inertness to the catalyst. These phenomena occur in the gas phase. It is clear that in liquid-phase reactions competitive sorption may even be much more important. It is also possible that a molecule which is for steric reasons excluded from the pores, for configurational and electronic reasons may have high affinity for the external surface, and behave differently. 2,2-Dimethylbutane will slowly diffuse as a molecular prop through the pores of ZSM-5 and thus retard the diffusion and conversion of n-octane. This effect of secondary shape selectivity will be dependent not only on the pore size of the zeolite but also on its dimensionality. This concept is based on the experimental data of Namba et al. [44].

A second example of secondary shape selectivity is found in the cracking of mixtures of hexane and hexadecane on SSZ-16 zeolite [43]. In open 12-MR zeolites (as Y), hexadecane will have disappeared completely before hexane is touched. In 10-MR ZSM-5 [43] this rate difference has decreased, while on the 8-MR SSZ-16 [43] the reactivity order is inversed (hexadecane is much less reactive than hexane). The molecular basis for the understanding of this kind of selectivity is lacking at the moment [45].

The interplay of primary shape selective phenomena such as transition state and diffusion shape selectivity is, as already mentioned, not easily enraveled, but has been dealt with on several occasions [45] with the help of diffusion/reaction kinetic models [27, 46]. It can easily be shown that if molecules A and B are reacted *separately*, the observed selectivity (S_O) is

$$S_O = \eta_A/\eta_B \, S_T \, (k'_A/k'_B)$$

in which η is the effectiveness factor, S_T the transition state shape selectivity and k' the reaction rate constants determined in the absence of diffusion or shape selectivity (i.e., in large pore zeolites). The method entirely neglects the existence of competitive sorption and secondary shape selectivity which might occur when mixtures of reagents are involved [45].

Cage and window effects in zeolite bifunctional catalysis of hydrocarbons have been reviewed recently [45]. Minima in the carbon number product distributions or bimodal product distributions with a minimum around carbon number 8 [47] were observed in the H-erionite cage. The original explanation was based on the preferential trapping of C_8 hydrocarbon fragments in the erionite cage thus undergoing intensive cracking, which explains the bimodal product distribution. The erionite cage effect is then a special case of product shape selectivity. Bimodal and even trimodal product distribution on zeolites ZSM-5, beta and twinned faujasites, respectively, are interpreted in terms of reduced central cracking [45]. Provided the adsorbed feed molecule stretches across two neighboring cages, separated by a window, the latter may become a steric barrier for alkyl shifts towards a centrally highly branched and energetically favored intermediate, thus reducing central cracking and giving rise to

Competitive sorption

C10 C10 C10
C10 C10

nC8 + nC10 → C8 C8 C8 C8 C8 →

C10 C10
C10 C10 C10

Primary shape selectivity

reactant shape selectivity

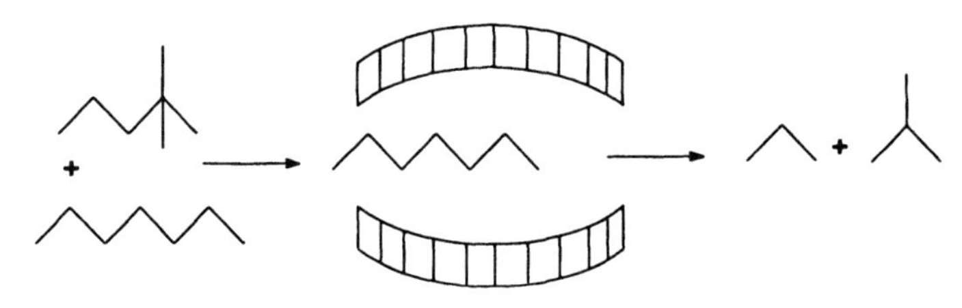

product shape selectivity

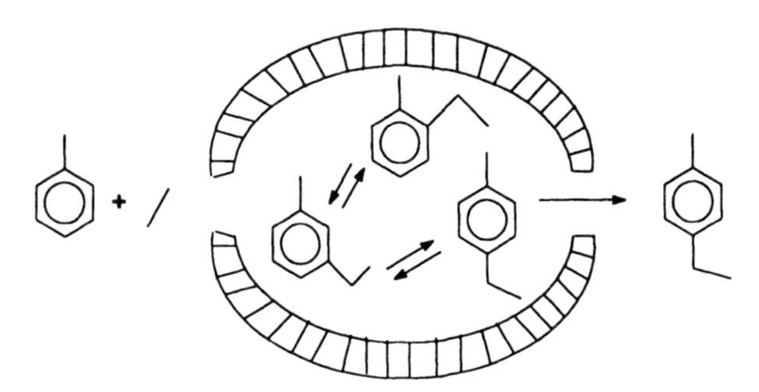

Fig. 6.2. Schematic representation of competitive sorption, primary and secondary shape selectivity in zeolites [nC_8 (n-octane), nC_{10} (n-decane)]

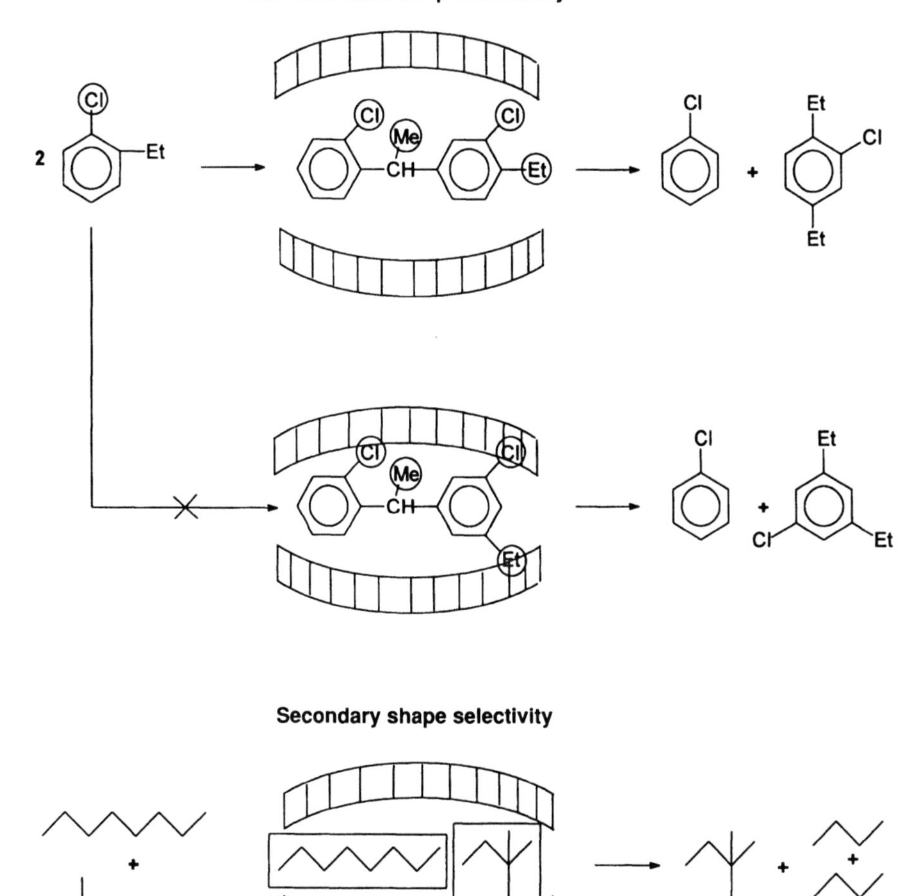

Fig. 6.2 (continued)

multimodal cracked product distributions (Fig. 6.3). A cage effect in the ketonization of carbonic acids will be reported later.

All mentioned types of molecular shape selectivity are intimately linked to reactions occurring inside the pores of the zeolite catalyst. The possibility of catalysis at the external surface was first mentioned by Venuto [48]. In cases of slow intracrystalline diffusion, only sites at the external surface may be responsible for catalysis. As counter-diffusion between product and reagent is slow in mordenite for the liquid-phase propylation of benzene, Katzer concluded this to be an example of *pore mouth catalysis* [49]. Fraenkel et al. [50] reported product discrimination between *o*- and *m*-ethyltoluene and cymene on internal sites and

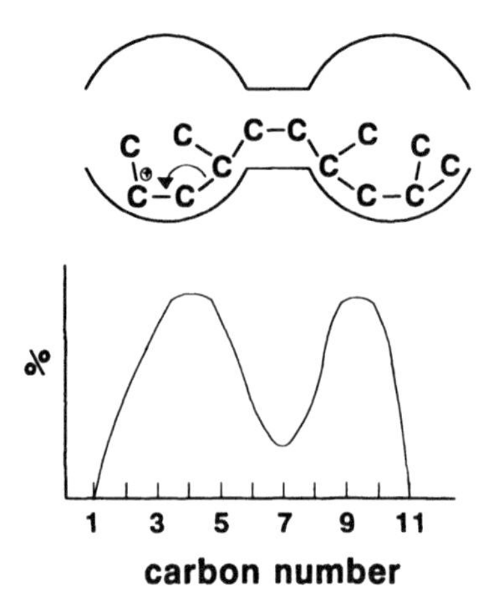

Fig. 6.3. C_{14} carbocation trapped in two neighboring zeolite cages and hypothetical product distribution resulting from restricted central cracking

sites present as half-channel intersections at the external surface. In the methylation of 2-methylnaphthalene [51,52] the "slim" isomers 2-methyl-, 2,6-dimethyl- and 2,7-dimethylnaphthalene were dominant over ZSM-5 and ZSM-11 in contrast to mordenite. Derouane et al. [53,54] proposed that the outer surface of zeolites be considered as an "ondulating micro-landscape". Molecules possessing a configuration which closely matches that of the zeolite outer surface (fitting in the half-cages of ZSM-5/-11 zeolites) are stabilized and their formation is favored during reaction. This basis of selectivity is called the "*nest effect*" and is expected for reactions over small crystals involving bulky molecules. However, no unambiguous evidence has been advanced to support this type of shape selectivity. Indeed, Neuber and Weitkamp [55, 56] questioned this concept recently based on a similar data set as Fraenkel et al. Over a whole series of 10-MR zeolites, the results of Fraenkel et al. were confirmed. Moreover, it was shown convincingly that 2-methylnaphthalene enters the pores of ZSM-5 already at low temperature and that the selectivity for slim isomers increases upon coking of the catalyst [55]. As coke is known to act as a selectivity enhancing parameter [57], the authors are convinced that selectivity is better explained by classic product shape selectivity rather than by nest effects in half-cavities at the external surface.

6.4 Generation of Active Sites in Zeolites

Traditionally, zeolites have been defined as (natural or synthetic) crystalline aluminosilicates with a tridimensional microporous framework made up by corner sharing of SiO_4 and AlO_4 tetrahedra [58]. They can be considered as originating from a SiO_2 lattice, in which Al^{3+} is partially but isomorphically substituted for Si^{4+}. This substitution generates an excess negative charge in the silicate lattice, which is compensated by cations. A general formula of an aluminosilicate zeolite can be written as

$$M^{n+}_{x/n}[(AlO_2^-)_x(SiO_2)_y]\,zH_2O$$

with n, the valence of the charge-compensating cation M and x/y smaller than or equal to 1, since AlO_4 tetrahedra can join only to SiO_4 tetrahedra according to the Loewenstein rule [59] and with z depending on the micropore volume. The charge-compensating cations to a variable degree can be exchanged for other cations, including protons. Water occluded in the intracrystalline channels may be removed by thermal treatment to leave an open framework.

Recently, this definition has become a matter of debate [60–62] as new developments in "zeolite chemistry" have expanded the field to other framework compositions and topologies. Framework compositions have been modified by:

(i) isomorphous substitution of other tetrahedral elements as Ge, Ga, B, ..., for Si [63],
(ii) the synthesis of pure silica microporous crystalline oxides as silicalite [64],
(iii) the discovery of the second generation of molecular sieves having aluminophosphate frameworks and isomorphously substituted forms thereof (SAPO's and MeAPO's) [65], and
(iv) the synthesis of a third generation of molecular sieves consisting of metallosulfide frameworks [66].

The number of framework topologies was expanded by the synthesis of non-fully connected frameworks (i.e., frameworks containing some remaining hydroxyls of the tetrahedral units interrupting the framework) or frameworks which contain as well as tetrahedrally, also pentahedrally and/or octahedrally coordinated framework cations [67]. Consequently, a more general definition may be used: "*Zeolites are crystalline microporous solids containing cavities and channels of molecular dimensions synonymously called molecular sieves*" [68].

Zeolites can be synthesized with different framework topologies, and with pores that vary in shape, size and dimensionality. These parameters determine the architecture of a pore system and are of crucial importance in zeolite catalysis (Table 6.3). The aperture sizes of the pores and cages range between 0.4 and 2 nm, depending on the number of tetrahedra in the rings that define them. The number of tetrahedra circumscribing the main zeolite channel is often used for classification purposes. Thus a distinction is made between 8-, 10-, 12-, 14-, 16-MR ... zeolites.

The broad range of structure types, framework compositions, exchangeable cations, supported organic or inorganic molecules provide a virtually unlimited structural and chemical diversity. Changing these parameters tailors the properties of the material.

Zeolites have been used as catalysts for almost al principal reactions encountered in homogeneous phase organic chemistry, either acid, base or redox catalyzed. Rather than enumerating these reactions in detail (see Table 6.2 and [7–23]), it is the purpose of this review to point to the aspects of zeolite-catalyzed reactions which distinguish them from their homogeneous counterparts.

Most of the large-scale catalytic applications are based on the Brønsted acidity of the zeolites (see Table 6.1), which is provided [68] either by:

(i) ion exchange in acid medium,
(ii) exchange of ammonium ions, followed by an activation step whereby ammonia is expelled,

(iii) hydrolysis of hydration water of polyvalent (e.g. trivalent) cations, or
(iv) hydrogen reduction of cations to a lower valency state.

The reactions involved in the respective procedures are represented in the following equations, where ZO^- represents the anionic zeolite lattice and M a cation or metal.

$$ZO^{n-}\,M^{n+} + n\,H^+ \longrightarrow ZO^{n-}\,(H^+)_n + M^{n+}$$

$$ZO^-\,NH_4^+ \longrightarrow ZO^-\,H^+ + NH_3$$

$$(ZO^-)_3\,M^{3+}\,(H_2O) \longrightarrow ZO^-\,H^+ + (ZO^-)_2(MOH)^{2+}$$

$$(ZO^-)_2\,M^{2+} + H_2 \longrightarrow 2\,ZO^-H^+ + M$$

Not all lattice hydroxyl groups in a zeolite exhibit Brønsted acid properties. Two types are distinguished: the terminal and bridging hydroxyls. The former terminate the framework at crystal boundaries or stacking faults and are characterized by an IR band around 3745 cm^{-1}. The latter are bound to an oxygen atom bridging an Al^{3+} and Si^{4+} and have an IR band in the range 3650 to 3550 cm^{-1}. They are thought to be the catalytically active sites in acid catalysis and can be dehydroxylated to form Lewis acid sites on high-temperature treatment [69].

In four-coordinated metallosilicates, the number of negative framework charges generated by isomorphous substitution of trivalent for tetravalent ions (Fig. 6.4, mechanism IS), and therefore the number of potential Brønsted sites, is proportional to the degree of substitution. For substituted aluminophosphate

Table 6.3. Topology and Pore Architecture of 12-MR Zeolites [6]

Zeolite	Structure type	Characteristics of the micropores	
		dimensionality[a]	channel dimension (nm) and orientation[b]
faujasite	FAU	3	⟨111⟩ **12** 7.4***
beta	BEA	3	[100] **12** 7.6 × 6.4* ⇄ ⟨100⟩**12** 5.5 × 5.5**
SAPO-37	FAU	3	⟨100⟩ **12** 7.4***
omega	MAZ	2	[100] **12** 7.4* | [100] **8** 3.4 x 5.6*
ZSM-5	MFI	2	{[010] **10** 5.3 × 5.6 ⇄ [100] **10** 5.1 x 5.5}***
mordenite	MOR	1	[100] **12** 6.5 × 7.0* ⇄ [010] **8** 2.6 × 5.7*
offretite	OFF	1	[001] **12** 6.7* ⇄ [001] **8** 3.6 × 4.9**
ZSM-12	MTW	1	[010] **12** 5.5 × 5.9*
zeolite L	LTL	1	[001] **12** 7.1*

[a] Represents the dimensionality of the main channels.
[b] The channel direction is represented as e.g., [001], parallel to the crystallographic c-direction; bold numbers represent the number of T-atoms in a ring; ⇄ and | represent interconnecting and non-interconnecting channels, respectively; the number of asterisks represents the dimensionality of the channel system.

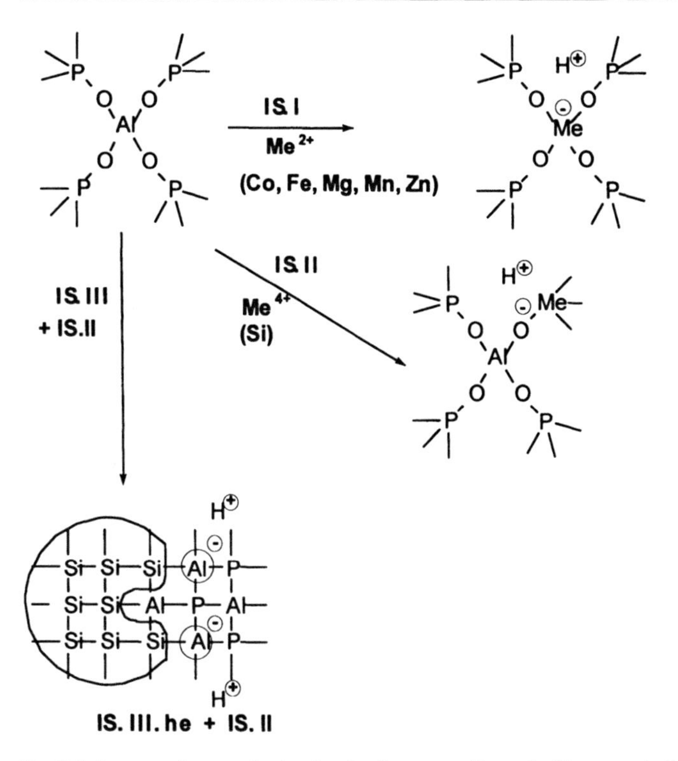

Fig. 6.4. Isomorphous substitution in four-coordinated silicate and aluminophosphate materials, generating negative framework charges

zeolites, the situation is more complex and the number of sites is not necessarily proportional to the chemical composition of the framework. As long as only substitution of divalent ions for aluminum (Fig. 6.4, mechanism IS.I) or tetravalent ones for phosphorus (mechanism IS.II) occurs the composition will reflect the Brønsted site density. When pairs of adjacent tri- and pentavalent ions are substituted (mechanism IS.III.he + IS.II) silica patches can be formed and new acidity is generated at the border of the silicate and aluminophosphate domain (Fig. 6.4). If the silica patches are large enough, substitution with trivalent ions will also generate "normal" zeolitic acidity. In the latter case, the relationship

IS

Me^{3+}

(Me = Al, B, Fe, Ga)

between composition and acidity is not straightforward and is dependent on the way the materials are synthesized. This has been reviewed recently [70]. The expectedly neutral gallophosphate cloverite framework was recently claimed to contain lattice terminating P-OH groups of high acid strength [71]. This would mean that the presence of heteroatoms in $AlPO_4$ and $GaPO_4$ frameworks is not a prerequisite for the existence of Brønsted acidity [71]. However, these physico-chemical predictions have not yet been confronted with catalytic reality.

For a complete characterization of zeolite acidity, it is necessary to determine the number and strength of Brønsted and Lewis acid sites. Several methods have been developed for this purpose. The most important ones are:

(i) titration with a base,
(ii) temperature-programmed desorption of bases,
(iii) IR spectroscopy of lattice OH groups before or after adsorption of weak bases, and
(iv) ^1H NMR spectroscopy [72].

Presently, there is increasing interest in also using zeolites as basic, oxidation or reduction catalysts. Basic catalysts are prepared by:

(i) exchange of the zeolite with basic cations (K^+, Rb^+, Cs^+),
(ii) deposition of basic oxides (MgO, ...), or by
(iii) synthesis of metallic clusters inside the zeolite cages.

Recently, the preparation of alkali metal clusters and particles in zeolites has been reviewed in detail [74].

Several types of redox functions were built in zeolites some of which possess unique properties due to the zeolitic environment. Transition metals, either as oxides supported on the zeolite [74], substituted in the framework [75-77], on exchangeable positions or chelated by organic ligands and encapsulated in the cages of zeolites [78], were reported to be active entities in zeolites for oxidation reactions. All of them can be prepared in several ways:

(i) transition metal oxides supported on zeolites are commonly prepared by an impregnation or cation exchange procedure combined with hydrolysis of the transition metal [74];
(ii) transition metals are incorporated into the framework of zeolite either during synthesis [75] or by post-synthesis modification using methods similar to dealumination (e.g., treatment with $TiCl_4$) [79];
(iii) cation exchange either in liquid phase or via solid state mixing [80] is the sole method to obtain transition metal ions in cation exchangeable positions; and
(iv) a wide variety of methods is applicable to synthesize complexes encapsulated in zeolites.

They all start from a transition metal containing zeolite and can be divided into three main categories. In the so-called *ship-in-a-bottle synthesis*, the organic ligand is synthesized *in situ* around the transition metal present, either as a cation [81] or as pre-adsorbed complex [82-84]. In the second method the ligand is adsorbed as such on zeolites with transition metal ions in cation exchangeable sites [85]. In the third method the zeolite is synthesized around the complex [86].

Metals supported on zeolites are the main hydrogenation-dehydrogenation compounds which enable reduction as well as isomerization reactions. They are prepared by first incorporating the transition metal either via a cation exchange or via an impregnation procedure [87]. Subsequently, the transition metal zeolite is oxidized and in the last step reduced, both occurring at high temperature. In rare cases hydrogen insertion reactions are catalyzed by transition metals chelated by organic ligands and encapsulated in zeolites. They are prepared by the same procedure as the encapsulated complexes which catalyze oxidation reactions [88, 89].

6.5 The Latest Visions on Zeolite Acidity

There is common agreement that acid sites, especially Brønsted sites, are the main source of the catalytic activity of zeolites [69, 90], although the role of Lewis sites is still subject to intensive discussions [91, 92]. Experimental as well as theoretical evidence was found for an enhancement of catalytic activity of zeolites by synergistic effects of conjugated Brønsted- and Lewis-type sites [93, 94]. Recent developments in theoretical modeling of zeolite catalysis indicate a similar effect between Brønsted acid sites and conjugated bridging oxygens, i.e. Lewis base sites, at least for reactions which do not proceed via ionic intermediates [95, 96]. However, the concept of acidity together with shape selectivity [97, 98] of zeolites alone are not sufficient to explain all selectivity effects observed in zeolite catalysis. This situation is very similar to the early 1960s when Pearson introduced the *concept of hard and soft acids* (HSAB) and bases [99] in order to quantify the strength of the interaction between different types of acids and bases.

First attempts to apply the HSAB principle to zeolite catalysis were undertaken in the early 1980s. Wendlandt et al. [100] found a close relationship between the softness of acid sites in large and medium pore zeolites and the selectivity in a variety of catalytic reactions, e.g., the *p*/*o*-selectivity in the alkylation of aromatics, the ratio between isomerization and dealkylation of alkylaromatics on zeolites. It was shown, that soft sites direct the alkylation towards *para*-products (an orbital-controlled reaction), while *ortho*-products are preferred on hard sites (a charge-controlled reaction). The site softness itself increases with increasing Si/Al ratio but is also dependent on the zeolite structure.

Very recently, these conclusions have received further support by quantum chemical calculations using the LUMO energy of small zeolite moieties as a criterion for the "hardness acidity" of the zeolites [101, 102]. However, although ab initio and semi-empirical SCF calculations on zeolite cluster models have proved to be successful in describing a number of import properties of zeolites [103], these techniques are subject to severe restrictions. Due to the limited number of atoms which can be treated with reasonable effort, neither the medium and long range structure of a zeolite nor its overall chemical composition are reflected properly. While information depending on the actual electron distribution such as electronegativity, electrostatic potential, electrostatic field etc. can be corrected satisfactorily by different embedding techniques [104–106], this is hardly possible for information depending on the mobility of the electrons, such as polarizability, hardness and softness. Thus, in order to understand the catalytic behavior of a zeolite with respect to terms like hardness and softness, it will be necessary to renounce some accuracy and resolution in favor of more extended zeolite models.

Well-developed tools to evaluate quantities such as polarizability, hardness and softness are the *electronegativity equalization method* (EEM) [107] and *charge sensitivity analysis* (CSA) [108, 109] at the atom-in-molecule (AIM) level of resolution. They rest on the rigorous definitions of electronegativity [110] and hardness [111] within the framework of the Density Functional Theory (DFT) [112, 113]. Both approaches allow the investigation of systems containing several hundreds of atoms with moderate computational effort, i.e., the full zeolitic environment of an absorbed molecule can be taken into account, even with a depth of at least 2 T-layers. Some intrinsic properties of zeolites can also be calculated for the infinite crystal [114].

The EEM approach has already been successfully applied to investigate *structure- and composition-depending properties* of zeolites and their interaction with adsorbed molecules [115–117]. More recently, CSA investigations were made on zeolites containing adsorbed molecules [118–120]. The latter calculations indicate that the global and local properties of a molecule adsorbed in a zeolite cavity are strongly influenced not only by the direct interaction with the adsorption site but also with the more distant parts of the surrounding cavity. Even in large-pore zeolites the interaction of an adsorbed molecule with the opposite walls of the cage will be strong enough to change the global hardness of the molecule (related to the activity in orbital-controlled reactions) and the local softnesses (Fukui function indices, related to the site selectivity) significantly as compared with small clusters or flat surfaces. The effect is of course more pronounced if the fit between molecule and zeolite is better and/or if there are other active sites present in the same cavity.

It follows that the carbocations conveniently written to rationalize product distributions from proton-catalyzed zeolites, in reality have properties which are influenced by many zeolite parameters, such as pore architecture (as defined earlier), and chemical composition. It is clear that the theoretical methodology is present which can quantify the parameters in terms of first principles. The accurate and systematic data sets required to establish the rules are, however, absent in most cases.

All product distributions in proton-catalyzed hydrocarbon rearrangements can be rationalized via classic *carbenium-carbonium-type chemistry*. On a formal basis there is a close similarity between the low-temperature chemistry in liquid superacids and the proton-catalyzed rearrangements in zeolitic cages at higher temperatures [121, 122]. Experimentally, the apparent activation energies of, e.g., 1,2-hydride shifts and 1,2-methyl-shifts catalyzed by zeolites at high and by superacids at low temperatures, are very similar [122]. This does not imply

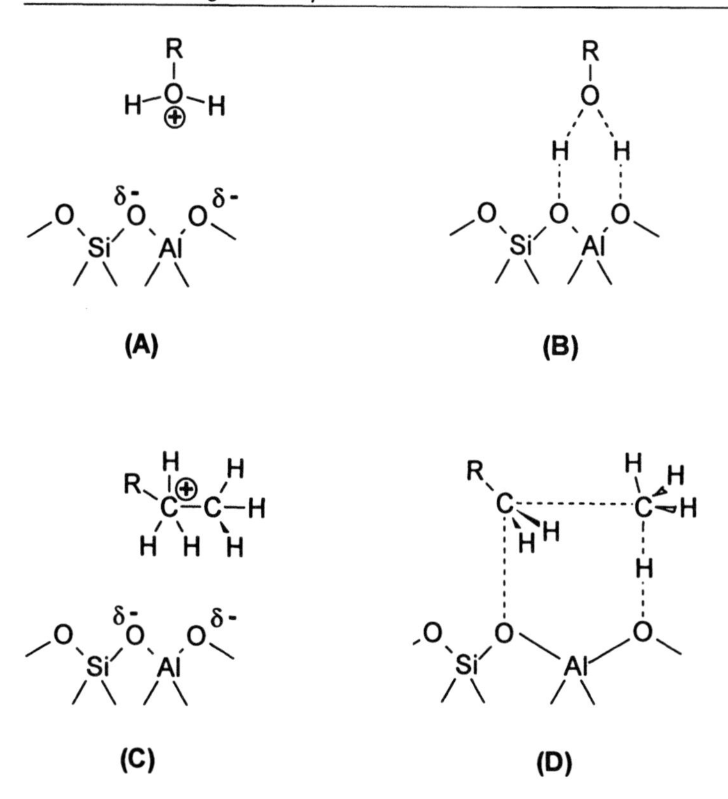

Fig. 6.5. Alcohol (ROH) (**A**) and alkane (R–CH_2–CH_3) (C) protonated by a Brønsted site of a zeolite in terms of the classical carbenium-carbonium ion theory or as a bidentate (**B**, **D**) species [123, 124] in terms of recently developed quantum chemical concepts

necessarily that carbenium-carbonium ions are the dominating surface species. Kazanski [123] and van Santen [124] have recently made the state of the art in this respect. According to Kazanski, the classic concepts of the carbonium-carbenium ion theory have to be refined in the following sense. The set of quantum chemical calculations, the NMR data of zeolite-adsorbed molecules, and the low temperature IR data using H_2 as probe, force a view of *carbocations in zeolites as transition states* and not as surface covering species. Carbocations in zeolites are electronically excited unstable ion pairs which stem from covalent ground states of surface alkoxy groups [123]. A classical oxionium ion formed by adsorption of e.g., an alcohol on an acid zeolite (see Fig. 6.5, species A) is not a surface-covering species, but becomes coordinated to the lattice as a bidentate species (Fig. 6.5, species B). Species A at best is an unstable transition state in the decomposition of a surface-covering bidentate intermediate into the reaction product, adsorbed water and an olefin.

Upon protonation of an olefinic substrate by a Brønsted site, via a surface π-complex and in excited state, an essentially covalent surface alkoxide is formed (a silyl-alkyl ether link with the lattice), which is decomposed into the reaction

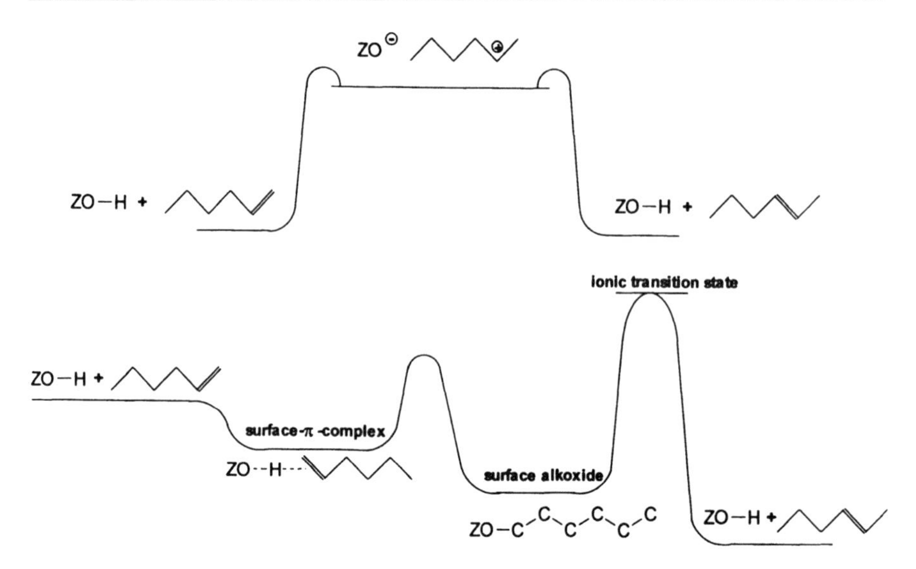

Fig. 6.6. Energy profile of alkene isomerization via carbenium ion theory (*upper part*) or via surface alkoxide intermediates (*lower part*)

products via an unstable ionic transition state (with carbocation resemblance) [124]. A pictorial representation of the energy profiles is given in Fig. 6.6 and compared with classic carbenium ion behavior. It should be stressed here that all authors that used the simplified carbenium-carbonium ion theory to rationalize their observations never intended to deny, e.g., adsorption, nor have they ever claimed that the carbocation transition states were surface-covering species. Therefore, one can safely continue to rationalize in terms of the carbenium-carbonium theory, as long as one remains conscious that reality is more complicated.

The intermediates formed (Fig. 6.5 B, D) are the result of the action of *Brønsted sites with bifunctional nature*: their Brønsted acid part protonates the adsorbed molecules, while interaction with the neighboring basic lattice oxygens converts the protonated intermediates into covalent species linked to the zeolite framework via bidentate coordination [123, 124]. Protonation is thus a *Brønsted acid Lewis base assisted coordination* with the zeolite framework. In this frame, acid/ catalysis becomes concerted and resembles an acid S_N-2-type mechanism [123]. Acid catalysis is thus controlled by the protonation chemistry as well as by the interaction with the stabilizing zeolites of adsorbed reagents and products [124]. The relative stability of the intermediates is dependent on deprotonation energy as well as on stabilization by the negative charge of lattice oxygens. The deprotonation energy is sensitive to local geometric constraints and to zeolite composition [124].

6.6 Zeolite Superacidity

Compared to amorphous silica-alumina catalysts, zeolites have similar aluminum concentration, but an intrinsic activity which is often several orders of magnitude higher. This difference is often attributed to the following parameters [125, 126]:

(i) the concentration of accessible tetrahedral Al, that is higher with zeolites;
(ii) the average intrinsic activity of these sites, that is (*inter alia* because of the regular microporous environment) much higher in zeolites; and
(iii) the reagent concentration that is higher in zeolites.

High reagent concentrations should contribute significantly to the superactivity of zeolites. It enhances not only the overall reaction rate but, more particularly, the rate of *bimolecular reactions. Intermolecular hydrogen transfer activity* is in this respect very much zeolite specific. In monomolecular alkene rearrangements, Y zeolites (e.g., proton-containing Y and rare earth exchanged Y) can be up to 30 times more active than silica-alumina [125], while in bimolecular cracking of alkanes, this rate enhancement can reach a factor of over 1,000 [125]. It seems that zeolites with monodimensional tubular channels such as mordenite, and in particular mazzite-type structures, have higher activity than zeolites with cages and intersecting channels [126]. Apparently, superacidity resulting from enhancement of bimolecular reaction rates also seems to have a geometric component.

The favorable interaction of a neighboring Lewis site [127, 128], of an extra-framework Al–O_X species [127, 129] and even of intrazeolitic amorphous silica-alumina [127, 130] has been invoked to be at the origin of superactivity. In the frame of the new concepts on the surface-covering sites (Brønsted acid sites are bifunctional in nature) (see Sect. 6.5) the first two explanations seem very acceptable. In this context, it seems that ultrastable Y zeolite below 373 K is also not functioning as a superacid as it is unable to protonate σ-bonds in alkanes [131]. It has acidity properties comparable to those of sulfuric acid, while those of ZSM-5 zeolite are comparable to a 70% solution of sulfuric acid (as determined from the ^{13}C chemical shift of adsorbed mesityl oxide) [131].

Intermolecular hydrogen transfer activity can easily be quantified from the determination of the isobutane formation rate (R_{iC4}) and isobutene ($[iC_4^=]$) and hexane ($[C_6]$) concentration (in the hexane cracking at 673 K [132]):

$$R_{iC4} = k_{HT}^{*}\,[C_6]\,[iC_4^=]$$

in which k_{HT}^{*} is proportional to the hydrogen transfer rate. Steric inhibition as well as site density can override this concentration effect as the hydrogen transfer activity in zeolite Y was found to be a factor of 6 higher than in ZSM-5 [132]. Intermolecular hydrogen transfer reactions involving adsorbed hydrocarbon heavy species transform them to hydrogen-deficient moieties that are at the origin of formation of "hard" coke [22, 30, 133].

Intermolecular hydride transfer also occurs with functionalized substrates at low temperature [134] as diphenylmethanol gave equimolecular amounts of phenylmethanes and phenyl ketones:

$$2\ \text{CHOH} \longrightarrow \text{C=O} + \text{CH}_2$$

This disproportion reaction is zeolite specific, as it is never observed in homogeneous catalysis, and depends on the density of the sites in zeolite Y [134].

6.7 Zeolite Specificity in Organic Catalysis with Functional Molecules: Zeolite Effects

As acidic, basic or redox catalysts, zeolites essentially act in a similar way to their homogeneous counterparts. However, due to their specific nature, these solid microporous materials sometimes show a behavior not observed in the homogeneous phase. It is the main purpose of this review to elaborate on these zeolite pecularities and to point out how zeolites succeed in changing and controlling reaction pathways. In Table 6.4, the main zeolite effects and their origin are enumerated. They are rationalized and illustrated further in this chapter.

6.7.1 Zeolite Effect I: Shape Selectivity

In zeolites, most of the active sites are located in the well-defined and molecularly sized pores and cages. During the reaction, the transforming molecules are continuously subjected to steric limitations imposed by the zeolite structure, possibly changing the course of reaction and finally resulting in product distributions deviating from those obtained in the homogeneous phase. The restrictions imposed on guest molecules by size and shape of the zeolite pores is called "molecular shape selectivity".

Investigations by Weisz et al. [28] and Csiscery et al. [29] distinguish three types of molecular shape selectivity, namely reactant, product, and transition state selectivity. Reactant and product selectivity are the manifestation of a diffusion controlled reaction. The former implies that those molecules with high diffusivity will react preferentially and selectively, while molecules which are excluded from the zeolite interior will only react on the external surface of the zeolite. The latter implies that products with high diffusivity will be preferentially desorbed while the bulkier molecules will be converted and equilibrated to

Table 6.4. Ranking and Origin of Zeolite Effects

Origin of zeolite effect	Zeolite effect
location of active site in pores and cages of molecular dimensions	*Shape selectivity* - reactant shape selectivity - product shape selectivity - transition state shape selectivity
zeolites as adsorbents: specific adsorption	*Concentration effect* Reagent selectivity Nest effect
location of active site in pores and cages of molecular dimensions + specific adsorption	*Functional selectivity*
zeolites as solid solvent	*Multifunctional synergy:* multi-step to single-step - hydrogenation + alkylation - hydrolysis + hydrogenation - hydration + dehydrogenation *Tuning of chemical properties:* easy - hydrogenation catalysts - oxidation catalysts - fine tuning of acid base properties *New chemistry by zeolites* - Ti-zeolites - encapsulated complexes

smaller molecules which will diffuse, or eventually react to larger species which will block the pores. Accordingly, the (observed) bulk product distribution does not reflect the intrinsic reactivity of the molecules. Transition state selectivity takes place when certain reactions are prevented as the transition state necessary for them to proceed is not reached due to space restrictions.

6.7.1.1
General Procedures

To change the shape-selective properties of zeolites, different approaches are possible. Metal cation exchange, deposition of inorganic oxides in the pores, increase of the crystal size of the catalyst, coverage of the outer surface, are some of the most common methods. Most of them not only change the shape-selective but also the acid properties. It is indeed almost impossible to modify these characteristics independently.

(i) cation exchange: small cations such as H^+ and Na^+ can be replaced by large, more voluminous cations such as K^+, Rb^+. Cs^+, using the ion-exchange method. This replacement causes narrowing of the pore diameter of the zeolite channels in addition to a change in the number and strength of the Brønsted sites. Acid sites are neutralized or even more they are converted to basic sites.

(ii) deposition of inorganic oxides: by deposition of inorganic oxides in the pores of the zeolite, the effective pore diameter is reduced and the steric constraints on reactants, transition state or products are enhanced. Deposition of these oxides also changes the acid/base properties of the zeolite.

(iii) crystal size: the crystal size of zeolite catalysts can be varied by adapting the synthetic conditions. By increasing crystal size, the fractional amount of sites on the outside of the crystal compared to the total amount of sites decreases and consequently the locus of catalytic activity shifts from extra- to intrazeolitic.

(iv) deactivation of the external surface: for small crystals, the active sites on the external surface largely contribute to the catalytic activity of the material. In the case of Brønsted acid sites, they can be deactivated by deposition of inorganic oxides or neutralized by chemical reaction with organosilanes which are too bulky to penetrate the intracrystalline voids. By silylation the hydrophobicity of the materials is simultaneously increased.

(v) deposition of organic bases: deposition of organic bases in the intracrystalline voids narrows the pore apertures and thus enhances the steric constraints on molecular diffusion.

6.7.1.2
Manifestation of Shape Selectivity in Organic Reactions

6.7.1.2.1
Reactant Shape Selectivity

Reactant shape selectivity is generally observed in all types of zeolite-catalyzed reactions when a mixture of at least two reactants is fed to a catalyst. The only prerequisite for this phenomenon to occur is that differences in size and shape between reactant molecules exist and that the zeolite pores are small enough to discriminate among them.

Reactant shape selectivity has been exploited in separation processes (e.g., in the ISOSIV process) [135] and in the selective conversion of one reactant out of a reactant mixture (e.g., in the selectoforming process; see Table 6.1). Since this phenomenon can be observed for all kinds of chemical reactions, we limit ourselves to one recent example.

Ample evidence is provided in the literature that the oxidation reactions of alkanes with peroxides or iodosylbenzene over metal complexes encapsulated in Y-type zeolites are controlled by the shape and size of the pore mouths. With iron-phthalocyanines (FePc) in solution, cyclododecane is oxidized three times faster than cyclohexane. For zeolite-encapsulated complexes this preference decreases by a factor of two. Post-synthesis modifications by exchanging the zeolite with cations of different size further improves this discrimination [139–141]. Such post-synthesis modifications can be categorized as fine-tuning of the reactant shape selectivity. Much stronger differences are obtained for zeolites

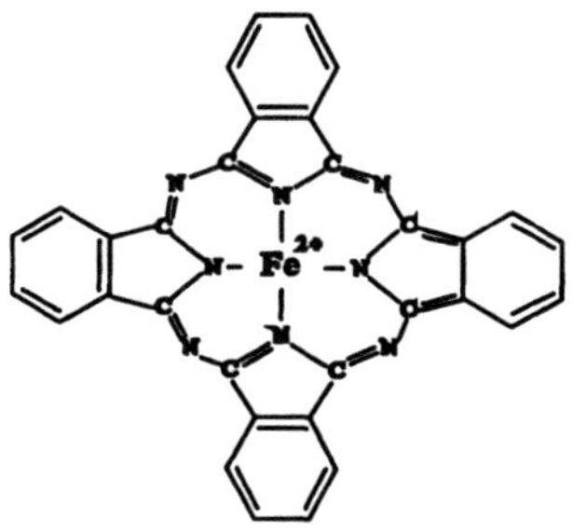

FePc 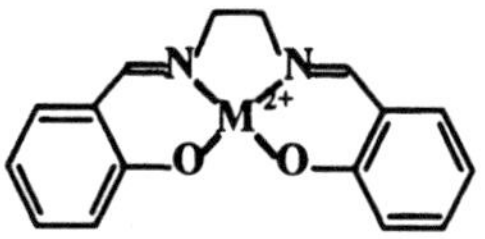 MSALEN 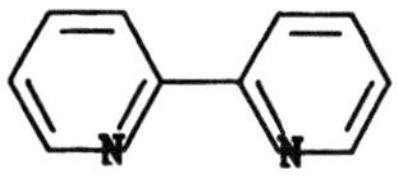 Bipy

with smaller pores. On a Fe/Pd-Y zeolite the activity ratio of octane to cyclohexane and of n-pentane to 2-methylpentane is 0.67 and 0.35, respectively [136–138]. On Fe/Pd-A, an 8-MR zeolite, however, this discrimination is reversed and ratios are now obtained of 200 and 180, respectively [139–141].

Similarly on Mnsalen encapsulated in zeolite-Y [142], cyclohexene is oxidized two or three times faster than styrene and *trans*-stilbene, respectively. In homogeneous solution, these ratios are 0.7 and 1, respectively. Pdsalen-Y prefers the hydrogenation of 1-hexene over cyclohexene [88]. Again the reverse order is observed homogeneously.

6.7.1.2.2
Product Shape Selectivity

Isomerization among products and differences in their diffusivity are the prerequisites for product shape selectivity. It is difficult to discriminate between product and transition state selectivity since both favor the slimmest isomer. Both types of shape selectivity are usually differentiated by applying different crystal sizes. Since the effectiveness of product shape selectivity depends on the path length necessary for a product molecule to escape from the zeolite after its formation, this type of shape selectivity is enhanced by increasing the crystal size. This reasoning can be critisized. Indeed, by increasing the crystal size, the ratio of the sites on the external compared to the internal surface decreases. Consequently, the number of non-selective reactions at the outer surface decreases also. Combination of different crystal sizes and deactivation of the sites at the external surface should indisputably demonstrate the impact of product shape selectivity.

The concept of product shape selectivity implies that isomerization of bulkier isomers to slimmer isomers occurs. Consequently, deactivation of the catalyst activity should not be severe. This deactivation occurs when only transition state selectivity exists as is the case when the zeolite possesses cages or intersecting pores. Under such circumstances bulky products are retained, preventing further access to the active sites. This explains why most of the industrial zeolite-based processes in Friedel-Crafts alkylation occur under conditions allowing consecutive isomerization.

a) Product Shape Selectivity in Friedel–Crafts Reactions

The synthesis of small and large zeolite crystals allows the intracrystalline product mixture from a catalytic transformation to be influenced by changing the intracrystalline diffusion characteristics. Regioselectivity (i.e., *para*-selectivity) in the electrophilic substitution of substituted arenes can be increased by using large crystals, as the less bulky *para*-isomer diffuses faster than the *ortho*-isomer which further isomerizes. A typical example is the very high selectivity for *p*-xylene by alkylation of toluene in the presence of medium pore pentasil zeolites [143].

Another example is found in the gas-phase methylation of chlorobenzene [144]. The selectivity for *para*-chlorotoluene increases significantly when the crystal size of the H-ZSM-5 catalyst is systematically increased (Table 6.5). The example convincingly proves the dominant effects of product shape selectivity.

b) Product Shape Selectivity in Methanol Conversion over Acid Medium Pore Zeolites

Product shape selectivity can also be enhanced by silylation of the zeolites. If the pore openings of the zeolite are smaller than the dimension of the organosilane used, only the hydroxyl groups of the outer surface of the zeolites are expected to react [145]. In contrast, if the pore dimensions are larger than the silylation agents, the silylation reaction takes place at the outer crystal surface as well as in the intracrystalline volume [14]. According to the proposal of Niwa et al. [146, 147], tetramethoxysilane reacts with the hydroxyl groups and forms a monolayer of silica first:

$$\text{Z–OH} + \text{Si(OCH}_3)_4 \longrightarrow \text{Z–O–Si(OCH}_3)_3 + \text{CH}_3\text{OH}$$

$$\text{Z–O–Si(OCH}_3)_3 + 3\,\text{H}_2\text{O} \longrightarrow \text{Z–O–Si(OH)}_3 + 3\,\text{CH}_3\text{OH}$$

The subsequent deposition of silica multilayers in narrow and medium pore size zeolites causes a decrease in the pore diameter at the pore mouth. Such a narrowing of the pore opening has the interesting ability to increase the shape-selective properties of zeolites.

H-ZSM-5 was treated with tetramethoxysilane and afterwards used in the conversion of methanol to hydrocarbons. The formation of elongated molecules is preferred; the yield of aromatics is decreased in favor of ethene and propene and in the xylene fraction, *p*-xylene formation is favored over the bulkier *meta*- and *ortho*-isomers.

Table 6.5. Percentage of *para*-Isomer in the Methylchlorobenzene Fraction from the Methylation of Chlorobenzene over H-ZSM-5 [144]

crystal size (μm)	2	3	7	10	12	220
p-selectivity	57	57	59	62	66	90

6.7.1.2.3
Transition State Shape Selectivity (TSSS)

With acid catalysis it is often difficult to distinguish transition state selectivity from product selectivity when secondary isomerization reactions occur. Both favor the formation of the slimmest, i.e., the *para*-isomer. However, for reactions performed at low temperature, which inhibit consecutive isomerization, product shape selectivity is excluded. These conditions are usually met in liquid-phase Friedel–Crafts alkylation reactions. In the literature some typical examples in the field of organic intermediates and fine chemicals have been disclosed, confirming the control of positional selectivity by transition state shape selectivity.

a) TSSS in Friedel–Crafts Alkylations and Acylations
Direct alkylation of polynuclear aromatics, such as biphenyl (BP), to produce a mixture of dialkylated polynuclear aromatic compounds enriched in *para*-isomers was reported recently [148, 149].

+ 2

In particular, the unique behavior of dealuminated mordenite zeolites was reported for the liquid-phase alkylation of biphenyl with propene. High yields of 4,4′-diisopropylbiphenyl (4,4′-DIPB) were obtained.

With a mineral acid-leached mordenite (with a SiO_2/Al_2O_3 molar ratio of 2600) a yield of 73.5% of the desired 4,4′-DIPB isomer was reached at 98% conversion [148, 149]. In contrast, with a homogeneous catalyst such as phosphoric acid, only 1.9% selectivity is obtained at 48% conversion. Other large pore zeolites such as Y, L and offretite combine high feed conversion with low selectivity (<15%). Possibly, the presence of mesopores (0.227 cm^3/g) resulting from the chosen dealumination procedure, allows fast product diffusion out of the selectivity-determining regions. The high SiO_2/Al_2O_3 ratio of the lattice should prevent deactivation and keep the selectivity high.

Next to dealuminated mordenite, zeolite ZSM-12 with Si/Al = 18, is a preferred catalyst for the propylation of biphenyl [150]. 77% of monoisopropylbiphenyl (IPBP) and 21% of diisopropylbiphenyl (DIPBP) containing 64% of 4,4′-DIPBP was obtained at 67% conversion.

Alkylation of Naphthalene to 2,6-Diisopropylnaphtalene. As in the biphenyl alkylation, it is difficult to obtain high yields of 2,6-dialkylnaphthalenes in the presence of conventional catalysts such as silica/alumina or Lewis acids. In order to control the substitution position of naphthalene, the shape-selective H-mordenite was again very much suited for the formation of 2,6-diisopropylnaphthalene (2,6-DIPN) [151]. In contrast to HY and HL zeolite, the favored formation of the 2,6- and 2,7-isomers was observed on dealuminated mordenite. If the tight fit of the molecular dimensions of these β,β-isomers with the intracrystalline space of mordenite is selectivity determining, the lack of attack in the α-position is explained. The selectivity of the 2,6- and the 2,7-isomer (the former is twice as high as the latter), which according to their size should be comparable, is then necessarily the consequence of TSSS, implying that the activated complex for the formation of 2,6-DIPN has a more linear structure than in case of 2,7-DIPN.

In contrast, the methylation of naphthalene with methanol seems possible in the presence of a 10-MR zeolite (*vide supra*). Preferred formation of 2-methylnaphthalene among the monoalkylates and of 2,6-dimethylnaphthalene among the dialkylates is possible with H-ZSM-5 [51, 152–156]. Although the initial interpretation of the data caused some controversial discussion in the literature (*vide supra*), there is now convincing evidence for intracrystalline and thus transition state shape selective catalysis.

Acylation of Diphenylether to 1,4-bis(4-phenoxybenzoyl)benzene. 1,4-Bis(4-phenoxybenzoyl)benzene (BPBB) is technically produced by the acylation of diphenyl ether with 1,4-benzenedicarbonyl chloride in the presence of Friedel–Crafts catalysts such as $AlCl_3$. In its presence, high *para*-isomer selectivity is not expected *a prioi.*

Cl—CO—C_6H_4—COCl + excess C_6H_5—O—C_6H_5 $\xrightarrow[190 - 250°C]{zeolite}$

C_6H_5—O—C_6H_4—CO—C_6H_4—CO—C_6H_4—O—C_6H_5 + 2 HCl

(BPBB)

In a new process for the manufacture of BPBB [157], the reaction of diphenyl ether with 1,4-benzenedicarbonyl chloride is carried out in the liquid phase in the presence of an acid large pore zeolite. At 523 K a 53:1 molar ratio of diphenyl ether and 1,4-benzenedicarbonyl chloride [157] is converted over HY-zeolite yielding 62% BPBB (based on the chloride). The purity of the BPBB product was greater than 99%.

b) TSSS in Condensation of Carboxylic Acids

Low carbon number carboxylic acids are known to undergo ketonization in homogeneous acids or on some oxides. Thus, condensation of acetic, propionic and butyric acid yields acetone, 3-pentanone and 4-heptanone, respectively [158]. The ketonization reaction is followed by acid-catalyzed self-condensation. On zeolites the chemistry is essentially identical [158]. In the ketonization of propionic and butyric acid over some acid zeolites of the offretite-erionite-chabasite family, and especially over zeolite T, consecutive condensation of the respective ketones (and therefore catalyst deactivation) is strongly suppressed [158]. When equimolar mixtures of propionic acid and butyric acid are fed to an

CH_3-CH_2-COOH
propionic acid

\+

CH_3-CH_2-CH_2-COOH
butyric acid

⟶ CH_3-CH_2-C(=O)-CH_2-CH_2-CH_3
3-hexanone

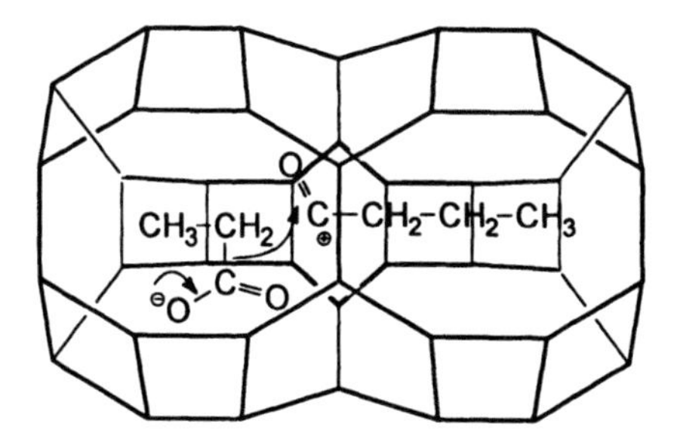

Fig. 6.7. Hetero-condensation in the erionite cage of ionic species from propionic and butyric acid into 3-hexanone (after Ref. [158])

acid zeolite T, the hetero-condensation and thus 3-hexanone formation is strongly favored over the homo-condensation. The propionic and butyric acid molecules seem to have the optimum size for a bimolecular ketonization reaction inside an erionite cage (Fig. 6.7). This clear preference for hetero-ketonization is an example of molecular recognition of the acid molecules by the erionite cage.

c) TSSS in the Halogenation of Aromatics

Whereas halogenation of aromatics is a completely different reaction type compared to a Friedel–Crafts reaction, the shape-selective effects on the orientation of substitution in both reactions are very similar. Formation of the *para*-isomer is enhanced when the bulkiness of the substituent (on the aromatic) and/or of the attacking species is increased. In the halogenation of monosubstituted benzenes on K-L zeolite [159], the increase in *para*-selectivity for the halogenation reactions parallels the size of the halogen atom: I > Br > Cl. K-L zeolites are able to chlorinate toluene up to 90%, almost without benzyl chloride formation, the yield of the *para*-isomer exceeding 75% [160]. Biphenyl is chlorinated on the same zeolite with a yield for the 4,4′-isomer of almost 90% [161, 162]. Since the reactions were performed under conditions preventing consecutive isomerization, as no *meta*-isomer was observed, transition state selectivity is the main reason for the high preference for halogenation at the *para*-position.

d) TSSS in the Alkylation of Ammonia

Alkylamines, in particular methylamines, are of considerable technical importance and are used as very valuable intermediates for the production of solvents such as dimethylformamide and dimethylacetamide, insecticides, herbicides, pharmaceuticals and detergents. The reaction of methanol with ammonia *via* a nucleophilic substitution is mostly used for the industrial production of methylamines. The reaction in the presence of amorphous oxide catalysts such as alumina yields an equilibrium mixture of mono-, di- and trimethylamine (MMA, DMA, TMA).

In many cases, however, a product mixture of MMA and DMA is desired. The employment of shape-selective zeolite catalysts is a very elegant concept in order to minimize the formation of the bulky TMA. Different zeolites have been used to meet this requirement. Large pore as well as small pore zeolites have been investigated [163–170].

Mordenite in the H-form displays a product mixture with equilibrium composition. A great increase in selectivity for dimethylamine can be achieved by careful adjustment of the alkali content of the mordenite, or by silanation with $SiCl_4$ [169], thus carefully narrowing the pore diameter.

e) TSSS in Oxidation Reactions

In oxidation reactions consecutive isomerization reactions are very unlikely and product shape selectivity effects will not intervene. Thus, it is evident that preferential formation of the slimmer isomers is the result of differences in shape and size of the transition state. Small differences in size of the transition state in the formation of secondary alcohols from n-alkanes can lead to the preferential formation of the smallest alkanol, namely the one with the alcohol function on a secondary carbon atom close to the end of the chain [82, 171, 172]. In this way, it should, in principle, be possible to overcome unfavorable energetics by a well-designed catalytic site and succeed in the synthesis of primary alkanols from n-alkanes.

The site selectivity for secondary alcohol and ketone formation is higher on 10-MR zeolites such as TS-1 [171, 172] than on 12-MR zeolites (e.g., FePc encapsulated in Y zeolite) [82]. Indeed, whereas for hexane oxidation there is no preference with FePcY catalysts, the ω-1 position is preferred over the ω-2 position on TS-1 [82, 171, 172]. For longer n-alkanes site selectivity that increases with the chain length of the substrate is also observed with FePcY. The site selectivity increases strongly with small pore 8-MR zeolites (Fe/Pd-A) which enable even limited oxidation at the unreactive primary carbon atoms of n-alkanes [139–141].

A special case of transition state selectivity is the oxidation of norbornane and methylcyclohexane observed for FePc-faujasite catalysts [136, 137]. In this case stereoselective preferences are induced by the steric constraints of the zeolite on the transition state of the reaction. Precisely, the preference of the *exo*- over the *endo*-hydroxy isomer of norbornane decreases from a molar ratio of 9.2 to 5.5 upon encapsulation. In the same way the *trans/cis* ratio of 4-methylcyclohexanol from methylcyclohexane increases from 1.1 to 2. Similar catalysts, $(Na^+)_4FePc^{4-}$-Y, also exhibit transition state selectivity in hydrogenation reactions [173]. In the hydrogenation of butadiene, but-1-ene formation is preferred over *trans*-but-2-ene, which, in turn, is preferred over *cis*-but-2-ene, as predicted by the size of the transition states [173].

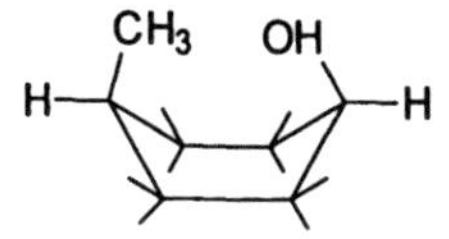

cis-4-methylcyclohexanol 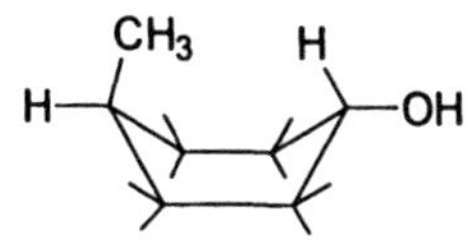 trans-4-methylcyclohexanol

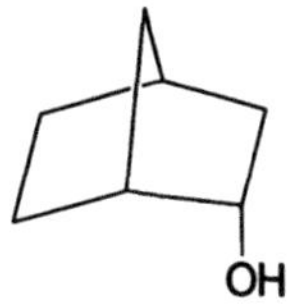

endo-norborneol 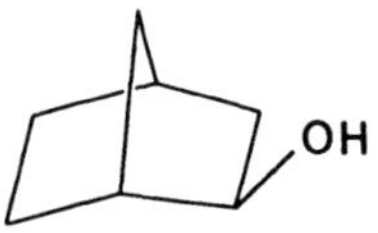 exo-norborneol

f) TSSS in the Meerwein–Ponndorf–Verley Reduction
Using alkali-X-zeolites, non-conjugated aldehydes such as citronellal can react with isopropanol in a Meerwein–Ponndorf–Verley reduction, either to citronellol or, as a result of ring closure to isopulegol, depending the nature of the doped metal [174].

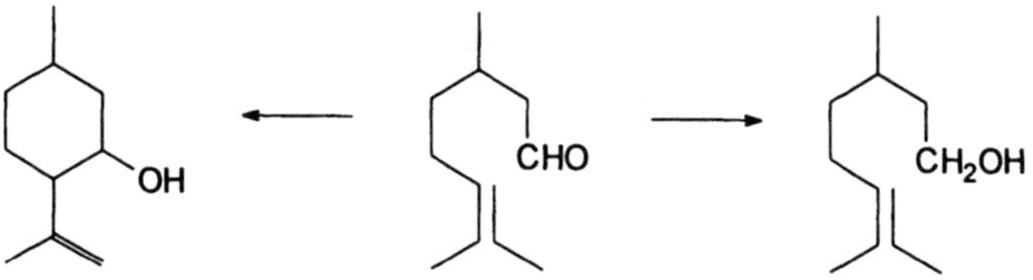

isopulegol ← citronellal → citronellol

With Cs-X-zeolite the reaction yields 92% of citronellol at 77% conversion. The authors [174] attribute these results to the difference in the steric requirements of the two competing reactions. Smaller pore sizes force the aldehyde to adopt a stretched conformation within the channel, thereby preventing cyclization and favoring citronellol formation. The difference in the Brønsted acid properties, however, has only a minor influence, if any at all.

g) TSSS in the Fischer Indole Synthesis
The Fischer indole synthesis involves acid-catalyzed rearrangement of arylhydrazones. Hydrazones originating from asymmetric ketones afford two isomeric indoles. Recently it was found that zeolites catalyze this transformation while exerting in some cases a profound effect on the regioselectivity [175]. An example is given in the equation.

The phenylhydrazone of 1-phenyl-2-butanone rearranges via homogeneous acid catalysis towards 2-ethyl-3-phenylindole. When applying zeolites (in refluxing isooctane or xylene) the other isomer is also formed and even becomes

2-ethyl-3-phenylindole 2-benzyl-3-methylindole

the predominant product with H-mordenite as the catalyst. This is again a clear case in which a tight fit of transition state in the intracrystalline space determines the reaction selectivity.

h) TSSS in Epoxide Rearrangement

Often the consecutive reaction which has to be suppressed by transition state selectivity is of the same type as the primary reaction, as has been demonstrated for monoalkylation and monohalogenation reactions (*vide supra*). However, transition state selectivity can also be used to suppress secondary reactions of a different type (also *vide infra*, TSSS in ketonization reactions). Brunel et al. reported the selective rearrangement of 1,2-alkene oxides to aldehydes over silanated offretites [176]. The consecutive aldol condensation of the aldehydes was avoided by the use of these zeolites, resulting in much higher selectivities than with non-zeolitic catalysts. More generally, Hoelderich and Goetz showed that transition state selectivity on MFI-zeolites suppresses the consecutive reactions in the rearrangements of styrene oxides, aliphatic epoxides and glycidic acid esters [177].

zeolites zeolites ALDOL CONDENSATION

Such types of shape selectivity are intimately linked to reactions occurring inside the pores of the catalyst. The concept of shape selectivity was expanded by Fraenkel and co-workers [50, 178] and Derouane et al. [53–54] to encompass also reactions occurring near and/or on the outer surface. Derouane proposed that the outer surface of zeolites can be considered as an undulating micro-landscape. Molecules possessing a configuration which closely matches that of the zeolite outer surface are stabilized and their formation is favored during reaction. The basis for selectivity is called the "nest effect" and is expected to occur with reactions over small crystals involving bulky molecules. However, no unambiguous evidence has been advanced to support this type of shape selectivity (*vide supra*).

i) TSSS in Diels-Alder Reactions
Zeolites accelerate Diels-Alder cycloaddition reactions and improve stereo-selectivities due to transition state selectivity. The combination of CeY-zeolite and a Lewis acid such as $ZnBr_2$ is found to be the most stereoselective catalyst; 100% of the *endo*-product is achieved in 96% yield in the cyclodimerization reaction of methylacrylate with cyclohexadiene or cyclopentadiene as diene [179].

+ $COOCH_3$ ⟶ $COOCH_3$ exo + $COOCH_3$ endo

6.7.2
Zeolite Effect II: Specific Adsorption

Adsorption equilibria determine the concentration of the reactants, products and solvents (in the case of liquid-phase reactions) on the surface of the catalyst in the absence of diffusion limitation and consequently determine reaction rate and product selectivity. As a result, concentrations inside and outside the zeolite can differ by several orders of magnitude. The hydrophilic/hydrophobic properties of zeolites can be tuned by changing the chemical composition of the zeolites. They will directly determine the nature of the molecules which are favorably adsorbed. Concentration effects also exist for gas-phase reactions where molecules are concentrated in the micropores of zeolites by adsorption.

6.7.2.1
Diels–Alder Cycloadditions

It is well known that Diels–Alder reactions are accelerated in aqueous media [180–181] or in conventional organic solvents under ultra-high pressure [182]. It is assumed that water enhances intimate contact between the dienophile and diene due to its high cohesive pressure. Similarly, by strong adsorption of both reactions inside zeolite pores, the rate of the cyclodimerization reaction of butadiene to vinylcyclohexene is dramatically enhanced. The rate enhancement caused by such concentration effects can be several orders of magnitude and is strongly dependent on reaction temperature [183–184].

6.7.2.2
Friedel–Crafts Alkylation

In the case study on Friedel–Crafts chemistry (*vide infra*), it was shown that the sorption properties of the aromatic substrate, alkylating agent and solvent in a liquid-phase alkylation completely control the reaction rate and product selectivity.

A pertinent example was provided by Venuto et al. as early as 1966 [185–187]. Although benzene could be used as a solvent in the ethylation of phenol, it was easier to ethylate benzene than phenol on acid Y-type zeolites. They showed that this paradox was based on the stronger adsorption of phenol compared to benzene, reducing the sorption of ethene and benzene.

6.7.2.3
Beckmann Rearrangement

In many cases reactions on zeolites are directed by adsorption phenomena. One is the Beckmann rearrangement of cyclohexanone oxime to ε-caprolactam. The most active and selective catalysts are found to be the highly silicious ZSM-5, which have a considerably better performance than non-zeolitic catalysts. It was shown that the rearrangement proceeds mainly on the outer surface or at the pore mouths of the zeolite on weakly acidic or even non-acidic sites [16, 188–191]. Temperatures higher than 573 K are required for the desorption of the desired lactam. However, at these high temperatures solid acid catalysts catalyze side reactions as decomposition of the reagent oxime to 5-cyanopent-1-ene or cyclohexanone and the polymerization reaction of the lactam. Therefore, tuning of the acidity is important and is realized on ZSM-5 at a partially deactivated external surface. The sites are still able to catalyze the rearrangement and product desorption occurs rapidly, thus avoiding side reactions. Therefore, fast desorption of the product combined with weak acidity seems to be the key-step in combining high activity with selectivity. Up to now this has only been realized on zeolitic catalysts. However, Corma et al. provide some contradictory results as they claim that medium and strong acid sites catalyze the reaction and that medium as well as large pore zeolites are suitable catalysts [192–194].

NOH
cyclohexane oxime
Beckmann rearrangement
O
NH
ε-caprolactam
fragmentation
C≡N
CH_2
+ H_2O
5-cyanopent-1-ene
hydrolysis
O
+ NH_2OH
cyclohexanone

6.7.3
Zeolite Effect III: Functional Selectivity

In the conversion of poly-functional reagents, the catalyst can select among different functions.

6.7.3.1
Hydrogenation of Unsaturated Aldehydes

Cinnamaldehyde possesses two functions, susceptible to hydrogenation, namely a double bond and an aldehyde function. On zeolites the latter can be hydrogenated in a selective way to an alcohol whereas the double bond remains intact.

$$C_6H_5\text{-}CH{=}CH\text{-}CHO \xrightarrow[\text{PtY, PtBeta}]{H_2} C_6H_5\text{-}CH{=}CH\text{-}CH_2OH$$

cinnamaldehyde **cinnamyl alcohol**

Large Pt or Rh clusters are formed in the cages or intersections of beta or Y-type zeolites [195–197]. Only the aldehyde function is capable of reaching these clusters and is consequently hydrogenated. Thus the functional selectivity originates in this case from transition state selectivity.

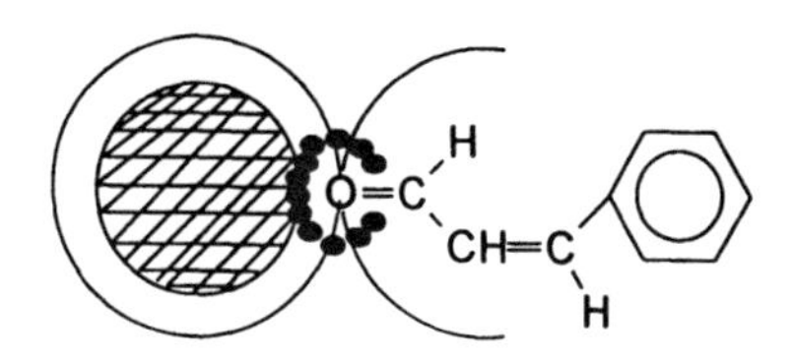

On large clusters, cinnamaldehyde molecules can adsorb only in an "end-on" fashion, i.e., via the C=O group on metal clusters entrapped in the zeolite supercage. However, if the clusters are too small and cinnamaldehyde can enter the supercage then C=C bond hydrogenation also takes place.

6.7.3.2
Preparation of Allyl-Substituted Aromatics by Friedel–Crafts Methods

Recently, we have shown [198] that allyl-substituted aromatics can be produced over acid 12-MR zeolites (such as Y, beta, mordenite) by electrophilic substitution of aromatics with allyl alcohol. In the homogeneous phase with sulfuric acid, the reaction of allyl alcohol with an aromatic substrate does not stop after the formation of the allyl-substituted aromatic but proceeds with the formation of di-aryl propanes. The main reason for the absence of functional selectivity in traditional Friedel–Crafts chemistry is related to the high activity of the acids used, failing to discriminate between different functional groups. The functional selectivity on zeolites is attributed to the suppression of bimolecular side reactions (transition state selectivity) and to the reduced isomerization rate of the double bond.

$$C_6H_5R + CH_2{=}CHCH_2OH \longrightarrow R{-}C_6H_4{-}CH_2CH{=}CH_2$$

6.7.4
Zeolite Effect IV: Multifunctional Synergy

The term multifunctional synergy denotes the cooperative action of at least two chemical functions that realizes a complex catalytic transformation of molecules. The two functions are usually present in a single zeolite and in close proximity. As a result of this multifunctional action, traditional multistep processes can be replaced by a "one-pot reaction".

The cation exchange capacity and crystallinity of zeolites make them particularly suitable for performing multifunctional catalysis. Bifunctionality is sometimes inherently linked to the method of preparation of the catalyst: hydrogenation of cations results in protons and metals. Therefore, the combination of acid catalysis with some hydrogenation capacity is the most common type of bifunctional catalysis on zeolites. Another example consists of the hydrolysis of metal ions resulting in protons and metal oxides for a combination of redox and acid catalysis. In other cases the bifunctionality is introduced in separate subsequent steps.

With the term bifunctional synergy we do not denote the inherently bifunctional action of a surface Brønsted acid site: the electrophiles are generated on acids, but neighboring basic lattice oxygens enhance the nucleophilicity of the compound on which the attack takes place.

6.7.4.1
Hydrogenation + Alkylation

A typical example of such bifunctional behavior is found for the hydrodimerization of benzene and alkylbenzene [199]. Pd or Ni metal supported on acid zeolites hydrogenate part of the aromatics to the corresponding cyclic olefins, which in a Friedel–Crafts alkylation step are substituted on the aromatic. The counterpart of the acid-hydrogenation reaction is probably the side-chain alkylation of toluene and other alkyl aromatics with methanol [200–204]. Methanol is dehydrogenated on occluded basic oxides to formaldehyde, while the side chain of toluene is activated on basic sites. Styrene is formed, as well as its hydrogenation product ethylbenzene.

6.7.4.2 Hydrolysis + Hydrogenation

Another type of reaction that combines acid sites with noble metals is Ru/HUSY, which enables the one-step conversion of polysaccharides (such as starch) to polyhydric alcohols such as D-glycitol [205]. This catalyst is superior to a combination of two separate monofunctional catalysts: (Ru/C with HUSY). The acid sites catalyze the hydrolysis of the polysaccharides to monosaccharides which are subsequently hydrogenated on the Ru metal.

6.7.4.3 Hydration + Dehydrogenation

Mochida et al. [206] studied the reaction of propene and water over Cu^{2+}-exchanged zeolite Y and found isopropanol and acetone as the main reaction products. Acetone is formed from propene by subsequent hydration and dehydrogenation. This implies that the Cu^{2+}-Y zeolite is highly acidic and plays a bifunctional catalytic role in the reaction. Protons in Cu^{2+}-Y can only be generated upon reduction of cupric ions with propene or by hydrolysis of Cu^{2+}, either during the ion exchange of the parent Na^{+}-Y zeolite with copper, as a result of an a-stoichiometric exchange process, or during reaction.

6.7.4.4 Isomerization + Dehydrogenation

The dehydroisomerization of α-limonene is an alternative way to obtain the industrially important *para*-cymene. α-Limonene can be transformed into isomer-free *para*-cymene via the zeolite specific shape selectivity. A yield higher than 80% *p*-cymene could be obtained by using a Ce- and Pd-modified ZSM-5-type zeolite. The reaction proceeds via a two-step mechanism, first the isomerization (double-bond shift into the cyclohexene ring) to α-terpinene or γ-terpinene and the subsequent dehydrogenation to *p*-cymene. The two steps occur simultaneously in the zeolite pore, thereby the reaction rates between isomerization and dehydrogenation have to be adjusted carefully to each other and a pentasil zeolite has to be used in order to maintain the *para*-position [15, 207].

6.7.4.5 Complete Process Changes by Zeolite Catalysts: ε-Caprolactam Production

ε-Carpolactam is a valuable intermediate for polyamide production. The classic synthetic route for ε-caprolactam uses benzene as feedstock and involves its hydrogenation to cyclohexane. The oxidation of the latter gives cyclohexylperoxide, the cleavage of which in the presence of transition metals [208], such as Co, Cr, Cu and Mn, leads to a mixture of cyclohexanol and cyclohexanone; its complete oxidation to cyclohexanone is followed by the oximation of this ketone with hydroxylamine sulfate and the subsequent Beckmann rearrangement of the obtained oxime. The last two reaction steps are carried out in fuming sulfuric acid, after which the reaction requires neutralization with ammonia, and thus forms large amounts of ammonium sulfate (2.5–4.0 ton/ton ε-caprolactam).

This traditional process technology has many drawbacks, e.g., oxidation with oxygen, use of highly toxic hydroxylamine, formation of ammonium sulfate as by-product, handling of a large amount of fuming sulfuric acid and use of corrosion-proof equipment.

Alternatively, titanium-silicalite (TS-1) can be used for the liquid-phase preparation of cyclohexanone oxime via ammoximation of cyclohexanone with ammonia and hydrogen peroxide [209–211]. At 353 K, 99.9% substrate conversion is obtained with a yield of 93.2% of oxime [209]. It has even been demonstrated that on TS-1 the ketoximes can be synthesized starting from the secondary alcohol without isolation of the intermediate ketone [212].

6.7.5 Zeolite Effect V: New Chemistry with Zeolites

Sometimes, active zeolite-specific sites are available to generate new, innovative chemistry.

6.7.5.1 Pseudo-Solid-Solvent Effect

A very specific property of zeolite topologies containing cages is their ability to isolate molecules by steric constraints [213–214]. Zeolites accomplish what solvents do in the homogeneous phase, realizing monomolecular dispersion of complexes. This, however, is not the result of solvation, but is due to steric constraints on the encapsulated complexes, preventing their aggregation. They are retained in the zeolite because the pore windows are smaller than the size of the complexes. Such zeolite-included complexes are easy to handle, and the problems of cluster formation and oxidative destruction of complexes by reactions between active complexes is avoided. Consequently, stability and turn-over numbers are enhanced upon encapsulation. These catalysts are called enzyme-mimics as the protein mantle is replaced by a zeolite framework and the active complex of the enzyme by a synthetic analogue. Recently, their catalytic performance has been considerably enhanced by embedding the mimic in a dense hydrophobic membrane [215]. This represents the ultimate mimic of the enzyme cytochrome P-450 on a formal basis. At the same time it was reported that mechanistic similarities are also present [215].

6.7.5.2 New Complexes Through Encapsulation

Monomeric *cis*-$[Mn(bpy)_2]^{2+}$ can be stabilized in the cages of faujasite zeolites under oxidative conditions, which is not possible in the homogeneous phase [216–218]. Moreover, zeolite-occluded monomeric *cis*-$[Mn(bpy)_2]^{2+}$ has unique oxidation properties [216–218]. Olefinic substrates are oxidized selectively with hydrogen peroxide to epoxides, diols or dicarboxylic acids depending on the ratio substrate/peroxide and the presence of acidity. In the absence of acid sites the reaction can be stopped at the epoxide stage. This is realized on X zeolites which have weak or no acid sites. On Y zeolites with stronger residual acid sites the epoxide is hydrated to diols and if enough peroxide is present it is further oxidized to dicarboxylic acids. Acid *cis*$[Mn(bpy)_2]^{2+}$-Y represents a bifunctional catalyst which produces the acids via oxidative, acid and again oxidative reactions. Conversion, selectivity and stability exceed the values reported for other oxidation catalysts.

6.7.5.3
Ti-Zeolites

Titanium-zeolites allow oxidation chemistry, previously unknown for Ti. Traditionally, titanium catalysts are recognized for the efficient and selective epoxidation of olefins. They are rather inert for saturated hydrocarbons.

Ti-silicalite (type 1 as well as type 2, probably as well as other Ti-containing topologies), however, catalyzes, besides epoxidation, a broad range of oxidation reactions with hydrogen peroxide as oxidant (Table 6.6).

The feature which makes this chemistry unique is probably related to the isolation of titanium and its specific coordination in the zeolite matrix. Although the discussion is not closed, and the exact coordination of titanium in zeolites is not fully elucidated, a convincing picture taking into account many XRD, EXAFS, XANES, UV-Vis and IR-data, has been advanced recently by Bellussi and Rigutto [75]. Ti seems isomorphously substituted in the zeolite structure in tetrahedral coordination, with two positions easily accessible for electron-donating ligands such a water, alcohols, peroxides and amines. This site is much more resistant against hydrolysis compared to isolated titanium species on amorphous silica.

Table 6.6. New Catalytic Chemitry with Ti-Silicates

Substrate	Product	Ref.
vinylbenzene compounds	β-phenylaldehydes	[219]
olefins and methanol	glycol monomethyl ethers	[219]
diolefins	monoepoxides	[220]
olefins	epoxides	[221-224]
phenol	hydroquinone and catechol	[225-227]
benzene	phenol	[228]
alkylbenzene	alcohols and ketones	[229]
primary alcohls	aldehydes	[229]
secondary alcohols	ketones	[229]
tertiary alcohols	alkyl hydroperoxides	[230]
ammoximation of cyclohexanone	ε-caprolactam	[231-233]
alkanes	alcohols and ketones	[171, 172, 234, 235]
N,N-dialkyl-amines	*N,N*-dialkylhydroxylamines	[236]
primary aliphatic amines	oximes	[237]
thioethers	sulfoxides	[238]

A key characteristic of these materials is their relatively strong hydrophobicity, resulting in the favorable adsorption of alkanes. Thus the low concentration of hydrogen peroxide present at all times in the catalyst favors its efficient use. The strong hydrophobicity also enables fast desorption of the product. Therefore, oxidations can occur up to high conversions with high selectivities and high efficiencies. Attempts to make comparably active titanium-zeolites with large pores as Ti-beta zeolites failed, partly because of their hydrophilic nature [239].

6.8 Case Study: Zeolites as Non-Corrosive, Environmentally Friendly Friedel–Crafts Alkylation Catalysts

6.8.1 Introduction

The term "Friedel–Crafts (FC) chemistry" has been used to cover an ever-increasing number of reactions related to the first aluminum chloride reaction discovered by Friedel and Crafts in 1877 [240]. They observed accidentally that, in the presence of anhydrous $AlCl_3$, an alkylhalogenide or acylhalogenide reacted with benzene with the formation of an alkyl- or acyl-substituted aromatic, respectively. In the continuation of the work they found that a whole class of Lewis acids could catalyze this and many other related reactions as dealkylation, polymerization, isomerization,... [240, 241]. It became generally accepted that any reaction combining two or more organic molecules through the formation of carbon to carbon bonds under the influence of anhydrous $AlCl_3$ or related catalysts, is a *Friedel–Crafts reaction.*

Later on, Brønsted acids and Brønsted–Lewis acid associations were found to also catalyze these reactions, extending the original scope of the FC reaction to any substitution, isomerization, elimination, cracking, polymerization or addition reaction taking place under the effect of Lewis or Brønsted acids [242]. The principal relationship between these different reactions is their electrophilic reaction mechanism. There is no fundamental reason to limit the scope of FC reactions to the formation of C–C bonds. The formation of many other bonds (C–N, C–O, C–S, C–X) conform to the general Friedel–Crafts mechanistic principle. However, as the original contributions of Friedel and Crafts were almost exclusively limited to the substitution of aromatic substrates, it is appropriate to use the name "Friedel–Crafts" for those reactions involving the formation of a new carbon–carbon bond by electrophilic substitution on an aromatic ring with any Lewis or Brønsted acid as catalyst. Consequently, FC reactions are considered as a subdivision in the class of electrophilic aromatic substitution (EAS) reactions. This definition is arbitrary, as it has no mechanistic basis.

6.8.2
Friedel–Crafts Chemistry over Zeolites from a Historical Perspective

Due to (i) the corrosive nature of the homogeneous Lewis and Brønsted acid catalysts giving acidic and salty waste waters, (ii) the difficult separation of products from the catalyst, (iii) the often stoichiometric consumption of the "catalyst", and (iv) the lack of elevated positional selectivity, a high interest for solid Lewis and Brønsted acid catalysts has been observed during the past two decennia. Acid-treated clays, acid zeolites and acid resins have emerged as the most promising substitutes for homogeneous Lewis and Brønsted acids.

A milestone in the development of zeolite-based FC catalysts was the generation of acid sites in a zeolite Y [243, 244]. Venuto et al. [7, 8, 185–187] indicated that these acid and rare earth exchanged Y zeolites are catalysts in the alkylation of monosubstituted aromatics with olefins, alkylhalogenides and alcohols in the liquid phase. In subsequent years, however, Venuto's studies became overshadowed by major developments in the field of petroleum refining. The discovery of ZSM-5 [245] gave a new impetus to the development of zeolite-based Friedel–Crafts catalysts. The employment of this highly acidic and thermally stable catalyst led to new industrial processes for the selective preparation of bulk chemicals as ethylbenzene (Mobil-Badger process, see Table 6.1), *p*-ethyltoluene, and *p*-xylene. The success of this catalyst delayed the evaluation of other zeolites as potential FC catalysts for the synthesis of organic fine chemicals. At the end of the 1980s, a renewed interest in all zeolites as FC catalysts emerged, the main reason being an increased environmental pressure on homogeneous FC methods.

6.8.3
Overview of Friedel–Crafts Literature with Zeolites

An overwhelming amount of data on FC alkylation reactions over zeolites have been accumulated in the patent and scientific literature. To avoid in this overview an enumeration of reactions, we propose a classification of zeolite FC chemistry based on the nature of the aromatic which is substituted, and on the properties of the alkylating agent. The division between alkyl aromatics and heteroatom-substituted aromatics basically reflects the boundary between relatively poorly and strongly adsorbing aromatics. Indeed, molecules such as phenol, anisole, thiophenol, thioanisole and aniline strongly adsorb on acid zeolites compared to alkyl aromatics [246]. The classification scheme is represented in Table 6.7.

6.8.3.1
Group 1 Reactions: Alkylation of Alkyl Aromatics with Olefins

Alkylation of alkyl aromatics with olefins has been almost exclusively performed in the gas phase over an acid form of ZSM-5 and related 10-ring zeolites using a high ratio of aromatic substrate to alkylating agent [143, 247, 248]. Appropriate ratios of aromatic substrate to alkylating agent are higher than 5.

Table 6.7. Classification of Friedel-Crafts Alkylation Reactions over Zeolites Based on the Adsorption Behavior of Aromatic Substrates and Alkylating Agents

Strength of adsorption of aromatic substrate	Strength of adsorption of alkylating agent	Examples
weak	weak	alkyl aromatic + olefin
weak	strong	alkyl aromatic + alcohols, aldehydes, ethers, amines ...
strong	weak	hydroxy, amino ... aromatics + olefin
strong	strong	hydroxy, amino ... aromatics + alcohols, aldehydes, ethers, amines...

The almost exclusive application of ZSM-5 originates from its specific pore size, which closely matches the size of many monosubstituted aromatics. Transition state shape selectivity inhibits polyalkylation and deactivation of the catalyst on one hand and diffusion selectivity enhances the yield of the *para*-isomer in the case of alkylation of monosubstituted aromatics on the other. *para*-Selectivity can be further enhanced by impregnation of ZSM-5 with boric acid [249], phosphoric acid [250] and deposition of oxides (MgO and others) [143]. These modifications reduce the pore size (or pore mouth) of the zeolite channel and the activity of the active sites on the outer surface [251]. High ratios of aromatic substrate to alkylating agent are required to suppress the bimolecular olefin polymerization.

There has been controversy over the precise reason for high *para*-selectivity in the ethylation of toluene over modified acid ZSM-5 catalysts. Keading et al. [252] reported that impregnation with P, Si, B, Mg, Ca and Mn oxides and zeolite coking caused enhanced *para*-selectivity. Indeed, as a result of product shape selectivity, the intracrystalline diffusivity of the *para*-isomer is much higher than that of the other two isomers. The same argument was advanced by Kolboe et al. [253] when explaining the enhanced *para*-selectivity with increasing size (30-50 μm) of ZSM-5 crystals. These authors tacitly assumed that the alkylation occurred inside the pores of the zeolite with a low preference for the *para*-isomer. The observed enhanced *para*-selectivity in the bulk medium was the result of a rate-controlling isomerization of the *meta*- and *ortho*-isomer to the *para*-isomer.

On the other hand, a number of authors have proposed that the primary alkylation reaction in the channels of ZSM-5 is inherently *para*-selective. To achieve high *para*-selectivity in the bulk medium, consecutive isomerization must be inhibited. According to Yashima et al. [254] and Paparatto et al. [251] this isomerization takes place essentially on sites located at the outer surface of the crystals. Their number can be reduced by impregnation with inorganic oxides or by increasing the crystal size. Alternatively, Lonyi et al. [255, 256] and Engelhardt et al. [257] proposed that both internal and external acid sites catalyzed isomerization. They ascribed the enhanced *para*-selectivity after impregnation or coke deposition to a reduction in the overall number of Brønsted acid sites, which decreased the rate of the side reactions more effectively than the formation of the primary product, *para*-ethyltoluene. A third explanation was advanced by Kim et al. [258–260]. They argued that it was not the decreased number but the lower strength of the acids sites after impregnation or coke deposition that reduced the extent of isomerization. This hypothesis was supported by the observation that acid ZSM-5 zeolites isomorphously substituted with Ga, Fe and B, and consequently exhibiting lower acid strength than Al-ZSM-5, resulted in higher *para*-selectivities [259]. However, the validity of this argument was contested by Parikh et al. [261, 262]. They reported that *para*-selectivity in reactions over isomorphously substituted ZSM-5 was only slightly dependent on the nature of the substituted cation and acid strength and strongly increased with increasing crystal size. They proposed, in agreement with Paparatto et al. [251], that with increasing crystal size, the relative number of sites on the external surface decreased which resulted in reduced isomerization of the primary product.

The alkylation of alkyl aromatics with olefins has also been performed over 12-MR zeolites. However, the extent of polyalkylation and coking is strongly enhanced compared to 10-MR zeolites [185]. *para*-Selectivity is also lower, especially for reactions with small alkylating agents. These disadvantages, and particularly the faster deactivation, have long prevented their application. However, operation in the liquid phase, combined with modification of the catalyst, e.g., by dealumination, and use of specific process technology, considerably enhance the stability of the catalyst. The isopropylation of benzene with propene is an example in this respect. Pradhan and Rao [263] performed this reaction in the continuous mode in the gas phase and at atmospheric pressure over a rare earth exchanged HY zeolite (493 K, WHSV of propene = 2.5 h^{-1} and ratio aromatic/alkylating agent = 8:1) and observed a stability in time of about 10 h. Performing this reaction at 2.5 MPa, the catalytic activity was prolonged to about 150 h. The extended stability was attributed to the operation in liquid phase. This avoided reactions like oligomerization and enabled the washing out of the coke precursors formed, if any [263, 264]. Meima et al. [265] used a dealuminated mordenite (Si/Al = 78) at temperatures between 423 and 443 K and a pressure of 3.6 MPa in a continuous liquid phase reactor (WHSV of propene = 1 h^{-1} and a ratio of aromatic/alkylating agent of 5.8). They observed no activity decay for 900 h. In addition no undesirable n-propylbenzene was produced due to the low reaction temperature. Performing this reaction in a catalytic distillation mode further prolonged the stability of the catalyst to about 3 months [266] and made this process an excellent substitute for the vapor-phase process based on a Kieselguhr catalyst impregnated with phosphoric acid.

Partially collapsed Y zeolites have been reported to be excellent catalysts for the alkylation of alkyl aromatics with long-chain α-olefins [267]. The reactions were run in a continuous flow reactor at temperatures from 338 and 423 K (hydrogen pressure of 1.2 MPa; WHSV of alkylating agent = 1 h^{-1}; ratio aromatic to alkylating agent of 5). Using ethylbenzene as aromatic substrate and 1-n-hexadecene as alkylating agent, selectivities to mono-alkylated aromatics higher than 99% were reported and no dimerization of the olefin occurred. The stability of the catalyst in time amounted to about 100 h. An analogous critical dependence of the catalyst stability on its preparation procedure was reported by Grey

[268] and Juguin et al. [269] for the same and related reactions over an acid mordenite. A strong dealumination of mordenite (Si/Al up to 93) was required. The shape-selective effect of the zeolite structure on the isomer distribution was in all cases not pronounced.

Acid zeolites have also been claimed to be good catalysts for the preparation of specialty chemicals in this group of reactions. Shape-selective isopropylation of biphenyl was reported (up to 90% of *p,p'*-dialkyl isomer) on a dealuminated mordenite [150, 270–272] and ZSM-12 [150]. Dealuminated mordenites were also suitable catalysts for di-isopropylation of naphthalene (88% of 2,6-isomer) and diphenylmethane (55% of *p,p'*-dialkyl-isomer) [270].

6.8.3.2 Group 2 Reactions: Alkylation of Alkyl Aromatics with Alcohols, Ethers, Aldehydes, Amines etc.

A separate discussion of the alkylation of alkyl aromatics with olefins on one hand and with all other alkylating agents on the other is justified because it generally holds that alcohols, ethers, amines, haloalkanes etc. (i) adsorb more strongly on acid zeolites, and (ii) are much less reactive than the corresponding olefins. Consequently, the required optimal reaction temperature and ratio of alkylating agent to aromatic are higher. The EAS reactions with alcohols, ethers, amines, haloalkanes are also accompanied by the formation of water, alcohols, ammonia or hydrogen halides, resulting possibly in a decrease in the rate of the catalytic reaction or destruction of the catalyst. The latter especially holds for reactions with haloalkanes. Positional selectivity is almost independent of the alkylating agent.

A typical reaction of this group is the preparation of *p*-xylene by direct alkylation of toluene with methanol over modified ZSM-5. On the original ZSM-5 zeolite, the isomer distribution is near thermodynamic equilibrium [143]. Enhanced *para*-selectivities are obtained by increasing the ratio of internal to external crystal surface area (by enlarging the crystal, by coating the outer surface or terminating the zeolite crystal with an Al-free surface), by enhancing the pore tortuosity or by partly blocking the channel structure with an inorganic filler.

As for the reaction of toluene with ethene, there is no agreement on the specific reason for the enhancement of *para*-selectivity on modified ZSM-5.

(i) The most obvious explanation was originally proposed by Kaeding et al. [273, 274], Young et al. [143] and Wie [275]. Based on kinetic experiments, it was advanced that transport limitation of the bulkier isomers (*m*-, *o*-) permitted the faster diffusing products (*para*-isomer) to escape from the catalyst at higher rate; thus, a higher selectivity towards the *para*-products was observed. In addition Kaeding et al. [273, 274] also suggested that restricted transition state selectivity could be operative in the initial alkylation reaction.
(ii) On the contrary, Seko et al. [276] proposed that variations in selectivity were caused by the different strength of adsorption of xylene isomers as they found that the more basic *m*- and *o*-xylenes adsorbed with higher strength than *p*-xylene.

(iii) Kim et al. [277, 278], Chandavar et al. [279], Bezouhanova et al. [280] and Vinek et al. [250] attributed the *para*-selective effect of modified ZSM-5 type zeolites to a decrease in the concentration of strong protic centers, the latter catalyzing the isomerization of the initially formed *para*-isomer.

(iv) Finally, Nayak and Riekert [281] also considered an initial *para*-selective formation of xylenes, but positional isomerization was according to these authors restricted to external surface sites. These sites were destroyed by impregnation with inorganic oxides. Recently, Mirth et al. [282] performed the alkylation of toluene with methanol in a continuously stirred tank reactor equipped with CaF_2 windows. This allowed simultaneous IR monitoring of the adsorption complexes under reaction conditions and analysis of the reactor effluent. The main isomer found in the reactor effluent was *p*-xylene, whereas *m*- and *o*-xylene were the dominating isomers adsorbed on the catalyst. These observations obviously support explanation (i).

In three hypotheses, the *initial* positional selectivity is thought to be determined by the intrinsic reactivity of the different carbon atoms in the aromatic substrate and by geometrical constraints imposed by the pores of the zeolite. The observed *para*-selectivity is either higher [explanation (i)] or lower [explanations (iii) and (iv)] than the initial selectivity, as a result of consecutive isomerization. It is also tacitly assumed that the chemical composition of the catalyst has no influence on the intrinsic reactivity of the aromatic positions or of the electrophile.

Yashima et al. [283, 284], however, observed that in the methylation of toluene over zeolite Y, *para*-selectivity was slightly enhanced when the strength of the acid sites was increased. In this zeolite no geometric constraints existed, neither for the transition state nor for product diffusion. The authors correlated selectivity with acid strength, but did not show the mechanistic background for it. Wendlandt and Bremer [100] reported similar observations but explained the selectivity dependence by properties of electrophilic species. They assumed that the alkylation reaction was orbitally controlled and that the catalyst modified the hardness/softness characteristics of the electrophile. With decreasing Al content of the zeolite the adsorbed electrophile becomes softer and prefers the *para*-position, in agreement with the knowledge that softer electrophiles prefer substitution in *para* [285]. In an attempt to verify such general statements, Corma et al. [286–289] calculated the energy of the LUMO orbital of a protonic zeolite cluster and of an adsorbed electrophile (as a measure of their hardness) with varying Si/Al ratios and found that with increasing Si content, the zeolitic proton and the carbenium ion formed upon protonation of the alkylating agent became softer. This agreement of theory and experiment does not unequivocally confirm this hypothesis, as in no study was the absence of consecutive isomerization proven! Possibly consecutive isomerization is reduced upon dealumination of the framework.

Besides methanol, other alcohols have been utilized as alkylating agents for monocyclic arenes: ethanol (e.g., [261, 262]) even commercially [290], 1- and 2-propanol (e.g., [248]), linear long-chain alcohols (e.g., [185]) etc. Aldehydes have been used to prepare styrenes [291] and diarylmethanes [187,292]. Haloalkanes have hardly been used due to their corrosive nature [185] and amines have not been employed at all. However, their use could be recommendable in the alkylation of strongly adsorbing aromatics.

Alkylation of oligocyclic arenes and especially naphthalene with methanol has been reported [293–298]. Neuber and Weitkamp [55, 56] demonstrated that H-ZSM-5 adsorbs 6.5 wt% of 2-methylnaphthalene at 373 K. As framework flexibility will increase at higher temperatures, such data strongly indicate that the catalytic alkylation is intraporous. A more detailed discussion of this reaction has been made in a previous section, as it has been the basis for first the introduction [51, 53, 54, 178] and subsequently the rejection [55, 56] of a new concept of shape selectivity, namely the "nest effect" or shape selectivity induced by the external surface of the catalyst.

6.8.3.3
Group 3 Reactions: Alkylation of Heteroatom-Substituted Aromatics with Olefins

Only the alkylation of the so-called "*activated aromatics*" which are more reactive than benzene in homogeneous EAS reactions is considered here; among them are phenol, anisole, thiophenol, thioanisole, aniline and alkylanilines. The reactivity of such substrates, in particular of phenol and aniline, has been extensively studied. Alkylation of so-called "*deactivated aromatics*" (e.g., haloaromatics, benzoic acid, benzonitirile) have occasionally been investigated. No solid catalysts yet have been investigated which are possible candidates for substitution of homogeneous acids.

An overview of the EAS reactions of phenol and aniline with olefins is presented in Tables 6.8 and 6.9. Alkylation reactions of these compounds are more complex than those of alkyl aromatics, as not only carbon but also heteroatom alkylation is possible. The ratio of heteroatom to carbon alkylation depends on reaction temperature, ratio of aromatic substrate to alkylating agent, acid strength of the catalyst, nature of aromatic and alkylating agent and contact time. Heteroatom alkylation is preferred on weak acid sites and at low reaction temperature (e.g., [299, 300]). High temperatures and contact times favor alkylation on the ring carbon atoms.

In contrast to the alkylation of alkyl aromatics, the generally stronger adsorption of heteroatom-substituted aromatics enables operation at lower ratios of aromatic substrate to alkylating agent. Venuto et al. [185, 246] reported that with equimolar amounts of aromatic substrate and alkylating agent, high selectivity for monoalkylation is obtained and that benzene is alkylated faster than phenol, although benzene is functioning as an inert solvent in the ethylation of phenol. Venuto and Wu [246] pointed out that this paradox was based on the stronger adsorption of phenol compared to benzene, thus reducing the sorption of benzene and ethene.

Positional selectivity in such reactions is strongly influenced by the presence of the heteroatom. However, it remains unsettled whether heteroatom alkylation is an intermediate step in carbon alkylation. It has been suggested that such a pathway could explain the high *ortho*-selectivity in the alkylation of

Table 6.8. Olefins Used in Phenol Alkylation on Zeolites Under Vapor- and Liquid-Phase Conditions

Type of olefin [a]	Mode of operation [b]	Zeolite type	Ref.
ethene	batch/liquid	REX [c]	[185]
propene	cont./gas	ZSM-5	[301]
long-chain	unspecified	pentasil family [d]	[302]
olefins	cont./liquid	Mordenite, ZSM-12, REY [e]	[303, 304]
	cont./liquid	Mordenite, ZSM-12, REY [e]	[299]
	batch/liquid	REX [c], HY, Beta, Mordenite	[185]
	batch/liquid	REY, HZSM-5	[305]
	cont./liquid	unspecified	[306]
specialty olefins	cont./liquid		[307]

[a] Specialty olefins: e.g., styrene, isoamylene, ...
[b] cont.: continuous.
[c] REX: zeolite X exchanged with rare earth metals.
[d] Pentasil family: family of zeolites exemplified by ZSM-5, ZSM-11, ZSM-12, ZSM-23, ZSM-35, ZSM-38, ZSM-48.
[e] REY: zeolite Y exchanged with rare earth metals.

Table 6.9. Olefins Used in Aniline Alkylation on Zeolites Under vapor- and Liquid-Physe Conditions

Type of olefin	Mode of operation [a]	Zeolite type	Ref.
propene	cont./liquid	REX, REY, HY, HMordenite	[310, 311] [300, 312–314]
	cont./liquid	HY, HMordenite	[315]
		REX, NaX	[315–318]
	cont./liquid	HUSY, AlY [b], SAPO-11,-37, AlPO-5	[319]
long-chain olefins	batch/liquid	HUSY, HMordenite	[310, 311]
specialty olefins	cont./liquid	REX, REY, HY, HMordenite	[312–314, 300]

[a] cont.: continuous.
[b] AlY: Zeolite Y exchanged with Al^{3+}.

phenol and aniline [308]. Burgoyne et al. went even further and claimed that the selective *ortho*-alkylation of aniline was based on a concerted cycloaddition of the alkene to the protonated arylamine [300]. This mechanism is represented in Fig. 6.8.

Shape-selective effects in the alkylation of phenol with long-chain olefins on mordenite and ZSM-12 have been reported [303, 309]. Transition state selectivity would especially favor the formation of the *para*-isomer. However, Campbell et al. [299] seemed to confirm these results and claimed that incorrect product identification was at the basis of the early claims.

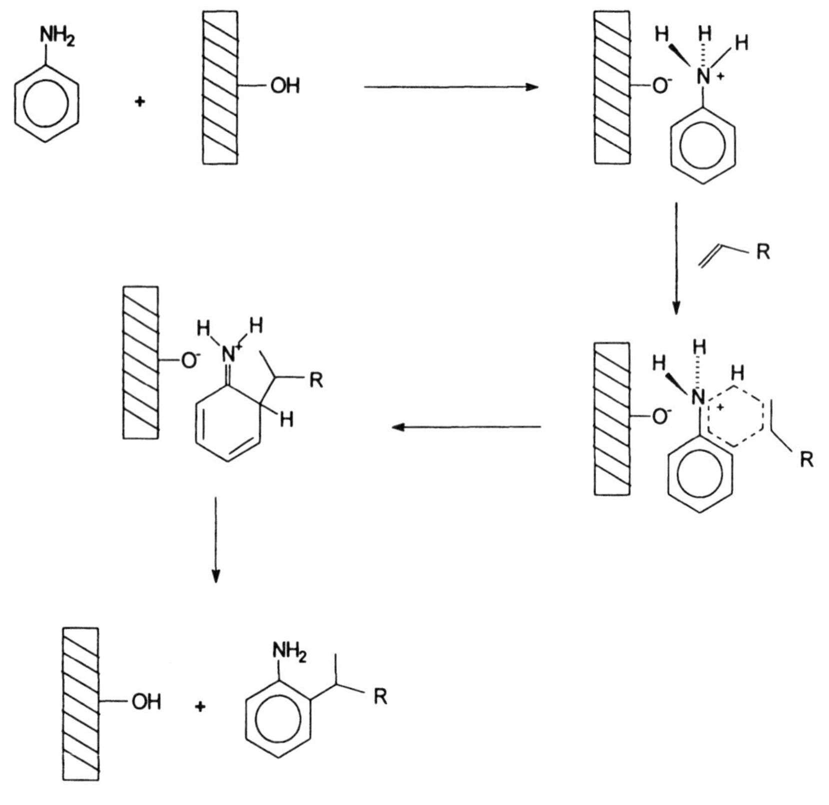

Fig. 6.8. Mechanism of the alkylation of aniline with olefins [300]

6.8.3.4
Group 4 Reactions: Alkylation of Heteroatom-Substituted Aromatics with Alcohols, Aldehydes, Haloalkanes etc.

In Table 6.10, an overview of the work in the area of the phenol alkylation with various (non-olefinic) alkylating agents is given (for reaction temperatures between 300 and 700 K and an aromatic/alkylating agent ratio of 5:1). The excess of substrate is required to suppress dimerization of the alkylating agent and polyalkylation.

As for the alkylation of phenol with olefins, there is a quite general belief among different authors that carbon alkylation requires stronger acid sites than oxygen alkylation [308, 316, 320, 321]. This dependence of carbon alkylation on acid strength has also been observed for metal phosphate catalysts, as a correlation between the selectivity of ring alkylation and the electronegativity of the metal ion has been established [322].

para-Selective substitution of phenol is hardly possible even in ZSM-5 samples modified with inorganic oxides [309]. The only exception was reported

Table 6.10. Different Alkylating Agents Used in Phenol Alkylation on Zeolites Under Vapor- and Liquid-Phase Conditions

Alkylating agent	Mode of operation[a]	Zeolite type	Ref.
methanol	cont./gas	HZSM-5, modified with H_3PO_4	[309]
	cont./gas	REX, REY, HY, HZSM-5	[332]
	cont./gas	HZSM-5	[333]
	cont./gas	HZSM-5	[334, 335]
	cont./gas	X	[336]
	cont./gas	AlPO-5, SAPO-5,-11,-34	[337]
	cont./gas	HY, exchanged with Na^+ and K^+	[320]
	cont./gas	HTsVM	[338]
	cont./gas	REY, REX, CaY, NaX	[339]
	cont./gas	Ti-ZSM-5	[340]
	cont./gas	HZSM-5, HUSY	[308]
	cont./gas	HY, HUSY	[327]
	cont./gas	unspecified	[341, 342]
	cont./gas	HZSM-5	[316]
higher alcohols	batch/liquid	HY	[321]
	cont./gas	P-ZSM-5	[302]
	cont./gas	HZSM-5	[301]
	batch/liquid	HMordenite, HZSM-12	[299]
	cont./liquid	HMordenite, REY, HZSM-12	[303]
methyliodide	cont./gas	HX	[343]
benzylchloride		HX, HY, HL	[344]
alkylchloride	batch/liquid	HX, HY	[185]
anisole	cont./gas	REX, REY, HY, HZSM-5	[324, 325]
trioxane	batch/liquid	REY, HZSM-5 HBeta, HMordenite	[305]
formaldehyde		REY, HZSM-5, HBeta, HMordenite	[305] [187]
dialkylcarbonate	cont./gas	unsepcified	[345]

[a] cont.: continuous.

by Corma et al. [321] who found *para*-selectivities up to 92% in the *tert*-butylation of phenol over an acid zeolite Y.

The exact reason for the low *para*-selectivity, especially in the methylation of phenol, has been the subject of much speculation [322]. It is, however, beyond question that rapid isomerization of the initially formed *para*-isomer to the *ortho*-isomer cannot be the reason, because low *para*-selectivities are obtained even in absence of *meta*-isomer formation [22].

Most hypotheses can be simplified to an interaction between the alkylating agent and the oxygen atom of phenol [324–327] which favors alkylation at the *ortho*-position. However, this explanation supposes a highly questionable interaction between an electrophilic oxygen of the alkyl donor (methanol) and a

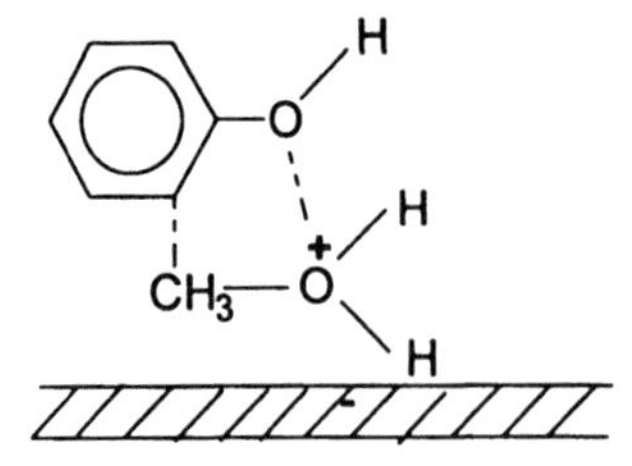

Fig. 6.9. Possible transition state of the alkylation of phenol with methanol according to Marczewski et al. [326]

nucleophilic oxygen atom of phenol as shown in Fig. 6.9. These models are also unable to explain the high *ortho*-selectivity in the alkylation of phenol with olefins.

Parton et al. [308] preferred another explanation. Based on the higher *para*-selectivity in the phenol methylation over zeolite Y compared to ZSM-5 (59 and 35%, respectively), they proposed that in 10-MR zeolites bimolecular formation of cresol isomers (reaction of phenol with methanol) was suppressed in favor of the intramolecular rearrangement of anisole to *o*-cresol. This suppression did not occur in 12-MR zeolites. This explanation requires the favorable formation of anisole, involving also a bimolecular reaction, on the strongly acidic H-ZSM-5. The theory is supported by results which show that the selectivity for anisole formation on zeolite Y is higher than on ZSM-5 [308], caused by the higher conversion rate of anisole to *o*-cresol on the pentasil zeolite, and the formation of three primary products, viz. phenol, *o*-cresol and *p*-methylanisole in the anisole conversion on pentasil zeolites [328]. According to the authors, the *para*-selective alkylation of phenol occurs and the *ortho*-selective alkylation especially is suppressed, but the rate of oxygen alkylation is higher than the rate of carbon alkylation. However, the intermediate anisole is less stable and transforms to the more stable *o*-cresol via a rearrangement mechanism. This theory, however, gives no general explanation for the relatively low yield of the *para*-isomer in the alkylation reactions of phenol over acid oxides as silica-alumina and zirconium sulfate, and acidic resins as Nafion-H and Amberlyst-15 [299]. A possible explanation for this phenomenon could be that on the weaker acid sites the rate of carbon to oxygen alkylation is strongly shifted to oxygen alkylation. Anisole again is not necessarily stable and can transform to *o*-cresol.

In Table 6.11, an overview of the alkylation of aniline with various (non-olefinic) alkylating agents is given.

Methylation of aniline to *N*-methylaniline and the toluidines is the most extensively studied reaction [329–331]. At low temperature (275 °C), *N*-alkylation seems to predominate, while at higher temperatures *C*-alkylation takes over. Regioselectivity in the *C*-alkylated fraction varies with temperature and space velocity. As the selectivity for *para*-toluidine decreases with WHSV, some of the *para*-toluidine seems to be formed via secondary reactions. Finally at 450 °C, the thermodynamic equilibrium composition is obtained, with a substantial production of the *meta*-isomer.

Table 6.11. Different Alkylating Agents Used in Aniline Alkylation on Zeolites Under Vapor- and Liquid-Phase Conditions

Alkylating agent	Mode of operation[a]	Zeolite type	Ref.
methanol	cont./gas	HZSM-5, modified ZSM-5	[346-348]
	cont./gas	HZSM-5	[349-350]
	cont./gas	HZSM-5, B-ZSM-5	[330]
	cont./gas	AlPO-5	[351]
	cont./gas	HOmega, HBeta, HL	[352]
	cont./gas	HZSM-5	[353]
	cont./gas	HZSM-5	[353]
higher alcohols	cont./liquid	SAPO-37	[317]
alkylbromide	batch/liquid	alkali cation-exchanged X, Y	[344, 354, 355]
trioxane	batch/liquid	HZSM-5, HY, Ti-ZSM-5	[356]

[a] cont.: continnous.

6.8.4 Recent Developments in Friedel-Crafts Alkylation: Solvent Effects

Solvent effects in heterogeneous catalysis that control chemo-, regio- and stereo-selectivity in a reaction have been examined recently by Gilbert and Mercier [357]. Factors that can be categorized as solvent effects in the absence of diffusion limitations are:

(i) physical effects of solvents, such as the solubility of the substrate in the solvent, and the hydrophilic-hydrophobic interactions between solvent, substrate, and catalyst; and
(ii) chemical effects of solvents; the solvent can be reactive (modification of the reaction intermediate), influence the acido-basicity of the medium or influence chemisorption.

Gilbert and Mercier view molecular sieves as polar solvents [357]. The Si/Al ratio is the main parameter affecting the polarity. Alternatively, the Si/Al ratio can be invoked as a parameter in the HSAB (hard and soft acids and bases) approach; an increase in the Si/Al ratio is known to increase the softness of the active sites. The maximum in the hydroxylation of fluorobenzene with N_2O observed for ZSM-5 zeolites with different Si/Al ratios is explained this way: For more siliceous zeolites, the hydrophobicity of the zeolite increases and consequently the solvation of the aromatic by the zeolite, while that of the polar oxidant decreases. In terms of Derouane's "*Confinement theory*", zeolite solvation is determined by "*surface curvature*", that is by the radius ratio of sorbate and zeolite cage (pore).

Next to this activity effect, selectivity can change as a consequence of the confinement in the zeolite cages. The attractive effects in this case can be molecular recognition, preorganization and preferential interactions. As selectivi-

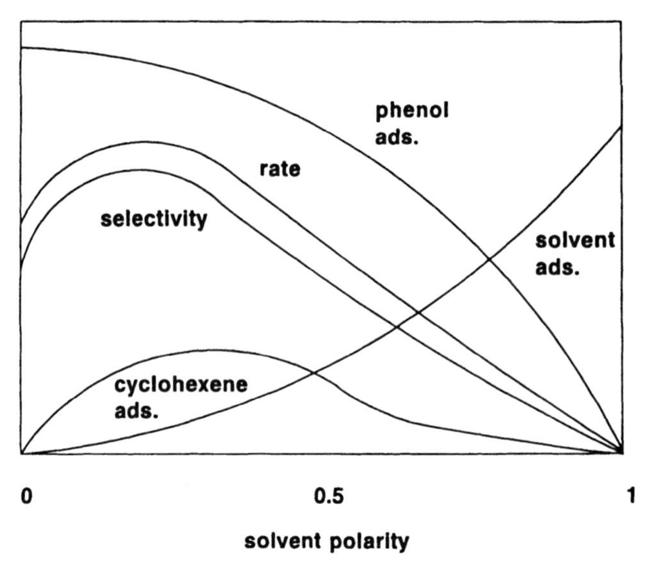

Fig. 6.10. Schematic representation of the dependence of reaction rate, selectivity, amount of solvent and reagent adsorbed in the cyclohexylation of phenol [358]

ties in the chlorination of biphenyl on zeolites [162] in a given topology depend on the solvent, Gilbert and Mercier [357] interpret the solvent influence on the selectivity in terms of this theory.

The cyclohexylation rate and selectivity of phenol over USY zeolite showed a volcano-shape relation with the solvent amount present in the zeolite and its polarity (Fig. 6.10) [358]. The selectivity for *C*- against *O*-alkylation also goes through a maximum. It was quantitatively demonstrated that the equilibrium distribution of reagents, products and solvents among bulk solution and zeolite cages, was at the basis of this behavior. Competitive sorption allowed to rationalize the existence of maxima in rate and selectivity without the intervention of a direct kinetic solvent effect.

We have recently demonstrated [359] that for a wide range of organic protic and aprotic solvents, a linear free energy relationship exists between the relative amounts of phenol adsorbed at saturation and the solvent properties.

Acknowledgments. This work was sponsored by a Belgian Federal IUAP programme. P.E. and R.P. acknowledge the Flemish NFWO for a fellowship. JM is grateful to NFWO for a research position and HT to EU for a postdoctoral fellowship in the Human Capital and Mobility Program. The authors appreciate the help of W. Souverijns and G. Vanbutsele for their assistance in the lay-out.

References

1. Misono M, Nojori N (1990) Appl Catal 64:1
2. Nojori N, Misono M (1993) Appl Catal 93:103
3. Armor JN (1991) Appl Catal 78:141
4. Enze M, Peiling Z (1993) Appl Catal 95:1
5. Chauvel A, Delmon B, Hölderich W (1994) Appl Catal 115:173
6. Meier WM, Olson DH, Baerlocher Ch (1996) Atlas of Zeolite Structure Types, 4th Ed, Elsevier
7. Venuto PB, Landis PS (1968) Adv Catal 18:259
8. Venuto PB (1971) Adv Chem Ser 102:260
9. Venuto PB (1994) Microporous Mater 2:297
10. Hölderich WF, Hesse M, Naumann F (1988) Angew Chem 27:226
11. Hölderich WF, van Bekkum H (1991) Stud Surf Sci Catal 58:701
12. Hölderich WF (1990) Stud Surf Sci Catal 67:257
13. Hölderich WF (1988) Proc Int Symp Acid-Base Catalysis, Saporro (eds). In: Tanabe K, Hattori H, Yamaguchi T, Tanaka T (1989) VCH & Kodansha, 1
14. Hölderich WF (1988) Stud Surf Sci Catal 41:83
15. Hölderich WF (1989) Stud Surf Sci Catal 49:69
16. Hölderich WF (1989) Stud Surf Sci Catal 46:193
17. Hölderich WF (1988) Stud Surf Sci Catal 58:631
18. Van Bekkum H, Kouwenhoven H (1988) Stud Surf Sci Catal 41:45
19. Van Bekkum H, Kouwenhoven H (1989) Recl Trav Chim 108:283
20. Maxwell IE (1982) Adv Catal 31:1
21. Parton RF, Jacobs JM, Huybrechts DR, Jacobs PA (1989) Stud Surf Sci Catal 46:163
22. Perot G, Guisnet M (1990) J Mol Catal 61:173
23. Ono Y (1981) Heterocycles 16:1755
24. Weisz PB, Frilette VJ (1960) J Phys Chem 64:382
25. Weisz PB, Frilette VJ, Maatman RW, Mower EB (1962) J Catal 1:307
26. Csicsery SM (1971) J Catal 23:124
27. Csicsery SM (1976) In: Rabo JA (ed) Zeolite Chemistry and Catalysis. ACS monograph 171:680
28. Weisz PB (1980) Pure Appl Chem 52:2091
29. Csicsery SM (1984) Zeolites 4:202
30. Guisnet M (1990) Acc Chem Res 23:392
31. Weitkamp J, Ernst S (1994) Catal Today 19:107
32. Anderson MW, Klinowsky J (1989) Nature 339:200
33. Csicsery SM (1986) Pure Appl Chem 58:841
34. Derouane EG (1982) Intercalation Chemistry. In: Whitinghton MS, Jacobsen AJ (eds) Academic Press, 101
35. Derouane E.G. (1984) Zeolite Science and Technology. In: Ribeiro RF et al. (eds) NATO ASI Ser E, Nijhoff, 347
36. Derouane EG (1980) Stud Surf Sci Catal 4:5
37. Unverricht S, Hunger M, Ernst S, Karge HG, Weitkamp J (1994) Stud Surf Sci Catal 84A:37
38. Corma A, Corell C, Lopis F, Martinez A, Perez-Pariente (1994) J Appl Catal 115:121
39. Souverijns W, Verrelst W, Vanbutsele G, Martens JA, Jacobs PA (1994) J Chem Soc, Chem Commun, 1671
40. Leonowicz ME, Lawton JA, Lawton SL, Rubin MK (1994) Science 264:1910
41. Dauns H, Weitkamp J (1986) Chem-Ing.-Tech 58:900
42. Martens JA, Jacobs PA, to be published
43. Santilli DS, Zones SI (1990) Catal Lett 7:383
44. Namba S, Sato K, Fujita K, Kim JH, Yashima T (1986) Stud Surf Sci Catal 28:661
45. Martens JA, Jacobs PA (1992) Zeolite Mircorporous Solids. In: Derouane EG, Lemos F, Naccache C, Ribeiro FR (eds) Nato ASI Ser C352, Kluwer, 511

46. Haag WO, Lago RM, Weisz PB (1982) Faraday Disc Chem Soc 72:317
47. Chen NY, Garwood WE (1971) Adv Chem Ser 121:575
48. Venuto PB (1977) ACS Symp Ser 40:260
49. Katzer JR (1969) Ph D Thesis, MIT, Cambridge, Mass
50. Fraenkel D, Cherniavsky M, Ittah B, Levy M (1984) Proc 8th Int Congr Catal 4, VCH Weinheim, 545
51. Fraenkel D, Cherniavsky M, Ittah B, Levy M (1986) J Catal 101:273
52. Fraenkel D (1989) J Mol Catal 51:L1
53. Derouane EG (1986) J Catal 100:541
54. Derouane EG, André JM, Lucas AA (1986) J Catal 110:58
55. Neuber M, Weitkamp J (1989) Stud Surf Sci Catal 49, Recent Research Reports, 425
56. Weitkamp J, Neuber M (1991) Stud Surf Sci Catal 60:291
57. Dessau RM (1984) US Patent 4,444,986
58. Smith JV (1963) Mineralogical Society of America, Special Paper No 1
59. Loewenstein W (1957) Am Mineral 39:92
60. Rees LVC (1982) Nature 296:491
61. Liebau F (1984) Zeolites 3:191
62. Smith JV (1984) Zeolites 4:309
63. Szostak R (1991) Stud Surf Sci Catal 58:153
64. Flanigen EM, Bennett JM, Grose RW, Cohen JP, Patton RL, Kirchner RM, Smith JV (1978) Nature 271:512
65. Lok BM, Messina CA, Patton RL, Gajek RT, Cannan TR, Flanigen EM (1984) US Patent 4,440,871; Lok BM, Messina CA, Patton RL, Gajek RT, Cannan TR, Flanigen EM (1984) J Am Chem Soc 106:6092; Flanigen EM, Lok BM, Cannan TR, Wilson ST (1986) Stud Surf Sci Catal 28:103
66. Bedard RL, Wilson ST, Vail LD, Bennet JM, Flanigen EM (1988) Stud Surf Sci Catal 49:375
67. Grobet PJ, Martens JA, Balakrishnan I, Mertens M, Jacobs PA (1989) Appl Catal 56:721
68. Vaughan DEW (1988) Chem Eng Progress 25
69. van Hooff JHC, Roelofson JW (1989) Stud Surf Sci Catal 58:242
70. Martens JA, Jacobs PA (1994) Stud Surf Sci Catal 85:653
71. Barr TL, Klinowski J, He H, Alberti K, Müller K, Lercher JA (1993) Nature 365:429
72. Engelhardt G (1989) Stud Surf Catal 58:285
73. Srdanov VI, Blake NP, Markgraber D, Metiu H, Stucky GD (1944) Stud Surf Sci Catal 85:115
74. Trifiro F, Jiru P (1988) Catal Today 3:519
75. Bellussi G, Rigutto MS (1994) Stud Surf Sci Catal 85:177
76. Huybrechts DRC, Parton RF, Jacobs PA (1991) Stud Surf Sci Catal 60:225
77. Parton RF, Huybrechts DRC, Buskens Ph, Jacobs PA (1991) Stud Surf Sci Catal 65:47
78. De Vos DE, Thibault-Starzyk F, Knops-Gerrits PP, Parton RF, Jacobs PA (1994) Macromol Symp 80:157
79. Kraushaar B, Van Hoof JHC (1988) Catal Lett 1:81
80. Karge GH, Beyer KH (1989) Stud Surf Sci Catal 49:43
81. Meyer G, Wöhrle D, Mohl M, Schulz-Ekloff (1984) Zeolites 4:30
82. Parton RF, Uytterhoeven L, Jacobs PA (1991) Stud Surf Sci Catal 59:395
83. Thibault-Starzyk F, Parton RF, Jacobs PA (1994) Surf Sci Catal 84:1419
84. Parton RF, Bezoukhanova CP, Thibault-Starzyk F, Reynders RA, Grobet PJ, Jacobs PA (1994) Stud Surf Sci Catal 84:813
85. De Vos DE, Thibault-Starzyk F, Jacobs PA (1994) Angew Chem Int Ed Engl 33:431
86. Balkus KJ Jr, Gabrielov AG, Bell SL, Bedioui F, Roué L, Devynck J (1994) Inorg Chem 33:67
87. Jacobs PA (1986) Stud Surf Sci Catal 29:357
88. De Vos DE, Jacobs PA (1992) Proc 96th Int Zeolite Conf. In: Ballmoos R, Higgins JB, Treacy MMJ (eds) Montreal, 615
89. Zakharov AN (1991) Mendeleev Commun, 80
90. Karge HG (1991) Stud Surf Sci Catal 69:133

91. Kustow LM, Mishin IV, Borowkow Vyu, Kazanski VB (1984) Kinetika i Katalisis 25:724
92. Zhobolenko VL, Kustow LM, Borowkow Vyu, Kazanski VB (1987) Kinetika i Katalisis 28:965
93. Haag WO, Lago RM, Mikovsky RJ, Olson D.H., Hellring SD, Schmitt KD, Kerr GT (1986) Stud Surf Sci Catal 28:677
94. Sauer J, Schirmer W (1988) Stud Surf Sci Catal 37:323
95. Kramer GJ, van Santen RA, Emels CA, Nowak AK (1993) Nature 363:529
96. Evleth EM, Kassab E, Sierra LR (1994) J Phys Chem 98:1421
97. Haag WO, Lago RM, Weisz PB (1981) Disc Farad Soc 72:317
98. Derouane EG, Dejaifve P, Gabelica Z (1981) Disc Farad Soc 72:330
99. Pearson RG (1963) J Am Chem Soc 85:3533
100. Wendlandt K-P, Bremer H (1984) Proc 86th Int Congr Catal, Berlin 1984, Verlag Chemie, Weinheim 4:507
101. Corma A, Lopis FP, Viruela P, Zicovich-Wilson C (1994) J Am Chem Soc 116:134
102. Corma A, Sastre G, Viruela P (1994) Stud Surf Sci Catal 84:2171
103. Sauer J (1989) Chem Rev 89:199
104. Rode BM, Reibnegger GJ (1979) JCS Faraday II 75:178
105. Allavena M, Seiti KE, Kassab E, Ferenczy Gy, Angyan JG (1990) Chem Phys Lett 168:461
106. Sauer J (1994) Stud Surf Sci Catal 84:2039
107. Mortier WJ (1987) Structure and Bonding 66:125
108. Baekelandt BG, Mortier WJ, Schoonheydt RA (1993) Structure and Bonding 80:187
109. Nalewajski RF (1993) Structure and Bonding 80:115
110. Parr RG, Donnelly RA, Levy M, Palke WE (1978) J Chem Phys 68:3801
111. Parr RG, Pearson RG (1983) J Am Chem Soc 105:7512
112. Hohenberg P, Kohn W (1964) Phys Rev B 136:864
113. Parr RG (1983) Ann Rev Phys Chem 34:631
114. Mortier WJ (1988) Stud Surf Sci Catal 37:253
115. Mortier WJ, Van Genechten K, Gasteiger J (1985) Am Chem Soc 107:829
116. Van Genechten KA, Mortier WJ, Geerlings P (1987) J Chem Phys 86:5063
117. Van Genechten KA, Mortier WJ (1988) Zeolites 8:273
118. Baekelandt BG, Mortier WJ Lievens JL, Schoonheydt RA (1991) J Am Chem Soc 113:6730
119. Nalewajski RF, Köster AM, Bredow T, Jug K (1993) J Mol Catal 82:407
120. Kazansky VB, Mortier WJ, Baekelandt BG, Lievens JL (1993) J Mol Catal 83:135
121. Martens JA, Jacobs PA (1990) Theoretical Aspects of Heterogeneous Catalysis. In: Moffat JB (ed) Van Nostrand Reinhold Catalysis Ser, 52
122. Jacobs PA, Martens JA (1991) Stud Surf Sci Catal 58:445
123. Kazanski VB (1994) Stud Surf Sci Catal 85:251
124. van Santen RA (1994) Stud Surf Sci Catal 85:273
125. Chen NY, Haag WO (1988) Hydrogen Effects in Catalysis. In: Paal Z, Menon PG (ed) Dekker, 695
126. Chen NY (1986) Stud Surf Sci Catal 28:653
127. Dwyer J (1987) Stud Surf Sci Catal 37:333
128. Lunsford JH (1986) J Phys Chem 90:4847
129. Mirodatos D, Barthomeuf D (1981) J Chem Soc, Chem Commun, 39
130. Corma A, Fornez F, Martinez A, Melo F, Pallato O (1988) Stud Surf Sci Catal 37:495
131. Sommer J, Hachoumy M, Garin F, Barthomeuf D (1994) J Am Chem Soc 116:5491
132. Lukyanov DB (1994) J Catal 145:54
133. Karge H (1991) Stud Surf Sci Catal 58:531
134. Climent MJ, Corma A, Garcia H, Iborra S, Primo J (1991) Recl Trav Chim 110:275
135. Maxwell IE, Stork WHJ (1991) Stud Surf Sci Catal 58:571
136. Herron N (1988) J Coord Chem 9:25
137. Herron N (1986) J Chem Soc, Chem Commun, 1521
138. Weitkamp J (1992) Proc 9th Int Zeolite Conf. In: von Ballmoos R, Higgins JB, Treacy MMJ (eds) Montreal 13
139. Herron N (1991) New J Chem 13:761

140. Herron N, Tolman CA (1987) ACS Prepr Div Petrol Chem 32:200
141. Herron N, Tolman CA (1987) J Am Chem Soc 109:2837
142. Bowers C, Dutta PK (1990) J Catal 122:271
143. Young LB, Butter SA, Kaeding W, Young W (1982) J Catal 76:418
144. Chen Fang Ren, Courdurier G, Naccache C (1986) Stud Surf Sci Catal 28:733
145. Sosztak R (1991) Stud Surf Sci Catal 58:153
146. Niwa M, Hidalgo CV, Hattori T, Murakami Y (1986) Stud Surf Sci Catal 28:297
147. Niwa M, Yamazaki K, Murakami Y (1989) Chem Lett 3:441
148. Lee GS, Maj JJ, Rocke SC, Garces JM (1989) Catal Lett 2:243
149. Lee GS, Burdett KA, Maj JJ (1989) US 4,996,353
150. Fellmann JD, Saxton RJ, Wentrcek PR, Derouane EG, Massiani P (1991) Int Patent WO 91/03443
151. Katayama A, Toba M, Takeuchi G, Mizukami F, Niwa S, Mitamura S (1991) J Chem Soc Chem Commun, 39
152. Eichler K, Leupold E (1985) DBP 3.334.084
153. Sampson RJ, Hanson ChB, Candlin JP (1986) Eur Pat 202.752
154. Takahashi O (1987) Jap Pat 62.29.536
155. Neuber M, Weitkamp J (1989) Proc 8th Int Zeolite Converence, Recent Research Reports, 425
156. Weitkamp J, Neuber M, Höltmann W, Collin G (1989) DEOS 37.23.103
157. Corbin DR, Kumpinsky E, Vidal A (1988) EP 316.133
158. Martens JA, Wydoodt M, Espeel P, Jacobs PA (1993) Stud Surf Sci Catal 78:527
159. Miyake T, Sekizawa K, Hironaka T, Nakano M, Fujii S, Tsutsumi Y (1986) Stud Surf Sci Catal 28:747
160. Botta A, Buysch H-J, Puppe L (1991) Ger Offen DE, 3.930.839
161. Botta A, Buysch H-J, Puppe L (1991) Ger Offen DE, 3.930.848
162. Botta A, Buysch H-J, Puppe L (1991) Angew Chem 103:1687
163. Shannon RD, Keane M, Abrams L, Staley RH, Gier TE, Sonnichsen GC (1988) Catal 113:367
164. Shannon RD, Keane M, Abrams L, Staley RH, Gier TE, Sonnichsen GC (1988) Catal 114:8
165. Shannon RD, Keane M, Abrams L, Staley RH, Gier TE, Sonnichsen GC (1989) Catal 115:79
166. Bergna HE, Keane M, Ralston D.H., Sonnichsen GC, Abrams L, Shannon RD (1989) J Catal 115:148
167. Bergna HE, Corbin DR, Sonnichsen GC (1987) US Pat 4,683,334
168. Bergna HE, Corbin DR, Sonnichsen GC (1988) US Pat 4,752,596
169. Segawa K, Sugima A, Sakaguchi M, Sakurai K (1989) Acid-Base Catal. In: Tanabe K et al. (eds) Kodansha
170. Mochida I, Yasutake A, Fujitsu H, Takeshita K (1983) J Catal 82:313
171. Huybrechts DRC, De Bruycker L, Jacobs PA (1990) Nature 345:240
172. Clerici MG (1991) Appl Catal 68:249
173. Kimura TA, Fukuoka A, Ichikawa M (1990) Lett 4:279
174. Shabtai J, Lazar R, Biron E (1984) J Mol Catal 27:35
175. Prochazka MP, Eklund L, Carlson R (1990) Acta Chem Scand 44:610
176. Brunel D, Chamoumi M, Geneste P, Moreau P (1993) J Mol Catal 79:297
177. Hoelderich WF, Goetz N (1992) Proc 9th Int Zeolite Conf. In: Ballmoos R, Higgins JB, Treacy MMJ (eds) Montreal, 309
178. Fraenkel D, Ittah B, Levy M (1986) Stud Surf Sci Catal 28:271
179. Narayana Murthy YVS, Pillai CN (1991) Synth Commun 21:783
180. Grieco PA, Nunes JJ, Gaul MD (1990) J Am Chem Soc 112:4595
181. Breslow R (1991) Acc Chem Res 24:159
182. Le Noble WJ, Asaro T (1978) Chem Rev 78:407
183. Dessau RM (1986) J Chem Soc, Chem Commun, 1167
184. Dessau RM (1987) US Pat, 4,665,247
185. Venuto PB, Hamilton LA, Landis PS, Wise JJ (1966) J Catal 4:81
186. Venuto PB, Hamilton LA, Vandis PS (1966) J Catal 5:484

187. Venuto PB, Landis PS (1966) J Catal 6:237
188. Sato H, Hirose K, Kitamura M, Nakamura Y (1989) Stud Surf Sci Catal 58:1213
189. Sato H, Ishii N, Hirose K, Nakamura S (1986) Stud Surf Sci Catal 28:755
190. Sato S, Takematsu K, Sodesawa T, Nozaki F (1992) Bull Chem Soc Jpn 65:1486
191. Takahashi T, Ueno K, Kai T (1991) J Chem Eng 69:1096
192. Aucejo A, Burguet MC, Corma A, Fornes V (1986) Appl Catal 22:187
193. Burguet MC, Aucejo A, Corma A (1987) Canad J Chem Eng 65:944
194. Corma A, Carcia H, Primo J, Sastre E (1991) Zeolites 11:593
195. Gallezot P, Giroir-Fendler A, Richard D (1990) Catal Lett 5
196. Blademond DG, Oukaci R, Blanc B, Gallezot P (1991) Catal 131:401
197. Gallezot P, Blanc B, Barthomeuf D, Païs da Silva (1994) Stud Surf Sci Catal 84:1433
198. Espeel PH, Janssen B, Jacobs PA (1993) J Org Chem 58:7688
199. Smirnitsky VI, Plakhotnik V.A., Lishchiner II, Mortikov ES (1994) Stud Surf Sci Catal 84:1813
200. Yashima T, Sato K, Hayasaka, Hara N (1972) J Catal 26:303
201. Lacroix C, Deluzarche, Kienneman A, Boyer A (1984) Zeolites 4:109
202. Garces JM, Vrieland GE, Bates SI, Scheidt FM (1985) Stud Surf Sci Catal 20:67
203. Miyamoto A, Iwamoto S, Agusa K, Inui T (1988) Proc Int Symp Acid-Base Catal, Sapporo. In: Tanabe K et al. (eds) Kodansha VCH, 497
204. Hathaway PE, Davis ME (1989) J Catal 119:497
205. Jacobs PA, Hinnekens H (1990) US Pat 4,950,812
206. Mochida I, Hayata S, Kato A, Seiyama T (1971) Bull Chem Soc Jpn 44:2282
207. Weyrich P, Hölderich WF (1995) PhD-thesis RWTH-Aachen
208. Hartig J, Stössel A, Hermann G (1982) DE-PS 3.222.144
209. Roffia P, Leofanti G, Cesana A, Mantegazza M, Padovan M (1989) Stud Surf Sci Catal 55:43
210. Reddy JS, Ravishankar R, Sivasanker S, Ratnasamy P (1993) Catal Lett 17:139
211. Hölderich WF (1993) Proc 10th Int Congr on Catal, Budapest, Hungary, 127
212. Roffia P, Paparatto G, Cesana A, Tauszik G (1989) EP 0314147
213. Parton RF, De Vos DE, Jacobs PA (1992) Zeolite Microporous Solids: Synthesis, Structure and Reactivity. In: Derouane EG, Lemos F, Naccache C, Ribeiro FR (eds) NATO ASI Ser C352, Kluwer Academic Publishers, 552
214. Romanovskii BV (1986) Proc 5th Int Symp on Relations between Homog Heterog Catal. In: Yermakov Yu, Likholobov V (eds) VNU Science Press, 343
215. Parton RF, Vankelecom IFJ, Casselman MJA, Bezoukhanova CP, Uytterhoeven JB, Jacobs PA (1994) Nature 370:541
216. Knops-Gerrits P-P, De Vos D, Thibault-Starzyk F, Jacobs PA (1994) Nature 369:543
217. Knops-Gerrits P-P, Thibault-Starzyk F, Jacobs PA (1994) Stud Surf Sci Catal 84:1411
218. Knops-Gerrits P-P, Jacobs PA, Li X-Y, Yu N-T (1994) XIVth Int Conf on Raman Spectroscopy. In: Yu N-T, Li X-Y (eds) Hong-Kong, John Wiley & Sons, A165
219. Neri C, Anfossi B (1984) Eur Pat 100 118 A1
220. Maspero F, Romano U (1986) Eur Pat 190 609 A2
121. Clerici MG, Romano U (1987) Eur Pat 230 949
222. Clerici MG, Romano U (1989) US Pat 4,824,976
223. Clerici MG, Bellussi G, Romano U (1991) J Catal 129:159
224. Clerici MG, Ingallina P (1993) J Mol Catal 140:71
225. Esposito A, Taramasso M, Neri C (1982) UK Pat 2 083 816
226. Esposito A, Taramasso M, Neri C (1984) DE 3135559 A1
227. Reddy JS, Sivasanker S, Ratnasamy P (1992) J Mol Catal 71:373
228. Thangaraj A, Kumar R, Ratnasamy P (1990) Appl Catal 57:L1
229. Esposito A, Neri C, Buonomo F (1984) US Pat 4,480,135
230. Maspero F, Romano U (1992) WO 90/01074
231. Roffia P, Padovan M, Moretti E, De Alberti G (1987) Eur Pat 208 311
232. Roffia P, Paparatto G, Cesana A, Tauszik G (1988) Eur Pat 301 486 A2
233. Reddy JS, Srivasanker S, Ratnasamy P (1991) Non-Cryst Solids 69:383

234. Huybrechts DRC (1990) WO 90/05126
235. Reddy JS, Sivasanker S, Ratnasamy P (1991) J Mol Catal 70:335
236. Tonti S, Roffia P, Cesana A, Mantegazza MA, Padovan M (1989) Eur Pat 314 147
237. Reddy JS, Jacobs PA (1993) J Chem Soc Perkin Trans 1:2665
238. Reddy RS, Reddy JS, Kumar R, Kumar P (1992) J Chem Soc, Chem Commun 84
239. Camblor MA, Corma A, Martinez A, Perrez-Pariente J (1992) J Chem Soc, Chem Commun 589
240. Friedel C, Crafts JM (1877) Compt Rend 84:1450
241. Nencki M (1897) Ber 20:1766
242. Olah GA (1974) Friedel Crafts Chemistry, Wiley N-Y, 29
243. Stamires DN, Turkevich J (1964) J Am Chem Soc 86:749
244. Uytterhoeven J-B, Christner LG, Hall WK (1965) J Phys Chem 69:2117
245. Argauer RJ, Landolt GR (1972) US 3,702,886
246. Venuto PB, Wu EL (1969) J Catal 15:205
247. Kaeding WW, Holland RE (1988) J Catal 109:212
248. Fraenkel D, Levy M (1979) J Catal 118:10
249. Sayed MB, Védrine JC (1986) J Catal 101:43
250. Vinek H, Rumplmayr G, Lercher J (1989) J Catal 115:291
251. Paparatto G, Moretti E, Leofanti G, Gatti F (1987) J Catal 105:227
252. Kaeding WW, Young LB, Chu C (1984) J Catal 89:267
253. Kolboe S, Larsen M, Anundskaas A (1988) Proc 96th ICC, Ottawa 1:468
254. Yashima T, Sakaguchi Y, Namba S (1981) Stud Surf Sci Catal 7:739
255. Lonyi F, Engelhardt J, Kallo D (1989) Stud Surf Sci Catal 58:1357
256. Lonyi F, Engelhardt J, Kallo D (1991) Zeolites 11:169
257. Engelhardt J, Kalló D, Zsinka I (1992) J Catal 135:321
258. Kim J-H, Namba S, Yashima T (1988) Bull Chem Soc Jpn 61:1051
259. Kim J-H, Yamaghishi K, Namba S, Yashima T (1990) J Chem Soc, Chem Commun, 1793
260. Kim J-H, Namba S, Yashima T (1991) Zeolites 11:169
261. Parikh PA, Subrahmanyam N, Bhat YS, Halgeri AB (1992) Ind Eng Chem Res 31:1012
262. Parikh PA, Subrahmanyam N, Bhat YS, Halgeri AB (1992) Catal Lett 14:107
263. Pradhan AR, Rao BS (1991) J Catal 132:79
264. Pradhan AR, Kotasthane AN, Rao BS (1991) Appl Catal 72:311
265. Meima GR, van der Aalst MJM, Samson MSU (1992) Proc 9th Int Zeolite Conf. In: Ballmoos R, Higgins JB, Treacy MMJ (eds) Montreal, 327
266. Dauzenberg F (1993) personal communication
267. Boucher HA, Cody IA, Somers HA (1985) Eur Patent 0160144
268. Grey RA (1988) US Patent 4,731,497
269. Juguin B, Raatz F, Travers C, Martino G (1992) Eur Patent 0466558A1
270. Taniguchi K, Tanaka M, Takahata K, Sakamoto N, Takai T, Kurano Y, Ishibashi M (1988) Eur Patent 0288582
271. Butruile J-R, Pinnavaia R (1992) Catal Lett 12:187
272. Sugi Y, Matsuzaki T, Hanoaka T, Kubota Y, Kim J-H (1994) Catal Lett 26:181
273. Kaeding WW, Chu C, Weinstein LB, Butter SA (1981) J Catal 67:159
274. Kaeding WW, Chu C, Young LB, Butter SA (1981) J Catal 69:392
275. Wei J (1982) J Catal 76:433
276. Seko M, Miyake T, Inada K (1979) The ACS/CJS Chemical Congress, Honolulu
277. Kim J-H, Namba S, Yashima T (1992) Appl Catal A 83:51
278. Kim J-H, Namba S, Yashima T (1989) Stud Surf Sci Catal 46:71
279. Chandavar KH, Hedge SG, Kulkarni SB, Ratanasamy P, Chitlangia G, Singh A, Deo AV (1983) Proc 6th IZC, Reno Butterworths, 325
280. Bezouhanova C, Dimitrov C, Nenova V, Spassov B, Lercher JA (1986) Appl Catal 21:149
281. Nayak VS, Riekert L (1986) Appl Catal 23:403
282. Mirth G, Teyka J, Lercher JA (1993) J Catal 139:24
283. Yashima T, Ahamad H, Yamazaki K, Katsuka M, Hara N (1970) J Catal 16:273
284. Yashima T, Ahamad H, Yamazaki K, Katsuka M, Hara N (1970) J Catal 17:151

285. Langenaeker W, Demel K, Geerlings P (1991) J Mol Struct (Theochem) 234:329
286. Corma A (1990) Guidelines for Mastering the properties of Molecular Sieves. In: Barthomeuf D (eds) Plenum Press, N-Y, 299
287. Corma A, Sastre G, Viruela R, Zicovich C (1992) J Catal 136:521
288. Corma A, Sastre E, Sastre G, Ciruela P, Zicovich C (1992) Proc 9th Int Zeolite Conf. In: von Ballmoos R, Higgins JB, Treacy MMJ (eds) Montreal, 521
289. Corma A, Lopis F, Viruela P, Zicovich-Wilson C (1994) J Am Chem Soc 116:134
290. a) Ratnasamy P (1989–1991) Nat Chem Laboratory, Pune, India, Biennial Report, 3; (b) Personnel Communication by Ratnasamy, P to Hoelderich, WF
291. Haag WO, Lago RM (1991) US Patente 5,008,482
292. Climent MJ, Corma A, García H, Prima J (1991) J Catal 130:138
293. Dimitrov C, Popova Z, Mai Tuyên (1978) React Kinet Catal Lett 8:101
294. Solinas V, Monaci R, Marongiu B, Forni L (1983) Appl Catal 5:171
295. Solinas V, Monaci R, Marongiu B, Forni L (1984) Appl Catal 9:109
296. Solinas V, Monaci R, Rombi E, Forni L (1987) Stud Surf Sci Catal 34:493
297. Matsuda T, Yogo K, Nagaura R, Kikuchi E (1989) Stud Surf Sci Catal 58:431
298. Komatsu T, Araki Y, Namba S, Yashima T (1994) Stud Surf Sci Catal 84:1821
299. Campbell CB, Onopchenko A, Santilli DS (1990) Bill Chem Soc Jpn 63:2555
300. Burgoyne WF, Dixon DD, Casey JP (1989) Chemtech, 690
301. Wu MM (1983) US Patent 4,391,998
302. Swanson BJ, Shubkin RL (1985) US Patent 4,532,368
303. Young LB (1981) Eur Patent 0029333
304. Le QN, Marler DO, McWilliams JP, Rubin MK, Shim J, Wong SS (1990) US Patent 4,962,256
305. Tobias MA (1973) US Patent 3,728,408
306. Kolesnichenko NV, Kurashev MV, Romanovskii BV, Menyailov AA (1981) Petrol Chem USSR 21:223
307. Sadykov SG, Rasulov CK, Zeinalova LB, Koshevnik AY, Shishkina MV, Khodzhayeva VL (1980) Petrol Chem USSR 20:141
308. Parton RF, Jacobs JM, Van Ooteghem H, Jacobs PA (1989) Stud Surf Sci Catal 46:211
309. Young LB (1983) US Patent 4,371,714
310. Dixon DD, Burgoyne WF (1987) Eur Patent 0226781
311. Dixon DD, Burgoyne WF (1990) Appl Catal 62:161
312. Burgoyne WF, Dixon DD (1986) Eur Patent 0202557
313. Burgoyne WF, Dixon DD (1990) Appl Catal 63:117
314. Burgoyne WF, Dixon DD (1990) J Mol Catal 62:61
315. Agrawal R, Auvil SR, Deeba M (1987) Eur Patent 0240018
316. Pierantozzi R, Nordquist AF (1986) Appl Catal 21:263
317. Pierantozzi R (1987) Eur Patent 0245797
318. Pierantozzi R (1988) Eur Patent 0286029
319. Bacskai R, Valeiras HA (1987) Eur Patent 0265932
320. Namba S, Yashima T, Itaba Y, Hara N (1980) Stud Surf Sci Catal 5:105
321. Corma A, Garcia H, Primo J (1988) J Chem Research (S), 40
322. Moffat JB (1978) Catal Rev-Sci Eng 18:199
323. Hoelderich WF (1990) Guidelines for Mastering the properties of Molecular Sieves. In: Barthomeuf D (eds) Plenum Press, N-Y, 319
324. Beltrame P, Beltrame PL, Carniti P, Catelli A, Forni L (116) Gaz Chim It 116:473
325. Beltrame P, Beltrame PL, Carniti P, Catelli A, Forni L (1987) Appl Catal 29:327
326. Marczewski M, Perot G, Guisnet M (1988) Stud Surf Sci Catal 41:273
327. Marczewski M, Bodibo J-P, Perot G, Guisnet M (1989) J Mol Catal 50:211
328. Jacobs JM, Parton RF, Boden A-M, Jacobs PA (1989) Stud Surf Sci Catal 41:228
329. Onaka M, Ishikawa K, Izumi Y (1984) J Incl Phenom 2:359
330. Ione KG, Kikhtyanin OV (1989) Stud Surf Sci Catal 49:1073
331. Chen PY, Yuan T, Chu SJ, Chu H, Chen MC, Diahn H, Chang NS, Lin WC, Chu H (1989) US Patent 4,801,752
332. Balsama S, Beltrame P, Beltrame PL, Craniti P, Forni L, Zuretti G (1984) Appl Catal 13:161

333. Santaceseria E, Grasso D, Gelosa D, Carrà S (1990) Appl Catal 64:101
334. Chantal P, Kaliaguine S, Grandmaison JL (1984) Stud Surf Sci Catal 19:93
335. Chantal P, Kaliaguine S, Grandmaison JL (1985) Appl Catal 18:133
336. Janardanarao M, Salvapati GS, Vaidyeswaran R (1978) Proc 4th Nat Symp Catal (Bombay), 51
337. Durgakumari V, Narayanan S, Guczi L (1990) Catal Lett 5:377
338. Agaev AA, Tagiev DB (1985) Z Prikl Kh 58:2734
339. Kozhevnikov SA, Motovilova NN, Sibarov DA, Vinogradov MV (1985) Z Prikl Kh 58:1539
340. Bezouhanova C, Al-Zihari MA, Lercher H (1992) Kinet Catal Lett 46:153
341. Churkin Y, Chvalyuk LA, Glazunova VI, Kirichenko GN (1985) Neftekhim 12:19
342. Churkin UV, Chvalyuk LA, Akhunova LV, Kirichenko GAN, Spivak SI (1985) React Kinet Catal Lett 29:145
343. Yamawaki J, Ando T (1979) Chem Letters, 755
344. Onaka M, Kawai M, Izumi Y (1983) 1101
345. Rhodes T, Nightingale P (1986) PCT Pat 03485
346. Chen KC, Tsuchyia T, MacKenzie JD (1986) J Non-Crystalline Solids 81:227
347. Chen PY, Chu SJ, Chang NS, Chuang TK (1989) Stud Surf Sci Catal 58:1105
348. Chen FR, Davis JG, Fripiat JJ (1992) J Catal 133:263
349. Weigert FJ (1986) US Patent 4,593,124
350. Weigert FJ (1989) US Patent 4,804,784
351. Prasad S, Rao BS (1990) J Mol Catal 62:L17
352. Mitchel RS (1989) US Patent 4,795,833
353. Buysch H-J, Pelster H, Puppe L, Wimmer P (1991) Eur Patent 0414077A1
354. Onaka M, Ishikawa K, Izumi Y (1982) Chem Lett 1783
355. Onaka M, Umezono A, Kawai L, Izumi Y (1985) Chem Soc, Chem Commun, 1202
356. Clerici MG, Bellussi G, Romano U (1987) Eur Patent 0264744
357. Gilbert L, Mercier C (1993) Stud Surf Sci Catal 78:51
358. Espeel PHJ, Vercruysse KA, Debaerdemaker M, Jacobs PA (1994) Stud Surf Sci Catal 84:1457
359. Espeel PHJ, Debaerdemaker M, Janssens B, Maes A, Jacobs PA (1994) J Phys Chem; in press

CHAPTER 7

Zeolites as Catalysts in Industrial Processes

P. M. M. Blauwhoff, J. W. Gosselink, E. P. Kieffer, S. T. Sie, and W. H. J. Stork

List of Abbreviations

ASA	amorphous silica-alumina
BP	British Petroleum Co.
bpsd	barrels per stream day
BTX	benzene toluene xylene
CCR	Conradson carbon residue
DHCD	dehydrocyclodimerization
DIPD	diisopropyldiphenyl
DME	dimethyl ether
EB	ethylbenzene
EBHP	ethylbenzene hydroperoxide
FCC	fluid catalytic cracking
FIMS	Field Ion Mass Spectrometry
GHSV	gas hourly space velocity
HDN	hydrodenitrogenation
LABS	linear alkylbenzene sulfonates
LCO	light cycle oil
LPG	liquified petroleum gases
LR	long residue
MDDW	Mobil Distillate Dewaxing
MEK	methyl ethyl ketone
MHC	Mild Hydrocracking
MOGD	Mobil Olefins to Gasoline and Distillate
MON	Motor octane number
MSTDP	Mobil Selective Toluene Disproportionation Process
MTBE	Methyl-tertiary-butyl ether
MTDP	Mobil Toluene Disproportionation Process
MTG	methanol to gasoline
MTO	methanol to olefins
PCAH	polycyclic aromatic hydrocarbons
PCP	protonated cyclopropane
PSD	pore size distribution or particle size distribution
RON	reseach octane number
RVD	Reid vapor pressure
SMDS	Shell Middle Distillate Synthesis

SPA	supported phosphoric acid
SPGK	Shell Poly-Gasoline and -Kero process
TAME	*tert*-amyl-methyl ether
TBHP	*tert*-butyl hydroperoxide
TIGAS	Topsøe Integrated Gasoline synthesis
TIP	Total Isomerization Process
UCS	unit cell size
UOP	Universal Oil Products Inc.
URBK	Union Rheinische Braunkohlen Kraftstoff AG
USY	ultrastable Y-zeolite

7.1 Introduction and General Overview

Zeolites find three main commercial applications: in detergents, because of their ion-exchange capacity, in separations and adsorption, because of their molecular sieve properties, and as catalysts, mainly because of their high acidity. In volume, the application in detergents is by far the largest [1]. However, with respect to financial market size, catalysts form the largest application of zeolites [2]. As an example, the market volume in US $/annum of zeolite catalysts (basis 1994 figures) constitutes about 60% of the total market, with fluid catalytic cracking (FCC) catalysts forming the greatest contribution, 48% of total market, followed by hydrocracking catalysts, 7% of total market. In catalysts currently mainly zeolite Y is used, with the biggest consumption being in catalytic cracking and hydrocracking, but ZSM-5, mordenite, and other zeolites have also found commercial application in catalysts. Continuously, new catalytic applications are being developed, or existing ones further fine-tuned, by making use of the many opportunities zeolites offer for tailoring a catalyst.

We will discuss the application of zeolites in the main refinery and petrochemical processes, such as fluid catalytic cracking (FCC) in Sect. 7.2, hydrocracking in Sect. 7.3 catalytic dewaxing in Sect. 7.4, upgrading of naphtha and tops in Sect. 7.5, syn-fuels production in Sect. 7.6 and several chemical processes, such as acid-catalyzed, oxidation and ammoxidation processes, in Sect. 7.7. For a better understanding of these applications and developments we will first give a brief introduction to oil refining.

7.1.1 Oil Refining: Basics

Oil refining is a major industry which processes world-wide some 80 million barrels (1 barrel = 159 l) of crude oil per day. Although presumably all refineries are different, the basic principles are always the same, see for an artist's impression Fig. 7.1. First the crude oil is separated in the primary distillation into various fractions, serving as the raw materials for the main products: C_5 until material boiling at about 180 °C, i.e., naphtha, for gasoline; a fraction boiling at 180–250 °C for kerosene, and a fraction boiling at 250–370 °C for diesel/gas oil

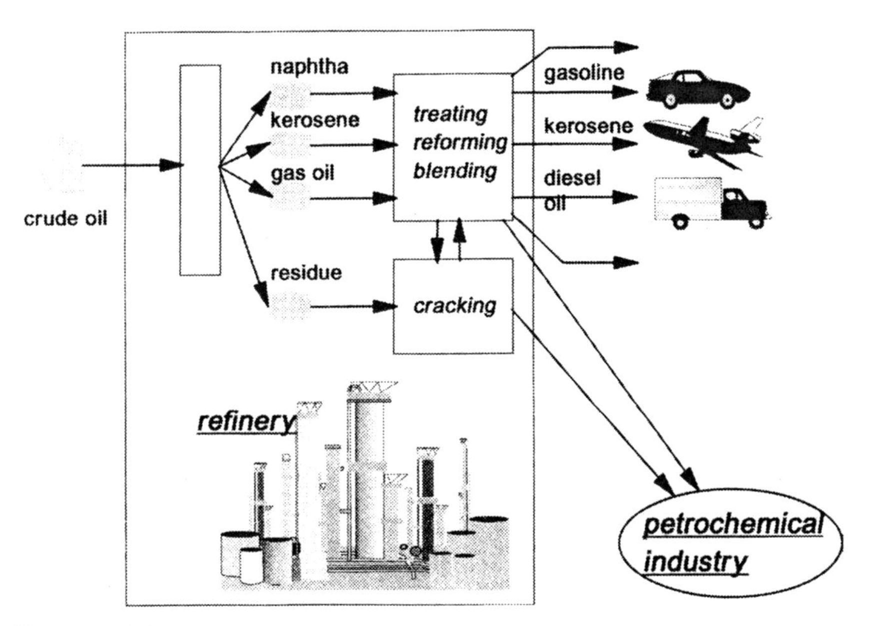

Fig. 7.1. Artist's impression of a refinery

(transportation and industrial fuel). In addition, base oils for lubricants can be recovered (Fig. 7.2 [3]). Refineries are being faced with more stringent specifications for the transportation fuels, see Table 7.1 [4], posing challenges to all refinery processes.

The fractions thus obtained must be treated or upgraded to arrive at on-specification fuels. Thus the naphtha, generally has to be desulfurized, for environmental reasons, but also often to allow the subsequent catalytic reforming which increases to octane number through aromatics formation. The tops fraction, C_{5-6}, is often isomerized to increase the octane quality. The middle distillates, i.e., the kerosene and diesel fractions, are desulfurized, again for environmental reasons, while in some cases gas oils are dewaxed (removal of normal paraffins) to improve the cold-flow properties (to meet the pour-point specification).

In this simple set-up (a "hydroskimming" refinery) the residue of the atmospheric distillation, a fraction with a boiling point above some 370 °C, is blended into a heavy industrial or marine fuel. Whether or not this fraction is a large part of the barrel depends on the crude oil type. Figure 7.3 clearly illustrates that, for instance, the amount of residue increases significantly on switching from North Sea to Middle East and further to South American crude. Since these heavy fuels often demand even lower prices than crude oil, there is a large incentive to convert the atmospheric residue into lower boiling fractions, in particular transportation fuels. This is done in a "conversion" refinery, according to various possible routes. An example of a hypothetical conversion refinery is given in Fig. 7.4, with zeolite applications given in solid black rectangles. Usually the atmospheric residue is separated again according to boiling point in a vacuum

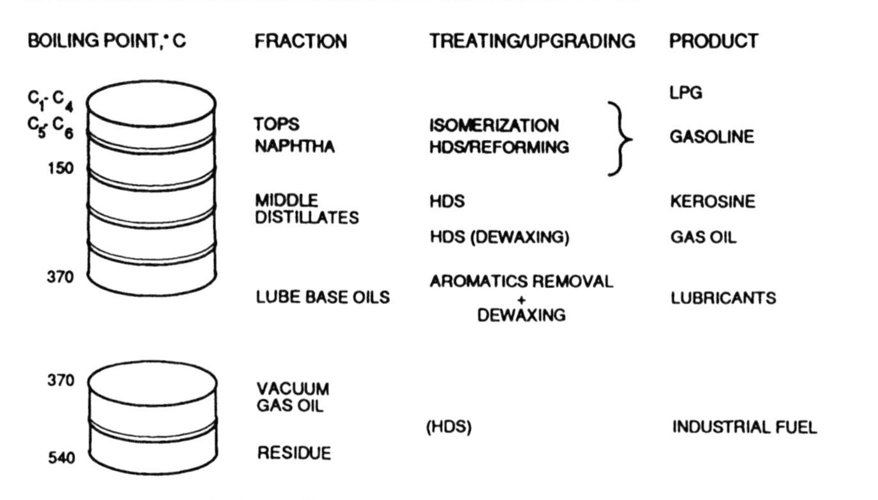

Fig. 7.2. A simple oil refinery (basics)

Table 7.1. Expected European Union (EU) motor gasoline (Mogas) and diesel fuel specifications (year 2000/2010) as Proposed by the EU Council of Environmental Ministers June 1997 [4] and Expected Specification of International Jet Al Kerosene Grade by the year 2000/2010

EU ULG95 Mogas spec	EN 228
RON	≥95
MON	≥85
RVP summer period, kpa	≤60.0
Aromatics, vol%	≤42.0 (possibly ≤35 by year 2010)
Olefins, vol%	≤18.0 (≤21 for ULG92)
Benzene, vol%	≤1.0
Oxygen, wt%	≤2.3
Sulphur, wt-ppm	≤150 (possibly ≤50 by year 2010)
MTBE, vol%	≤15
E100 min, vol%	≥46.0
E150 min, vol%	≥75
International Jet A-1/AVTUR Kerosene spec	AFQRJS Checklist, meeting ASTM D1655/DEF STAN spec and IATA Guidance Material
Smoke Point, mm	≥25
OR Smoke point, mm, and Naphthalenes, vol%	≥19 (mm) and ≥3.0 (vol%)
OR Luminometer number	≥45
Copper corrosion (2 h/100 °C)	≤1
Silver corrosion	will be deleted
Density 15 °C, kg/m³	775–840
End Point, °C	≤300

Table 7.1 (continued)

EU ULG95 Mogas spec	EN 228
Aromatics, vol%	≤ 22.0
Sulfur, wt%	possibly to be reduced from ≤ 0.3 to ≤ 0.2
EU Diesel spec	EU 590
Cetane number	≥ 51.0
Density, kg/m³	≤ 845
T95, °C	≤ 360
PCAH, wt%	≤ 11
Sulphur, wt-ppm	≤ 350

flasher (cut point about 520°C), after which the resulting flashed distillate or vacuum gas oil can be cracked into lighter fractions relatively easily by established processes such as catalytic cracking (preferably to gasoline range fractions) and hydrocracking (preferably to middle distillates). This then only leaves the vacuum residue, which can be used for fuel (possibly after desulfurization), or which can be thermally cracked in a coker or visbreaker, deasphalted (by extraction with propane, butane or pentane to yield a relatively clean deasphalted oil that is suitable for further processing, e.g., for lubricating base oil) or gasified to produce hydrogen [5,6]. In addition, the vacuum residue can be hydroconverted [7], which is basically a more severe form of hydrodesulfurization, producing apart from a low-sulphur residue significant amounts of lower range boiling point products, which can be further processed in the other conversion and hydrotreating processes to eventually yield transportation fuels. An example of a commercial residue hydroconversion process is the HYCON process, present in Shell's Pernis refinery near Rotterdam, The Netherlands [8]. A similar range

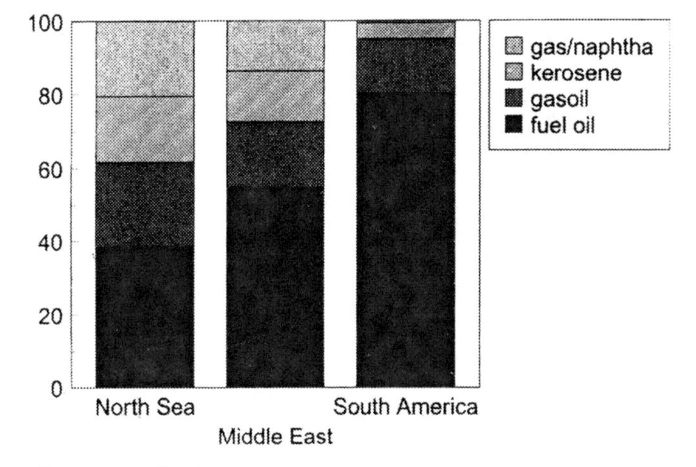

Fig. 7.3. Crude oil variety

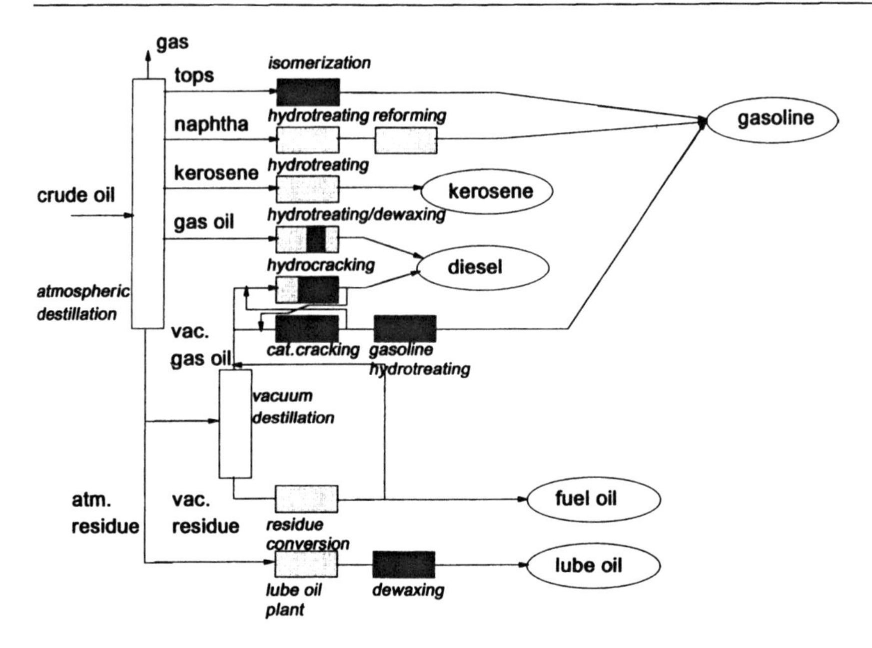

Fig. 7.4. A hypothetical "conversion" refinery

of options exists if the vacuum flashing is omitted: the entire atmospheric residue can be sent to a catalytic cracker (residue catalytic cracking), can be desulfurized or hydroconverted, thermally cracked, etc.; combinations of first desulfurizing the atmospheric residue and then catalytically cracking the product are also practised.

These "conversion" refineries are more profitable than the simple ones, and the conversion processes, mainly thermal cracking, catalytic cracking and hydrocracking, are operated on a large scale. In some cases the products require more treating than the "straight-run" products in a hydroskimming refinery, in other cases very high quality products are directly produced; moreover, the conversion processes produce quite some gaseous hydrocarbons, which one would like to convert to liquid transportation fuel components. This is for instance done in processes such as paraffin/olefin alkylation, oligomerization and paraffin aromatization (Fig. 7.5). The synthesis of liquid hydrocarbons from small molecules is also important in the context of synthetic fuels. Thus processes have been developed for the conversion of methane into gasoline, e.g., Mobil's methanol to gasoline (MTG) process, or into middle distillates, e.g., the Shell Middle Distillate Synthesis (SMDS) process, while the conversion of synthesis gas stemming from coal gasification into hydrocarbons has already been practised for a considerable time in South Africa.

Finally, the interrelation between oil refining and the chemical industry is receiving increasing attention nowadays. This of course was already clear in the aromatics production in catalytic reforming, but the partial conversions in hydrocrackers to make hydrowax for ethylene manufacture, the production of

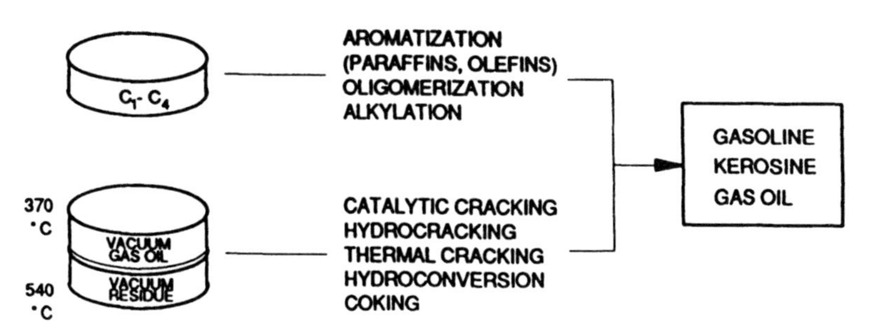

Fig. 7.5. Conversion options in a complex refinery

additional olefins in a cat. cracker for polypropylene manufacture, and the production of ethers for octane enhancement are also examples of the growing importance of the oil/chemical interface.

7.1.2 The Petrochemical Industry

The petrochemical industry differs from oil refining in a number of aspects. Petrochemistry is more varied than oil refining; apart from all the reactions involved in oil refining, oxidation and reduction reactions of oxygenates are important in the chemical industry. In addition, the polymerization reactions have been refined much further than in refining (oligomeriziation) because of the important product properties. Another characteristic difference is that in refining one always deals with feedstocks and products consisting of many different chemical compounds characterized by average properties (multi-component systems), while in the chemical industry this is hardly ever the case.

Because the chemical industry is smaller in volume than the refining industry, and at the same time more diverse, the projects in the chemical industry are usually smaller, and hence can bear lower research costs. Neglecting a few obvious exceptions such as steam cracking, aromatics processing and polypropylene manufacture, the subjects have therefore generally been less extensively studied than in refining, and still larger steps can be made and, indeed, have to be made to compensate for the research effort at the lower throughputs.

At present zeolitic catalysts are used extensively in the processing of aromatics, using (shape-selective) acid catalysis, while oxidation catalysis is also applied commercially with molecular sieves. On top of this a number of applications in the fine chemicals business have been developed, which have been reviewed by Espeel et al. [9].

Despite their differences, oil refinery and the petrochemical industry are inter-linked in a sometimes very complex way. For example, building blocks of the chemical industry, such as ethylene and propylene, are manufactured by thermal steam crackers (and catalytic crackers) which are fed by a wide variety of refinery streams. Also, some compounds produced in the petrochemical complexes can be 'sent back' to the refinery, for instance as components for the gasoline pool.

7.2 Fluid Catalytic Cracking

Fluid Catalytic Cracking (FCC) is a process for the conversion of residual hydrocarbon material vacuum distillates and residues – into more valuable, olefinic gases, as well as high-octane gasoline and diesel-boiling-range products.

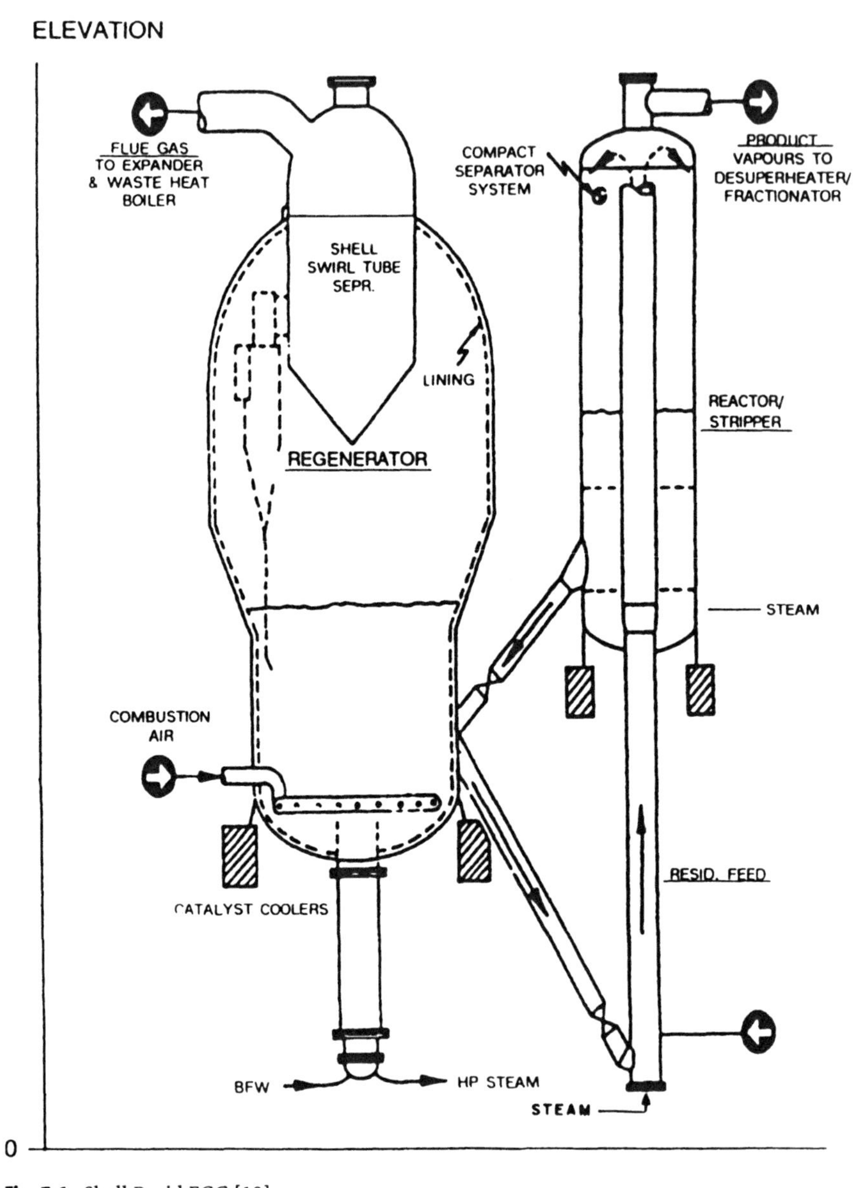

Fig. 7.6. Shell Resid FCC [10]

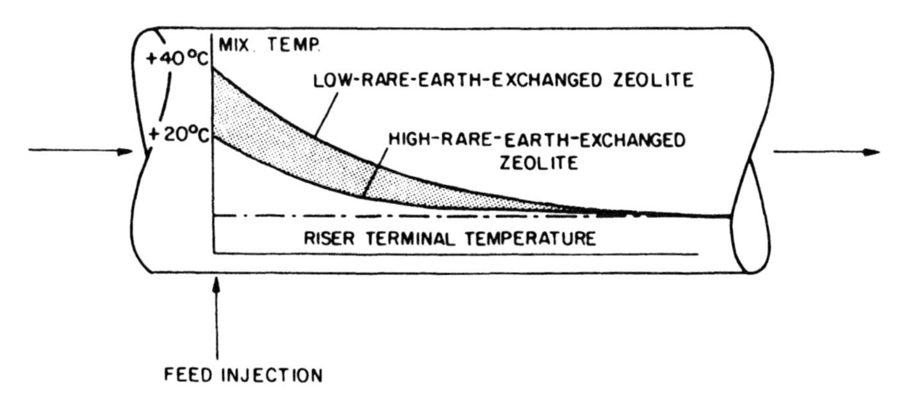

Fig. 7.7. Typical riser reactor temperature profile

The process employs an acidic catalyst at high reaction temperatures, typically 480–550°C, at pressures close to ambient (2–3 bar). The hydrocarbon feed is preheated to 200–300°C, dispersed at the riser lift pot and mixed with hot (680–750°C), fluidized catalyst (Fig. 7.6 [10]). The resulting temperature is typically some 500–570°C and causes the feed to be cracked over the acid catalyst. The volume expansion induced by vaporization and cracking of the feed, combined with the steam injected at the lift pot, entrains the gaseous hydrocarbon/catalyst mixture to the top of the riser reactor. The cracking reactions are endothermic and, consequently, the reaction temperature drops significantly over the riser height in a matter of a few seconds (Fig. 7.7). During its journey to the riser top, coke is deposited on the catalyst as a byproduct of the cracking reactions.

At the top of the riser the catalyst is separated from the gaseous reaction products in a series of cyclones, and heavy hydrocarbons are steam-stripped from the catalyst. The stripped, spent catalyst typically contains 0.8–1.3 wt% of coke – depending on the operating conditions and feedstock composition – and is transported to the regenerator. Here the coke is burnt from the catalyst to generate the heat required for the endothermic cracking reactions in the riser and also to restore the catalytic activity. More details on the heat balance in conventional FCCs can be found in [11].

At many points in the FCC unit, a fluidizing medium (steam, fuel gas, air) is injected to keep the catalyst in a fluidized state. The catalyst circulation flow rate, and hence the riser temperature, are controlled by means of slide valves located in the stripper and regenerator standpipes (Figs. 7.6 [10] and 7.8 [11]). The above hardware description refers to state-of-the-art catalytic cracking equipment. Since its introduction in the late 1930s, however, the process has lived through a more or less continuous stage of development, in terms of both hardware and catalyst [10, 12, 13]:

- improved metallurgy enabling higher temperature levels to be achieved while at the same time improving unit integrity, advanced reactor technology, starting from fixed beds (Houdry technology), through fluidized beds now

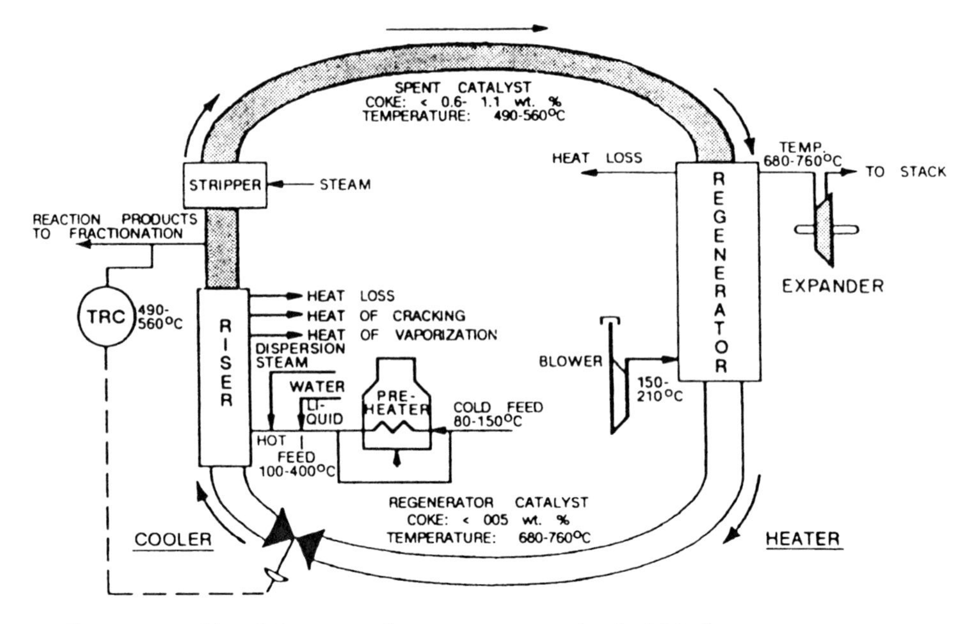

Fig. 7.8. Typical heat balance contributions in a conventional FCC [11]

developing in more and more sophisticated lean riser technology, better dispersing feed nozzle technology,

- increasingly efficient gas/solids separation technology,
- improved energy efficiency, e.g., through the application of power recovery systems and sophisticated heat integration,
- the switch from amorphous to zeolite-Y-based FCC catalysts and the rapid development of the latter,
- increasing metals resistance of the catalysts,
- maximizing gasoline octane via catalyst formulation of the use of octane-boosting additives.

As a result, the severity of the FCC operation has continuously increased and consequently the FCC feed diet has widened considerably, now ranging from vacuum gas oils to atmospheric residues. Moreover, the product slate has moved significantly towards gasoline production. At the moment, FCC is the most efficient high-tech conversion process for the production of gasoline and is the cornerstone of most refineries.

7.2.1 Feedstocks and Products

With the drop in fuel oil demand the conversion of heavy residual oil fractions into transportation fuels has become a prerequisite for economic survival in the refining industry (Fig. 7.9). As a consequence of the economic incentive, FCC feedstocks have become heavier and heavier, as can be seen in, e.g., Fig. 7.10 [10],

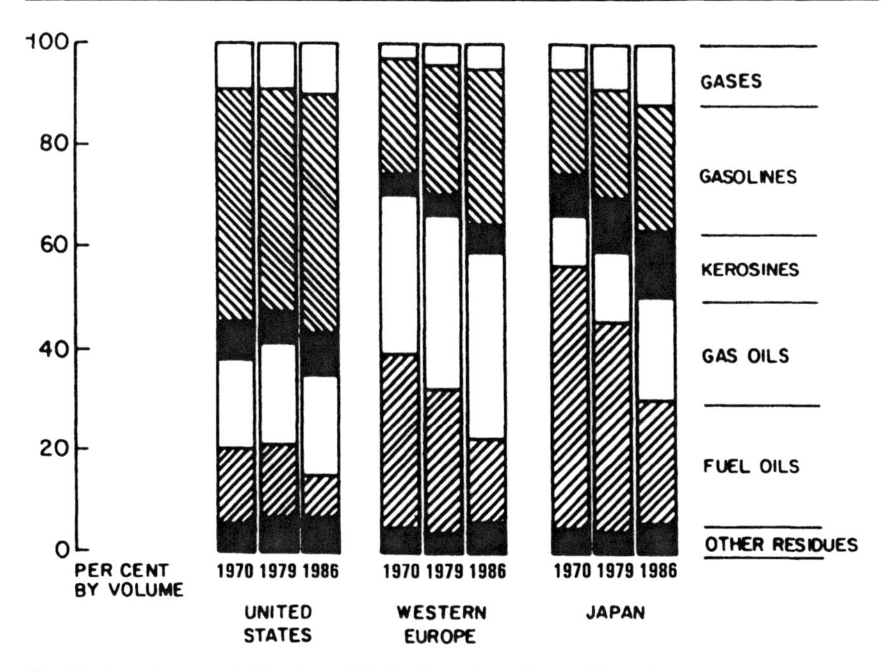

Fig. 7.9. Development/whitening of the hydrocarbon demand barrel

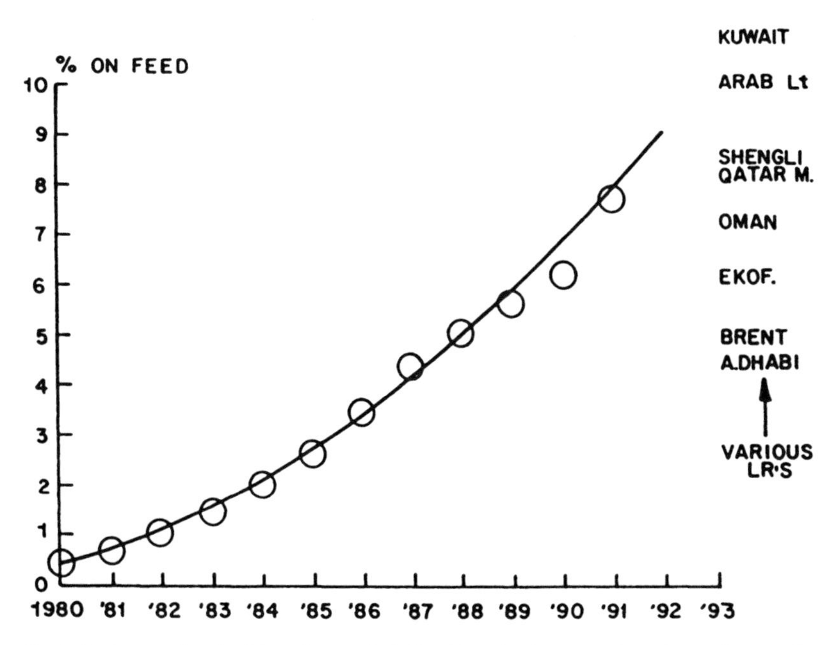

Fig. 7.10. Increasing feed heaviness of Shell FCC units (expressed as Conradson carbon content of the feed) [10]

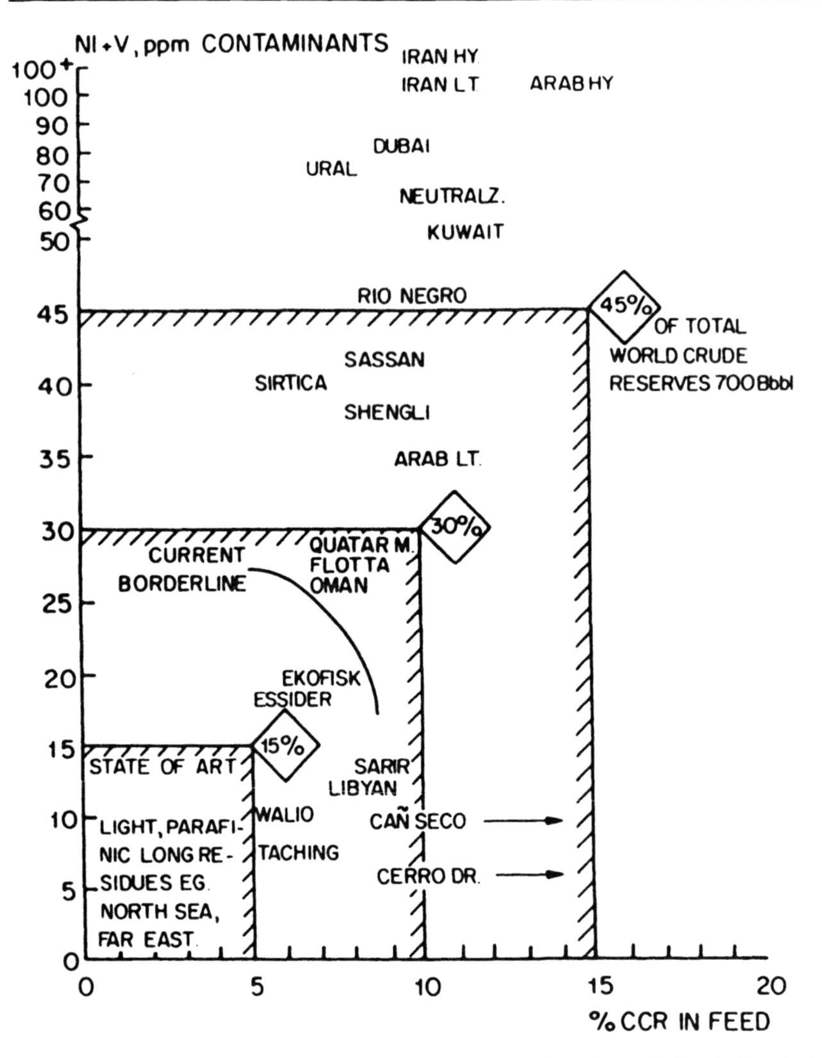

Fig. 7.11. Long-residue Conradson carbon residue (CCR) and metals levels in relation to processability [10]

and have reached the stage where full straight-run long residue (LR, >370 °C) processing is feasible. In the initial phase, FCCs have been drastically revamped to increase their ability to process heavier feedstocks but since the early 1980s dedicated LR FCCs have been designed and constructed which are now in successful operation [10, 14]. These LR FCC designs feature a number of aspects which are of crucial importance for processing residual feedstocks [10, 14]:

- high-performance feed injection system to realize a good dispersion of the heavy feed and to effectively contact it with the passing catalyst,

- a quick catalyst/hydrocarbon vapor separator at the riser end to minimize aftercracking,
- an efficient and fast stripping of adsorbed heavy hydrocarbons to minimize coke make,
- adequate catalyst cooling in the regenerator to remove the surplus heat associated with heavy feed processing; alternatively, two-stage regeneration has been proposed to cope with these phenomena [15],
- use of FCC catalysts capable of operating at high nickel and vanadium levels whilst maintaining sufficient activity to crack the heavy, metal-rich feed (e.g., [16]).

Present-day reactor and catalyst technology allows feedstocks to be processed with Conradson carbon contents up to ~7 wt% and metals contents (Ni + V) up to ~20 wt-ppm (Fig. 7.11 [10]). Figure 7.12 shows the progression in FCC designs with time [17]. It should be realized, however, that whereas in the recent past the emphasis has been on developing LR FCC technology proper, the (near) future limitations will be mainly on the environmental impact of LR FCC processing, e.g.:

- SO_x and NO_x emission control/reduction,
- spent, high-metal catalyst disposal/regeneration/rejuvenation [18],
- particulate emission control,
- product quality improvement, sulfur removal in particular.

The FCC product slate evolution has to a large extent been induced by catalyst developments and is clearly geared towards gasoline production (Fig. 7.13). By

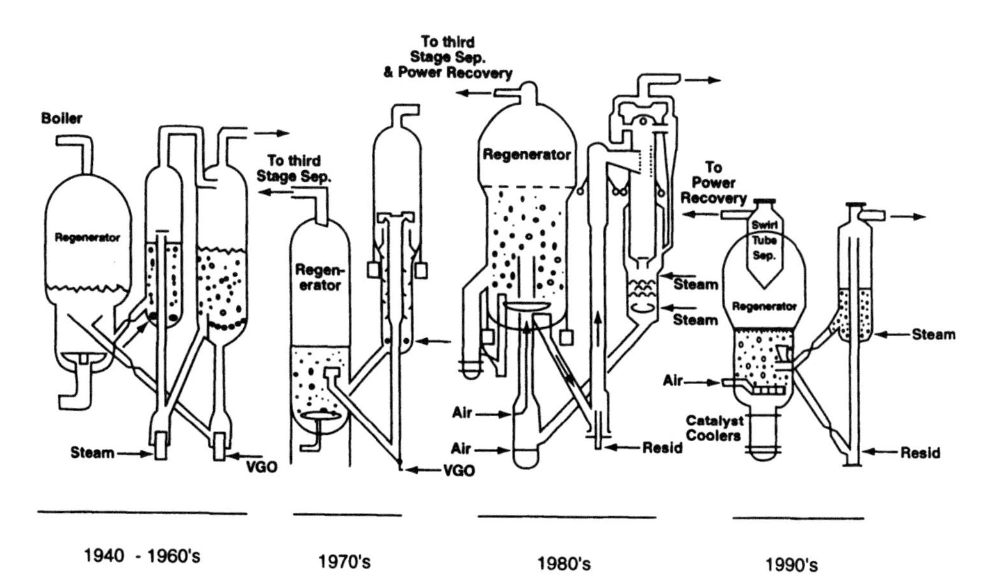

Fig. 7.12. Development of residue FCC designs [17]

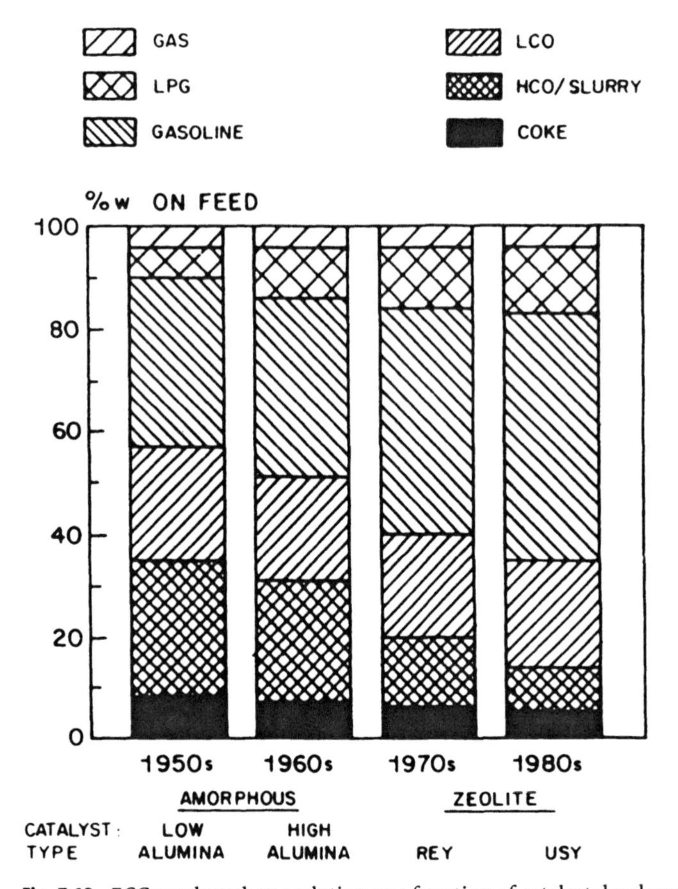

Fig. 7.13. FCC product slate evolution as a function of catalyst developments

now, 40–50 wt% of the products is in the gasoline range, even with residual feedstocks. Also a trend towards more LPG can be discerned. FCC gasoline is a desired product because of its high octane quality. Typical octane distributions over the gasoline boiling range are shown in Fig. 7.14 [19]. FCC gasoline is rich in olefins and aromatics, which contribute significantly to octane numbers, respectively, but as a consequence FCC gasoline has a relatively high density. Usually, the olefin-rich light fraction (C_5 – 100 °C) can be blended straight into the gasoline pool. The octane quality of the heart-cut fraction (100–150 °C), however, is often too low for direct use and further octane boosting may be effected by catalytic reforming after proper hydrotreating. The heavy gasoline has a good octane quality because of its aromatics content but cannot be used directly for blending in today's gasolines because of its high sulfur levels. This aspect becomes increasingly important in view of new legislation, which may lead to complete reformulation of the gasoline pool (see below, but also, e.g., [20]).

The cycle oils consist mainly of dealkylated aromatic nuclei and, therefore, their density is high and light cycle oil cetane is low (typically 10–20). As the

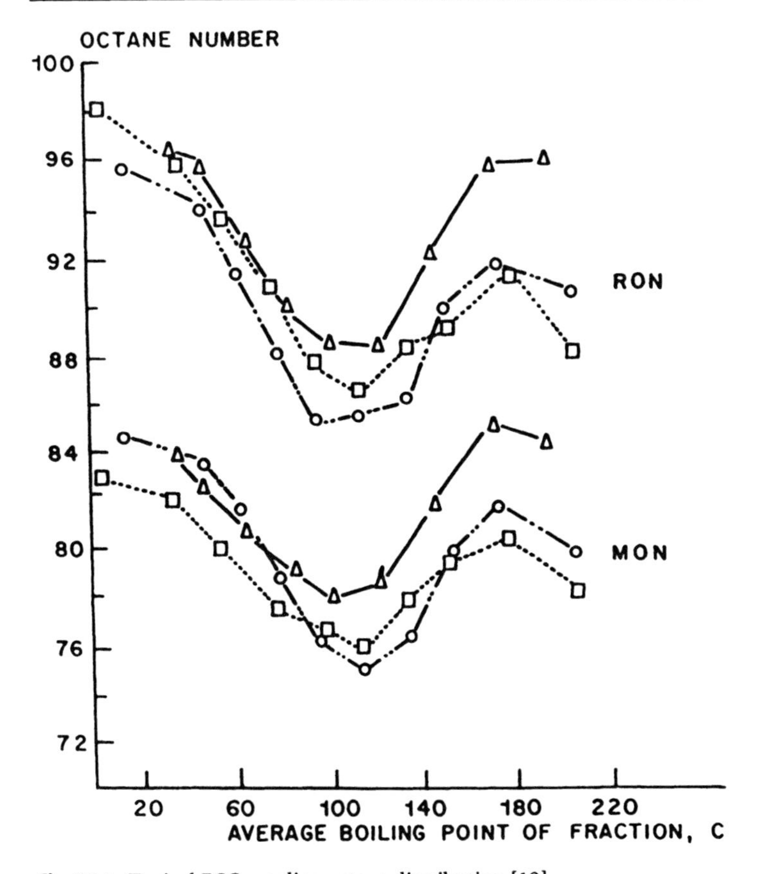

Fig. 7.14. Typical FCC gasoline octane distribution [19]

FCC process hardly desulfurizes, the sulfur content of cycle oils is usually high and these oils have to a large extent to be treated or used in the fuel pool. The slurry oil normally contains some 500–1500 wt-ppm of catalyst fines, and is high in sulfur, which makes it unsuitable for any outlet except refinery fuel. Some outlets exist as carbon black feedstock, however.

7.2.2
Application of Fluid Catalytic Cracking (FCC)

FCC is the most widespread conversion process in the world and represents the single largest market for refining catalysts (Table 7.2, from [21]). These FCC catalysts are consumed in some 350 FCC units world-wide, representing a total conversion capacity of well over 10 million tons per day. FCC processes are licensed through a few major licensers: UOP, Shell, Stone & Webster, and Kellogg.

To date the number of different FCC catalysts available on the market exceeds the number of FCC units (Fig. 7.15 gives the growth during the eighties, a recent

Table 7.2. Market for Solid Refining Catalysts [21]

Process	Production (t/a)	Value (10^6 French Francs per year)
Catalytic cracking	440,000	5,500
Hydrotreating	80,000	2,800
Claus	20,000	200
Hydrocracking	8,000	600
Reforming + isomerization	6,000	600 [a]
Other	12,000	200
Total	566,000	9,900 [a]

[a] Excluding precious metals costs.

figure quoted [21] being 700 different grades), which makes the selection of an appropriate catalyst a difficult task. These catalysts are manufactured and marketed by: AKZO/Filtrol, CCIC, Engelhard, and Grace Davison. FCC catalysts are certainly the largest catalyst application of zeolites (about 100,000 ton zeolite/a, vs. hydrocracking at best about 5,000 t zeolite/a, [21]); certainly it is the area where the application of zeolites had the largest economic impact. It has been estimated [22] that the application of zeolites vs. the older amorphous catalysts gives a yield improvement with an economic value of 40 billion US dollars a year.

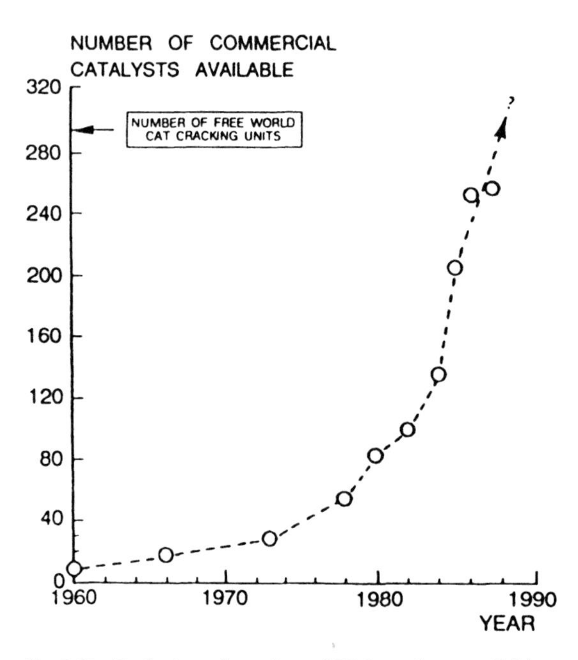

Fig. 7.15. Evolution of number of FCC catalysts available on the market [16]

As will be clear from the detailed process description below, the laboratory simulation of FCC is particularly interesting, not only because of high gas velocities etc., but also because the catalyst has only a short contact time, within this time a marked catalyst deactivation, but also a marked (cumulative) age distribution. This age clearly reflects the state and the performance of the catalyst both due to the hydrothermal deactivation and due to deposition of metals such as Ni and V (whose representative deactivation is an issue in itself). This aspect is discussed in [23, 24].

7.2.3
Reaction Mechanism

Although at the temperatures prevailing in modern FCCs some thermal cracking of the feed may take place, we will focus here on the acid-catalyzed route, which consists of a number of rather complex reaction pathways. These are complicated even further as primary reaction products undergo secondary reactions. It is generally accepted that catalytic cracking of hydrocarbons occurs via a carbocation ion mechanism in the vapor phase and that the reactions can be catalyzed by any material with Brønsted or Lewis acidity [16, 25, 26]. Two reaction mechanisms are known to date to play a role in catalytic cracking of hydrocarbons. The first mechanism proceeds via carbenium ions, whereas the second mechanism proceeds via the formation of carbonium ions.

Carbenium ions are formed either by protonation of unsaturated bonds, olefins or aromatics, at Brønsted acid sites, or by hydride abstraction from an alkane or alkyl aromatic molecule at Lewis acid sites [25, 27–29]. Carbenium ions can react further by a number of simultaneously occurring primary reaction mechanisms [25, 27–29] (Fig. 7.16):

- proton loss to the catalyst resulting again in an olefin, a termination reaction,
- intermolecular hydride transfer from a carbenium ion to a paraffin, a propagation reaction,
- C–C bond cleavage of the carbenium ion at the beta-position, the most important reaction in catalytic cracking. This mechanism proceeds via beta-scission and yields another, smaller carbenium ion and a smaller olefin.

The above classical carbenium ion mechanism has been widely accepted but has some shortcomings in explaining a number of experimental facts. More recently, an alternative reaction mechanism has been proposed [30]. At temperatures over 500 °C, a reaction path via penta-coordinated carbonium ion intermediates is proposed to exist [31]. The carbonium ions, created at Brønsted acid sites, are cracked via beta-scission to smaller paraffins and carbenium ions or are converted directly into carbenium ions through the loss of hydrogen (Fig. 7.17). As these reactions are monomolecular, they are favored in, e.g., zeolites with high constraint indexes (ZSM-5). In general, the cracking reaction rates decrease in the following order: olefins > isoparaffins/naphthenes > n-paraffins > aromatics.

As pointed out above, various secondary cracking reactions proceed during the cracking process. Hydrogen transfer and acid catalysis play important roles

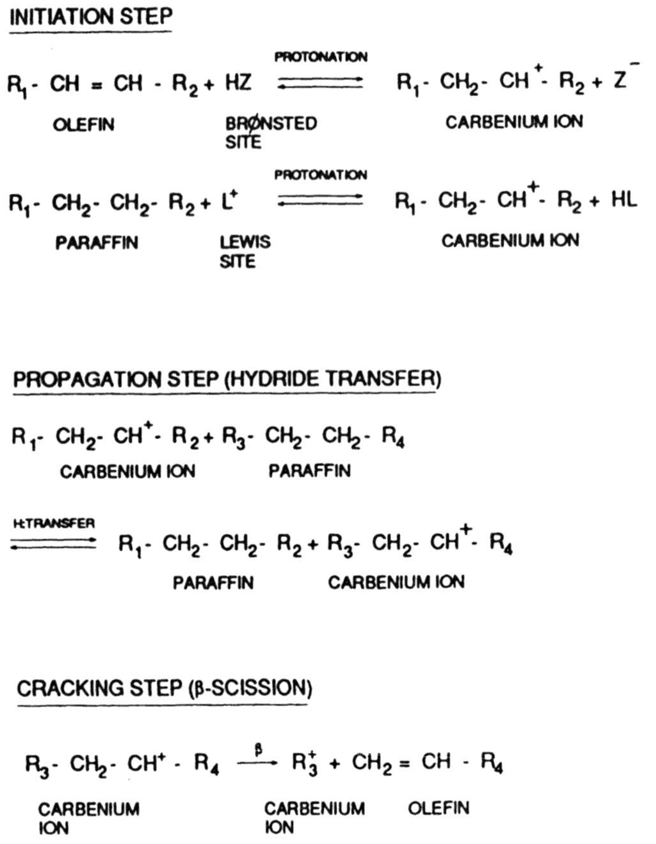

INITIATION STEP

$$R_1 - CH = CH - R_2 + HZ \xrightleftharpoons{\text{PROTONATION}} R_1 - CH_2 - CH^+ - R_2 + Z^-$$

OLEFIN — BRØNSTED SITE — CARBENIUM ION

$$R_1 - CH_2 - CH_2 - R_2 + L^+ \xrightleftharpoons{\text{PROTONATION}} R_1 - CH_2 - CH^+ - R_2 + HL$$

PARAFFIN — LEWIS SITE — CARBENIUM ION

PROPAGATION STEP (HYDRIDE TRANSFER)

$$R_1 - CH_2 - CH^+ - R_2 + R_3 - CH_2 - CH_2 - R_4$$

CARBENIUM ION — PARAFFIN

$$\xrightleftharpoons{\text{H-TRANSFER}} R_1 - CH_2 - CH_2 - R_2 + R_3 - CH_2 - CH^+ - R_4$$

PARAFFIN — CARBENIUM ION

CRACKING STEP (β-SCISSION)

$$R_3 - CH_2 - CH^+ - R_4 \xrightarrow{\beta} R_3^+ + CH_2 = CH - R_4$$

CARBENIUM ION — CARBENIUM ION — OLEFIN

Fig. 7.16. Primary reaction pathways involving carbenium ions [25]

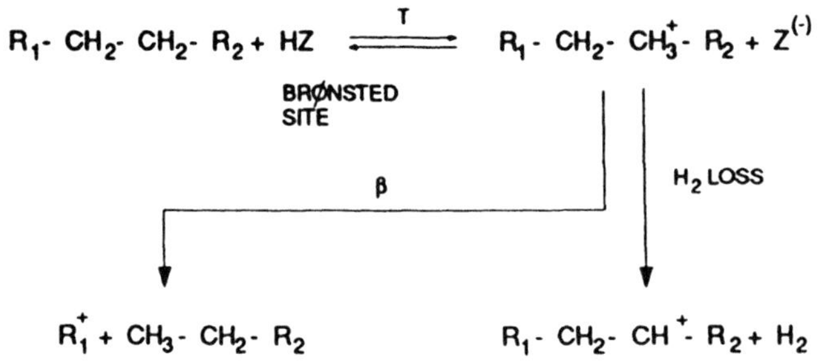

$$R_1 - CH_2 - CH_2 - R_2 + HZ \xrightleftharpoons{T} R_1 - CH_2 - CH_3^+ - R_2 + Z^{(-)}$$

BRØNSTED SITE

β: $R_1^+ + CH_3 - CH_2 - R_2$

H_2 LOSS: $R_1 - CH_2 - CH^+ - R_2 + H_2$

Fig. 7.17. Primary reaction pathways involving carbonium ions [25]

in most of these reactions: cyclization, aromatization, condensation, and double-bond and skeletal isomerization. Intermolecular hydrogen transfer from naphthenes to olefins, yielding aromatics and paraffins, is a bimolecular reaction which reduces the amount of olefins in the product. The aromatics content also increases, and upon further hydrogen transfer reactions these species may condense to polycyclic aromatics and finally to coke.

7.2.4 The FCC Catalyst

7.2.4.1 Catalyst Constituents

In the early 1940s, synthetic silica-alumina was introduced as an FCC catalyst [12]. From 1960 onwards, Y-zeolites found their way more and more in FCC applications, amongst other factors, because of:

- their high intrinsic activity,
- good selectivity towards gasoline combined with low coke and gas make,
- good activity retention.

The introduction of the high-activity zeolite-based catalysts was a breakthrough in fluid catalytic cracking as it enabled the introduction of lean phase, full riser cracking at contact times in the order of a few seconds only. As a result, virtually all FCC catalysts are at present zeolite-Y based, a development which has been further enhanced by a continuous series of zeolite and catalyst improvements since the 1960s [16].

Apart from the 5–40 wt% Y-zeolite, finished FCC catalysts also contain an active matrix and a binder material. The amorphous alumina-, silica- or silica-alumina-based matrix in fact determines the catalyst particle shape, it acts as an attrition resistant carrier for the (1–5 μm) zeolite crystals and provides a porous structure for hydrocarbon diffusion. Depending on the formation, the matrix may also display catalytic activity, which can significantly contribute to the cracking reactions. The latter aspect is particularly relevant for cracking of the heavier hydrocarbon molecules in residual feedstocks, which diffuse faster through the wide pore (50–150 Å) matrix than through the small pore (7.4 Å) openings of the zeolite. In fact, heavy hydrocarbons are "precracked" on the matrix to smaller molecules, which can subsequently enter the zeolite pores for further reaction. As a consequence, active matrices can significantly contribute to the FCC catalysts' bottoms upgrading and are indispensable in resid processing. A clear example illustrating this effect is given in Table 7.3 [32].

Modern matrix systems are designed to assist in binding the feed metals (in particular nickel and vanadium), thereby improving the metals resistance of the catalysts. Moreover, active matrices react with the feed nitrogen and prevent it from deactivating the zeolite through strong adsorption. Active matrices are, however, less coke selective and produce more dry gas and C_3/C_4 olefins than zeolites. They also exhibit a lower hydrogen transfer than zeolites and, therefore, matrix cracking, e.g., contributes to gasoline octane.

Table 7.3. Effect of Matrix Activity: Product Distribution of 100–482 °C fraction [32]

Yields, wt%	Low matrix surface area catalyst	High matrix surface area catalyst
Paraffinic Feed		
Coke	3.2	4.2
C_2	2.1	2.2
$C_3 + C_4$	21.8	21.3
Gasoline	59.4	60.7
LCO	12.0	9.5
640 °C + bottoms	1.5	2.1
LCO/bottoms	8.0	3.4
Conversion	86.5	88.4
Aromatic Feed		
Coke	2.6	6.3
C_{2-}	1.1	2.3
$C_3 + C_4$	6.3	10.6
Gasoline	34.7	37.0
LCO	16.9	17.3
640 °C + bottoms	38.4	26.5
LCO/bottoms	0.4	0.7
Conversion	44.7	56.2

In an operating FCC unit, the freshly delivered catalyst is aged mainly in the high-temperature (680–750 °C), high-steam partial-pressure (typically 0.2 bar) section of the regenerator, where it spends 80–90% of its life. Here, the zeolite component dealuminates to a smaller equilibrium unit cell size (UCS), which is in the range of 24.22–24.35 Å, and consequently loses cracking activity. For a normal HY zeolite this would simultaneously lead to massive crystal destruction and finally the collapse of the zeolite structure. If zeolite Y dealumination occurs in the manufacturing process rather than in the FCC, the zeolite stability improves and higher Si/Al ratios can be obtained, with higher crystallinity. Both (hydro)thermal and chemical treatment (silicon enrichment) steps are known for zeolite dealumination and stabilization [16, 25]. Potential side effects of this dealumination comprise (i) the removal of alumina debris from the zeolite framework, which reduces gas and coke make and increases the zeolite accessibility, (ii) creation of hydroxyl nests or defect sites, (iii) increase in mesoporosity which may be helpful in residue processing, and (iv) crystallinity loss. During the production, the zeolites' sodium content can also be reduced to the lowest possible level, and rare-earth metals can be incorporated via ion exchange (REY) either on the pure zeolite or on the zeolite/matrix blend. These factors have a positive effect on zeolite stability. The effect of rare-earth level on UCS is illustrated in Fig. 7.18.

Reducing the (framework) alumina content of zeolite Y, either during the zeolite preparation (ultra-stable Y zeolite, USY) or in the FCC operation itself, proportionally reduces the number of acid sites as well as the zeolites' intrinsic cracking activity and unit cell size (Figs. 7.19 [13] and 7.20). As the acid sites,

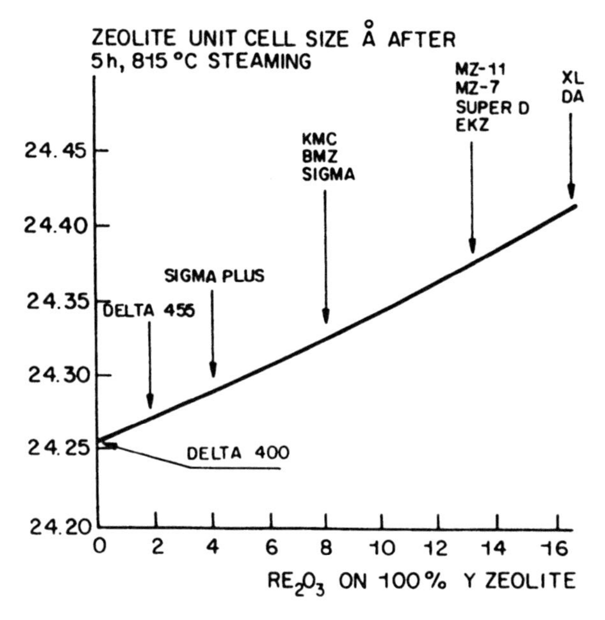

Fig. 7.18. Effect of rare-earth level on equilibrated unit cell size

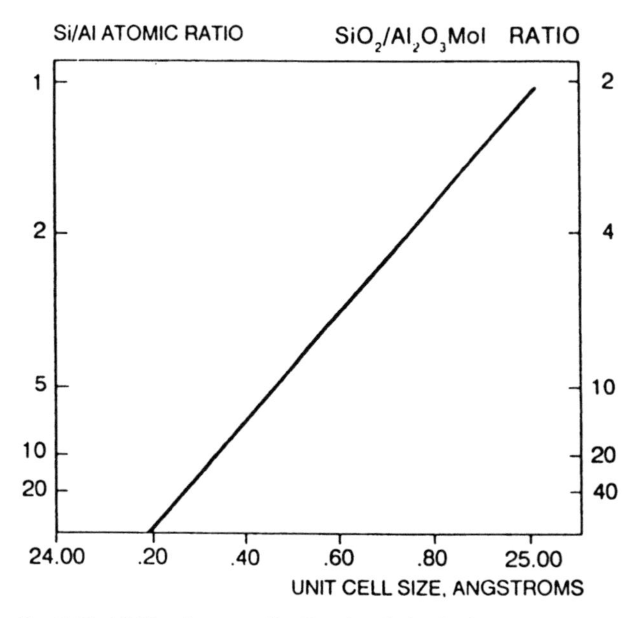

Fig. 7.19. Si/Al ratio vs. zeolite Y unit cell size [13]

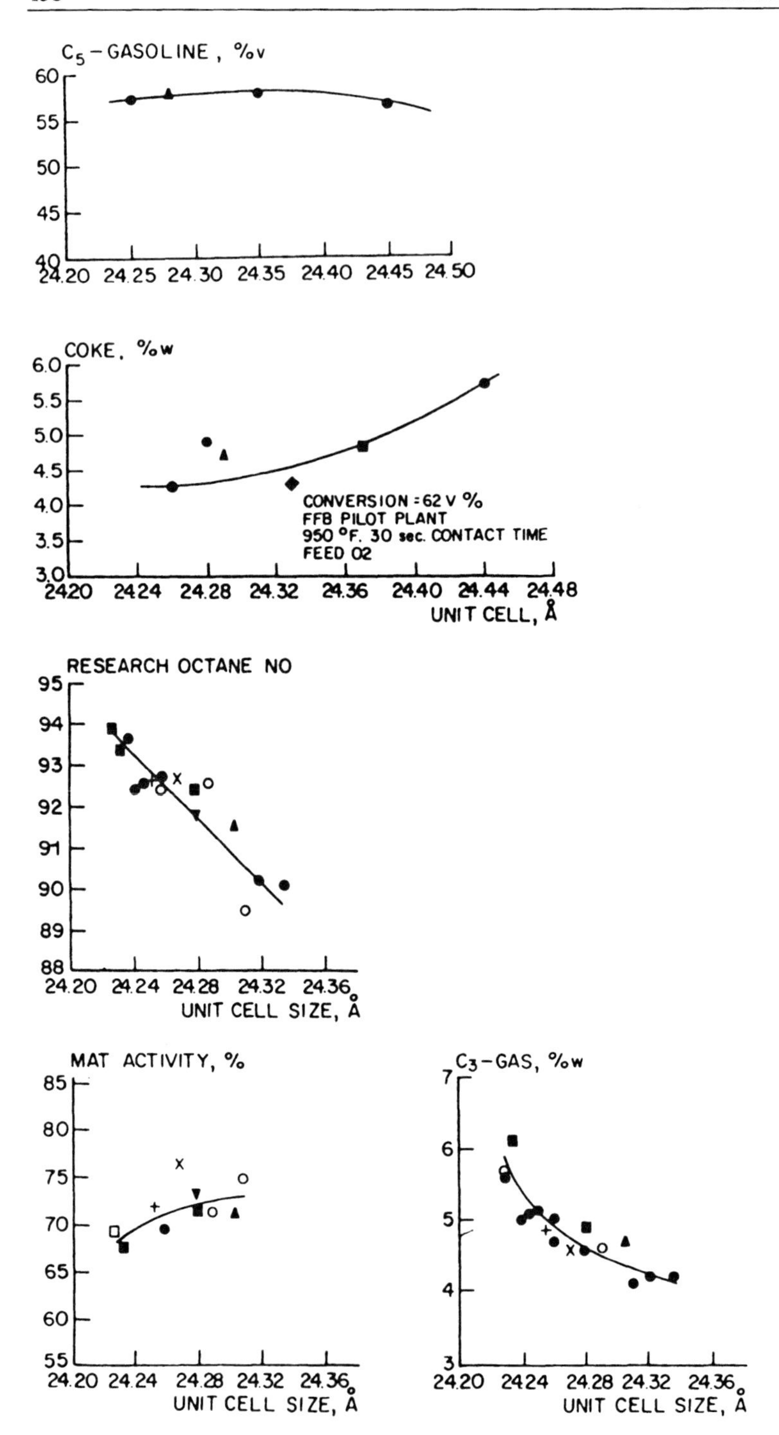

Fig. 7.20. Effects of unit cell size on catalyst performance

e.g., in USY zeolites also become more isolated, the bimolecular hydrogen-transfer reactions are reduced: the FCC products become more olefinic, thereby reducing aromatics formation and condensation, which in turn reduces coke make (Fig. 7.20) with, as a net result, an increase in the gasoline octane quality. Low UCS USY increases LPG make, mainly at the expense of gasoline yield, because the olefin-rich gasoline is very reactive and tends to be overcracked on the more isolated, highly acidic sites of the USY zeolite (Fig. 7.20). The effects of low UCS are more complex than this simplified interpretation suggests; they are described in more detail in, e.g., [16, 25].

7.2.4.2
Effect of Metals on FCC Catalyst Behavior

Most FCC feedstocks and in particular residual ones contain metals of which nickel and vanadium are the most relevant for FCC catalyst behavior. Nickel deposits on the catalyst surface and acts as a dehydrogenation catalyst at the prevailing FCC conditions. It enhances the coke formation tendency of the catalyst at the cost of gasoline make, and it increases the production of gas as well as hydrogen. This means that it changes the selectivity pattern but does not affect the stability of the catalyst. Vanadium, on the other hand, also deposits on the catalyst but has a detrimental effect on its stability as it migrates through the catalyst. It forms vanadates and – in the presence of sodium – a low-melting eutectic, both of which destroy the zeolite [16].

The detrimental effect of nickel on the selectivity of the catalysts can be suppressed by adding antimony to the feed. It has been suggested that antimony forms a complex with nickel, thereby reducing the dehydrogenation effect over nickel. In view of the environmental problems related with antimony, alternatives like bismuth compounds have been developed [33, 34]. Vanadium poisoning is usually combated by trapping the V_2O_5, which is formed at regenerator conditions, with basic alkali earth oxides, e.g., MgO, or metals like strontium [16]. The high-melting product prevents vanadium migration to the zeolite. Tin additives have also been reported to be effective in vanadium passivation [35, 36].

7.2.4.3
Novel Zeolites in FCC Catalysts

As FCC is usually the refinery's money-maker, there is a continuous drive towards catalyst improvements, however small. ZSM-5 is a well-known octane booster, which is already widely applied in combination with normal Y-based catalysts. The mechanism according to which ZSM-5 acts on the FCC products and enhances the gasoline octane quality, while increasing in particular propylene and butylene make, is extensively discussed in the literature (see, e.g., [16, 25]). It appears that depending on the activity of the zeolite (i.e., the state of framework dealumination due to the hydrothermal regenerations) ZSM-5, in order of increasing activity, will isomerize the olefins in the naphtha range, crack these olefins and, finally, crack the paraffins. This is illustrated graphical-

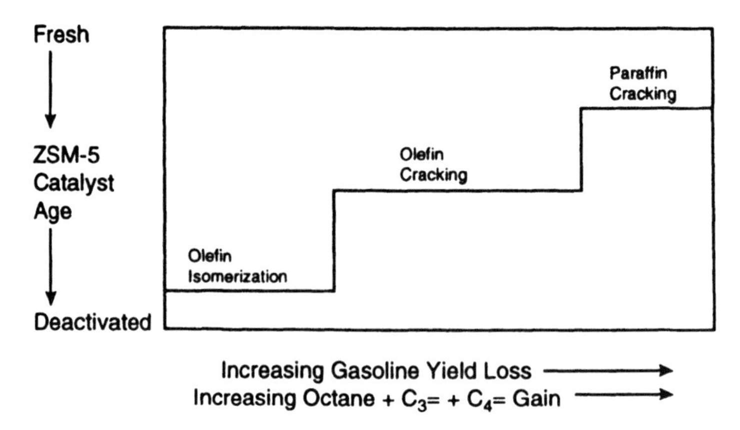

Fig. 7.21. Activity regimes of ZSM-5 [37]

ly in Fig. 7.21 [37]. Other members of the ZSM family (ZSM-11, -12, -23, -35, -38) have been reported to boost gasoline octane [33]. Novel zeolites have been proposed in the patent and the open literature, e.g., mordenite, silicalite, ZSM-20, MCM-22 [38], and beta, which is claimed to produce more high-octane C_6 and C_7 branched alkanes that contribute to the gasoline octane quality [25]. In addition, ALPOs, SAPOs, MeAPOs and MeAPSOs are being studied to assess their potential in the FCC environment [25].

Nevertheless, a recent review concludes that "the reign of zeolite Y over FCC appears to be in little danger of being overthrown [23, 24]".

7.2.4.4
Physical Catalyst Parameters

Several physical parameters play an important role in FCC catalyst performance:

- The pore volume and pore size distribution are important factors in the performance of the catalyst. Too many small pores (<100 Å) or a low pore volume may lead to pore blockage by coke and limit the diffusional transport of feed and product molecules. Moreover, the hydrothermal stability of the matrix may be negatively affected. Too many large pores (>200 Å), on the other hand, lower the surface area and consequently reduce the catalytic contribution of the matrix. In the extreme case they also deteriorate the attrition resistance of the catalyst. An optimized FCC catalyst has a pore size distribution which is balanced for a specific feed type at specific operating conditions. Pore volume and pore size distribution are controlled by, amongst other factors, the composition of the silica-alumina gel, reaction temperature, slurry concentration and pH, ageing and calcination conditions.
- The thermal and hydrothermal stability of both zeolite and matrix constituents during the catalyst regeneration is required for good activity and selectivity maintenance. To this end, zeolites are stabilized by ion exchange with rare-earth metals (REY), while a high Si/Al ratio, a low residual sodium

content, a high crystallinity and large crystallites are also favorable (USY). Matrices should not contain too large a fraction of micropores if they are to retain their activity.

7.2.4.5 Mechanical Aspects

Mechanical aspects of FCC catalysts form, probably more than in any other catalytic process, an integral part of the overall performance of the catalyst. Two important aspects are commented on below [25, 26].

- The attrition resistance of the finished catalyst at the severe FCC conditions determines to a large extent the losses of catalyst to the environment, either directly via the flue gases or indirectly via the slurry. It can also largely determine the rate of addition of fresh catalyst and the particle size distribution of the equilibrium catalyst. The latter parameter influences the fluidizing behavior of the catalyst in the FCC unit and is hence an important operational factor. The attrition resistance usually drops with increasing zeolite content, but up to 30–35% can be incorporated with retention of sufficient attrition resistance. For in situ crystallized zeolites, however, the attrition resistance is sometimes claimed to increase with the zeolite content, and up to 70% can be accommodated without sacrificing attrition resistance. Furthermore, zeolite crystal size, zeolite dispersion and morphology, binder content and type, as well as spray drier conditions have an impact on attrition resistance.
- The particle size distribution (PSD) of the FCC catalyst determines its fluidizing behavior and transport properties. The average particle size is in the order of 60–80 μm, and the inventory should contain some fines (5–10% <40 μm) to enable stable fluidization. The PSD of the fresh catalyst is determined by the spray drier conditions. For equilibrium catalysts, however, the PSD is a dynamic property, influenced of course by the PSD of the fresh catalyst but also by the loss of fines through the cyclone systems and the production of fines as a result of attrition.

7.3 Hydrocracking

Besides fluid catalytic cracking, hydrocracking is the other major catalytic process for converting residual feedstocks into distillate products. Whereas in catalytic cracking the rapid catalyst deactivation is coped with by a process set-up in which the catalyst is regenerated after a reaction time of only a few seconds, in hydrocracking high hydrogen pressures prevent rapid catalyst deactivation, allowing catalyst lives of five years and more. At the same time the feedstock is markedly hydrogenated, giving products that are rich in hydrogen and low in heteroatom and aromatics contents.

7.3.1
Process Configurations

Various hydrocracking process configurations are applied commercially. In general the configurations are governed by the marked sensitivity of the dedicated cracking catalysts to poisoning by the (basic) organic nitrogen and polyaromatic compounds in the feedstock [39]. The effect of poisoning by organic nitrogen-containing molecules is illustrated in Fig. 7.22 [40]. The emphasis in the "first stage" of the process is generally on "hydrotreating" (denitrogenation, aromatics saturation together with desulfurization) of the feedstock [41]. This can be realized in a dedicated way, using selective hydrodenitrogenation (HDN) catalysts at relatively low temperatures giving little cracking ("pretreat" stage); alternatively, using higher temperatures, hydrocracking catalysts can be used without a pretreat step, but they will then necessarily have a low cracking activity at the top of the reactor. In the second stage of the process the hydrotreated feedstock is cracked to distillates over dedicated hydrocracking catalysts.

The simplest process configuration (A), shown in Fig. 7.23 [42], is the so-called single-stage hydrocracker. Here the feedstock is passed in a trickle-flow operation, together with hydrogen, over a single catalyst, or over a stacked bed of a hydrotreating and a hydrocracking catalyst. The unconverted feedstock, which is considerably enriched in hydrogen, can either be recycled or used as ethylene cracker or catalytic cracker feedstock, as lubricating base oil, or as low-sulfur fuel. This simple configuration is the basis of many modern hydrocrackers. For instance, recently a new hydrocracker has been built and put into operation at the Shell refinery in Pernis near Rotterdam as part of a large modernization program [39, 43]. A simplified scheme is given in Fig. 7.24. The design throughput of this hydrocracker is 8,000 tons per stream day, and it is as high as 34 m. It converts, among others, vacuum gas oil to distillates at a conversion level

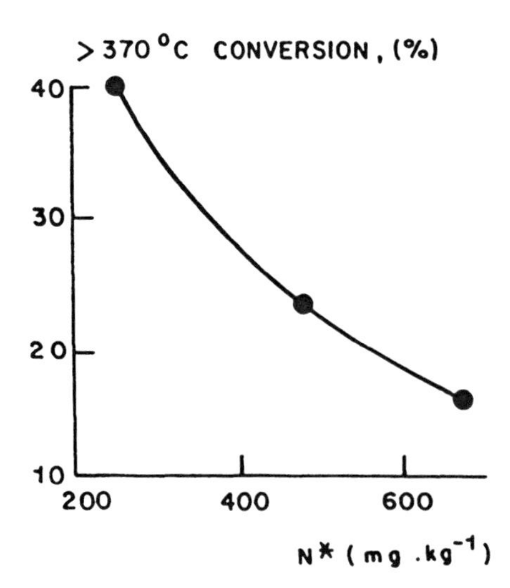

Fig. 7.22. The effect of organic nitrogen in the feed on the cracking over a zeolitic catalyst (MHC conditions) [40]

A SINGLE STAGE

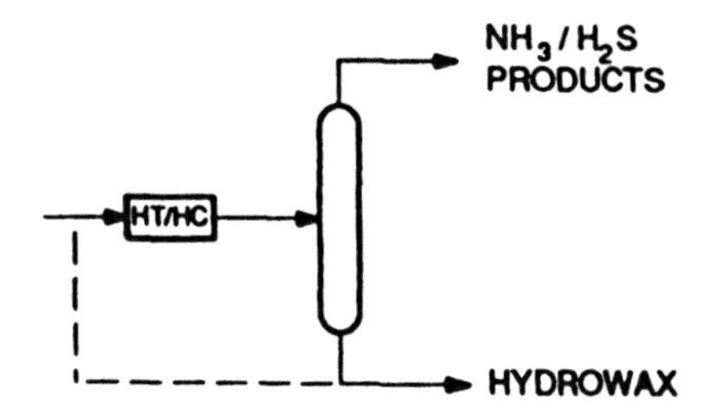

B SERIES FLOW

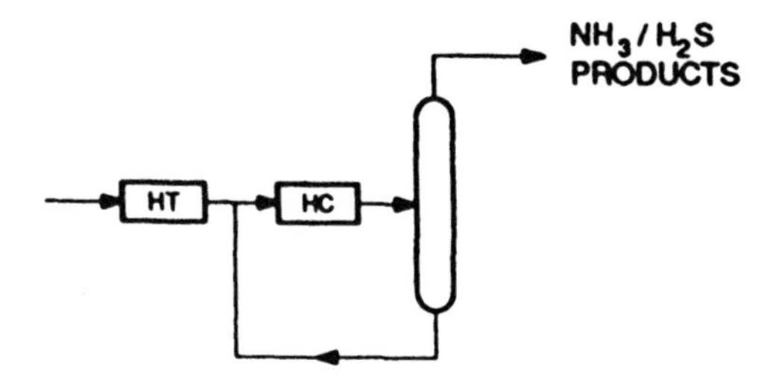

C TWO STAGE

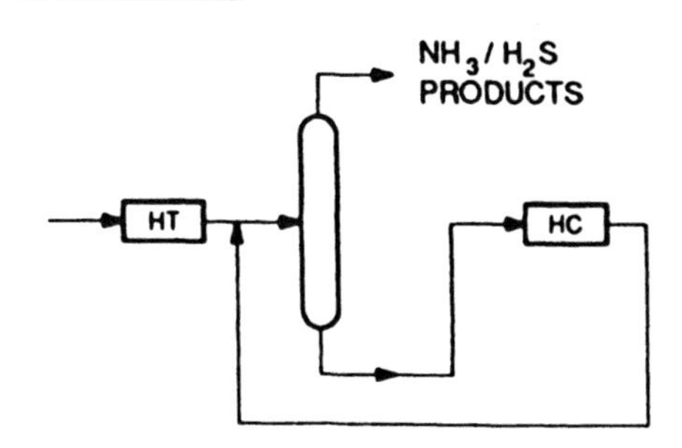

Fig. 7.23. Modes of hydrocracking operation [42]

of 80 wt%. The unconverted material is subsequently processed by the catalytic crackers. The single stage (without recycle of unconverted product) is also the basic scheme for mild hydrocrackers, which operate within a lower pressure window, viz. 30–90 bar. These low-pressure hydrocrackers are from origin revamped vacuum gas oil hydrodesulfurization units, but have also been proposed as a first phase in a step-by-step introduction of conversion capacity in (simple) refineries [7, 44].

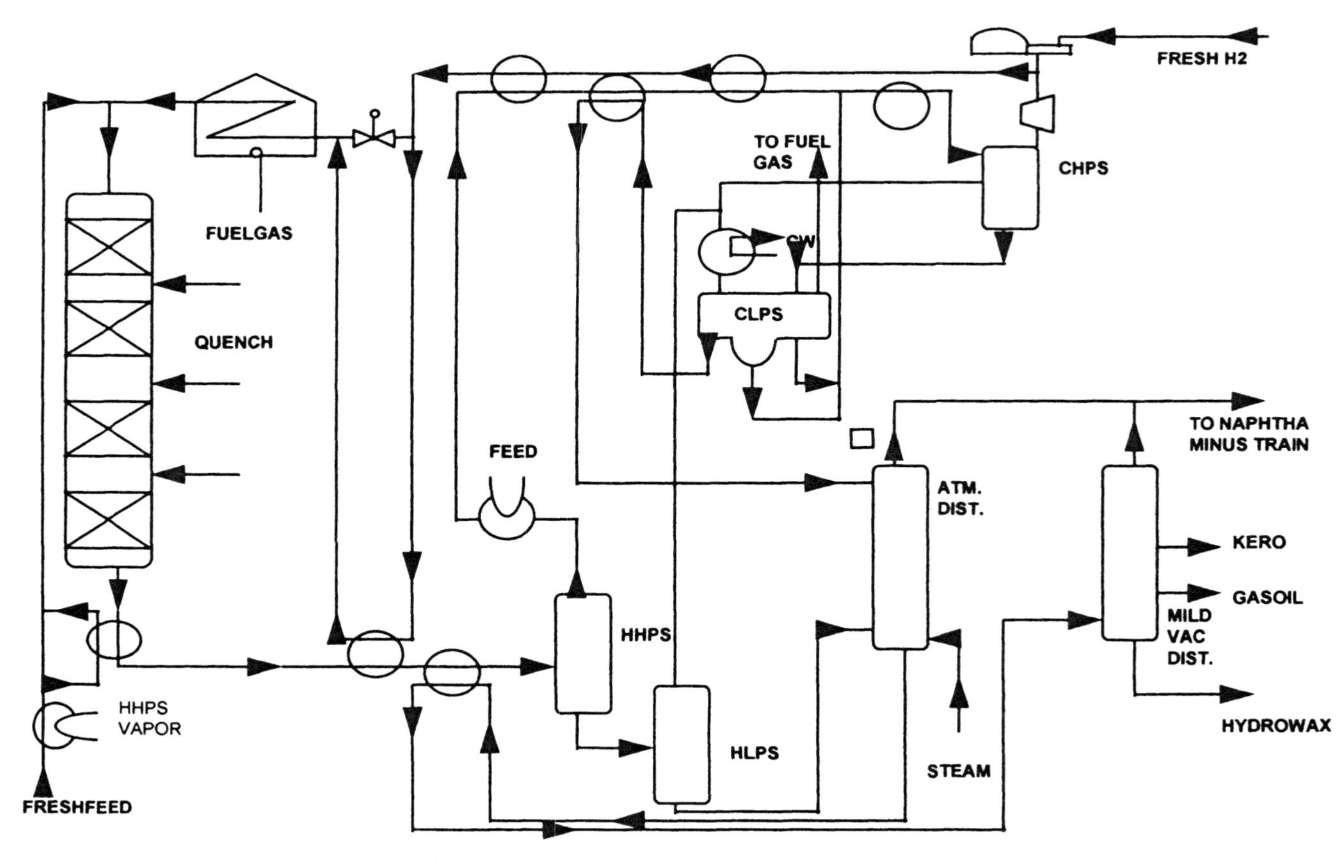

Fig. 7.24. Simplified flow scheme of the Shell Pernis hydrocracker [39, 43]

The "series flow" configuration (B) is a variation on this scheme, with two reactors, and a recycle to either the first or the second reactor, such recycle can obviously be effective if high conversions are desired. The "two-stage" configuration (C) differs from series flow in that the products from the first reactor(s) are first separated, and that only the heavy fraction, without cracked products, hydrogen sulfide or ammonia are passed over the second reactor. This certainly favors the activity (ammonia reduces the acidic cracking activity); effects on selectivity can be complex: removal of light products from the first stage favors middle distillate selectivity, but the series flow conditions in themselves often lead to higher middle distillate selectivities (see below). Obviously, in the two-stage mode the second-stage catalyst does not have to be resistant to high concentrations of hydrogen sulfide. The costs of the two-stage configuration are clearly higher, however (compression!). In addition to these schemes, one finds combinations in which the feed is first passed over a pretreating catalyst and a cracking catalyst, and only then, after product separation, over the second stage with recycle. In these schemes the split of the conversion over the stages, the recycle ratio (i.e., the conversion per pass), and of course the choice of the catalyst(s) are important parameters to vary the selectivity pattern. Slight upgrading of the hydrocracker products is sometimes required, such as catalytic prevention of mercaptans formation via recombination of olefins and H_2S and deep hydrogenation in the case of lubricating base oils production [45].

7.3.2 Feedstocks and Products

The conversion of heavy residual oil fractions into transportation fuels is economically attractive, and therefore in hydrocracking, as in catalytic cracking, the feedstocks are becoming ever heavier in terms of boiling fraction; often also cracked fractions are included. As an illustration, Fig. 7.25 shows the increasing heaviness of feedstocks processed in Shell designed hydrocracking units in the period 1970–1990 [46]. The latest increase is due to the start-up of a hydrocracking unit processing a mixture of vacuum gas oil and deasphalted oil (a butane extract of vacuum residue). In this application the residual feedstocks are converted into naphtha, kerosene and gas-oil fractions, where for the latter two their high hydrogen content is a very favorable point for hydrocracking. Figure 7.26 shows a modern combination of a high vacuum unit (to split atmospheric residue into vacuum distillate and a vacuum residue), a deasphalter (to remove asphalt from the vacuum residue), a gasifier (to gasify asphalt), a hydrodemetallization unit (to remove metals from and upgrade the deasphalted oil) and a single-stage hydrocracker (to convert the blend of vacuum distillate and upgraded deasphalted oil into premium middle distillates) [47].

In the United States with the large number of catalytic cracking units and coking units, on the one hand, and the large demand for gasoline, on the other, hydrocracking is often applied to convert highly aromatic products from catalytic crackers or cokers boiling in the middle distillate range, because they are

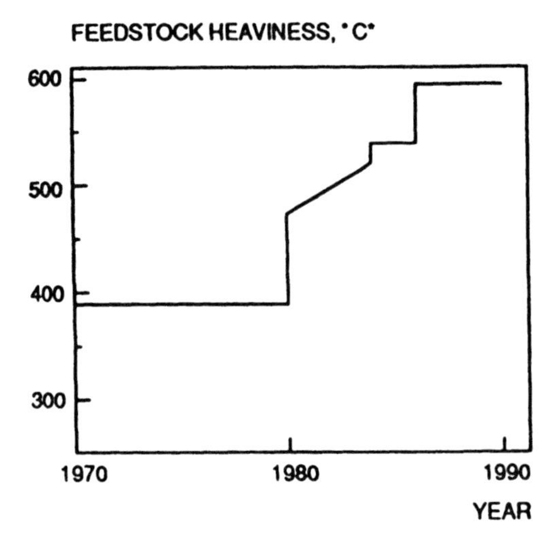

Fig. 7.25. Heaviness of feedstocks processed in Shell hydrocrackers [40]

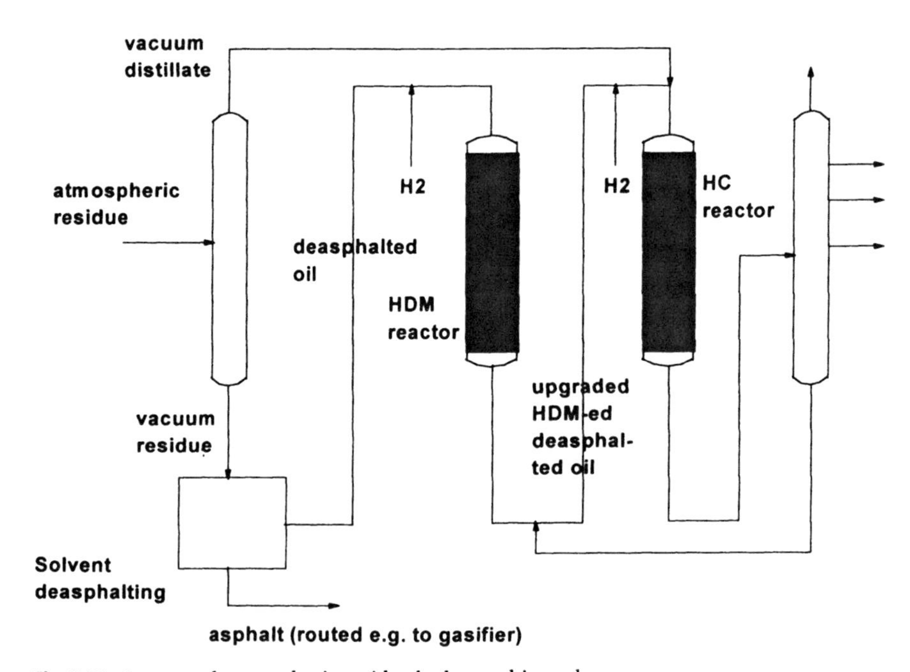

Fig. 7.26. Integrated atmospheric residue hydrocracking scheme

Table 7.4. Typical Properties of Middle Distillate Products ex-Catalytic Cracking and Hydrocracking [46]

	Carbon rejection (FCC)	Hydrogen addition (Hydrocracking)
Kerosene (150–250 °C)		
H, wt%	10.4	14.0
S, wt-ppm	1,660	20
Smoke point/mm	8	23
Gas oil (250–370 °C)		
H, wt%	9.4	13.6
S, wt-ppm	10,000	40
Cetane No.	27	55

difficult to crack further without hydrogen addition. Thus the American units produce from relatively low-boiling but very unreactive feedstocks additional naphtha, and often some kerosene.

A few remarks on product properties have already been made. Table 7.4 shows typical properties in comparison with catalytically cracked products, illustrating the low aromatics and sulfur contents. This makes the hydrocracked products environmentally attractive and, also in terms of performance specification, excellent kerosene and diesel components. Although the C_5/C_6 fraction can have a high degree of branching, giving a high octane number, in general the naphtha is low in octane because of the absence of olefins and aromatics. To cope with this, Chevron have studied the Paragon process [48] in which the naphtha is passed at low pressures over ZSM-5-type catalysts, giving selective conversion of low-octane components. To our knowledge, however, this has not yet found any commercial application.

Because of the hydrogenation of the oil and the selective hydrocracking of aromatics, the feedstock part that has not been cracked to the desired boiling range (in a hydrocracking unit with incomplete conversion) is an excellent feedstock for a catalytic cracker or for an ethylene cracker; it is often also an attractive lubricating base oil. Obviously, the product quality still depends on factors such as catalyst choice, feedstock origin, process conditions, and in particular temperature and hydrogen partial pressure. Thus products from mild hydrocracking, operating at low pressure and high temperature, are rather aromatic because of the hydrogenation/dehydrogenation equilibria [49].

7.3.3 Application of Hydrocracking

Hydrocracking is a wide-spread conversion process, and currently some 120 units are operational world-wide, with an estimated total capacity of about 420,000 tons/day. About 40% of these units are in North America, generally converting

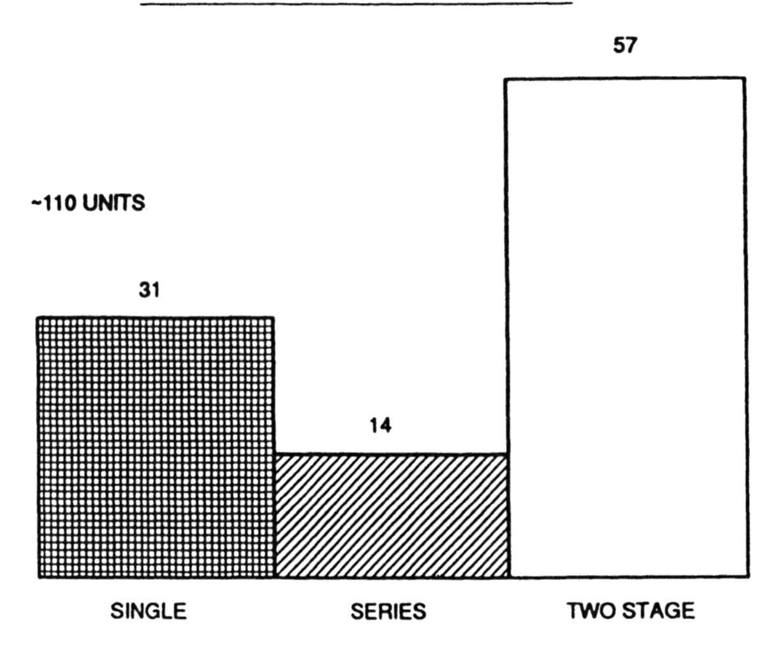

Fig. 7.27. Split of hydrocracking units according to configuration (1990)

lighter feedstocks to naphtha, while in the rest the emphasis is more on processing heavier feedstocks. Figure 7.27 indicates the split over unit configurations, with most of the two-stage units in the United States due to the high demands the cracking of typical US feedstocks places on catalyst activity.

The hydrocracking process and also catalyst development are dominated by a few major licensers: Unocal/UOP, Chevron, and Shell. Unocal and UOP used to be clear competitors, but first UOP acquired the molecular sieves division of Union Carbide which manufactured the zeolitic catalysts for Unocal, and after that a close cooperation was established between UOP and the Unocal Science and Technology division. As a result, the UOP/Unocal combination is now number one in regard to hydrocracking capacity in operation. Of course, other companies are involved as well, but with a clearly smaller market share. As referred to earlier, the technologies can also be quite different, ranging from multistage units with a clear pretreat step to single-stage units operating at relatively high temperature.

The main players are all involved in developing and producing dedicated catalysts, sometimes in close cooperation with dedicated catalyst companies such as AKZO, Criterion, Procatalyse and Zeolyst International. The annual consumption of hydrocracking catalysts can be estimated at roughly 5,000 to 10,000 tons, of which about half are zeolitic catalysts. This application of (generally) zeolite Y makes hydrocracking, after catalytic cracking, volume-wise the largest catalytic application of zeolites.

7.3.4
Catalytic Aspects

7.3.4.1
Hydrocracking Mechanism and "Ideal Hydrocracking"

Hycdrocracking is a typical example of bifunctional catalysis, in which both the hydrogenation/dehydrogenation and the cracking function play crucial roles. Extensive work has been carried out recently, among others by Weitkamp et al. [50, 51] and Jacobs et al. [52, 53], to unravel the mechanism, in particular for the hydrocracking of higher paraffins over noble metal/zeolite catalysts. From these and earlier studies [54] the mechanism shown in Figs. 7.28 and 7.29 emerges. The paraffins are first dehydrogenated over the metallic function (or a metal sulfide) producing olefins that form carbenium ions or possibly alkoxides at the acid sites. These undergo the usual acid-catalyzed reactions, that is they are first isomerized and, at a prolonged reaction time or higher temperature, cracked. The isomerized or cracked species can desorb from the acid sites and be hydrogenated again to the corresponding paraffins. In this mechanism one usually finds that a substantial degree of isomerization can be achieved with only little cracking, and that per cracked parent paraffin molecule two cracked molecules are formed. This is referred to as "ideal hydrocracking" and means that the con-

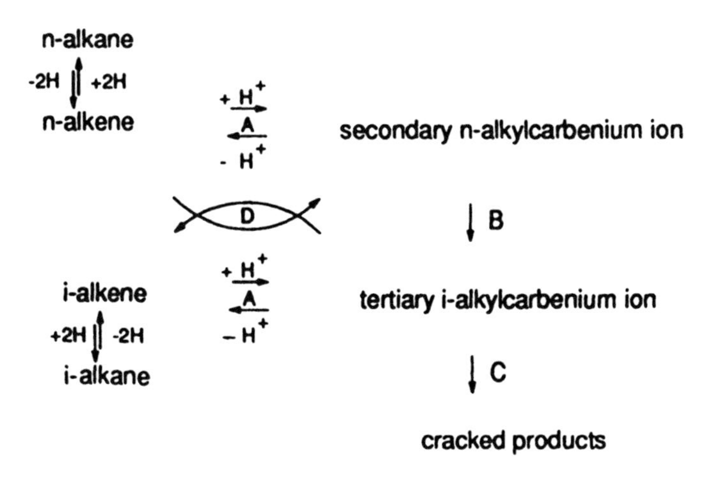

A: protonation - deprotonation reaction via catalyst acid sites
B: rearrangements of carbenium ions via protonated cyclopropane (pcp) intermediates
C: cracking of carbenium ions via β- scission
D: competitive sorption - desorption of alkenes and carbenium ions

Fig. 7.28. Classical bifunctional conversion of n-alkanes [52, 53]

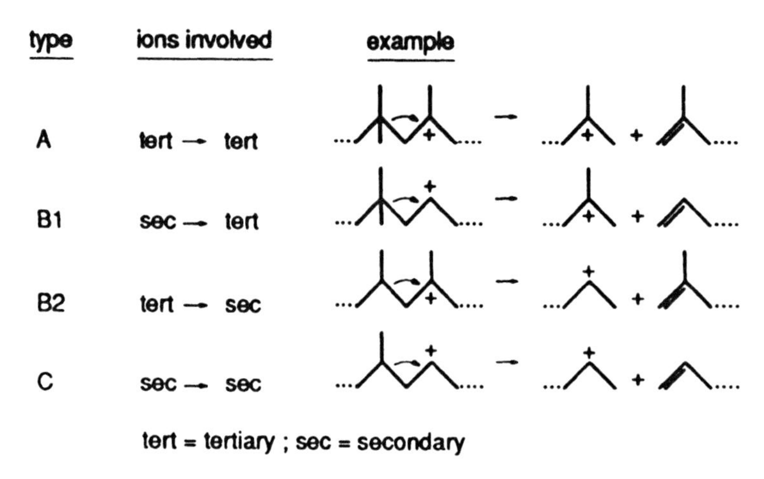

Fig. 7.29. Possible beta-scission mechanisms of secondary and tertiary carbocations [52, 53]

secutive reactions (cracking over isomerization, secondary cracking after primary cracking) play a minor role (see also Sect. 7.5.3 and [30]).

The prime requirement for "ideal hydrocracking" is a highly active (de)hydrogenation function; this is shown for instance in Fig. 7.30 [50, 51]. A high steady state concentration of olefins is generated, which enhances the desorption by competitive chemisorption. This shortens the residence time of the carbenium ions and thus decreases the chance of secondary cracking reactions [41, 55]. In short, the hydrogenation and the cracking function should be "in balance"; the hydrogenation function must be able to maintain the equilibrium composition of olefins notwithstanding their interconversions at acid sites. This of course also requires a sufficient proximity of the two functions, but application of the criterion by Weisz [56]indicates that this can be achieved relatively easily.

The ideal hydrocracking mechanism has a further interesting consequence, namely that a high hydrogen pressure, resulting in lower equilibrium olefins concentrations, leads to lower reaction rates. Indeed this can be observed with very pure feedstocks [57]; in general, however, with the large amounts of catalyst poisons present in oil fractions, higher hydrogen pressures help to convert these poisons and hence maintain the catalyst in a more active state. The active hydrogenation function also serves a related purpose [58]: it prevents coking up of acidic sites by timely hydrogenation of coke precursors. Coke is formed either from precursors in the feed or from precursors generated through condensation reactions as part of the hydrocracking process itself [59–61] and may affect the hydrogenation and the cracking function differently [62]. Again the activity of the hydrogenation function is crucial, this time not for selectivity but for stability reasons.

Generally, the hydrocracking catalysts employ mixed sulfides or Group VIII metals as hydrogenation function, while the acidic function is provided by (fluorided) alumina, amorphous silica-alumina or zeolites. In this chapter we will first discuss the hydrogenation function and then the acidic function.

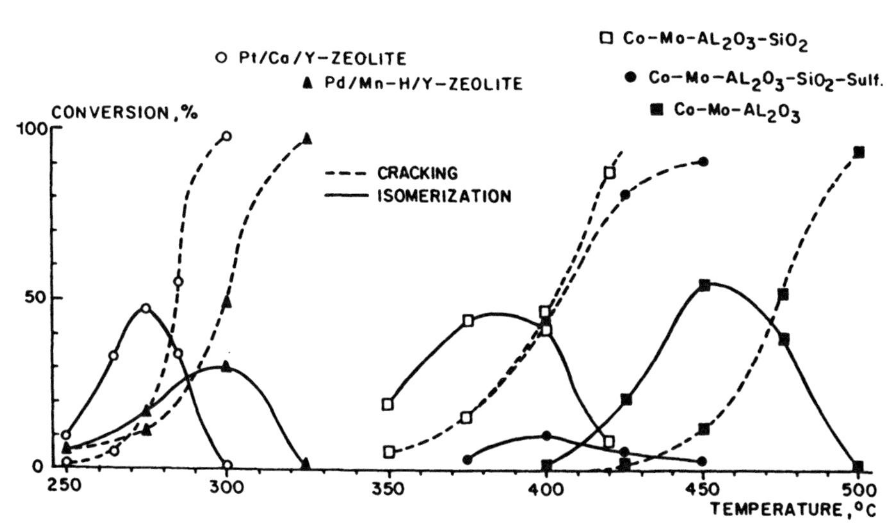

Fig. 7.30. Hydrocracking and hydroisomerization conversion. Dependence on reaction temperature for different bifunctional catalysts [50, 51]

7.3.4.2 Hydrogenation Function

Mixed Sulphides. The Ni/Mo and Ni/W mixed-sulfide hydrogenation functions are generally the most effective for denitrogenation and aromatics hydrogenation in a sulfur-rich environment such as the one that exists at the early stages of the hydrocracking unit; Co/Mo is more selective to desulfurization. So far alumina is the most suitable carrier for these systems; addition of phosphorus promotes the HDN activity. The Ni/W system offers a higher hydrogenation activity than the Ni/Mo system, but only at low sulfur levels. As a result, the pretreat catalysts usually consist of Ni/Mo/P/alumina, and for reaction temperatures below some 380 °C, in the presence of sulfur, no systems with higher HDN activity have been found.

Although alumina appears the best dispersing agent for the mixed sulfides, it offers of course only a moderate cracking activity, and the activity of mixed sulfides dispersed on more acidic materials such as amorphous silica-alumina and zeolites is of interest. Zeolites of course are particularly poor dispersing agents for mixed sulfides since the large anionic precursor species (e.g., $Mo_7O_{24}^{6-}$) do not enter the zeolite crystals during preparation. In commercial zeolitic catalysts the extrudates often contain a binder such as alumina, apart from the zeolite, however, and though the mixed sulfides should be primarily on the alumina, the (indirect?) effect of the zeolite on the hydrogenation activity can still be considered.

Several studies have now been published that show that such mixed sulfides systems on an acidic support (zeolitic catalysts or amorphous silica-alumina) do indeed give a lower activity than when they are supported on alumina, unless the operating temperature is quite high, e.g., about 400 °C [49, 63–65]. Obviously,

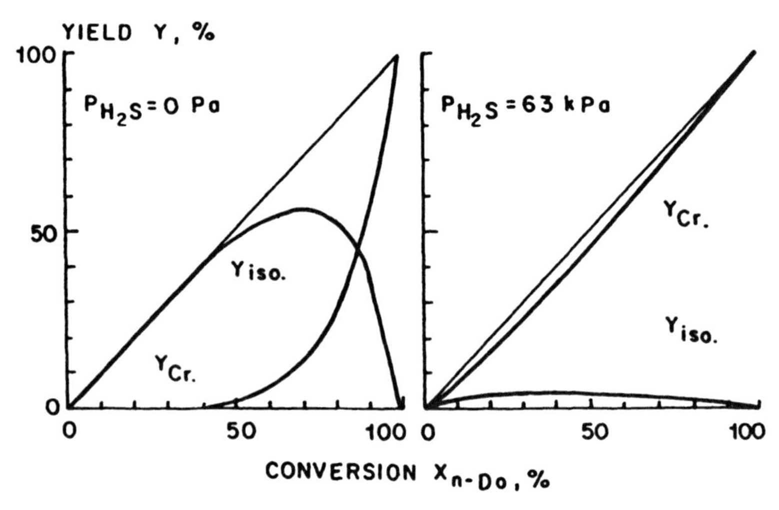

Fig. 7.31. Hydroconversion of n-dodecane on Pd/LaY. Yields of iso-dodecanes and hydrocracked products in the absence and presence of hydrogen sulfide [67]

these systems exhibit a higher activation energy; this has been ascribed to a stronger adsorption energy of the nitrogen compounds to these acid systems, but this has not been fully explained yet.

(Noble) Metals. In the presence of only low concentrations of sulfur, Group VIII metals are of course very powerful hydrogenation catalysts, and therefore metals such as Ni and Pd provide an excellent hydrogenation function for second-stage hydrocracking catalysts that operate under two-stage (low hydrogen sulfide and ammonia conditions; pre-treated feedstock) conditions. Catalysts promoted with noble metals instead of mixed sulfides generally give a higher activity, while the products are of course also very well hydrogenated. Noble metals can be placed inside the pores of the zeolites, close to the acidic sites, in contrast to the mixed sulfides which are generally placed on the catalyst binder [66]. These differences, however, are enormously reduced if the hydrogen sulfide concentration increases, and hence the activity of the metal function is suppressed. The effect of hydrogen sulfide on dodecane hydrocracking over Pd/Y sieve catalysts was studied by Weitkamp et al. [67]; in the presence of hydrogen sulfide the activity of the (sulfide) Pd function is quite insufficient to achieve "ideal" hydrocracking, as shown in Fig. 7.31. Of course Ni catalysts are generally even more sensitive to sulfur. The desirability of low hydrogen sulfide pressure obviously adds to the operational costs when such catalysts are used.

7.3.4.3
Acidic Function

The main acidic functions applied in hydrocracking catalysts are alumina (possibly fluorinated [68]), amorphous mixed oxides such as amorphous silica-alumina, silica-magnesia, etc., and zeolites.

Obviously, alumina is the least active cracking function of these, even after fluoridation (see also [68]). Since alumina is also an excellent support for the hydrotreating function, these catalysts do have merits in hydrocracking of, e.g., nitrogen-rich feeds, such as in first-stage or mild hydrocracking.

Amorphous silica-alumina (ASA) has been a well-known base material since the early days of fluid catalytic cracking. These materials can be prepared in many ways, and also over a range of alumina contents; they usually have a rather high surface area. Many materials can be purchased commercially. The low-alumina grade, with 13% alumina, is the most acidic. Increasing the alumina content to 25% (high alumina), or beyond leads to a material with a lower cracking activity, as was shown by Ward [69]. Since with varying alumina content the dispersing power of the hydrogenation function also changes, it is not easy to define from first principles the optimum ASA. Often ASAs are considered to have a rather small number of acid sites of high strength; this makes them less active than zeolites, and also more prone to poisoning by, e.g., nitrogen compounds. Altogether the field of ASA-based hydrocracking catalysts is quite complex and is somewhat outside the scope of this book.

Clays, too, have been used in hydrocracking catalysts, e.g., pillared clays have attracted much interest, without so far emerging as commercial catalysts [70]. Given the molecular sieve effect with zeolites (see below), with a drive to heavier feedstocks, interest in clay-containing catalysts will certainly increase further.

Zeolites of course exhibit by far the highest cracking activity [71], which is very attractive. For application in hydrocracking catalysts a high thermal and

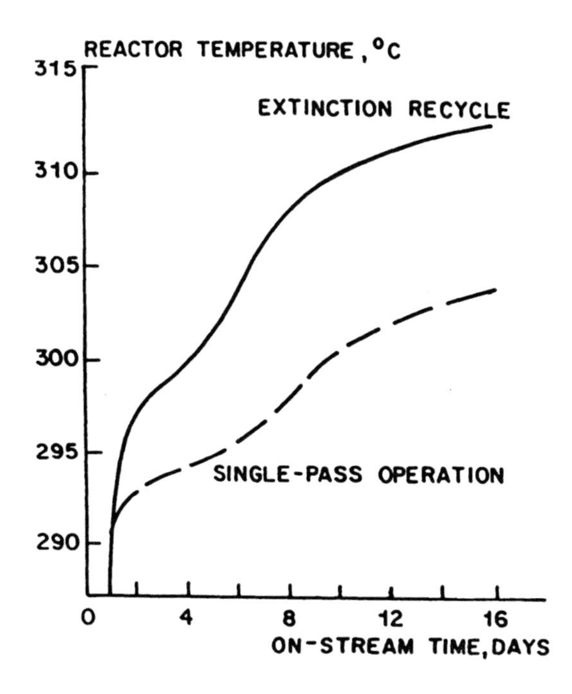

Fig. 7.32. Reactor temperatur required for 60% conversion vs. on-stream time [72]

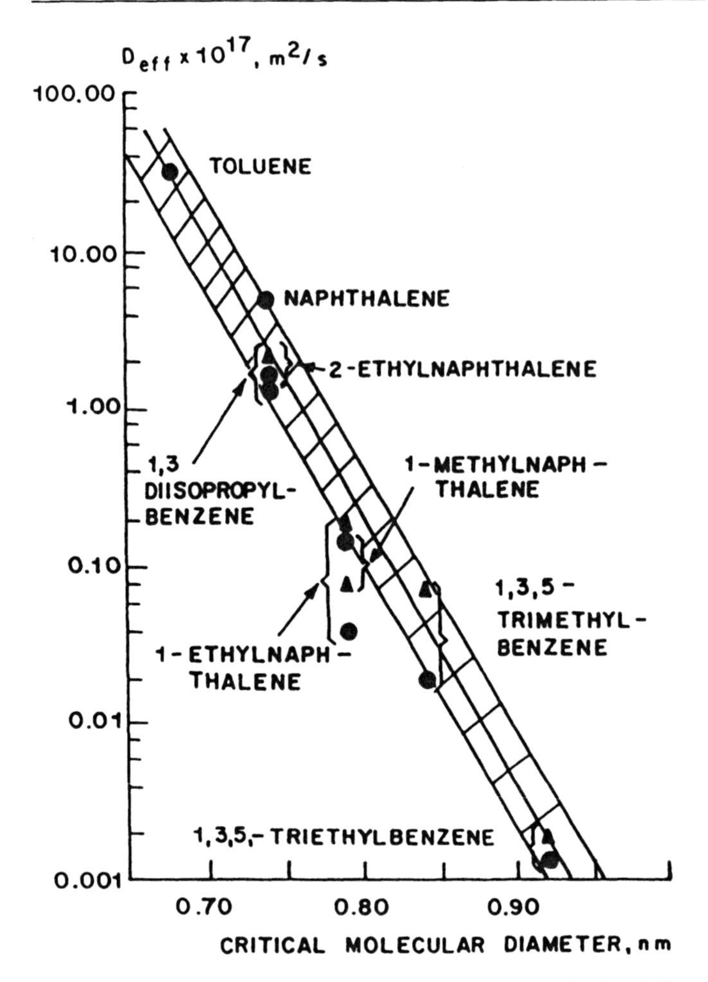

Fig. 7.33. Effective diffusivity in zeolite Y as a function of critical diameter for a variety of alkyl-substituted benzene and naphthalene molecules [74]

hydrothermal activity is required, while in addition the zeolite pore system must be sufficiently wide to allow access to the reactant molecules.

With regard to the zeolite structure, this means that wide-pore zeolites with 12-(Si, Al atoms) membered pore openings are preferred for hydrocracking of, e.g., vacuum gas oils. Even with zeolite Y, with its open structure and three-dimensional channel system, the largest molecules are converted much less effectively, easily resulting in a disproportionate build-up of a heavy fraction in recycle operation, and in a lower overall activity and selectivity in a recycle operation to total extinction. This has been clearly shown [72, 73] and is illustrated in Fig. 7.32. Indeed one can easily show that the larger molecules diffuse quite slowly into the zeolite (Fig. 7.33, from [74]), and one can also model this build-up phenomenon. Moreover, large polynuclear aromatics enhance both the catalyst

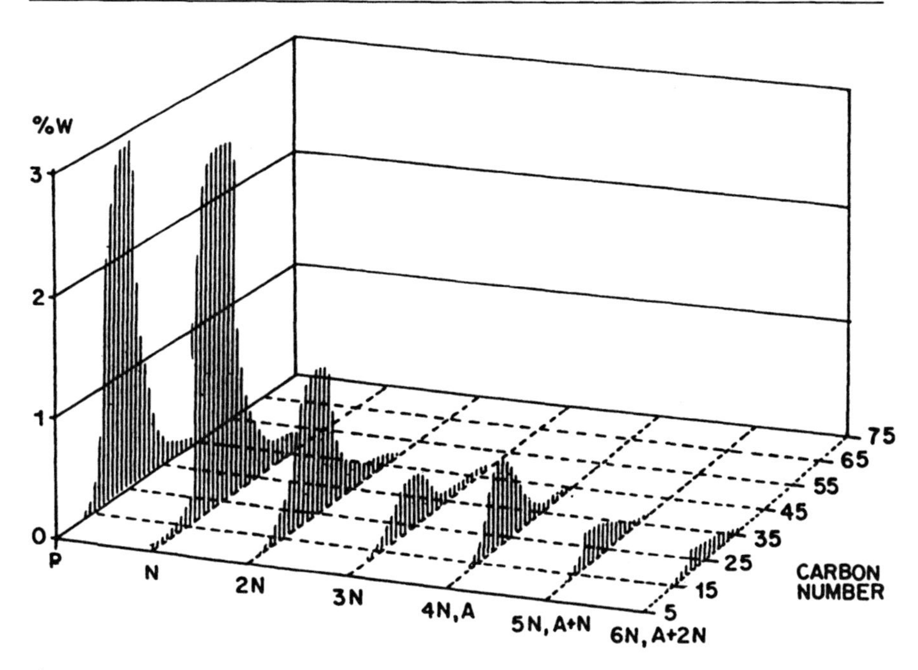

Fig. 7.34. FIMS characterization of unconverted feed in product after processing treated vacuum gas oil over an amorphous hydrocracking catalyst (two-stage operation) [42]

deactivation and the fouling of downstream heat exchangers leading to severe reduction of the heat exchanger heat transfer coefficients, eventually shortening the hydrocracker cycle [59]; this phenomenon is known as the "red death". The size exclusion effect also emerges from a field ion mass spectrometric (FIMS) analysis of the unconverted fractions obtained over a zeolitic and an amorphous catalyst. As Figs. 7.34 and 7.35 show, the amorphous catalyst is clearly superior in the conversion of multi-ring naphthenes [42]. It is consistent with this that zeolites with smaller pores than faujasite, such as mordenite, ZSM-5 and erionite, are used as a rule not in high conversion hydrocracking, but mainly in applications in which the less branched molecules are to be selectivity converted, such as catalytic dewaxing and Selectoforming (see below). In high conversion of vacuum gas oils these zeolites are less selective, overcracking the lighter components and giving high gas yields. By the same token, the newer very wide pore systems, such as MCM-41 [75], VPI-5 and cloverite [76], are extremely interesting, but have not yet led to commercial catalysts. The wide pore zeolites with 14-(Si, Al atoms) membered pore openings also hold some promise, but have only now been reported for oxidation catalysis [77]. Such materials would add scope to the conversion of heavier oil fractions, with larger molecules, over molecular sieves. Another option is the use of catalyst systems that contain both zeolites and ASA as acidic components, the ASA serving to convert the species the zeolite cannot cope with. Optimization of the synergy between both acidic cracking components in one catalyst is obviously of prime performance. An

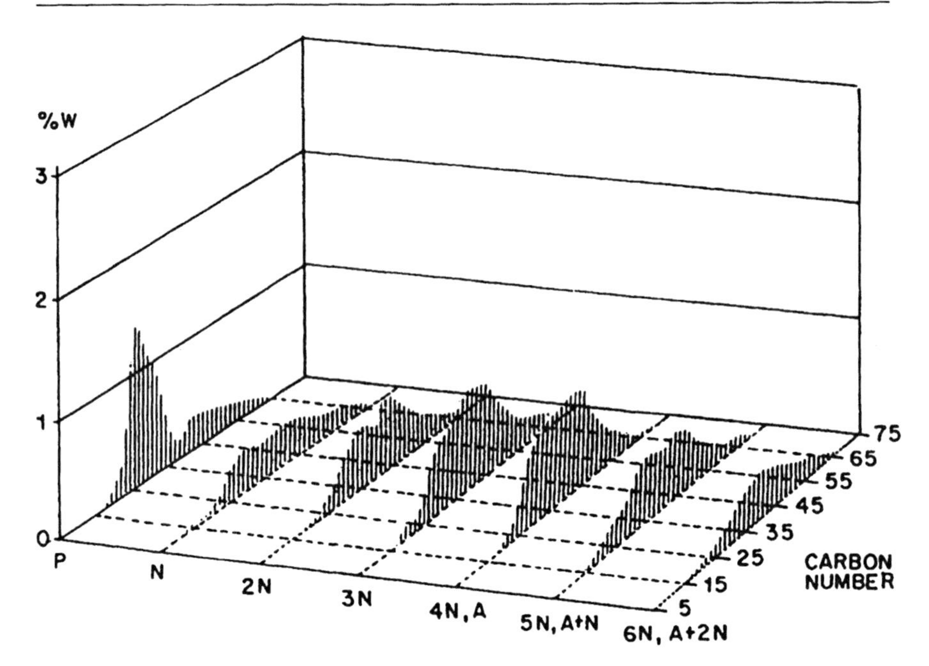

Fig. 7.35. FIMS characterization of unconverted feed in product after processing treated vacuum gas oil over a zeolitic hydrocracking catalyst (two-stage operation) [42]

illustration of this concept is given [78], while others have demonstrated [42, 79] that such catalysts can be designed with different performances.

Within a given structure the site density can often still be varied through changes in synthetic conditions or by subsequent modification (ion exchange or dealumination). The aluminum content of the framework in faujasite can be varied over a large range and can be conveniently monitored by the unit cell size. At high aluminum contents the acid sites are so close that their intrinsic acidity is reduced; at lower values, however, the reduction of the aluminum content just leads to a reduction in the number of acid sites without changing the intrinsic activity. The performance of hydrocracking catalysts with Y sieves stabilized to different degrees has been studied for heavy feedstocks [46]. Surprisingly, it was found that the lower aluminum content did not lead to a reduced activity, but that selectivity to middle distillates increased markedly. At the same time, the lower unit cell size catalyst was more stable to coke deactivation. This may be part of the explanation for the unchanged activity upon reduction in acid site density: with the low unit cell size sieve a smaller fraction of sites is deactivated by coke, both as a result of the decrease of the acid site density and an increase of the hydrogenation over cracking activity. In evaluating these effects one must stress, however, that the change in unit cell size is accompanied by other changes, e.g., in mesopore volume, extra-lattice aluminum, hydrophobicity, etc.

7.3.4.4
Hydrocracking Catalysts

The various aspects can now be integrated into a discussion of the design of hydrocracking catalysts for optimal performance. Again we are faced with the balance of the two functions in the hydrocracking catalyst under the working conditions, i. e., passivated or poisoned by the species present such as hydrogen sulfide (in particular affecting (noble) metal activity, but also acid activity [67]), ammonia (passivating acid sites), and organic nitrogen compounds and polyaromatics (both primarily poisoning acidity) [59]. A number of illustrative examples have been given by Ward [69]. He reports an example of two zeolite catalysts with different amounts of zeolite which have equal activity for second-stage hydrocracking in the absence of ammonia (hydrogenation function limited); in the presence of 2000 ppm of ammonia the catalyst with the higher zeolite content is 10 °C more active than the other. In the presence of ammonia the catalyst is therefore acidity limited. Another example quoted deals with zeolite catalysts with two levels of hydrogenation: in the presence of ammonia there is not much influence (acidity limited), but in the absence of ammonia the catalyst with the higher hydrogenation activity is again some 10 °C more active (hydrogenation limited). Thus the Pd content of a catalyst will mainly play a role in the two-stage operation. In this context [80] is also of interest.

So far the arguments have concentrated on activity. In selectivity, however, large effects can also be observed. Thus the middle distillate selectivity of a catalyst is much higher in the presence than in the absence of ammonia [81]. This can be generalized following the concepts of "ideal hydrocracking": for a high middle distillate selectivity the catalyst must have a high hydrogenation activity relative to the cracking activity, so that intermediate species can be hydrogenated and desorb quickly before further cracking. This can be achieved either by using amorphous catalysts, or by using zeolite catalysts in series flow, with ammonia moderating the zeolite activity. For naphtha production of course the reverse holds. Indeed these arguments are nicely illustrated by plotting the middle distillate selectivity for a number of catalysts operated under different ammonia pressures against the temperature requirement for a desired conversion (activity). In general a plot such as Fig. 7.36 [81] is found, showing that middle distillate selectivity increases with decreasing activity, though of course this is not a unique relationship, and a well-designed catalyst can have a higher selectivity at the same activity. Nevertheless such a plot seems to give a striking illustration of the principles outlined above. One should be aware, though, that these effects may be extra pronounced by evaporation phenomena: at high temperatures the middle distillate products will largely be in the gas phase, with the liquid phase enriched in the heavy feedstock molecules, while at low temperatures this occurs to a much lesser extent.

Another aspect of the hydrogenation function is of course the hydrogenation of the product itself. Here one observes that a catalyst with a more effective hydrogenation function gives many more hydrogenated products. Examples are noble metal catalysts in two-stage operations where the presence of hydrogen sulfide can help the quality of the gasoline components produced, or also

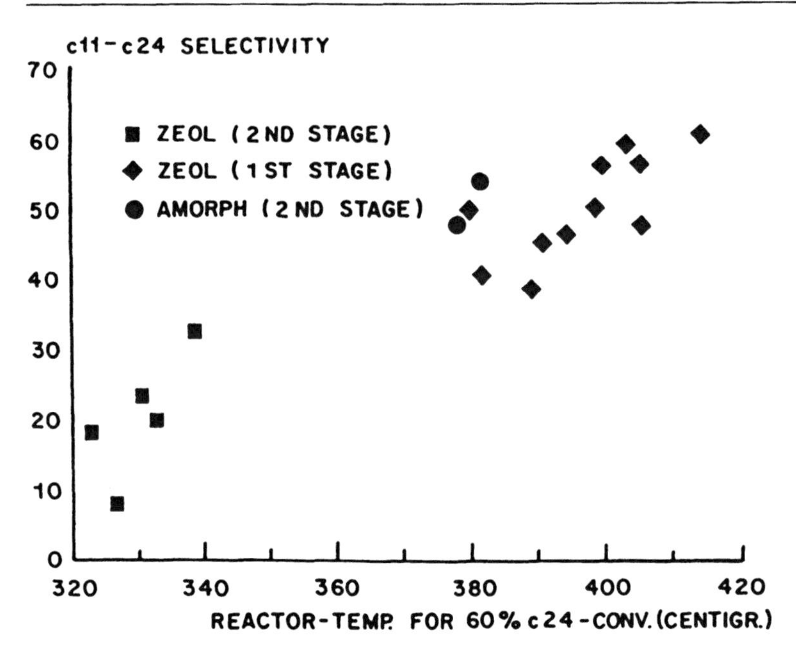

Fig. 7.36. Middle distillate selectivity vs. activity [81]

mixed-sulfide catalysts where a better hydrogenation efficiency leads to lower aromatics contents in the products. These aspects can be valuable not only from a product quality point of view (e.g., for hydrowax, [82]) but also for retention of catalytic activity.

Finally, it is of interest to compare zeolite catalysts with amorphous catalysts. The big advantage of zeolites is their high activity and nitrogen resistance, at the expense, however, of selectivity. Although amorphous silica-alumina has the reputation of being unable to tolerate high amounts of ammonia and nitrogen this is not completely true, as their application in (mild) hydrocracking demonstrates. Amorphous catalysts can be applied in the presence of nitrogen compounds, but only at high temperatures, to reduce the adsorption of the poisons. As shown above, at these conditions such catalysts can be highly selective towards middle distillate.

7.3.5 Challenges in Hydrocracking

The challenge in hydrocracking is to produce in the most cost-efficient way high quality transportation fuels. This implies that the hydrocracker feedstock slate needs to be as cheap and flexible as possible, the hydrocracking process should have low investment and operational costs and the product slate should be of high flexibility with high quality naphtha, kerosene and gas oil as prime products; gas production should be minimized and production of hydrowax only as outlets as feedstocks for either base oil, ethylene cracker and catalytic cracker plants.

Cheaper feedstocks are generally heavier and less reactive. They often contain higher levels of polyaromatics and combined with their lower reactivities requiring higher operation temperatures, this results in higher tendencies to both catalyst deactivation by coke formation and fouling of the downstream heat exchangers by, amongst others, polycyclic aromatics, as a result of higher levels of coke precursors in the feed and a less attractive naphthenes/aromatics equilibrium. Both catalyst deactivation and fouling decrease the hydrocracker cycle length. The result of catalyst deactivation is not only a lower activity, which is compensated by increasing operation temperature (until the design end of run operating temperature is reached), but also gradual changes in product yields and properties [59]. Future hydrocracking catalysts should be even more resistant to these deactivation phenomena than the current ones. The high activity of the zeolites is in this respect advantageous, but size exclusion effects clearly form a disadvantage. Dedicated composite catalysts (zeolites combined with amorphous cracking components) and highly active molecular sieves with larger pore diameters might also be a way forward.

By simplifying the hydrocracking process, e.g., large-scale single-stage (once-through) hydrocrackers and improved versions based on alternative reactor concepts [7, 83–90], capital investment can be decreased. Dedicated highly active (e.g., zeolite-based) catalysts are required for these modern hydrocrackers. In addition to improved catalysts, increasing operation pressure also suppresses catalyst deactivation and fouling, but this makes the process significantly more expensive. Close integration of the hydrocracker with other refinery processes, like catalytic cracking and residue hydroconversions, also contributes to the overall refinery efficiency.

The hydrocracking process by nature (cracking plus hydrogen addition) is excellently suited for the production of high quality middle distillates and feedstocks for other conversion units (catalytic cracking, ethylene cracking, base oil manufacturing). Further improvement in the properties of these products might be accomplished by further increasing the hydrogenating over the cracking power of the catalysts or dedicated multiple catalyst combinations. Producing at the same time better naphtha is quite a challenge. Options might be an increase of the level of isomerization in the naphtha range (dedicated catalysts or multiple catalyst combinations) and increasing the level of olefins by subsequent processes.

7.4 Catalytic Dewaxing

7.4.1 Introduction

In "normal" hydrocracking the aim is, in general, total conversion of the feedstock, but in a number of applications the aim is only selective conversion of part of the feedstock, i.e., a selective group of feed molecules. The latter applies mostly to the conversion of normal paraffins, either by selective cracking or by isomerization. Long-chain linear and slightly branched paraffins have the highest melting points of any class of hydrocarbons of equivalent molecular weight. Low concentrations of these paraffins adversely affect the cold flow properties of the oil fractions, e.g., diesel oil and lubricating base oil, by crystallization of wax [91]. Lubricating base oils are manufactured from straight-run oil fractions in the vacuum gas oil range by first removing aromatics, by extraction or by hydroprocessing, and then removing normal (and slightly branched) paraffins (to avoid the wax crystallization). An ideal lube oil molecule is an only slightly branched paraffin (or even one containing an aromatic ring) to arrive at a low temperature dependency of the viscosity (i.e., a high viscosity index) [41]. In the light of the increasing demand for high quality base oil stocks, production from the hydrowax, bottoms of the hydrocracker, is being considered as an advantageous alternative for a full conversion hydrocracker for the production of transportation fuels [92].

The dewaxing step can be carried out by several routes [92–94]: By crystallization from a solvent, producing wax as a by-product; by selective catalytic (hydro)cracking the wax molecules, producing lighter components such as naphtha and gas; by isomerization of wax molecules, not affecting the boiling range. Catalytic dewaxing can also be applied to atmospheric gas oils, where the reduced pour point that is achieved allows an extension of the boiling range of the oil fraction blended into on-spec. gas oil, and so increasing the gas oil yield. Also in the naphtha range selective conversion of normal paraffins by isomerization (e.g., Hysomer) or cracking (Selectoforming, etc.) can be desirable for reasons of gasoline octane quality, but these processes are discussed later.

To illustrate the importance of dewaxing technology, the global base oil production capacity up to 1997 is close to one million barrels per day [95] which corresponds to about 50 million tons per year. The hydrodewaxing process uses trickle flow operation, under hydrogen as in normal hydrocracking; the technology is dominated by Mobil [96]. The Mobil technology is centered around ZSM-5, e.g., their lube oil and gas oil dewaxing processes, and protected by an extensive patent portfolio. They have also recently introduced processes with higher selectivity to isomerization with full integration with hydrotreating processes [97]. Chevron [98, 99] also has dewaxing technology with, as a new development, the commercialization in 1993 of an "Isodewaxing" process presumably based on molecular sieves such as SAPO-11 [41, 100]. AKZO and Fina have jointly developed a gas oil cold flow improvement process [101, 102]. BP has been active in this area, with a mordenite-based catalyst [103]. The next process is the isomerization of the wax fraction produced by solvent dewaxing to very high viscosity index base oils [93, 95].

7.4.2
Principles of Catalytic Dewaxing

The basic principle of catalytic dewaxing is shape-selective conversion of wax which is nicely illustrated by Fig. 7.37 [104], in which the chromatogram clearly indicates the selective conversion of the normal paraffins, either by cracking or by isomerization. Indeed, Fig. 7.38 [105] clearly demonstrates that ZSM-5 cracks

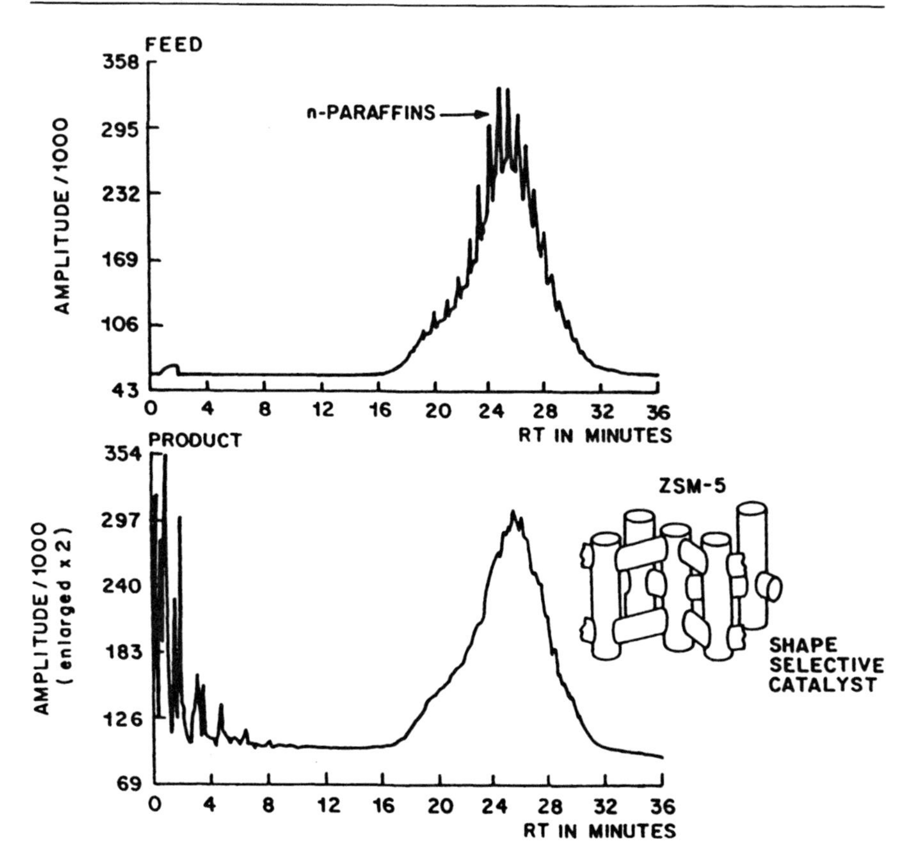

Fig. 7.37. Shape-selective catalytic dewaxing [104]

normal paraffins faster than monobranched ones, which in turn are cracked much faster than dibranched ones. The principles of shape selectivity are discussed elsewhere in this book, and aspects such as reactant shape selectivity and transition state shape selectivity need not be dealt with here.

Economically, in dewaxing, several aspects play a role in selectivity:

- selectivity in conversion of normal vs. branched paraffins,
- selectivity in cracking vs. isomerization,
- selectivity in product pattern of cracking: e.g., naphtha vs., e.g., propane.

In principle the zeolite should convert normal and monobranched molecules, but not the more branched ones (this would only result in yield loss, if conversion proceeds by cracking). In the selective removal of n-paraffins from gasoline-range hydrocarbons (as in the Selectoforming process of Mobil) this can be achieved by an 8-ring zeolite such as erionite. However, for heavier feedstocks such as gas oils and lubricating basestocks, the wax constituents are not necessarily purely normal only. Therefore the 8-ring zeolites such as erionite are not suitable for the larger molecules involved in dewaxing [106], whereas 12-ring

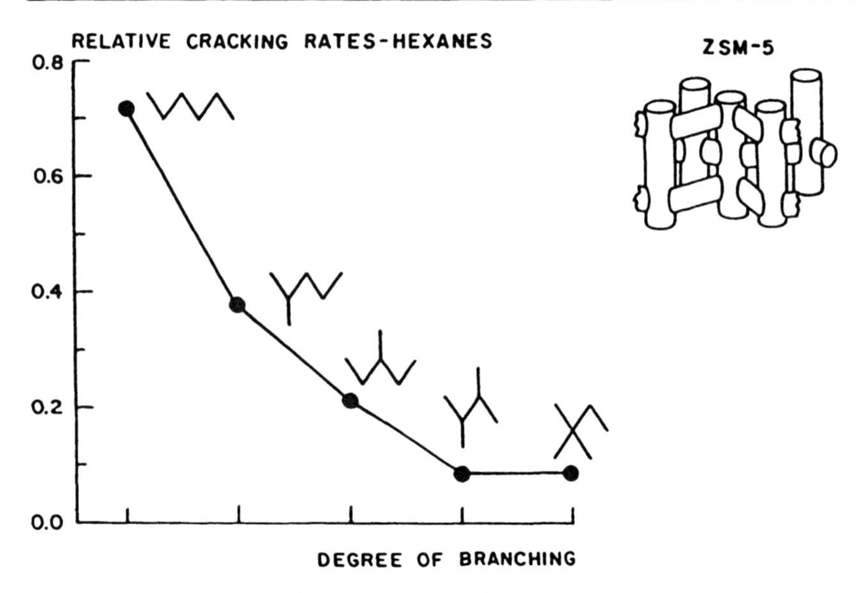

Fig. 7.38. Shape-selective catalytic dewaxing [105]

zeolites such as mordenite have pores that ore too wide (see below). Another advantage of the 10-ring zeolites (over the 12-ring ones) is their low coking tendency, which reduces the need for a hydrogenation function such as Pt or Pd [107]. Within the 10-ring zeolites the precise channel structure of the zeolite catalyst is crucial for its performance. This is shown in Tables 7.5 and 7.6 [108], comparing the performance of mordenite, ZSM-5 and ZSM-23. For the feedstock and the conditions studied, the narrow-pore ZSM-23 is clearly the most selective [108], while the yield loss with mordenite is clear. Further examples are available [3].

Table 7.5. Dewaxing with ZSM-5 vs. Mordenite [108]

Catalyst	Feed	ZSM-5	ZSM-5	Pt/mordenite	Pt/mordenite
Conversion, wt%	0	20	23	27	36
Product pour point, °C	+35	-12	-26	-12	-26
Viscosity index	-	94.3	89.4	89.6	77.6

Table 7.6. Dewaxing over ZSM-5 and ZSM-23 [108]

	ZSM-5	ZSM-23
Pore size (nm)	0.54 × 0.56	0.45 × 0.56
Conversion, wt%	15	11
Product pour point °C	-12	-12
Viscosity index	101.0	108.7

ZSM-5 based catalysts, applying reactant shape selectivity, are excellent dewaxing catalysts, but their drawback is that the paraffins are cracked, instead of isomerized, producing light paraffins. For gas oil dewaxing upgrading of the selectivity to naphtha range light paraffins (instead of gases) as dewaxing by-product can be a solution, but for lube oil dewaxing decrease of the high value lube oil product by cracking is economically very unattractive [41]. Chevron has disclosed selection criteria for structures of molecular sieves for optimal isomerization performance: The pore diameter should be less than or equal to 7.1 Å, while at least one pore dimension should not be lower than 4.8 Å; the crystallite size should be lower than or equal to 0.5 µm; acidity should be sufficiently high to convert at least 50% of n-hexadecane at specific conditions, with 40% isomerization selectivity at 96% n-hexadecane conversions, applying a group VIII metal as hydrogenation function. The minimum pore size is to enable methyl branching. By applying small crystallites the escape of the initially formed isomers from the zeolite is ensured. The crystallite size is optimized with respect to the number of active sites in the zeolites, whereby the length of the crystallites in the direction of the pores is important. The most preferred sieves are those with 10-membered rings, although 12-membered rings can also be used as long as the pores are non-circular, to meet the cross dimension criterion of ≤7.1 Å. The molecular difference between dewaxing by cracking and dewaxing by isomerization can be nicely illustrated for a large paraffin like normal-uneicosane (n-C_{21}) [94]. Conversion of this molecule by conventional dewaxing (cracking) results in the formation of propane (2 C_3), normal-butane (n-C_4) and normal-undecane (n-C_{11}), while isodewaxing (isomerization) yields 3,8,14-trimethyl-octadecane (i-C_{21}).

Obviously, the most preferred situation with respect to wax removal while suppressing yield losses, is one of selective isomerization of the wax molecules only, while selective cracking to naphtha products is the second best option. The Mobil and Fina processes appear quite successful in cracking to naphtha rather than gas, but isomerization rather than cracking of course remains preferable. For the latter option zeolites with somewhat wider pores than ZSM-5 are attractive, and Mobil has indeed filed a number of patents using zeolite beta. Chevron has introduced increased isomerization using catalysts comprising molecular sieves with optimal structures for isomerization, like, e.g., SAPO-11, ZSM-22 and ZSM-23 [109, 110]. These catalysts combine a high isomerization activity with a low cracking activity, by effectively suppressing the formation of multi-branched molecules. These latter molecules could be readily cracked to light paraffins [41] Judging from results [111] reporting improved selectivities for boron and iron silicates (MFI structure) relative to (Al)-ZSM-5, this is presumably due to the lower acidity of the sites. Combinations of isomerization and subsequent cracking ("cascade dewaxing") have also been described.

The feedstocks to be dewaxed can vary significantly in heteroatom content. Lubricating base oils are often reasonably clean (particularly after hydroprocessing to remove aromatics), but gas oils are usually straight-run with considerable amounts of sulfur and nitrogen. Generally, the heteroatom content is important for catalyst life: it appears that whereas with "clean" feedstocks life can be a year or longer, depending on catalyst characteristics, feedstock and conditions, with dirty feedstocks a cycle between hot hydrogen strips frequently lasts a month, and between oxidative regenerations six months [96]. Often catalytic dewaxing is combined with a subsequent hydrotreating step. With clean lubricating base oil feedstocks this is a very mild step to saturate olefins; with dirty straight-run feedstocks, such as gas oils, a more severe step is needed to achieve the desired sulfur specification or also to bring about additional cracking (e.g., [101]).

7.5 Upgrading of Naphtha and Tops

Apart from feedstock for the production of petrochemicals, naphtha and tops are mainly used as gasoline components, together with the products with the same boiling range obtained from cracking of heavier fractions (ex-fluid catalytic cracking or thermal cracking), and from condensation of lighter molecules (olefins, etc.). In this section the use of zeolites in upgrading straight-run or cracked naphtha, and in the production of naphtha components from light gases, is considered.

7.5.1 Cracking of Normal Paraffins

In naphtha the normal paraffins are less desirable components, because of their unfavorable ignition properties (low octane number), as shown in Table 7.7 [3, 112]. Several processes have therefore been developed to increase the octane number of naphtha for meeting product specifications (see Table 7.1). The most prominent is of course catalytic reforming, where aromatics are formed from paraffins through dehydrocyclization, but where also a significant fraction of the paraffins (contrary to the naphthenes) is cracked. Mobil has developed processes in which the reformer operates at lower severity, and selective cracking is carried out over zeolitic catalysts [113]. The oldest process, Selectoforming, uses

Table 7.7. Paraffin isomerization: Research Octane Numbers of C_5 and C_6 Paraffins [3, 112]

	RON-0
Pentanes	
n-Pentane	62
2-Methylbutane	93
2,2-Dimethylpropane	83
Hexanes	
n-Hexane	29
2-Methylpentane	78
3-Methylpentane	76
2,2-Dimethylbutane	92
2,3-Dimethylbutane	104

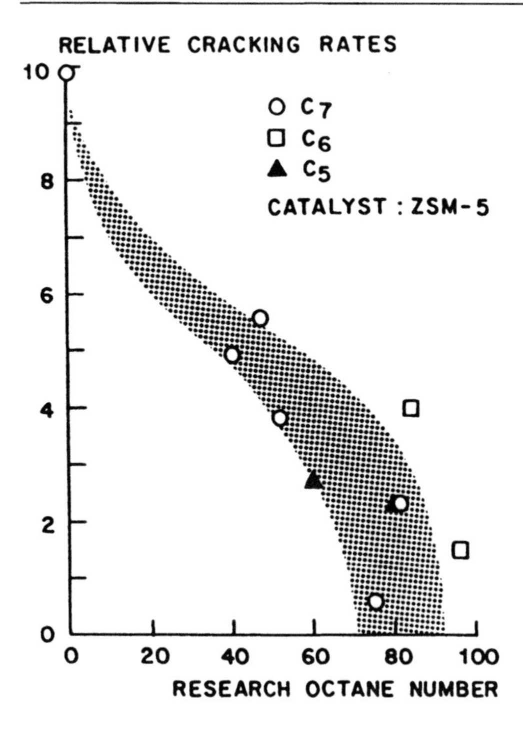

Fig. 7.39. Octane rating vs. cracking rate of paraffins [120, 121]

the 8-ring zeolite erionite promoted with a weak hydrogenation function; it is operated under hydrogen, and selectively cracks normal paraffins to propane [114–117]. An improvement to this process is the so-called M-Forming process developed in the 1970s at Mobil [118, 119]. Here the catalyst is based on ZSM-5; it selectively cracks the lower octane components, as can be seen from Fig. 7.39 [113, 120, 121], where the relative cracking rates for various C_5–C_7 paraffins are plotted vs. their octane numbers. In addition, the catalyst achieves aromatics alkylation by the light gases from cracking, thereby increasing the liquid yields. The process is therefore optimal in combination with catalytic reforming, and the optimum severity of the reformer in this combination depends on factors such as feedstock composition [122]. Compared to conventional reformates at the same octane level, the M-Forming products are low in aromatics and high in isoparaffins [113]. To some extent, however, both Selectoforming and M-Forming suffer from the yield loss that is a consequence of cracking undesired species, so isomerization to produce high-octane components would be preferable. Another process is ZEOFORMING, based on a ZSM-5 catalyzed upgrading process (presumably isomerization and aromatization) for low-octane feedstocks. It is suitable for gasoline production from light paraffinic hydrocarbons (e.g., gas condensate) and has been shown to be economic even at small capacity, making it an attractive gasoline production option (small scale) in areas with a less-developed logistic infrastructure. The process was developed by Ione at al. (Novosibirsk), and there is now a collaboration with KTI (Zoetermeer, The Netherlands).

7.5.2
Desulfurization of Naphtha ex Fluid Catalytic Cracking

Naphtha obtained from an FCC unit is an import component for the gasoline pool, both as to volume and octane quality. With the increasingly tight sulfur specifications for gasoline, however, this stream also has to have a low sulfur content, and achieving this in an economical way presents a challenge. The most straightforward route is direct hydrodesulfurization using well-known CoMo on alumina HDS catalysts, but unfortunately a deep desulfurization is then accompanied by a significant hydrogenation of olefins, resulting in a marked reduction in the octane number. An alternative would be the desulfurization of the feedstock of the FCC unit, but this requires high hydrogen pressures and hence a significant capital investment, and even then may not yield the desired low sulfur level [123]. For this reason processes have been commercialized in which the FCC gasoline is first desulfurized over a conventional CoMo-HDS catalyst, with the above-mentioned octane drop, but then directly in a second step converted over (presumably) ZSM-5-type catalysts, also containing a metal function such as Ni. In this way the normal paraffins and (remaining) olefins are cracked into light gases, and a higher octane stream is obtained, although at a loss in yield (see also Sect. 7.5.1). The overall yield might then be partially restored by processing these light gases in an alkylation step [124]. Processes of this nature as developed by Mobil [OCTGAIN] and Intevep/UOP [ISAL] have found commercial application, and are described in some detail [125, 126].

7.5.3
Isomerization of Light Paraffins

The isomerization of paraffins is an equilibrium reaction, acid-catalyzed, with the equilibrium at higher temperatures shifting to the lower octane components as shown in Fig. 7.40 [112]. The reaction is generally carried out in the presence of hydrogen, with a hydrogenation component (Pt) on the catalyst to achieve high catalyst stability. With the lead phase-out, and reduction of benzene in gasoline, paraffin isomerization has become very important for generating the high-octane components, and capacity has increased enormously in recent years. Thus in 1988 some 800,000 barrels per stream day (corresponding to about 40 million tons per year) world-wide licensed isomerization capacity was reported [127]; the growth in capacity of the Shell Hysomer isomerization process is shown in Fig. 7.41. A recent figure (1997) is 403,000 bpsd combined intake capacity (corresponding to ca. 20 million tons per year) [128].

The mechanism is somewhat similar to that of hydrocracking: again both functions of the catalyst are involved, the hydrogenation function controlling the olefin/paraffin equilibrium, and the acidic sites inducing the olefins to react by a carbenium ion mechanism (Fig. 42 [112]). In fact, there is strong evidence that the isomerization of the carbenium ion occurs via a protonated cyclopropane (PCP) intermediate [30, 112], while it is assumed that the cracking occurs via beta-scission. Recently this mechanism, in relation to that of hydrocracking of paraffins, was revisited by Sie [30], who suggested that also in

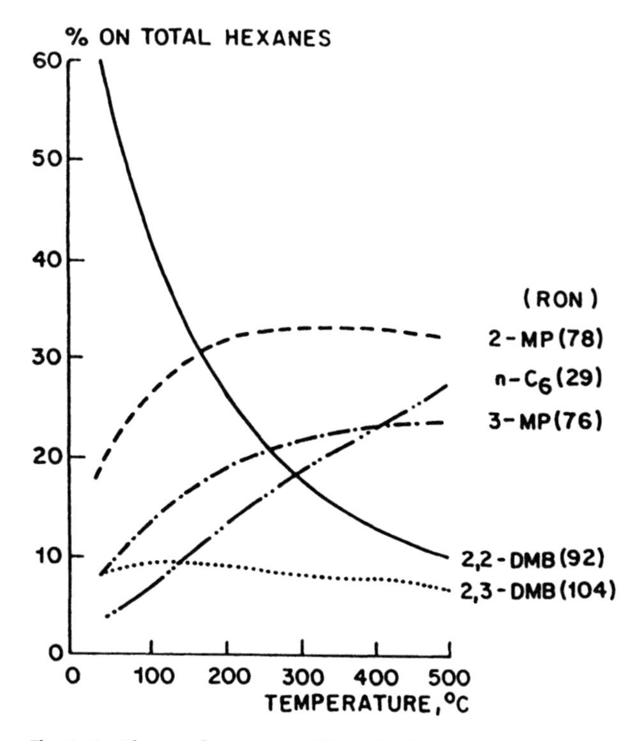

Fig. 7.40. Thermodynamic equilibria for hexane isomerization [112]

cracking the PCP intermediates are usually involved; isomerization due to the PCP ring opening is generally the faster reaction, however.

On this basis he then explains why the hydrocracking rate increases so strongly with carbon number from C_6 through C_7 to C_8. This dependence is very important when the isomerization of naphtha components (C_{7+}) instead of tops (C_5-C_6) is considered, because it implies that the selectivity to isomerization is much lower; this is also borne out by other results [30]. In general, of course, the reactivities of the paraffins increase with chain length, and this is the reason why co-processing of a wide range of paraffins gives such poor results: at the high temperatures required for the isomerization of the shorter molecules, the longer ones already crack, as shown in Fig. 7.43 [129]. The PCP mechanism cannot apply to the isomerization of butane (which indeed operates at clearly higher temperatures) and hence here one can either propose a monomolecular mechanism involving a primary carbenium ion (as suggested in [130]) which clearly is energetically an unfavorable intermediate, or, alternatively, assume that the reaction proceeds through a dimer, as proposed most recently by Sachtler and others [131, 132]. Interestingly enough, for C_7 isomerization also the importance of dimers has recently been identified and quantified by Blomsma et al. [133–135].

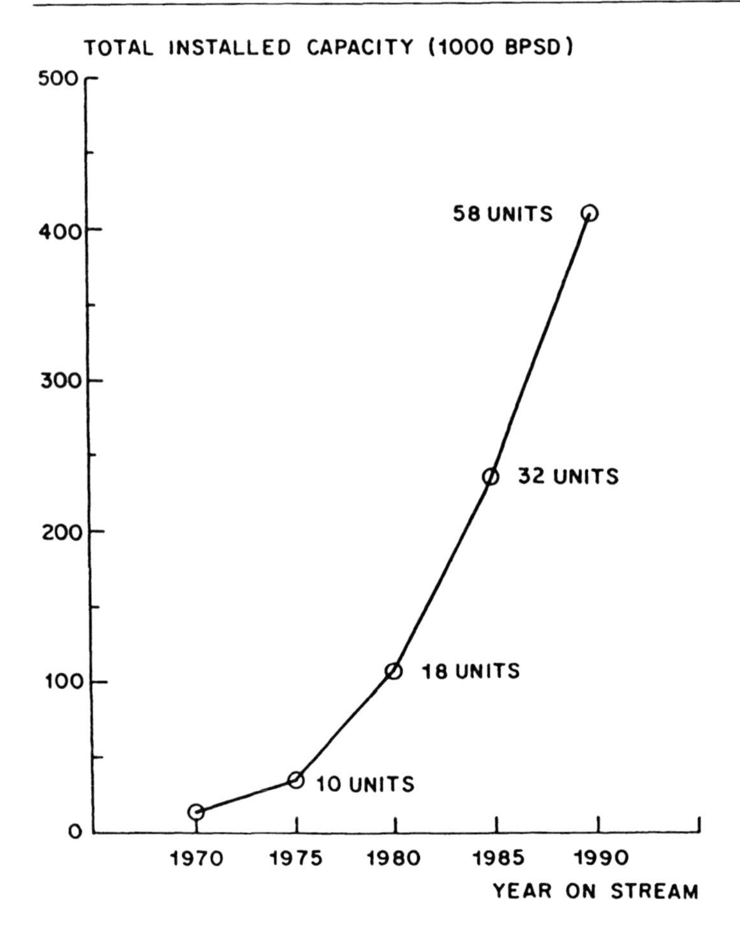

Fig. 7.41. Growth of total installed Hysomer capacity

7.5.3.1
Isomerization over Amorphous Catalysts

Paraffin isomerization over amorphous catalysts has been practised for a long time [127, 136], and at present the amorphous catalysts based on Pt/Cl alumina also remain the most active ones available. In the current UOP Penex process the I-8 catalyst is used which [136] can isomerize the C_5–C_6 stream at 115–150 °C (often two reactors are used in series, with the second one at the lower temperature). As outlined earlier, this low temperature is advantageous because of thermodynamics, and allows in a once-through operation an increase in Research Octane Number (RON) form about 69 to 83.

It might of course be attractive to recycle the normal paraffins, and this can be done by combining the Penex process with a UOP Molex separation step [136]. In this way the octane number can be boosted to 88–90. The Molex pro-

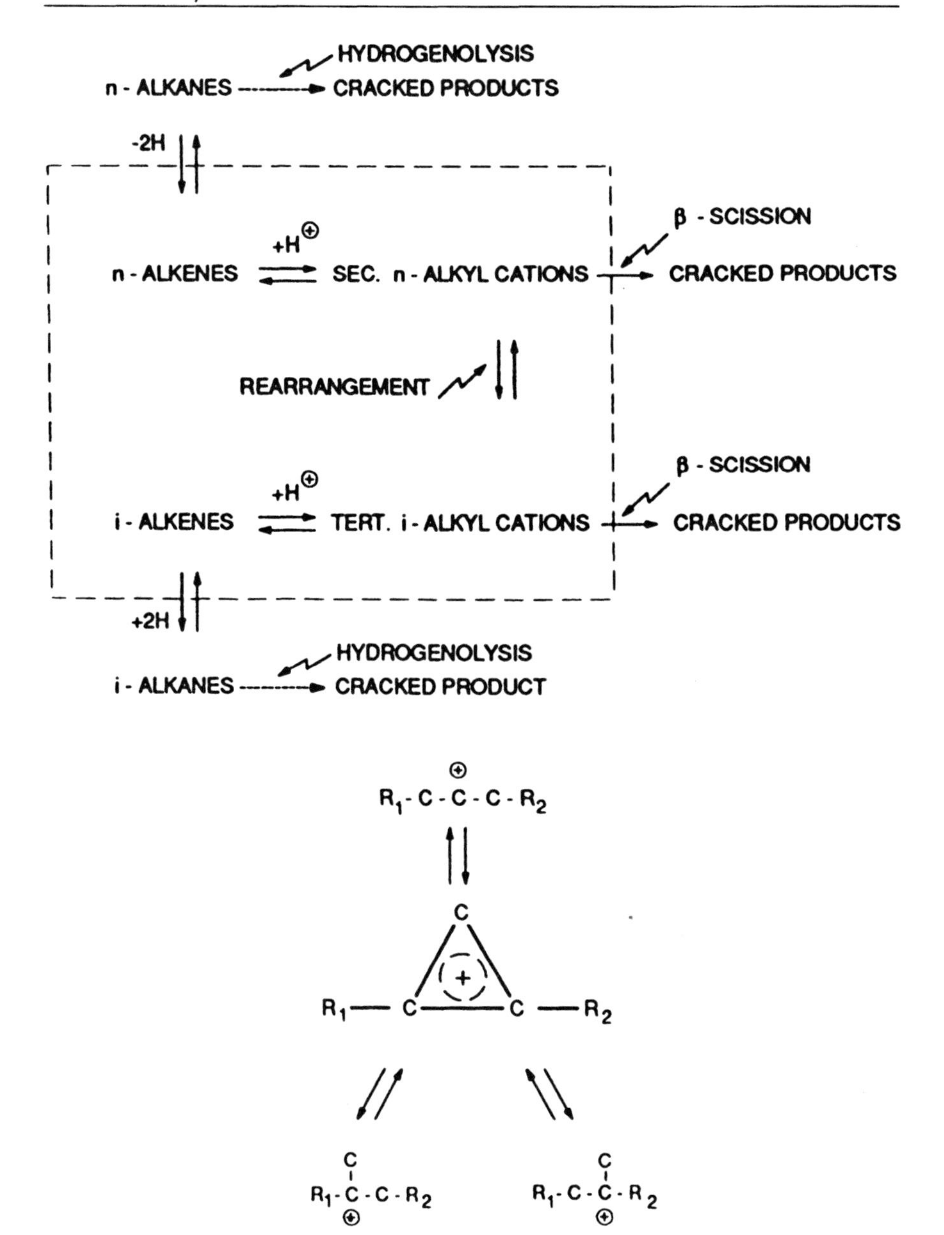

Fig. 7.42. Mechanism for n-alkane hydroisomerization, with protonated cyclopropane intermediate [112]

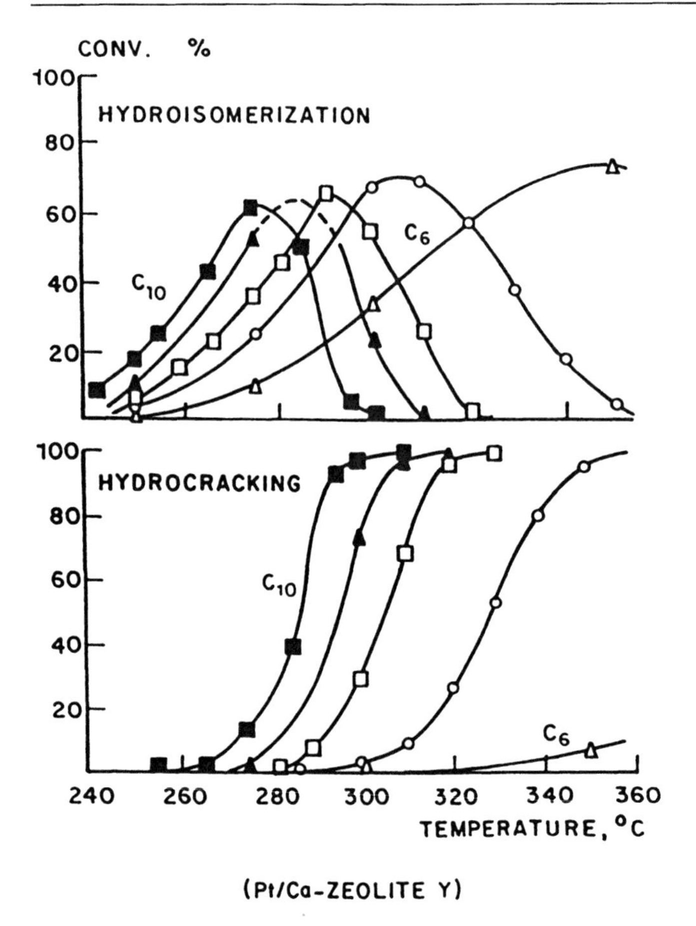

Fig. 7.43. Comparison of hydroisomerization and hydrocracking for C_6–C_{10} paraffins [129]

cess uses a molecular sieve adsorbent and operates at low temperature in the liquid phase. The latest development in the Penex process is to use once-through hydrogen while ensuring a minimal partial pressure at the end of the reactor, thus enabling a reduction in costs.

The I-8 type catalysts, which are also used in the Butamer butane isomerization process, are generally manufactured by additional chlorination of a Pt/alumina catalyst. Such catalysts had already been patented in 1961. Sublimation of $AlCl_3$ was used to increase the chlorine level on the catalyst [137]; another attractive chlorinating agent is CCl_4 [138]. Altogether the chlorination and activation of such catalysts contain a number of intricacies, which are referred to in the literature [139–142]; however, in the context of this book we will not go into any details. The high acidity of the chlorinated alumina catalyst is of course responsible for the benefits of the low-temperature operation, but also constitutes its weakness: these catalysts are sensitive to poisons such as sulfur and water and therefore require a rigorous feedstock pretreatment [127].

Moreover, they are less robust than the zeolitic catalysts operating at higher temperature (see below). Very recently, it was announced that LPI-100, a sulfated metal oxide (non-chlorinated) catalyst, has been commercialized by UOP (basis earlier work at Cosmo Oil and Mitsubishi Heavy Industries), which is about 80 °C more active than the zeolitic ones but does not have the contaminant sensitivity of chlorinated catalysts, and has already been in commercial operation since December 1996 [128, 143].

7.5.3.2
Isomerization over Zeolitic Catalysts

Zeolitic paraffin isomerization catalysts are less active than the amorphous ones, and as a consequence have to be operated at higher temperatures, i.e., at about 250 °C, with the inherent disadvantages of a lower maximum conversion per pass and of a lower content of dibranched species that can be obtained. These disadvantages are, however, offset by their much lower sensitivity to poisons, and hence these catalysts have found wide use, both in the dedicated Shell-developed Hysomer process, and in the UOP process, with the I-7 catalyst. These processes are often coupled with an i/n-separation process, such as Isosiv developed by Union Carbide, with recycle of the normal paraffins. Thus the Hysomer/Isosiv combination is called the Total Isomerization Process (TIP) (see Fig. 7.44 [112]); while Hysomer alone can raise the research octane number to about 80, with the Isosiv process values of 90 are achieved [112]. The preferred catalysts are based on mordenite, which is first converted into the acid form and then acid-leached to increase the silica/alumina ratio [112, 144, 145]. Indeed, the value of the silica/alumina ratio which gives maximum activity observed by Guisnet et al. (about 20) is close to the value expected for isolated aluminum atoms;

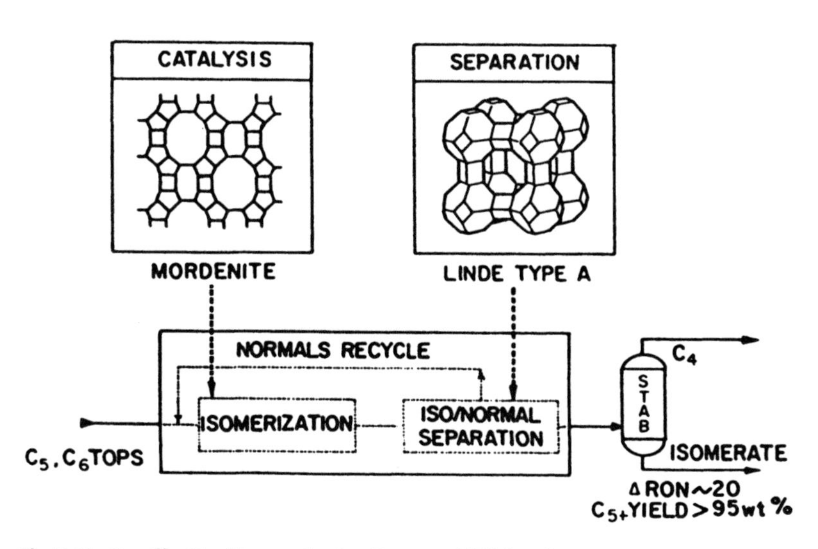

Fig. 7.44. Paraffin Total Isomerization Process (TIP) [112]

a higher silica/alumina ratio merely leads to a decreased number of sites, whilst a lower ratio leads to next nearest neighboring Al atoms, which are known to be less acidic [146]. These authors also report higher stability at the higher silica/alumina ratio, with high-alumina catalysts even benefiting from increased hydrogen pressures [147], and some speculations on the origin of the effects have been made. The final calcination temperature of the mordenite catalyst was reported to be important [148], which is not surprising. Some studies on isomerization catalysts based on other zeolites than mordenite have been published recently, dealing with Pd/USY (Pd particle size effects) [149, 150], with zeolite omega [151, 152] and with zeolite beta [133–135, 153]; a detailed discussion is beyond the scope of this review.

One can expect the isomerization area to remain important, and the challenge is to isomerize also the higher paraffins. In the conventional bifunctional catalysis, mechanistic considerations indicate, however, that the selectivities that can be attained will be clearly lower than with C_5–C_6 isomerization. Possibly the high i/n ratios that are obtained in hydrocracking can be of value to obtain suitable C_{7+} streams.

7.5.4
Isomerization of Light Olefins

This is a good place to discuss the skeletal isomerization of olefins, a process that in recent years has attracted a lot of attention, also from a mechanistic point of view. With the reduction of the concentration of high octane components such as olefins and benzene in gasoline, a need exists to produce other high octane components. The ethers such as MTBE and TAME have high octane numbers and are therefore of clear interest. With the anticipated rapid increase in demand for these ethers alternative manufacturing routes are of immediate interest. One attractive route for MTBE is the manufacture from isobutene and methanol, and this makes isobutene an attractive raw material. Therefore a selective skeletal isomerization of n-butene into isobutene is clearly attractive, as discussed in [154]. A number of catalyst systems for this reaction are known, such as those based on alumina or silica-alumina (see, e.g., [155]), but in general these exhibit a very short lifetime. In particular the system developed at Shell, Amsterdam, using a particular ferrierite, is very attractive in that it shows both a high selectivity and stability [156], and is the basis of a process developed at Lyondell [157]. The catalytic and mechanistic issues are similar to those in paraffin isomerization: an equilibrium which is better at lower temperatures, the optimum aluminum content of the zeolite, and the issue whether the reaction follows a monomolecular or bimolecular mechanism (a recent reference to this issue, debated by many groups, is given in [158]).

7.5.5
Paraffin/Olefin Alkylation

Paraffin/olefin alkylation is a major process for the manufacture of gasoline components, since the formed "C_8" compounds have a high degree of branching,

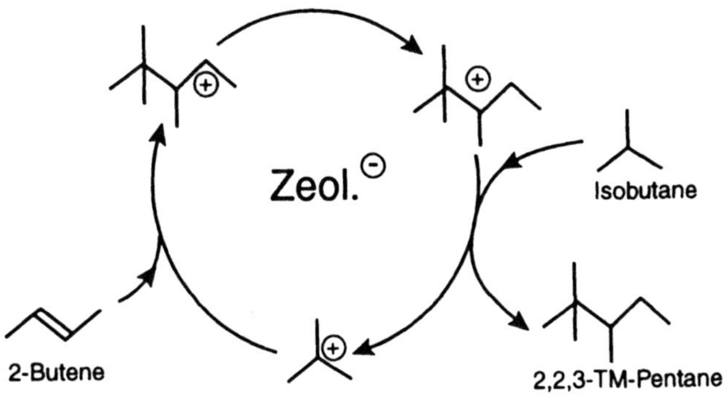

Zeol.
Isobutane
2-Butene
2,2,3-TM-Pentane

Fig. 7.45. Mechanism of alkylation [159]

and therefore a high octane number. Given the fact that for every olefin molecule used one (often less valuable) isobutane molecule is also used, it is not surprising that this process can be more attractive than olefin oligomerization. Worldwide capacity is high, and has been estimated at 40 million t/a (HF process) and 30 million t/a (H_2SO_4 process) [159]. These two existing processes both have their disadvantages as to undesirable chemicals (HF), or waste streams (spent acid, H_2SO_4). Therefore for a long time a solid, non-toxic, non-corrosive catalyst has been sought. To understand the issues involved it is useful to consider the mechanisms of alkylation (e.g., Fig. 7.45) and olefin oligomerization (Fig. 7.46), as given in [159]. Starting with an adsorbed carbenium ion (e.g., from butene or from isobutane) an olefin is added yielding a branched C_8 molecule. This molecule can now be desorbed through a hydrogen transfer reaction with isobutane, creating an alkylation cycle. Failure to do this timely will result in oligomerization, instead of alkylation, (the general phenomenon for a deactivated alkylation catalyst) and will also lead to heavy paraffins responsible for

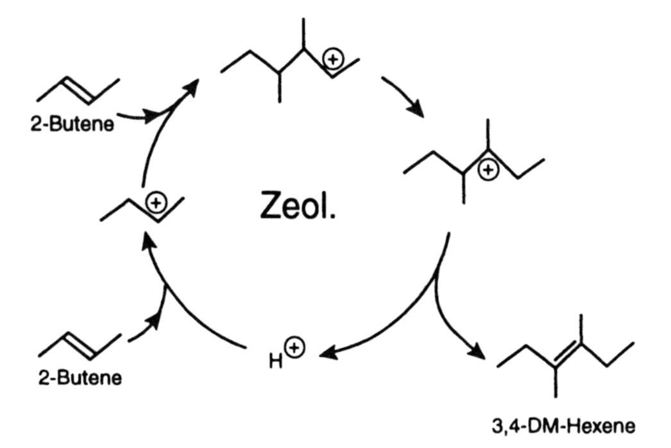

Fig. 7.46. Mechanism of olefin oligomerization [159]

the deactivation of the catalyst. More detailed descriptions of the chemistry are available [159–161]. The main zeolites studied for alkylation are Y-sieves and beta [162]; in general a high number of acid sites is positive for catalyst performance. The beta systems, as described by Mesters et al., in combination with a slurry reactor allowed high selectivities for alkylation and a very long life of 200 hours; regeneration by a hydrogen treatment of the noble metal promoted catalyst was also demonstrated. Nevertheless at present the front runners in heterogeneous alkylation systems, as far as commercial development is concerned, are not using zeolites, but strong acids carried on a support. An example is supported triflic acid (Haldor Topsøe), which has reached a pilot plant (55 kg/d) stage [163].

7.5.6 Zeolite-Supported (De)Hydrogenation Catalysts

The previous text always discussed catalysts where the zeolite performed the main duty, usually acid catalysis, sometimes supported by a hydrogenation function. Catalysts are emerging, however, in which the zeolite is the optimal support for a hydrogenation function such as Pt or Pd for specific applications.

7.5.6.1 Aromatization Catalysts

As stated above, aromatization is generally achieved in catalytic reforming using monometallic (Pt) or bimetallic catalysts, with chlorinated alumina as a support. It has been found, however, at several laboratories, that Pt supported on, e.g., zeolite BaK-L gives very high aromatization selectivities, especially with light paraffins, as is illustrated in Fig. 7.47 [164]. In particular with light paraffinic feedstocks high activities and selectivities have been reported [165]. To understand this system, one must primarily recognize that all effective acidity of the zeolite has been neutralized to avoid cracking. Exchanging K-L with Ba ions, in combination with high temperature calcination, has been found most effective because this locates the (non-acidic) K ions at accessible sites, and the (potentially acidic, because of hydrolysis) Ba ions at sites that are inaccessible to reactants (as witnessed by IR spectroscopy of adsorbed pyridine) [165]. In this non-acidic system the ring closure presumably occurs directly from linear paraffinic chains (as shown in Fig. 7.48 [166]), so the mechanism is different from that in "conventional" bifunctional reforming over Pt(/X)/Cl/alumina catalysts [166]. The aromatization properties have been ascribed to the combination of the platinum clusters and the shape-selective properties of the zeolite [112]. Lane et al. [167] conclude from a comparison between Pt/K-L zeolite and Pt/K-Y zeolite that indeed the high selectivity in zeolite L is consistent with a confinement model [168], in which n-hexane is adsorbed as a 6-ring pseudo-cycle. Derouane, however, reported that similar aromatization properties were shown by Pt supported on stabilized magnesia [169], which would suggest that shape selectivity is not an important factor. Interestingly the catalyst deactivation rate decreases

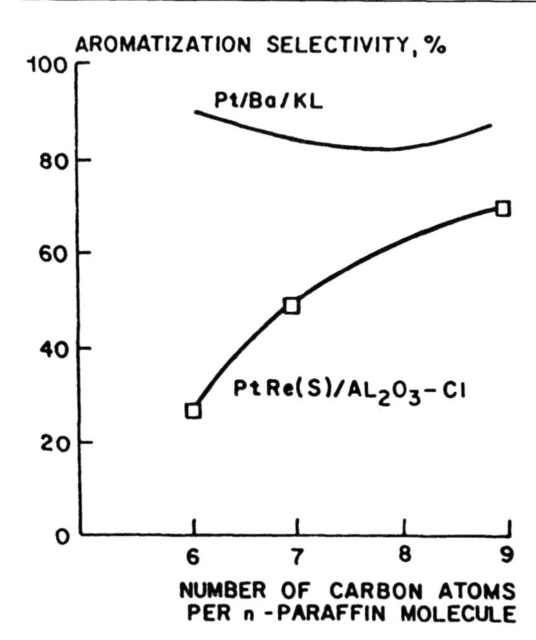

Fig. 7.47. Catalytic reforming with Pt on zeolite L [164]

with decreasing pressure. This unusual phenomenon is attributed to the zeolite structure (which leads to a low coking tendency), in combination with the lower temperature requirement at lower pressure [170].

These findings are the basis for the AROMAX process being developed by Chevron [171]. A drawback of the process is the extremely high sensitivity of these catalysts to sulfur, although this can be coped with by using proper sulfur guard beds [170]. It should be mentioned that Pt/BaK-L is not the only zeolitic reforming catalyst described in the literature (and neither is Chevron the only company active in this area). Chevron in particular, though, has reported a number of alternative systems such as Pt/Cs/B/(Al)/beta (or "SSZ-X"), but it is not clear whether these have (already) found commercial application [172, 173]. Interestingly, the issue of acidity emerges here again [174], and it is not immediately clear whether in all cases the same mechanism prevails.

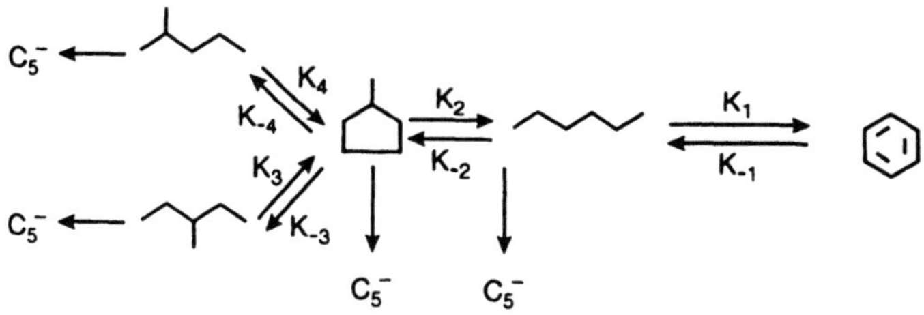

Fig. 7.48. Reaction network for hexane aromatization with AROMAX™ catalyst [166]

7.5.6.2
Sulfur-Tolerant Hydrogenation Catalysts for Production of Low Aromatics Diesel

Clearly a trend exists towards cleaner oil products, such as gas oil, with lower aromatics contents [175], see also Table 7.1. The hydrogenation reaction should, because of thermodynamic constraints, preferably be carried out at low temperatures, and hence with a noble metal catalyst whose activity does not suffer too much from the traces of sulfur and other impurities still present in the feed. Recently, such catalysts have been announced in which the active metals are supported on specific zeolites (or also on amorphous supports), resulting in much better activity with sulfur-containing feedstocks [79, 176, 177]. Interestingly enough, for the sulfur resistance a high degree of acidity of the support appears beneficial, completely in contrast to the situation in AROMAX (Sect. 7.5.6.1), where the Pt on a basic support was very sulfur-sensitive. A precise explanation for these effects is not yet available. These sulfur-tolerant hydrogenation catalysts have found commercial application in various process configurations [178] such as in a two-stage process line up (where fresh hydrogen gas is supplied to the noble metal catalyst), [177], and in countercurrent operations, where the gaseous catalyst poisons are already quickly removed at the top of the reactor [179]. Given the importance of this area, further developments may be expected.

7.6
Zeolites in Synfuels Production

7.6.1
Conversion of Methanol to Gasoline (MTG)

The conversion of methanol over ZSM-5 zeolite to mixtures of olefinic and aromatic hydrocarbons offers a means to produce synthetic gasoline of high octane quality from methanol. Since methanol can be produced from synthesis gas ($CO + H_2$), which in turn can be made from natural gas or coal, a route is thus offered for producing synthetic gasoline from the latter resources. A commercial plant to produce gasoline from natural gas along this route was constructed at Motonui, Taranaki in New Zealand and has been operated by the New Zealand Synthetic Fuels Corp. Ltd. (now Methanex New Zealand Ltd.) since 1985. It has a design capacity of 570,000 tons per year (14,450 bpsd) of gasoline [180–183].

7.6.1.1
Reaction Mechanism

Since the initial publications on the conversion of methanol to high octane gasoline [184, 185] considerable research effort has been devoted worldwide to elucidate the underlying reaction mechanism. Extensive reviews of earlier work have been published by Chang [186–188]. More recently, a review on the mechanism of the initial steps in methanol conversion has been published by Hutchins and Hunter [189]. In the reaction path for converting methanol to hydrocarbons the

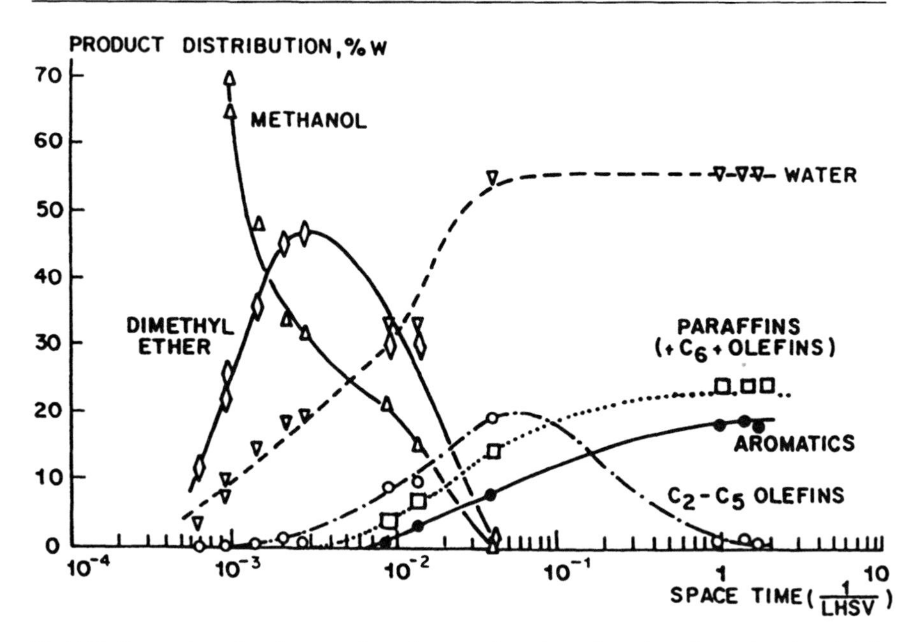

Fig. 7.49. Formation of various products from methanol as a function of reaction time [185]

initial step is the dehydration of methanol to dimethyl ether, followed by the formation of light olefins. These olefins are converted to olefins with a longer carbon chain, which are eventually transformed by C/H disproportionation into aromatics and paraffins. The sequential character of these transformations is clearly manifested in the product spectrum as a function of the reaction time as shown in Fig. 7.49. Whereas the sequence of the formation of dimethyl ether, light and heavier olefins, and aromatics plus paraffins seems quite clear, there is less certainty about the crucial step in which a C–C bond is formed in the transformation of methanol or dimethyl ether to ethene or in the conversion of an olefin into another olefin with one more carbon atom. Mechanisms which have been proposed include the formation of a carbene from methanol, a radical pathway, an intermolecular reaction of dimethyloxonium methylide, and deprotonation of a surface-bonded methyloxonium ion to give a surface-bonded oxonium methylide. These mechanisms are discussed in detail in the article by Hutchins and Hunter [189].

7.6.1.2 The Fixed-Bed MTG Process

The Methanol To Gasoline (MTG) process is the result of research and development efforts by the Mobil Research and Development Corporation including chemistry, process development and scale-up studies, which have been well documented [190–196]. Important features, from the process point of view, of the methanol-to-gasoline reaction are the high exothermicity of the overall

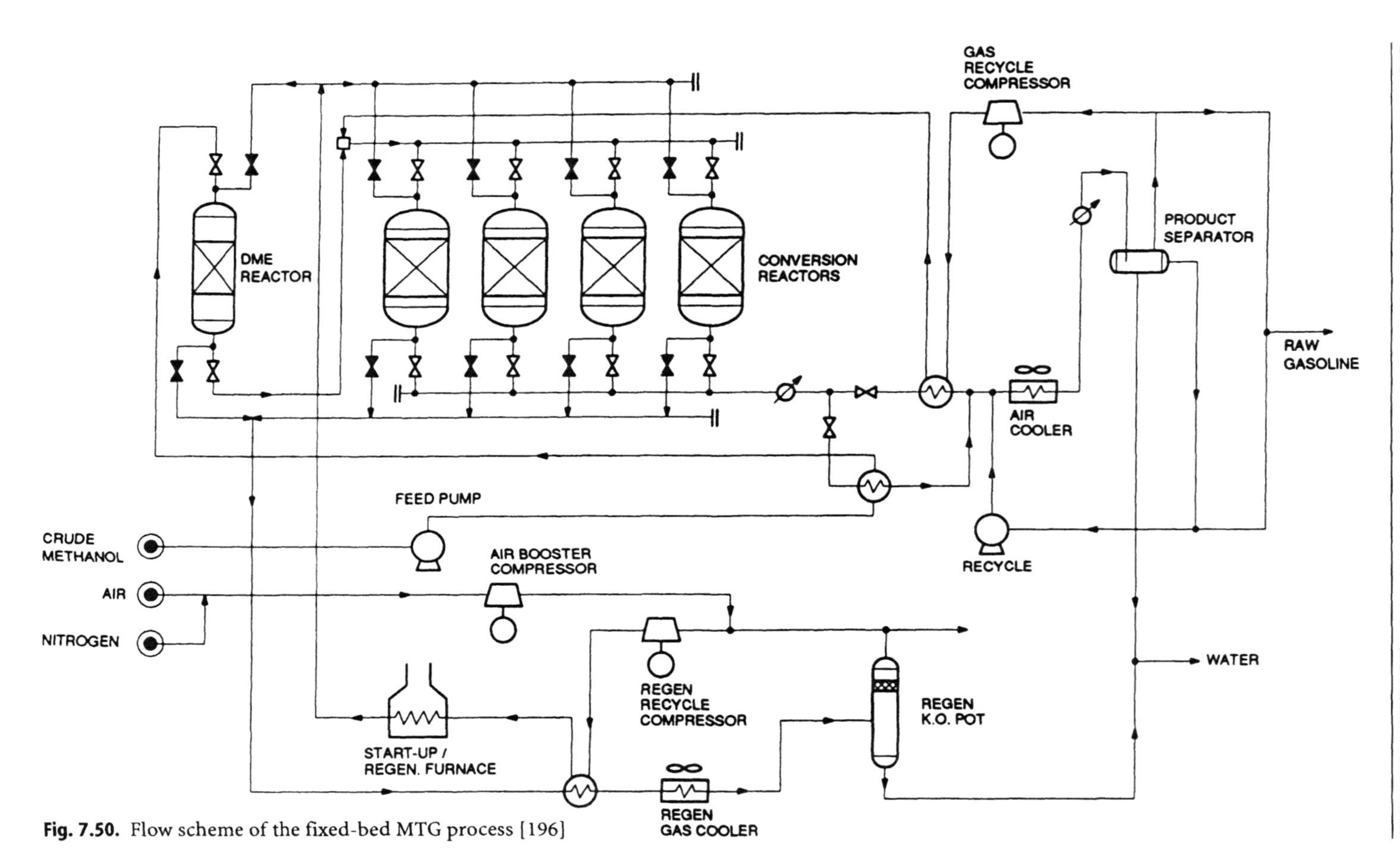

Fig. 7.50. Flow scheme of the fixed-bed MTG process [196]

reaction and the deactivation of the catalyst, corresponding with life cycles of not more than a few hundred hours. In the fixed-bed version of the process, where adiabatic reactors are applied, temperature rises are kept within acceptable limits by dividing the overall reaction in two stages, viz. a first stage in which methanol is dehydrated to an equilibrium mixture of dimethyl ether, water and residual methanol, and a second stage in which this mixture is converted to hydrocarbons and water. The temperature rise in the latter-stage reactor is further reduced by applying a large recycle of gas. A simplified flow diagram of the fixed-bed MTG process is depicted in Fig. 7.50 [196]. A typical feed to the process is crude methanol containing some 20 wt% of water. Temperatures vary from about 315 °C (dehydration reactor inlet) to about 415 °C (conversion reactor outlet). The process is operated at about 20 bar and uses alumina as dehydration catalyst in the first stage and ZSM-5 as catalyst in the second stage. The gas recycle ratio is typically about 9 mol recycle gas per mol methanol feed. A typical space velocity amounts to 2 kg crude methanol per kg catalyst per hour. As a result of deactivation of the ZSM-5 catalyst the composition of the second-stage reactor effluent is not constant but varies during a cycle as a deactivation front progresses through the bed. This causes the main conversion zone to move through the bed, as evidenced by the temperature profiles within the bed measured at different times (Fig. 7.51 [196]). The end of a cycle is reached

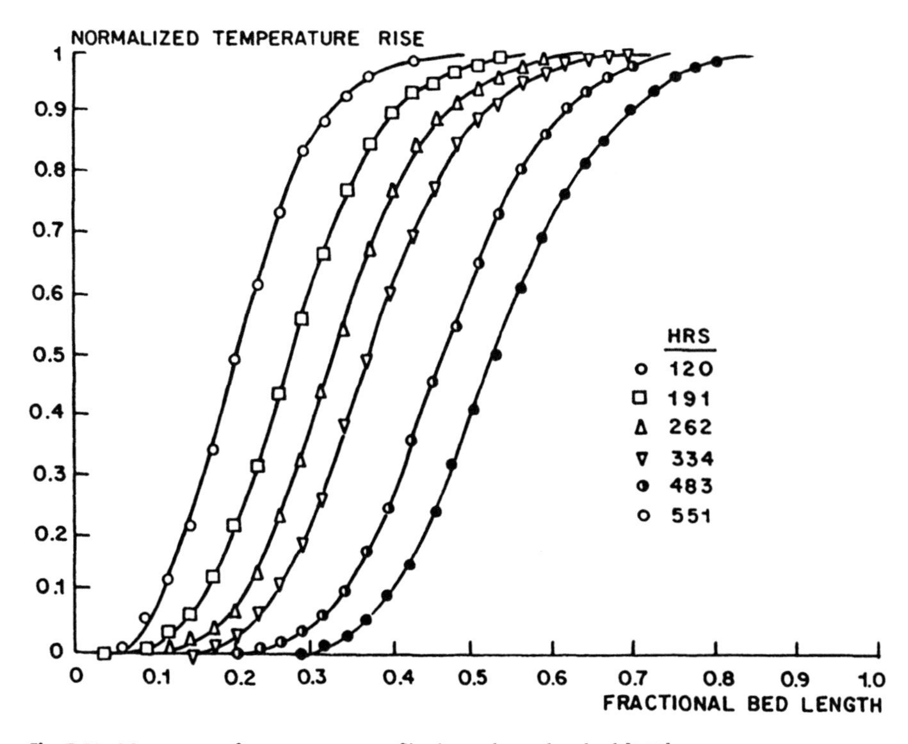

Fig. 7.51. Movement of temperature profile through catalyst bed [196]

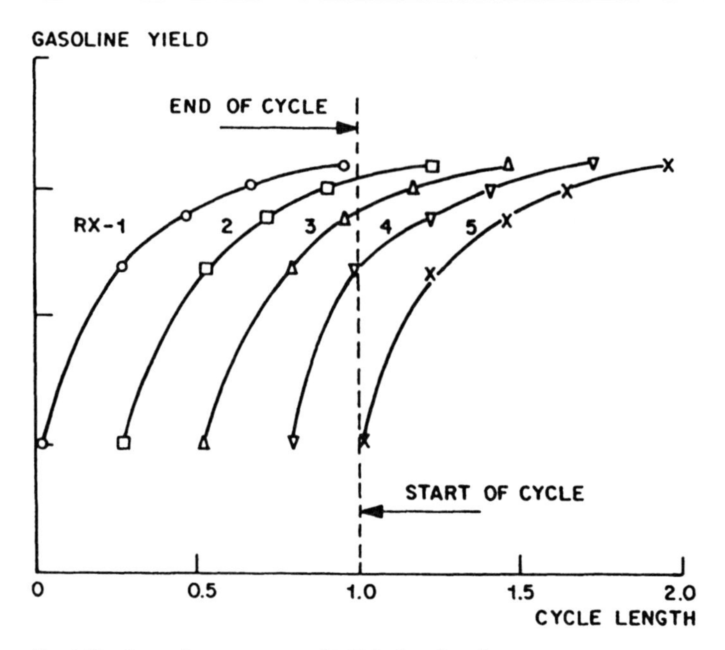

Fig. 7.52. Operating sequence of MTG plant [195]

when methanol starts to break through. To cope with the varying reactor effluent composition and to satisfy the need for frequent regenerations without interrupting the process, several second-stage reactors are used in a parallel arrangement, allowing any one reactor to be taken out for regeneration (see Fig. 7.50 [196]). A staggered reactor operating sequence is applied to smoothen out the effect of catalyst deactivation in each single reactor on the total plant stream (see Fig. 7.52 [117]). Deactivated catalyst is regenerated by burning off deposited coke. A typical product break-down of the fixed-bed MTG process is shown in Table 7.8 [197]. Data on gasoline quality, as produced in the commercial MTG plant at Motonui in New Zealand, are shown in Table 7.9 [198].

Table 7.8. Yield Breakdown of Fixed-Bed MTG Processes [198]

Yield on methanol charged, wt%		Hydrocarbon product breakdown, wt%	
Methanol + dimethyl ether	0.0	Light gas	1.3
Water	56.15	Propane	4.6
CO, CO_2	0.04	Propene	0.2
Coke, oxygenates	0.15	Isobutane	8.8
Hydrocarbons	43.66	n-Butane	2.7
		Butenes	1.1
		C_{5+}	81.3
	100.0		100.0

Table 7.9. Quality of Synthetic Gasoline Produced in the MTG Plant in Motonui [199]

	Average value
Density, kg/m^3 at 15°C	730
Reid vapor pressure, kPa	86.1
RON	92.2
MON	82.6
Durene content, wt%	2.0
Induction period, min	325
Distillation	
% evaporation at 70°C	31.5
% evaporation at 100°C	53.2
% evaporation at 180°C	94.9
End point, °C	204.5

7.6.1.3 The Fluid-Bed MTG Process

The exothermic character and the necessity of frequent catalyst regeneration in the conversion of methanol over ZSM-5 zeolite catalyst makes a fluid-bed reactor much more attractive than a fixed-bed reactor in view of the more efficient heat removal and the possibility of removing deactivated catalyst from the reactor during operation and replacing it with a fresh or regenerated one. Studies on the MTG reaction in a laboratory-scale fluid-bed reactor have been reported by Liederman et al. [199] and have been followed by development studies in a pilot plant (capacity 4 barrels per stream day or ca. 0.65 m^3/d) reported by Lee et al. [200].

Apart from the heat removal and catalyst replenishment possibilities mentioned above, the fluid-bed reactor differs from the fixed-bed reactor in that the degree of staging is much less due to backmixing. For a series of reactions as shown in Fig. 7.49, this implies that, at a similar degree of conversion of primary feedstock, the reactor effluent will contain higher levels of intermediate products. This is evident from a comparison of the MTG hydrocarbon yields as shown in Table 7.10 [197].

It is clear that less C_{5+} hydrocarbons are obtained in the fluid-bed process and that the fluid-bed product contains more light olefins. However, the lower yield of primary gasoline in the fluid-bed process can be compensated for when use is made of the alkylation process to convert the light olefins together with the isobutane also produced in higher yields. When alkylate is included, the fluid-bed process actually yields more gasoline of higher octane quality (due to the excellent octane quality of alkylate) than the fixed-bed process, as shown in Table 7.10 [197]. With the aid of a conversion model developed from the process development studies [201], the Mobil fluid-bed MTG process was scaled up from the 0.04-m diameter pilot reactor with a capacity of 0.65 m^3/d to a 0.6-m diameter demonstration plant.

Table 7.10. Comparison of Hydrocarbon Yields Obtained in Fixed- and Fluid-Bed Processes [198]

	Fixed-bed	Fluid-bed
Hydrocarbon product, wt%		
Light gas	1.3	4.3
Propane	4.6	4.4
Propene	0.2	4.3
Isobutane	8.8	11.0
n-Butane	2.7	2.0
Butenes	1.1	5.8
C_{5+}	81.3	68.2
	100.0	100.0
C_{5+} gasoline, including alkylate	83.9	91.2
Gasoline octane numbers		
RON	93	95
MON	83	85

The commercial feasibility of the fluid-bed MTG process was further studied in a demonstration project with three industrial participants (Uhde GmbH, Union Rheinische Braunkohlen Kraftstoff AG, and Mobil Research and Development Corporation) partly financed by the governments of the USA and the Federal Republic of Germany. A demonstration plant with a capacity of 100 barrels per stream day or ca. 16 m^3/d was constructed at the URBK facility at Wesseling, Germany, and was operated from the end of 1982 to the end of 1985 [202]. In these demonstration studies product yields and quality as observed in the pilot plant investigations were essentially confirmed and the feasibility of cooling inside the reactor as well as outside (with catalyst circulating through a catalyst cooler) was demonstrated. Figure 7.53 [197] shows a flow diagram of the fluid-bed MTG demonstration unit. Some considerations on a commercial-scale fluid-bed MTG plant, based on the results of the above demonstration project, have been published by Grimmer at al. [203].

7.6.2 Integration of Methanol Synthesis and Methanol Conversion (TIGAS Process)

In the Topsøe Integrated Gasoline Synthesis (TIGAS) process as developed by Haldor-Topsøe A/S [204, 205] a reduction in capital investment of a natural gas to syngasoline plant as has been realized in New Zealand is aimed at by combining the methanol synthesis and methanol conversion steps in a single process. The reactions are carried out in separate reactors using similar catalysts to those employed in individual methanol synthesis and methanol-to-gasoline processes, viz. a Cu/ZnO-type catalyst for methanol synthesis and a ZSM-5-type

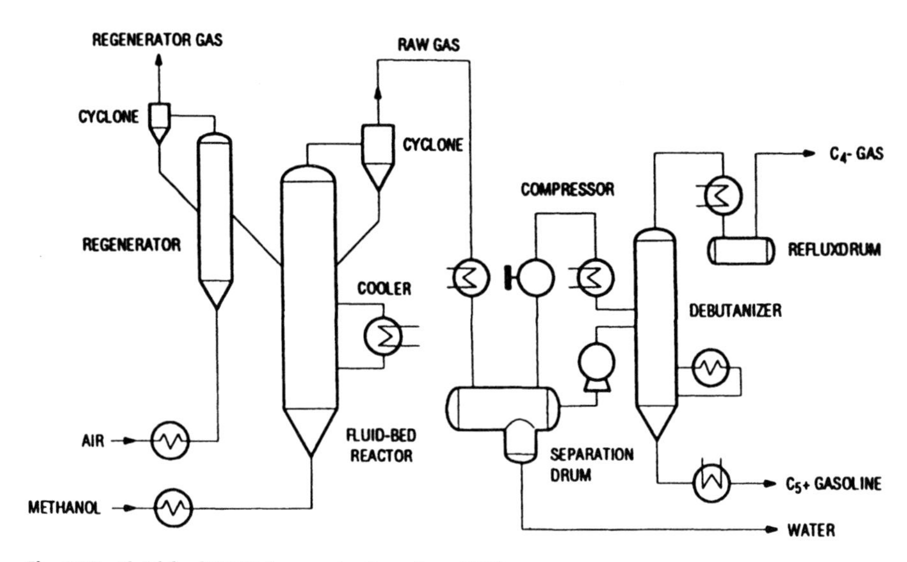

Fig. 7.53. Fluid-bed MTG demonstration plant [198]

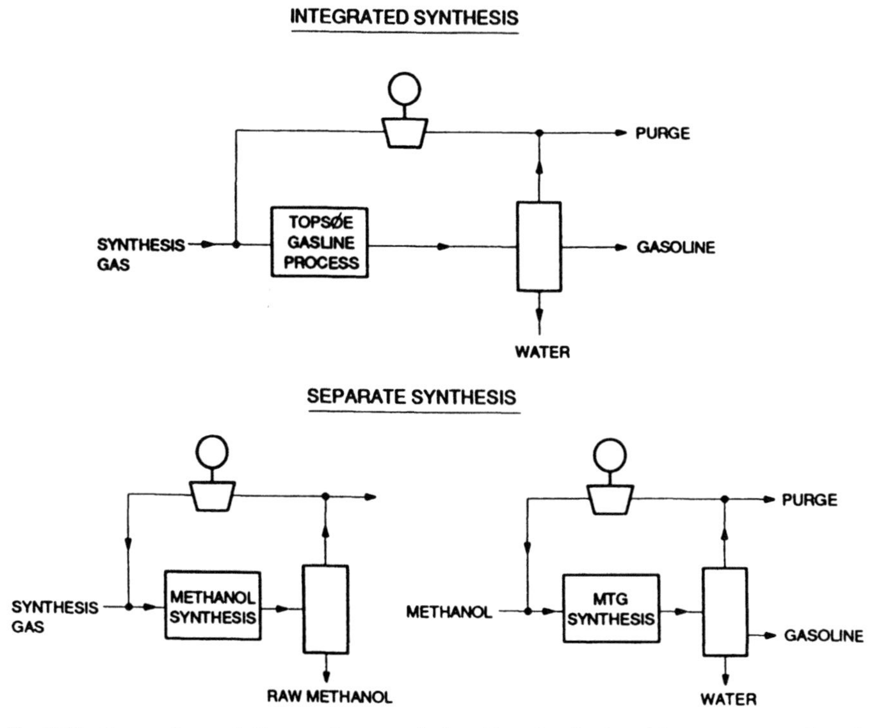

Fig. 7.54. Comparison of Topsøe Integrated Gasoline Synthesis with separate methanol synthesis and methanol conversion [204]

catalyst for methanol conversion, but the two reactors are integrated into a single gas recycle loop. A simplified scheme of the TIGAS principle is shown in Fig. 7.54 [204].

The main advantage of the integrated process is the saving in energy and equipment necessary for heating and cooling, and compression of gases. Additional advantages may be derived from the lesser requirements on methanol purity in the synthesis step, since oxygenates other than methanol can also be converted over the ZSM-5 zeolite. Thus a more favorable thermodynamic equilibrium conversion of synthesis gas may be obtained. This is illustrated in Fig. 7.55 [205].

A problem in the integration of the methanol synthesis and the MTG reaction is the difference in pressure: methanol synthesis is usually carried out at 50–100 bar, whereas the MTG reaction is normally performed at 15–25 bar. However, by a proper choice of catalyst and operating parameters, a process can be obtained in which the two reactions are carried out at about the same pressure level, which is also close to the pressure level at which natural gas can be converted to synthesis gas.

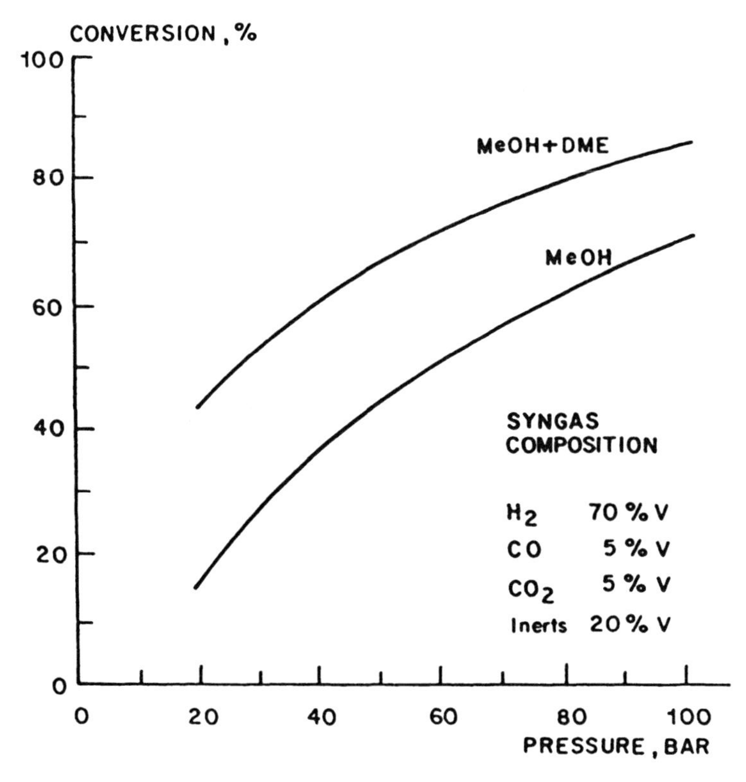

Fig. 7.55. Maximum conversion of synthesis gas as a function of pressure at 250 °C, for methanol synthesis and for combined synthesis of methanol and dimethyl ether [205]

After exploratory studies on the integrated gasoline synthesis in bench-scale reactors, process studies and catalyst ageing tests have been carried out by Haldor Topsøe in a 15 kg/day pilot plant at their laboratory in Lyngby, Denmark. On the basis of the results obtained in the pilot plant, it was decided to build a process demonstration unit of 1 ton/day capacity. This unit was constructed at a location in Houston, Texas. Using a methanol synthesis catalyst developed by Haldor Topsøe and a ZSM-5 catalyst supplied by Mobil, the viability of the TIGAS process was demonstrated in more than 10,000 hours of operation [205].

7.6.3
Direct Conversion of Synthesis Gas into Gasoline

A further reduction in the number of process units is possible by carrying out the synthesis of methanol or hydrocarbons from synthesis gas in the same reactor as the production of gasoline from these intermediate products. This requires a bifunctional catalyst, i.e., one which combines a synthesis gas conversion function with the gasoline formation function provided by the ZSM-5 zeolite. The synthesis gas conversion function may be provided by a Fischer–Tropsch metal such as iron, cobalt or ruthenium. Alternatively, oxides such as zirconium or thorium oxide which convert synthesis gas in the classical isosynthesis process or a methanol synthesis catalyst may be used.

Composite catalysts consisting of a combination of ZSM-5 with either iron, cobalt or zirconium oxide have been tested for synthesis gas conversion to gasoline by Chang et al. [206, 207]. Although the direct synthesis of gasoline from synthesis gas has been demonstrated in these experiments, the technical as well as economic feasibility of a direct gasoline synthesis process still seems rather remote.

Catalyst ageing, excessive carbon formation and poor regenerability of the composite catalyst systems are problems which still have to be overcome. Another problem is that at the relatively high temperature of synthesis (compared to the usual temperature in Fischer–Tropsch synthesis, which is necessary for the aromatization reactions over ZSM-5) rather large amounts of methane are produced when a transition metal such as iron is used as the CO reduction component (see Table 7.11 [207]).

The large amounts of methane represent a great economic penalty since they adversely affect the yield of gasoline from syngas. As the production of synthesis gas represents the greatest cost factor in the conversion of coal or natural gas to liquid hydrocarbons, this non-optimal utilization of costly synthesis gas represents an economic drawback which is generally not compensated for by the advantage of a single-step operation. In addition, the formation of methane represents a technical problem in that it builds up in the recycle gas loop.

Formation of methane is reduced when an oxide such as ZrO_2 or a methanol synthesis catalyst is used instead of a Fischer–Tropsch metal as synthesis gas conversion component.

Data on the combination of a Zn/Cr methanol synthesis catalyst with a ZSM-5-type catalyst reported by Chang et al. are also given in Table 7.11 [207], and these show a reduced methane production.

Table 7.11. Synthesis Gas Conversion over Composite Catalysts [207]

Catalyst:			
Metal component	Fe	Zr	Zn/Cr
Zeolite	ZSM-5	ZSM-5	ZSM-5
Feed gas H_2/CO molar ratio	1	1	1
Reaction conditions			
T, °C	371	427	427
P, bar	35	83	83
GHSV, h^{-1}	3000	720	1780
$(H_2 + CO)$ conversion, mol%	12.0	13.8	44.1
Product distribution, wt%			
C_1	33.4	1.6	3.9
C_2	10.8	6.8	13.1
C_3	17.3	6.0	22.9
C_4	19.3	0.8	15.5
C_{5+}	19.2	84.8	44.6
Oxygenates	0	0	< 0.1
Percentage olefins in product	5.1	0.5	0.6
Percentage aromatics in C_{5+}	25.2	99.8	75.6

The direct synthesis of gasoline from synthesis gas has been extensively studied by Sie et al. at the Shell Laboratory in Amsterdam [208]. Following exploratory studies in micro- and bench-scale fixed-bed reactors, the process was developed in fluid-bed pilot reactors. Although the process performance proved to be quite attractive in that gasoline of high octane quality was produced from CO-rich synthesis gas at very high selectivity the process was not commercialized because production of synthetic gasoline from coal turned out not to be competitive with normal gasoline production at current and foreseeable crude oil prices.

7.6.4
Conversion of Methanol to Synfuels via Light Olefins

From the sequential reaction scheme of methanol conversion to aromatic gasoline as illustrated in Fig. 7.49 it follows that the reaction can be carried out in two distinct process steps, viz. a first step in which methanol is converted to mainly light olefins and a following step in which the olefins are converted to liquid hydrocarbons. The advantage of this scheme over a single-step MTG process is that the second step offers greater flexibility to vary the product formed. Thus, by oligomerization of light olefins with little hydrogen transfer in the second step it is possible to produce a more aliphatic product which is suitable as middle distillate fuel (kerosine, diesel fuel). Along these lines Mobil Research and Development Corporation have developed two processes, viz. the Methanol to Olefins (MTO) and Mobil Olefins to Gasoline and Distillate (MOGD) processes. Both processes use as catalyst a ZSM-5 zeolite.

7.6.4.1
Methanol to Light Olefins (MTO) Process

The MTO process of Mobil was developed after the MTG process when it became apparent to them that a need existed for a parallel route leading to diesel oil rather than gasoline. The MTO process, in essence, stops the MTG reaction, leading to a product which contains 75–80% light olefins. The relationship between these two processes can be depicted as follows:

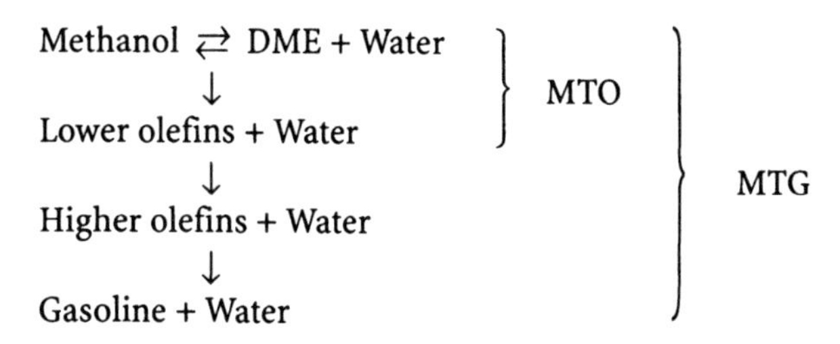

Like the MTG process the MTO process is based on a ZSM-5 zeolite as catalyst. In its preferred form the MTO process is carried out in a fluid-bed reactor. Process development work on this process has been carried out at the Paulsboro laboratory of Mobil Research and Development Corporation in a pilot plant with a capacity of 4 barrels per stream day or 0.65 m^3/d [209, 210].

Hydrocarbon selectivities of the MTO process as observed in the development studies are given in Table 7.12 [209]. These data show that the product consists predominantly of C_3–C_5 olefins. The ratio of the individual olefins can be varied within a certain range, depending upon the process severity, see Fig. 7.56 [197]. As a follow-up to the process development work in the laboratory, demonstration of the MTO process was undertaken in the same fluid-bed unit (capacity 16 m^3/d) at Wesseling, FRG, used before for the demonstration of the fluid-bed MTG process. Results which in essence confirm the data obtained in the 0.65 m^3/d pilot plant have been reported by Keim et al. [211].

Table 7.12. Hydrocarbon Selectivities in the MTO Process [210]

Hydrocarbon	wt%
C_1	2.0
C_2	0.4
$C_2^=$	5.2
C_3	2.1
$C_3^=$	32.9
i-C_4	2.3
n-C_4	0.7
$C_4^=$	19.1
C_{5+} (total)	19.7
C_{5+} PNA	15.6
Total	100.0

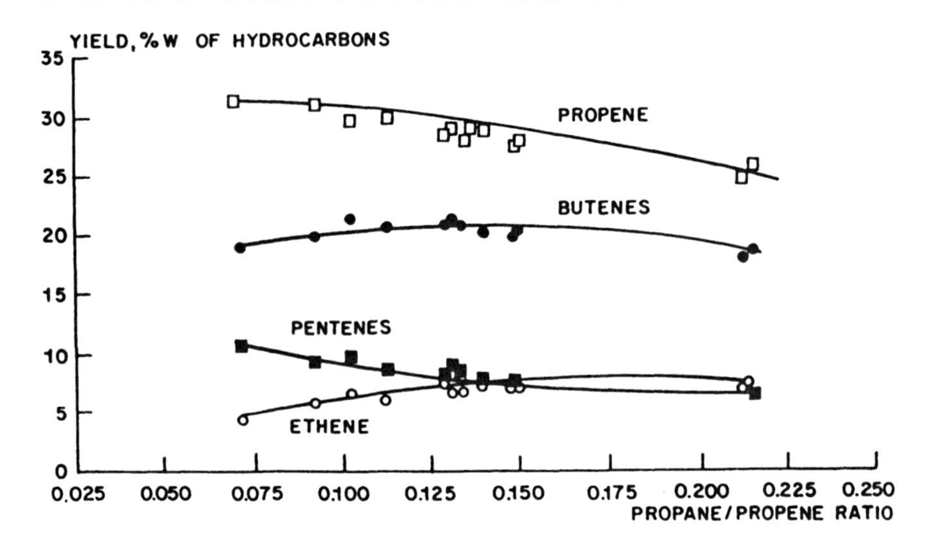

Fig. 7.56. Variation of light olefins distribution in the MTO process [198]

The MTO process cannot only be considered as an element in the conversion of methanol to synthetic liquid hydrocarbon fuels, but may also be seen as a means to produce light olefins in the context of chemicals manufacture (vide infra).

The newest development is an MTO process co-developed by UOP and Norsk Hydro ASA as part of a two-step process to convert natural gas to ethylene and propylene [212]. The process is being tested out on a pilot scale (20 kg/h), see Table 7.13 [212]. Although an olefins plant based on natural gas may require a larger investment than a conventional naphtha cracker, there may also be a break-even point as a result of the lower feedstock costs [212]. Nevertheless, conventional steam cracking of gas and naphtha will remain for the time being the dominant technology, while alternative routes to olefins like methane oxydimerization, oxydehydrogenation and MTO will play no more than a niche role [212].

Table 7.13. UOP/Norsk Hydro MTO process: product yields (excluding water) [212]

Product	Yield, wt%
Ethene	48.0
Propene	33.1
Butenes	9.6
C_{5+}	2.4
Hydrogen + ethane + propane	3.6
CO_x + coke	3.4
Total	100.0

7.6.4.2
Light Olefins to Gasoline and Distillates (MOGD) Process

The conversion of light olefins over ZSM-5 can lead to higher olefins up to C_{30} and higher by acid-catalyzed oligomerization and isomerization-cracking to intermediate carbon number olefins which can also be (co-)oligomerized to yield a continuous boiling product containing all carbon numbers within a certain range [213]. Thus, aliphatic products in the middle distillate or the lubricating oil range can be produced. As in the MTG process, cyclization and hydrogen transfer reactions leading to aromatics and light paraffins can also take place so that aromatic gasoline is co-produced. The ratio between distillate and gasoline production depends on operating conditions and can be varied within a certain range.

The Mobil Olefin to Gasoline and Distillate (MOGD) process can be used to convert light olefins from a catalytic cracker to liquid transport fuels, but as an adjunct to the MTO process it provides a route from methanol to hydrocarbon transport fuels. The process and its underlying chemistry have been described in a number of publications [197, 210, 211, 213–215].

As part of the MOGD development effort the process was demonstrated in a 33 m^3/d test in a Mobil refinery for the conversion of a mixed C_3/C_4 stream of an FCC unit. The reactor applied in the MOGD test run was a commercial wax hydrofinisher [215]. MOGD process yields are shown in Table 7.14 for two modes of operation, emphasizing distillate and gasoline production. Properties of the gasoline product are given in Table 7.15, while Table 7.16 shows the properties of the distillate fuels after hydrotreating. The catalyst in the MOGD process is based on ZSM-5: its shape-selective properties lead to products with predominantly methyl side chains (the degree of branching decreases with the pore size in the sequence omega > HY > mordenite>ZSM-5 [216]), while in addition the degree of coke deactivation is small. The ZSM-5 catalyst can be optimized by, e.g., steaming, or poisoning of the outer surface [217, 218]. A ZSM-5-type catalyst also appears to have been used in the NESCO process by Neste where light olefins are converted to high molecular weight products such as diesel, solvents and lube oils [219]. Exxon, on the other hand, has commercialized a process for propene that uses ZSM-22 as a catalyst [220].

The Shell Poly-Gasoline and -Kero process (SPGK) is another process for converting light olefins ($C_2^=$–$C_5^=$, e.g., from FCC units) into liquid transporta-

Table 7.14. MOGD Process Yields (in wt%) with C_3–C_6 Olefins as Feed [216]

	Max. Distillate Mode	Gasoline Mode
C_1–C_3	1	4
C_4	2	5
C_5–165 °C Gasoline	15	–
165 °C$^+$ Distillate	82	–
C_5–200 °C Gasoline	–	84
200 °C$^+$ Distillate	–	7

Table 7.15. Properties of Gasoline from the MOGD Process [216]

Density, g/cm^3	0.738
Octane numbers:	
Research clear	93.0
Motor clear	85.0
Reid vapor pressure, kPa	57.2
Sulfur, ppm	<5
Distillation ASTM D86, °C	
10%	48
30%	92
50%	105
70%	120
90%	135
EP	178

Table 7.16. Properties of Hydrotreated Distillate Fuels from the MOGD Process [216]

	Jet Fuel	Diesel Fuel
Volume, % of total distillate	30	70
Density, g/cm^3	0.774	0.800
Freeze point, °C	-60	-
Flash point, °C	50	100
Cetane No.	-	52
Smoke point, mm	25	-
Aromatics, vol%	5	-
Viscosity at 40%, m^2/s	-	$3.1 \cdot 10^{-6}$
Distillation, ASTM D86, °C		
10 %	205	258
30 %	218	261
50 %	226	270
70%	231	284
90%	235	302
EP	238	338

tions fuels. It uses a novel zeolitic catalyst system which, in contrast to the ZSM-5 catalyst employed in the MOGD process, allows conversion of ethylene under very much the same conditions as the higher olefins. The process is characterized by a very high flexibility towards selective production of either gasoline or middle distillates of attractive quality [221, 222].

7.6.5 Upgrading of Fischer-Tropsch Products with Zeolites

The MOGD process may also be applied for the conversion of light olefins produced in the Fischer-Tropsch synthesis. Such olefins are, for instance, produced

Table 7.17. Processing of Fischer–Tropsch Aqueous Product over ZSM-5 [223]

Temperature, °C	500
Pressure, kPa	276
Conversion, %	71
Product distribution, wt%	
C_1	8.3
$C_2^=$	50.4
$C_3^=$	25.3
$C_4^=$	5.4
LPG	7.4
C_{5+}	3.2
Aromatics	-

in relatively high proportions in the Fischer–Tropsch synthesis at relatively high temperature with an iron catalyst, as practised by SASOL in South Africa in the Synthol units. Conversion of these olefins over a ZSM-5-type catalyst may thus be an alternative to the currently applied polymerization over a phosphoric acid catalyst.

An advantage of the use of ZSM-5 catalysts in the conversion of Fischer–Tropsch products over the latter polymerization is that apart from higher olefins aromatics may also be formed in the production of gasoline.

Another feature of the use of ZSM-5 as catalyst is that not only olefins may be converted, but oxygenates can be processed as well. These oxygenates are generally obtained as an aqueous solution in the condensed Fischer–Tropsch product, and it is in principle possible to process the oxygenates over ZSM-5 in the presence of water, as shown by Stowe and Murchison of Dow Chemical Co. [223].

Table 7.17 [223] shows the product obtained by processing the aqueous phase of a Fischer–Tropsch product containing methanol, ethanol and propanol as main oxygenated products, with minor quantities of other alcohols, acetaldehyde, acetone and methyl ethyl ketone (MEK). It can be seen that this Fischer–Tropsch product fraction can be upgraded to a product consisting mainly of ethene and propene.

Under appropriate conditions ZSM-5 is also able to convert paraffins as produced in the Fischer–Tropsch reaction to aromatics plus light paraffins, analogous to what occurs in the M-Forming process to upgrade low-octane reformate, and the Mobil Distillate or Lube-Oil Dewaxing processes. A disadvantage of this route is that the coproduction of LPG implies a reduced liquid yield.

The omnivorous character of ZSM-5 zeolite as catalyst allows the upgrading of Fischer–Tropsch products containing paraffins, olefins as well as oxygenates without prior separation into the constituent types. This is illustrated by the data in Table 7.18 [223], pertaining to the conversion of C_{5+} Fischer–Tropsch oil over ZSM-5. In particular at high temperatures the formation of aromatics is evident, which is, however, accompanied by the production of LPG:

The upgrading of Fischer–Tropsch product over ZSM-5 has been considered by Mobil in combination with a slurry-phase Fischer–Tropsch process as developed by Koelbel at Rheinpreussen in Germany before and during World War II.

Table 7.18. Distribution of Products (in wt%) Obtained by Processing Fischer–Tropsch Oil[a] over ZSM-5 Catalysts [223]

Temperature, °C	400	450	500	525
Ethene	1.5	2.8	1.7	2.5
Propene	6.6	6.3	1.3	2.4
LPG	11.6	24.3	36.8	34.0
Butenes	5.0	3.6	0.5	0.7
Methane	–	0.2	2.3	3.0
Aliphatic/Oil	61.8	37.1	6.6	7.8
Aromatics	13.6	25.6	50.8	49.7

[a] A C_{5+} product consisting of 24.9 wt% paraffins, 62.3 wt% olefins, 2.5 wt% oxygenates (10.3 wt% not accounted for).

An advantage of the latter synthesis is that it can convert CO-rich synthesis gas as produced by modern coal gasifiers without the need for an expensive pre-adjustment of the H_2/CO ratio of the gas. Mobil Research and Development Corporation are reported to have been engaged in studies of the slurry-phase Fischer–Tropsch synthesis within this context [224].

7.6.6 Aromatics from Light Paraffins (Cyclar Process)

The shape-selective acidic properties of ZSM-5 zeolite as catalyst also allow the conversion of light paraffins such as propane and butane into aromatic hydrocarbons and ethane plus methane by reshuffling of hydrogen and carbon in a thermodynamically allowed sense at high temperatures and relatively low pressures. This process, also called dehydrocyclodimerization (DHCD), involves dehydrogenation, dimerization, cyclization and hydrogen transfer. A review of the chemistry and catalysis of this process has recently been published by Seddon [225].

The catalyst in the DHCD is generally a bifunctional one on which the shape-selective acidic zeolite is combined with a dehydrogenation metal. In early work by Csicsery [226] platinum was used as the dehydrogenation metal; later on workers at Mobil extensively investigated zinc as promoter for ZSM-5. However, these catalysts generally suffered from poor stability due to high coking rates and evaporation of zinc at the high operating temperatures (typically 500 to 600 °C). Better catalysts were obtained by using gallium instead of zinc as promoter on ZSM-5 or a gallium silicate of ZSM-5 structure in which gallium (at least partly) replaced the usual aluminum atoms in tetrahedral positions in the zeolite framework. Such gallium-containing catalysts have been investigated by, e.g., Davies and Kolombos of British Petroleum [227].

The Cyclar process [228] is a DHCD process resulting from a joint development of British Petroleum Co. PLC (BP) and Universal Oil Products Inc. (UOP). The process converts LPG to aromatics and dry hydrocarbon gas using a BP-developed catalyst. The process is carried out in a similar reactor system to

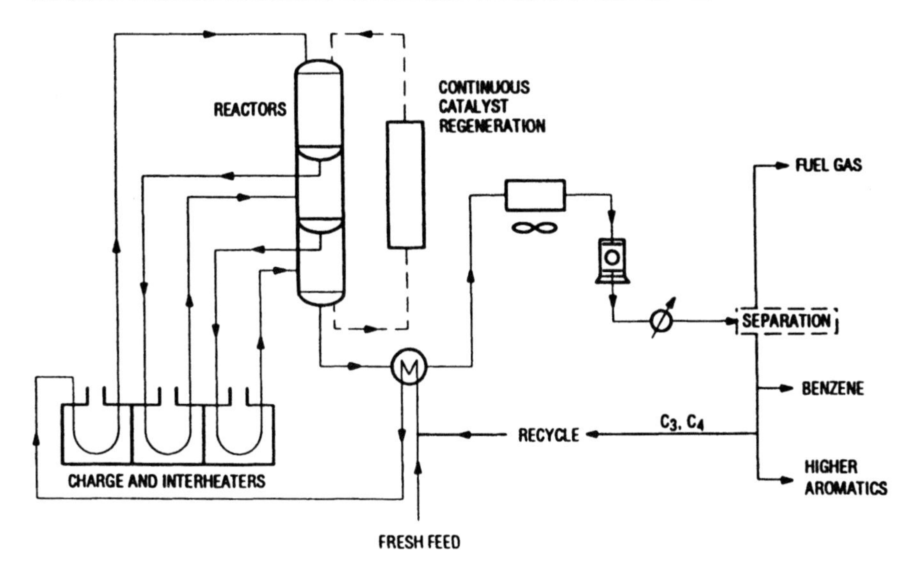

Fig. 7.57. Reactor section of the DHCD process for aromatization of LPG

that applied in UOP's continuous catalyst regeneration platforming process, in which the reaction takes place in a number of vertically stacked moving-bed radial-flow reactors, with interheaters in between to supply the heat for the endothermic conversion. Catalyst is withdrawn from the bottom reactor, regenerated and supplied to the top reactor without interrupting the process flow. A simplified scheme of the process is shown in Fig. 7.57.

The liquid product of the DHCD process, which is obtained in some 50–65 wt% yield, consists largely of aromatic hydrocarbons in the BTX range. It can be used as a high-octane gasoline blending component but it also constitutes a source of aromatics for petrochemical applications.

A demonstration unit for the Cyclar process has been erected at Grangemouth, Scotland, and has been in operation for more than a year [229]. An improved catalyst for this process has recently been announced [230].

7.7 Application of Zeolites in the Chemical Industry

7.7.1 Introduction

In this part of the chapter we report on the progress that has been made and is being made in the application of zeolites in chemical processes. As an introduction to this area of zeolite application, it may be worthwhile itemizing those aspects which distinguish chemical processes from oil processes in their most general terms.

Generally, chemical processing is associated with the reaction of a well-defined feedstock containing only a few components into a (single) well-defined product of relatively high value in which the reaction selectivity to the (single) desired product is of prime importance. We are normally dealing with molecules which contain heteroatoms and which are often unstable and feature polyfunctionality. These molecules usually have critical diameters for diffusion in excess of the pore diameters encountered in zeolites. Oil processes, on the other hand, almost invariably only involve transformations of hydrocarbons and frequently require only the production of a broad range of similar products from a complex reactant mixture. In such a case the requirements on reaction selectivity, although far from trivial, are less stringent than for the typical chemical process.

Two main findings emerge from a study on the application of zeolites as catalysts in chemical processes. The first finding is that between 1970 and 1985 the only processes that have been commercialized are in the largest class of reactions identified, namely in the class where both acid catalysis and shape selectivity are utilized for the production of pure hydrocarbons. Not surprisingly, reactions of this class lie closest to the reactions exploited with success in the oil processing industry. The second finding is that there is at present a lively interest in studying zeolite catalysis of a wide range of chemical reactions [231, 232]. Here, however, it is not always clear just what specific property of the zeolite is expected to give it an advantage over a more conventional catalyst. On the other hand, that zeolites offer equally exciting and profitable applications for chemicals outside the commodity chemicals business as those for the oil and petrochemical industry is illustrated, for example, by commercial processes which have recently been developed for selective oxidation and ammoxidation making use of a new class of "redox" zeolites.

7.7.2 Acid-Catalyzed Reactions Giving Hydrocarbon Products

7.7.2.1 *Ethylbenzene from Benzene plus Ethylene*

The conversion of benzene and ethylene to ethylbenzene over ZSM-5 catalyst constitutes the heart of the Mobil/Badger process [233]. This process was first brought on stream in 1980 by American Hoechst, Bayport, Texas. This high-temperature vapor-phase alkylation process, which is an alternative to liquid-phase processes, (by Monsanto or UOP) based on $AlCl_3$- or BF_3-type catalysts, offers various advantages such as greater simplicity, higher energy efficiency, elimination of corrosion-resistant materials, and the production of non-polluting effluents. With the ZSM-5 catalyst an excellent selectivity of higher than 98% is obtained at a benzene conversion level of about 20%. Since 1981, about 90% of the new plant capacity worldwide has used the Mobil-Badger ethylbenzene technology, and currently the combined capacity of these plants is estimated at nearly 7,000,000 t/a [234].

The Chemistry. ZSM-5 shows a relatively high activity in this process: a weight hourly space velocity of up to 400 kg feed per kg of catalyst per hour is achieved at an optimum temperature of about 400 °C, an outlet pressure of 15–30 bar and a benzene/ethylene molar ratio between 5 and 20. Although other zeolites (e.g., faujasites) have been tested for this reaction [235], ZSM-5 was the first catalyst to be used in practice on account of its much lower coke production and thus much lower rate of deactivation. These effects are due to the three-dimensional channel structure of ZSM-5, as opposed to the cage and window structure or pore dimension/distribution of the other zeolites; probably they are also due to the greater spacing of the acid sites in ZSM-5. Owing to the low coke-forming tendency of ZSM-5, cycle times of up to 40–60 d between regenerations have been attained.

The reaction involves carbenium ion chemistry and gives a broader range of products than the Lewis acid catalyzed processes; the products include propylene, butylene and hexene from ethylene oligomerization/cracking and the corresponding alkylated species. To reduce the polymerization reactions, a high ratio of benzene to ethylene is maintained in the process. The operating conditions are sufficiently different from those for the xylene isomerization process (see below) for disproportionation to be avoided [236].

The uniform channels of the ZSM-5 zeolite permit the entrance of the feed molecules as well as the leaving of the product molecules ethylbenzene and diethylbenzene, while higher alkylated products are restricted from leaving. Any of the latter products formed within the channels of the zeolite are forced to undergo transalkylation or dealkylation to facilitate their diffusion back to the bulk. Hence, in this case, the formation of heavier by-products from the desired primary product is a reversible process. Steric hindrance of the necessary transition states is also believed to inhibit the formation of polyalkylated products.

The Process. A process flow scheme for the Mobil/Badger ethylbenzene process is given in Fig. 7.58 [236]. To control the exothermic reaction (reaction enthalpy = -105.8 kJ/mol ethylbenzene), the reaction is carried out in an adiabatic multistage reactor. For an optimal temperature control and for an optimal ethylene to benzene ratio over the reactor, the feedstocks are injected between the catalyst beds. The by-product diethylbenzene can be converted with benzene to ethylbenzene in the same reactor or in a separate transalkylation reactor.

The process is operated in two reactions in a swing mode of operation, in which one reactor is in operation while the catalyst is regenerated in the other.

New Developments. Recently, a novel zeolite-Y-catalyzed ethylbenzene process was announced by CD-Tech. The process employs catalytic distillation. The reaction is carried out at lower temperature and pressure than the Mobil/Badger process. The claimed advantage of the CD-Tech process is a higher catalyst stability, which implies a saving on the capital investment required for the parallel reactor for regeneration needed in the Mobil/Badger process. The CD-Tech process is in operation at Nippon Steel Chemical Co. (Japan) in a 200,000 tons/year styrene production plant [237].

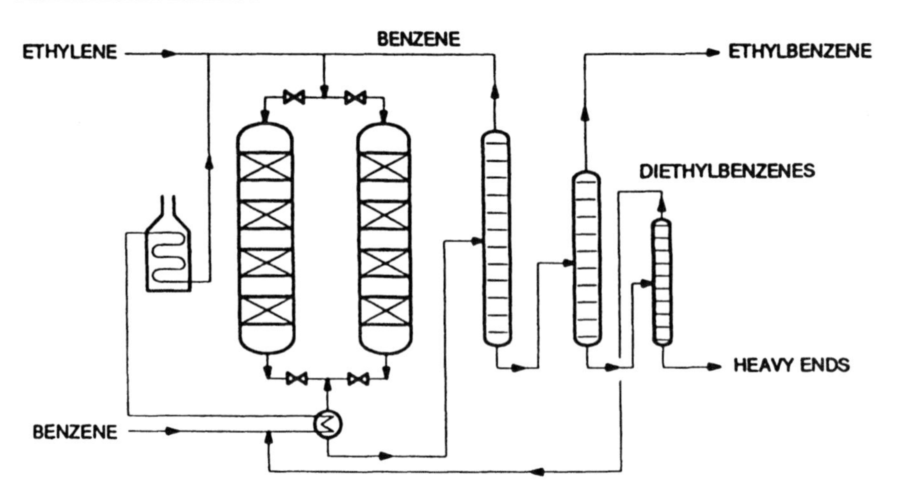

Fig. 7.58. The Mobil Badger ethylbenzene process [236]

Other alternatives to the vapor-phase EB technology involve the liquid-phase alkylation. Such processes also use zeolitic catalysts, and several units are in commercial operation [234, 238]; these have the advantage of decreased by-products yields.

7.7.2.2
Isopropylbenzene (Cumene) from Benzene plus Propylene

Cumene has for a long time been manufactured by the UOP process (already developed during World War II), which operates at 200–250 °C, and uses supported phosphoric acid (SPA) [238, 239]. The yield of cumene is typically some 96 %, with an added transalkylation reactor 98 %. In the mid 1980s Monsanto developed a process based on $AlCl_3$, similar to their EB process, but also with its disadvantages.

Application of the Mobil/Badger EB catalyst ZSM-5 for cumene manufacture has drawbacks since the (relatively) narrow pore zeolite favors isomerization to n-propylamine, and inhibits diffusion of the polyalkylated species [234]. Therefore 12-ring zeolites have been selected: modified mordenite zeolites by Dow, because of their activity and transalkylation capability [240], zeolite beta by Enichem [241], with a very high selectivity, and, last but not least, in a new Mobil process [234], with a new proprietary catalyst (supposedly MCM-22, [239]), which also features alkylation and transalkylation sections, and reportedly gives high yields of cumene of a high purity. Mention should also be made of the CD-Tech process based on catalytic distillation, where long catalyst life and high purity cumene are claimed [239, 242], and of the UOP Q-max process of which no details on the catalyst have been published [238, 239].

7.7.2.3
Higher Alkylbenzenes

Manufacture of higher alkylbenzenes, such as those with 10–14 carbon atoms in the alkyl chain, are important intermediates for alkylbenzene sulfonates used as detergents. For reasons of biodegradability nowadays linear alkenes are used (e.g., from ethylene polymerization) generating linear alkylbenzene sulfonates (LABS). Conventional processes achieved this alkylation step with liquid HF as a catalyst, but more recently processes using solid catalysts [238, 243] including zeolites have been developed. Interestingly, with medium pore zeolites, one finds a preferential attachment of the phenyl ring to the end of the alkyl chain, for kinetic reasons, while with larger pore zeolite catalysts such as REY, a broader mixture of isomers is obtained, as with liquid HF.

7.7.2.4
p-Ethyltoluene from Toluene plus Ethylene

Direct alkylation of toluene with ethylene can be steered to give *p*-ethyltoluene selectively (see Table 7.19, [113]). For this purpose molecular shape selectivity has been used to produce a different substitution pattern from that normally found with conventional electrophilic substitution. The effect of the modifications of ZSM-5 on the selectivity can be traced back to changing the diffusion properties (especially of the slowest diffusing species) [244]. Mobil has commercialized a process to produce *p*-ethyltoluene using these catalysts [234]. *p*-Ethyltoluene can be dehydrogenated, in an analogous manner to ethylbenzene, to yield *p*-methylstyrene, which has potential as a new monomer.

Table 7.19. Alkylation of Toluene with Ethylene over Modified ZSM-5 Catalyst (reproduced from [113])

	Product composition, wt%
Light gas and benzene	0.9
Toluene	86.2
Ethylbenzene and xylenes	0.5
p-Ethyltoluene	11.9
m-Ethyltoluene	0.4
o-Ethyltoluene	0
C_{10+} Aromatics	0.1
Total	100
Ethyltoluene isomers, %:	
para	96.7
meta	3.3
ortho	0
Total	100

7.7.2.5
Alkylation of Binuclear Aromatics

The alkylation of higher aromatics is of considerable interest: alkylnaphthalenes are precursors to fine chemicals such as vitamin K, and to speciality polymers [245, 246]. Zeolites, especially ZSM-5, modified mordenites and ZSM-12 have been studied; a brief review with references is available [234]. High-silica mordenites have also been studied for the propylation of biphenyl, to yield preferably 4,4′-DIPD [247].

7.7.2.6
Xylenes Production: Isomerization (Including Ethylbenzene) and Toluene Disproportionation

Xylenes can be produced directly, for example from reformate or from pyrolysis gasoline, or indirectly by alkylation of benzene and toluene or by toluene disproportionation. In general, as the supply pattern differs from the demand pattern there is a need for various inter-conversion processes which are almost invariably acid-catalyzed. A great deal of work has been reviewed, principally by Mobil using variously modified zeolites, mostly based on ZSM-5, for carrying out many of these reactions [113, 119, 248, 249].

Xylene Isomerization. The principal industrial interest is in *para-* and *ortho-*xylenes, and various isomerization processes have been developed to increase the yield of these isomers. Basically, two types of xylene isomerization processes exist, viz. processes that operate at moderate pressure with hydrogen circulation and processes that operate at low pressure without hydrogen circulation. Currently, about 80% of the xylene isomerization units throughout the world operate with hydrogen circulation. The total world capacity for xylene isomerization is about 2.5 million tons per annum.

Apart from the use of hydrogen, another characteristic which distinguishes the various xylene isomerization processes is their ability to convert ethylbenzene, which is present in the virgin C_8-aromatics feed in quantities between 10 and 40%. In order to avoid the build-up ethylbenzene in the recycle loop, the ethylbenzene is either recovered via distillation or extraction (UOP-Ebex process), or it is converted in the isomerization reactor. In the majority of commercial processes ethylbenzene is converted. Depending on the catalyst type and the process conditions the ethylbenzene is isomerized to xylenes, disproportionated to benzene and C_{8+}-aromatics, or dealkylated to benzene and ethylene.

Commercial xylene isomerization units which can also effect the isomerization of ethylbenzene into xylene isomers utilize bifunctional catalysts having hydrogenation/dehydrogenation capacity and acidic properties, e.g., platinum on silica-alumina. In the presence of hydrogen, these catalysts allow the isomerization of ethylbenzene to xylenes via hydrogenated intermediates, as shown in Fig. 7.59 [113]. However, with these catalysts non-selective reactions such as disproportionation to toluene and trimethylbenzenes and hydrogenolysis occur, which may cause an important loss in xylene selectivity. The contribution

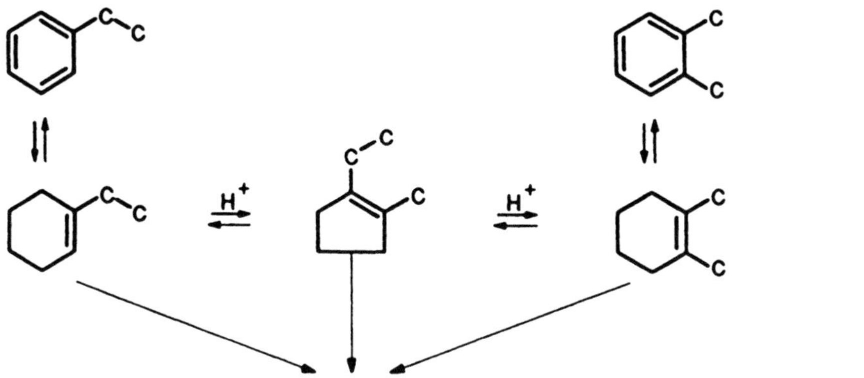

C_1 - C_7 PARAFFINS - NAPHTHENES

Fig. 7.59. Isomerization of ethylbenzene to xylenes over bifunctional catalysts [113]

of disproportionation to the loss of xylenes can be reduced either by diluting the feed stream with toluene [235] or by applying zeolites in the bifunctional catalyst. With specific zeolites such as mordenite [250, 251] the ratio of the rates of xylene isomerization and disproportionation is dramatically increased compared to that found with amorphous acids. Currently, most commercial processes use zeolites as acidic components in their catalysts. Leading processes capable of isomerizing ethylbenzene to xylenes are the Isomar (UOP) and the Octafining (Engelhard) process, with a total market share of about 40% worldwide. Several companies, including Mobil, ICI, IFP and Zeolyst, seem active in designing dual functional catalysts based on zeolites for xylene isomerization, which could also convert ethylbenzene to xylene. Mobil, for example, have claimed [252] the use of ZSM-23 modified with Pt whilst ICI [253] use zeolite FU-1 modified with Ru and Pt, Ir or Ni.

As can be seen in Table 7.20 [248], with ZSM-5, disproportionation of xylenes occurs at a very low rate even compared to that found with other zeolites. The smaller diameter of ZSM-5 means that the formation of the bulkier, bimolecular transition state for disproportionation is inhibited while the smaller intermediate for the 1,2-methyl shift for isomerization is unaffected. Mobil have commercialized three different xylene isomerization processes using ZSM-5-type catalysts. Acceptane of Mobil xylene isomerization technology has been extensive. Currently, about 40% of the xylene isomerization capacity in the world uses Mobil technology.

In the Mobil Vapor Phase Isomerization process, commercialized in 1973, the isomerization is carried out between 315 and 375°C with a monofunctional ZSM-5 catalyst. Although pure ZSM-5 is itself inactive for the conversion of ethylbenzene to xylenes it does convert this material to benzene and C_{8+} aromatics, and thus prevent its build-up in recycle streams. A highly selective reaction to xylenes over ZSM-5 requires higher transalkylation rates for ethylbenzene than for xylene. Due to the combined effect of configurational diffusion and restrict-

Table 7.20. Xylene Isomerization with ZSM-5 Catalyst; Mobil Vapor-Phase Isomerization Process (reproduced from [249])

	Feed composition, wt%	Product composition, wt%
C_1–C_5	0	0.2
Benzene	0	1.6
Toluene	0	1.0
Ethylbenzene	14.0	
Xylene		
para	10.0	20.1
meta	64.0	45.0
ortho	12.0	18.4
C_{9+} Aromatics	0	3.7

ed transition state selectivity, the rate of transalkylation is more than 100 times higher for ethylbenzene than for xylenes [113].

For un-extracted feeds Mobil have developed the Mobil High-Temperature Isomerization process, applying a low acidity ZSM-5 catalyst which contains a trace concentration of platinum [68]. The process operates at temperatures above 425 °C to hydrocrack the C_{8+} aliphatics to lighter products and hydrodealkylate ethylbenzene to benzene and ethane. Here, too, the xylene losses due to disproportionation are claimed to be low.

The Mobil Low-Pressure Isomerization process is an example of a process where no hydrogen is applied. The process was first commercialized in 1978. Because of the absence of hydrogen, catalyst stability is lower than in the other processes, limiting operation cycles to about six months.

The purpose of all xylene isomerization processes is identical, viz. to take a mixture of *ortho-*, *para-* and *meta*-xylene, and to selectively isomerize this mix to the desired ratio of isomers. The xylenes feed includes ethylbenzene, which must also be converted in the isomerization reactor, because fractionation is complex as a consequence of the close boiling points of ethylbenzene and *para*-xylene. Although normally the *para*-isomer is the desired product, *ortho-*, and *meta*-xylene are co-produced in many cases. All the processes employ a process set-up (see Fig. 7.60, [254]) which includes a *para*-xylene separation unit, a xylene isomerization reactor and a distillation equipment. The *para*-xylene separation is effected by adsorption or crystallization.

Toluene Disproportionation. Toluene disproportionation is a further reaction of alkylbenzenes that shows improved selectivity over ZSM-5. Due to the steric hindrance for the production of polymethylated species, disproportionation is limited to the formation of benzene and xylene. Two process versions for toluene disproportionation were developed based on ZSM-5 as catalyst, viz. the Mobil Toluene Disproportionation (MTDP) and the Mobil Selective Toluene Disproportionation (MSTDP) process.

The MTDP process has been in operation at Mobil's Naples petrochemical complex in Italy since 1975. The process operates at a significantly higher tem-

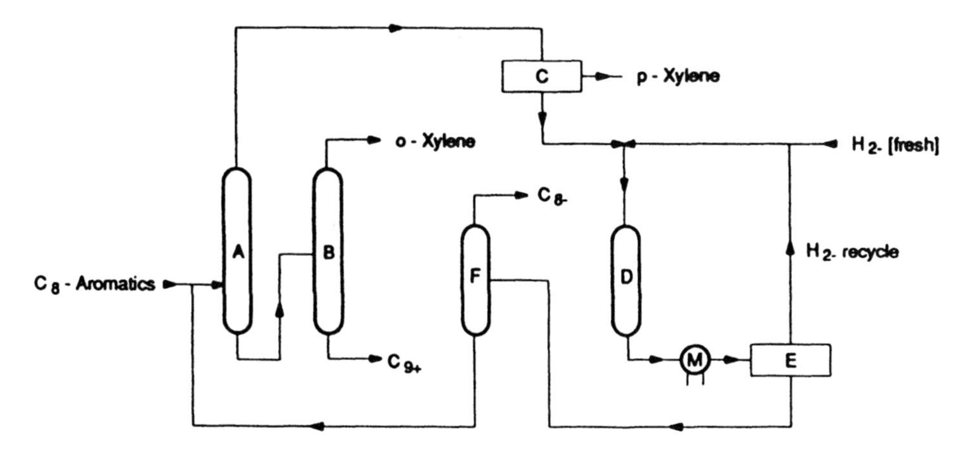

Fig. 7.60. Isomerization of C_8-aromatics [255]

perature than the isomerization processes, because of the much lower activity of the catalyst for toluene disproportionation than for xylenes isomerization. However, at temperatures above 450 °C toluene is disproportionated selectively to an almost equilibrium mixture of benzene and xylenes. Because the steric constraints in pure ZSM-5 are greatly reduced at the higher operating temperature, the xylene isomer distribution is about at equilibrium [119]. Another process for the disproportionation of toluene is the Tatoray process developed by Toray Industries.

As illustrated in Fig. 7.61 [244], when ZSM-5 is modified (with phosphorus, magnesium, silica or, especially, coke) the selectivity to *para*-xylene is considerably enhanced compared to the equlibrium mixture obtained with the unmodified catalyst. This is related to the differences in diffusional effects, as is implied in Fig. 7.61, and elucidated elsewhere [244]. For the commercial MSTDP process selectivities to *para*-xylene between 85 and 95% are claimed [255, 256]. The Enichem refinery at Gela, Italy, gave the first commercial demonstration of the technology in 1988–1990, with a cycle life of 500 days, and an excellent regenerability. The technology has since been commercialized in several other plants [234].

An analogue of selective toluene disproportionation is selective disproportionation of ethylbenzene; the selective manufacture of *p*-di-ethylbenzene is now practised commercially with a modified ZSM-5 catalyst [234].

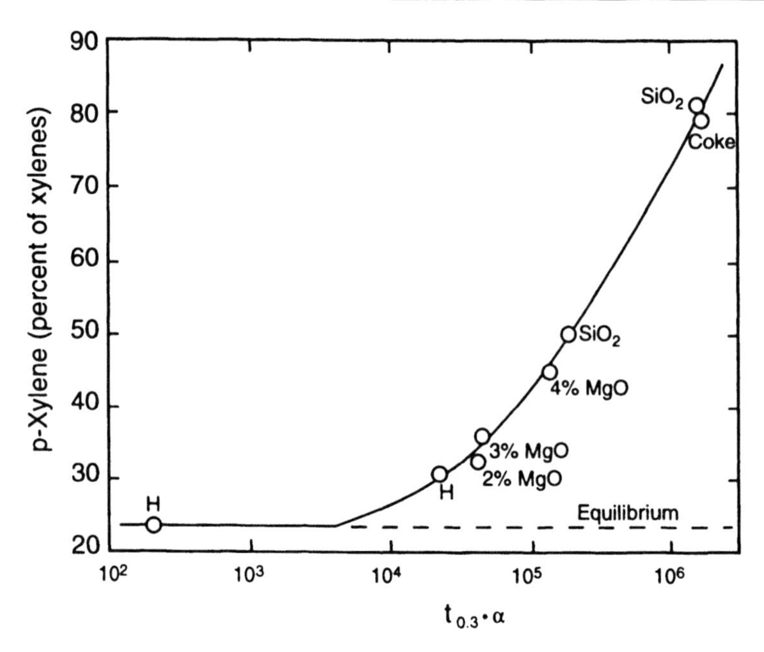

Fig. 7.61. Dependence of *para*-xylene selectivity on pseudo-Thiele modulus. Toluene disproportionation at 550 °C, 20% conversion [240]

7.7.3 Oxidation and Ammoximation Processes

In recent years, the use of zeolites in the area of selective oxidation and ammoxidation reactions has been studied extensively [231, 232, 257]. The discovery of titanium silicalite (ZSM-5 type, MFI structure [258]), TS-1, as a selective oxidation/ammoxidation catalyst highlights the progress in this area of importance to the chemical industry. TS-1 has been reported to catalyze a wide variety of oxidation reactions; worthwhile mentioning are ketone ammoximation, phenol hydroxylation, olefin epoxidation and alcohol oxidation with 30% H_2O_2 under mild conditions (typically 20–150 °C) [257, 259, 260]. Essential to the selectivity of this catalyst appears to be the optimal dispersion of the titanium sites. Isolated tetrahedrally coordinated Ti^{4+} ions have been reported to be the active centers [241, 257, 258]. More generally the concept implies isolation of redox metals in silicalites, zeolites, aluminophosphates (AlPOs) and silicoaluminophosphates (SAPOs) [257]. The zeolites have a preferred SiO_2/TiO_2 molar ratio between 35 and 65. Due to the rigid pore geometry, this type of catalyst provides high shape selectivity towards substrates, intermediates or transient complexes and products, which may even offer opportunities for selective oxidation of single compounds in complex mixtures, but hamper their applications for oxidation of bulky molecules, of interest for fine chemicals.

Enichem have already taken the use of Ti silicalites for aromatics hydroxylation to commercialization. Since 1985, Enichem have operated a 10,000 tons/year plant at Ravenna, Italy, for the production of hydroquinone and catechol from phenol [261]. In this process hydrogen peroxide is used as an oxidant. TS-1 is furthermore effective in the oxidation of a wide range of organic molecules. In a new Enichem process cyclohexanone oxime is directly processed by ammoximation of cyclohexanone with ammonia and hydrogen peroxide [262, 263]. Furthermore, a process for the production of propylene oxide making use of this catalyst has been developed [264].

A large variety of oxidation catalysts have been explored since the breakthrough of the TS-1-type catalysts, for instance, Ti and V silicalites having the MEL structure, TS-2 (inferior performance compared with TS-1 [258]) and VS-2 (this catalyst also oxidizes the side chains of alkyl-substituted aromatic hydrocarbons [265, 266]), vanadium-modified ZSM-48 [258]. CrAPO-5 and CrAPO-11 have also been reported to be efficient oxidation catalysts (benzylic, allylic and secondary alcohol oxidation [257]. With respect to the efficiency of the oxidant H_2O_2 the following ranking has been reported for the metals in metal-silicates: Ti- > V- ≈ Sn- ≈ As-silicates [267]. More alternative catalysts will be discussed in Sect. 7.7.3.1 (hydroxylation).

The range of molecules that can be oxidized over TS-1 and TS-2 is limited as a result of its relatively small pore size (about 0.6 nm). In fine chemistry, oxidation of larger molecules is important [268]. Ti-modified beta with larger pores showed less steric hindrance in the oxidation of cycloalkanes and cycloalkenes, although a real step forward is the synthesis of mesoporous Ti-silicate (able to catalyze the oxidation of 2,6-di-*tert*-butyl phenol [268–270], Sn-modified MCM-41 (for the oxidation of large molecules like 2-methylnaphthalene to naphthaldehyde) [271].

7.7.3.1
Hydroxylation of Phenol with Hydrogen Peroxide

By 1990 the total world production of hydroquinone (used for the production of photographic developers, polymerization inhibitors and dyes [272]), and catechol amounted to 50,000 tons per year, of which about 50% was produced via H_2O_2 processes. As illustrated in Fig. 7.62 [273], three different processes were developed for the hydroxylation of phenol with hydrogen peroxide, each representing a different reaction mechanism. In the Rhône-Poulenc process strong homogeneous acid catalysts such as $HClO_4$ or H_3PO_4 are employed, in the Brichima process a Fenton's reagent made of Co(II) and Fe(II) is used, and in the Enichem process the titanium silicalite is likely to be active through a titanium peroxo complex. The features of the different processes are shown in Table 7.21 [261]. Compared to the homogeneously catalyzed processes, the zeolite-catalyzed process can operate at substantially higher conversion levels of phenol, while maintaining an acceptable level of tar formation and a high selectivity on both H_2O_2 and phenol. The higher phenol conversion per pass significantly improves the process economics because of a reduction of the phenol recycle. Moreover, the zeolite-catalyzed process produces the *para*-isomer at significantly higher selectivity (catechol/hydroquinone ratio equals about 1), suggesting shape selectivity of the TS-1 zeolite for this reaction. In their review on the

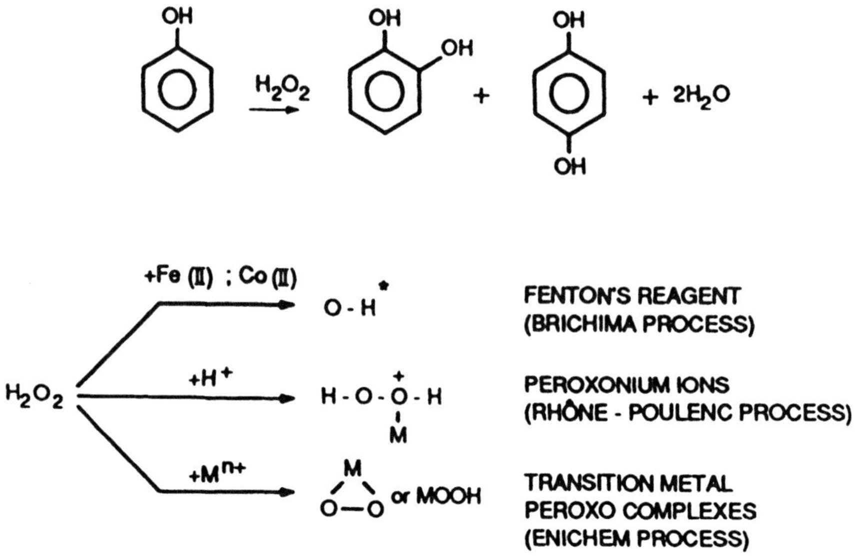

Fig. 7.62. Hydroxylation of phenol with H_2O_2: schematic representation of the mechanism

Table 7.21. Phenol Hydroxylation with Processes Using Hydrogen Peroxide as Oxidant

Process	Rhône-Poulenc ($HClO_4$, HP_3O_4)	Brichima [Fe(II), Co(II)]	Enichem (TS-1)
Phenol conversion, %	5	10	25
H_2O_2 yield, %	70	50	70
Phenol selectivity, %	90	80	90
Ratio catechol/hydroquinone	1.4	2.3	1
100* tars/(tars + diphenols)	10	20	12

hydroxylation of phenol, Parton et al. [273] conclude that both the reaction mechanism and the pore geometry are selectivity-determining parameters. Detailed studies showed that in TS-1, external sites give catechol and coking, while internal sites give hydroquinone [274].

Following the success with TS-1 as hydroxylation catalyst a great variety of other molecular sieves have been explored. Metal complexes (e.g., iron(II)-phenanthroline) supported by mesoporous MCM-41 open new routes for phenol hydroxylation to hydroquinones and catechol [275]. The Sn^{4+} centers in Sn-modified MCM-41 are accessible for the hydroxylation of large molecules like 1-naphthol [271], whereas Ti-modified MCM-41 gave surprisingly high hydroquinone yields [272]. Transition metal aluminophosphates, such as MeAPO-11 (Me = Fe, Co, Mn and FeMn) have also been synthesized and evaluated for the hydroxylation of phenol to catechol and hydroquinone, although not leading to significantly better performances [276].

7.7.3.2
Epoxidation of Propylene

Propylene oxide is produced commercially via chlorohydrin or via hydroperoxide processes with a world-wide production of about 3 million tons per year. About 50% is produced via the hydroperoxide routes. Commercial hydroperoxide processes based on *tert*-butyl hydroperoxide (TBHP) and on ethylbenzene hydroperoxide (EBHP) have been developed on the basis of homogeneous and amorphous heterogeneous catalysts. Currently, as an alternative process, the epoxidation of propylene with hydrogen peroxide using zeolite TS-1 as a catalyst is being developed by Enichem.

A potential advantage of the hydrogen peroxide process is that in principle no co-product is formed. This is in contrast with the hydroperoxide processes where economics are influenced by co-product demand. (In the hydroperoxide processes *tert*-butyl alcohol or methylphenylcarbinol are co-produced, which are generally converted to ethyl *tert*-butyl ether and styrene, respectively). From this point of view, titanium silicates seem to be very promising catalysts for the epoxidation of propylene; however, the attractiveness of a large-scale commercial utilization of this process for propylene oxide production will strongly depend on the cost of hydrogen peroxide.

For the epoxidation reaction, the zeolites are claimed to be effective with low concentrations of H_2O_2 in water. Selectivities are reported to be about 95% based on H_2O_2 converted at about 40 °C and using a feed of H_2O_2/olefin with a molar ratio of typically 1/2. Similar values were found in both batch and continuous experiments. In the latter case productivities of typically 4 kg/(l · h) were reported.

The TS-1 catalyst has roughly the same elementary composition as the catalyst used in the commercial Shell process for the epoxidation of propylene with EBHP as an oxidant. This similarity in catalyst composition suggests that similar mechanisms are operative. Notari [261] demonstrated that even the effects on the selectivity on silylation of the catalysts are similar. For the epoxidation reaction it turns out that it is crucial for a good overall selectivity that the silica-bound titanyl (Ti = O) species are isolated from each other [277, 278]. On the other hand, an important role of the zeolite structure in terms of shape selectivity is considered unlikely.

Titano-silicates with MCM-41 and MCM-48 structure enable through their mesoposity the catalytic oxidation of bulky olefins such as norbornene in the presence of bulky oxidants like *tert*-butyl hydroperoxides [279]. These reactions cannot be catalyzed by TS-1 due to its small pores. The pores of the MCM-41-type molecular sieves are large enough for the placement inside of Ti-based organometallic complexes, which then serve as oxidation catalysts [279]. An alternative approach might by the so-called "ship-in-a-bottle" catalysts, e.g., Fe-phthalocyanine encapsulated in faujasites (Y and X) and the 18-membered ring aluminophosphate VPI-5 [257]. A new complex catalytic system (mimicking in some way enzymes) are zeolites embedded in a PDMS membrane. An encaged *cis*-Mn-bis-2,2′-bipyridyl complex has been reported to catalyze the epoxidation of olefins with *tert*-butyl hydroperoxide, while making the use of a solvent obsolete [280].

7.7.3.3
Ammoximation of Cyclohexanone

Cyclohexanone is an intermediate in the caprolactam process. The current technology to caprolactam involves the cyclohexanone oximation with hydroxylamine sulfate, leading to the co-production of ammonium and sodium sulfate. Cyclohexanone oxime is directly synthesized in the new Enichem process via direct synthesis, ammoximation of cyclohexanone with ammonia and H_2O_2 on TS-1 [281]. Via this route several oximes can be produced from their corresponding ketones. The ammoximation of cyclohexanone to cyclohexanone oxime with TS-1 avoids the co-production of these salts and may therefore be of industrial interest for environmental reasons. Moreover, high conversions and high selectivities can be obtained with this process under mild conditions. The reaction is optimally carried out in a water/*tert*-butanol liquid phase at a temperature between 80 and 95 °C. At conversion levels of about 90%, selectivities between 96 and 99% of cyclohexanone to cyclohexanone oxime and between 89 and 95% on hydrogen peroxide were obtained [262]. The unique features of the TS-1 catalyst for this reaction were illustrated by comparing conversion and selectivities with thermal reaction and Ti on amorphous SiO_2, as shown in Table 7.22 [262].

7.7.4
Amination

Methylamines, which are of considerable commercial interest, can be produced from methanol and ammonia by means of nucleophilic substitution. They are important intermediates in the product of solvents, insecticides, herbicides, pharmaceuticals and detergents [282]. In 1992 a production capacity of 500,000 t/a was reported [282]. The aim in most cases is to obtain a reaction product having a composition different from equilibrium distribution and con-

Table 7.22. The Ammoximation of Cyclohexanone to Cyclohexanone Oxime with H_2O_2 and NH_3; the Influence of the Catalyst (reproduced from [262])

Catalyst	Ti, %	H_2O_2/cyclohexanone, molar ratio	Cyclohexanone conversion, %	Oxime, %	Oxime yield on H_2O_2, %
None	–	1.07	53.7	0.6	0.3
SiO_2 amorphous	0	1.03	55.7	1.3	0.7
Silicalite	0	1.09	59.4	0.5	0.3
TiO_2/SiO_2	1.5	1.04	49.3	9.3	4.4
TiO_2/SiO_2*	9.8	1.06	66.8	85.9	54.0
Ti-silicalite	1.5	1.05	99.9	98.2	93.2

Catalyst concentration: 2 wt%
Temperature: 80 °C
NH_3/H_2O_2 molar ratio: 2.0
Reaction time: 1.5 h except for experiment indicated with star (*): reaction time 5 h.

taining as much mono- or dimethylamine as possible. In particular the dimethylamines are the more favored products [282]. For instance ethylene diamine has a major industrial importance for the production of fungicides, chelating agents, epoxy curatives, hot melt adhesives, corrosion inhibitors and lubricating oil and fuel additives [283].

BASF uses the direct amination of olefins over zeolites to produce *tert*-butylamine [284]. For the production of alkylamines with 2 to 10 C-atoms, e.g., *tert*-butylamine, dealuminated Y-zeolites have been found to be stable catalysts [285]. Lower aliphatic amines are important intermediates for the synthesis of drugs, and also, e.g., for bactericides, herbicides, rubber accelerators, corrosion inhibitors, extraction agents in the production of penicillin and surface agents [285]. Du Pont have described [286] the use of NaH-mordenite (containing 2–4.3% Na) and similar zeolites for the production of monomethylamine in high selectivity, compared with di- and trimethylamine at atmospheric pressure and ca. 400 °C. At methanol conversions of about 40% and with C/N feed ratios of 1.0, selectivities above 80% to monomethylamine were obtained in contrast with values of ca. 30% over amorphous silica-alumina. Monomethylamine so produced may be disproportionated over the same catalyst, if desired, to give dimethylamine in high selectivity [287]. The zeolite in this instance is found to be both active and selective, irrespective of whether water (a byproduct of the above reaction) is present or not; this provides added processing simplicity. Silica-alumina was found to be non-selective, whereas alumina was only selective for the desired disproportionation in the absence of water. Selectivities to dimethylamine of >90% were found for conversions of monomethylamine of >70% at atmospheric pressure and 400 °C.

It has been established [288] that the elimination of trimethylamine depends on the size of the zeolite. It was demonstrated that small pore zeolites such as H-Rho, H-ZK-5 and chabazite yield higher dimethylamine selectivities than other zeolites. Hence, in the reaction both the acidic and the shape-selective properties of the zeolite catalyst are being utilized. With respect to the acidity both the number and the strength of the acidic sites were found to be the controlling factors on the amination of 2-methylpropene, in particular H-ZSM-5 zeolite with a silica/alumina ratio of 81 was established to provide the optimal activity [289]. In the production of ethylene diamine by amination of ethanolamine with ammonia the selectivity depends on the degree of dealumination of the H-mordenite catalyst, moderate dealumination giving the best selectivity [283]. The shape selectivity of for instance H-ZK-5 can be enhanced by deactivation of the outer surface with tetramethoxysilane [282].

Japanese workers have also studied the direct reaction of methanol with ammonia to dimethylamine [290, 291]. They also found that mordenite is effective, especially when protons are partially exchanged with lanthanum or alkali-metal ions. In 1985, Nitto Chemical commercialized a process in Japan for the selective production of dimetyhlamine from methanol and ammonia, using steam-modified ion-exchanged mordenite [273, 292]. For the amination of bulky olefins to their corresponding amines, zeolites of the type SSZ-26, SSZ-33, SSZ-37, CIT-5 [293, 294], aluminophosphates with FAU structure [295] and pillared clays have recently been reported [296].

7.8 Concluding Remarks

From the foregoing discussion it will be clear that zeolites have found important applications as catalysts in a large number of oil-refining, synfuel-production and petrochemical processes. The most important catalytic function used in these processes is the acidic function. The attractiveness of zeolites as solid acidic catalysts derives for a large part from the fact that the acid strength can be quite high and that it can also be tailored to suit a particular process; e.g., the density of acidic sites can be varied by dealumination. Another feature of zeolites, in particular zeolites of medium pore size, is their ability to discriminate between feedstock, intermediate and product molecules of different size, which forms the basis of shape-selective catalysis. This is particularly true for the selective conversion of a particular compound or a particular type of compounds in complex mixtures (Selectoforming, dewaxing) or the production of monoaromatics from such different feedstocks as methanol (MTG) or light paraffins (Cyclar). The variability of the pore structure offered by the large number of zeolites and related microporous crystalline substances such as the AlPOs offer many possibilities in this context.

A third feature of zeolites is the possibility to substitute other elements for Al in the zeolite framework. This may offer possibilities for types of catalysis other than acidic catalysis, e.g., oxidation catalysis.

The application of zeolites is now well entrenched in oil refining, see Fig. 7.63. The economic benefit derived from the use of zeolites in oil refining, especially catalytic cracking and hydrocracking, is very large and can be estimated at dollar figures of ten to eleven digits per annum.

In the scene of production of transportation fuels, there have been several developments over the past few years. The most important one is undoubtedly the rapidly increasing tightening of the product (gasoline, etc.) specifications for environmental reasons. The gasoline specifications for instance aim at a complete ban on lead, a reduction of the emission of volatile organic compounds, and a decrease of the concentrations of sulfur, benzene and possibly aromatics. Specifications on olefin content are still under debate. Many of these changes will tend to lower the octane number of the gasoline, and hence other components (ethers) to increase the octane number must be manufactured (via alkylation, isomerization). In automotive gas oil the aromatics and sulfur contents will have to decrease, requiring additional hydroprocessing capacity, usually at a higher severity than that currently practised. It is certain that these tightening standards will force the oil industry to make major new investments. A second development observed over the years is the increasing importance of catalytic processing relative to thermal processes or physical separations. In many cases the catalysts used contain zeolites, as shown in Fig. 7.63. Thus, in paraffin isomerization zeolites have become an attractive alternative to the amorphous catalysts; in dewaxing the catalytic dewaxing with shape-selective zeolites is a preferred alternative to solvent dewaxing for newly built equipment, etc. Often the catalytic route also yields environmentally cleaner products, as is illustrated by comparing thermal cracking of residues with hydroconversion.

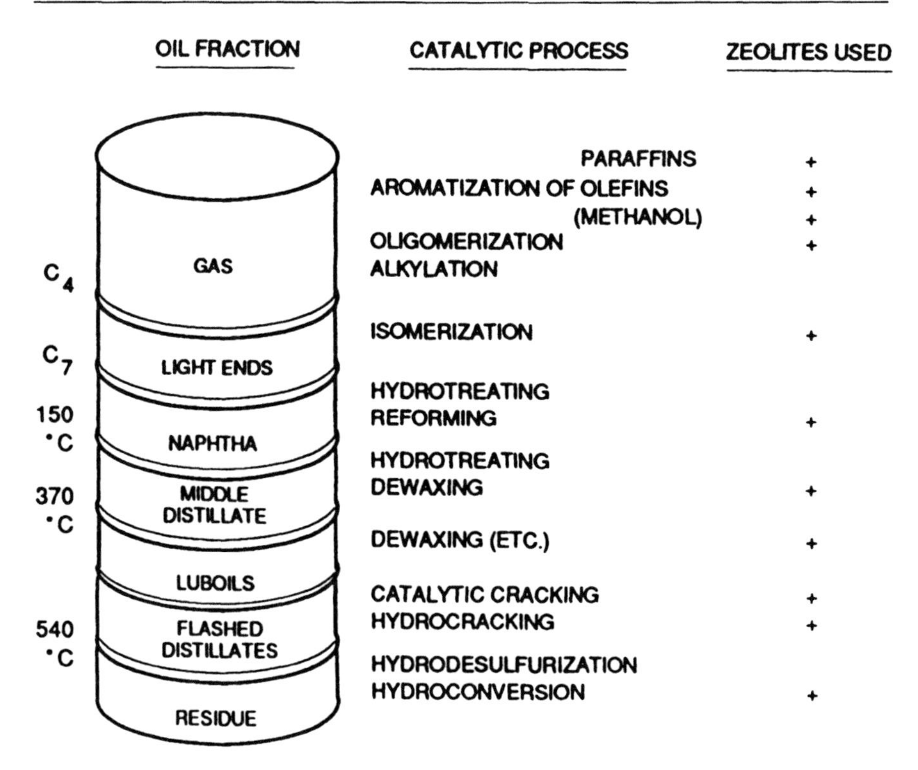

Fig. 7.63. Zeolite catalysis in oil refining

In the scene of chemicals production, processes such as light olefins production from methanol over zeolitic catalysts may increase in importance. For the production of speciality chemicals, zeolites may find increasing application, making use of the possibilities offered by tailorable acidity, shape selectivity and isomorphous substitution. An important area of growth of zeolite catalysis may be their use as solid acids replacing currently used acidic materials such as sulfuric acid, $AlCl_3$, BF_3 etc., which quite often present a problem from an environmental point of view. This is already exemplified by the Mobil/Badger technology for the production of ethylbenzene. Metallosilicates, in particular titanosilicates (TS-1) have become important in oxidation catalysis, while the introduction of metallosilicates with mesoporous structures will find application in fine chemicals. Finally, an area which is rapidly growing in importance is the cleaning of waste streams for environmental conservation reasons. Zeolite catalysis in this area is still in a state of emergence but may well become quite important in the future.

Acknowledgements. The authors are indebted to Messrs. J.P. van den Berg, W. Bosch, P.R. Grandvallet, J.K. Minderhoud, T. Terlouw and J.A.R. van Veen for critically reading parts of the manuscript and to Mrs. A. Derking for literature searches.

References

1. Moscou L (1991) In: van Bekkum H, Flanigen EM and Jansen JC (eds) Introduction to zeolite science and practice. Elsevier, Amsterdam, p 1
2. Naber JE, De Jong KP, Stork WHJ, Kuipers HPCE, Post MFM (1994) In: Zeolites and Related Microporous Materials: State of the Art 1994, Weitkamp J, Karge HG, Pfeifer H, Hölderich W (eds) Elsevier Science BV, Amsterdam, p 2197–2219
3. Maxwell IE, Stork WHJ (1991) In: Van Bekkum H, Flanigen EM and Jansen JC (eds) Introduction to zeolite science and practice, Elsevier, Amsterdam, p 571
4. EU Environmental specifications for market fuels for year 2000, proposed by the Council of Environmental Ministers 19.6.97
5. Van Zijll Langhout WC, Ouwerkerk C, Pronk KMA (1980) Paper presented at the 88th National Meeting of the American Institute of Chemical Engineers, Philadelphia, Pennsylvania, June 8–12
6. Röbschläger KW, Deelen WJ, Naber JE (1992) Proceedings International Symposium on Heavy Oil and Residue Upgrading and Utilization, Chongren H and Chu H (ed) International Academic Publishers Beijing, 249–254
7. Gosselink JW (1998) In: Weber T et al (eds) Transition Metal Sulphides, Kluwer Academic Publishers, Dordrecht, The Netherlands, p 311–355
8. Scheffer B, Van Koten MA, Röbschläger KW, de Boks FC (1998) Catal Today 43:217–224
9. Espeel P, Parton R, Toufar H, Martens J, Hölderich W, Jacos P, this book, chap 6
10. Naber JE, Barnes PH, Akbar M (1988) The Shell residue fluid catalytic cracking process. Commercial experience and future developments, Japan Petroleum Institute, Petroleum Refining Conference, Tokyo, October 19–21
11. Mauleon JL, Courcelle JC (1985) FCC heat balance considerations with heavy feeds, 6th Katalistiks FCC Symposium, Munich, May 22–23
12. Avidan AA, Edwards M, Owen H (1990) Innovative improvements highlight FCCs past and future, Oil Gas J, Jan 8, p 33
13. Advances in fluid catalytic cracking (1987) Catalytica study number 4186 CC
14. Dean RR, Hibble PW, Brown GW (1987) Crude oil upgrading utilizing residual oil fluid catalytic cracking, 8th Katalistiks FCC Symposium, Hungary, June 1–4
15. Shaffer Jr AG, Hemler CL (1990) Seven years of operation proves RCC capability, Oil Gas J, May 28, p 62
16. Biswas J, Maxwell IE (1990) Appl Catal 58: 1–19
17. Naber JE, Barnes PH, Akbar M, Japan Petrol Inst Petr Ref Conf, Tokyo, Japan, 19–21 October 1988
18. Elvin FJ (1991) Fluid cracking catalyst demetallization. Commercial results. NPRA Annual Meeting, March 17–19, San Antonio, TX, paper AM-91-40
19. Akbar M, Borley M, Claverin B, Otto H (1986) Some experiences with FCC octane enhancement, AKZO Catalysts Symposium, May 25–28, paper F-3
20. Young GW, Suarez W, Roberie TG, Cheng WC (1991) Reformulated gasoline: the role of current and future FCC catalysts, NPRA Annual Meeting, march 17–19, San Antonio, TX, paper AM-91-34
21. Martino G, Courty, P, Marcilly (1997) In: Ertl G, Knoezinger H, Weitkamp J (eds) Handbook of Heterogeneous Catalysis, Vol 4, VCH-Wiley, Weinheim, p 1805
22. Naber JE, de Jong KP, Stork WHJ, Kuipers HPCE, Post MFM (1994) In: Weitkamp J, Karge HG, Pfeifer H, Hoelderich (eds) Zeolites and Related Microporous Materials, Stud Surf Sci Catal 84. Elsevier, Amsterdam, p 2197
23. von Ballmoos R, Harris DH, Magee JS (1997) In: Ertl G, Knoezinger H, Weitkamp J (eds) Handbook of Heterogeneous Catalysis, VCH-Wiley, Weinheim, Vol 4, p 1955
24. Helmsing MP, Makkee M, Moulijn JA (1995) In: O'Connor P, Takatsuka T, Woolery GL (eds) Deactivation and testing of hydrocarbon processing catalysts, ACS Symposium Series 364, 1995, p 322
25. Scherzer J (1989) Octane-enhancing, zeolitic FCC catalysts: scientific and technical aspects, Catal Rev Sci Eng 31 (3):215

26. Venuto PB, Habib ET (1979) Fluid catalytic cracking with zeolite catalysts. Dekker, New York, NY
27. Ward JW (1984) In: Leach BE et al. (eds) Applied industrial catalysis. Academic Press, New York, NY, vol 4, p 272
28. Wojciechowski BW, Corma A (1986) Catalytic cracking. Dekker, New York, NY
29. Keyworth DA, Nieman J, O'Connor P, Stanger CW (1987) NPRA Annual Meeting, San Atnonio, TX, paper AM-87-62
30. Sie ST (1992) Ind Eng Chem Res 31:1881; (1993) 32:403
31. Haag W, Dessau MR (1984) Proc 8th Int Congress Catal, Berlin, vol 2, p 305
32. Davison Catalagram, No 76, 1987
33. Ramamoorthy P et al. (1988) A new metals passivator in fluid catalytic cracking, NPRA Annual Meeting, San Antonio, TX, March 20–22
34. Heite RS, English AR, Smith GA (1990) Bismuth nickel passivation effective in FCCU, Oil Gas J, June 4, p 81
35. English AR, Kowalzyk DC (1984) Oil Gas J, July 16, 127
36. Groenenboom CJ, Van Houtert FW, Van Maare J (1984) Metals in cracking catalysts, Ketjen Catalysts Symposium, Amsterdam, Mai 27–30
37. Dwyer FG, Degnan F (1993) In: Magee JS, Mitchell MM jr (eds) Fluid Catalytic Cracking: Science and Technology, Stud Surf Sci Cat 76, Elsevier, Amsterdam, p 499
38. Corma A, Gonzalez-Alfaro V, Orchilles AV (1995) Appl Catal 129: 203
39. Van Wechem HMH (1996) paper presented at The Institute of petroleum. Hydrogen – The refinery green light, London, April 25, 1996
40. Gosselink JW, Stork WHJ, De Vries AF, Smit CH (1987) In: Delmon B, Froment GF (eds) Catalyst Deactivation 1987. Elsevier, Amsterdam, p 279
41. Maxwell IE, Minderhoud JK, Stork WHJ, Van Veen JAR (1997) In: Ertl G, Knoezinger, Weitkamp J (eds) Handbook of Heterogeneous Catalysis, Vol 4, VCH-Wiley, Weinheim, pp 2017–2038
42. Van Dijk A, De Vries AF, Van Veen JAR, Stork WHJ, Blauwhoff PMM (1991) Catal Today 11:129
43. Ladeur P, Bijwaard H (1993) Oil Gas J, April 25, 64
44. Basta N, Reno M (1988) Chem Eng V93 N 32–33 (1/6/88)
45. Stork WHJ (1997) In: Froment GF, Delmon B, Grange P (eds) Hydrotreatment and Hydrocracking of Oil Fractions, Elsevier, Amsterdam, pp 41–67
46. Hoek A, Huizinga T, Esener AA, Maxwell IE, Stork WHJ, Van de Meerakker FJ, Sy O (1991) Oil Gas J, April 22, 77
47. European Patent Specificaton 0 683 218
48. O'Rear DJ (1987) Ind Eng Chem Res 26:2337
49. Gosselink JW, Van de Paverd A, Stork WHJ (1990) In: Trimm DL et al. (eds) Catalysts in Petroleum Refining 1989, Elsevier, Amsterdam, p 385
50. Schulz H, Weitkamp J (1972) Ind Eng Chem Prod Res Develop 11:46
51. Weitkamp J, Dauns H (1987) Erdöl, Kohle – Erdgas – Petrochem 40:111
52. Martens JA, Jacobs PA, Weitkamp J (1986) Appl Catal 20:239, 283
53. Jacobs PA, Martens JA (1991) In: van Bekkum H, Flanigen EM, Jansen JC (eds) Introduction to Zeolite Science and Practice, Elsevier, Amsterdam, p 445
54. Coonradt ML, Garwood WE (1964) Ind Eng Chem Prod Res Dev 3:38
55. Minderhoud JK, Van Veen JAR (1993) First-stage hydrocracking: Process and catalytic aspects, Fuel Process Technol 35:87
56. Weisz PB (1962) Adv Catal 13:137
57. Nat PJ, Schoonhoven JWFM, Plantenga FL (1990) In: Trimm DL et al. (eds) Catalysts in Petroleum Refining 1989. Elsevier, Amsterdam, p 399
58. Beuther H, Larson OA (1965) Ind Eng Chem Proc Des Dev 4:177
59. Gosselink JW, Stork WHJ (1997) Ind Eng Chem Res 36:3354
60. Absi-Halabi M, Stanislaus A (1991) Appl Catal 72:193
61. de Jong KP, Reinalda D, Emeis CA (1994) In: Delmon B, Froment GF (eds) Catalyst Deactivation. Stud Surf Sci Catal. Elsevier, Amsterdam 88:155

62. Hadjiloizou GC, Butt JB, Dranoff JS (1992) J Catal 135:481
63. Esener AA, Maxwell IE (1989) In: Occelli ML, Anthony RG (eds) Advances in Hydrotreating Catalysts. Elsevier, Amsterdam, p 263
64. Maxwell IE, Van de Griend JA (1986) In: Murakami Y et al. (eds) New Developments in Zeolite Science and Technology, Proceedings of the 7th International Zeolite Conference. Elsevier, Amsterdam, p 795
65. Van Veen JAR et al., to be published
66. Yan TY (1990) Ind Eng Chem Res 29:1995
67. Dauns H, Ernst S, Weitkamp J (1986) Stud Surf Sci Catal 28 (New Developments Zeolite Science and Technol). Elsevier, Amsterdam, 787-794
68. Douwes CT, 't Hart M (1968) Erdöl, Kohle - Erdgas - Petrochem 21:202
69. Ward JW (1983) In: Poncelet G, Grange P, Jacobs PA (eds) Preparation of catalysts III. Elsevier, Amsterdam, p 587
70. Occelli ML (1997) US Patent 5,614,453
71. Miale JN, Chen NY, Weisz PB (1966) J Catal 6:278
72. Yan T (1983) Ind Eng Chem Proc Des Dev 22:154
73. Maxwell IE (1987) Catal Today 1:385
74. Nace DM (1970) Ind Eng Chem Prod Res Dev 9:203
75. Corma A, Kan Q, Navarro MT, Perez-Pariente J, Rey F (1997) Chem Mater 9: 2123-2126
76. Martens JA, Jacobs PA, this book, chap 2
77. Balkus KJ jr, Khanmamedova A, Gabrielov AG, Zones SI (1996) In: Hightower JW, Delgass WN, Iglesia I, Bell AT, Studies in Surface Scinece and Catalysis, vol 101, 11th International Congress on Catalysis, 40th Anniversary. Elsevier, Amsterdam, p 1341-1348
78. US Patent Specification (1972) 3,702,818
79. Naber JE, Stork WHJ (1991) paper presented at the JECAT'91 Tokyo, Japan, 2-4 December
80. Foley RM, George SE (1991) NPRA Annual Meeting, March 17-19, paper AM 91-13
81. Nat PJ (1989) Erdöl, Kohle - Erdgas - Petrochem 42:447
82. Goossens AG (1986) Hydrocarbon Processing, November, p 84
83. Ondrey G, Kim I, Parkinson G (1996) Chemical Engineering, June, 39
84. Doherty MF, Buzad G (1992) Trans Ichem 70:448
85. Barchas R, Samarth R, Gildert G (1993) Fuels Reformulation 3:44
86. Hearn D, Putman HM (1997) PCT Patent WO 97/03149
87. Hickley TP, Adams JR (1994) US Patent 5,321,163
88. Sie ST (March 1995) NPT Processtechnologie, 9-13 (in Dutch)
89. Sie ST (April 1995) NPT Processtechnologie, 9-12 (in Dutch)
90. Krishna R, Sie ST (1994) Chemical Engineering Science 49:4029-4065
91. Chen NY, Degnan TF (1988) Chem Eng Prog 84 (February 1988), 32-41
92. Andre J-P, Hahn S-K, Min W (1996) NPRA Annual Meeting, March 17-19, San Antonio, Texas, USA, AM-96-38
93. Everett GL, Suchanek A (1996) NPRA Annual Meeting, March 17-19, 1996, Conventional Center, San Antonio, Texas, USA, AM-96-37
94. Qureshi WR, Howell RL, Hung C, Xiao J (1996) 2nd Annual Fuels and Lubes Asia Conference, Singapore, January 29-31
95. Rhodes AK 81997) Oil Gas J, Sept 1, 66-70
96. Appetiti A, Simola F, Binaco B (1990) NPRA Annual Meeting, San Antonio, TX, paper AM 90-43
97. Bryan FM, Chen NY, LaPierre RB, Pappal DA, Patridge RD, Wong SF (1995) HTI Quartely Refining, Autumn, 14-19
98. Farrell TR, Zakarian JA (1986) Oil Gas J 84(20):47
99. Miller SJ, Shippey, MA, Massada, NPRA Fuels and Lubriants National Meeting (Houston 11/5-6/1992), paper N.FL-92-109
100. Law DV (1993, October 19) Presentation at the Institute of Petroleum Economics of Refining Conference, London
101. Grootjans J, Sonnemans JWM (1991) Ketjan Catalysts Symposium, paper H-5

102. Homan Free HW, Schockaert T, Sonnemans JWM (1993) Fuel Processing Technology 35:111–117
103. Hargrove JD, Elkes GJ, Richardson AH (1979) Oil Gas J 77(3):103
104. Chen NY, Gorring RL, Ireland HR, Stein TR (1977) Oil Gas J 75, June 6, p 165
105. Smith KW, Starr WC, Chen NY (1980) Oil Gas J 78, May 26, p 75
106. Chen NY, Schlenker JL, Garwood WE, Kokotailo GT (1984) J Catal 86:24
107. Rollmann LD, Walsh DE (1979) J Catal 56:139
108. Bendoraitis JG, Chester AW, Dwyer FG, Garwood WE (1986) In: Murakami Y, Iijima A, Ward JW (eds) Proceedings 7th International Zeolites Conference, Kodansha/Elsevier, Tokyo, p 669
109. Patent WO 92/01657
110. US Patent 5,135,638
111. Sivasanker S, Waghmare KJ, Reddy M, Ratnasamy P (1988) In: Proceedings 9th International Congress on Catalysis, Calgary, Vol 1, Phillips MJ, Ternan M (eds) The Chemical Institute of Canada, Ottawa, Ontario, p 120
112. Maxwell IE, Stork WHJ (1999) In: Van Bekkum H, Flanigen EM, Jansen JC (eds) Introduction to Zeolite Science and Practice 2nd edition, Elsevier, Amsterdam, in preparation
113. Chen NY, Garwood WE (1986) Catal Rev-Sci Eng 28 (2–3):185
114. Chen NY, Maziuk J, Schwartz AB, Weisz PB (1968) Oil Gas J 66(47):154
115. Chen NY, Garwood WE, US Patent specification 3.379.640
116. Chen NY, Rosinski EJ, US Patent Specification 3.630.966
117. Roselius RR, Gibson KR, Orriiston RM, Maziuk J, Smith FM (1973) NPRA Annual Meeting, San Antonio, TX, April
118. Chen NY, US Patent Specification 3.729.409
119. Heinemann H (1977) Catal Rev-Sci Eng 15:53
120. Chen NY, Garwood WE, Haag WO, Schwartz AB (1979) paper presented at AIChE 72nd Annual Meeting, San Francisco, CA, November 25–29
121. Weisz PB (1980) Pure Appl Chem 52:2091
122. Bonacci JC, Patterson JR, US Patent Specification 4.292.167
123. Gonzalez RG (1996) Hart's fuel technology and management, November/December, p 56
124. WO 93/04146, world patent application
125. Hilbert TL, Durand PP, Sarli MS (1996) Hydrocarbon Asia, November/December, p 86
126. Salazar JA, Badra C, Perez JA, Palmisano E, Garcia W, Solari B (1996) Symposium on the removal of aromatics, sulfur and olefins from gasoline and diesel, presented at 212th national ACS Meeting, Division of petroleum chemistry, Orlando, FL, August 25–29
127. Fernandez PD, Schmidt RJ, Johnson BH, Rice LH (1988) NPRA Annual Meeting, San Antonio, TX, March 20–22, paper AM-88-45
128. Gosling C, Rosin R, Bullen P, Shimizu T, Imai T (1997) paper presented at the World Conference on Refining, Vehicle Technology and Fuel Quality, Brussels (Belgium), May 21–23. In addition, taking into account information obtained at SRTCA
129. Weitkamp J (1975) In: Ward JW (ed) Hydrotreating and Hydrocracking, ACS Symposium Series 20, p 1
130. Kuchar PJ, Bricker JC, Reno ME, Haizman RS (1993) Fuels Processing Technology, 35: 183
131. Sachtler WMH, to be published in Topics in Catalysis
132. Liu H, Lei GD, Sachtler WMH (1996) Applied Catalysis, A, 146(1):165
133. Blomsma E, Martens JA, Jacobs PA (1997) J Catal 165:241
134. Blomsma E, Martens JA, Jacobs PA (1996) J Catal 159:323
135. Blomsma E, Martens JA, Jacobs PA (1995) J Catal 155:141
136. Bloch HS, Haensel V, US Patent Specification 2.999.074
137. Raghuram S, Haizmann RS, Lowry DR, Schiferli WJ (1990) Oil Gas J, Dec 3, p 66
138. Stork WHJ, personal communication
139. Myers JW (1977) Ind Eng Chem Prod Res Develop 10(2):200

140. Franck JP, Le Page J-F, Roumegous A (1981) In: Seiyama T, Tanabe K (eds) New Horizons in Catalysis, Studies Surf Sci Catal, vol 7. Elsevier, Amsterdam, p 1018
141. Melchor A, Garbowski E, Mathieu MV, Primet M (1986) J Chem Soc Faraday Trans 1, 82:1893
142. Ollendorf RL, Boskovic G, Butt JB (1990) Appl Catal 62:85 and references cited therein
143. Oil Gas Journal, April 7, 1997, p 41
144. Koradia PB, Kiovsky JR, Asim MY (1980) J Catal 66:290
145. Hays GR, Van Erp WA, Alma NCM, Couperus PA, Huis R, Wilson AE (1984) Zeolites 4:377
146. Guisnet M, Fouche V, Belloum M, Bournonville JP, Travers C (1991) Appl Catal 71:283
147. Guisnet M, Fouche V, Belloum M, Bournonville JP, Travers C (1991) Appl Catal 71:295
148. Gray JA, Cobb JT (1975) J Catal 36:125
149. Leglise J, Chambellan A, Cornet D (1991) Appl Catal 69:15
150. Leglise J, Goupil JM, Cornet D (1991) Appl Catal 69:33
151. Fajula F, Boulet M, Coq B, Rajaofanova V, Figueras F, des Courieres T (1993) In: Guczi L et al. (eds) New Frontiers in Catalysis, Proc 10th Int Congress on Catalysis, Juli 19–24, 1992, Budapest. Elsevier, Amsterdam
152. Allain JF, Magnoux P, Schulz Ph, Guisnet M (1996) Proceedings DGMK Meeting, Berlin, March 14–15, pp 219–226
153. Leu L-J, Iion L-Y, Kang B-C, Li C, Wu S-T, Wu J-C (1991) Appl Catal 69:49
154. Kumar Kunchal S, Dwyer FG, Ambler CP (1993) Paper AM 93-45 at the 1993 NPRA Annual Meeting, march 21–23, San Antonio, Texas, USA
155. Gialella RM, Mank L, Burzynski J-P, Duee D, Travers C (1993) Paper AM 93-15, presented at the NPRA Annal meeting, March 21–23, San Antonio, Texas, USA
156. Mooiweer HH, de Jong KP, Kraushaar-Czarnetski B, Stork WHJ, Krutzen BCH (1994) In: Weitkamp J, Karge HG, Pfeifer H, Hölderich W (eds) Zeolites and Related Microporous Materials: State of the Art 1994. Stud Surf Sci Cat, vol 84, pp 2327–2334. Elsevier, Amsterdam, 1994
157. Powers DH (1993) Presentation O-1 at the 1993 DEWITT Petrochemical review, Houston, Texas, USA, March 23–25
158. Houzvicka J, Diefenbach O, Ponec V (1996) J Catal 164:288
159. Weitkamp J, Traa Y (1997) In: Ertl G, Knoezinger H, Weitkamp J (eds) Handbook of Heterogeneous Catalysis, vol 4, Wiley-VCH, Weinheim, p 2039
160. Corma A, Martinez A (1993) Catal Rev-Sci Eng 35:483
161. Guisnet M, Gnep NS (1996) Appl Catal A 146:33
162. Mesters CMAM, Peferoen DGR, Gilson JP, de Groot C, van Brugge PTM, de Jong KP (1996) In: Weitkamp J, Luecke B (eds) Proceedings of the DGMK Conference on Solid Acids and Bases, Berlin, March 14–15, DGMK, Hamburg, pp 57–58
163. Oil Gas J, 94(14):69
164. Hughes TR, Buss WC, Tamm PW, Jacobson RL (1986) In: Murakami Y, Iijima A, Ward JW (eds) Proceedings 7th International Zeolites Conference, Kodansha/Elsevier, Tokyo, p 725
165. Bernard JR (1980) in Rees LVC (eds), Proc Fifth Intern Conf Zeolites, Heyden, London, p 686; see also ref. 170
166. Tamm PW, Mohr DH, Wilson CR (1988) Stud Surf Sci Catal 38 (Catalysis 1987) Elsevier, Amsterdam 335–353
167. Lane GS, Modica FS, Miller JT (1991) J Catal 129:142
168. Derouane EG, Vanderkerken DJ (1988) Appl Catal 45:L15
169. Davis RJ, Derouane EG (1991) Nature 349:313
170. Law DV, Tamm PW, Detz CM (1987) Energy Prog 7: 215–222
171. Law DV, Tamm PW, Detz CM (1987) Spring Natinal meeting, AIChE, March 29–April 2, Houston, Texas
172. PCT WO 91/11501
173. Zones SS (1996) Paper S7b.2, Symposium Advanced Catalytic Materials, MRS Fall Meeting, Boston, USA, December 2–5
174. US Patent 5,328,595

175. Naber JE, Stork WHJ, Blauwhoff PMM, Groeneveld KJW (1991) Oil Gas European Magazine 1:26
176. Milan SN, Winquist BHC, Murray BD, Minderhoud JK, Del Paggio AA (1991) Reduction of aromatics in diesel fuel, a low pressure, two stage approach, Paper presented at the 1991 AIChE Meeting in Houston, TX, April 8–12
177. Van den Berg JP, Lucien JP, Germaine G, Thielemans GLB (1993) Fuels Process Technol 35:119
178. Stork WHJ (1997) Stud Surf Sci Catal 106:41
179. Dave D, Gupta A, Karlson K, Suchanek AJ, van Stralen H (1993) Paper AM 93–24 in the 1993 Spring Annual NPRA Meeting
180. Maiden CJ (1988) ChemTech, January, p 38
181. Maiden CJ (1988) In: Bibby DM, Chang CD, Howe RF, Yurchak S (eds) Methane Conversion. Elsevier, Amsterdam, p 1
182. Zenn JZ (1988) ChemTech, January, p 42
183. Zenn JZ (1988) In: Bibby DM, Chang CD, Howe RF, Yurchak S (eds) Methane Conversion. Elsevier, Amsterdam, p 663
184. Meisel SL, McCullough JP, Lechthaler CH, Weisz PB (1976) ChemTech, February, p 86
185. Chang CD, Silvestri AJ (1977) J Catal 47:249
186. Chang CD (1983) Hydrocarbons from Methanol. Dekker, New York
187. Chang CD (1983) Catal Rev Sci Eng 25:1
188. Chang CD (1988) Stud Surf Sci Catal 36:127
189. Hutchins GJ, Hunter R (1990) Catal Today 6:279
190. Yurchak S, Voltz SE, Warner JP (1979) Ind Eng Chem Process Des Dev 18:527
191. Kam AY, Schreiner M, Yurchak S (1984) In: Meyers RA (ed) Handbook of Synfuels Technology. McGraw-Hill, New York, chaps 2–3
192. Penick JE, Lee W, Maziuk J (1983) In: Wei J, Georgakis C (eds) Chemical reaction engineering – plenary lectures, ACS Symposium Series 226, ACS, Washington DC, p 19
193. Yurchak S (1988) Stud Surf Sci Catal 36:251
194. Yurchak S (1988) In: Bibby DM, Chang CD, Howe RF, Yurchak S (eds) Methane Conversion. Elsevier, Amsterdam, p 251
195. Krohn DE, Melconian MG (1988) In: Bibby DM, Chang CD, Howe RF, Yurchak S (eds) Methane Conversion. Elsevier, Amsterdam, p 679
196. Lee W, Yurchak S, Daviduk N, Maziuk J (1980) A fixed-bed process for the conversion of methanol-to-gasoline, Paper presented at the NPRA Annual Meeting, New Orleans, LA, March 23–25
197. Tabak SA, Yurchak S (1990) Catal Today 6:307
198. Allum KG, Williams AR (1988) In: Bibby DM, Chang CD, Howe RF, Yurchak S (eds) Methane Conversion. Elsevier, Amsterdam, p 691
199. Liederman D, Jacob SM, Voltz SE, Wise JJ (1978) Ind Eng Chem Process Des Dev 17(3):340
200. Lee W, Maziuk J, Thiemann WK (1978) Ein neuer Prozeß zur Umwandlung von Kohle in Ottokraftstoff, Paper presented at the 26th DGMK Meeting, Berlin, October 4–6
201. Edwards M, Avidan A (1986) Chem Eng Science 41(4):729
202. Maziuk J, Gould R, Keim KH, Gierlich R (1984) The methanol-to-gasoline (MTG) process, Paper presented at the PetroPacific '84 Converence, Melbourne, September 16–19; Keim KH, Maziuk J, Tonnesmann A (1984) Erdöl und Kohle, Erdgas, Petrochemie 37:558
203. Grimmer HR, Thiagarajan N, Nitschke E (1988) In: Bibby DM, Chang CD, Howe RF, Yurchak S (eds) Methane Conversion. Elsevier, Amsterdam, p 273
204. Joenseni F, Kiilerich Hansen JR, Rostrup-Nielsen J, Skov A, Topp-Jorgensen J (1985) Conversion of synthesis gas to high octane gasoline, Paper presented at a Conference on large chemical plants, Antwerp, October 9–11
205. Topp-Jorgensen J (1988) In: Bibby DM, Chang CD, Howe RF, Yurchak S (eds) Methane Conversion. Elsevier, Amsterdam, p 293
206. Chang CD, Kuo JCW, Lang WH, Jacob SM, Wise JJ, Silvestri AJ (1978) Ind Eng Chem Process Res Dev 17:255

207. Chang CD, Lang WH, Silvestri AJ (1979) J Catal 56:268
208. Schaper L, Sie ST (1977) Netherlands Patent Application 7.711.719 of 26/10/1977, assigned to Shell International Research Mij BV
209. Tabak S, Weiss GJ (1984) New synthetic fuel routes for production of gasoline and distillate, Paper presented at the Synfuels 4th Worldwide Symposium, Washington DC, November 7-9
210. Tabak SA (1984) New synthetic fuel routes for production of gasoline and distillate, Paper presented at the symposium Nonpetroleum Vehicular Fuels IV, Arlington, VA, April 16-18
211. Keim K-H, Krambeck FJ, Maziuk J, Tonnesmann A (1987) Erdöl Erdgas Kohle 103(2):82
212. Piciotti M (1997) Oil Gas J, June 30, 71
213. Garwood WE (1983) In: Stucky GD, Dwyer FG (eds) ACS Symposium Series 218, American Chemical Society, Washington DC, p 23
214. Tabak SA, Krambeck FJ, Garwood WE (1984) Conversion of propylene and butylene over ZSM-5 catalyst, Paper presented at the AICHE Meeting, San Francisco, CA, November 25-30
215. Avidan AA (1988) In: Bibby DM, Chang CD, Howe RF, Yurchak S (eds) Methane Conversion, Elsevier, Amsterdam, p 307
216. Miller SJ (1988) In: Ward JW (ed) Catalysis 1987. Elsevier, Amsterdam, pp 187-197
217. O'Connor, CT (1997) In: Ertl G, Knözinger H, Weitkamp J (eds) Handbook of Heterogeneous Catalysis, vol 5. Wiley-VCH Weinheim, pp 2380-2387
218. Chen CSH, Bridger RF (1996) J Catal 161(2):687
219. Keskinen KM (1996) VTT Symposium, 163, 111-116; see also WO 9620988
220. Martens LR, Verduyn J, Mathijs GM (1997) Catal Today 36(4):51
221. Van den Berg JP, Roebschlager KHW, Maxwell IE (1989) The Shell Polygasoline and Kero (SPGK) process, Paper presented at the 11th North American Catalysis Society Meeting, Dearborn, MI, May 7-11
222. van den Berg JP, Blauwhoff PMM (1991) The Shell Polygasoline and Kero process, Paper presented at the American Chemical Society Meeting, petroleum Chemistry Division, New York, NY, August 25-30
223. Stowe RA, Murchison CB (1982) Hydrocarbon Processing, January, p 147
224. Synfuls (1982) April 16, p I
225. Seddon D (1990) Catal Today 6:361
226. Csicsery JM (1970) J Catal 17:207, 216, 315, 323; 18:30
227. Davies EE, Kolombos AJ (1976) British Patent Specification 53012/76, December 20, assigned to British Petroleum Co Ltd
228. Johnson JA, Weiszmann JA, Hilder GK, Hall AHP (1984) NPRA Annual Meeting, an Antonio, TX, March 25-27, paper AM 84-45
229. European Chemical News (1991) January 21
230. European Chemical News (1991) July 15, p 39
231. Hoelderich WF (1989) Stud Surf Sci Catal 49:69
232. Hoelderich WF (1988) Angew Chem Int Ed Engl 27:226
233. Haag WO, Olson DH, Weisz PB (1984) Chem Future, Proceedings 29th IUPAC Congress, 1983, p 327
234. Beck JS, Haag WO (1997) In: Ertl G, Knözinger H, Weitkamp J (eds) Handbook of Heterogenous Catalysis, vol 5, Wiley-VCH, Weinheim, p 2123-2136
235. Bolton AP (1976) In: Rabo JA (ed) Hydrocracking, Isomerisation and other Industrial Processes, Zeolite Chemistry and Catalysis. ACS Monograph 171, American Chemcal Society, Washington DC, p 714
236. Dwyer FG, Lewis PJ, Schneider FM (1976) Chem Eng, January 5, p 90
237. Industrial News (1991) Appl Catal 72, 2, N14; Chem Eng (1990) September, p 19
238. Benthem MF, Gajda GJ, Jensen RH, Zinman HA (1996) Proceed. DGMK meeting, Berlin, March 14-15, pp 155-166
239. Meima GR, van der Aalst MJM, Samson MSU, Garces JM, Lee JG (1996) Proceed. DGMK Meeting Berlin, March 14-15, pp 125-138

240. Meima GR, van der Aalst MJM, Samson MSU, Garces JM, Lee JG (1992) In: von Ballmoos R, Higgins JB, Treacy MMJ (eds) Proceed 9th Int Zeolite Conf Montreal, Butterworth-Heinemann, Boston, p 327
241. Chem Eng News 1995, December 18, p 12
242. Portela JAO, Hildreth JM (1994) de Witt Petrochem Rev, Houston, March 22–24, E1–22
243. Vora BV, Pujado PR, Imai AT, Fritsch TR (1990) Recent advances in the production of detergent olefins and linear alkylbenzenes. Soc of Chem Ind, University of Cambridge, England
244. Olson DH, Haag WO (1984) ACS Symposium Series, 89, p 267
245. Song C, Schobert HH (1993) Fuels Processing Technology 34:157
246. Song C, Kirby S (1993) Prepr Am Chem Soc Div Petr Chem 38:783
247. Lee GS, Maj JJ, Rocke SC, Garces JM (1989) Catal Letters 2(4):243
248. Keading WW, Barile GC, Wu MM (1984) Catal Rev-Sci Eng 26:597
249. Bhatia S (1990) Zeolite Catalysis: Principles and Applications. CRC Press, Boca Raton, FL
250. See e.g., GB Patent Specification 9.011.412
251. Sie ST, de Vries AF, Mesters CMAM, Boon AQM, Bottenberg K, Trautrims B (1996) DGMK Conference "Catalysis on solid acids and bases". March 14, 15, Berlin (Germany)
252. European Patent Specification 136.133
253. UK Patent Specification 2.042.490
254. Hoelderich W, Gallei E (1984) Chem-Ing-Tech 56(12):908
255. The Catalyst Review Newsletter (1990) 3(3):3
256. Gorra F, Breckenridge LL, Guy WWM, Sailor RA (1992) Oil Gas J, October 12, 60
257. Sheldon RA (1996) J Mol Catal A: Chemical 107:75
258. Tuel A, Ben Taârit Y (1993) Appl Catal A: General 102:201
259. European Patent Specification 102.655
260. US Patent Specification 4.396.783
261. Notari B (1988) Stud Surf Sci Catal 37:413
262. Roffla P, Leofanti G, Cesana A, Mantegazza M, Padovan M, Petrini G, Tonti S, Gervasutti V, Veragnolo R (1990) La Chimica & L'Industria 72:598
263. Mantegazza MA, Cesana A, Pastori M (1996) Topics in Catal 3327–335
264. ECN Speciality Chemicals Suppl (1991) February, p 14
265. Ramaswamy AV, Sivasanker S (1993) Catal Lett 22:239
266. Ramaswamy AV, Sivasanker S, Ratnasamy P (1994) Microporous Materials 2:451
267. Ratnasamy P, Kumar R (1995) In: Bonneviot L, Kaliaguine S (eds) Zeolites: A refined tool for designing catalytic sites, pp 367–376
268. Corma A, Navaro MT, Perez-Pariente J, Sanchez F (1994) In: Weitkamp J, Karge HG, Pfeifer H, Hölderich W (eds) Zeolites and Related Microporous Materials: State of the Art 1994. Stud Surf Sci Cat 84, Elsevier, pp 69–75
269. Tanev PT, Chibwe M, Pinnavala TJ (1994) Nature 368:321
270. Corma A, Navarro MT, Peres Pariente J (1994) Chem Soc, Chem Commun 147
271. Das TKR, Chaudhari K, Chandwadkar AJ, Sivasanker S (1995) J Chem Soc Chem Commun 2495
272. Kulawik K, Schulz-Ekloff G, Rathousky J, Zukal A, Had J (1995) Collect Czech Chem Commun 60:451
273. Parton RF, Jacobs JM, Huybrechts DR, Jacobs PA (1988) Stud Surf Sci Catal 46:163
274. Tuel A, Moussa-Khouzami S, Ben Taarit Y, Naccache C (1991) J Mol Catal 68:45
275. Liu C, Ye X, Wu Y (1996) Catal Lett 35:263
276. Dai P-SE, Petty RH, Ingram CW, Szostak R (1996) Appl Catal A 143:101
277. Sheldon RA (1981) In: Ugo R (ed) Aspects of Homogeneous Catalysis. Reidel, Dordrecht, 4, p 1
278. Sheldon RA (1991) Chemtech, September, p 566
279. Murugavel R, Roesky HW (1997) Angew Chem Int Ed Engl 36:477
280. Knops-Gerrits PP, Vankelecom IFJ, Béatse E, Jacobs PA (1996) Catal Today 32:63–70
281. Mantegazza MA, Cesana A, Pastori M (1996) Topics in Catal 3:327–335
282. Fetting F, Dingerdissen U (1992) Chem Eng Technol 15:202–212

283. Deeba M, Ford ME, Johnson TA, Premecz JE (1990) J Mol Catal 60:11
284. Lequitte M, Figueras F, Moreau C, Hub S (1996) J Catal 163:255
285. Benac BL, Knifton JF, Dai E, EP 0.587.424 A1 (8-9-1993)
286. US Patent Specification 4.254.061
287. European Patent Specification 0.26.071
288. Keane M, Sonnichsen GC, Abrams L, Corbin DR, Gier TE, Shannon RD (1987) Appl Catal 32:361
289. Mizuno N, Tabata M, Uematsu T, Iwamoto M (1993) J Chem Soc Faraday Trans 89:3513
290. Mochida I, Yasutake A, Fujitsu H, Takeshita K (1983) J Catal 82:313
291. European Patent Specification 125.616
292. Misono M, Nojiri N (1990) Appl Catal 64:1
293. Eller K, Kummer R, Stops P, PCT Patent WO 97/21661 (19-6-1997)
294. Eller K, Kummer R, Stops P, EP 0.778.259 A2 (11-6-1997)
295. Eller K, Kummer R, Stops P, EP 0.785.185 A1 (23-7-1997)
296. Dingerdissen U, Eller K, EP 0.752.411 A2

Subject Index

C

Printing: Saladruck, Berlin
Binding: H. Stürtz AG, Würzburg

Molecular Sieves

Science and Technology

Editors:
H.G. Karge
J. Weitkamp

Volume 1
Synthesis

1998. IX, 302 pp.
113 figs., 37 tabs.
Hardcover DM 298
ISBN 3-540-63622-6

Subscription price valid for subscribers to the complete work: DM 238

This book covers, in a comprehensive manner, the science and technology of zeolites and all related microporous and mesoporous materials. Authored by renowned experts, the contributions are grouped together topically in such a way that each volume of the book series is dealing with a specific sub-field.

Contents:

R.W. Thompson:
Recent Advances in the Understanding of Zeolite Synthesis.

H. Gies:
Synthesis of Clathrasils.

S. Ernst:
Synthesis of more Recent Aluminosilicates with a Potential in Catalysis and Adsorption.

J.C. Vartuli, W.J. Roth, J.S. Beck, S.B. McCullen and C.T. Kresge:
The Synthesis and Properties of M41S and Related Mesoporous Materials.

E.N. Coker and J.C. Jansen:
Approaches for the Synthesis of Ultra-large and Ultra-small
Zeolite Crystals.

R. Szostak:
Synthesis of Molecular Sieve Phosphates.

G. Perego, R. Millini and G. Bellussi:
Synthesis and Characterization of Molecular Sieves Containing Transition Metals in the Framework.

S.A. Schunk and F. Schüth:
Synthesis of Zeolite-Like Inorganic Compounds.

P. Cool and E.F. Vansant:
Pillared Clays: Preparation, Characterization and Applications.

Please order from
Springer-Verlag Berlin
Fax: + 49 / 30 / 8 27 87- 301
e-mail: orders@springer.de
or through your bookseller

Errors and omissions excepted. Prices subject to change without notice.
In EU countries the local VAT is effective. Gha.

Printed in the United States
131824LV00001B/25/A

9 783540 636502